Soils for Management of Organic Wastes and Waste Waters

Soils for Management of Organic Wastes and Waste Waters

Proceedings of a symposium held 11–13 March 1975 at the TVA National Fertilizer Development Center, Muscle Shoals, Alabama, and cosponsored by the Soil Science Society of America, American Society of Agronomy, Crop Science Society of America, Tennessee Valley Authority, American Society of Agricultural Engineers, The Institute of Ecology, Society of American Foresters, Society for Range Management, and Soil Conservation Society of America.

Editorial Committee: L. F. ELLIOTT, co-editor
                       F. J. STEVENSON, co-editor
                       C. R. FRINK
                       R. R. HILL
                       C. C. HORTENSTINE
                       V. J. KILMER
                       E. J. MONKE
                       F. G. VIETS, JR.

Organizing Committee: L. F. ELLIOTT, chairman
                        C. R. FRINK
                        C. C. HORTENSTINE
                        W. R. GARDNER
                        T. M. MC CALLA
                        F. G. VIETS, JR.

Coordinating Editor: MATTHIAS STELLY

Managing Editor: RICHARD C. DINAUER

Assistant Editors: JEAN M. PADRUTT
                    JUDITH H. NAUSEEF

Published by

Soil Science Society of America
American Society of Agronomy
Crop Science Society of America
Madison, Wisconsin USA
1977

American Society of Agronomy, Inc.
Crop Science Society of America, Inc.
Soil Science Society of America, Inc.
677 South Segoe Road, Madison, Wisconsin 53711 USA

Second Printing 1986

Library of Congress Catalog Card Number: 77-73344

Standard Book Number: 0-89118-049-4

Printed in the United States of America

CONTENTS

15 Transportation and Application of Organic Wastes to Land

J. R. MINER and T. E. HAZEN

19 Land Utilization and Disposal of Organic Wastes in Hot, Humid Regions

GRANT W. THOMAS

20 Special Problems and Opportunities in Use of Waste Heat for Soil Warming

DAVID A. MAYS

21 Odors and Emissions from Organic Wastes

A. R. MOSIER, S. M. MORRISON, and G. K. ELMUND

22 Pathogen Considerations for Land Application of Human and Domestic Animal Wastes

23 Problems and Need for High Utilization Rates of Organic Wastes

W. E. LARSON and G. E. SCHUMAN

FOREWORD

The practice of applying organic wastes and waste water to soils is not a recent innovation. The current intensity of study of the effects of wastes on soils has developed rather recently, however. Previous practices of disposing of wastes into water, or by stockpiling or incinerating them are no longer adequate or acceptable. Our increasing concern for the quality of our environment and the need to utilize wisely all of our resources have stimulated projects to utilize organic wastes and waste water in the production of food, feed, and fiber for the ever-increasing population of the world.

Agricultural scientists, representing many areas of specialization, have played a major role in studying the fate of organic wastes applied to soils, the effects of such wastes on soils and the role of soils in the transformation processes involved. This publication is a compilation of symposium papers that were presented by scientists having a wide diversity of training and experience.

The American Society of Agronomy, the Soil Science Society of America, and the Crop Science Society of America are proud to have been among the cosponsors of this symposium and to be the publishers of this book. Appreciation is expressed to all other cosponsors and especially to the Tennessee Valley Authority at Muscle Shoals, Alabama who acted both as a host and cosponsor of the symposium.

March 1977

MARLOWE D. THORNE, *president*
American Society of Agronomy

DALE N. MOSS, *president*
Crop Science Society of America

VICTOR J. KILMER, *president*
Soil Science Society of America

PREFACE

This volume, *Soils for Management and Utilization of Organic Wastes and Waste Waters,* comprises the contributions of outstanding authorities on subjects related to land utilization and disposal of organic wastes. The papers were presented at a symposium held at Muscle Shoals, Alabama, 11–13 March 1975, cosponsored by the Soil Science Society of America, American Society of Agronomy, Crop Science Society of America, Tennessee Valley Authority, American Society of Agricultural Engineers, Institute of Ecology, Soil Conservation Society of America, Society of American Foresters, and The American Society of Range Management.

The initial chapter provides a brief resume of sources and quantities of organic wastes, present and potential future regulations, and economic and social implications of utilization and disposal on agricultural lands. The next 23 chapters can be grouped into the following categories: (i) properties of wastes and waste waters; (ii) chemical, physical, and biological properties of soils; (iii) effect of waste application on nutrient cycles; (iv) site selection, design, and transportation; (v) special utilization and disposal problems occurring within various climatic regions; and (vi) special environmental considerations (odors, pathogens, etc.). The final chapter deals with the future direction of waste utilization.

Tremendous quantities of organic wastes are produced each year from domestic and municipal sewage, animal and poultry enterprises, food processing industries, pulp and paper mills, municipal garbage, and others. It is imperative to have safe and economic practices for the utilization and disposal of these wastes. With increased fertilizer and energy costs, some organic wastes (e.g., animal manures and selected municipal organic wastes) are again being considered resources for crop growth and as beneficial soil amendments. Use of some products, especially municipal and domestic organic wastes, incurs special problems, such as heavy metals, pathogens, nutrient imbalances, odors, and social acceptance. These areas of concern have resulted in a voluminous literature on the use of soil for disposal of organic wastes and waste waters. It was the purpose of this book to synthesize current literature on the subject into a comprehensive treatise.

Benefits and problems associated with utilization and disposal of organic wastes on agricultural land are discussed herein. Both the theoretical and practical aspects are covered. The various chapters, which were assembled by knowledgeable scientists from many disciplines, represent an evaluation of existing facts, and they point out future research needs.

No attempt was made to arrive at a consensus on controversial subjects, and the reader will discover variation in personal philosophies among authors. This reflects, to a large extent, a paucity of information on some key aspects of the disposal problem. Until such time as complete information becomes available, some confusion will exist as to how soils can, or should be, managed for proper and effective waste utilization. It is felt that the material presented will be of value to layman and agricultural scientist alike and that the volume will be of value to administrators and others involved in policy decisions regarding organic waste application to soil.

The editorial committee is grateful to the authors and the organizations they represent. Acknowledgement is given to Richard C. Dinauer, Dr. Matthias Stelly, and other members of the Headquarters Office of the Soil Science Society of America for editing and preparing the manuscripts for publication. Assistance by anonymous reviewers is greatly appreciated.

L. F. ELLIOTT, *co-editor*
USDA-ARS
Washington State University
Pullman, Washington

F. J. STEVENSON, *co-editor*
Department of Agronomy
University of Illinois
Urbana, Illinois

CONTRIBUTORS

William H. Allaway

Soil Scientist, U. S. Plant, Soil and Nutrition Laboratory, Agricultural Research Service, U. S. Department of Agriculture, Ithaca, New York

Stanley A. Barber

Professor of Soil Fertility and Chemistry, Department of Agronomy, Purdue University, West Lafayette, Indiana

Eric G. Beauchamp

Associate Professor, Department of Land Resource Science, University of Guelph, Guelph, Ontario, Canada

Sidney M. Beck

Professor of Bacteriology, Department of Bacteriology and Biochemistry, University of Idaho, Moscow, Idaho

Hinrich L. Bohn

Professor of Soil Chemistry, Department of Soils, Water and Engineering, University of Arizona, Tucson, Arizona

F. E. Broadbent

Professor of Soil Microbiology, Department of Land, Air, and Water Resources, University of California, Davis, California

Rufus L. Chaney

Plant Physiologist, Biological Waste Management and Soil Nitrogen Laboratory, Agricultural Research Service, U. S. Department of Agriculture, Beltsville, Maryland

G. K. Elmund

Graduate Research Assistant, Department of Microbiology, Colorado State University, Fort Collins, Colorado

James O. Evans

Research Hydrologist, Forest Service, U. S. Department of Agriculture, Washington, D. C.

Klaus W. Flach

Assistant Administrator for Soil Survey, Soil Conservation Service, U. S. Department of Agriculture, Washington, D. C.

Dennis D. Focht

Associate Professor of Soil Microbiology, Department of Soil Science and Agricultural Engineering, University of California, Riverside, California

Wallace H. Fuller

Biochemist, Department of Soils, Water and Engineering, University of Arizona, Tucson, Arizona

Campbell M. Gilmour

Professor of Bacteriology and Biochemistry, Department of Bacteriology and Biochemistry, University of Idaho, Moscow, Idaho.

Paul M. Giordano

Research Soil Chemist, Soils and Fertilizer Research Branch, Tennessee Valley Authority, Muscle Shoals, Alabama

Thamon E. Hazen

Assistant Director, Iowa Agriculture and Home Economics Experiment Station, Ames, Iowa

Louis T. Kardos

Professor of Soil Physics, Department of Agronomy, The Pennsylvania State University, University Park, Pennsylvania

Dennis R. Keeney   Professor, Department of Soil Science, University of Wisconsin, Madison, Wisconsin

William E. Larson   Soil Scientist, North Central Region, Agricultural Research Service, U. S. Department of Agriculture, University of Minnesota, St. Paul, Minnesota

James P. Law, Jr.   Chief, Agricultural Wastes Section, Robert S. Kerr Environmental Research Laboratory, U. S. Environmental Protection Agency, Ada, Oklahoma

John Letey, Jr.   Professor of Soil Physics, Department of Soil Science and Agricultural Engineering, University of California, Riverside, California

Cecil Lue-Hing   Director, Research and Development, Metropolitan Sanitary District of Greater Chicago, Chicago, Illinois

Jesse Lunin   Staff Scientist, National Program Staff, Agricultural Research Service, U. S. Department of Agriculture, Beltsville, Maryland

T. M. McCalla   Microbiologist, North Central Region, Agricultural Research Service, U. S. Department of Agriculture, University of Nebraska, Lincoln, Nebraska

James P. Martin   Professor of Soil Science, Department of Soil Science and Agricultural Engineering, University of California, Riverside, California

David A. Mays   Agronomist, Soils and Fertilizer Research Branch, National Fertilizer Development Center, Tennessee Valley Authority, Muscle Shoals, Alabama

Burl D. Meek   Research Leader, Imperial Valley Conservation Research Center, Agricultural Research Service, U. S. Department of Agriculture, Brawley, California

James D. Menzies   Research Microbiologist, Agricultural Research Service, U. S. Department of Agriculture, Beltsville, Maryland (now retired and located at Fox Island, Washington)

J. Ronald Miner   Professor and Head, Department of Agricultural Engineering, Oregon State University, Corvallis, Oregon

Sumner M. Morrison   Professor, Department of Microbiology, Colorado State University, Fort Collins, Colorado

Arvin R. Mosier   Research Chemist, Western Region, Agricultural Research Service, U. S. Department of Agriculture, Fort Collins, Colorado

Fred A. Norstadt   Soil Scientist, Western Region, Agricultural Research Service, U. S. Department of Agriculture, Fort Collins, Colorado

Sterling R. Olsen   Research Leader, Western Region, Agricultural Research Service, U. S. Department of Agriculture, Fort Collins, Colorado

James R. Peterson

Soil Scientist III, Metropolitan Sanitary District of Greater Chicago, Cicero, Illinois

Parker F. Pratt

Professor of Soil Science, Department of Soil Science and Agricultural Engineering, University of California, Riverside, California

Burns R. Sabey

Professor of Soil Science, Agronomy Department, Colorado State University, Fort Collins, Colorado

Clarence E. Scarsbrook

Professor of Soils, Agronomy and Soils Department, Auburn University, Auburn, Alabama

Gerald E. Schuman

Soil Scientist, High Plains Grasslands Research Station, Agricultural Research Service, U. S. Department of Agriculture, Cheyenne, Wyoming

Wayne H. Smith

Professor of Forest Nutrition and Assistant Director School of Forest Resources and Conservation, University of Florida, Gainesville, Florida

B. A. Stewart

Research Leader, Southwestern Great Plains Research Center, Agricultural Research Service, U. S. Department of Agriculture, Bushland, Texas

Norris P. Swanson

Agricultural Engineer, North Central Region, Agricultural Research Service, U. S. Department of Agriculture, University of Nebraska, Lincoln, Nebraska

Grant W. Thomas

Professor of Agronomy, Department of Agronomy, University of Kentucky, Lexington, Kentucky

Richard E. Thomas

Research Soil Scientist, Robert S. Kerr Environmental Research Laboratory, U. S. Environmental Protection Agency, Ada, Oklahoma

Marlowe D. Thorne

Professor of Agronomy, Department of Agronomy, University of Illinois, Urbana, Illinois

Thomas C. Tucker

Soil Scientist, Department of Soils, Water and Engineering, University of Arizona, Tucson, Arizona

V. Van Volk

Associate Professor of Soil Science, Department of Soil Science, Oregon State University, Corvallis, Oregon

Len R. Webber

Professor, Department of Land Resource Science, University of Guelph, Guelph, Ontario, Canada

Frank Wiersma

Agricultural Engineer, Department of Soils, Water and Engineering, University of Arizona, Tucson, Arizona

Raymond E. Wildung

Manager, Environmental Chemistry, Battelle-Northwest, Richland, Washington

John E. Witty

Soil Correlator, Technical Service Center, Soil Conservation Service, U. S. Department of Agriculture, Broomall, Pennsylvania

CONVERSION FACTORS FOR U. S. AND METRIC UNITS

To convert column 1 into column 2, multiply by	Column 1	Column 2	To convert column 2 into column 1, multiply by
Length			
0.621	kilometer, km	mile, mi	1.609
1.094	meter, m	yard, yd	0.914
0.394	centimeter, cm	inch, in	2.54
Area			
0.386	kilometer^2, km^2	mile^2, mi^2	2.590
247.1	kilometer^2, km^2	acre, acre	0.00405
2.471	hectare, ha	acre, acre	0.405
Volume			
0.00973	meter^3, m^3	acre-inch	102.8
3.532	hectoliter, hl	cubic foot, ft^3	0.2832
2.838	hectoliter, hl	bushel, bu	0.352
0.0284	liter	bushel, bu	35.24
1.057	liter	quart (liquid), qt	0.946
Mass			
1.102	ton (metric)	ton (U.S.)	0.9072
2.205	quintal, q	hundredweight, cwt (short)	0.454
2.205	kilogram, kg	pound, lb	0.454
0.035	gram, g	ounce (avdp), oz	28.35
Pressure			
14.50	bar	lb/inch^2, psi	0.06895
0.9869	bar	atmosphere, atm	1.013
0.9678	kg(weight)/cm^2	atmosphere, atm	1.033
14.22	kg(weight)/cm^2	lb/inch^2, psi	0.07031
14.70	atmosphere, atm	lb/inch^2, psi	0.06805
Yield or Rate			
0.446	ton (metric)/hectare	ton (U.S.)/acre	2.24
0.892	kg/ha	lb/acre	1.12
0.892	quintal/hectare	hundredweight/acre	1.12
Temperature			
$\left(\dfrac{9}{5}\,°C\right) + 32$	Celsius $-17.8C$ 0C 100C	Fahrenheit 0F 32F 212F	$\dfrac{5}{9}\,(°F - 32)$
Water Measurement			
8.108	hectare-meters, ha-m	acre-feet	0.1233
97.29	hectare-meters, ha-m	acre-inches	0.01028
0.08108	hectare-centimeters, ha-cm	acre-feet	12.33
0.973	hectare-centimeters, ha-cm	acre-inches	1.028
0.00973	meters^3, m^3	acre-inches	102.8
0.981	hectare-centimeters/hour, ha-cm/hour	feet^3/sec	1.0194
440.3	hectare-centimeters/hour, ha-cm/hour	U.S. gallons/min	0.00227
0.00981	meters^3/hour, m^3/hour	feet^3/sec	101.94
4.403	meters^3/hour, m^3/hour	U.S. gallons/min	0.227

Plant Nutrition Conversion—P and K

P (phosphorus) $\times$ 2.29 = P_2O_5
K (potassium) $\times$ 1.20 = K_2O

chapter 1

Huge machines windrow and aereate sludge and woodchips at USDA's test site in Belts-ville, Md. Scientists are experimenting with digested and undigested municipal sewage sludge of varying wood-to-sludge ratios (Photo courtesy of USDA).

Introduction

JESSE LUNIN, USDA-ARS National Programs Staff, Beltsville, Maryland

During the past few years, a succession of symposia and conferences have been oriented toward waste management. Some have dealt rather narrowly with one specific waste management problem; others have addressed the subject in rather broad terms. One common thread runs through most of these meetings—a strong emphasis on the role of land for recycling organic wastes and waste waters. Many municipal, State and Federal agencies, as well as certain industries and private operations, are already involved in this effort. All are looking for advice and guidelines to most effectively use these wastes. Since soil is the recipient in any land-application system, it is logical that the expertise available within the Soil Science Society of America be called upon to provide the most current information available. The organizing committee is to be commended for initiating this symposium.

When attending conferences on waste management or while reading reports on this subject, one is often overwhelmed by the recital of statistics to indicate the magnitude of the problem. Numbers are cited for total tonnages produced annually, total nutrient content of various wastes on an annual basis, population equivalents, and total organic loadings. For any given category, the numbers vary widely because it is hard to make a reasonable estimate when there is such wide variability in waste characteristics and in the data available.

Steffgen[1] recently stated that, "The materials we refer to as organic wastes are merely those which are not put to use in our existing technological system. Once we begin to use them, they will no longer be called wastes, and if they are in demand, we may even seek to increase their production. Organic wastes are really resources out of place." This should be an important consideration in this symposium. Although we will continue to refer to organic wastes as a matter of convenience, we must continually be aware that these are potential resources.

Farmers historically have applied animal manure to the soil. Human wastes, both treated and untreated, also have been applied to the land for crop production. We know that incorporation of crop residues can be beneficial. During the last decade or so, the outlook on waste management has changed drastically. Concentrated animal production units result in large amounts of wastes in relatively small areas, which complicates handling those wastes. Dense populations in metropolitan and suburban areas are producing huge quantities of sewage and garbage. Recent environmental constraints have dictated that much of these wastes will have to be recycled through the soil.

Three major current concerns must be considered in assessing waste management problems and solutions. The first deals with environmental quality. Wastes must be managed so that they do not pollute air or soil-water resources, nor should they be used in any way that might introduce a toxic or pathological component into the food chain.

The energy crisis has added a second dimension. Organic wastes are now

---

[1]F. W. Steffgen. 1974. Energy from agricultural products. p. 23–35. *In* A new look at energy sources. ASA Special Publication No. 22. Am. Soc. of Agron., Madison, Wis.

regarded as potential energy sources. Waste-management practices are now viewed in terms of energy consumption as well as economic and social considerations. For example, until recently, incineration was considered a viable alternative to land application for municipal sludge disposal. Shortages and high costs of fuel have reversed this situation and, again, land application appears to be the most logical option. The energy crisis has contributed to a shortage of fertilizer, particularly nitrogen. Many are now asking to what extent organic wastes can replace chemical nitrogen requirements.

We have recently added the "food crisis" to our list of concerns. We are concerned both with current food shortages in the developing countries and future worldwide shortages brought about by a rapidly increasing population. There is a challenge to more effectively use organic wastes in food production and to conserve the soil resource that produces our food and fiber.

We have a good knowledge base related to management of organic wastes and waste waters. We need to build on this base to provide the necessary information for developing efficient management systems compatible with current and projected future constraints.

The burst of research effort on waste management started with animal wastes. This was followed by a major effort on sewage effluents and sludges. Some research is being done on solid wastes and effluents from the processing of agricultural products. This is an area that needs additional support. Considerable information is available on the management of crop residues, but many researchable problems remain. For example, increased mechanical harvesting and a trend toward preliminary processing of agricultural products in the field have introduced new aspects to the management of plant residues. The use of organic wastes in no-till and minimum-till planting and multiple and double cropping systems presents many researchable problems. Finally, we may have to give greater consideration to land application of garbage from large communities and cities, either unprocessed or as a compost.

The research effort on land application of organic wastes and waste waters is sizeable. A printout from the Current Research Information System (CRIS) was obtained on 2 Jan. 1975, for all projects dealing with this subject. This covers research within the U. S. Department of Agriculture and the State Agricultural Experiment Stations that is in progress or recently terminated. In the area of wastes produced from processing of agricultural products, there are 10 projects in 8 states. For municipal wastes, the USDA has 7 projects in 5 states and the State Agricultural Experiment Stations have 54 projects in 22 states. In all, research on land application of municipal wastes is being carried on in 26 states. The effort on animal wastes is much greater. The USDA has 26 projects in 12 states, while the State Agricultural Experiment Stations have 194 projects in 39 states. This is indeed a sizeable effort in a very important problem area. When the total effort on projects related to this symposium is considered, two things become apparent: first, we are accumulating a large body of data (and hopefully knowledge), and second, there is a need for improved research coordination. Knowledge gained will depend upon the synthesis of ideas and concepts from these data.

We must look at land application systems for organic wastes and waste waters for several purposes. Any one system may be designed for a single objective or a combination of objectives. One objective is the use of organic wastes solely for soil improvement: to improve soil structure, provide nutrients, create more favorable soil water regimes, or otherwise provide a more favorable medium for plant growth. Crop residues and other wastes have been used effectively as a mulch to prevent soil erosion. Rates of application must be determined with this objective in mind. In contrast to this is the use of the soil as a disposal medium. In this situation, maximum loading rates need to be determined, based on the capabiltiy of the soil to absorb these wastes without creating a pollution hazard or destroying the soils future productive capacity. These wastes have another use in the reclamation of drastically disturbed areas. Of principal interest are construction sites and strip mine spoils. In these situations a different set of criteria may be necessary for application rates and incorporation methods.

Waste waters require different considerations. These may be used as a valuable source of irrigation water in certain water-short areas. In more humid areas, effluents may be used for supplemental irrigation for crop production, but they must be stored or disposed of during wet periods. Land application also may be regarded solely as a disposal system. On the other hand, soil systems can be used to renovate waste waters. The use of rapid infiltration ponds for ground-water recharge is very successful under certain conditions. Renovation through overland flow through grassed areas is a possibility that merits further investigation. Hydrologic considerations are important in most land application systems involving waste waters.

This symposium deals largely with the soil as a medium for disposal or use of wastes. We recognize the tremendous variability in soils and their related topographic and geologic factors throughout the United States. Consequently we must consider the great variability in physical and chemical characteristics of the wastes, the large differences in climatic factors, and the many vegetative alternatives. An effective management system needs to integrate all of these variables to optimize use of our resources and meet the stated objectives. Most of the speakers at this symposium are research scientists and specialists in their specific subject matter areas. Multidisciplinary teams will be needed, which must include engineering and economic expertise, to develop these management systems.

Certain constraints must be taken into account in developing waste management systems. One of the major concerns relates to the Federal Water Pollution Control Act Amendments of 1972 (PL 92-500). This act calls for the development of management practices to control water pollution from nonpoint sources such as agriculture. This includes, of course, practices for land application of organic waste waters. Sediment, nitrate, phosphates, soluble salts, and dissolved organics are of primary concern here. Although this act deals largely with surface waters, consideration also must be given to minimizing pollution hazards to ground water. With regard to municipal wastes, however, the act specifically encourages the development of facilities that will recycle potential sewage pollutants by using sewage sludge in the production of agricultural products.

Other constraints deal with pathogen hazards, the introduction of heavy metals into the food chain, and adverse effects on plant growth and quality.

Economic factors must be taken into account, and these change from time to time. For example, manure from feedlots was not in great demand for a long time and disposal was a problem. The energy crisis created a shortage of nitrogen fertilizer and prices rose sharply in 1974. As a result, many feedlots now have a ready market for the manure. Numerous inquiries are received regarding the use of organic wastes to replace chemical fertilizers; perhaps we can learn to use them more effectively. The fuel problem has altered the sludge picture. Energy requirements have made dehydration and incineration of sludge much less desirable options than disposal on land. Logistics, pretreatment, and application problems should also be considered. Costs of land application systems must be evaluated both in terms of benefits derived and in comparison with other feasible disposal or utilization practices.

We must not overlook sociological factors. While food products grown on soils supplemented with farm manure have always been readily accepted, there is some aversion in the urban sector to the use of sewage for crop production. Some may even object to the use of sewage effluents to produce fiber crops. Then, there are those that favor land application of these wastes as long as they are not applied in their neighborhood. One major concern in the management of any waste is odor. While some of the public's apprehensions are unjustified, odor problems do exist. In developing practices for land application of wastes, we must remember that public acceptance is an important factor.

Legal constraints, such as zoning restrictions and the concept of "private nuisance," could complicate the development of land application systems. When dealing with waste waters, both surface and ground-water rights must be considered. Practices that adversely affect the quality, quantity, or natural flow patterns of existing water resources could cause legal problems. Adequate initial planning can avoid some of these. This emphasizes the need for good research information that will enable valid predictive capabilities in the initial planning stages.

Nevertheless, the Environmental Protection Agency has stated that land application is indeed a viable alternative municipal waste management system. In fact, that agency believes that if land treatment can be demonstrated to be the most cost effective alternative, is consistent with the environmental assessment, and other aspects satisfy applicable tests, it should be used in lieu of other systems of waste treatment.

Many aspects of the topics I have mentioned will be discussed in the presentations to follow. However, a few points should be stressed and kept in mind. First, these presentations represent the current state of the art in the various related subjects. As the social and economic considerations change, research must be flexible enough to provide information to meet changing needs. Second, we tend to be discipline-oriented. We need to provide a multidisciplinary team approach that can produce integrated management systems designed to meet specific objectives. Along these same lines, great emphasis is being placed on technology transfer. Research scientists have an obligation to see that research results are packaged in a usable form and made available to potential users. Improved communications and coordination of efforts are needed among research, action, and regulatory agencies. This is obvious from the statistics presented earlier on current research efforts. This symposium is an effective means of communications for a specific subject. Coordination can be greatly improved. In one or two instances, a Regional Research Committee has been quite effective, but there should be a greater effort made on a national basis. Finally, we must bear in mind that many Federal and State regulatory agencies will be requiring recommendations and guidelines for design criteria and control measures. It is unrealistic to think that we can ever reach a consensus among a large group of scientists. We should be in a position, however, to provide a sound rationale for decision making.

chapter 2

Sludge destined for composting rolls off the vacuum filters and is checked for consistency at the Blue Plains wastewater treatment plant near Washington, D. C. Samples are taken daily and analyzed for acidity, chemical content, and bacteria in compliance with local, State, and Federal laws (Photo courtesy of USDA).

Properties of Agricultural and Municipal Wastes

T. M. MC CALLA, USDA-ARS, Lincoln, Nebraska

J. R. PETERSON and C. LUE-HING
Metropolitan Sanitary District of Greater Chicago, Chicago, Illinois

I. INTRODUCTION

Waste has been defined as "useless, unwanted, or discarded material" (USEPA, 1972a). However, the soil has enormous capacity to adsorb and transform waste by microbial activity into useable nutrients for plant growth.

Solid waste can be classified as: (i) *residential*–waste normally originating in residential areas (domestic solid waste); (ii) *municipal*–residential and commercial solid waste generated by a community; (iii) *industrial*–waste from industrial processes and manufacturing; (iv) *commercial*–waste generated by stores, offices, and other activities that do not actually produce a product; (v) *institutional*–waste from educational, health-care, and research facilities; and (vi) *agricultural*–waste from raising, slaughtering, and processing animal products, forest waste, orchard and field crop residues, and food-processing wastes (USEPA, 1965, 1972b; Hart and McGaughey, 1966; Studdard, 1973).

Agricultural wastes produced annually in the United States include (in approximate metric tons): animal, 1.8 billion; agricultural food-processing waste and crop residues, 0.45 billion; and logging debris, 22.6 million.

Annual production of sewage sludge solids in the United States is currently estimated at 16 million metric tons, a major portion of which is from large metropolitan areas; e.g., Chicago produces 300,000 metric tons/year (Peterson et al., 1971).

Annual production of municipal refuse in the United States is 113 million metric tons (USEPA, 1974). Considerable separation and processing are essential before municipal refuse can be incorporated into cropped soils.

Solid waste is deposited on land and in water, sometimes with serious environmental problems. However, waste may be utilized as a resource. Major concerns of waste include problems and benefits of land disposal.

The purpose of this chapter is to discuss the properties of agricultural and municipal wastes as they relate to land application.

II. PROPERTIES OF AGRICULTURAL AND INDUSTRIAL PROCESSING WASTE

A. Animal Wastes

1. CHEMICAL

Sources of animal waste are beef and dairy cattle, chickens, swine, turkeys, ducks, and other types of animals. Major sources are beef feedlots and confined housing for dairy cattle, chickens, and swine. The mean and range of chemical analyses of samples taken from 23 outdoor beef cattle feedlots in Texas are shown in Table 1. The concentration of plant nutrient elements is highly variable, depending upon ration, collection, storage, and handling (Gilbertson et al., 1970, 1971). The total amount of all mineral elements, except C and N will remain constant with time. As manure is de-

Table 1—Chemical analyses of manure samples taken from 23 feedlots
(Mathers et al., 1973)

Nutrient	Range	Average	Amount in 10 metric tons
	— % —		kg
N	1.16 - 1.96	1.34	134
P	0.32 - 0.85	0.53	53
K	0.75 - 2.35	1.50	150
Na	0.29 - 1.43	0.74	74
Ca	0.81 - 1.75	1.30	130
Mg	0.32 - 0.66	0.50	50
Fe	0.09 - 0.55	0.21	21
Zn	0.005- 0.012	0.009	0.9
H_2O	20.9 -54.5	34.5	3,450

composed, part of the C is given off as CO_2, and some N is mineralized. To understand N, let us look at the nature of the nitrogenous compounds which exist chiefly in two forms: (i) protein aqueous substances that have resisted attack by digestive juices and intestinal bacteria and, presumably, are resistant to further biological decomposition in the manure heap or soil; and (ii) microbial protein in the cells of living and dead bacteria (Salter & Schollenberger, 1939). One-half to two-thirds of the N is present as microbial protein which has been shown to be readily attacked by soil microorganisms and has an N availability similar to that of dried blood and cottonseed meal. However, animal feces contain a rather large amount of lignin or lignin-like substances that have apparently combined with protein to form complexes resistant to biological decomposition and, hence, possess low N availability for plants. Crops seldom recover more than 2% of the soil N from humus in a single season. The low availability of N in solid manure is probably partly the result of the combination of proteinaceous substances with lignin or lignin-like substances. Availability of other nutrients also depends upon mineralization rate.

Humic substances may also complex with mineral nutrients, thereby keeping them available to plants (McCalla, 1972, 1974a, 1974b; McCalla & Norstadt, 1974; McCalla & Viets, 1969; McCalla et al., 1972).

Table 2 lists the estimated quantities of nutrients and energy in kg/ 1 metric ton of dry beef animal waste. Some of the trace elements and materials, including crude protein, fat, and fiber, are energy sources for microbial activity (Gilbertson et al., 1974). Live microorganisms and the humus in manure may improve soil structure, thereby increasing aeration, water intake, and stability of soil aggregates (Azevedo & Stout, 1974).

Of all the nuisances related to manures, odor is perhaps the most readily noticeable but least definable and most difficult to control (Azevedo & Stout, 1974). Although manure odors can be characterized chemically, odor nuisance judgments are often subjective.

The odor of freshly voided or dried livestock manure is inoffensive to most people, but odors produced by anaerobic bacterial activity during

Table 2—Estimated quantity of nutrients and energy in 1 metric ton of animal waste†
(Gilbertson et al., 1974)

	Ration roughage level				
	High feces‡	Medium feces‡	Low feces‡	Low manure‡	Low waste‡
	kg/metric ton (dry basis)				
Total solids	1,000	1,000	1,000	1,000	1,000
Volatile solids	869	894	939	858	245
Digestible solids	135	172	577	544	--
Energy (M calories)§	1,971	2,054	2,113	1,972	445
Crude protein	82	150	154	200	--
Crude fat	26	33	31	22	--
Crude fiber	303	241	121	124	--
Total N	13.10	24	24.70	32	10.50
Total P	10.20	12	6.60	7.70	0.80
Total C	329	293	326	404	--
K	5.27	7.94	3.64	14.60	2.89
Ca	4.67	5.07	3.67	5.43	1.66
Na	1.64	0.89	1.87	3.29	1.10
Mg	1.46	2.84	3.54	2.91	0.89
Mn	0.27	0.09	0.17	0.12	0.05
Fe	7.90	0.80	1.78	1.39	1.32
Zn	0.12	0.10	0.34	0.14	0.01
Cu	0.05	0.02	0.03	0.01	0.00

† One metric ton of dry solids was estimated as equivalent to 4.2, 5.7, 4.1, 5.1, and
2.4 metric tons wet weight for high-, medium-, and low-roughage ration feces,
housed feedlot manure, and outdoor feedlot waste, respectively.
‡ Feces were collected from animals confined in metabolism crates. Manure was feces,
urine, and dilution water collected from animals confined in a housed feedlot. Waste
was removed from outdoor, unpaved feedlots during cleaning.
§ M calories = calories $\times 10^6$ = 1 megacalorie.

fermentation of wet manure can be astoundingly offensive. Using chromatographic techniques, White et al., (1971) tentatively identified hydrogen sulfide, methanethiol, dimethyl sulfide, diethyl sulfide, propyl acetate, n-butyl
acetate, trimethylamine, and ethylamine as fermentative decomposition
products from dairy manure. Dimethyl sulfide caused the principal sensory
odor in anaerobic dairy wastes, but aeration reduced or eliminated the odorous sulfide compounds. Amines cause the major odorous compounds around
outdoor cattle feedlots (Azevedo & Stout, 1974). When applying animal
waste to land, practices must be used to minimize odor.

As shown in Table 3, animal waste contains considerable amounts of
energy, ranging from 8,900 to 18,500 Btu/kg manure, most of which is available for microbial activity. In some countries, manure is used as fuel.

2. BIOLOGICAL

Animal wastes have great numbers of different microorganisms. In
ruminant animals, like beef cattle, the feed is digested by microorganisms.
Table 4 shows the changes in the number of microorganisms in beef waste

Table 3—Heat value of animal manures (moisture-free basis) (Azevedo & Stout, 1974)

Type of manure	Btu/kg†	Ash content, %
Chicken	13,836	29.3
	13,655	29.3
	10,930	36.1
	12,813	35.9
Dairy	16,544	36.0
	16,355	17.0
	9,887	49.9
	17,780	14.2
	17,367	18.0
Beef	15,006	27.4
	14,606	34.5
	8,974	51.9
	18,548	11.3
	13,543	32.3
Turkey	12,738	44.0
	12,797	43.6
Horse	15,365	24.6
Swine	16,078	27.7
Sheep	16,865	26.3

† For comparison purposes, coal has a Btu/kg rating of 17,820 to 33,000; oakwood has a heat value of 18,304 Btu/kg.

Table 4—Number of microorganisms/g (dry basis) in beef cattle wastes before and after incubation in the absence of soil (McCalla & Viets, 1969)

Days of incubation	Bacteria $\times 10^7$	Fungi $\times 10^3$	Bacilli $\times 10^4$	Clostridium $\times 10^4$	E. coli $\times 10^8$	Enterococci $\times 10^4$
0	13	1	0.07	--	5.4	100
14	2,000	252	145	--	3.4	198
28	301	403	393	--	0.5	38
86	14	65	353	26	<0.01	7
119	14	77	127	2	<0.01	6
179	1	200	173	0.6	<0.01	1

after incubation in the absence of soil (McCalla & Viets, 1969). Bacteria, fungi, and bacilli increased greatly after 14 days of incubation and then decreased, whereas *Escherichia coli* and the coliforms decreased rather rapidly, but enterococci increased after 14 days, then decreased rapidly.

Rhodes and Hrubant (1972) found that beef feedlot waste contained, in terms of viable count per gram dry weight, 10^{10} total organisms, 10^9 anaerobes, 10^8 gram-negative bacteria, 10^7 coliforms, 10^6 sporeformers, and 10^5 yeasts, fungi, and streptomycetes. Hrubant et al. (1972) indicated that a wide spectrum of enterobacteria in feedlot waste (10^7 enterobacteria/g dry feedlot waste) may be a hazard if unsterilized waste is refed. The specific numbers and distribution pattern of these groups of organisms varied only slightly with time during the study in spite of wide variation in weather.

Besides the large number of saprophytic organisms that may occur in animal waste, several disease organisms may be present and transmitted by animal waste material, as shown in Table 5. The disease organism can be a bacterium, virus, protozoan, rickettsial fungus, or larger parasite. While the list is long and formidable, actually many of the disease organisms do not persist for very long periods because of the unfavorable soil environment. However, some disease organisms do persist. Thus, care should be observed in handling and utilizing manure on land. Composting manure may raise the temperature sufficiently for pasteurization to occur. (Further discussion on pathogens is given in Chapter 21.)

3. PHYSICAL

The particle-size analysis of typical feedlot manure is shown in Table 6. This indicates that about 70% of the manure particles are $> 2\ \mu m$. Table 7 shows the particle-size analyses of animal waste from high-roughage rations (HRR) and high-concentrate rations (HCR) feces. The volatile solids (materials lost in ignition at 600C) retained on each sieve decreased with particle size. Solids from the HRR feces were less volatile than the HCR feces at each sieve level.

Table 8 (Frecks & Gilbertson, 1974) shows engineering parameters for feces from cattle fed high-roughage rations (HRR) and high-concentrate rations (HCR). Parameters include shrinkage limit, shrinkage ratio, volumetric change, lineal shrinkage, porosity, void ratio, air voids, liquid limit, particle density, and wet bulk density. The shrinkage limit of the HRR feces occurred at a moisture content 1.5 times greater than that of the HCR feces. Particle and wet bulk densities of the feces were not affected by the ration. A linear regression equation was developed for estimating slump (S) as a function of total solids content (TS) wet-weight basis. The equations were $S = 85.98 - 2.40$ TS for the HCR feces and $S = 62.60 - 3.23$ TS for the HRR feces. The equation for HRR feces was linear within the range of 14 to 18% total solid content. The liquid limit was 78.1% (wet basis) for the HRR feces and 69.9% (wet basis) for the HCR feces, similar to organic clays and highly elastic silts and silty clays (Frecks & Gilbertson, 1974).

This type of information (Tables 6 and 8) is valuable in designing materials handling equipment for collecting, storing, and applying animal waste to the land.

B. Crop Plant Wastes

The approximate quantities of nutrients contained in various crops are shown in Table 9. This indicates that nutrient concentrations vary widely, depending upon the type of material. In utilizing plant materials on land, careful consideration must be given to the crop nutrient needs. Supplemental

Table 5—Diseases potentially transmitted by animal manures (Azevedo & Stout, 1974)

Type of disease	Etiologic agent	Common domestic reservoirs	Method of entry
		Bacterial	
Salmonellosis	*Salmonella* sp.	Most domestic animals & man	Ingestion
Leptospirosis	*Leptospira pomona*	Most domestic animals & man (urine)	Cutaneous & facial membranes
Anthrax	*Bacillus anthracis*	Most domestic animals & man (urine)	Cutaneous, inhalation, ingestion
Tuberculosis	*Mycobacterium tuberculosis*	Man, swine	Inhalation, ingestion
	M. avium	Poultry, swine	
	M. bovis	Cattle, swine, man	
Johnes Disease (Paratuberculosis)	*Mycobacterium tuberculosis*	Cattle, sheep, goats	Ingestion
Brucellosis	*Brucella abortus*	Cattle, swine, goats, man	Ingestion, inhalation, body openings
	B. melitensis	Goats, swine, cattle, man	
	B. suis	Swine, goats, cattle, man	
Listeriosis	*Listeria monocytogenes*	Cattle, sheep, swine, chickens, man, horse	Ingestion (?)
Tetanus	*Clostridium tetani*	Horse, sheep, man	Deep cutaneous
Tularemia	*Pasturella tularensis*	Sheep, rabbits, man	Ingestion, inhalation, biting anthropods
Erysipelas	*Erysipelothrix rhusiopathiae*	Swine, turkeys, man, decaying organic matter	Ingestion, cutaneous
Colibacilosis	*E. coli* (some serotypes)	Most domestic animals & man	Ingestion, especially in newborn
Coliform mastitis-metritis	*E. coli* (some serotypes)	Cattle	Body openings

(continued on next page)

Table 5. Continued

Type of disease	Etiologic agent	Common domestic reservoirs	Method of entry
		Rickettsial	
Q fever	*Coxiella burneti*	Cattle, sheep, goats, man (vaginal discharges)	Dust-inhalation, ingestion
		Viral	
Newcastle	Virus	Avian species, man	Contact, dust
Hog cholera	Virus	Swine	Contact, ingestion
Foot and mouth	Virus	Cloven-foot animals, rarely man	Contact, ingestion
Psittacosis (ornothosis)	Virus	Avian species and man	Contact, dust-inhalation
		Fungal	
Coccidioidomycosis	*Coccidioides immitus*	Cow, hog, horse, sheep, man, free-living	Dust-inhalation, cutaneous
Histoplasmosis	*Histoplasma capsulatum*	Cow, horse, hen, man, free-living	Dust-inhalation, some ingestion
Ringworm	Various *Microsporum & Trichophyton*	Most domestic animals & man	Contact
		Protozoal	
Coccidiosis	*Eimeria* sp.	Most domestic animals	Ingestion of oocytes
Balantidiasis	*Balantidium coli*	Swine (normal), man	Ingestion of oocytes
Toxoplasmosis	*Toxoplasma* sp.	Many domestic animals & man	Uncertain
		Parasitic	
Ascariasis	*Ascaris lumbricoides*	Swine, man	Ingestion
Sarcocystiasis	*Sarcocystis* sp.	Most domestic animals & man	Ingestion

Table 6—Particle-size analysis of fresh manure (oven-dry weight)
(McCalla, unpublished data)

Particle size	Percent of total
1.12 mm or greater	2.45
<1.12 mm but >2,000 μm	25.09
<2,000 μm but >500 μm	36.69
<500 μm but >210 μm	4.75
<210 μm but >2 μm	1.01
<2 μm	30.02

Table 7—Volatile solids retained at each sieve level for feces from cattle fed high-concentrate and high-roughage rations (Frecks & Gilbertson, 1974)

Sieve size	High-roughage ration feces					High-concentrate ration feces			
	Mean†	Range		Standard deviation		Mean‡	Range		Standard deviation
		Low	High				Low	High	
mm					%				
2.0	94.7	93.5	95.7	0.91		99.3	99.2	99.4	0.07
1.0	95.0	94.8	95.2	0.14		98.7	98.7	98.8	0.03
0.50	94.0	93.4	94.5	0.47		97.9	97.8	97.9	0.06
0.25	90.6	89.7	91.2	0.67		93.3	92.7	93.9	0.58
0.105	86.0	85.7	86.2	0.20		96.3	95.9	96.7	0.41
0.053	77.5	74.8	81.3	2.57		90.6	88.9	92.2	1.66
<0.053	76.5	75.9	77.4	0.52		85.4	85.1	86.0	0.47

† n = 6.
‡ n = 3.

fertilization, especially N, may be desirable and necessary to decompose crop residues and to release the minerals for plant growth.

The decomposition of soil organic matter depends mainly on the activities of microorganisms. Plant residues returned to the soil, naturally or artificially, furnish the main source of energy and food for microorganisms, and when those activities are the source of nutrients for plants, microbial activity is frequently the determining factor in crop production.

The C/N ratio of plant residues varies from about 80:1 in straw to about 20:1 in legume materials; the C/N ratio of microorganisms is frequently < 10:1 (Table 10). To carry on their life processes, microorganisms must break down and utilize plant residues. This breaking down process tends to continue until most of the degradable portion of plant residue is utilized in the growth and reproduction of these organisms. If the point is reached when all crop residues returned to the soil are utilized by microorganisms, the C/N ratio of soil organic matter would be similar to that of microorganisms. This is the narrowest ratio at which C and N could exist in the soil and, therefore, the C/N ratio where soil organic matter becomes comparatively stable. This narrow ratio is rarely, if ever, reached in virgin soils because of the continuous return of plant residues with a wider C/N ratio, and soil conditions are

Table 8—Calculated engineering parameters for feces from cattle fed high-roughage and high-concentrate rations (Frecks & Gilbertson, 1974)

Parameter	High-roughage feces				High-concentrate feces			
	Mean	Range		Standard deviation	Mean	Range		Standard deviation
		Low	High			Low	High	
Shrinkage limit, % dry basis†	364.5	263.1	366.6	1.86	245.6	238.0	250.0	6.62
Shrinkage ratio†	0.30	0.28	0.31	0.015	0.48	0.46	0.50	0.021
Volumetric change, %†	7.4	8.8	5.7	1.72	55.5	61.9	50.0	5.99
Lineal shrinkage, %	2.3	2.8	1.8	0.52	13.7	14.8	12.6	1.10
Porosity, %	81.6	77.1	86.7	3.35	78.8	74.8	80.5	2.35
Void ratio	4.3	3.4	4.9	0.67	3.8	3.0	4.1	0.48
Air voids, %	79.5	77.9	81.7	2.02	69.0	68.0	70.8	1.51
Liquid limit, % wet basis	78.1	--	--	--	69.9	--	--	--
Particle density, g/cm³	1.53	1.49	1.57	0.022	1.50	1.46	1.54	0.231
Wet bulk density, g/cm³†	1.09	1.05	1.13	0.033	1.10	1.05	1.12	0.036

† Based on moisture content of the raw feces which was 72.07% for the high-concentrate feces samples and 74.15% (wet basis) for the high-roughage ration feces samples used in this experiment.

Table 9—Approximate nutrients contained in portion of crop of the size shown (Garman et al., 1962)†

Crop	Yield	N	P	K	Ca	Mg	S	Cu	Mn	Zn
	ha					kg/ha				
Barley (grain)	2,150	39	7	9	1	2	3	0.03	0.03	0.07
Barley (straw)	2,240	17	2	28	9	2	4	0.01	0.36	0.06
Corn (grain)	9,408	151	26	37	18	22	16	0.07	0.10	0.17
Corn (stover)	10,080	112	18	135	31	19	11	0.06	1.68	0.34
Oats (grain)	2,867	56	10	14	2	3	6	0.03	0.13	0.06
Oats (straw)	4,480	28	7	74	9	9	10	0.03	--	0.32
Rice (rough)	4,032	56	10	9	3	4	3	0.01	0.09	0.08
Rice (straw)	5,600	34	5	65	10	6	--	--	1.77	--
Rye (grain)	1,882	39	5	9	2	3	8	0.02	0.25	0.03
Rye (straw)	3,360	17	4	23	9	2	3	0.01	0.16	0.08
Sorghum (grain)	3,763	56	12	14	4	6	6	0.01	0.04	0.04
Sorghum (stover)	6,720	73	10	88	32	20	--	--	--	--
Wheat (grain)	2,688	56	12	14	1	7	3	0.03	0.10	0.16
Wheat (straw)	3,360	22	2	33	7	3	6	0.01	0.18	0.06
Hay - alfalfa	8,960	202	20	167	125	24	21	0.07	0.49	0.47
bluegrass	4,480	67	10	56	18	8	6	0.02	0.34	0.09
coastal bermuda	17,920	207	34	251	66	27	--	0.24	--	--
cowpea	4,480	134	12	74	62	17	15	--	0.73	--
peanut	5,040	118	12	88	50	19	18	--	0.26	--
red clover	5,600	112	12	93	77	19	8	0.04	0.60	0.40
soybean	4,480	101	10	46	45	20	11	0.04	0.52	0.17
timothy	5,600	67	12	88	20	7	6	0.03	0.35	0.22

(continued on next page)

Table 9. Continued

Crop	Yield	N	P	K	Ca	Mg	S	Cu	Mn	Zn
Fruits & Vegetables										
Apples	26,320	34	5	42	9	6	11	0.03	0.03	0.03
Beans, dry	2,016	84	12	23	2	2	6	0.02	0.03	0.07
Cabbage	44,800	146	17	121	22	9	49	0.04	0.11	0.09
Onions	16,800	50	10	37	12	2	20	0.03	0.09	0.35
Oranges	62,720	95	15	130	37	13	10	0.22	0.07	0.27
Potatoes (tubers)	26,880	90	15	139	3	7	7	0.04	0.10	0.06
Spinach	11,200	56	7	28	13	6	4	0.02	0.11	0.11
Sweet potatoes	18,480	50	7	70	4	10	7	0.03	0.07	0.03
Tomatoes (fruit)	44,800	134	20	149	8	12	16	0.08	0.15	0.18
Turnips (root)	22,400	50	10	84	13	7	--	--	--	--
Other Crops										
Cotton (seed & lint)	1,680	45	10	14	2	4	2	0.07	0.12	0.36
Cotton (stalks, leaves & burs)	2,240	39	5	33	31	9	--	--	--	--
Peanuts (nuts)	2,800	101	5	14	1	3	7	0.02	0.01	--
Soybean (grain)	2,688	168	17	51	8	8	4	0.04	0.06	0.04
Sugarbeets (roots)	33,600	67	10	46	37	27	11	0.03	0.84	--
Sugarcane	67,200	107	27	251	31	27	27	--	--	--
Tobacco (leaves)	2,240	84	7	112	84	20	16	0.03	0.62	0.08
Tobacco (stalks)	--	39	7	46	--	--	--	--	--	--

† These figures will vary with soil type, season, and soil fertility.

Table 10—Classification of common crop residues on the basis of their C/N ratios
(Sievers & Holtz, 1926)

Residues	C	N	C/N ratio
		%	
High in both C and N			
Legume residues			
Legume green manures			
Young cereal green manures	40	2.5	16/1
Low C and high N			
Composted strawy manure			
Animal feces	25	3.0	8.3/1
High C and low N			
Straw			
Stubble			
Strawy manure			
Mature cereal green manures			
Leaves, sawdust, or wood shavings	40	0.5	80/1
Low C and low N			
Composted straw	25	0.8	31.3/1

rarely optimum for decomposition for any considerable time. When little effort is made to return crop residues to the soil and conditions are optimum for decomposition of organic matter, this C/N ratio can be narrowed quite rapidly and, thus, approach the theoretical minimum (Sievers & Holtz, 1926).

Other substances that should be considered in land application of plant residues are lignin and cellulose and substances in plant material which inhibit microbial and plant growth (Rice, 1974). Sweetclover residues (*Melilotus* sp.) contain appreciable quantities of water-soluble substances, primarily coumarin, that depressed corn germination and seedling growth (McCalla & Norstadt, 1974). Most crop residues contain water-soluble substances that inhibited germination and seedling growth of wheat, corn, and sorghum. Several phenolic acids—ferulic, *p*-coumaric, syringic, vanillic, and *p*-hydroxy-benzoic—were found in corn (*Zea mays* L.), wheat (*Triticum aestivum* L.), sorghum (*Sorghum vulgare* Pers.), and oat (*Avena sativa* L.) residues. Concentrations of *p*-coumaric acid were greater than those of the other acids (Guenzi & McCalla, 1966).

C. Forest Waste

The mineral content of different tree parts may vary greatly, with the stems being low and the leaves much higher (Table 11). The ash composition of different parts of trees, as shown in Table 12, varies considerably. Application of woody material to the land will require considerable attention to chemical composition, depending on whether the woody material is being used as a plant nutrient source, soil conditioner, energy source for microbial activity, or to influence physical properties of the soil.

Allison (1965), in a decomposition study, showed that most softwood

Table 11—Ash content of parts of various trees (Lutz & Chandler, 1947)†

	Tree species			
Tree part	Aspen (Populus tremula)	Beech (Fagus sylvatica)	Ash (Fraxinus excelsior)	English oak (Quercus robur)
	% dry matter basis			
Stem wood	0.398	0.355	0.361	0.311
Branches	2.275	1.590	1.831	2.426
Bark	3.334	5.857	4.114	5.575
Leaves	8.865	5.142	7.001	4.508

† Reproduced by permission of publisher: John Wiley & Sons, Inc., New York.

Table 12—Composition of ash from parts of various trees (Lutz & Chandler, 1947)†

		Tree species			
Chemical constituent	Tree part	Aspen (Populus tremula)	Beech (Fagus sylvatica)	Ash (Fraxinus excelsior)	English oak (Quercus robur)
		% (based on ash content)			
K	Stem	9.8	12.0	10.9	8.1
	Branches	7.5	9.8	14.6	11.8
	Bark	6.4	4.2	6.9	2.4
	Leaves	15.3	18.1	15.5	18.6
Ca	Stem	50.8	43.0	44.3	54.4
	Branches	55.5	47.5	46.0	49.4
	Bark	52.0	59.5	57.3	62.8
	Leaves	35.4	31.6	28.1	33.6
Mg	Stem	2.3	2.7	3.5	2.2
	Branches	3.2	3.2	1.9	3.8
	Bark	4.3	2.2	1.4	1.2
	Leaves	2.4	4.4	4.9	1.7
Fe	Stem	0.7	1.6	1.3	1.1
	Branches	0.8	1.1	1.1	0.4
	Bark	2.1	0.5	0.8	0.7
	Leaves	1.4	1.6	0.8	2.2
P	Stem	1.9	1.2	3.0	1.6
	Branches	1.4	2.7	3.7	2.0
	Bark	1.4	0.9	1.7	1.3
	Leaves	3.8	3.4	9.9	5.4
Si	Stem	1.3	4.7	1.0	1.1
	Branches	1.0	2.6	0.8	0.8
	Bark	1.0	1.7	0.7	0.4
	Leaves	3.5	4.9	1.2	2.4

† Reproduced with permission of publisher, John Wiley & Sons, Inc., New York.

Table 13—Carbon and N composition of woods and barks of some tree species (Allison, 1965)

Species	Source	Wood			Bark		
		C	N	C/N ratio	C	N	C/N ratio
		%			%		
Softwoods							
California incense cedar (*Libocedrus decurrens*)	California	51.1	0.097	526:1	51.8	0.038	1,363:1
Redcedar (*Juniperus virginiana*)	Virginia	50.8	0.139	366:1	46.0	0.206	223:1
Cypress (*Toxidium distichum*)	North Carolina	50.3	0.057	883:1	47.3	0.324	146:1
Redwood (*Sequoia sempervirens*)	California	49.9	0.060	832:1	48.3	0.060	805:1
Western larch (*Larix occidentalis*)	Montana	48.6	0.180	270:1	49.9	0.161	310:1
Eastern hemlock (*Tsuga canadensis*)	Pennsylvania	48.5	0.106	458:1	51.1	0.060	852:1
Red fir (*Abies magnifica*)	California	48.2	0.227	212:1	49.1	0.259	190:1
White fir (*Abies concolor*)	California	44.8	0.045	996:1	51.8	0.135	384:1
Douglas fir (*Pseudotsuga menziesii*)	Oregon	48.1	0.051	943:1	52.7	0.041	1,285:1
Engelman spruce (*Picea engelmannii*)	Montana	48.5	0.118	411:1	51.2	0.390	131:1
White pine (*Pinus strobus*)	Maine	48.3	0.087	555:1	51.5	0.101	510:1
Shortleaf pine (*Pinus echinata*)	Maryland	45.0	0.130	346:1	51.3	0.128	401:1
Loblolly pine (*Pinus taeda*)	South Carolina	48.7	0.068	716:1	50.9	0.082	621:1
Slash pine (*Pinus elliottii*)	Florida	49.2	0.050	984:1	52.1	0.056	930:1
Longleaf pine (*Pinus palustris*)	Louisiana	49.9	0.038	1,313:1	50.2	0.092	546:1
Ponderosa pine (*Pinus ponderosa*)	Oregon	45.1	0.052	867:1	51.8	0.048	1,079:1
Western white pine (*Pinus monticola*)	Idaho	48.9	0.113	433:1	49.5	0.171	290:1
Lodgepole pine (*Pinus contorta*)	Montana	46.9	0.071	661:1	49.3	0.179	275:1
Sugar pine (*Pinus lambertiana*)	California	50.1	0.124	404:1	51.7	0.166	311:1
Hardwoods							
Black oak (*Quercus velutina*)	Illinois	47.3	0.070	676:1	44.5	0.102	436:1
White oak (*Quercus alba*)	Illinois	46.9	0.104	451:1	41.6	0.129	323:1
Red oak (*Quercus falcata*)	Louisiana	47.4	0.099	479:1	46.2	0.284	163:1
Post oak (*Quercus stellata*)	Louisiana	47.2	0.096	492:1	41.8	0.270	155:1
Hickory (*Carya* sp.)	Arkansas	46.8	0.100	468:1	48.1	0.413	117:1
Red gum (*Liquidambar styraciflua*)	Arkansas	46.7	0.057	819:1	45.1	0.177	255:1
Yellow poplar (*Liriodendron tulipifera*)	North Carolina	47.1	0.088	535:1	47.6	0.351	136:1
Chestnut (*Castanea dentata*)	North Carolina	47.1	0.072	654:1	47.3	0.273	173:1
Black walnut (*Juglans nigra*)	South Carolina†	47.0	0.100	470:1	45.1	0.177	255:1
Averages		48.0	0.093	615:1	48.7	0.174	452:1

† Bark from Maryland.

species have high C/N ratios (Table 13). However, addition of N significantly increased the decomposition of only shortleaf pine and western white pine (refer to Table 13 for scientific names of trees). Likewise, N additions did not significantly increase decomposition of any of the softwood barks. In a few instances, notably white pine and loblolly pine, adding fertilizer N produced a marked decrease in CO_2 evolution. Woods of hardwood species decomposed much more readily than did the woods of the softwood species, except for shortleaf pine. The decomposition of all hardwoods was accelerated by fertilizer N. Hardwood barks were more resistant to microbial attack than were the woods, but their decomposition was nearly three times as great as softwood barks. Some woods, like redcedar and Ponderosa pine, which decomposed to only a limited extent during the first 60 days, yielded considerable CO_2 later; others, like sugar pine, were considerably more resistant. White oak, which decomposed comparatively rapidly initially, showed limited CO_2 release later, as expected, but the total CO_2 released was greater than that for any of the woods or barks studied. Delay in CO_2 release from some woods, especially redcedar, followed by more rapid oxidation later, may indicate initial toxicity of the woods to the microflora. The chief factor controlling decomposition, however, is probably chemical composition, including arrangement of individual molecules. Cellulose, for example, may be attacked readily when in a pure state, but if present in intimate relation with lignin or resinous material, it may decompose slowly and only as these other wood constituents break down.

Allison (1965), in a study with garden peas (*Pisum sativum* L.) grown in soil-sawdust mixtures with adequate nutrients present, showed no significant toxic effects on 22 of the 28 species tested of the woods and barks. Allison (1965) found that California incense cedar wood was very toxic to pea germination and growth, even at the 1 and 2% rates of addition. The bark retarded growth slightly at the 2 and 4% rates. White pine bark was somewhat harmful to germination and very injurious to pea seedlings at the 1 and 2% rates. Redcedar wood and bark had no effect on germination but were slightly inhibitory to pea seedlings at the 2% rate. Yellow poplar bark and Ponderosa and loblolly pine woods slightly injured growth. The harmful effects were less evident on a second crop of peas than on the first, but they usually did not disappear entirely (Allison, 1965). Caution must be exercised in applying forest waste to land.

D. Industrial Processing Wastes

Analysis of the mineral nutrients in various industrial and miscellaneous wastes has shown their wide range in concentration of different mineral nutrients (Table 14). These data indicate that composition of material applied to the land must be obtained before application, and care must be exercised in the amount applied, depending upon plant nutrient needs and the possibility of toxic concentrations of heavy metals and other substances.

Table 14—Moisture and main nutrients in industrial and miscellaneous wastes†‡ (Loehr, 1974)

Material	Details	Moisture	N	P	K	Mg	Ca	No. of samples
					%			
Brewer's waste		6-94	0.9-3.2	0.04-0.18	0.83 (1)	--	0.02 (1)	2
Carbon (activated) waste		48	0.9	0.18	0.02	--	--	1
Castor meal		11	5.5	1.06	1.16	--	--	1
Chicory waste		--	1.5	--	--	--	--	1
Cockleshell		2	0.3	0.04	0.08	--	--	1
Cocoa waste		6-71	0.8-2.9	0.26 -0.66	--	--	10.2 (2)	8
	M		1.8	0.44				
Coconutfiber/matting waste	Dust	48	0.2-0.9 (3)	0.04	0.33 -0.58	--	(1)	3
	M		0.4		0.50			
	Fiber	18 (1)	0.1	(1)	--	--		1
	M							
Coffee waste		42-70	0.7-3.1	0.01 -5.50	0.01 -0.75	--	0.1	6
	M	62	1.5	0.09				
Coffee-chicory residue		66-67	0.8-1.7	0.26 (1)	0.01 (2)	0.04 (1)	(1)	2
Cotton cake		11	4.1-6.9	0.75	1.33 -2.32 (1)	--	--	1
	M		4.8 (3)		-- (2)			
Felt waste		--	13.6	--	--	--	--	1
Flax	Cavings & cleanings	6	--	1.58	7.64	--	--	1
	Residues from retting tanks (dried)	7	0.8	0.09	0.58	--	--	1
	Flax & jute dust	9	1.2	--	--	--	--	1

(continued on next page)

Table 14. Continued

Material	Details		Moisture	N	P	K	Mg	Ca	No. of samples
Ginger root			--	1.8	--	--	--	--	1
Glue factory waste			28–64	1.4–4.7	0.004–0.48	0.03 –0.50	--	--	6
		M	41	2.1	0.09	0.21	--	--	
Hemp waste			66–68	0.7–1.5	--	--	--	--	4
		M	(2)	0.9					
Hops	Spent		9–87	0.6–5.7	0.09 –1.50	0.005–2.16	0.12–0.16	--	9
		M	73	1.1	0.13	0.08	(2)	--	
Leather	Chamois dust		--	8.3	--	--	--	--	1
	Hide meal, ground leather		19–68	0.4–7.5	0.04 –0.17	0.02 –0.25	--	--	4
		M	52	6.1 (7)	0.04	0.12	--	--	
Malt fiber			7 (1)	4.4–5.2	0.70 –0.92	0.91 –1.83	--	--	2
Paper waste			85	0.1	--	--	--	--	1
Sawdust			4–66	0.1–0.9	0.004–0.22	0.03 –1.16	--	--	4
		M	48	0.2	0.03	0.06	--	--	
Tannery waste			6–85	0.1–14.1	--	--	--	--	10
Tobacco waste		M	55	5.3	--	--	--	--	
Woodshavings			71	0.4	0.04	0.75	--	--	1

† Percent in material as received.
‡ The range in composition and the median value M are given where possible; where the number of determinations differs from the total number of samples, the number is given in parentheses below the analysis. (Reprinted with permission of Academic Press, New York, N. Y.)

III. PROPERTIES OF MUNICIPAL WASTES

A. Sewage Sludge

Sewage sludge is obtained from the processing of waste waters from domestic and/or industrial areas. The organic and inorganic matter in waste water are separated by a variety of waste water treatment processes. The waste water usually undergoes a primary treatment, where settleable solids are removed by some form of physical separation, e.g., gravity settling. Next, the waste water is subjected to processes designed to remove dissolved and colloidal material via some form of chemical or biological treatment process. Further treatment of the resultant waste water may include various combinations of physical and chemical treatment. The solids removed during these processes are usually high in organic matter and are biologically unstable. They may be further stabilized by aerobic or anaerobic digestion. Extended storage of the digested sludge also reduces its volatile solids. More information on engineering and design of various sludge handling systems is discussed by Metcalf and Eddy, Inc. (1972).

1. CHEMICAL

Table 15 presents data which illustrate the chemical variability of sewage sludge from select cities in five midwestern and two eastern states of the United States. These results indicate high total N content in sludges from Wisconsin, Minnesota, and Ohio; sludges from Wisconsin and Minnesota were high in P. The higher N and P contents in these sludges indicate that either the raw sewage was high in these elements or the treatment plants were efficient in removing N and P from these waste waters and concentrating them in the sludges.

A state survey of 24 Illinois communities by the Metropolitan Sanitary District of Greater Chicago (MSDGC) is summarized in Table 16. These were the larger sanitary districts in Illinois excluding the MSDGC. Most cities have some industrial wastes, which is indicated by the maximum metal contents reported in Table 16.

An intensive chemical monitoring of sewage sludge was conducted from 1973 to 1974 at the West-Southwest Sewage Treatment Plant (WSW) of MSDGC. Three grab samples were composited daily from the anaerobic digester drawoff. A composite sample was analyzed daily for total solids, pH, Kjeldahl N, NH_3-N, and total P, according to *Standard Methods for the Examination of Water and Wastewater* (American Public Health Association, 1971). Approximately every fourth day, the daily composite sludge sample was analyzed for metal. Results of 13 months of daily composited sampling of the anaerobic digester drawoff are summarized in Table 17. The contents of the sewage sludge are a function of the waste water composition and of the treatment processes employed. To show the operating efficiency of the

WSW plant during this sampling period, the final plant effluent biochemical oxygen demand (BOD) and total suspended solids averaged 8.3 and 8.4 mg/ liter, respectively.

No seasonal differences were detectable in any of the constituents reported in Table 17. The daily variation was such that any possible longer term trends were not detectable in this sewage treatment plant with a daily flow of about 3.8 gigaliters (10^9 gallons) of waste water.

The presence of many essential and nonessential metals in sewage sludge is not surprising. A frequent criticism of the use of sewage sludge on land is that toxic levels of metals are present in many sludges from the larger industrial cities. (This toxicity is discussed in Chapter 10.) To illustrate the concentration variability among sludges with high metal contents, the data shown in Table 15 for Zn and Cd have been used to plot the cumulative frequency curves (Fig. 1 and 2, respectively), which illustrate the concentration differences between sludges from different states, many with Cd levels > 50 μg/g.

An ordinance has been in effect since September 1969 at the MSDGC to regulate the maximum allowable metal concentration which an industrial facility can discharge to the sewer system. To evaluate possible effects of this ordinance, metal content of an anaerobically digested sewage sludge, which was placed in holding lagoons before 1969, has been compared with sludge from this same sewage treatment plant in summer 1974 (Table 18). Effects of changes in industrial processing methods and nature of industrial changes were not evaluated individually; however, this ordinance did greatly reduce the metal concentrations in the sewage. These data also indicate that presently operating a sanitary district without having some heavy metals in the sewage sludge is virtually impossible. Further reductions in allowable metal content of the incoming sewage would mean additional extensive pretreatment of all industrial wastes. Available data show that often metal inputs from nonindustrial sewer users and street stormwater runoff are significant. While regulations limiting the maximum concentrations of metals delivered to the sewer system are helpful in reducing metal content, this method alone will never totally eliminate the metals in municipal sludges.

Sometimes the question of pesticide residues in sewage sludges is raised by regulatory agencies. There is a paucity of data on the pesticide content of sewage sludges. The contents of organochlorine insecticides in sludges from four Ontario (Canada) sewage treatment plants were reported by Chawla et al. (1974) (Table 19). Table 20 presents the mean monthly levels of polychlorinated biphenyls (PCB) in chemical sewage sludge for the same four plants. Pesticide contents of five sewage sludge sources in the MSDGC were: lindane, < 0.5 mg/liter; DDT, toxaphene, or organophosphates and carbamate insecticides, < 0.1 mg/liter; heptachlor or heptachlor epoxide, < 0.2 mg/liter; aldrin, dieldrin, chlordane, or lindane, < 0.01 mg/liter; 2,4-D, < 1 mg/liter; 2,4,5-T, < 0.005 mg/liter; and 2,4,5-TP, < 0.03 mg/liter. Organic matter and clay minerals in soil are effective in inactivating most pesticides (Adams, 1972).

Table 15—Summary of the chemical properties of sewage sludge from select cities in seven states as compiled by the NC-118 North-Central Regional Agricultural Committee on Utilization and Disposal of Municipal, Industrial, and Agricultural Processing Wastes on Land (Unpublished data courtesy of NC-118)

Constituent		Wisconsin (38)†					Michigan (47)†				
		Minimum	Maximum	Median	Mean	Coefficient of variability, %	Minimum	Maximum	Median	Mean	Coefficient of variability, %
pH		7.8	8.1				5.3	11.7			
Organic C	%	25.7	38.8	35.8	33.5	19	6.5	47.4	31.7	30.5	29
Total N	%	2.5	17.6	5.4	7.3	67	0.0	3.3	1.6	1.4	63
NH₄-N	%	0.14	6.76	2.71	2.55	81					
NO₃-N	µg/g	9.0	2,958	187.5	679.3	151					
Total P	%	2.2	6.1	2.7	3.2	41	0.0	3.3	1.6	1.5	66
Total S	%										
Ca	%	3.3	18.0	3.9	6.0	84	0.1	24.8	3.5	4.9	91
Fe	%	0.7	7.9	1.1	2.1	118	0.0	0.6	0.3	0.3	72
Al	%	0.4	1.2	0.5	0.6	45					
Na	%	0.56	2.19	0.98	1.07	45	0.04	0.96	0.12	0.16	89
K	%	0.80	1.92	1.13	1.20	33	0.05	0.87	0.14	0.18	89
As	µg/g						6.0	18.0	9.0	10.7	46
Mg	µg/g	7,600	12,100	8,680	9,171	18	1,400	19,700	5,100	6,502	58
Ba	µg/g	285	1,340	634	676	58					
B	µg/g	154	757	226	301	68					
Zn	µg/g	490	12,100	2,400	3,012	74	1,200	9,400	5,200	4,900	61
Cu	µg/g	140	10,000	825	1,168	138	580	10,400	900	2,423	148
Ni	µg/g	10	2,300	140	372	152	240	2,800	620	1,040	85
Cr	µg/g	50	28,850	1,250	3,593	179	22	30,000	3,200	8,086	130
Mn	µg/g	100	1,130	294	420	90					
Cd	µg/g	5	460	19	77	146	4	520	18	163	143
Pb	µg/g	100	4,600	585	868	106	340	12,400	880	2,940	147
Hg	µg/g	1	31	6	7.4	85	2	3	3	2.6	21
Cd/Zn	%	0.2	32	0.9	3.7	189	0.2	7.6	0.7	2.5	119

(continued on next page)

Table 15. Continued

Constituent		Indiana (14)†					Minnesota (19)†				
		Minimum	Maximum	Median	Mean	Coefficient of variability, %	Minimum	Maximum	Median	Mean	Coefficient of variability, %
pH							6.0	8.0			
Organic C	%	19.2	36.6	22.7	24.1	19	27.2	37.0	29.9	31.3	13
Total N	%	1.5	13.7	3.1	4.3	74	0.5	7.7	5.3	5.1	30
NH₄-N	%	0.01	2.4	0.72	0.93	84	0.01	2.48	0.06	0.19	299
NO₃-N	µg/g	22	1,257	134	237	144					
Total P	%	0.9	3.0	1.9	2.0	29	1.4	4.6	3.5	3.4	24
Total S	%	0.8	1.5	1.1	1.1	15	0.6	1.4	0.9	1.0	25
Ca	%	2.6	14.9	6.6	6.5	45	2.6	7.5	4.0	4.4	30
Fe	%	1.3	4.0	2.1	2.3	35	0.3	1.4	0.8	0.8	31
Al	%						0.2	1.6	0.5	0.7	63
Na	%						0.38	2.66	1.03	1.19	53
K	%	0.14	1.4	0.33	0.41	78	0.08	0.75	0.30	0.36	58
As	µg/g										
Ng	µg/g	6,500	19,200	10,700	11,214	30	3,300	9,700	5,300	5,807	33
Ba	µg/g										
B	µg/g						23	66	35.5	36.4	30
Zn	µg/g	1,550	27,800	4,115	8,107	100	942	5,600	1,800	2,368	63
Cu	µg/g	422	10,125	1,580	2,846	109	281	3,680	1,290	1,521	69
Ni	µg/g	44	3,515	317	993	136	32	419	240	231	48
Cr	µg/g	65	13,000	1,033	3,195	134	17	3,470	906	931	94
Mn	µg/g						145	1,050	396	487	49
Cd	µg/g	16	846	54	163	146	13	435	25	131	129
Pb	µg/g	517	19,730	832	2,970	175	164	4,095	626	1,190	103
Cd/Zn	%	0.1	6.4	1.9	2.4	83	0.3	20.1	1.7	6.9	125

(continued on next page)

Table 15. Continued

Constituent		New Jersey (13)†					New Hampshire (28)†				
		Minimum	Maximum	Median	Mean	Coefficient of variability, %	Minimum	Maximum	Median	Mean	Coefficient of variability, %
pH											
Organic C	%	0.8	7.5	2.6	2.9	66	18.0	48.0	38.0	35.4	23
Total N	%	0.01	0.45	0.06	0.15	108	1.4	5.0	2.5	2.6	38
NH_4-N	%										
NO_3-N	µg/g	130	4,880	410	1,043	129					
Total P	%	0.5	3.2	1.7	1.7	41	0.5	2.0	0.9	1.1	46
Total S	%										
Ca	%	0.6	10.9	2.0	3.7	99	0.3	6.0	3.0	3.2	52
Fe	%	0.4	4.0	0.9	1.2	84	0.1	0.2	0.1	0.1	18
Al	%	0.4	3.5	0.6	1.2	88	0.1	0.1	0.1	0.1	9
Na	%	0.02	0.18	0.08	0.09	58	0.01	0.15	0.06	0.06	67
K	%	0.02	0.54	0.16	0.17	85	0.20	0.52	0.30	0.30	17
As	µg/g	1,400	5,150	3,000	2,979.2	34	300	2,900	700	948	92
Mg	µg/g	148	8,980	230	956.6	253	21	112	80.5	73.8	38
Ba	µg/g						4	253	16.5	26.1	176
B	µg/g	735	6,775	1,850	2,205.6	77	101	240	108	120.6	28
Zn	µg/g	480	2,643	1,200	1,399.9	43	84	100	85	85.7	4
Cu	µg/g	20	860	89	155.8	151	23	190	51	63.2	66
Ni	µg/g	86.0	7,502	994	1,606.4	132	30	99,000	140	6,762.9	377
Cr	µg/g	55	1,380	75	250.8	146	18	482	107.5	124.3	77
Mn	µg/g	5.0	82	11	29.3	99					
Cd	µg/g	115	1,354	181	326.7	104	5	45	5	10.3	113
Pb	µg/g						13	15,000	650	3,346.7	142
Cd/Zn	%	0.1	7.5	1.1	1.6	115	4.6	41.7	4.6	9.6	113

(continued on next page)

Table 15. Continued

Constituent		Ohio (15)†					All seven states†					
		Minimum	Maximum	Median	Mean	Coefficient of variability, %	n	Minimum	Maximum	Median	Mean	Coefficient of variability, %
pH							61	5.3	11.7			
Organic C	%	0.5	12.7	3.8	5.0	83	101	6.5	48.0	30.4	31.0	27
Total N	%						134	0.0	17.6	2.5	3.2	85
NH$_4$-N	%						56	0.0	6.76	0.13	0.74	171
NO$_3$-N	µg/g						34	9	4,880	194	646	158
Total P	%						128	0.0	6.1	1.8	1.9	61
Total S	%						27	0.6	1.5	1.1	1.1	21
Ca	%	2.2	25.0	4.9	9.1	86	143	0.1	25.0	3.8	5.0	87
Fe	%	1.3	15.3	2.3	3.7	101	104	0.0	15.3	0.8	1.4	148
Al	%	0.1	1.3	0.4	0.5	70	83	0.1	3.5	0.4	0.5	119
Na	%	0.02	1.20	0.21	0.25	122	126	0.01	2.66	0.12	0.35	146
K	%	0.02	0.31	0.15	0.15	60	142	0.02	1.92	0.24	0.30	99
As	µg/g						10	6.0	230.0	10.0	43.1	171
Mg	µg/g	400	7,100	4,000	3,549	63	139	300	19,700	4,600	5,414	75
Ba	µg/g	380	4,400	1,000	1,311	82	60	21	8,980	162	576	225
B	µg/g	100	700	460	437	45	60	4	757	28	109	162
Zn	µg/g	520	15,000	1,800	4,153	116	134	101	27,800	1,800	2,997	134
Cu	µg/g	300	4,700	1,000	1,392	89	131	84	10,400	850	1,308	138
Ni	µg/g	120	2,800	560	710	98	109	10	3,515	190	440	162
Cr	µg/g	100	2,800	1,360	1,281	78	119	17	99,000	906	3,280	309
Mn	µg/g	40	7,100	240	869	205	82	18	7,100	200	390	209
Cd	µg/g	12	700	200	198	98	115	4	846	20	101	157
Pb	µg/g	200	5,000	1,200	1,634	82	116	13	19,730	652	1,656	177
Hg	µg/g	3.0	10,600	4,900	4,370	77	53	1.0	10,600	6.0	1,077	232
Co	µg/g						15	1.0	18	6.0	5.3	83
Mo	µg/g						29	5.0	39	30	27.7	26
Cd/Zn	%	0.5	46.2	4.7	14.3	117	115	0.1	46	1.7	5.7	163

† Number of sewage treatment plants surveyed in each state. This was the maximum n value, but some of the constituents analyzed had lower n values, as noted in the first column of the "All seven states" section.

Table 16—Summary of the chemical properties of sewage sludge from Illinois communities
with populations ranging from 6,000 to 200,000 and excluding all MSDGC sewage
treatment plants. One grab sample was collected from each plant between
23 Mar. and 22 Apr. 1974†

Constituent	Samples analyzed	Minimum	Maximum	Median
		——————— % dry-weight basis ———————		
Kjeldahl N	11	2.6	9.8	4.1
NH₃-N	11	0.14	6.1	1.2
Total P	11	0.7	4.9	2.4
Fe	11	1.3	3.8	2.0
		——————— µg/g dry-weight basis ———————		
Zn	24	338	14,900	1,430
Cd	24	3	3,410	12
Cu	11	117	4,060	716
Ni	24	0	1,650	70
Pb	24	58	1,300	267
Hg	24	0.5	15.6	2.5

† Based on a survey of the chemical characteristics of sludges produced by publicly
owned sewage treatment plants in Illinois (unpublished data, Dept. of Research and
Development, The Metropolitan Sanitary District of Greater Chicago).

Table 17—Summary of the chemical properties of daily composite primary and waste-
activated, anaerobically digested sewage sludge from the West-Southwest Sewage
Treatment Plant of MSDGC, Cicero, Illinois, from August 1973 to
September 1974 (unpublished data, MSDGC)

Constituent	Samples analyzed	Minimum	Maximum	Mean	Median
	No.	——————— % dry-weight basis ———————			
Kjeldahl N	349	4.2	9.65	6.93	6.77
NH₃-N	349	1.52	5.00	3.26	3.09
Total P	344	1.08	8.08	2.80	2.79
Total S	12	0.35	1.28	0.90	0.98
K	79	0.22	0.79	0.47	0.43
Ca	79	0.27	4.77	2.16	2.14
Fe	79	1.54	4.90	3.63	3.54
Al	79	0.60	1.49	1.05	1.05
Na	79	0.16	0.66	0.29	0.25
		——————— µg/g dry-weight basis ———————			
Mg	79	6,140	17,400	11,400	10,800
Zn	79	1,670	4,850	2,770	2,630
Cd	79	120	312	205	197
Cu	79	679	2,270	1,370	1,380
Cr	79	11	4,120	2,330	2,400
Ni	78	186	840	355	343
Mn	79	120	550	370	386
Pb	79	304	1,160	699	680
Hg	79	0.8	7.5	3.2	3.2
pH	347	6.7	9.1		

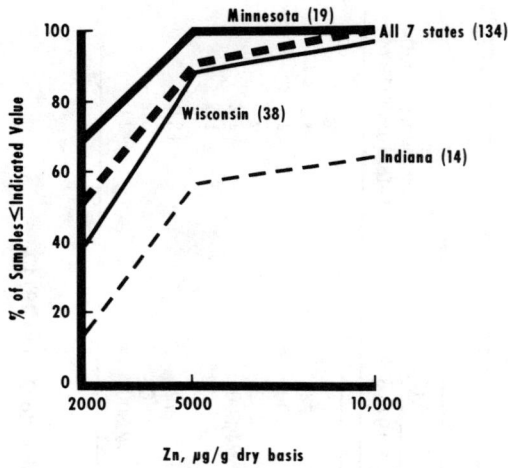

Fig. 1—Cumulative frequency of Zn in sludge samples from seven states. Values in parentheses show the number of samples analyzed. (Unpublished data courtesy of NC-118).

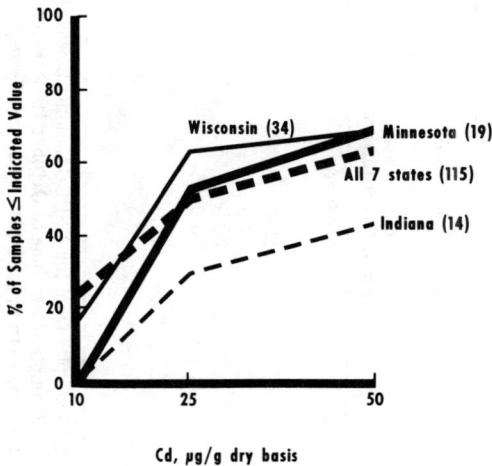

Fig. 2—Cumulative frequency of Cd in sludge samples from seven states. Values in parentheses show the number of samples analyzed. (Unpublished data courtesy of NC-118).

Table 18—Comparison of metal content of the Calumet Sewage Treatment Plant's sludge before and after the 1969 Metropolitan Sanitary District's Sewage and Waste Control Ordinance on maximum allowable metal concentration to the sewer system

Metal	Maximum allowable concentration in incoming sewage by MSDGC ordinance	Sludge source			Metal reduction ratio before 1969:1974
		Lagoons	Anaerobic digester		
		Before 1969†	1972‡	1974§	
	mg/liter	—— μg/g dry-weight basis ——			
Cd	2.0	190	100	54	3.5
Cr (total)	25.0	2,100	1,100	790	2.7
Cu	3.0	1,500	900	282	5.3
Fe	50.0	53,700	36,800	24,200	2.2
Hg	0.0005	3.3	3.0	2.15	1.5
Ni	10.0	1,000	200	77	13.0
Pb	0.5	1,800	1,800	486	3.7
Zn	15.0	5,500	3,500	2,800	2.0

† Mean of 10 samples from the MSDGC Calumet Sewage Treatment Plant sludge lagoon.
‡ Mean of 22 samples from June to October 1972 (Peterson et al., 1973).
§ Mean of 6 samples from the MSDGC Calumet Sewage Treatment Plant anaerobic digesters from June to October 1974.

Table 19—Mean levels of organochlorine pesticides in digested chemical sludges
(Chawla et al., 1974)

Organochlorine pesticides	Source of sludge and treatment†			
	North Toronto iron	Pt. Edward alum	Newmarket lime	Sarnia iron
	µg/liter			
P, P′-DDT	1.33	0.77	1.12	2.70
O, P′-DDT	6.87	4.11	1.15	5.25
P, P′-DDE	7.18	3.68	2.18	20.77
P, P′-DDD	4.31	3.98	1.60	6.41
Total DDD/DDE/DDT	19.69	12.54	6.68	35.13
a-chlordane	15.79	12.51	5.69	16.99
γ-chlordane	12.56	11.41	2.84	12.92
Total chlordane isomers	28.35	23.92	8.53	29.91
Heptachlor	5.37	2.29	0.70	14.95
Heptachlor epoxide	3.04	1.77	0.58	6.78
Total heptachlor	8.41	4.06	1.28	21.73
Aldrin	2.03	1.09	0.70	9.40
Dieldrin	3.03	1.69	1.65	3.23
Lindane	1.48	1.16	0.63	3.22
Total organochlorine pesticides	63	45	20	103

† Chemicals used for phosphorus precipitation.

Table 20—Mean monthly levels of PCB's (Arochlor 1254) in chemical sewage sludges
(Chawla et al., 1974)

Month	Source of sludge and treatment†			
	North Toronto iron	Pt. Edward alum	Newmarket lime	Sarnia iron
	µg/liter			
June 1973	206.2	148.6	45.2	975.0
July 1973	170.7	128.2	106.1	1,730.0
August 1973	224.0	97.3	133.5	1,477.8
September 1973	286.7	145.3	105.2	1,511.7
October 1973	205.5	87.5	84.5	1,241.0
November 1973	228.3	88.8	73.5	954.7
December 1973	216.8	66.3	45.7	822.0
January 1974	200.6	62.9	48.5	864.7
February 1974	195.2	118.0	53.0	830.2
March 1974	207.8	135.1	43.3	811.8
Mean	214	108	74	1,122

† Chemicals used for phosphorus precipitation.

2. BIOLOGICAL

Stabilizing sludge by anaerobic digestion greatly reduces the possibility of sewage sludges emitting odors. Anaerobic digestion also effectively reduces the coliform and virus population of the sewage sludge. Research with the porcine enterovirus (ECPO-1), with germ-free, 10-day-old piglets to test virus viability, indicated that no porcine enterovirus survived in the sewage sludge digester after 5 days (Meyer et al., 1971). Reed, Fenters, and Lue-Hing (*In* Abstr., 1975 Annual Meeting, Am. Soc. Microbiol.) reported on an incubation study with digested sludge supernatant where all inoculated Echoviruses were inactivated in 2 days, Coxsackie B4 in 3 days, and poliovirus type 1 in 5 days. Other experiments with poliovirus type 1, coxsackievirus type A-9, coxsackievirus type B-4, and Echovirus type 11 have shown an average inactivation of 93.8, 97.5, 89.5, and 58%, respectively, after 24 hours. After 48 hours of digestion, the respective inactivations were 98.5, 99.7, 98.6, and 92.5%, respectively, for the viruses listed (Bertucci et al., unpublished report no. 74–19, MSDGC).

In most farming programs where sewage sludge is utilized as a nutrient source, the sludge often must be stored. During storage, further reduction in virus and bacterial counts can be expected. Berg (1966) determined the time in days required for 99.9% reduction in the number of viruses and bacteria by storage of untreated sewage at different temperatures (Table 21).

Adding $FeCl_3$ or Fe_2SO_4 and lime to aid in dewatering raw sewage sludge from about 7 to 26% total solids reduced the aerobic bacteria population by two to three logs and the enterobacteriaceae population by two to four logs (Kampelmacher & Van Noorle Jensen, 1972). The *Salmonella* population of the dry sludge was significantly less than that of wet sludge.

3. PHYSICAL

Physical characteristics of sewage sludge are affected by quality of the waste water, type and extent of waste water treatment, and method of sludge stabilization. A granular, dense sludge may be produced from primary waste water treatment. The waste-activated sludge from secondary waste water treatment results in a sludge containing mostly bacterial cells which may be viscous and difficult to dewater. The distribution of water in sewage sludge was estimated by Bjorkman (1969) to be 70% between cells, 22% adhesion and capillary water, and 8% adsorption and intracellular fluids.

Particle size of digested waste-activated sludge has been reported as 99% < 9 μm and 60% < 3 μm (Peterson et al., 1973).

The density of sludge is a function of moisture content. Heat-dried, waste-activated sludge has a density of 0.58 g/cm^3, liquid digested sludge is 1.01 g/cm^3, lagooned digested sludge is 1.08 g/cm^3, and aged Imhoff sludge (25 to 60% moisture) is approximately 1.2 g/cm^3 (unpublished data, MSDGC).

Table 21—Effects of storage and temperature on days required for 99.9% reduction of
viruses and bacteria in untreated sewage (Berg, 1966)

Organism	Temperature, °C		
	4	20	28
	—Days required to reduce numbers 99.9% —		
Poliovirus 1	110	23	17
Echovirus 7	130	41	28
Echovirus 12	60	32	20
Coxsackievirus A9	12	--	6
Aerobacter aerogenes	56	21	10
Escherichia coli	48	20	12
Streptococcus faecalis	48	26	14

An engineering study was conducted by Rimkus and Heil (1975) on a
sludge lagoon, constructed 15 years ago, which had been filled periodically
almost every year since its construction with a small amount of Zimpro ash
and mostly waste-activated, anaerobically digested sewage sludge. Physical
properties of the stored sludge are: total solids, 13%; density, 1.08 g/cm^3;
plastic viscosity, 0.79 poise; yield stress, 232 dynes/cm^2; Reynold's number,
163; and pressure loss, 6.10^6 dynes/cm^2. This sludge had non-Newtonian
flow when pumped and behaved as a thixotropic pseudoplastic. Mechanical
mixing fluidized the sludge. Plastic viscosity and yield stress varied as an exponential function of the solids content of the sludge.

The method of handling sewage sludges depends on solids content.
Sludge of up to 10% total solids can be pumped with special techniques.
Slurries with up to 5 to 6% total solids may be applied with field sprinklers.
The sludge solids content must be at least 25 to 30% before it can be handled
as a solid (e.g., by shovel or fork).

B. Municipal Refuse

Municipal refuse may include any waste generated by the domestic and
industrial sector of a municipality. Usually, sewage sludge is not included.
The composition of this refuse may vary greatly. Composition of a repre-

Table 22—Composition of representative municipal refuse (McCalla, 1974b)

Constituent	Percent by weight
Paper products	58.79
Food scraps	9.24
Wood and yard plants refuse	10.08
Metals	7.52
Glass, ceramics, and ash	8.49
Miscellaneous: plastic, rags, leather, paints, oils, and dirt	5.88

sentative municipal refuse is presented in Table 22 (McCalla, 1974b). The paper, food, and plant refuse fraction accounts for 80% of municipal refuse. The other 20% should be separated to utilize the refuse on land for its fertility or organic matter. This separation requires intensive labor and/or machinery inputs. The inert fraction which is useable, e.g., metal and glass, may be recycled. The decomposable fraction can be mixed with coal and used as an energy source or shredded as a mulch, or composted for use on land. The remaining inert fraction often goes to landfills.

There are about 25 different mechanical methods for composting, but they can be summarized as either a static or a dynamic process. For the static method, refuse is stacked in well-ventilated layers and the layered piles turned once or twice during the composting process. In the static compost method, development of fungi is intense.

The dynamic process requires continuous mixing of the compost; therefore, the flora is mainly bacteria. This is followed by a static phase where the compost matures (Stickelberger, 1974). This mixing phase is important if liquids, e.g., sewage sludge or liquid manures, are to be incorporated with the shredded refuse. Using hog manure or undigested sewage sludge as the moisture source causes an odorous compost when anaerobic pockets develop in the pile. If undigested, waste-activated sludge is composted with a bulking agent, e.g., woodchips, odors are likely. But if the waste-activated sludge is

Table 23—Analyses of composted refuse and freshly shredded refuse

Constituent	Japan[†]	Florida[‡]	Tennessee[§]	Washington[¶]	Ontario[#]
			% dry-weight basis		
Total N	1.0	1.2	1.3	0.78	0.57
P	0.24	0.26	0.26	0.16	0.08
K	0.98	0.38	0.97	0.27	0.314
Ca	5.1	1.30	4.6	1.97	0.850
Mg		0.07	0.5	0.11	0.209
Na			0.6	0.26	0.187
Fe				0.51	
S			0.5		
Total C	28		27	48	
Organic C					37
			μg/g dry-weight basis		
Mn		130		242	250
B		25		58–62	
Cu		125	100	96	28
Zn		250	1,500	715	400
Pb					200
Cd					7
Cr					31

† Egawa, 1974. Composted refuse.
‡ Hortenstine and Rothwell, 1968. Composted refuse.
§ Terman et al., 1973. Compost may contain up to 20% sewage sludge.
¶ Volk and Ullery, 1972 (unpublished report). Freshly shredded refuse > 2-mm fraction.
# King et al., 1974. Freshly shredded unsorted refuse.

first digested, the odor problem is greatly reduced. However, the maximum compost temperature is then 60C as compared with 80C using undigested sludge.

A slow rise in the compost temperature is desirable to force spore-forming pathogenic organisms to germinate. In their vegetative stages, spore-forming organisms can be killed at temperatures of 50 to 60C. Worm eggs and parasites also were reportedly destroyed by· composting (Stickelberger, 1974).

Composted refuse, free of glass, wire, and other sharp inert matter, is a useful, low-grade fertilizer and soil conditioner. Data presented in Table 23 give some of the chemical properties of uncomposted and composted refuse only. The composts from Japan and Florida were composted, sorted refuse only. The compost from Johnson City, Tennessee was sorted garbage with up to 20% sewage sludge added during composting, which resulted in high N and Zn concentrations in the final compost (Terman et al., 1973). The fresh-ly shredded refuse from Vancouver, Washington was screened to exclude the < 2-mm fraction and was handsorted to remove metal, glass, and large pieces of wood (unpublished report of Boeing Co. by Volk and Ullery, 1972). Volk and Ullery found that most of the B was from glue in the paper and cor-rugated cardboard. King et al., (1974) analyzed unsorted, shredded munici-pal refuse from St. Catherines, Ontario which contained 71.2% paper, 5.2% plastic, 8.3% metal, 5.0% glass and dust, 1.3% miscellaneous, and 48.6% moisture.

LITERATURE CITED

Adams, R. S., Jr. 1972. Effect of soil organic matter on the movement and activity of pesticides in the environment. p. 81–93. In D. Hemphill (ed.) Trace substances in the environment.-V. University of Missouri Press, Columbia, Mo.

Allison, F. E. 1965. Decomposition of wood and bark sawdusts in soil, nitrogen require-ments, and effects on plants. USDA-ARS Tech. Bull. 1332.

American Public Health Association. 1971. Standard methods for the examination of water and wastewater. 13th ed. Am. Public Health Assoc., New York.

Azevedo, J., and P. R. Stout. 1974. Farm animal manures: an overview of their role in the agricultural environment. California Agric. Exp. Sta. Ext. Serv. Manual 44.

Berg, G. 1966. Virus transmission by the water vehicle. II. Virus removal by sewage treatment procedures. Health Libr. Sci. 3:90.

Bjorkman, A. 1969. Heat processing of sewage sludge. 4th Congress International Re-search Group on Refuse Disposal. Basle, Switzerland, 2–5 June. p. 670–686.

Chawla, V. K., J. P. Stephenson, and D. Liu. 1974. Biological characteristics of digested chemical sewage sludges. In Proc. Sludge Handling and Disposal Seminar, Environ-ment Canada, Toronto, Ont. 18–19 Sept.

Egawa, T. 1974. The use of organic fertilizer in Japan. FAO/SIDA Expert consultation on organic materials as fertilizers, Rome, Italy. 2–6 Dec. 1974. p. 253–274.

Frecks, G. A., and C. B. Gilbertson. 1974. The effect of ration on engineering properties of beef cattle manure. Am. Soc. Agric. Eng. Trans. 17:383–387.

Garman, W. H., D. A. Williams, A. W. Tenny, and E. T. York, Jr. 1962. Our land and its care. The Fertilizer Institute, Washington, D. C.

Gilbertson, C. B., T. M. McCalla, J. R. Ellis, O. E. Cross, and W. R. Woods. 1970. The effect of animal density and surface slope on characteristics of runoff, solid wastes and nitrate movement on unpaved beef feedlots. Nebraska Agric. Exp. Sta. Bull. 508.

Gilbertson, C. B., T. M. McCalla, J. R. Ellis, and W. R. Woods. 1971. Characteristics of manure accumulations removed from outdoor, unpaved, beef cattle feedlots. p. 132-134. *In* Livestock waste management and pollution abatement. Am. Soc. Agric. Eng., St. Joseph, Mich.

Gilbertson, C. B., J. A. Nienaber, J. R. Ellis, T. M. McCalla, T. J. Klopfenstein, and S. D. Farlin. 1974. Nutrient and energy composition of beef cattle feedlot waste fractions. Nebraska Agric. Exp. Sta. Bull. 262.

Guenzi, W. D., and T. M. McCalla. 1966. Phenolic acids in oats, wheat, sorghum, and corn residues and their phytotoxicity. Agron. J. 58:303-304.

Hart, S. A., and P. H. McGaughey (chairmen). 1966. Solid wastes management. Proc., Nat. Conf. on Solid Wastes Management. Univ. of California, Davis. 4-5 Apr. 1966.

Hortenstine, C. C., and D. F. Rothwell. 1968. Garbage compost as a source of plant nutrients for oats and radishes. Compost Sci. 9(2):23-25.

Hrubant, G. R., R. V. Daugherty, and R. A. Rhodes. 1972. Enterobacteria in feedlot waste and runoff. Appl. Microbiol. 24:378-383.

Kampelmacher, E. H., and L. M. Van Noorle Jensen. 1972. Reduction of bacteria in sludge treatment. J. Water Pollut. Contr. Fed. 44:309-313.

King, L. D., L. A. Rudgers, and L. R. Webber. 1974. Application of municipal refuse and liquid sewage sludge to agricultural land. 1. Field study. J. Environ. Qual. 3:361-366.

Loehr, R. C. 1974. Agricultural waste management: problems, processes and approaches. Academic Press, New York. p. 551-555.

Lutz, H. J., and R. F. Chandler, Jr. 1947. The organic matter of forest soils. p. 140-197. *In* Forest soils. John Wiley and Sons, Inc., New York.

Mathers, A. C., B. A. Stewart, J. D. Thomas, and B. J. Blair. 1973. Effects of cattle feedlot manure on crop yields and soil conditions. p. 1-13. *In* Proc. Symp. on Animal Waste Management. 18 Jan. 1973. USDA Southwestern Great Plains Research Center, Bushland, Tex. Tech. Rep. 11.

McCalla, T. M. 1972. Think of manure as a resource, not a waste. Feedlot Manage. 14(5).

McCalla, T. M. 1974a. Use of animal wastes as a soil amendment. J. Soil Water Conserv. 29:213-216.

McCalla, T. M. 1974b. Waste management problems. p. 121-127. *In* Land use: persuasion or regulation? Soil Conservation Society of America, Ankeny, Iowa.

McCalla, T. M., and L. F. Elliott. 1974. Municipal and animal wastes as fertilizers. p. 179-180. *In* 1974 McGraw-Hill yearbook on science and technology. McGraw-Hill Book Co., New York.

McCalla, T. M., J. R. Ellis, C. B. Gilbertson, and W. R. Woods. 1972. Chemical studies of solids, runoff, soil profile and groundwater from beef cattle feedlots at Mead, Nebraska. p. 211-223. *In* Waste management research. Proc. 1972 Cornell Agric. Waste Manage. Conf., Syracuse, N. Y.

McCalla, T. M., and F. A. Norstadt. 1974. Toxicity problems in mulch tillage. Agric. Environ. 1:153-174.

McCalla, T. M., and F. G. Viets, Jr. 1969. Chemical and microbial studies of wastes from beef cattle feedlots. p. 1-24. *In* Proc. Pollution Research Symp., Lincoln, Nebraska. 23 May 1969.

Metcalf and Eddy, Inc. 1972. p. 575-632. *In* Wastewater engineering collection treatment disposal. McGraw-Hill Book Co., New York.

Meyer, R. C., F. C. Hines, H. R. Isaacson, and T. D. Hinesly. 1971. Porcine enterovirus survival and anaerobic sludge digestion. p. 183-189. *In* Livestock waste management and pollution abatement. Am. Soc. Agric. Eng., St. Joseph, Mich.

Peterson, J. R., C. Lue-Hing, and D. R. Zenz. 1973. Chemical and biological quality of municipal sludge. p. 26-37. *In* W. E. Sopper and L. T. Kardos (ed.) Recycling treated municipal wastewater and sludge through forest and cropland. Penn State University Press, University Park, Pa.

Peterson, J. R., T. M. McCalla, and G. E. Smith. 1971. Human and animal wastes as fertilizers. p. 557-596. *In* Fertilizer technology and use. 2nd ed. Soil Sci. Soc. Am., Madison, Wis.

Rhodes, R. A., and G. R. Hrubant. 1972. Microbial population of feedlot waste and associated sites. Appl. Microbiol. 2 :369-377.

Rice, E. L. 1974. Allelopathy. Academic Press, New York. 353 p.

Rimkus, R. R., and R. W. Heil. 1975. The rheology of plastic sewage sludge. *In* Proc. 2nd Annu. Conf. on Complete Wastewater Reuse. Am. Inst. Chem. Eng., New York, 8 May.

Salter, R. M., and C. J. Schollenberger. 1939. Farm manure. Ohio (Wooster) Agric. Exp. Sta. Bull. 605.

Sievers, F. J., and H. F. Holtz. 1926. The significance of nitrogen in soil organic matter relationships. Washington Agric. Exp. Sta. Bull. 206.

Stickelberger, D. 1974. Survey of city refuse compost. FAO/SIDA Expert consultation on organic materials as fertilizers, Rome, Italy. 2-6 Dec. 1974. p. 185-209.

Studdard, G. J. (ed.). 1973. Common environmental terms—a glossary. U. S. Environmental Protection Agency, Washington, D. C.

Terman, G. L., J. M. Toileau, and S. E. Allen. 1973. Municipal waste compost: effects on crop yields and nutrient content in greenhouse pot experiments. J. Environ. Qual. 2:84-89.

U. S. Environmental Protection Agency. 1965. Restoring the quality of our environment. Report of Environmental Pollution Panel. President's Science Advisory Committee, Washington, D. C.

U. S. Environmental Protection Agency. 1972a. Glossary of solid waste management. Publ. GP-1972-3. National Center for Resource Recovery, Inc., Washington, D. C.

U. S. Environmental Protection Agency. 1972b. Solid waste management glossary. USEPA Publ. SW-108tf. U. S. Government Printing Office, Washington, D. C.

U. S. Environmental Protection Agency. 1974. Resource recovery and source reduction. 2nd Report to Congress. Publ. SW-122. U. S. Government Printing Office, Washington, D. C.

White, R. K., E. P. Taiganides, and G. D. Cole. 1971. Chromatographic identification of malodors from dairy animal waste. p. 110-113. *In* Livestock waste management and pollution abatement. Am. Soc. Agric. Eng., St. Joseph, Mich.

chapter 3

Lime is removed from effluent in a recarbination basin at the Durham Facilities of the Unified Sewerage Agency of Washington County, Oreg. (Photo courtesy of Stevens, Thompson & Runyan, Inc., Engineers/Planners).

Properties of Waste Waters

RICHARD THOMAS and JAMES P. LAW
Robert S. Kerr Water Research Center, EPA, Ada, Oklahoma

I. INTRODUCTION

The planning, design, and implementation of land-based waste water management projects should include assessment of the properties of waste water at an early stage. The objective of this early assessment is to weigh the properties of the waste water relative to stated or implied criteria and to make sound judgments leading to a successful project. The important key to early assessment is the desire to avoid circumstances that may result in disruptive and costly revisions during construction of facilities or to facility operation of a newly constructed system. It is equally important to avoid costly overdesign of a facility. The purpose of this chapter is to provide information on waste water properties which will be of value to those planning and designing projects.

Two approaches were considered during collection and review of information. Presentation of extensive quantitative data was considered and rejected. The approach adopted was to present selected quantitative data in combination with other information to depict ways for assessing factors which have strong influence on project planning, design, and management. The categories covered include waste water flow rates and the properties of raw waste waters and effluents from selected treatment processes. Waste water sources include municipalities, animal feedlots, and irrigation return flows. Waste water source is used as the principal heading. Waste water flow rates and changes in properties induced by treatment processes are addressed separately for each source. The treatment processes covered include primary processes and the conventional biological processes referred to as secondary treatment processes.

II. MUNICIPAL WASTE WATERS

The flow and composition of municipal waste waters are determined by factors related to properties of the water, per capita water consumption, industrial contributions, and amount of used water reaching the sewers. Properties of the water are important because the water supply can be the principal source of dissolved constituents which will appear in raw waste water collected in the sewerage system. Durfor and Becker's (1964) survey of water supplies for the 100 largest cities in the United States is a good source of information on the properties of water used for public supplies. Per capita water consumption is one of the principal factors used to estimate the flow of waste water to be discharged into the sewerage system. Industrial flows can cause unusual variability in flow rate as well as contribute substantial quantities of specific constituents. There is established methodology to estimate waste water flows and sewage composition for the general case, however, the final determination is obviously site specific and must be based on appropriate collection of onsite data.

A. Waste Water Flows

Many reference books on waste water engineering, such as Fair et al. (1971), Metcalf and Eddy, Inc. (1972), and Bond and Straub (1974), discuss the determination of waste water flows for design of waste water treatment facilities. They describe how data on population projections, per capita water use, and industrial flows are used to predict waste water flows. A per capita water use of 0.56 m^3 per day (147 gallons per day [gpd]) and a value of 60% to 80% for the amount of used water reaching the sewers are the basis for the commonly accepted value of 0.38 m^3 per capita per day (100 gpcd) for estimating municipal waste water flows. An in depth understanding of how waste water flows are estimated is readily obtainable by reference to textbooks such as those cited. The average flow rate which can be estimated by use of the 0.38 m^3 per capita per day value or a more accurate value for a specific site is important, but it needs to be supplemented by information on fluctuations in flow rate.

Fluctuations in flow rate usually occur in daily, weekly, and seasonal patterns. Each of these patterns can have a strong influence on the design and management of land-based waste water management systems. Factors which are the primary cause of fluctuations in flow rate include domestic use patterns, frequency and quantity of industrial discharges, and wet weather contributions. Domestic use patterns are fairly predictable and are covered in textbooks on waste water engineering. Typical patterns depicted by Metcalf and Eddy, Inc. (1972) estimate the 1-day maximum flow to be 180% of the average daily flow while the daily flow during the maximum week is estimated to be 140% of the average daily flow for the year. Prediction of fluctuations in domestic use patterns induced by seasonal recreational use requires more onsite investigation. Affected municipalities may experience substantial variations in flow with the high flow occurring in either summer or winter, depending on the predominant recreational season. Fluctuations associated with daily, weekly, or seasonal changes in industrial operations are diverse and may be very abrupt. In extreme situations the influence of a seasonally operated industry may require special designs including supplemental components for intermittent operation. Incorporation of seasonal food processing waste waters is an example of one situation which may require additional facilities for seasonal operation. Wet weather contributions to separate sanitary sewerage systems are associated with infiltration into sewer lines. The amount of infiltration is dependent on local conditions but it is not uncommon for wet weather flows to be double dry weather flows for a comparable measuring period. This brief account of factors which cause fluctuations in flow of waste water in municipal sewerage systems highlights some of the more prominent factors. Those desiring deeper inquiry will find an ample literature ready for thorough exploration. References cited in waste water engineering textbooks provide ready access to the technical literature on this subject.

B. Properties of Municipal Waste Waters

As alluded to in the brief coverage of waste water flows, municipal waste waters are usually a mixture of discharges coming from residences and a diverse assortment of industries. The properties of such mixtures are obviously source dependent and can be expected to vary over wide ranges of values. For example, the properties of waste water coming from a strictly residential community with a population of 10,000 is not comparable to the same quantity of flow from a residential population of 5,000 plus flow from a tannery and metal plating industry. Coverage of the properties of municipal waste waters reflects compositional changes which are associated with the rapid industrialization of the United States and the very recent surge of interest in the quality of our environment. Both of these factors are contributing to changes in water use patterns and to changes in the composition of waste discharges to the sewer systems. In an attempt to provide a flexible base which can be readily adjusted to these factors, data on the properties of waste waters is presented in a comparative format. The discussion begins with the properties of raw waste water and progresses through typical treatment processes to the effluents which will be the source water for most land-based waste water management systems.

1. PROPERTIES OF RAW MUNICIPAL WASTE WATERS

Historically, the properties of raw municipal waste waters have been determined with a relatively small list of physical, chemical, and biological properties. Thousands of samples have been analyzed and are frequently depicted in generalized presentations such as shown in Table 1 for physical and chemical properties. Two entries in Table 1 warrant specific notation. The values for total P which represent a recent change in the composition of

Table 1—Typical properties of raw municipal waste waters†

Constituent	Concentration, mg/liter		
	Strong	Medium	Weak
Solids			
Total	1,200	700	350
Dissolved	850	500	250
Suspended	350	200	100
Biochemical oxygen demand, 5-day at 20C	300	200	100
Total organic carbon	300	200	100
Chemical oxygen demand	1,000	500	250
Nitrogen			
Total	85	40	20
Free ammonia	50	25	12
Phosphorus, total	20	10	6
Chlorides (added by a use cycle)	100	50	30

† Adapted from Metcalf and Eddy, Inc. (1972).

municipal waste waters, represent a downward adjustment from those of the 1950's and reflect the influence of recent changes in the formulation of cleaning agents. This change in P concentration is an example of how recent interest in environmental quality has influenced the properties of municipal waste waters. Further changes can be expected in the future. The second entry warranting specific notation is the Cl concentration. Values for Cl are indicative of those added during a municipal use cycle. They do not include the Cl already present in the water source. In addition to the physical and chemical parameters entered in Table 1, the total coliform test and more recently the fecal coliform test are used in the routine assessment of waste water properties. Use of the fecal coliform test is particularly important when dealing with land-based systems since the total coliform test will be influenced by coliform subgroups common to the soil (National Academy of Sciences, 1973, p. 57–58).

It is helpful to indicate the reasonableness of typical values such as those shown in Table 1 with comparative data from actual case histories. Data collected by Barsom (1973) were selected to illustrate the properties of raw waste water from smaller domestic sources (Table 2). These data are taken from reported tests at 141 lagoons in 22 states and should be relatively free of industrial influences. Parameters selected were total suspended solids (147 determinations), and BOD (191 determinations). Three groupings (weak, medium, and strong) show an example of the range of actual values typified in Table 1. Median values for suspended solids and for BOD in Table 2 are comparable to the typical values in Table 1; content of suspended solids was 25% less than the typical value for strong waste water and BOD was 7% greater than the typical value for strong waste water. This comparison of case history data to typical values highlights some aspects of data variability and substantiates the usefulness of typical values like those in Table 1 for planning purposes.

The metal content of municipal waste water is attracting much attention and is an important factor to consider in the planning of a land-based waste management system. Influences on crop quality, soil productivity, and ground water quality are principal concerns for decision makers. Final decisions for specific situations will usually require some onsite determinations, whereas an assessment of available literature may be adequate for planning

Table 2—Raw waste water data for 141 lagoon systems†

Constituent	Concentration, mg/liter		
	Strong	Medium	Weak
Total suspended solids			
Range	219–730	117–204	22–116
Median	260	161	83
Biochemical oxygen demand			
Range	266–667	171–266	13–170
Median	322	230	115

† Adapted from Barsom (1973).

decisions. Surveys of the metal content of municipal waste waters in highly industrialized areas are presented to provide information on the upper range of metal concentrations occurring in raw municipal waste waters. Data summarized in Table 3 are taken from Mytelka et al. (1973) and Konrad and Kleinert (1974). These data comprise the results of 148 to 481 individual determinations for each constituent on samples from some 140 waste water treatment facilities in Connecticut, New Jersey, New York, and Wisconsin. The National Academy of Sciences (1973) recommended maximums are included as appropriate criteria for making decisions about pretreatment needs. This reference source lists two sets of recommended maximums for irrigation use. The more restrictive values for continuous use on all soils are listed in Table 3 and will be used throughout this section on municipal waste waters. The format of using the 50th and 90th percentiles, hereafter referred to as the 50% value and 90% value, respectively, was chosen to facilitate comparison of raw waste water to the reference values. This summary of data from about 140 waste treatment facilities expected to have comparatively high metal content shows several trends of importance to decision makers. Cobalt, Fe, Pb, Ni, and Zn show little tendency to require reduction by pretreatment prior to use of even raw waste water for irrigation while Cd, Cr, Cu, and Mn would require up to 67% removal in order to make the 90% value conform to the recommended maximum values for irrigation use. Copper and Zn concentrations would seldom require removal during the conditioning of raw waste water for use as a public water supply. The remaining metals would require removals of up to 83% to bring the 90% value into conformance with the criteria for public water supplies. Data in Table 3 on metal concentrations in raw waste water will be used as a comparative base in the subsequent discussion of metal removal by primary and secondary treatment processes.

2. PROPERTIES OF PRIMARY EFFLUENTS

The efficiency of primary treatment has been judged, historically, through the measurement of relatively few parameters. The principal design objective of primary treatment is removal of 50% to 65% of the suspended solids and 25% to 40% of the BOD by providing a brief sedimentation period. The sedimentation period may range from 30 to 150 minutes and the settled solids are removed as primary sludge. The shorter detention periods of 30 to 60 minutes are usually used as preliminary treatment ahead of a secondary process. The generalized values listed in Table 4 are indicative of the expected concentrations of constituents frequently included in an assessment of the properties of primary effluents. The comparatively high BOD of primary effluents has been a deciding factor in their use for crop irrigation and other land application approaches. The potential for development of septicity and associated obnoxious odors in holding ponds or distribution systems has been a frequent cause for abandonment of the practice of using primary effluents for irrigation.

Table 3—Metals in raw municipal waste waters†

| Constituent | Recommended maximum values Nat. Acad. Sci. (1973) | | Concentration, mg/liter | | | |
| | Continuous irrigation | Public water supply | Wisconsin | | Connecticut, New Jersey, and New York | |
			50% value	90% value	50% value	90% value
Cadmium	0.01	0.01	--	0.03	--	0.02
Chromium	0.10	0.05	0.2	3.6	0.05	0.30
Cobalt	0.05	--	--	--	--	0.05
Copper	0.20	1.0	0.1	0.3	0.1	0.4
Iron	5.0	0.3	--	--	0.9	1.9
Lead	5.0	0.05	0.1	0.3	--	0.2
Manganese	0.20	0.05	--	--	0.14	0.30
Mercury	--	--	0.001	0.006	0.0013	0.0045
Nickel	0.20	--	0.08	0.30	--	0.2
Silver	--	--	--	--	--	0.05
Zinc	2.0	5.0	0.38	1.0	0.18	0.76

† Data from Konrad and Kleinert (1974) and Mytelka et al. (1973).

Table 4—Typical properties of primary effluent

Constituent	Concentration, mg/liter	
	Range	Median
Solids		
Total dissolved	200–1,500	500
Total suspended	50– 150	100
Biochemical oxygen demand	65– 200	135
Chemical oxygen demand	150– 750	335
Nitrogen		
Total	10– 60	40
Free ammonia	7– 40	30
As nitrate	–	<0.1
Phosphorus, total	5– 17	8

The metal content of primary effluents is of interest because it gives an indication of the amount of metals tied up in suspended solids removed as primary sludge. The data of Mytelka et al. (1973) have been selected to estimate the metals removal observed at operating primary treatment plants. The survey included more than 70 primary treatment plants in Connecticut, New Jersey, and New York. This assessment of the metal content of raw waste waters and primary effluents is presented in Table 5. Cadmium, Co, Pb, and Ag were omitted because concentrations in both the raw waste waters and the primary effluents were consistently below the detection limit for analytical procedures used for the survey. The differences between the raw waste water and the primary effluent at both the 50% and 90% frequencies suggest that primary treatment results in low to moderate reduction for most of the metals listed in Table 5. Primary treatment deserves little credit for reducing metal concentrations to meet criteria for use as irrigation water or for further conditioning as a source for public water supplies.

3. PROPERTIES OF SECONDARY EFFLUENTS

Secondary treatment processes regularly in use include several types of activated sludge processes, trickling filters, and treatment pond systems. Buzzell (1972) projects typical properties of secondary effluents for nine separate processes. Driver et al. (1972) and Menzies and Chaney (1974) give ranges for a typical secondary effluent. For the purpose of this chapter, it seems appropriate to list typical values such as these for comparison to one another and for comparison to some data collected at operating waste treatment facilities with secondary treatment processes. The values listed in Table 6 are the tabulations of typical values or ranges of values. Typical values and ranges similar to those given in Table 6 for the suspended solids, oxygen demand measurements, N, and P are well accepted in the waste water management profession as reliable estimates for activated sludge or trickling filter treatment. There is little purpose to tabulate actual data to confirm the reliability of the estimated values and ranges for these treatment processes. The capability of lagoon systems to produce effluents with com-

Table 5—Metals removal by primary treatment†

Constituent	Concentration, mg/liter for 50% values		Percent difference for 50% values	Concentration, mg/liter for 90% values		Percent difference for 90% values
	Raw waste water	Primary effluent		Raw waste water	Primary effluent	
Chromium	--	--	--	0.30	0.25	17
Copper	0.1	0.1	0	0.55	0.40	27
Iron	1.0	0.8	20	3.4	3.0	13
Manganese	0.16	0.16	0	0.30	0.32	--
Mercury	0.0012	0.0009	31	0.0040	0.0040	0
Nickel	--	--	--	0.3	0.2	50
Zinc	0.20	0.18	10	0.74	1.02	27

† Adapted from Mytelka et al. (1973).

parable levels of suspended solids and oxygen demanding substances has been the subject of considerable recent debate (McKinney, 1970). Two studies including a total of 154 lagoon systems of various designs will be used to illustrate the quality of effluents from this type of secondary treatment. Barsom (1973) conducted a survey involving 105 lagoon systems of different types scattered throughout the United States. Pierce (1974) conducted an intensive study of 49 multicell lagoon systems in Michigan with semiannual discharge. Data collected in these two studies are summarized in Table 7. Since terminology for describing lagoon designs is not consistent, it is appropriate to elaborate further regarding terms in the "type of lagoon" column. The facultative term includes lagoons which are usually 1 to 2 m deep, are loaded with 10- to 100-kg of BOD_5 ha^{-1} day^{-1}, have O_2 stratification resulting in aerobic and anaerobic zones, and have retention times of about 40 days. Mechanically aerated lagoons are similar in physical design to facultative lagoons but they are loaded at 10- to 300-kg of BOD_5 ha^{-1} day^{-1}. Their oxygen source is mechanical aeration; their retention time is 2 to 30 days and O_2 stratification may or may not occur depending on the level of mechanical aeration. As included in Table 7, polishing (tertiary) ponds receive secondary effluents from other treatment processes at a loading of 10- to 50-kg of BOD_5 ha^{-1} day^{-1} with detention times of a few hours to a few days. Polishing ponds may be designed with several months of detention time but the data in Table 7 represent the 18 short term polishing ponds included in the study by Barsom (1973). Oxidation ditches are shaped like a race track, are loaded at 25- to 200-kg of BOD_5 ha^{-1} day^{-1}, have mechanical brush aerators to supply O_2, have a detention time of about 1 day, and are usually followed by a clarifier for solids separation. Multicell lagoons with semiannual discharge are usually a combination of a facultative lagoon and one or more polishing (tertiary) ponds which provide a total detention time of 180 days. This brief description of the terms in the "type of lagoon" column helps to clarify the observed trends in the suspended solids and BOD properties of the various lagoon effluents. The facultative lagoon, which is the simplest of the designs, produces an effluent with the greatest concentration of suspended solids and BOD. The quality of the lagoon effluent improves steadily with aeration and time of detention. The multicell, long term detention systems produce the best quality effluent with respect to suspended solids and BOD. By comparing the median values and the ranges for the lagoon systems (Table 7) to estimates for the properties of secondary effluents (Table 6), it can be seen that facultative and artificially aerated lagoons as defined for Table 7 do not produce effluents with properties comparable to the estimates in Table 6. The other four lagoon systems described for entry in Table 7 do produce effluents with qualities comparable to the estimates in Table 6.

Fecal coliform counts are an important indicator of the biological properties of treated municipal waste waters. The fecal coliform count has not been included in the tabular presentations because the raw waste water, primary effluents, and even most secondary effluents contain tens of

Table 6—Three estimates of the properties of secondary effluents

| | | | Concentration, mg/liter | | |
| | Buzzell (1972)† | Driver et al. (1972) | Menzies and Chaney (1974) | |
Constituent	Range	Typical value or range	Typical value	Range
Suspended solids	13-62	25	–	–
Biochemical oxygen demand	13-75	25	–	–
Chemical oxygen demand	50-160	70	–	–
Nitrogen, total	–	20	25	15-40
Phosphorus, total	7-10	10	10	0.5-40
Trace metals				
Cadmium	–	0.015	<0.005	<0.005-6.4
Chromium	0.01-2.5	0.02-0.14	0.025	<0.05-6.8
Cobalt	–	–	<0.05	<0.05-0.05
Copper	0.10-1.4	0.07-0.14	0.10	<0.02-5.9
Iron	0.10-3.0	0.10-4.3	–	–
Lead	0.01-1.0	0.01-0.03	0.05	<0.02-6.0
Manganese	–	0.20	–	–
Mercury	<0.005	0.01	0.001	<0.0001-0.125
Nickel	0.02-2.0	0.03-0.20	0.02	<0.02-5.4
Zinc	0.10-1.10	0.20-0.44	0.15	<0.02-20
Other parameters				
Boron	0.5-1.0	1.0	–	–
Calcium	1-40	20	–	–
Magnesium	1-10	17	–	–
Potassium	7-10	14	–	–
Sodium	40-100	50	–	–
Chloride	40-100	45	–	–
Sulfate	12-52	–	–	–
Oil	0-10	–	–	–
Phenol	0-1	–	–	–

† Range given encompasses seven different activated sludge processes and trickling filters.

Table 7—Suspended solids and biochemical oxygen demand (BOD) of lagoon effluents†

| | Concentration, mg/liter | | | | | |
| | Suspended solids | | | Biochemical oxygen demand | | |
Type of lagoon	Number‡	Median	Range §	Number‡	Median	Range §
Facultative	41- 72	83	32-222	94-130	40	15-100
Mechanically aerated	29- 57	66	21-135	29- 59	40	17-160
Polishing (tertiary)	14- 14	42	13- 57	18- 18	20	9- 29
Oxidation ditch	21- 43	37	12-151	21- 51	25	7- 90
Two-cell semiannual discharge	28-789	30	15- 45	28-789	15	5- 25
Three- or more-cell semiannual discharge	21-686	25	10- 45	21-686	12	5- 25

† Adapted from Barsom (1973).
‡ 41-72 equals 41 lagoon systems with 72 samples taken, etc.
§ Range given encompasses the 80% of the values falling between the 10% and 90% values.

thousands to millions of these organisms as indicators of recent human pollution. Disinfection through the use of Cl or to a lesser degree through the use of other disinfectants has been adopted as the principal approach for reducing the fecal coliform count in treated municipal effluents. Use of such disinfection practices can, at least momentarily, reduce the fecal coliform count to a few hundred counts per 100 ml. Lagoon systems with long retention times also reduce fecal coliform counts to a few hundred counts per 100 ml without addition of disinfectants. Pierce (1974) reported on intensive fecal coliform monitoring of unchlorinated discharges from the 49 lagoon systems included in his study of Michigan lagoons. Data showing fecal coliform counts for 163 discharge periods were assessed for conformance with the EPA secondary treatment requirements that the geometric mean of the fecal coliform counts be < 200 for samples taken in a period of 30 consecutive days and < 400 for samples taken in a period of 7 consecutive days. A total of 147 or 90% of these unchlorinated discharges conformed to the EPA criteria for secondary treatment. Releases which did not conform occurred during discharges from ice-covered ponds or when ponds were being drawn down to very low levels.

The metals content of secondary effluents is of particular interest because secondary effluents are most apt to be considered for land-based management systems oriented to food production or recharge of ground waters which may be used for public water supplies. Municipal effluents from rural communities or smaller urbanized areas with little industrial development are expected to have comparatively low metal content. Secondary effluents from highly industrialized urban areas are most apt to have high metal content and it is this type of source which will be addressed in the following discussion. The data used were reported by Blakeslee (1973), Konrad and Kleinert (1974), and Mytelka et al. (1973). These data are summarized in Table 8 to facilitate several comparisons regarding the properties of secondary

Table 8—Metals removal through secondary treatment and properties of secondary effluents for up to 200 plants

Constituent	Maximum recommended values in mg/liter for Nat. Acad. Sci. (1973)		Effluent properties†		Metals removal by secondary treatment in %‡	
	Continuous irrigation	Public water supply	50% value	90% value	Konrad and Kleinert (1974)	Mytelka et al. (1973)
Arsenic	0.10	0.10	-	0.01	-	-
Cadmium	0.01	0.01	-	<0.02	33	-
Chromium	0.10	0.05	<0.05	0.17	58	67
Cobalt	0.05	-	-	<0.05	-	-
Copper	0.20	1.0	0.05	0.22	50	28
Iron	5.0	0.30	0.50	2.0	-	47
Lead	5.0	0.05	-	<0.20	47	-
Manganese	0.20	0.05	-	0.38	-	13
Mercury	-	-	0.0005	0.0015	83	26
Nickel	0.20	-	<0.10	0.20	33	-
Silver	-	-	-	<0.05	-	-
Zinc	2.0	5.0	0.12	0.35	50	47

† Percentiles interpolated from data collected by Blakeslee (1973), Konrad and Kleinert (1974), and Mytelka et al. (1973).
‡ Degree of metals removal is estimated by comparison of 90% values on the influent and the effluent.

effluents with respect to metal content. It should be noted that the entries for effluent properties in Table 8 are derived from varying numbers of individual analyses ranging from 137 for Hg to 336 for Cr.

The National Academy of Sciences (1973) recommendation for maximum levels in waters used continuously for irrigation on any soil is included for direct comparison to the values for effluent properties which are the 50% and 90% values for all of the individual analyses. This comparison shows that metal content of these secondary effluents from highly industrialized areas are less than the recommended levels in most instances. No recommended levels are exceeded at the 50% value for the effluents. Chromium, Cu, Mn, and Zn exceed the recommended value at the 90% level and the comparison for Cd is indeterminant because the recommended level is below the detection limit of the analytical procedure used for most of the individual analyses. This comparison shows planners that the metal content of secondary effluents from highly industrialized areas usually meet the National Academy of Sciences (1973) recommendations, yet it alerts planners to be aware that concentrations of some metals may exceed these same recommendations in 10% or more of the cases. Reduction of metal concentration by pretreatment of industrial waste streams would seem appropriate in these cases.

Table 8 also includes the National Academy of Sciences (1973) recommendation for maximum levels in waters used for public water supplies. These levels are of particular interest in land-based systems designed to recharge ground waters. The comparison between the recommended maximums for public water supplies and the effluent concentrations shows that Fe exceeds the recommended values at both the 50% and 90% values. Chromium, Fe, Pb, and Mn exceed the recommended values at the 90% value. This comparison alerts planners to the importance of metals in secondary effluents when planning recharge systems using secondary effluents from highly industrialized areas. Pretreatment of industrialized waste water streams to reduce metal concentrations and the degree of control over the local hydrologic cycle are important considerations in planning land-based systems for recharge purposes. Both the 50% and 90% values for the metals data in Table 8 fall within the typical values or ranges for secondary effluents presented in Table 6. In using estimates such as those reproduced in Table 6 one must remember that individual cases or segments of a population frequently deviate from median values or ranges set forth to be representative of a whole population. Difference between the estimated high values in Table 6 and the 90% values in Table 8 shows the divergence that may occur in the top 10% of these distributions. The high value of 6.8 mg/liter for Cr is 40 times greater than the 0.17 mg/liter concentration at the 90% level for all data combined as given in Table 8 and is 34 times greater than the 0.20 mg/liter concentration which is the 90% value for the data of Mytelka et al. (1973).

The other comparison detailed in Table 8 is an estimate of the degree of metals removal achieved by secondary treatment processes. Results of a study of three treatment plants conducted by the U. S. Public Health Service (1975) showed that 37% to 40% of the Cr was removed from the waste water as it passed through a secondary treatment plant. Similarly, determined values for the same study were 16% to 73% for Cu, 53% to 85% for Zn, and 8% to 78% for Ni. Data collected by Konrad and Kleinert (1974) and by Mytelka et al. (1973) offered an opportunity to estimate the degree of removal for a large population sample of treatment plants. The percentage reductions shown in Table 8 are derived by dividing the 90% value for the influent into the value obtained by subtracting the 90% value for the effluent from the 90% value for the influent. The validity of this approach is subject to many factors and the Zn data of Konrad and Kleinert (1974) were treated in an alternate manner as a check of the reasonableness of estimates with this approach. The alternate method was a determination of percent removal for each individual analysis for the 33 pairs of analyses included in their data. The median value for these 33 data pairs was 50%, which is the same as the value obtained by using the 90% values for influent and effluent. The exact agreement for this specific case lends credence to the method used to obtain the percent removals as shown in Table 8. The range for the 33 individual determinations was minus 85% to plus 90% with 27 of the 33 determinations showing removals. The assessment presented in Table 8 shows that secondary treatment does reduce metal concentrations by some 15% to 80% from the values in the raw waste water. Reduction of Cr and Zn are quite comparable for the different sources of data, whereas the removal of Cu and Hg are quite different for the two sources. Regardless of their absolute accuracy, the values in Table 8 help one to estimate the degree of metals removal to be expected during the secondary treatment process.

The foregoing discussion of the properties of municipal waste waters has been directed to presentation of information useful to those considering application of municipal effluents to land. It is based on the premise that selected information presented in a specific manner would be more pertinent for inclusion than a comprehensive review of hundreds of reports discussing treatment of municipal waste waters and including intensively collected data for a single site or experimental trial. To this end, the references utilized have been those which included data collected at numerous sites and provided an opportunity to make generalized conclusions about certain properties of raw waste water and the effluents produced by several primary and secondary processes in common usage.

III. AGRICULTURAL WASTE WATERS

Two major sources of agricultural waste waters are irrigation return flows and animal production wastes. Irrigation return flows constitute the greater volume of the two sources, but they do not carry the high organic loading that animal wastes do and, therefore, do not pose the same treatment, disposal, and/or utilization problems. Runoff from animal production areas carries high loadings of oxygen demanding organic materials which require treatment, disposal, or control measures to protect the integrity of any possible receiving water resources. The quantity and quality of irrigation return flows will be discussed only briefly and in general terms while the major emphasis will be given to animal production wastes which are amenable to land-based waste water management systems.

A. Irrigation Waste Waters

Since 1890 the total irrigated land in the United States has increased 1,200%, from about 1.6 million ha to over 19.4 million ha (Skogerboe & Law, 1971). Projections of agricultural irrigation growth in the United States (Pavelis, 1967) indicate that by 1980 there will be 20.2 million ha under irrigation; by the year 2000, 23 million ha and by 2020, 25.5 million ha. If realized, these projections will mean increased withdrawals, consumptive use, and return flows, but the magnitude of such increases would be difficult to predict.

The major pollutional factor associated with irrigation is the increased concentration of soluble mineral constituents in the drainage water (Law & Bernard, 1970). Water is transpired by growing plants and evaporated from the soil, resulting in a concentration of the dissolved mineral salts that are present in all natural water resources. Irrigation may increase the pollutant load of receiving waters by leaching natural salts from weathered minerals occurring in the soil profile or deposited below. Irrigation return flows provide the vehicle for conveying the concentrated salts and other pollutants out of the crop root zone and to a receiving stream or ground-water reservoir.

Irrigation causes substantial water quality changes due to varying exposure conditions (Law & Skogerboe, 1972). Surface return flows (tailwater), experiencing limited contact and exposure to the soil surface may undergo the following changes in quality between application and runoff: slight increase in dissolved solids; addition of variable amounts of fertilizer elements and pesticides; an increase in sediments and other colloidal material; increase in organic matter (crop residues, etc.); and increased bacterial content. Drainage water that has moved through the soil profile will experience different changes in quality. Because of its more intimate contact with the dynamic soil-plant-water system, the following changes in quality are predictable: considerable increase in dissolved solids concentration; the distribution of various cations and anions may be quite different from the water supply; variation in the total salt load, depending on whether there has been deposition or dissolution; little or no sediment or colloidal material; possible increase in NO_3-N content; very low P content; and reduction of oxidizable organic substances and pathogenic organisms, if present. Either type of return flow will affect the receiving water in proportion to the respective discharges and relative quality of the receiving water.

Many factors affect the quality of both surface return flows and subsurface drainage from irrigated lands. Among these are quality of applied water, salinity status of the soil, climatic variables, and water demand of the crop being grown. These and other factors combine to preclude setting "typical" values for the quality of irrigation waste waters, which may be very site-specific. A field investigation in southwestern Oklahoma (Law et al., 1970) found total dissolved solids (TDS) of surface return flows to range from 1,700 to 2,100 mg/liter during the first irrigation with 1,400 mg/liter TDS irrigation water after an over-winter fallow season. Subsequent irrigations produced tailwater with from 1,500 to 1,600 mg/liter TDS. Percolating soil water extracted with vacuum ceramic samplers at the 45 cm depth ranged from 6,000 mg/liter following the first irrigation to 12,000 mg/liter after the third irrigation. At 75 cm, soil water extracts were as high as 16,000 mg/liter TDS after the third irrigation of the season. The crop was cotton (*Gossypium hirsutum* L.) and the area is in an average rainfall belt of 63 cm where supplemental irrigation is required.

Extensive studies have been conducted in the Grand Valley of western Colorado to evaluate certain salinity control technology (Skogerboe & Walker, 1972; Skogerboe et al., 1974). Irrigation in this arid region is with Colorado River water ranging from 500 to 600 mg/liter TDS and by furrow and flood application methods. The area is underlain with saline Manchos shale which results in excessive salt pickup from overirrigation and flushing the local ground water into the river system. The ground-water quality is consistently found to range from 6,000 mg/liter to over 8,000 mg/liter TDS. In contrast to this high salt-loading situation are the results of irrigation return flow quality studies reported by Carter et al. (1971) from southern Idaho. Irrigation there is with Snake River water with an average TDS content of 295 mg/liter. Surface return flow had essentially the same TDS content as the applied water and subsurface drainage measured over a large area and at numerous sites had TDS values ranging from 555 to 735 mg/liter with a mean value of 665 mg/liter. These cases serve to illustrate the wide variability found in irrigation waste water quality and its dependence on local conditions. Only salinity concentrations have been cited; however, similar variances will be found in nutrient content, sediment loading, and other minor pollutants.

In summary, pollutants associated with irrigation waste waters are dissolved mineral salts, suspended solids (sediment), plant nutrient elements, and pesticides or other minor organics. Since these waste waters do not carry excessive loadings of biodegradable organic materials and are not amenable to land-based waste water management systems, further discussion will not be pursued.

B. Animal Production Waste Waters

During recent decades, animal production methods in the United States have undergone dramatic changes. The trend has seen animal production changing from small, individual farm operations into large-scale commercial enterprises (Loehr, 1968). Small animals—i.e., poultry and swine—are being confined within smaller areas or in buildings where environmental conditions are controlled to produce the greatest weight gain in the shortest possible time. Beef cattle are being produced in confined feedlots rather than on pasture and range, and the more recent trend is toward controlled areas under roof for better environmental control. The efficiency of animal production of all types has greatly increased under these newer methods. However, there has been a concomitant increase in a variety of aesthetic and environmental quality problems associated with these changes in animal production methods. Not the least of these problems is the potential for severe degradation of our water resources that may receive runoff or direct discharges of wastes from these concentrated animal production areas. The thrust of this discussion will be confined to the consideration of the properties of those waste waters that are important from the standpoint of land-based waste water management systems. The most obvious solution to severe water quality problems arising from animal waste areas is the collection, retention, and ultimate land disposal or crop utilization of the liquid wastes. For protection of water quality, many states have enacted legislation requiring that all such waste waters be collected and not allowed to enter freshwater streams and reservoirs. In addition, the U. S. Environmental Protection Agency (1974) has promulgated effluent guidelines for the point source category of animal feedlots which recommend that there be no discharge of waste water pollutants to navigable water bodies after 1 July 1977. Other factors also enhance and support the wisdom of such efficacious practices. For example, the recent history of high cost and shortages of commercial fertilizers has stimulated a new interest in animal manures and water-borne wastes. By crude estimates, 16 dairy cows, 32 growing beef cattle, or 1,700 laying hens will excrete approximately 1 metric ton of N in a year. Since it requires about 1,200 m^3 of natural gas to produce 1 metric ton of N as anhydrous NH^3, resource conservation by on-the-farm utilization of animal wastes becomes a significant factor. These and other considerations provide ample reason for a greatly increased interest in controlling these waste waters and effectively utilizing them for crop production and/or efficiently designed and operated land-based waste water management systems.

1. RAINFALL-RUNOFF RELATIONSHIPS

There are many factors which exert a direct influence on the quantity of runoff produced by any given rainfall event. Among these are the size and slope of the drainage area, duration and intensity of rainfall events, type of vegetative cover, soil type and surface roughness, and antecedent soil moisture. The condition of feedlot surfaces will also influence runoff quantity and quality from those areas. Clark et al. (1975) analyzed runoff from a large feedlot in the Southern Great Plains near Bushland, Texas. An area about 4 ha in size, stocking about 3,000 cattle, was instrumented and data were collected and analyzed for a 3-year period. Rainfall totaled 46.0, 45.3, and 37.4 cm during 1971, 1972, and 1973, respectively. Runoff totaled 9.4,

5.2, and 2.0 cm, respectively, for the same years. The regression analysis of these data showed a linear relationship between rainfall and runoff and that about one-third of the rainfall ended up as runoff when the rainfall event exceeded 1.0 cm. In general, rainstorms of < 1.0 cm did not produce runoff. It was noted that similar storms produced different runoff patterns. One storm, following a 4-month dry period, produced three times more runoff than resulted from a similar rainfall after a relatively wet period. The observation that less runoff occurred when the feedlot surface was wet was due to feedlot surface conditions. The wet surface is roughed by the animals and numerous tracks and depressions create more surface storage; whereas, a dry lot is packed smooth with very little surface storage. The authors compared the results of the Bushland study with rainfall-runoff data from other locations. The regression curves showed the Bushland runoff to be about 60% of that produced by similar rainfall in western Kansas, about 50% of that for two locations in Nebraska, and only 40% of that for a southeast Texas location. Higher stocking densities, producing a thicker manure pack to absorb more water, would also influence the runoff pattern.

Studies conducted at Mead, Nebraska (McCalla et al., 1972) have shown that about one-third of the 71 cm annual precipitation will run off the feedlots. They found also that from 3% to 6% of the material deposited on a feedlot will be transported in the rainfall runoff and that snowmelt runoff will transport even greater amounts of pollutant materials. Slopes of 3%, 6%, and 9% were employed in the runoff studies. An increase in the slope of the feedlot surface increased the quantity of total solids, volatile solids, N and P removed with the runoff, but the variability was such that it was not possible to determine the exact effect of slope on runoff quantity and quality. A major conclusion from this study was that cattle feedlot runoff transports an excess of nutrients and, because of its poor quality, must not be discharged into streams.

Wells et al. (1971) reported on a study which seemed to show that ration had some influence on the quantity of runoff from a concrete-surfaced feedlot. Over a 2-month period, 46.5 cm of rainfall were measured. The lot holding cattle being fed an all-concentrate ration produced 56.5% of the rainfall as runoff, while the lot on which cattle were being fed a 12% roughage ration produced only 43% as runoff. Similar dirt-surfaced lots receiving the same rainfall over the same period produced 25.8% and 23.1% runoff from the all-concentrate and 12% roughage rations, respectively. The greater difference, of course, is observed between concrete- and dirt-surfaced feed pens.

Kreis et al. (1972) measured rainfall and runoff from a large dirt-surfaced feedlot in north Texas from July 1969 to April 1970. During this period, runoff from the feedlot area totaled 39% of the 66 cm of rainfall recorded.

Other rainfall-runoff data have been reported from west central Kansas (Manges et al., 1975). An area of 25 feedlot pens covering 11.1 ha received 17.4 cm of rainfall which produced 6.7 cm (38.5%) of runoff. A smaller area

(0.82 ha) received 13.1 cm of rainfall and produced only 3.5 cm (26.5%) of runoff.

Rainfall and runoff were measured over a 5-year period from a feedlot in Nebraska (Porter et al., 1975). The pen area was approximately 37 m wide and 91 m long with a 6% slope in the long direction. From July 1968 through December 1972, precipitation totaled 314 cm and occurred on 323 days. There were 89 runoff events and 104 cm (34.5%) of runoff. It was estimated that about 23 cm (32.5%) of runoff can be expected annually from a sloping lot with 71 cm of average annual precipitation at that location.

It soon becomes obvious that there is no magic formula by which one may accurately predict the runoff to be expected from animal feedlots under a wide range of climatic and site conditions. For the design of settling basins, collection lagoons, and runoff retention structures, one must resort to historical rainfall data, evaporation data, local runoff records (if available), and other pertinent information that will assist in adequate design parameters. Detailed discussion of these and other design criteria are beyond the scope of this chapter.

2. PROPERTIES OF ANIMAL PRODUCTION WASTE WATERS

Numerous data sources are found in the literature which indicate the quality of runoff from cattle feedlots to be highly variable (Miner et al., 1966; Loehr, 1968; Powers et al., 1975), depending upon such factors as rainfall intensity, feedlot surface, temperature, antecedent moisture, manure accumulation, etc. For example, organic content, measured as COD, in cattle feedlot runoff can range from 3 to 11 times that of untreated domestic sewage (Miner & Willrich, 1969). Historically, where animals graze on vegetated range or pasture, there has been little potential for serious water quality problems. Manure is distributed in light applications, liquids are absorbed by the soil, and the vegetative cover utilizes the added nutrients and inhibits erosion. Low-intensity rainfall is absorbed by the soil and high-intensity rainfall provides ample dilution to minimize the concentration of potential pollutants in the runoff. On the other hand, animals produced in open feedlots in such concentrations as to remove or prevent the growth of all vegetative cover present pollution hazards quite unlike the pasture systems. During and immediately following rainfall and spring thaws, water flows over the manure-covered surfaces, carrying with it heavy loads of both particulate solids and soluble manure constituents. The action of animal hooves on the manure-covered surface creates an area void of vegetation and one through which infiltration rates are greatly reduced; however, considerable surface storage is available on feeding areas in the hoof depressions (Miner & Willrich, 1969).

In assessing the significance of animal production waste waters within a drainage basin, both quantity and quality of the waste water source must be considered. Assuming an earthen lot with 2% slope, about 28–30 cm of annual runoff might be expected from 76 cm of annual rainfall, with runoff occurring during 30 days of the year. At an average of 1,000 mg/liter of

Table 9—Concentrations of chemical constituents measured in direct runoff from beef cattle feedpens†

	No. of samples	Mean	Minimum	Maximum
			mg/liter	
Total solids	8	11,429	3,110	28,882
Total suspended solids	8	5,912	745	17,202
Volatile suspended solids	7	3,426	475	9,286
Total dissolved solids	8	5,526	882	22,372
Chloride	7	450	97	648
Total PO_4-P	16	69.2	21	223
NO_3-N	15	0.64	<0.05	2.3
NH_3-N	15	108	4	173
Total organic N	15	228	31	493
COD	15	7,210	1,439	16,320
BOD_5	4	2,201	1,075	3,450
TOC	15	2,010	150	4,400
Ca	6	698	194	1,619
Mg	6	69	28	89
Na	6	408	130	655
K	6	761	226	1,352

† From Kreis et al. (1972).

BOD_5, the runoff from a feedlot on each of these 30 days would be equivalent to the untreated sewage from about 1,250 people for each hectare of feedlot surface (Miner & Willrich, 1969). Admittedly, such an average is of little value in actual situations, but it does indicate that runoff from cattle feedlots is a significant source of organic waste waters.

Several data sources are available in which investigators have provided rather complete analyses of the quality of animal production waste waters. Table 9 provides the minimum, maximum, and mean for concentrations of chemical constituents measured in direct runoff from a cattle feedlot in north Texas (Kreis et al., 1972). Collection of the feedlot runoff in holding ponds had a significant impact on the quality of the waste water. Reduction in pollutant concentrations was a result of dilution by direct rainfall on the holding pond surface and, probably more importantly, several days' detention in the ponds settled out most of the particular matter and suspended solids. Detention in the holding ponds resulted in a reduction in suspended solids concentrations ranging from 60% to 90%. Mean nutrient concentrations were reduced from 40% to 80%. Mean concentrations of the organic constituents in the form of COD, BOD, and TOC were reduced by approximately 70%. It was concluded that a large proportion of all pollutants was associated with the suspended solids carried in the feedlot runoff.

Long term studies reported by Porter et al. (1975) have produced detailed analyses of the characteristics and chemical values of runoff from beef cattle feedlots in Nebraska. A comparison of the runoff from rainfall and snowmelt is shown in Table 10. It was found from these studies that from 3% to 6% of the material deposited on a feedlot will be transported in the rainfall runoff. The quantity of material removed by snowmelt was even

Table 10—Ranges in the characteristics and chemical values of runoff from beef cattle feedlots, Mead, Nebraska 1968–1972†

	Snowmelt runoff			Rainstorm runoff		
	Low	High	Mean	Low	High	Mean
pH	4.1	9.0	6.3	4.8	9.4	7.0
Conductivity, mmhos/cm	3.0	19.8	7.1	0.9	5.3	3.2
Total solids, %	0.8	21.8	7.7	0.24	3.3	1.93
Volatile solids, %	0.6	14.3	3.9	0.12	1.5	0.82
Ash, %	0.2	9.2	3.8	0.12	2.8	1.11
COD, mg/liter	14,100	77,100	41,000	1,300	8,200	3,100
P, ppm	5	917	292	4	5,200	300
NH_4-N, ppm	6.0	2,028	780	2	1,425	151
NO_3-N, ppm	0	280	17.5	0	217	10
Total nitrogen, ppm	190	6,528	2,105	11	8,593	854

† From Porter et al. (1975).

Table 11—Chemical analyses of feedlot runoff effluent applied to grass and clover plots, Springfield, Nebraska, 1970–1972†

	1970		1971		1972	
	Average	Standard deviation	Average	Standard deviation	Average	Standard deviation
Total solids, ppm	3,800	914	3,000	717	2,200	443
Volatile solids, ppm	2,000	542	1,300	282	1,000	169
Total nitrogen, ppm	188	92	91	23	96	88
NH_4-N, ppm	54	21	42	22	29.9	16.6
NO_3-N, ppm	1.4	0.5	0.6	0.8	9.4	12.7
Total P, ppm	9.4	4.6	42	26	21.1	5.2
pH	8.1	0.8	8.6	0.8	8.4	0.5
Electrical conductivity, mmhos/cm	2.76	0.55	1.47	0.85	1.30	0.18
No. of observations	6		18		16	

† Adapted from Swanson et al. (1974).

greater. When thaws occurred, it was observed that a slurry of undecomposed manure flowed from the lot. The snowmelt runoff that contained high solids content came from lots with cattle that were on high-concentrate rations. Total N in the winter runoff was as high as 6,500 ppm, while NO_3-N varied from 0 to 280 ppm in the runoff from rain.

Another example of feedlot runoff quality following detention in holding ponds is provided by Swanson et al. (1974). Plots of tall fescue (*Festuca elatior*) and perennial ryegrass (*Lolium perenne*), both overseeded with Ladino clover (*Trifolium repens*), were irrigated with feedlot runoff holding pond effluent. The data of Table 11 were derived from chemical analyses performed at each irrigation application. There were 6 irrigations reported in 1970, 18 in 1971, and 16 in 1972. Averages and standard deviations for each year are shown. The standard deviations reported give a good indication of the variability of the data.

Another long term study has been reported by Manges et al. (1975) which was carried out in west central Kansas. The study commenced in 1969 and continued through 1973. Two areas of a large feedlot were instrumented to measure runoff volume and at the same time collect samples for chemical analyses. The resultant data are shown in Table 12 for the two feedlot areas. The range, mean, and standard deviation are shown for each chemical parameter measured. An added aspect of the study dealt with the effect of anaerobic lagoon storage on the quality of the feedlot runoff. After a period of rains in May and early June, the COD in the lagoon was about 6,000 mg/liter. With no additional runoff for over 2 months, the COD value had dropped to 2,000 mg/liter by mid-August. Evaporation during this period resulted in a concentrating effect on total solids and some of the inorganic constituents.

Emphasis has been given to studies on waste water from beef cattle feedlots. From a rainfall runoff standpoint, this is by far the greatest source

Table 12—Characteristics of rainfall runoff from beef cattle feedlots†

Parameter	Range	Mean	Standard deviation	No. of observations
Area 119 (0.82 ha)				
COD, mg/liter	16,100–861	7,596	3,255	72
Total N, mg/liter	1,580–165	675	364	55
Total P, mg/liter	242– 9	79	42	56
Total solids, mg/liter	19,252–214	8,442	5,190	48
Volatile solids, mg/liter	9,552– 36	3,888	2,680	48
NH_4^+-N, mg/liter	580– 0	159	112	81
NO_3^--N, mg/liter	48– 0	10	18	81
Na, mg/liter	2,970– 31	560	551	81
K, mg/liter	2,990– 29	796	586	81
Ca, mg/liter	402– 31	181	83	81
Mg, mg/liter	183– 35	98	29	81
EC, mmhos/cm	15– 1	7	3	78
Area 2 (11.1 ha)				
COD, mg/liter	14,309–1,514	6,111	2,631	85
Total N, mg/liter	962– 85	494	211	51
Total P, mg/liter	482– 19	87	89	86
Total solids, mg/liter	17,669–2,971	7,528	2,622	63
Volatile solids, mg/liter	11,437–1,429	3,891	1,627	63
NH_4^+-N, mg/liter	285– 14	141	57	99
NO_3^--N, mg/liter	45– 0	5	6	99
Na, mg/liter	735– 67	334	132	99
K, mg/liter	2,150– 134	851	480	99
Ca, mg/liter	1,040– 40	187	124	99
Mg, mg/liter	228– 22	86	40	99
EC, mmhos/cm	11– 1	5	2	99

† From Manges et al. (1975).

of animal production waste water. Other species and operations do produce waste water that cannot be discharged without creating severe water quality problems. Dairy operations often employ washwaters in the stalls and milking areas that produce slurry-type wastes that are definitely liquid in nature. An area which is experiencing much difficulty due to the high density of dairy animals is the Chino-Corona area of the upper Santa Ana River Basin in southern California (Grant & Brommenschenkel, 1974). Because of the unavailability of adequate land area for waste water disposal, studies have been conducted to find alternate treatment and/or disposal methods. The problem has not been fully solved. During their investigations, Grant and Brommenschenkel (1974) characterized fresh dairy manure slurry. These data are summarized in Table 13, for comparison with beef feedlot runoff data. Being a slurry, the organic parameters have much greater values than those previously given for feedlot waste waters.

Swine are often grown in housing with slotted floors which results in a slurry pit below that can be flushed with water. Robbins et al. (1971) reported values for swine waste lagoon influent. Typical values for the diluted fresh swine waste are shown in Table 14. Factors such as the amount of

Table 13—Experimental characterization of fresh dairy cow manure (slurry)†

Component	Average	Range	Standard deviation	Coefficient of variation	No. of samples
Total solids, % of manure	15.4	12.9 - 19.8	2.17	0.14	21
Volatile solids, % of solids	86.1	76.7 - 91.8	3.0	0.035	21
Total COD, g/liter	149	81 -284	57	0.38	17
Soluble COD, g/liter	33	19 - 53	10	0.30	16
Total BOD_5, g/liter	16.1	8.6 - 21.5	3.5	0.22	12
Soluble BOD_5, g/liter	9.3	4.6 - 14.4	3.0	0.32	11
pH	6.2	5.2 - 6.8	0.5	0.08	12
Total nitrogen, % TS	2.8	2.6 - 2.9			2
Soluble phosphorus, % TS	0.25	0.17- 0.32			4
Total potassium, % TS		0.5 - 5			7

† From Grant and Brommenschenkel (1974).

Table 14—Typical values for diluted fresh swine waste†

Total coliform	20 $\times 10^6$ colonies/100 ml
Fecal coliform	6 $\times 10^6$ colonies/100 ml
Fecal Streptococci	1.5 $\times 10^6$ colonies/100 ml
BOD_5	725 mg/liter
TOC	680 mg/liter
COD	1,400 mg/liter
Total solids	1,700 mg/liter
Volatile solids	1,000 mg/liter
Total N	200 mg/liter
NH_3-N	100 mg/liter
NO_3-N	1 mg/liter
Total PO_4	85 mg/liter
Ortho PO_4	75 mg/liter

† From Robbins et al. (1971).

washwater used per animal per day can markedly affect the concentrations of pollutants in lagoons receiving the wastes. Very little data are found in the literature indicating the bacterial content of animal waste waters. However, values for swine waste (Table 14) show these to be significant parameters that must be considered where health hazards are to be avoided.

3. GENERAL CONSIDERATIONS

The data amply illustrate the extreme variability in both quantity and quality of waste water derived from animal production areas. Add to this the variance of climatic, edaphic, and other agricultural factors from region to region and the obvious conclusion is readily reached that precise design and operational criteria for land-based waste water management systems are not easily established. There are, however, several important recommendations to be considered (Powers et al., 1975).

Animal production waste waters must be analyzed in order to determine the concentrations of pollutants to be managed. Seasonal fluctuations of the quality of lagoon waters must also be known. This is especially important if land applications are to be managed for optimum crop utilization of the nutrient content. Total dissolved solids (salinity) concentrations are important in order to minimize their toxic effects on land, crop, and ambient water quality. In extreme cases, lagoon water dilution may be required to minimize the adverse effects on the soil-plant-water regime. Animal production waste waters carry high organic loads and nutrient value that can be beneficial to crop production under well-managed land-based systems.

LITERATURE CITED

Barsom, G. 1973. Lagoon performance and the state of lagoon performance. Environ. Protect. Technol. Ser. EPA-R2-73-144, EPA, Washington, D. C. 214 p.

Blakeslee, P. A. 1973. Monitoring considerations for municipal wastewater effluent and sludge application to the land. p. 183–198. *In* Recycling municipal sludges and effluents on land. National Association of State Universities and Land-Grant Universities. Washington, D. C.

Bond, R. G., and C. P. Straub (ed.). 1974. Wastewater treatment and disposal, handbook of environmental control. Vol. IV. CRC Press, Cleveland, Ohio 905 p.

Buzzell, T. 1972. Secondary treatment processes. *In* Wastewater management by disposal on the land. Cold Regions Res. and Eng. Lab., Corps of Engineers, U. S. Army. Special Rep. No. 171. p. 35–47.

Carter, D. L., J. A. Bandurant, and C. W. Robbins. 1971. Water-soluble NO_3-nitrogen, PO_4-phosphorus, and total salt balances on a large, irrigation tract. Soil Sci. Soc. Am. Proc. 35:331–335.

Clark, R. N., A. D. Schneider, and B. A. Stewart. 1975. Analysis of runoff from southern great plains feedlots. Trans. ASAE 18:319–322.

Driver, C. H., et al. 1972. Assessment of the effectiveness and effects of land disposal methodologies of wastewater management. Department of the Army, Corps of Engineers. Wastewater Manage. Rep. 72-1. 147 p.

Durfor, C. N., and E. Becker. 1964. Public water supplies of the 100 largest cities in the United States, 1962. USDI, Geological Survey-Water Supply Paper 1812. 364 p.

Fair, G. M., J. C. Geyer, and D. A. Okun. 1971. Elements of water supply and wastewater disposal. Wiley and Sons, New York. 752 p.

Grant, F., and F. Brommenschenkel, Jr. 1974. Liquid aerobic composting of cattle wastes and evaluation of by-products. Environ. Protect. Technol. Ser. EPA-660/2-74-034, EPA, Washington, D. C. 50 p.

Konrad, J. G., and S. J. Kleinert. 1974. Surveys of toxic metals in Wisconsin. Department of Natural Resources, Madison, Wis., Tech. Bull. No. 74. p. 2–7.

Kreis, R. D., M. R. Scalf, and J. F. McNabb. 1972. Characteristics of rainfall runoff from a beef cattle feedlot. Environ. Protect. Technol. Ser. EPA-R2-72-061, EPA, Washington, D. C. 43 p.

Law, J. P., Jr., and H. Bernard. 1970. Impact of agricultural pollutants on water users. Trans. ASAE 13:474–478.

Law, J. P., Jr., J. M. Davidson, and L. M. Reed. 1970. Degradation of water quality in irrigation return flows. Oklahoma State Univ., Agric. Exp. Sta., Bull. B-684. 26 p.

Law, J. P., Jr., and G. V. Skogerboe. 1972. Potential for controlling quality of irrigation return flows. J. Environ. Qual. 1:140–145.

Loehr, R. C. 1968. Pollution implication of animal wastes—a forward oriented review. Water Pollut. Control Res. Ser. 13040—07/68, EPA, Washington, D. C. 148 p.

McCalla, T. M., J. R. Ellis, C. B. Gilbertson, and W. R. Woods. 1972. Chemical studies of solids, runoff, soil profile and groundwater from beef cattle feedlots at Mead, Nebraska. Proc. Cornell Agric. Waste Manage. Conf., 1972. p. 211–223.

McKinney, R. E., ed. Second international symposium for waste treatment lagoons. 1970. University of Kansas, Lawrence, Kans. 404 p.

Manges, H. L., R. I. Lipper, L. S. Murphy, W. L. Powers, and L. A. Schmid. 1975. Treatment and ultimate disposal of cattle feedlot wastes. Environ. Protect. Technol. Ser. EPA-660/2-75-013, EPA, Washington, D. C. 136 p.

Menzies, J. D., and R. L. Chaney. 1974. Waste characteristics in factors involved in land application of agricultural and municipal wastes. USDA-ARS. National Program Staff, Soil, Water, and Air Sciences, Beltsville, Md. 200 p.

Metcalf and Eddy, Inc. 1972. Wastewater engineering. McGraw-Hill Book Company, New York. 782 p.

Miner, J. R., R. I. Lipper, L. R. Fina, and J. W. Funk. 1966. Cattle feedlot runoff: its nature and variation. J. Water Pollut. Control Fed. 48:1582-1591.

Miner, J. R., and T. L. Willrich. 1969. Livestock operations and field-spread manure as sources of pollutants. p. 231-240. In T. L. Willrich and G. E. Smith (ed.) Agricultural practices. Water Pollut. Control Res. Ser. 13040EYX11/69, FWPCA (presently EPA), Washington, D. C.

Mytelka, A. I., J. S. Czachor, W. B. Guggino, and H. Golub. 1973. Heavy metals in wastewater and treatment plant effluents. J. Water Pollut. Control Fed. 45:1859-1864.

National Academy of Sciences. 1973. Water quality criteria, 1972. Ecological research EPA-R3-73-033, EPA, Washington, D. C. 564 p.

Pavelis, G. A. 1967. Regional irrigation trends and projective growth functions. Water Resour. Br., Natur. Resour. Econ. Div., Econ. Res. Serv., USDA, Washington, D. C.

Pierce, D. M. 1974. Performance of raw waste stabilization lagoons in Michigan with long period storage before discharge. p. 89-135. In E. J. Middlebrooks et al. (ed.) Upgrading wastewater stabilization ponds to meet new discharge standards. Utah State University, Logan, Utah.

Porter, L. K., F. G. Viets, Jr., T. M. McCalla, L. F. Elliott, F. A. Norstadt, H. R. Duke, N. P. Swanson, L. N. Mielke, G. L. Hutchinson, A. R. Mosier, and G. E. Schuman. 1975. Pollution abatement from cattle feedlots in northeastern Colorado and Nebraska. Environ. Protect. Technol. Ser., EPA-660/2-75-015, EPA, Washington, D. C. 120 p.

Powers, W. L., G. W. Wallingford, and L. S. Murphy. 1975. Research status on effects of land application of animal wastes. Environ. Protect. Technol. Ser. EPA-660/2-75-010, EPA, Washington, D. C. 96 p.

Robbins, J. W. D., D. H. Howells, and G. J. Kriz. 1971. Role of animal wastes in agricultural land runoff. Water Pollut. Control Res. Ser. 13020DGX08/71, EPA, Washington, D. C. 114 p.

Skogerboe, G. V., and J. P. Law, Jr. 1971. Research needs for irrigation return flow quality control. Water Pollut. Control Res. Ser. 13030—11/71, EPA, Washington, D. C. 98 p.

Skogerboe, G. V., and W. R. Walker. 1972. Evaluation of canal lining for salinity control in Grand Valley. Environ. Protect. Technol. Ser. EPA-R2-72-047, EPA, Washington, D. C. 197 p.

Skogerboe, G. V., W. R. Walker, R. S. Bennett, J. E. Ayars, and J. H. Taylor. 1974. Evaluation of drainage for salinity control in Grand Valley. Environ. Protect. Technol. Ser. EPA-660/2-74-084, EPA, Washington, D. C. 100 p.

Swanson, N. P., C. L. Linderman, and J. R. Ellis. 1974. Irrigation of perennial forage crops with feedlot runoff. Trans. ASAE 17:144-147.

U. S. Environmental Protection Agency. 1974. Development document for effluent limitations guidelines and new source performance standards—feedlots point source category. EPA-440/1-74-004a, Washington, D. C. 318 p.

U. S. Public Health Service. 1965. Interactions of heavy metals and biological sewage treatment processes. USDHEW, Environ. Health Ser. Pap. 999-WP-22. 201 p.

Wells, D. M., R. C. Albin, W. Grub, E. A. Coleman, and G. F. Meenaghan. 1971. Characteristics of wastes from southwestern cattle feedlots. Water Pollut. Control Res. Ser. 13040DEM01/71, EPA, Washington, D. C. 87 p.

chapter 4

Laboratory technician uses a diffusion porometer to measure water uptake of grasses and soybeans fertilized with composted sludge of varying rates of application. Scales measure water loss from both soils and plants. These test results will enable scientists to determine the proper amounts of composted sludge for fertilizing various crops (Photo courtesy of USDA).

Chemical Properties of Soils

D. R. KEENEY, University of Wisconsin, Madison, Wisconsin

R. E. WILDUNG, Battelle, Pacific Northwest Laboratories, Richland, Washington

I. INTRODUCTION

Consideration of the role of soils for management and utilization of organic wastes and waste waters must take into account the chemical reactions which may occur with components of the wastes. These reactions can be grouped conveniently into (i) ion exchange, (ii) adsorption and precipitation, and (iii) complexation. The mechanisms and rates of most, if not all, of these reactions are dependent upon the type and amounts of clay, hydrous oxide, and organic matter, as well as more dynamic properties including solute composition and concentration, exchangeable cations, pH, and oxidation-reduction status. The latter reactions are often profoundly affected by the physical and biological properties of soils and any comprehensive consideration of soil chemical reactions must consider the entire soil continuum.

The purpose of this chapter is to provide the framework for ensuing chapters on the effects of waste applications on the mobility and plant availability of specific soil and waste components. This essentially involves a condensed discussion of the vast amount of information on soil chemical reactions. The discussion will be limited to soils which have not been subject to heavy applications of organic wastes and emphasis will be placed on chemical reactions occurring in soils which affect the mobility of P, S, exchangeable cations, and metals.

II. EXCHANGE REACTIONS

A. Cation Exchange

The cations associated with negative charge sites on soil solids through largely electrostatic bonding and subject to interchange with cations in the soil solution with little or no alteration of the solids are termed *exchangeable cations*. The replacement process is referred to as *cation exchange*. The dominant exchangeable cations are Ca^{2+}, Mg^{2+}, K^+, Na^+, Al, and H. Under certain conditions, Mn^{2+} or Fe^{2+} may occupy a significant proportion of the exchange complex, and these ions will be considered later. The first four of these cations are commonly termed *exchangeable bases*. In acid soils, Ca^{2+}, Mg^{2+}, K^+, and Al (Al^{3+} and $Al[OH]^{2+}$) dominate, with their relative proportions being dependent on the relative abundance and on the types of exchange sites and pH. In calcareous soils, Ca^{2+} and Mg^{2+} occupy most of the exchange sites; in salt-affected soils, Na^+ can comprise a significant proportion of the exchangeable cations (Black, 1968).

Ions may be bound to soil solids by a combination of forces ranging from electrostatic to covalent, with corresponding increases in bonding energy. When covalent bonding dominates, a property of specificity (often termed *specific sorption*) is observed for certain cations and many anions, and the reversible ion exchange for these ions decreases. Retention by true ion exchange is less important. This phenomenon has been observed with clays, hydrous oxides, and organic matter.

Cation exchange and exchangeable cations in soils are usually quantified in terms of milligram equivalents (meq) per 100 g of soil. The total exchange capacity of arable surface soils ranges from 0.5 to 50 (greater in organic soils). Cation exchange capacity is determined, commonly, by displacing the exchangeable cations (or an index cation which was previously used to saturate the exchange sites) with neutral, $1N$ NH_4OAc, and estimation of the amount of NH_4^+ (or other index cation) held on the exchange sites. Exchangeable bases are determined from analysis of the extract. The relative quantities of exchangeable bases in soils commonly follow the order: Ca^{2+} $> Mg^{2+} > K^+$.

The net negative charge of soil colloids arises from isomorphous substitution (e.g., Al^{3+} for Si^{4+}) in many of the layer silicate minerals, from $SiOH$ and $AlOH$ groups of exposed edge surfaces of clay minerals, from amorphous surfaces, and to a lesser degree, crystalline hydrous oxides and hydrated oxides of Fe, Al, and Mn, and from the disassociation of acidic functional groups of organic matter. Some of these exchange sites vary in reactivity with soil pH; as pH declines, the disassociation of $SiOH$, $AlOH$, and organic matter functional groups is suppressed. Hydrous oxides and silicate edges develop positive charges at the pH below the isoelectric point of the oxide in question.

Due to the net negative charge of soil colloids, anions are repelled from water in the vicinity of these colloids. This has been termed *anion exclusion* (Thomas & Swoboda, 1970).

A considerable portion of the cations in soil solids are in a form such that they do not interchange with soil solution cations. These are termed *nonexchangeable cations*. Bear et al. (1945), for example, found that the nonexchangeable Ca^{2+}, Mg^{2+}, and K^+ constituted 76.9, 96.2, and 99.6%, respectively, of the total concentration of these elements. Nonexchangeable cations will be released slowly to solution from soil constituents through mineral weathering and organic matter decomposition.

B. Anion Exchange

While cation exchange is the dominant exchange process occurring in soils, some soils also have the ability to retain anions in truly exchangeable form (i.e., extractable with a replacing anion). Of the anions commonly found in soils, NO_3^-, Cl^-, and to a larger extent, SO_4^{2-}, exhibit this type of exchange. Anion exchange is especially important in acid-weathered soils high in hydrous oxides and kaolinite. Singh and Kanehiro (1969) reported that two Hawaiian soils sorbed from 1.3 to 2.6 meq/100 g of NO_3^- at pH 5.0 with the amount decreasing as pH increased. It is unlikely, however, that this type of sorption would be significant except in areas where hydrous oxides and kaolinite constitute a significant fraction of the soil matrix, and the soil is acid.

C. The Nature of Soil Acidity

True soil acidity involves the measured pH (active acidity), and cation exchange capacity (CEC), and the degree of saturation of the exchange complex with Al (reserve acidity). It is this property which provides the soil with a buffering capacity, i.e., a soil with a high CEC will require more limestone or neutralizing waste to raise the pH to a given level than one of low CEC, even though both had the same percentage of Al on the exchange sites. This acidity is of paramount importance in dealing with research and practical problems. Aluminum probably exists in solution as hydrated species (Richburg & Adams, 1970), and can be in any number of polymeric forms. As soil pH is raised, H_3O^+ from the hydrolysis of hydrated Al forms is neutralized. Evidence exists that Al polymers precipitate in interlayer positions in montmorillonitic and vermiculite minerals, resulting in partial blockage of exchange sites.

Soil acidity may arise from (i) crop removal of more cations than anions, (ii) oxidation of FeS or FeS_2 which has accumulated during anaerobic conditions, (iii) oxidation of reduced forms of N, (iv) increases in the partial pressure of CO_2 in soil solution from biological decomposition of organic wastes, and (v) leaching of bases from soils under conditions where rainfall exceeds evapotranspiration. Processes (ii), (iii), and (iv) are by far the most important on a short term basis and several workers have reported that N- and S-containing fertilizers and carbonaceous wastes will lower soil pH.

D. Effect of Excess Sodium and Soluble Salts

Excess soluble salts in soils result in adverse effects on water availability to plants. Excess Na is particularly detrimental and leads to unfavorable soil structure and decreased infiltration. The sodium-adsorption ratio (SAR), originally developed by Gapon and cited in Richards (1954), has long been used to evaluate the possible effects of salts in irrigation waters:

$$SAR = Na^+/[(Ca^{2+} + Mg^{2+})/2]^{1/2} .$$

Ellis (1973) has reviewed the possible use of the Gapon equation for evaluating waste water applications to soils. He also points out other factors which must be considered to properly evaluate the SAR (e.g., soil texture, associated anion, and K^+ concentration) and concludes that most effluents would not lead to excessive levels of exchangeable Na^+ in most soils. However, excess Na^+ could be a major deterrent to waste water application in semiarid and arid regions, especially if the quality of the irrigation water was low.

E. Exchange Models

A number of models, in addition to the Gapon-type approach, have been used to estimate the relative amount of a given cation on the exchange complex. Most of these were summarized by Ellis (1973). However, no single model has been found applicable to widely varying soils or to all possible cation combinations.

There are a number of complications in devising models, including selective fixation of K^+ or NH_4^+ (Shawney, 1972), precipitation of Ca and Mg carbonates, ion pair formation (Nakayama, 1968), and differences in the affinities of mono- and divalent cations for exchange sites. Nevertheless, these models may prove useful as predictors for specific situations. An example is the application of the NH_4^+ adsorption ratio (analogous to the SAR) (Lance, 1972), for estimation of the exchangeable NH_4^+ in an anaerobic soil column receiving septic tank effluent (Magdoff et al., 1974).

F. Exchange of Metals

The extent of participation of metals in true cation exchange reactions varies depending on the metal, metal concentrations, soil constituents and their corresponding properties, pH, and presence of chelating agents. In general, the metals tend to form coordinate covalent bonds. At high concentrations, oxides and hydroxides (or sulfides under reduced conditions) of low solubility may be formed, particularly at higher pH values.

Lagerwerff and Brower (1972, 1973) found that at low concentrations the exchange behavior of Cd^{2+} and Pb^{2+} follow mass-action principles with kaolinitic, montmorillonitic, and illitic soils. Bittell and Miller (1974) found that at very high concentrations, Pb^{2+} and Cd^{2+} competed with common divalent cations (such as Ca^{2+}) for clay absorption sites. Tiller et al. (1963), and McLaren and Crawford (1973), noted that Co^{2+} and Cu^{2+} sorption was dominated by more specific exchange sites (see also Hodgson et al., 1966). Cation exchange has been found to be important in controlling water soluble Mn^{2+} and Fe^{2+} in waterlogged soils (Gotoh & Patrick, 1972, 1974).

III. SORPTION AND PRECIPITATION REACTIONS

A. Mechanisms and Equations

Ellis (1973) discussed in some detail sorption and precipitation processes of anions and heavy metals in soils. *Adsorption* (the more general term, *sorption*, will be used here) generally is defined as adhesion of gas molecules, dissolved substances, or liquids to the surface of solids with which they are in contact. Precipitation, on the other hand, denotes formation of a sparingly soluble solid phase. The two reactions often are in competition, with precipitation dominating at relatively high concentrations of reactants. Oftentimes, it is difficult to discern which mechanism is dominating.

Several types of mechanisms can be postulated for the process of removing an ion from solution and bonding it to a solid surface. For convenience, these can be grouped into physical sorption, chemisorption, and penetration into the solid mineral phase. However, it is doubtful if they ever operate independently.

Physical sorption involves the attachment of the sorbent and sorbate through weak atomic and molecular interaction forces (van der Waal forces) which operate when the electron clouds of the atoms do not overlap sufficiently to cause strong attractive (ionic) forces. The activation energy for this type of attraction is characteristically low, much lower than is normally observed for ions (e.g., orthophosphate) which are bound more strongly. However, this may be the initial sorption mechanism, followed by other processes such as chemisorption.

Chemisorption involves chemical bonds similar to those holding atoms in a molecule and can be considered a special case of precipitation in which one ion remains a constituent of the solid phase to which the sorbate is attached. The process is thought to be the dominant mechanism for phosphate retention in acid soils (Syers et al., 1973; Mattingly, 1975). Penetration of ions into the solid mineral phase can also occur, probably concurrently with chemisorption. This reaction tends to be time-dependent, i.e., more of the sorbate becomes irreversibly retained with time, in contrast to the initial rapid chemisorption reaction.

Two equations are most often used to quantitatively describe sorption of ions from solution onto solids, namely the Freundlich and the Langmuir isotherms. The Freundlich isotherm has the form

$$x = kc^n \text{ or } \log x = \log k + n \log c$$

where x is the amount sorbed per unit sorbent at equilibrium concentration, c, of sorbate and k and n are constants.

The disadvantages of this equation are that it predicts unlimited sorption, the derived constants are often not applicable to soils, and it does not give information on sorption mechanisms.

The Langmuir equation has the form

$$M/(x/m) = 1/Kb + M/b$$

where M is the activity (moles/liter) of the ion, x/m is the amount of ion, M, sorbed per unit of sorbate (usually meq/100 g), K is a constant related to the bonding energy, and b is the maximum amount of the ion (usually meq/100 g) that will be sorbed by a given sorbate.

The advantages of this equation are that laboratory measurements can be made to predict a sorption maximum, and that, using K and b, predictions can be made of the quantity of ion sorbed at a given input concentration. The disadvantages are that little information is provided on mechanisms of sorption and it assumes uniformity of sorption sites and noninteraction laterally of sorbed molecules. Further, the soil data often deviates from the model at low and high ion concentrations, likely due to different types of sorption sites or mechanisms. However, the literature contains numerous examples of close agreement of experimental data with this isotherm, and it has often been used, especially for studies of P sorption in soils.

B. Phosphate Retention

The P in soils is present in inorganic and organic combinations and the relative distribution of P in these forms differs widely among soils. The major portion of P in soils is associated with the solid phase and P concentrations in the soil solution seldom exceed 1 ppm.

Black (1970) suggests that the P retention capacities of soils can be categorized into three general types. The first is the sorption capacity at low levels of P. The second is the capacity of the soil to react rapidly with high P concentrations, and the third is the ultimate capacity of the soil to react with P. Black (1970) points out that the theoretical ultimate capacity is achieved only when the anions of original carbonate, hydrous oxide, and silicate minerals are converted to phosphates. This would be an extremely large value, even compared to the second category.

Several workers have applied the Langmuir equation to obtain an estimate of the capacity of soils to sorb P. An excellent example is given by Ellis (1973). However, even these estimates are probably conservative, since Ellis (1973) points out that the sorption capacity of a soil initially saturated with phosphate may recover some sorption capacity in time. This is in line with the conceptual view of P sorption capacities as expressed by Black (1970). In contrast, Barrow (1974) found that previous addition of phosphate reduced the capacity of soils to sorb additional phosphate, even 3 years after initial application. Shah et al. (1975) combined the Langmuir with Darcy's law for one-dimensional water movement to model the movement of P under an animal waste disposal system. Novak et al. (1975) evaluated this model and found that, for the soils tested, the S-shaped P movement front (shock layer) was < 4 cm thick.

The accepted view at present is that the labile inorganic P in acid, unfertilized soils is sorbed on the surface of Fe- and Al-containing minerals to form surface compounds (Syers et al., 1972; Mattingly, 1975). This is based on the fact that, at the low levels of phosphate usually found in soil solution, variscite ($Al[OH]_2 H_2PO_4$) and strengite ($Fe[OH]_2 H_2PO_4$) are not stable, and solubilized P is sorbed on amorphous Fe and Al hydrous oxide coatings on the solid phase. In calcareous soils, sorption on $CaCO_3$ occurs at low P concentrations, while Ca phosphate minerals form at high P levels (Mattingly, 1975). The latter can maintain relatively high (about 1 ppm) solution P concentrations (Ellis, 1973).

In addition to the sorbed forms of P which are in equilibrium with the soil solution, there are phosphates of secondary origin bound to Fe and Al, but occluded in crystalline minerals as soil concretions or by Fe and Al hydrous oxide coatings on minerals. These will be slowly mobilized. Stable apatite also has been demonstrated to occur in some soils.

When P is added to soils in high concentrations, either as fertilizer or as a waste product, coprecipitation of P with Ca, Mg, Fe, or Al could result. However, unless additional P is added, these reaction products dissolve or revert to less available occluded or precipitated forms. The dissolved P is believed to be sorbed by the aforementioned mechanisms.

Under reduced conditions (lowland rice [*Oryza sativa* L.], swamps, etc.) and low pH ($<$ 5), strengite is partially solubilized (Patrick et al., 1973). However, P release and sorption under anaerobic conditions is apparently a function of solution P concentration. Patrick and Khalid (1974) found that anaerobic soils released more P to soil solutions low in soluble P, but sorbed more P from solutions high in soluble P, than corresponding aerobic soils. This was attributed to formation of gel-like reduced ferrous compounds. Phosphate resorption by clay and hydrous oxides of Al also appear important, at least in lake sediments (Williams et al., 1970; Shukla et al., 1971).

C. Retention of Other Anions

1. ARSENIC

In arable soils, the chemistry of As closely parallels that of P, being essentially the chemistry of arsenate. Hingston et al. (1968) obtained similar maxima for P and As sorption by goethite over the pH range of 3 to 13, although other workers have obtained evidence that P is more strongly sorbed by soils than As (Dean & Rubins, 1947; Swenson et al., 1949). Jacobs et al. (1970) found that retention of As by Wisconsin soils increased as the sesquioxide and free Fe_2O_3 content increased and that amorphous Fe and Al compounds preferentially sorbed added As. Jacobs and Keeney (1970) observed that phosphate apparently displaced sorbed arsenate enhancing As toxicity to corn (*Zea mays* L.).

2. SULFUR AND SELENIUM

Most soils exhibit the capacity to retain SO_4^{2-}, although sorption is generally much less than with phosphate (Harward & Reisenauer, 1966). As with phosphate, Fe and Al hydrous oxides show a marked tendency to retain SO_4^{2-}, and retention increases as pH decreases. The Langmuir-type equation has been used with some success to predict sorption (Aylmore et al., 1967), although Chao et al. (1962) found that sulfate sorption was more adequately described by the Freundlich isotherm. Sorption of sulfate usually results in an increase in the pH of the system, presumably through replacement of Al-bound hydroxyls (Harward & Reisenauer, 1966).

Selenium occurs in soils mainly in the selenite (SeO_3^{2-}) and selenate (SeO_4^{2-}) forms, with the latter predominating in alkaline soils (Lakin, 1972). Geering et al. (1968) obtained evidence that a ferric oxide-ferric selenite sorption complex existed in acid and neutral soils. In alkaline soils, $CaSeO_4$ is quite soluble, and Se can readily leach from such soils. Sulfate has been shown to be quite effective in displacing Se from the sorption sites in the alkaline soils (Brown & Carter, 1969). However, most of the Se in organic wastes probably occurs as the stable metal selenites.

3. MOLYBDENUM

The toxic effects of Mo on livestock have prompted studies on the transport mechanisms of Mo (Chappell, 1973; Katz & Runnells, 1974).[1,2] Below pH 6 Mo occurs chiefly as the bimolybdate ion ($HMoO_4^-$), while above pH 6 the principle ion is molybdate (MoO_4^{2-}). A number of workers have found that Mo sorption by soils and clays decreases with increasing pH, with very limited sorption of pH 7.5 (Davies, 1956; Katz & Runnells, 1974[2]). As with sulfate, sorption of molybdate increases the pH of the system. Katz and Runnells (1974)[2] found that the Langmuir equation predicted closely the sorption of Mo by soils, and that organic matter and hydrous Al oxides are of primary importance in sorption of Mo.

4. BORON

Ellis and Knezek (1972) and Ellis (1973) have reviewed the extensive literature on B mobility in soils. Boron exists as H_3BO_3 and, in neutral and alkaline soils (concentrations $<$ about $0.1M$), H_3BO_4 partially hydrates to form $B(OH)_4^-$. Boron is sorbed more strongly than Cl^- and NO_3^-, and sorption on clay surfaces is probably more similar to that of heavy metals than to other anions (Hodgson, 1963). Ellis (1973) considered four mechanisms for B sorption by soils, including interactions with (i) hydrous Fe and Al coatings, (ii) Fe and Al oxides, (iii) the interlattice structure of clays, particularly micaceous-type minerals, and (iv) Mg hydroxy coatings or clusters that exist on the ferro-magnesium minerals. It is believed that much of the B in soils is associated with the organic matter. Little is known of the mechanisms involved in bonding of B to organic matter but diols may be involved (Hodgson, 1963).

The Langmuir equation appears to hold only over limited B concentrations (Ellis, 1973). Rhoades et al. (1970) found that Langmuir equation underestimated B desorption from soils while Tanji (1970) found good agreement with the Langmuir model.

---

[1]W. R. Chappel. 1973. Transport and the biological effects of molybdenum in the environment. Presented at Heavy Metals in the Aquatic Environment Conf., Vanderbilt Univ., Nashville. 4–7 Dec.

[2]B. G. Katz and D. D. Runnells. 1974. Experimental study of sorption of Mo by desert agricultural and alpine soils. Presented at Eighth Annu. Conf. on Trace Substances in Environmental Health, Columbia, Mo. June 1974.

D. Retention of Metals

Several mechanisms for the retention of metals (viz. Cu, Zn, Cr, Ni, Cd, Hg, Pb, Co) in soils have been proposed. Hodgson (1963) lists these reactions as (i) association with soil surfaces, (ii) precipitates, (iii) occluded in other precipitates, (iv) native constituents of soil minerals, (v) solid-state diffusion into soil minerals, and (vi) incorporation into biological systems or residues. While all of these reactions undoubtedly operate to some extent in surface soils, differences in the relative importance, and rates, of these mechanisms exist depending on the metal, soil properties, and environmental conditions. This is an extremely complex subject, and the literature is difficult to evaluate with respect to the relatively high concentrations of metals added via wastes, particularly sewage sludges. However, these mechanisms must be understood if meaningful management systems for waste disposal are to be developed.

The factors controlling the relative dominance of precipitation over surface sorption reactions are solution metal concentration (sorption usually dominates at concentrations found in unamended soils), pH, ion pair formation, and possibly existence of organo-metal complexes (Hodgson, 1963; Jenne, 1968; Ellis & Knezek, 1972; Jurinak & Santillan-Medrano, 1974). Even if precipitates are formed, it is likely that dissolution and surface sorption will occur as the soil solution concentration of the metal decreases due to plant uptake or leaching.

Several workers have applied mineral solubility data to predict metal ion activities in soil solution (see reviews by Jenne, 1968, and Lindsay, 1972). These reviews point out that, with the possible exception of Fe and Mn, many difficulties are involved. These difficulties arise because of the diverse nature of the amorphous solids, complex ion formation, and dynamic changes in Eh, pH, and solute concentrations of soil systems. Ellis (1973) lists solubility product constants for a number of metal carbonates, hydroxides, and sulfides. Of the metals listed, Pb, Fe, and Mn are most likely to be controlled by precipitation reactions.

A number of workers regard ion exchange as one of the important mechanisms controlling metal availability. Hodgson (1963) and Ellis and Knezek (1972) state that in addition to ion exchange, specific adsorption processes, involving covalent bonding to certain functional groups on clay surfaces, are involved also. Chaney (1973) regarded cation exchange and organic matter chelation as the most important factors controlling metal availability. Leeper (1972) concluded that allowable metal additions to soil should be based solely on soil CEC, and that soil pH be maintained at 6.5 or greater. In contrast, Jenne (1968), in a review of factors controlling metal concentrations in soil and aquatic systems, felt that sorption of metals is not related to cation exchange capacity. Tiller et al. (1963) found that specific sorption of Co was essentially the same for 15 widely divergent soils. McLaren and Crawford (1973), using essentially the same approach as Tiller et al. (1963) for evaluation of Cu sorption, concluded that with the relatively small amounts of Cu normally present in soil solution, specific sorption was more important than cation exchange. For example, in the case of montmorillonite, specific sorption was greater than nonspecific (exchange) sorption up to 0.4 ppm Cu. At Cd concentrations ranging from 1 to 100 ppm, John (1972) found no significant relationship between Cd sorption and soil CEC for 30 diverse soils. However, the exchange behavior of Cd^{2+} has been shown to correspond to the principles of mass action in kaolinitic, montmorillonitic, and illitic soils pretreated with Al^{3+} or Ca^{2+} and maintained at several Cd^{2+} concentration levels (Lagerwerff & Brower, 1972). Dolar and Keeney (1974) found only 3.2 to 4.7% of the Cu and 0.2 to 1.1% of the Zn added to soil at rates of 50 to 100 ppm in a complexed form was exchangeable with $1N$ $MgCl_2$ (see also MacLean, 1974). Udo et al. (1970) found that when the specific sorption sites for Zn in calcareous soils were saturated, Zn concentration was governed by the solubility of the hydroxide or carbonates rather than by classical exchange reactions.

Organic matter is often regarded as a major factor in the sorption of metals (Ellis, 1973; Ellis & Knezek, 1972; Hodgson, 1963; Stevenson & Ardakani, 1972). This is based on (i) the known chelating ability of soil organic constituents, such as humic and fulvic acid extracts, (ii) high correlations between metal sorption or micronutrient deficiencies and organic matter content of soils; and (iii) reduction in sorption or increase in extractability of metals after treatment of soils with H_2O_2 to destroy organic matter. There is firm evidence that the humic and fulvic acids have relatively high stability constants for metals (Stevenson & Ardakani, 1972). In organic soils, it is likely that organic matter reactions will be dominant simply due to the overwhelming quantity present. Also, considerable evidence exists (Hodgson et al., 1966; Stevenson & Ardakani, 1972) that Zn, Mn, and Cu in the soil solution are largely complexed by soluble organics (see Section IV).

Considerable evidence has accumulated indicating that the hydrous metal oxides play a major role in the sorption of heavy metals in mineral soils (Jenne, 1968). Kinniburgh (1974)[3] provides literature citations and experimental evidence on the role of hydrous metal oxides. These oxides, particularly those of Fe, Mn, and Al, are common in soils. They may occur as discrete crystalline minerals or as coatings on other soil minerals. Many are of indefinite structure (amorphous) and composition (Ponnamperuma et al., 1969). They have high surface areas in relation to their weight, are highly reactive, and the Fe and Mn oxides are quite labile since they are formed and dissolved in oxidizing and reducing conditions, respectively, in the soil (Ponnamperuma, 1972). The latter is of importance since many of the observations on heavy metal mobilities as related to organic matter content of soils, pH, flooded conditions, and H_2O_2 treatment can be explained by consideration of Fe and Mn oxide reactions in response to pH and Eh (Jenne, 1968).

Jenne (1968) hypothesizes that the role of organic matter is often

---

[3]D. G. Kinniburgh. 1974. Cation adsorption by hydrous metal oxides. Ph.D. Thesis, Univ. of Wisconsin–Madison.

secondary, producing the periodic reducing environment necessary to maintain oxides in their reactive hydrous microcrystalline condition, and that H_2O_2 treatments (i) release metals already present in hydrous Mn oxides, (ii) dissolve the Mn oxides, and (iii) decrease the surface area of the Fe oxides. This hypothesis is supported by (i) sorption studies where organic matter was removed by heating from 300 to 400C (Sokoloff, 1953), (ii) investigations showing that soluble organics can rapidly reduce Fe^{3+} and Mn^{3+} chemically (Bloomfield, 1954, 1955), and (iii) indirectly through the low pH and Eh conditions often occurring during anaerobic respiration (Ponnamperuma, 1972).

This hypothesis at least partially explains the commonly observed finding that on waterlogging of soils containing sufficient organic matter to effect a decrease in Eh, the availability of many metals is increased. Increased mobilization of Cr, Ni, Cu, Zn, and Co have been observed (Kee & Bloomfield, 1962; Adams & Honeysett, 1964; Leeper, 1947), while waterlogging of low organic matter content soils has shown little effect on increased mobilization of metals. However, since waterlogging also increases soluble Fe^{2+} and Mn^{2+}, and these metals may displace other metals from sorption sites, a simplistic mechanism does not seem likely.

The hydrous metal oxides exhibit weakly acid characteristics similar to other polyelectrolytes. Their acidity can be characterized as a two-step reaction:

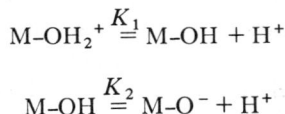

$$M\text{-}OH_2^+ \overset{K_1}{=} M\text{-}OH + H^+$$

$$M\text{-}OH \overset{K_2}{=} M\text{-}O^- + H^+$$

K_1 and K_2 are not constants, but depend on the surface charge. Also, the oxides exhibit a point of zero charge pH and an isoelectric point. The isoelectric point is the pH at which the potential at some plane in the double layer is zero. However, estimation of surface charge density-pH relationships is complicated by electrolyte effects and the slow attainment of equilibrium (Kinniburgh, 1974)[3], and models to quantitatively describe these types of curves have not been very successful. Recently, improvements in models of the electrical double layer of oxide surfaces have been made by taking ion penetration into account (Kinniburgh, 1974).[3] While sorption on metal oxides is a rapid process, several workers have found that sorption will continue slowly for a long period of time (Jenne, 1968).

It is commonly (but not universally) found that availability of metals decreases or sorption increases with an increase in soil pH. This can be attributed to (i) precipitation of Mn oxides or heavy metal hydroxides, (ii) precipitation-dissolution of Fe and Mn oxides, (iii) secondary effects on the concentration of carbonate and phosphate in soil solution, and (iv) rate and extent of sorption and desorption of metals by hydrous oxides and organic matter. Lindsay (1972) points out that the solubility of Cu and Zn in soils is several orders of magnitude less than would be predicted by hydroxide or carbonate solubility curves (pH 4 to 9); at high pH values, the concentration of these elements in solution is so low that very little would be on the cation exchange complex. Sorption on hydrous oxides, formation of insoluble phosphates, and complexation complicate these conclusions (Ellis & Knezek, 1972). MacLean (1974) found that P additions increased extractable Zn but decreased plant uptake of Zn in Zn-amended soils.

The rapid increase in specific sorption of multivalent cations on hydrous oxides with increasing pH has led many workers to envisage this as a proton-exchange reaction with the protons derived from weakly acidic surface-OH groups (Kinniburgh, 1974).[3] For divalent cations, the number of H^+ released (n) is pH dependent, and can vary between 1 and 2, with $n \cong 1$ more often observed while for multivalent cations, evidence suggests that sorption of an n-valent cation leads to release of $(n-1)$ H^+. This has led to the postulation that a $(MOH)^+_{n-1}$ hydrolyzed species is involved. Kinniburgh (1974)[3] points out that, due to the close similarity between sorption, coprecipitation, and precipitation, sorption isotherms might be expected to show no inflexion points when these different mechanisms come into play.

Jenne (1968) discusses the observation that pH $\sim$ 6 and above is often the point at which micronutrient deficiencies begin to occur. The recurring consistency of this value might indicate that the "average" isoelectric point of the soil sorbing complex is about pH 6. However, the marked differences in hydrolysis constants, solubilities, sorption energies, and selectivities indicate that each metal must also be considered separately. For example, Jurinak and Santillan-Medrano (1974) indicate that the solubility of $Pb(OH)_2$ and $Pb_5(PO_4)_3OH$ controls Pb mobility in soils and that Cd is 100 times more soluble, and correspondingly more mobile, than Pb in the pH range of 5 to 9. Hahne and Kroontje (1973) point out that hydrolysis of Hg^{2+}, Pb^{2+}, Zn^{2+}, and Cd^{2+} becomes significant above pH 1, 5, 7, and 8, respectively. The chemistry of Cd and Zn are not analogous as Cd^{2+} remains in the divalent form to pH 7.8, and only 50% is present as $Cd(OH)_2$ at pH 11 whereas $Zn(OH)_2$ represents 50% of the total Zn at pH 7.5. This suggests that soil amendments to raise pH and overcome Cd toxicities or minimize plant uptake may not be nearly as successful as for Zn.

Kinniburgh (1974)[3] evaluated divalent metal sorption by Fe and Al gels. He found the following selectivity sequences (figures in parentheses indicate pH at 50% sorption)

Fe gel: $Pb(3.1) > Cu(4.4) > Zn(5.4) > Ni(5.6) > Cd(5.8) > Co(6.0)$.

Al gel: $Cu(4.8) > Pb(5.2) > Zn(5.6) > Ni(6.3) > Co(6.5) > Cd(6.6)$.

The Langmuir isotherm has been used for analysis of sorption data for Cd and Zn by soils. The sorption maxima for Zn obtained with calcareous Arizona soils ranged between 2.7 and 15.6 meq/100 g soil (Udo et al., 1970); Cd sorption maxima for the Canadian soils varied very little and averaged about 28 meq/100 g soil (John, 1972). However, the previously stated limitations on use of the Langmuir model would hold for these metals.

IV. METAL COMPLEXATION

In addition to potential inorganic reactions of metals in soil, metals may be subject to complexation, chelation, and biological transformation. This section will deal separately with the principles of complex formation, the difficulties inherent in the study of these phenomena in soil, and metal complexation by soil organic matter.

A. Principles of Complex Formation

The interactions of metals with functional groups of soil organic matter can be treated theoretically from a knowledge of metal interactions with model organic compounds. Except in the case of well-defined biochemicals in soil, accurate extrapolation from studies with model compounds to soil is not possible due to a lack of specific knowledge of soil solution composition and pH in the vicinity of the colloid and the structure and chemistry of organic moieties in soil. However, the basic chemical principles apply to soil organic colloids and have proven useful in explaining the general behavior of metals in soil. A number of good textbooks are available on the chemistry of metal complexes (Chaberek & Martell, 1959; Martell & Calvin, 1952) and on metal ions in biological systems (Sigel, 1974). Succinct reviews of the principles of complex and chelation chemistry have been presented by Lehman (1963) and Martell (1971).

Classes of organic compounds which have the greatest potential for serving as electron donors in metal complexes include enolates, alkoxides, carboxylates, phenoxides, alkyl amino, heterocyclic N, mercaptides, phosphates, and phosphonates. Functional groups with lower metal affinities which may serve as auxiliary donors include hydroxyalkyls, carbonyls, ethers, esters, amides, and thioethers (Martell, 1971). The bond types formed with metals may range from nearly ionic to nearly covalent in character.

Stabilities of complexes are normally measured by the equilibrium constants for complex formation and expressed by stability constants. For the simplest case, where A represents a ligand with a single donor group and M is the metal ion, the equilibrium would be expressed as

$$M + A \rightleftharpoons MA$$

and the stability-constant defined as

$$K = [MA]/[M][A].$$

Where a number of donor groups are present on the same compound, a series of successive or stepwise constants are derived and the overall stability constant is equal to the product of the stepwise constants.

A number of factors influence the stabilities of complexes including (i) charge and ionic dimensions of the metal ion, (ii) electronegativity of the donor group on the ligand, (iii) steric and electrostatic repulsions between ligand donor groups, and (iv) other physical and chemical factors which may be classed generally as environmental factors.

The most stable complexes are those derived from transition metals or metals immediately following the transition metals in the periodic table. The coordination number, or the number of water molecules or donor groups which may be associated with the metal, is dependent on the charge, electron configuration, and dimension of the metal ion. In general, the stability of complexes increases with increased charge and decreased metal ion size. Most ions or molecules containing an unshared electron pair may act as a coordinating agent or donor group. However, increases in the number and electronegativity of donor groups on a ligand generally result in increased affinity for metal ions provided a complex is formed which is sterically stable. Increased resonance or delocalization of electrons in a molecule generally results in increased stabilization of the complex. For example, phenols exhibit high acidity relative to alcohols because delocalization of electrons through the aromatic ring is responsible for increased stabilization of the conjugate base. The position and type of substituent groups on organic moities strongly influence resonance stabilization as well as the electronegativity of functional groups.

Chelation, a form of complexation, occurs when an equilibrium reaction between a metal ion and an organic ligand results in more than one bond between the metal and a molecule of the complexing agent with the formation of a heterocyclic ring which includes the metal ion. Chelates have the potential for forming complexes of greater stability than analogous complexes in which the metal is coordinated in a noncyclic fashion with water, other organic ligands, or both. This effect is most pronounced in dilute solutions and appears to be related to entropy considerations.

Chelate stability generally increases with the number of chelate rings formed with the metal ion and with decreased ring size, but in addition, the stability of a chelate is a function of those factors influencing the stability of simple complexes. Lehman (1963) outlined environmental factors which may influence chelate stability. Recently, Norvell (1972) reviewed the factors influencing the stability and equilibria of specific synthetic metal chelates in soil.

Increased temperature generally results in a decrease in the chelate stability constant. An increase in the dielectric constant of the solvent or in ionic strength generally results in a decrease in the stability constant. Other ions in solution will compete for the donor groups and an indication of the final equilibrium of the system can be obtained from the stability constant, with the metal ion having the highest stability constant being the first chelated.

For most donor groups, H^+ is a highly efficient competitor with the metals and pH is probably the single most important parameter influencing chelate stability in solution. Where the stability constant with a given metal is low, the H^+ ion is a strong competitor for the donor site. In contrast, if the stability constant is high, the complex may be stable at relatively low pH values. In the case of most metals, pH also markedly influences solubility, and at pH values above 8, insoluble hydroxides may effectively remove those metals which do not have soluble OH products from solution, thus limiting severely the likelihood of complex formation.

B. Difficulties in Characterization of Soil Organic Complexes

Although the chemistry of metals of great agronomic significance has been extensively studied, a complete understanding of the reactions of these elements with soil organic matter and the role of the soil microbiota in influencing the chemical fate of these elements has not been attained. Less information is available regarding metals nonessential to higher plant life (e.g., Cd, Cr, Hg, and Ni) which have the potential for entering soil in fertilizers and in industrial and municipal wastes, and which may be sufficiently soluble under certain soil conditions to be taken up by plants. In general, the principal difficulties associated with the study of metal complexation and biotransformation in surface soils result from the complexity of the soil organic fraction and the low concentrations of most metals in soils.

The soil organic fraction has been studied for nearly 100 years with considerable progress (Kononova, 1966). Yet, due to its complexity, it has been concluded that the evidence substantiating metal complexation and chelation in soils is largely circumstantial (Mortensen, 1963). To verify complexation requires extraction and positive identification of the intact compounds. To accomplish this requires separation of the compound from the soil matrix without chemical modification during extraction, separation of the compound from other coextractable materials, and positive chemical identification of the unaltered isolate. Some definitive and highly useful investigations have recently been conducted and these will be summarized in a subsequent section, but unequivocal identification of specific, intact organometal complexes in soil remains to be accomplished for surface soils. However, by inference from other systems, from the results of studies of metal exchange as a function of soil properties, and from the application of a broad range of characterization methods (such as successive extraction with chelating agents, flocculation, peptization, and precipitation reactions, aqueous and nonaqueous titration, absorption spectrophotometry, electrophoresis, and chromatography) in studies of soils and model compounds, it would appear likely that metal complexation does occur in surface soils. Application of these chemical characterization techniques has resulted in identification of a number of organic ligands with potential for metal complexation in soils (Stevenson & Ardakani, 1972). Furthermore, the study of podzol B horizons where complexes of principally Fe and Al leached from surface soils are present in simpler associations than in surface soils (Schnitzer, 1969), has provided a great deal of evidence useful in evaluation of the forms of organometal complexes in surface soils.

The necessary use of extraction and chromatographic methods in which metal complexes are purified by successive partitioning requires that the concentration of the substance in soil be sufficiently high to allow for losses which occur during isolation and characterization. However, with the exception of Fe and Al, the concentration of metals in mineral soils not subject to recent pollution is low with total concentrations of most trace metals usually < 100 ppm (Swaine, 1955). Until the advent of neutron activation (which is an expensive technique for a single element), flameless atomic absorption spectroscopy, and the graphite furnace for atomic absorption spectroscopy, it was generally not possible to remove metals from soils with mild extractants in sufficient quantities for detection and subsequent characterization using standard techniques of membrane and gel filtration, electrophoresis, and chromatography.

The necessity for use of soil extractants and the relatively low concentrations of metals in nonpolluted soils has led to the study of "spiked" soils in which sufficient radio-labeled or unlabeled metal in ionic form is "equilibrated" with the soil, followed by extraction and characterization of isolated components. This method, although essential in the study of some metals, likely does not allow adequate assessment or prediction of the form in which many metals will be stable in soil over the long term. As discussed previously, metal exchange may not occur, particularly where bond types approach covalency. Furthermore, the type and kinetics of exchange will be highly dependent upon the conditions under which equilibration is attempted. In effect, three potential metal pools may result when addition of metals to soil and equilibration is attempted. These include the endogenous metal, the unlabeled metal (or carrier metals in the case of radiotracer studies), and the labeled metal in radiotracer studies. Similar problems are encountered in microbial studies when organisms must be isolated from soil and metabolic products studied in vitro.

While the diversity of organic materials present in soil in conjunction with the dynamic nature of the soil system will continue to serve as major deterrents to the complete elucidation of the physiochemical properties of soil organometal complexes, much progress in the characterization of metal complexes in soil may be expected in the next several years. This will result in large measure from recent developments in (i) atomic absorption allowing routine detection of endogenous metals present at relatively low concentrations (< 1 ppb for many metals) in soil extracts and purified eluates, and (ii) ultrafiltration, gel and thin-layer electrophoresis and thin-layer, column, and liquid chromatography of complexes over a wide molecular weight range.

C. Metal Complexation by Soil Organic Matter

The soil chemistry of micronutrients of greatest agronomic importance (B, Co, Cu, Fe, Mn, Mo, Se, Zn) and their interactions with the soil organic fraction have been the subject of a number of excellent reviews over the last two decades (Mitchell, 1964, 1972; Mortensen, 1963; Hodgson, 1963; Stevenson & Ardakani, 1972). In general, earlier studies emphasized metal interactions with intact soil or with the higher molecular weight humic components of soil whereas more recent studies have emphasized the more soluble components of soil.

It is practical to categorize metal complexes in soil in terms of their solubility since, in general, it is this factor which most influences their mobility and plant availability. Three principal categories have been proposed (Hodgson, 1963) although the complexity of the soil system results in considerable overlap between categories. These include (i) the relatively high molecular-weight humic substances containing condensed aromatic nuclei in complex polymers derived from secondary syntheses which have a high affinity for metals but are largely insoluble in soil, (ii) low molecular weight organic acids and bases, classified as nonhumic substances, and derived largely from microbial cells and metabolism which demonstrate relatively high solubility in association with metals, and (iii) soluble ligands which are precipitated on reaction with metals.

1. HUMIC SUBSTANCES

Humic substances are generally divided into three categories based on their solubilities (Felbeck, 1965). The humin (alkali and acid insoluble) fraction is soluble only under drastic conditions and is apparently of the highest molecular weight. The humate (alkali soluble, acid insoluble) and fulvate (alkali and acid soluble) fractions of soil may constitute up to 90% of the soil organic fraction (Kononova, 1966). The humates and fulvates are characterized, in part, by a high charge density due to acidic functional groups (Stevenson & Ardakani, 1972; Felbeck, 1965). This property leads to a high degree of reactivity and these materials exhibit a strong pH-dependent affinity for cations in solution and are likely strongly bound to soil minerals and other organic constituents in soil (Greenland, 1965). The acidic functional groups consist principally (in general order of acidity) of carboxyl, hydroxyl (phenolic, alcoholic), enolic, and carbonyl groups (Broadbent & Bradford, 1952; Felbeck, 1965; Schnitzer et al., 1959). Total acidity has been estimated to range between 500 to 900 and 900 to 1,400 meq/100 g for humic acids and fulvic acids, respectively (Stevenson & Butler, 1959). The acidic H of humic acids was differentiated by Thompson (1965)[4] into three groups at 100 to 200, 500 to 700, and 1,000 to 1,200 meq/100 g using nonaqueous titration methods. Basic functional groups, likely amides and heterocyclic nitrogen compounds (Bremner, 1965), probably also contribute to retention of metals but are of much less importance than acidic groups at most soil pH values.

Although humic and fulvic acids likely account for most of the metal immobilization attributed to the soil organic matter, (e.g., Hodgson, 1963;

---

[4]S. O. Thompson. 1965. Comparative properties of plant lignins and soil humic materials. Ph.D. Thesis, Univ. of Wisconsin–Madison.

Stevenson & Ardakani, 1972), they have the potential for formation of soluble complexes with metals, particularly in dilute solutions. Small quantities of metal fulvates thought to be of lower molecular weight than the humates, may be present in soil solution. A nondialyzable material with infrared absorption spectra and elemental analyses similar to fulvic acids was isolated from a dilute salt (0.01N KBr) extract of a mineral soil by Geering and Hodgson (1969). The material exhibited a concentration equivalent to 2.5% of a dialyzable fraction but was more effective in complexing Cu and Zn.

2. NONHUMIC SUBSTANCES WITH POTENTIAL FOR METAL COMPLEXATION

Lower molecular weight biochemicals of recent origin have been implicated in metal complexation and solubilization in soil. These materials represent (i) components of living cells of microorganisms and plant roots and their exudates and (ii) the entire spectrum of potential degradation products which ultimately serve as the building units of the soil humic fraction. The quantity and composition of these materials will vary with soil, vegetation, and environmental conditions (Alexander, 1961). Readily decomposable wastes disposed to soil under conditions appropriate for microbial growth may, for example, result in immediate and marked increases in organic materials identified in category (i) and longer term increases of materials in category (ii). Conversely, toxic materials may have the opposite effects. The specific compounds produced will be dependent upon the properties of the waste and soil environmental conditions after disposal (Routson & Wildung, 1969).

Although the concentration of metals soluble in the soil solution or in mild extractants is low, often near minimum detectable levels, the major portion of Cu and Zn have been shown to be associated with low molecular weight components. Most of the titratable acidity of this fraction has been attributed (Geering & Hodgson, 1969) to aliphatic acids ($<$ pH 7.0) and amino acids ($>$ pH 7.0).

The production, distribution, and action of organic acids in soil has been reviewed by Stevenson (1967). A wide range of organic acids are produced by microorganisms known to be present in soil. These include (i) simple acids such as acetic, propionic, and butyric, produced in largest quantities by bacteria under anaerobic conditions, (ii) carboxylic acids derived from monosaccharides, such as gluconic, glucuronic, and α-ketogluconic acids produced by both bacteria and fungi, (iii) products of the citric acid cycle such as succinic, fumaric, malic, and citric acid, which are common metabolic excretory products of fungi, and (iv) aromatic acids such as p-hydroxybenzoic, vanillic, and syringic acids thought to be fungal decomposition products of plant lignins. A variety of organic acids have also been reported in root exudates.

The other important group of compounds identified in significant quantities in the soil solution by Geering and Hodgson (1969) which may be expected to exhibit strong affinity for metals are the amino acids. The qualitative and quantitative aspects of amino acids and other nitrogenous components in soils have been reviewed by Bremner (1967). It was concluded that soil acid hydrolysates do not differ greatly in amino acid composition but quantitative differences may occur with differences in soil, climatic, and cultural practices. A number of acidic and basic amino acids have been reported in soil. However, it appears that the major portion of amino acid-N that is present in hydrolysates is in the neutral amino acids glycine, alanine, serine, threonine, valine, leucine, isoleucine, proline, aspartic acid, glutamic acid, lysine, and arginine. Most of the amino acids detected in soil hydrolysates have also been shown to exist free in small quantities in soils with levels seldom exceeding 2 ppm. In the soil solution (Geering & Hodgson, 1969), neutral amino acids also appeared to predominate. Basic amino acids were not detected although two acidic amino acids (aspartic and glutamic acids) were present.

Stevenson and Ardakani (1972) concluded that organic acids and amino acids, while present only in small quantities in soil, were present in sufficient quantities in water-soluble forms to play a significant role in solubilization of mineral matter in soil. Small quantities of a number of other complexing agents, such as adenosine phosphates, polyphenols, phytic acid, prophyrins, and auxins, also exist in soil (pertinent references have been summarized by Mortensen, 1963). However, it appears unlikely that these materials would be present in significant quantities in the soil solution under most soil conditions.

V. APPLICATIONS AND RESEARCH NEEDS

This review has given only cursory treatment to the vast amount of knowledge available on the chemical reactions occurring in soils. However, in many instances, sufficient information is available to predict for practical purposes the fate of waste components in the soil-plant-water continuum. In view of this, those involved in soil chemistry research should carefully evaluate the need of instigating new investigations when a different soil type or waste is encountered.

Little attention has been given in this review to (i) microbial processes which may directly and indirectly influence the chemical behavior and fate of wastes in soil, (ii) soil physical phenomena, or (iii) soil-plant interactions. These phenomena will receive detailed attention in other chapters. However, it should be emphasized that decisions regarding waste applications require an integrated view encompassing soil and plant aspects of the problem. In this respect, increased interaction is required between the disciplines of soil chemistry, biochemistry, microbiology, physics, plant physiology, and genetics. These disciplines have much to offer each other regarding soil-plant interactions and innovative research approaches.

Among the soil chemistry research needs which appear to be most pressing is that of elucidating the retention mechanisms and factors influencing the form and long term behavior of metals in soils. This information is essential to development of better recommendations and management techniques for application of high metal wastes to soils.

ACKNOWLEDGMENT

Contribution from the College of Agricultural and Life Sciences, University of Wisconsin, and from Battelle, Pacific Northwest Laboratories, Richland, Washington. This review was supported, in part, by the United States Energy Research and Development Administration, under Contract AT(45-1)-1830. The comments and suggestions by Drs. R. B. Corey, R. C. Routson, and S. S. Shukla are much appreciated.

LITERATURE CITED

Adams, S. N., and J. L. Honeysett. 1964. Some effects of soil waterlogging on the cobalt and copper status of pasture plants grown in pots. Aust. J. Agric. Res. 15:357-367.
Alexander, M. 1961. Introduction to soil microbiology. John Wiley, New York.
Aylmore, L. A. G., M. Karim, and J. P. Quirk. 1967. Adsorption and desorption of sulfate ions by soil constituents. Soil Sci. 103:10-15.
Barrow, N. J. 1974. Effect of previous additions of phosphate on phosphate adsorption by soils. Soil Sci. 117:28-33.
Bear, F. E., A. L. Prince, and J. L. Malcolm. 1945. Potassium needs of New Jersey soils. New Jersey Agric. Exp. Sta. Bull. 721, Rutgers, N. J. 42 p.
Bittell, J. E., and J. Miller. 1974. Lead, cadmium, and calcium selectivity coefficients on a montmorillonite, illite, and kaolinite. J. Environ. Qual. 3:250-253.
Black, C. A. 1968. Soil-plant relationships. John Wiley, New York.
Black, C. A. 1970. Behavior of soil and fertilizer phosphorus in relation to water pollution. p. 72-93. In T. L. Willrich and G. E. Smith (ed.) Agricultural practices and water quality. Iowa State Univ. Press, Ames.
Bloomfield, C. 1954. A study of podzolization. V. The mobilization of iron and aluminum by aspen and ash leaves. J. Soil Sci. 5:50-56.
Bloomfield, C. 1955. Leaf leachates as a factor in pedogenesis. J. Sci. Food Agric. 6: 641-651.
Bremner, J. M. 1965. Organic nitrogen in soils. p. 93-149. In W. V. Bartholomew and F. E. Clark (ed.) Soil nitrogen. Am. Soc. of Agron., Madison, Wis.
Bremner, J. M. 1967. Nitrogenous compounds. p. 19-66. In A. D. McLaren and G. H. Peterson (ed.) Soil biochemistry. Marcel Dekker, Inc., New York.
Broadbent, F. E., and G. R. Bradford. 1952. Cation-exchange groupings in the soil organic fraction. Soil Sci. 74:447-457.
Brown, M. J., and D. L. Carter. 1969. Leaching of added selenium from alkaline soils as influenced by sulfate. Soil Sci. Soc. Am. Proc. 33:563-565.
Chaberek, S., and A. E. Martell. 1959. Organic sequestering agents. John Wiley, New York.
Chaney, R. L. 1973. Crop and food chain effects of toxic elements in sludges and effluents. p. 129-141. In Recycling municipal sludges and effluents on land. Nat. Assoc. State Universities and Land-Grant Colleges, Washington, D. C.
Chao, T. T., M. E. Harward, and S. C. Fang. 1962. Adsorption and desorption phenomena of sulfate ions in soils. Soil Sci. Soc. Am. Proc. 26:234-237.
Davies, E. B. 1956. Factors affecting molybdenum availability in soils. Soil Sci. 81: 209-223.
Dean, L. A., and E. J. Rubins. 1947. Anion exchange in soils: I. Exchangeable phosphorus and the anion-exchange capacity. Soil Sci. 63:377-387.

Dolar, S. G., and D. R. Keeney. 1974. Availability of Cu, Zn, and Mn as influenced by application of Cu-polyflavonoid. Can. J. Soil Sci. 54:225–233.

Ellis, B. G. 1973. The soil as a chemical filter. p. 46–70. In W. E. Sopper and L. T. Kardos (ed.) Recycling treated municipal wastewater and sludge through forest and cropland. Penn State Press, University Park, PA.

Ellis, B. G., and B. D. Knezek. 1972. Adsorption reactions of micronutrients in soils. p. 59–68. In J. J. Mortvedt, P. M. Giordano, and W. L. Lindsay (ed.) Micronutrients in agriculture. Soil Sci. Soc. Am., Madison, Wis.

Felbeck, G. T., Jr. 1965. Structural chemistry of soil humic substances. Adv. Agron. 17:327–368.

Geering, H. R., E. E. Carey, L. H. P. Jones, and W. H. Allaway. 1968. Solubility and redox criteria for the possible forms of selenium in soils. Soil Sci. Soc. Am. Proc. 32:35–40.

Geering, H. R., and J. F. Hodgson. 1969. Micronutrient cation complexes in soil solution: III. Characterization of soil solution ligands and their complexes with Zn^{2+} and Cu^{2+}. Soil Sci. Soc. Am. Proc. 33:54–59.

Gotoh, S., and W. H. Patrick, Jr. 1972. Transformation of manganese in a waterlogged soil as affected by redox potential and pH. Soil Sci. Soc. Am. Proc. 36:738–742.

Gotoh, S., and W. H. Patrick, Jr. 1974. Transformation of iron in a waterlogged soil as influenced by redox potential and pH. Soil Sci. Soc. Am. Proc. 38:66–71.

Greenland, D. J. 1965. Interaction between clays and organic compounds in soils. Part 1. Mechanisms of interaction between clays and defined organic compounds. Soils Fert. 28:415–425.

Hahne, H. C., and W. Kroontje. 1973. Significance of pH and chloride concentration on behavior of heavy metal pollutants; mercury(II), cadmium(II), zinc(II), and lead(II). J. Environ. Qual. 2:444–451.

Harward, M. E., and H. M. Reisenauer. 1966. Reactions and movement of inorganic soil sulfur. Soil Sci. 101:326–335.

Hingston, F. J., R. J. Atkinson, A. M. Posner, and J. P. Quirk. 1968. Specific adsorption of anions on geothite. Int. Congr. Soil Sci. Trans. 9th (Adelaide, Aust.) 1:669–678.

Hodgson, J. F. 1963. Chemistry of the micronutrient elements in soils. Adv. Agron. 15:119–159.

Hodgson, J. F., W. L. Lindsay, and J. F. Trierweiler. 1966. Micronutrient cation complexing in soil solution. II. Complexing of zinc and copper in displaced solution from calcareous soils. Soil Sci. Soc. Am. Proc. 30:723–726.

Jacobs, L. W., and D. R. Keeney. 1970. Arsenic-phosphorus interactions on corn. Commun. Soil Sci. Plant Anal. 1:85–93.

Jacobs, L. W., J. K. Syers, and D. R. Kenney. 1970. Arsenic sorption by soils. Soil Sci. Soc. Am. Proc. 34:750–754.

Jenne, E. A. 1968. Controls on Mn, Fe, Co, Ni, Cu, and Zn concentrations in soils and water: The significant role of hydrous Mn and Fe oxides. In Trace inorganics in water. Adv. Chem. Ser. 73:337–387. Am. Chem. Soc., Washington, D. C.

John, M. K. 1972. Cadmium adsorption maxima of soils as measured by the Langmuir isotherm. Can. J. Soil Sci. 52:343–350.

Jurinak, J. J., and J. Santillan-Medrano. 1974. The chemistry and transport of lead and cadmium in soils. Res. Rep. 18. Agric. Exp. Sta., Utah State Univ., Logan. 121 p.

Kee, N. S., and C. Bloomfield. 1962. The effect of flooding and aeration on the mobility of certain trace elements in soils. Plant Soil 16:108–135.

Kononova, M. M. 1966. Soil organic matter. Pergamon Press, New York.

Lagerwerff, J. V., and D. L. Brower. 1972. Exchange adsorption of trace quantities of cadmium in soils treated with chlorides of aluminum, calcium, and sodium. Soil Sci. Soc. Am. Proc. 36:734–737.

Lagerwerff, J. V., and D. L. Brower. 1973. Exchange adsorption as precipitation of lead in soils treated with chloride of aluminum, calcium, and sodium. Soil Sci. Soc. Am. Proc. 37:11–13.

Lakin, H. W. 1972. Selenium accumulation in soils and its adsorption by plants and animals. Geol. Soc. Am. Bull. 83:181–190.

Lance, J. C. 1972. Nitrogen removal by soil mechanisms. J. Water Pollut. Control Fed. 44:1352–1361.

Leeper, G. W. 1947. The forms and reactions of manganese in the soil. Soil Sci. 63:79–94.

Leeper, G. W. 1972. Reactions of heavy metals with soils with special regard to their application in sewage wastes. Dep. of Army, Corps of Engineers, DACW 73-73-C-0026.

Lehman, D. S. 1963. Some principles of chelation chemistry. Soil Sci. Soc. Am. Proc. 27:167-170.

Lindsay, W. L. 1972. Inorganic phase equilibria of micronutrients in soils. p. 41-57. In J. J. Mortvedt, P. M. Giordano, and W. L. Lindsay (ed.) Micronutrients in agriculture. Soil Sci. Soc. Am., Madison, Wis.

McLaren, R. G., and D. V. Crawgord. 1973. Studies on soil copper. II. The specific adsorption of copper by soils. J. Soil Sci. 24:443-452.

MacLean, A. J. 1974. Effects of soil properties and amendments on the availability of zinc in soils. Can. J. Soil Sci. 54:369-378.

Magdoff, F. R., D. R. Keeney, J. Bouma, and W. A. Ziebell. 1974. Columns representing mound-type disposal systems for septic tank effluent. II. Nutrient transformations and bacterial populations. J. Environ. Qual. 3:228-234.

Martell, A. E. 1971. Principles of complex formation. p. 239-263. In S. D. Vaust and J. V. Hunter (ed.) Organic compounds in aquatic environments. Marcel Dekker, Inc., New York.

Martell, A. E., and M. Calvin. 1952. Chemistry of the metal chelate compounds. Prentice-Hall, New York.

Mattingly, G. E. G. 1975. Labile phosphates in soils. Soil Sci. 119:369-375.

Mitchell, R. L. 1964. Trace elements in soils. p. 320-368. In F. E. Bear (ed.) Chemistry of the soil. Reinhold Publishing Corp., New York.

Mitchell, R. L. 1972. Trace elements in soils and factors that affect their availability. Geol. Soc. Am. Bull. 83:1069-1076.

Mortensen, J. L. 1963. Complexing of metals by soil organic matter. Soil Sci. Soc. Am. Proc. 27:179-186.

Nakayama, F. S. 1968. Calcium activity, complex, and ion pair in saturated $CaCO_3$ solutions. Soil Sci. 106:429-434.

Norvell, W. A. 1972. Equilibria of metal chelates in soil solution. p. 115-138. In J. J. Mortvedt, P. M. Giordano, and W. L. Lindsay (ed.) Micronutrients in agriculture. Soil Sci. Soc. Am., Madison, Wis.

Novak, L. T., D. C. Adriano, G. A. Coulman, and D. B. Shah. 1975. Phosphorus movement in soils: Theoretical aspects. J. Environ. Qual. 4:93-98.

Patrick, W. H., Jr., S. Gotoh, and B. G. Williams. 1973. Strengite dissolution in flooded soils and other natural aqueous solutions from specific conductance. Soil Sci. 102:408-413.

Patrick, W. H., Jr., and R. A. Khalid. 1974. Phosphate release and sorption by soils and sediments: Effect of aerobic and anaerobic conditions. Science 186:53-55.

Ponnamperuma, F. N. 1972. The chemistry of submerged soils. Adv. Agron. 24:29-96.

Ponnamperuma, F. N., T. A. Loy, and E. M. Tianco. 1969. Redox equilibria in flooded soils: II. The manganese oxide systems. Soil Sci. 108:48-57.

Rhoades, J. D., R. D. Ingvalson, and J. T. Hatcher. 1970. Adsorption of boron by ferromagnesium minerals and magnesium hydroxide. Soil Sci. Soc. Am. Proc. 34:938-941.

Richards, L. A., ed. 1954. Diagnosis and improvement of saline and alkali soils. USDA Agric. Handb. No. 60.

Richburg, J. S., and F. Adams. 1970. Solubility and hydrolysis of aluninum in soil solutions and saturated-paste extracts. Soil Sci. Soc. Am. Proc. 34:728-734.

Routson, R. C., and R. E. Wildung. 1969. Ultimate disposal of wastes to soil. p. 19-25. In L. Cecil (ed.) Water-1969, Chem. Eng. Progr. Symp. Ser. Am. Inst. Chem. Eng., New York.

Schnitzer, M. 1969. Reactions between fulvic acid, a soil humic compound and inorganic soil constituents. Soil Sci. Soc. Am. Proc. 33:75-81.

Schnitzer, M., D. A. Shearer, and J. R. Wright. 1959. A study in the infrared of high-molecular weight organic matter extracted by various reagents from a podzolic B horizon. Soil Sci. 87:252-257.

Shah, D. B., G. A Coulman, L. T. Novak, and B. G. Ellis. 1975. A mathematical model for phosphorus movement in soils. J. Environ. Qual. 4:87-92.

Shawney, B. L. 1972. Selective sorption and fixation of cations by clay minerals: A review. Clays Clay Miner. 20:93-100.

Shukla, S. S., J. K. Syers, J. D. H. Williams, D. E. Armstrong, and R. F. Harris. 1971. Sorption of inorganic phosphate by lake sediments. Soil Sci. Soc. Am. Proc. 35: 244–249.

Sigel, H. 1974. Metal ions in biological systems: Simple complexes. Marcel Dekker, Inc., New York.

Singh, B. R., and Y. Kanehiro. 1969. Adsorption of nitrate in amorphous and kaolinitic Hawaiian soils. Soil Sci. Soc. Am. Proc. 33:601–683.

Sokoloff, V. P. 1953. Occlusion of copper and zinc by some soil materials of lower Mississippi River area. Science 118:296–297.

Stevenson, F. J. 1967. Organic acids in soil. p. 119–146. In A. D. McLaren and G. H. Peterson (ed.) Soil biochemistry. Marcel Dekker, New York.

Stevenson, F. J., and M. S. Ardakani. 1972. Organic matter reactions involving micronutrients in soils. p. 79–114. In J. J. Mortvedt, P. M. Giordano, and W. L. Lindsay (ed.) Micronutrients in agriculture. Soil Sci. Soc. Am., Madison, Wis.

Stevenson, F. J., and J. H. A. Butler. 1969. Chemistry of humic acids and related pigments. p. 534–557. In G. Englinton and Sister M. T. J. Murphy (ed.) Organic geochemistry. Springer-Verlag, Berlin.

Swaine, D. J. 1955. The trace-element content of soils. Tech. Commun. No. 48, Commonwealth Bur. of Soil Sci., Rothamsted Exp. Sta., Harpenden, England.

Swenson, R. M., C. V. Cole, and D. H. Sieling. 1949. Fixation of phosphate by iron and aluminum and replacement by organic and inorganic ions. Soil Sci. 67:3–22.

Syers, J. K., M. Browman, G. W. Smillie, and R. B. Corey. 1973. Phosphate sorption by soils evaluated by the Langmuir adsorption equation. Soil Sci. Soc. Am. Proc. 37: 358–363.

Tanji, K. K. 1970. A computer analysis on the leaching of boron from stratified soil columns. Soil Sci. 110:44–51.

Thomas, G. W., and A. R. Swoboda. 1970. Anion exclusion effects on chloride movement in soils. Soil Sci. 110:163–166.

Tiller, K. G., J. F. Hodgson, and M. Peech. 1963. Specific sorption of cobalt by soil clays. Soil Sci. 95:392–399.

Udo, E. J., H. L. Bohn, and T. C. Tucker. 1970. Zinc adsorption by calcareous soils. Soil Sci. Soc. Am. Proc. 34:405–407.

Williams, J. D. H., J. K. Syers, R. F. Harris, and D. E. Armstrong. 1970. Adsorption and desorption of inorganic phosphorus by lake sediments in a 0.1M NaCl system. Environ. Sci. Technol. 4:517–519.

chapter 5

Physical Properties of Soils

JOHN LETEY, JR., University of California, Riverside, California

I. INTRODUCTION

The soil matrix consists primarily of mineral particles of various size, shape, and mineralogy. Most soils also contain small percentages of organic matter which coats the mineral particles. Exceptions are where relatively large quantities of organic debris or solid organic wastes are incorporated into the soil.

The content and arrangement of mineral particles and organic matter constitute the physical framework within which a variety of physical-chemical-biological interactions occur. Most properties and reactions vary with respect to both time and position. This chapter describes those physical processes occurring within the soil matrix which can significantly affect disposal of organic wastes in the soil.

The schmatic diagram shown in Fig. 1 gives the interrelationships between various physical processes. An arrow leading from one soil property or process to another indicates that the first significantly affects the value of the second. For example, an arrow from water retention to gas diffusion indicates that the gas diffusion rate is dependent upon water retention. Soil texture, in the upper left hand corner, is the origin for all arrows. The arrows terminate at the three functions that soil serves: (i) soil as a medium for plant growth, (ii) soil as a material for construction, and (iii) soil as a moderator of environmental pollution.

II. SOIL STRUCTURE

Soil structure refers to the mineral and organic component arrangement. Mineral particles are not usually randomly and uniformly distributed in the soil but are grouped into subunits referred to as aggregates or peds. Structure is not static. Chemical composition of the water significantly affects the degree of soil aggregation. Polyvalent cations tend to cause clay flocculation leading to aggregate formation; on the other hand, sodium contributes to dispersion and breakdown of aggregates. High electrolyte concentration contributes to floculation, but low electrolyte concentration contributes to dispersion. Also decomposition of organic matter by organisms contribute to development and stability of aggregates.

III. POROSITY AND PORE SIZE DISTRIBUTION

Porosity (fraction of the volume which is not solid) and pore size distribution are important parameters of the physical matrix. Pore size distribution generally has a greater effect than total porosity on physical processes occurring in the soil. Coarse-textured soils have large pores with a relatively small range of sizes. Also nonaggregated fine-textured soils have small pores with a small size range. However, any combination of intermediate situations

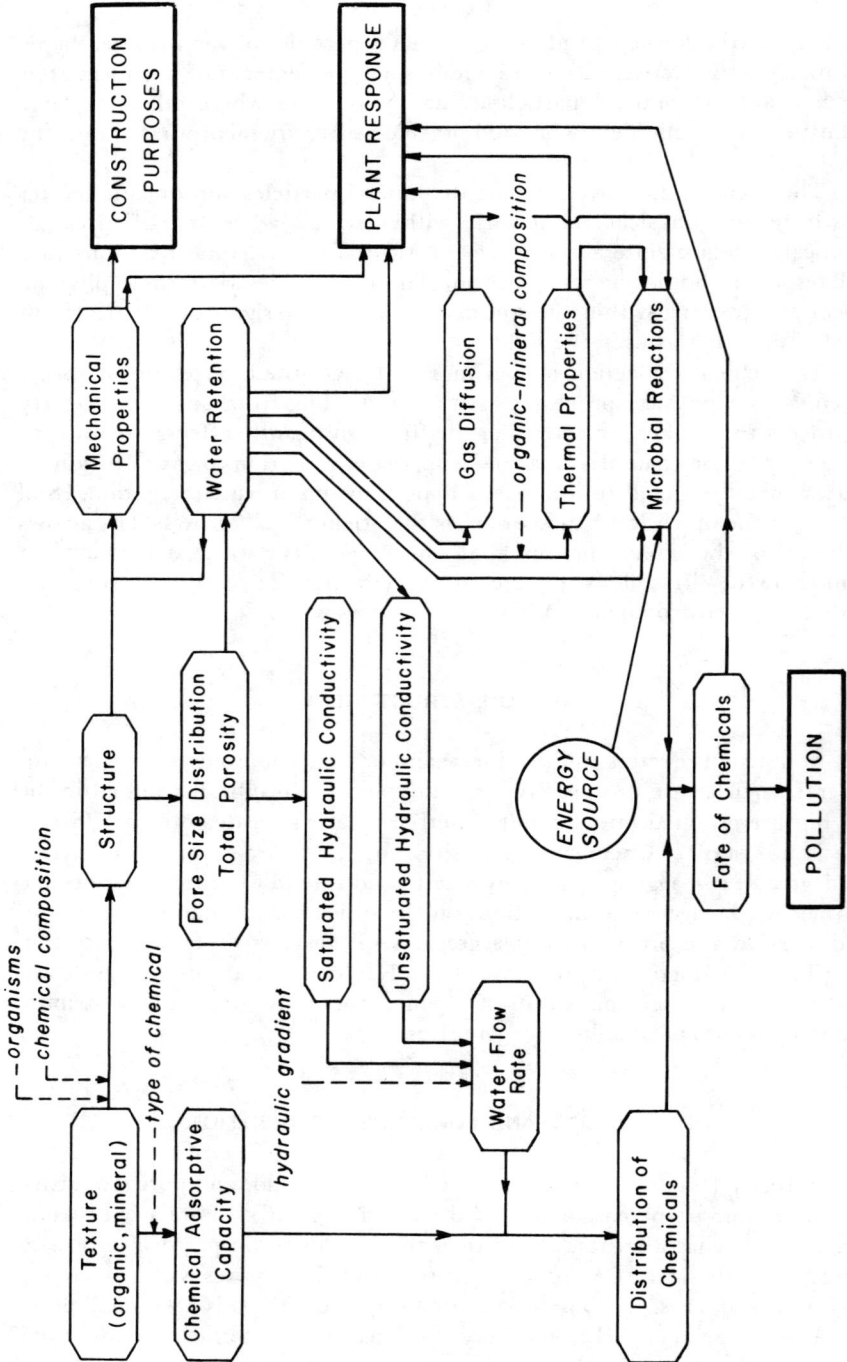

Fig. 1—Interrelationships between soil-physical processes.

can occur in aggregated soils. Soil may have small pores within and large pores between aggregates.

Disposal of organic wastes in soil may affect the pore size distribution. If large quantities of solid waste are incorporated into the soil, they become a significant component of the matrix and can alter pore sizes. In plant nurseries and landscaping, wood chips or other organic components are incorporated into the soil to provide large pores. The addition of organic chemicals to soil may affect aggregation thus also contributing to alteration in pore size distribution.

IV. FLOW PROCESSES

Various flow processes occur within the soil matrix. Although the movement of components such as gas, water, and heat have unique characteristics, the steady-state flow of each is expressed by the same basic relation

$$\text{flow rate} = \text{transmission coefficient} \times \text{driving force} \qquad [1]$$

where flow rate is quantity unit area^{-1} unit time^{-1}. The steady-state flow calculations for any component, therefore, involves proper characterization and measurement of the appropriate driving force and transmission coefficient for that component in soil.

Steady-state flow is characterized by a constant flow rate in which the quantity of flowing material remains constant at a given position. As examples, temperature at a given point remains constant at all times during steady-state heat flow or water content remains constant during steady-state flow. If the quantity at a given point changes with time, a transient flow process is operative. Mathematically a transient flow equation is developed by combining the steady-state flow equation with the equation of continuity. The equation of continuity is a mathematical expression for the conservation of matter.

Transient flow processes are quite complex and will not be treated extensively in this chapter. However, a transmission coefficient is involved in transient flow processes which usually equals the steady-state transmission coefficient divided by a "capacity" factor. The implications of the "capacity" factor for transient flow of various components will be discussed in the following sections dealing with specific flow processes.

V. WATER FLOW

The steady-state equation describing water flow through soil is known as Darcy's law. The appropriate driving force is the hydraulic head gradient and the transmission coefficient is the hydraulic conductivity. Water flows under either saturated or unsaturated states. (The term *saturated* will denote pores completely filled with water.)

With a uniform soil, the saturated hydraulic conductivity is dependent primarily upon the largest pore size because the velocity is proportional to the pore radius squared. In a soil consisting of a variety of pore sizes most of the flow will occur through the largest pores. Saturated water flow has generally been found to decrease with time as water flows through the soil. The decrease in hydraulic conductivity may be associated with gradual deterioration of aggregation leading to loss of the large pores. However, a decrease in saturated conductivity has also been measured in quartz sand particles in which the pore sizes do not change with time. The decrease in flow in coarse sands has been attributed primarily to development of microbial products.

Decrease in hydraulic conductivity with time has significant implications to organic waste disposal. For example, sewage effluent may be passed through a soil for further purification by ponding the water on the surface. Because the hydraulic conductivity decreases with time, it is advisable to allow the soil surface to dry periodically to re-establish a higher conductivity. Over a long period of time more water can be passed through soil by intermittent drying as compared to continuous ponding.

If saturated water flow occurs through a layered soil of differing hydraulic conductivity, the effective hydraulic conductivity for the layered profile can be calculated from

$$\Sigma \ l_i/K_a = \Sigma \ (l_i/K_i) \qquad [2]$$

where K_a is the effective hydraulic conductivity, K_i is the hydraulic conductivity of the ith layer, and l_i is the thickness of the ith layer. The effective hydraulic conductivity is not an arithmetic mean of the hydraulic conductivities of the various layers. The average hydraulic conductivity is greatly influenced by the layer of the lowest hydraulic conductivity and will approach that value. For example, if there are two layers of equal thickness, one layer having a hydraulic conductivity of 100 and the other of 1 cm hour^{-1}, the effective hydraulic conductivity for both layers would be 2 cm hour^{-1}. A very thin layer of material with low hydraulic conductivity can significantly reduce the overall flow rate through soil. A thin layer of low hydraulic conductivity material may be formed at the soil surface if the infiltrating liquid contains fine particles which are screened at the surface. A layer can also be caused by applying water in a manner which disturbs the soil surface, breaks down aggregates, and suspends clay in solution. Any system designed to pond water on the surface should have water free from turbidity before application and arrangements to periodically disrupt the fine layer which develops at the surface.

As soil is allowed to drain, the large pore sizes drain first. Indeed the amount of water retained is a direct function of pore size distribution. Obviously, sand with predominantly large pores retains very little water as compared to a poorly aggregated fine-textured soil with predominantly small pores. Aggregated soil with a variety of pore sizes have intermediate water retention properties.

Unsaturated hydraulic conductivity is strongly dependent upon water retention properties of the soil. As the large pores are drained they no longer conduct water and water is only transmitted through the small pores containing water. The unsaturated conductivity is much lower than the saturated conductivity and decreases as the water content decreases.

A soil with a high saturated hydraulic conductivity may have a lower unsaturated conductivity than a soil with a low saturated conductivity. For example, sand with large pores and high saturated conductivity retains very little water and has an extremely low unsaturated conductivity. On the other hand, a poorly aggregated fine-textured soil has low saturated hydraulic conductivity but retains more water and therefore has a higher unsaturated conductivity than sand. A pile of wet sand tends to dry very rapidly on the surface but remains wet for an extremely long period of time in the interior. This is a direct consequence of the low unsaturated conductivity of sand as it becomes dry and no longer can conduct water to the surface for evaporation.

Water flow rate equals the conductivity times the hydraulic head gradient. Obviously, a soil with high conductivity has a higher flow rate for a given hydraulic head gradient than soil with a low conductivity. On the other hand, it must be clearly recognized that water flow rate is dependent upon both the conductivity and the hydraulic head gradient.

Ponding water on the soil surface causes the soil to become nearly saturated. Undersaturated soil conditions can be maintained by using sprinklers which apply water at a rate lower than the saturated conductivity of the soil.

VI. CHEMICAL TRANSPORT

Water flowing through soil has the ability to transport chemicals within the profile. Referring again to Fig. 1 note that the chemical distribution depends on water flow and chemical adsorption by the soil. Adsorption is dependent upon properties of both the soil and chemical. Generally finer-textured soils have a higher adsorptive capacity than coarse-textured soils. Soil organic matter also has relatively high adsorptive capacity for most organic chemicals. If a chemical has a charge, adsorption will be highly dependent upon the type and amount of clay minerals present in the soil. For example montmorillonitic-type clays have relatively many negative charges and, therefore, can adsorb large amounts of positively charged ions or molecules.

Qualitatively, the depth to which a chemical moves for a given amount of water flow is inversely proportional to the adsorption coefficient of the chemical.

Movement of organic pesticides through soil has been reviewed by Letey and Farmer (1974) and Leistra (1973). The general principles discussed may also apply to other organic chemicals.

For chemicals which have linear adsorption isotherms with soil, the depth of maximum concentration of the organic can be estimated by dividing the depth of water movement by $\beta K'/\theta$, where β is the soil bulk density (g cm^{-3}), θ is the volumetric water content, and K' is the adsorption coefficient between the organic chemical and soil. K' is calculated by measuring the quantity of organic adsorbed (mass of chemical/mass of soil) at equilibrium with a given concentration of organic in solution (mass of solutes/volume of solution). Values of K' range from near zero to several hundred depending upon the chemical and soil. Hamaker and Thompson (1972) have reviewed adsorption of pesticides by soil and give many reported data.

The above analysis assumes no hysteresis in the adsorption isotherm. In other words, organic chemicals are reversibly adsorbed and the shape of the adsorption isotherm is independent of whether pesticide is being added to or removed from soil. Organic chemicals commonly have a nonsingular, adsorption-desorption value. For a given equilibrium solution concentration, a higher concentration of the organic chemical is adsorbed when the chemical is being removed from the soil than when it is added to the soil. Lindstrom et al. (1971), van Genuchten et al. (1974), Hornsby and Davidson (1973), Davidson et al. (1973) presented approaches to incorporating nonsingular adsorption relationships in the analysis of pesticide transport.

Another complicating factor in estimating movement of an organic chemical through soil is that the adsorption coefficient is usually measured in the laboratory where there is complete mixing between soil and solution. On the other hand, most water flow in the field may be in the large pores and the transported organic chemical only exposed to the external aggregate surfaces. Thus, the effective adsorption coefficient is much lower than measured in the laboratory, and movement in the soil profile is deeper than calculated. Estimation of organic chemical movement through soil is more difficult in aggregated as compared to dispersed single grain soils.

VII. GAS FLOW

Gas transfer through soil is important for a number of situations. Organisms in soil respire and, therefore, require gas exchange with the atmosphere. Also, most plants take up oxygen through their root systems. Diffusion is a significant mechanism for gas exchange in the soil. Organic compounds added to soil have various degrees of biodegradability which is partially dependent upon the type and activity of microbes. Thus, there is direct linkage between gas transfer, microbial reaction, and degradation of organic chemical compounds.

Another significant effect of gas diffusion is the escape of gases produced within the soil. For example, various gases are produced within sanitary landfills. The concentration of these gases occurring within the soil pores is dependent upon diffusion from the soil into the atmosphere. Some organic chemicals have a measurable vapor pressure and, as such, can volatilize. Disposal of these chemicals within the soil may ultimately lead to their presence in the aerial atmosphere and widespread distribution.

Steady-state gas diffusion through soil can be calculated by inserting the appropriate driving force and transmission coefficient into Eq. [1]. The appropriate driving force is the gas concentration gradient and the transmission coefficient is an effective diffusion coefficient, D'. D' is related to soil and vapor properties by

$$D' = D_v P^{10/3}/P_T{}^2 \qquad [3]$$

where D_v is the diffusion coefficient of gas in air, and P and P_T are the air-filled and total porosity of the soil, respectively. The porosity terms are included to account for geometric effects of the soil. Clearly, gas diffusion through soil is highly dependent upon the porosity of the soil and the water retained. Increasing the air-filled porosity (decrease in water content) causes a high rate of gas diffusion through soil.

Equation [3] applies for an unsaturated soil. In a saturated system

$$D'' = D_o P_T{}^{4/3} \qquad [4]$$

where D_o is the gas diffusion coefficient through water. Because the diffusion coefficient through air is so much greater than through water, diffusion through water can be neglected in an unsaturated state.

The gas concentration within the soil is often of greater interest than the rate of gas diffusion through soil. Oxygen concentration is of prime importance for biological reactions. The concentration of oxygen within pores can be calculated from

$$C_x = C_o - [\alpha x (2L-x)]/[2D'/P] \qquad [5]$$

where C_x is the concentration at depth x, C_o is the concentration in the atmosphere, α is the oxygen consumption rate within the soil (g cm^{-3} second^{-1}), and L is the depth at which the gas concentration gradient equals 0. Physically L can be related to a position in the soil profile which completely obstructs oxygen diffusion such as a saturated layer or it could represent the depth at which the oxygen concentration goes to 0 and, therefore, the gradient below that depth equals 0.

Solution of Eq. [5] for the case where $C_x = 0$ when $x = L$ gives

$$x = (2D'C_o/\alpha P)^{\frac{1}{2}}. \qquad [6]$$

This equation can be used to calculate the depth at which the concentration within the pores will be 0. Both Eq. [5] and [6] assumes α constant throughout the soil.

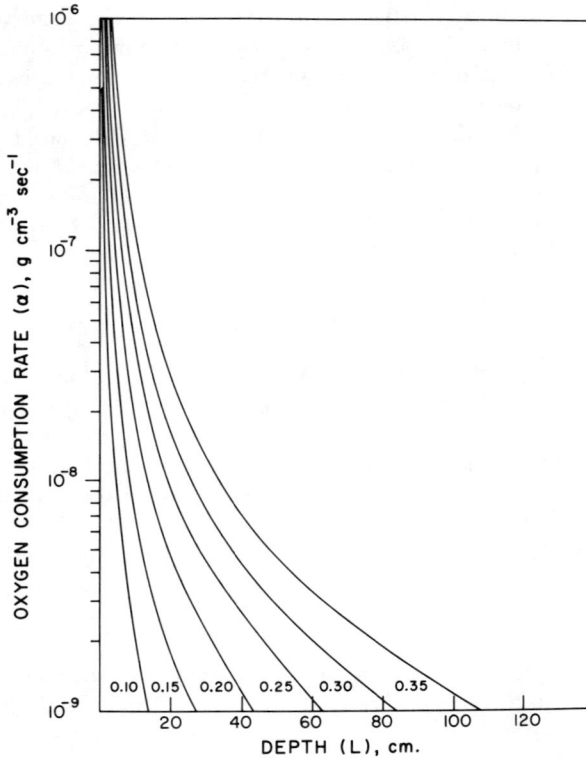

Fig. 2—The depth, L, at which the oxygen concentration reaches 0 for various oxygen consumption rates, α, and various air-filled porosities, P. The total porosity is 0.50 cm^3 cm^{-3}. Each curve represents an indicated air-filled porosity.

The depth at which the concentration equals 0 is plotted in Fig. 2 for a variety of air-filled porosities and oxygen consumption rates (a total porosity of 0.5 was assumed). Data presented in Fig. 2 indicate that the concentration drops to 0 only at reasonably great depths except for cases of low air-filled porosity or extremely high oxygen consumption rates. The lower α values plotted in Fig. 2 are probably more typical of natural soils except when an energy source is incorporated. Application of a readily decomposed energy source allows α to become relatively high for a short time. As the energy source is consumed, α decreases. Energy sources are usually incorporated in the upper soil horizons so α is not large below the application zone. The data presented in Fig. 2 tends to underestimate the depth at which oxygen occurs in soil.

Equation [5] can be used to calculate the concentration of a gas other than oxygen when this gas is generated within the soil profile by allowing α to equal the rate of gas production and changing the sign preceeding α in Eq. [5] from minus to plus. In this case it is assumed that the atmospheric concentration is negligibly small so that $C_o = 0$. Note that the gas concentration is a function of both production rate and diffusion through soil.

Data presented in Fig. 2 indicate that anaerobic conditions would not be frequently expected in soils with reasonable air porosity and biological activity. However, microorganisms are not necessarily at the air-water interface where the oxygen supply is adequate. Organisms within an aggregate must have oxygen supplied by diffusion from open pores through water. Thus it is appropriate to calculate the oxygen distribution within aggregates. For purposes of calculation, the aggregates will be considered to have spherical geometries.

For the condition of atmospheric oxygen concentration surrounding a spherical saturated aggregate, the sphere radius necessary for oxygen concentration to equal 0 at the center of sphere is

$$a = (6D''C_o'/\alpha)^{1/2} \qquad [7]$$

where a equals the sphere radius and C_o' is the oxygen concentration in water at equilibrium with atmospheric oxygen.

Figure 3 represents the spherical radius causing the concentration to equal 0 at the center of the spherical aggregate for a variety of oxygen consumption rates and aggregate porosities. The aggregates were assumed to be saturated with water. If the aggregates are not saturated with water, the diffusion coefficient for oxygen is sufficiently high that anaerobic conditions would not be expected.

If the aggregate is sufficiently dense (low porosity) and the oxygen consumption rate rapid, anaerobic conditions may exist within soil aggregates even though the concentration of oxygen within an open pore is equal to atmospheric concentration. Thus, positions of both aerobic and anaerobic conditions within the same profile are possible at the same time. Addition of organic matter to the soil may stimulate microbiological activity and increase the consumption rate, thus contributing towards anaerobic conditions. In general, anaerobic conditions are not expected except for soils with (i) very high oxygen consumption rates, (ii) extremely high water content, or (iii) large, dense aggregates which remain saturated.

Burial has been suggested as a disposal technique for waste organic compounds which are volatile. Vapor will diffuse away from the organic chemical position. Initially, transient state diffusion occurs and the transmission coefficient equals the diffusion coefficient of vapor through soil (D') divided by the adsorptive capacity of the soil for the chemical. The transmission coefficient can be quite small for soils with high adsorptive capacities for the chemical. Eventually, however, the organic will reach the soil surface and vaporize into the atmosphere under steady-state diffusion conditions. At this time the transmission coefficient is no longer affected by the adsorptive capacity of the soil.

Thus, there are two distinct phases to consider in estimating atmospheric contamination from buried volatile organic compounds. The first is

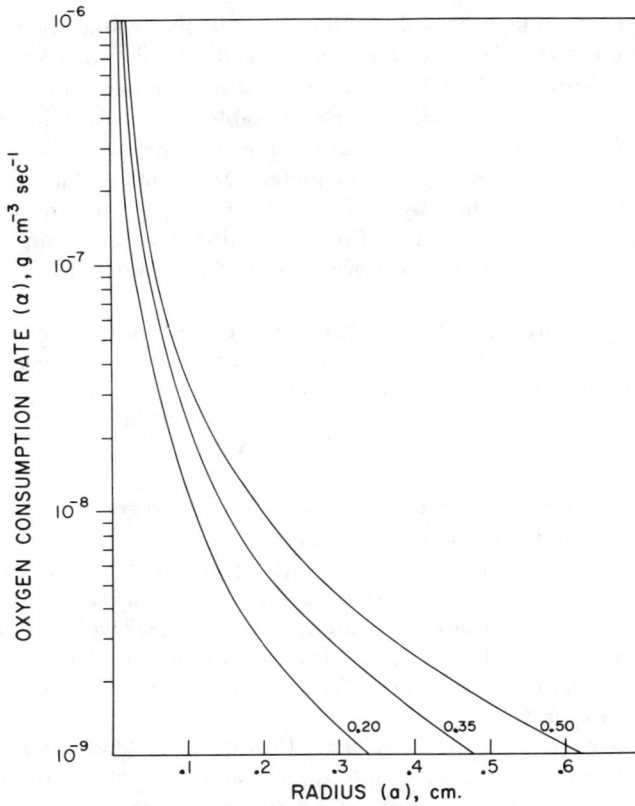

Fig. 3—The radius, *a*, of a saturated spherical aggregate when 0 oxygen concentration exists at its center for various oxygen consumption rates, α, and aggregate porosities. Each curve represents an indicated aggregate porosity.

transient-state diffusion and adsorption by the soil is very important. This phase determines the time before any organic appears at the soil surface and escapes into the atmosphere. The second phase is steady-state diffusion through the soil into the atmosphere and is proportional to D' and inversely proportional to the placement depth of the organic chemical.

VIII. HEAT TRANSFER

Biological and chemical reactions are usually temperature dependent. Heat transfer through soil is dependent upon the thermal conductivity of the soil components. Change in temperature is dependent both on thermal conductivity and heat capacity of the soil. Thermal conductivity values for water, glass, and air are 1.43×10^{-3}, 2.0×10^{-3}, and 5.7×10^{-5} calorie cm^{-1} second^{-1} degree^{-1}, respectively. Thus the thermal conductivity of a wet soil is greater than a dry soil.

One is generally not as interested in the heat transfer rate as in temperature changes occurring within the profile. Transient-state heat flow resulting in soil temperature changes is related to the thermal diffusivity. Thermal diffusivity equals the thermal conductivity divided by heat capacity (specific heat times density) of the component in question. Some representative specific heat values in units of calorie degree^{-1} g^{-1} are 0.194 for quartz sand, 0.229 for kaolin, 0.450 for humus, 1.0 for water, and 0.315×10^{-3} for air. When water is applied to an air-dried soil, thermal diffusivity increases to a maximum and then may decrease as the soil becomes saturated. Thermal conductivity is dominant at first, but then the heat capacity increases at a faster rate. One of the benefits of subsurface drainage in removing gravitational water is to increase the thermal diffusivity, thus aiding in warming the soil.

Soil temperature is dependent upon incoming radiation and thermal diffusivity. The surface color of soil affects the amount of energy absorbed from incoming radiation. A white-colored surface reflects, whereas a dark-colored surface absorbs energy. Soil temperatures can be significantly altered by manipulating soil surface color.

Although heat flow equations are well known, calculations for soil systems are complicated. One complication is that temperature gradients also cause water vapor density gradients. Water flows in the vapor phase from a zone of high temperature to a zone of low temperature. Water vapor carries latent heat and thus contributes to energy transfer in addition to normal thermal conduction.

IX. MICROBIAL REACTIONS

According to Fig. 1, microbial reactions depend upon temperature, oxygen, and water as well as an energy source. Microbial reactions may significantly affect the fate of organic chemicals. In turn, the fate of these chemicals may be related to pollution and in some cases plant response. The reader is referred to Chapter 6 for detailed information on this topic.

X. MECHANICAL PROPERTIES

Each segment of the schematic representation of Fig. 1 has been described except for mechanical properties of the soil. Mechanical properties depend on structure, water content, and organic matter. Generally, an increase in organic matter tends to decrease the strength of a soil which is detrimental to construction, but generally good for plant root growth. A detailed discussion of mechanical properties is not within the scope of this chapter, however.

XI. SUMMARY

The physical matrix of the soil consists of mineral and organic particles which are arranged into structural units which give rise to various porosities and pore size distributions. Pore characteristics significantly affect water retention and water flow. Water as a key constituent affects almost every physical process within the soil. Although the physical matrix of a soil significantly affects water retention and transport, water is largely controlled through management practices. Thus, through manipulation of water, one can significantly change other physical processes to achieve a desired result. This chapter has briefly reviewed each of the physical processes and their interrelationships. Several excellent books (Baver et al., 1972; Hillel, 1971; Kirkham & Powers, 1972; Taylor & Ashcroft, 1972) are available if the reader wants more detail than given in this chapter.

LITERATURE CITED

Baver, L. D., W. H. Gardner, and W. R. Gardner. 1972. Soil physics, 4th ed. John Wiley & Sons, Inc., New York. 498 p.

Davidson, J. M., R. S. Mansell, and D. R. Baker. 1973. Herbicide distributions within a soil profile and their dependence upon adsorption-desorption. Soil Crop Sci. Soc. Fla. Proc. 32:36–41.

Hamaker, J. W., and J. M. Thompson. 1972. Adsorption. p. 49–144. In C. A. I. Goring and J. W. Hamaker (ed.) Organic chemicals in the soil environment. Marcel Dekker, Inc., New York.

Hillel, D. 1971. Soil and water. Academic Press, New York. 288 p.

Hornsby, A. G., and J. M. Davidson. 1973. Solution and adsorbed fluometuron concentration distribution in a water-saturated soil: Experimental and predicted evaluation. Soil Sci. Soc. Am. Proc. 37:823–828.

Kirkham, D., and W. L. Powers. 1972. Advanced soil physics. John Wiley & Sons, Inc., New York. 435 p.

Leistra, M. 1973. Computation models for the transport of pesticides in soil. Residue Rev. 49:87–130.

Letey, J., and W. J. Farmer. 1974. Movement of pesticides in soil. p. 67–97. In W. D. Guenzi (ed.) Pesticides in soil and water. Soil Sci. Soc. Am., Madison, Wis.

Lindstrom, F. T., L. Boersma, and D. Stockard. 1971. A theory on the mass transport of previously distributed chemicals in a water saturated porous media: Isothermal cases. Soil Sci. 112:291–300.

Taylor, S. A., and G. L. Ashcroft. 1972. Physical edaphology. W. J. Freeman & Co., San Francisco, Calif. 533 p.

van Genuchten, M. Th., J. M. Davidson, and P. J. Wierenga. 1974. An evaluation of kinetic and equilibrium equations for the prediction of pesticide movement through porous media. Soil Sci. Soc. Am. Proc. 38:29–35.

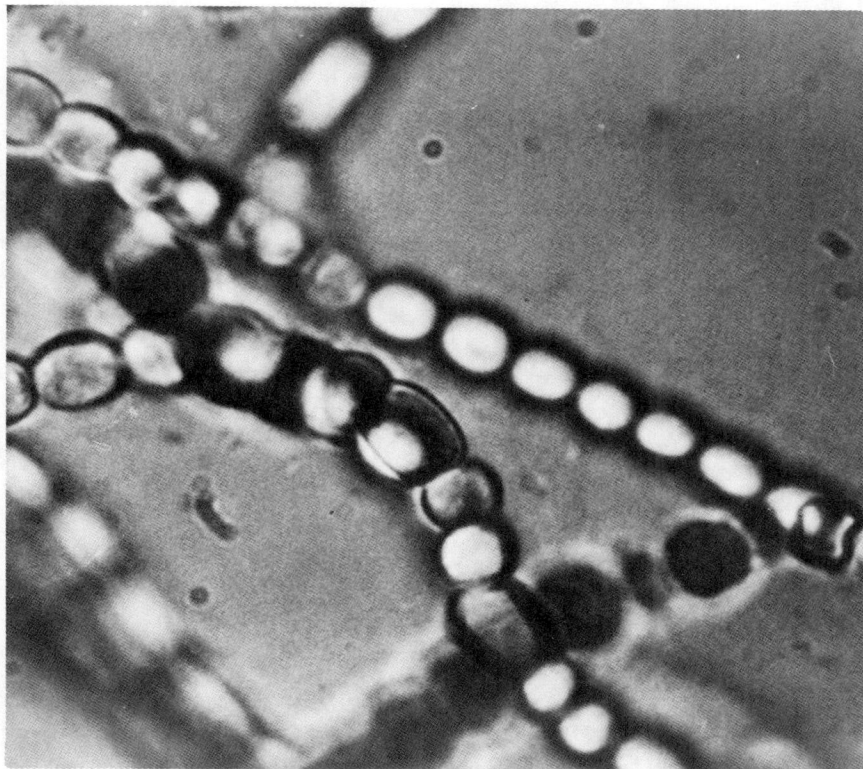

chapter 6

Blue-green algae found in soil (Photo courtesy of V. H. Goodman, Riverside, Calif.).

Biological Properties of Soils

JAMES P. MARTIN and DENNIS D. FOCHT
University of California, Riverside, California

I. INTRODUCTION

A good soil is a living system. It is the domain of myriads of microorganisms and small animals which are intimately associated with the soil organic fraction which represents a carbon, energy, and nutrient source for most forms. The numbers, kinds, and activities of these organisms are influenced by the food material available or organic matter content of the soil, plant species, amounts and kinds of organic amendments applied, soil texture, pH, moisture, aeration, amounts of salts, and other factors. They live according to the rule of the survival of the fittest. The organisms active in any particular microenvironment are best adapted to the conditions present or represent the most successful competitors.

Although in some soils a few organisms may parasitize or injure plant roots, or even on occasion infest animals or man, the vast majority perform beneficial functions which are important for the soil, the plant, and even for most living things upon the earth. Through decomposition of organic residues and soil humus, soil organisms release carbon, nitrogen, and other nutrients which are thereby made available for new generations of living things. Toxic organic substances from plants and applied organic pesticides are destroyed or utilized as a carbon and energy source. Soil aggregation or tilth is improved through their activities during organic residue decomposition. Insoluble inorganic nutrients are solubilized and soluble compounds of nitrogen and sulfur are oxidized or reduced depending upon environmental conditions.

Approximate numbers of organisms commonly found in soils are listed in Table 1. The estimates for bacteria, actinomycetes, fungi, and yeasts are based on plate counts and are conservative because there is no accurate method for determining all viable microbial types. Numbers based on direct microscopic counts and biomass estimates based on adenosine triphosphate determinations (MacLeod et al., 1969) indicate that actual numbers are much greater. It has been estimated that the live weight of the soil microflora may vary from 0.5 to over 4 metric tons or more in the surface 15 cm of 1 ha (Clark, 1967). Although the bacteria outnumber the fungi they are much smaller and in most soils the live weight of the fungi may surpass that of the bacteria. It has also been estimated that the live weight of the micro- and macrofauna of the soil may equal that of the microflora (Timonin, 1965).

II. THE SOIL POPULATION

A. Bacteria and Actinomycetes

1. BACTERIA

With the probable exception of the viruses, the bacteria represent the most numerous of the soil microinhabitants. They vary in shape and size from tiny spheres, ovals, and rods with dimensions under 1 μm to larger cells

Table 1—Approximate numbers of organisms commonly found in soils†

Organism‡	Estimated numbers/g
Bacteria	3,000,000 to 500,000,000
Actinomycetes	1,000,000 to 20,000,000
Fungi	5,000 to 900,000
Yeasts	1,000 to 100,000
Algae	1,000 to 500,000
Protozoa	1,000 to 500,000
Nematodes	50 to 200

† The figures for bacteria, actinomycetes, fungi, and yeasts are based on plate counts and refer to viable propogules able to grow on the plating media.
‡ In addition to these there are large numbers of slime molds (Myxomycetes), viruses or phages of bacteria, algae, fungi, insects, and plants, arthropods, earthworms, mycoplasmas, and other organisms.

several microns wide or long (Fig. 1). Both autotrophic and heterotrohpic forms are present. The chemoautotrophs are capable of obtaining energy from the oxidation of simple inorganic substances such as S, H_2S, NH_3, NO_2^-, Fe^{2+}, and Mn^{2+} (Stanier et al., 1970). As in the green plants C is obtained from CO_2 and H from H_2O. In wet soils, a few photoautotrophic bacteria may be present. Some of these obtain H from H_2O but others require H_2S.

Strictly aerobic and anaerobic bacteria are found in the soil. There are other forms which are capable of developing in both the presence and absence of O_2, and still other normally aerobic bacteria are able to carry on their usual activities in the absence of O_2 if nitrate, sulfates, and other oxi-

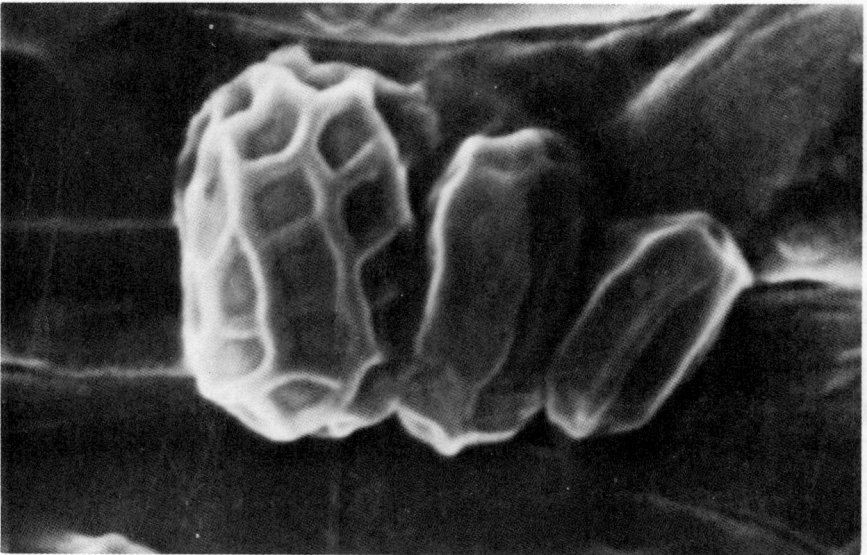

Fig. 1—Three bacterial cells growing on nylon gauze which had been placed in natural soil. Magnification is 26,000 times under a scanning electron microscope. (Courtesy of R. Honour and P. Tsao, Riverside, Calif.)

dized substances are present to act as electron or H acceptors. Transformations of inorganic elements by autotrophic and other bacteria will be discussed in more detail later.

The majority of the soil bacteria are heterotrophic and require preformed organic substances for C and energy. One of the more abundant types are small coccoid or rod-shaped forms with variable morphology and belong to the genus, *Arthrobacter*. These bacteria are well adapted to survival in soil. The sphere forms contain up to 40% of a glycogen-like material and have a half-life of 100 or more days under starvation conditions. The aerobic (*Bacillus*) and anaerobic (*Clostridium*) spore forming bacilli constitute other important soil forms. Additional common forms include *Pseudomonas, Nitrobacter, Rhizobium,* and *Azotobacter* (Clark, 1967; Clark & Paul, 1970; Hagedorn & Holt, 1975).

The numbers and activity of the bacteria in the soil depend on the available food supply and other environmental factors, such as pH, moisture, oxygen supply, temperature, salt concentration, soil texture, and available nutrients such as N. Any practice which increases the organic food supply including proper fertilization for crop production and application of organic wastes will increase their numbers (Table 2) provided other factors such as soil pH or salt concentrations are not adversely affected. They are especially abundant in the rhizosphere or near or on the surfaces of plant roots where food is available in the form of excreted organic substances such as amino acids and sloughed off plant cell debris (Rovira & McDougal, 1967).

Table 2—Numbers† of bacteria and fungi in soils of several orange tree (*Citrus* sp.) fertilizer plots at Riverside, Calif. (September 1951)

Treatment	Bacteria	Fungi
None	13,000,000	24,000
Calcium nitrate fertilizer	18,000,000	48,000
Winter cover crop	19,000,000	46,000
Winter cover crop and calcium nitrate	35,000,000	69,000

† Numbers in 1 g dry soil estimated by the dilution plate technique. Figures are averages of six, 0 to 30.5 cm samples from each of four plots. Sampled about 20 years after annual treatments commenced.

Table 3—Numbers of bacteria, actinomycetes, and fungi in soil fertility plots at Riverside, Calif., in relation to soil pH

Soil pH	Bacteria and actinomycetes	Fungi
	millions/g	thousands/g
7.5	95	180
7.2	58	190
6.9	57	235
4.7	41	966
3.7	3	280
3.4	1	200

Most bacteria grow best under neutral to slightly alkaline conditions. At pH 5 to 5.5 numbers begin to decrease sharply and below pH 4.0 numbers and activity sharply decline (Table 3). Many fungi are better adapted to activity at lower reactions and become more dominant in acid soils. In general the microflora are relatively tolerant of high salt concentrations and extremes of temperature although their species diversity and activity decline with extreme conditions (Johnson & Guenzi, 1963).

2. ACTINOMYCETES

The actinomycetes are unicellular, mostly aerobic organisms which form a branched mycelium and reproduce by fragmentation or asexual spore formation (Fig. 2). They may constitute from 1 to 50% of the colonies developing on soil dilution plates but generally average about 10%. The actinomycetes were first believed to be fungi but numerous studies have now proved them to be more closely related to the bacteria. Their cell walls are polymers of sugars, amino sugars, and amino acids similar to those of gram positive bacteria. Their high sensitivity to phages and antibiotics, cell diameter, spores, sensitivity to an acid environment, procaryotic nucleus, and chemical nature of their DNA all suggest a close relationship to bacteria (Stanier et al., 1970).

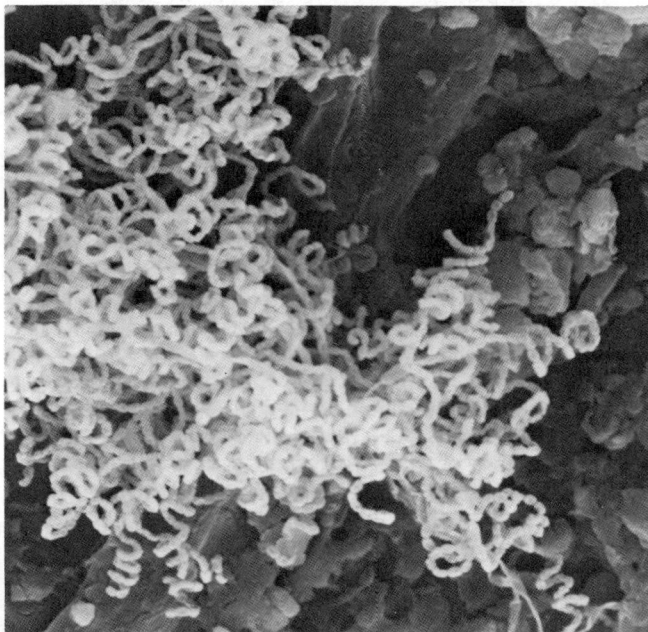

Fig. 2—Fruiting structures of an actinomycete (*Streptomyces* sp.) growing in soil as viewed by the electron scanning microscope. Magnification is 2,000 times. (Courtesy of R. Honour and P. Tsao, Riverside, Calif.)

In general, the numbers and activities of the soil actinomycetes are dependent upon the same factors affecting bacteria. Many forms, however, are favored over some bacteria at higher temperatures, at relatively lower moisture percentages, and at higher salt concentrations. They function in organic residue degradation and many produce antibiotics (Küster, 1967; Wong & Griffin, 1974).

Probably the two most typical soil genera of the actinomycetes are *Nocardia* and *Streptomyces*. The *Nocardia* form a mycelium or branching filaments which later fragment into rods or spheres. The colonies have a paste-like consistency as do many bacteria. *Streptomyces* mycelia do not fragment but the conidia are formed on sporophores or at the terminal ends of aerial hyphae (Waksman, 1961).

B. Fungi, Myxomycetes, and Yeasts

1. FUNGI

The fungi consist of a large heterogeneous and widespread group of heterotrophic microorganisms (Ainsworth et al., 1973). They produce a rela-

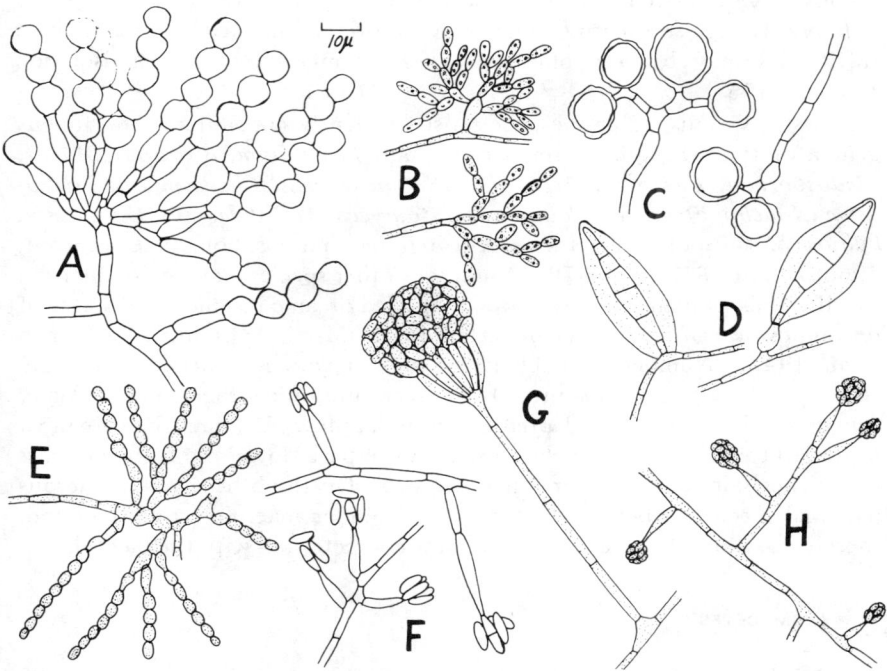

Fig. 3—Fruiting structures and mycelium of some fungi isolated from southern California soils. The fungi are: *(A) Scopulariopsis brevicaulis*, *(B) Cladosporium* sp., *(C) Sepedonium* sp., *(D) Murogenella* sp., *(E) Scopulariopsis* sp., *(F) Fusarium* sp., *(G) Stachybotrys chartarum*, and *(H) Gliomastix* sp.

tively large diameter mycelium which is generally divided by crosswalls into nucleate cells (Fig. 3). The fungi are not as numerous as the bacteria but with respect to total biomass probably represent the most important soil microbial group. As is true with bacteria and actinomycetes, extracellular enzymes are excreted and smaller organic molecules are absorbed over the entire mycelium cell surface (Ainsworth et al., 1965). They reproduce by forming asexual and/or sexual spores and the cell walls contain chitin, chitosan, or cellulose.

It is difficult to determine the most active types of fungi in any particular soil. Probably the most common plating media presently in use are adjusted to pH 5.0 to 7.0, contain dextrose and peptone as C, energy and N sources, and an antibiotic and rose bengal to prevent growth of bacteria and actinomycetes (Martin, 1950; Tsao, 1964). However, no single medium is suitable for all types, and special baits or specific organic substrates and pH values are used to isolate many species (Warcup, 1967). Some fungi such as the Endogone have never been cultured on the usual artificial media but their mycelium and large chlamydospores can be readily obtained from most soil and plant roots by sieving techniques which suggests that they are common soil fungi (Gerdeman, 1968). It is difficult to interpret the significance of the numerous species which do thrive on the common plating media. A colony may originate from a spore, a small fragment of mycelium, or a whole colony. Highly sporulating types may be overemphasized. Some of these difficulties may be overcome by a tedious method of hyphal isolation (Warcup, 1967).

Using various techniques, the most frequently encountered soil fungus genera or types have been reported to be: Penicillium, Rhizopus, Mucor, Cladosporium, Fusarium, Aspergillus, Scopulariopsis, Chaetomium, Trichoderma, Stachybotrys, Gliocladium, Doratomyces, Acrostalagmus, Gliomastix, Pullularia, nonsporulating forms, Rhizoctonia, and basidiomycetes (Barron, 1968; Domsch & Gams, 1972). Numerous other species may be dominant in specific soils as plant species, types of applied organic residues, and stages of decomposition of residues favor certain forms (Martin & Ervin, 1958; Martin et al., 1942). Numbers and activity of the fungi are influenced by the available C, soil reaction, moisture, O_2, temperature, and other factors. Many fungus species are better adapted for growth at low pH values than are most bacteria (Table 3). Many species also grow better at lower moisture contents and higher salt concentrations than bacteria. On the other hand, numerous bacterial species are better adapted to thrive under anaerobic conditions and, therefore, at low O_2 tensions fungi do not compete well with the bacteria.

2. MYXOMYCETES

Myxomycetes, or slime molds, are saprophytic organisms common in damp woodlands, lawns, and other surface environments where decaying organic residues are present in abundance (Ainsworth et al., 1965). They have two life stages. In the amoeba stage they consist of a mass of amoeba-like cells not bounded by a cell wall. The amoeba mass or cells digest living and dead bacteria and other organic constituents of decaying residues. At a certain stage the amoeba stream together and form fungus-like fruiting structures. Most form spores within a sporangium. The myxomycetes are difficult to culture on artificial media and their true importance in soils has not been fully evaluated.

3. YEASTS

Yeasts are closely related to the fungi and are considered to be fungi in which the unicellular form predominates (Rose & Harrison, 1969). Vegetative reproduction is usually by budding. Reproduction also may be by fission or ascospore formation. Under certain conditions some forms produce a true mycelium. Most yeasts prefer simple organic substances such as sugars as a carbon and energy source. They are abundant on plant leaves, stems and fruits, but most do not compete well with other microorganisms in soils.

C. Algae and Lichens

1. ALGAE

Algae are the most abundant photosynthetic microorganisms in soil (Fig. 4), and most soils have a characteristic algal flora. Many species belonging to *Botrydium, Chlorococcum, Zygnema, Oedogonium, Microcoleus,* and other genera are found only in soil (Shields & Durrell, 1964); species belonging to *Anabaena, Oscillatoria, Nostoc, Protococcus, Stichococcus,* and many other genera are found in both soils and waters. The soil algae have received relatively little attention and some microbiologists have considered them to be of minor importance. Many studies, however, have emphasized their role in N fixation and in the synthesis of organic materials which represent source material for soil humus formation (Fogg et al., 1973; Fuller & Rogers, 1952).

Most soil algae belong to the blue-green (*Cyanophycaea*), the green (*Chlorophyceae*), or the diatom (*Bacillariophyceae*) groups (Alexander, 1961). The blue-green algae are the least advanced forms. They are procaryotic and contain chlorophyl, carotenoids, and phycocyanin pigments which are diffused throughout the cytoplasm. They are unicellular or filamentous and acid sensitive. Very few are found when the pH of the soil or water is below 5.2. The green algae contain chlorophyl, xanthophyll, and carotene pigments in chloroplasts and are more acid tolerant. The diatoms are acid sensitive and have a highly silicified cell wall consisting of two overlaping plates which fit together like a petri plate (Lewin, 1962).

The algae, being photoautotrophic, are found in greatest abundance in the surface 2 to 4 cm of soils (Shields & Durrell, 1964). Some forms, however, can live heterotrophically on simple organic compounds and a few are found in the subsoil. Since they do not compete well with other hetero-

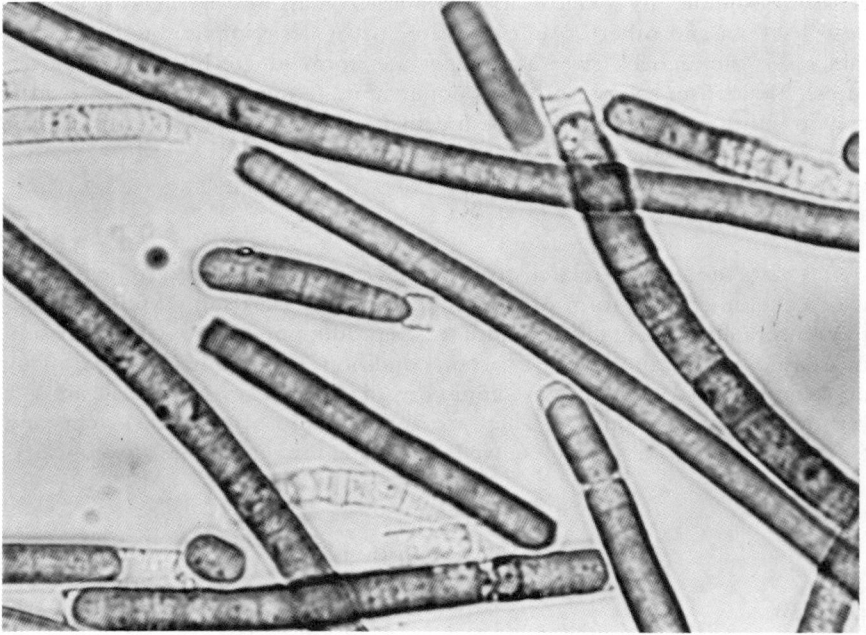

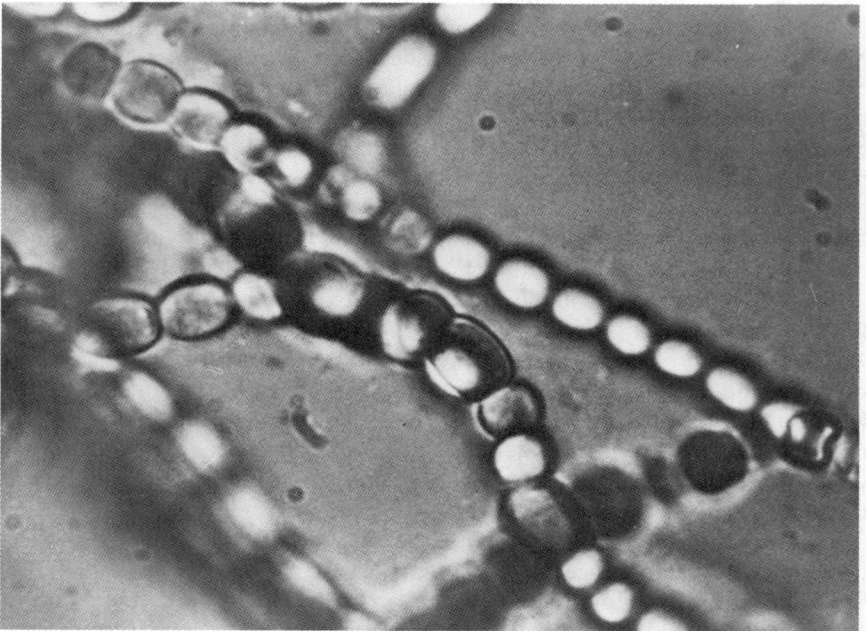

Fig. 4—Two blue-green algae found in soil: *(top)* *Lyngbia* sp. and *(bottom)* *Anabaena* sp. (Courtesy of G. Wagner, Columbia, Mo. and V. H. Goodman, Riverside, Calif., respectively.)

trophic forms in the dark, it is commonly believed that the subsurface algae were taken down by insects, by cultivation, or by other means and are present only as dormant forms.

Soil blue-green algae are tolerant to great extremes in temperature. They grow in arctic ice and in hot springs. Their abundance in the soil is most closely related to the available inorganic nutrient supply and moisture. The nutrients generally required include C, N, P, K, S, Fe, Mg, and Ca. Sodium and Co are required by some species and Si is essential for diatoms (Lewin, 1962). In liquid culture, in ponds, or in streams, algae will generally develop more rapidly in the presence of other organisms especially if a readily available energy source is present. This may suggest a growth requirement by some forms but probably the most important factor is an increase in the supply of available C as CO_2 which is produced by the bacteria and other microbes utilizing the organic energy sources (Wolk, 1973). Bubbling higher concentrations of CO_2 through axenic algal cultures will usually stimulate growth (Lang, 1971). Fertilizing ponds with manures or sewage will also stimulate algal growth.

The important functions of the soil algae include (i) nitrogen fixation (blue-greens), (ii) colonization of new rock or barren surfaces, (iii) supplying organic matter and nitrogen for soil humus formation, (iv) weathering of rocks and minerals, and (v) binding of soil surfaces which may aid in soil erosion control.

2. LICHENS

The lichens consist of two distinct organisms, a fungus and an alga. Together they can survive in some of the harshest environments (Ahmadjian & Hale, 1973). The algal partner may belong to the *Cyanophyceae* (blue-green) or *Chlorophyceae* (green) and may be found living independently in nature. The fungal component is usually an ascomycete but occasionally may be a basidiomycete, and the forms involved have not been found growing independently in nature. The fungus forms a clamp or other cell which attaches to the algal cell. The algal component furnishes carbohydrates formed through photosynthesis and the fungus partner furnishes nutrient elements from the rock and soil surfaces. Lichens grow very slowly and develop tough, gelatinous tissues which envelope the cells and allow the cells to adsorb enough water to survive long dry periods. Their functions are similar to those of algae on harsh, barren rock and soil surfaces.

D. Viruses and Mycoplasmas

1. VIRUSES

The viruses are the smallest of all biological entities and can develop only within the cells of host organisms (Stanier et al., 1970). They have a diameter of 0.02 to 0.3 μm and consist of a protein envelope which encloses a single type of nucleic acid. They are largely devoid of enzymes, thus their activity is dependent upon the enzymes produced by the host cell. After penetration the virus particle directs the metabolic activity of the cell to the production of the virus particle. The process usually terminates with the death and lysis of the host cell. The liberated particles are then free to infest another host cell. If the viruses may be viewed as an organism they are probably the most numerous and primitive of all soil forms. They have been found for nearly all bacterial species tested and for most other soil organisms including actinomycetes, algae, and fungi (Safferman & Morris, 1964; Schisler et al., 1967). Some plant viruses may survive in soil for short periods of time or may be carried in fungus or nematode cells (Cadman, 1965).

The importance or function of viruses in soil processes is difficult to assess. They may hasten the destruction of susceptible microbial cells which is probably of little significance. It has been suggested that they may reduce growth of legumes by destruction of *Rhizobium* species but good evidence of this possible effect is lacking.

2. MYCOPLASMAS

Mycoplasmas are small, primitive organisms which do not have cell walls and grow as spheres or in various pleomorphic forms. They are generally associated with animal or plant diseases or disorders (Smith, 1971). Recently, however, a mycoplasma-type organism was isolated from a hot refuse coal pile (Darland et al., 1970). The organism grows over a pH range of 0.96 to 3.5 and a temperature range of 45 to 62C. The existence of this saprophytic species suggests that saprophytic forms may possibly exist in soils or composts.

E. Soil Animals

Most soils harbor large numbers and a great variety of soil animals which play a very important role in soil processes and organic residue degradation (Kevan, 1955). Most soil scientists, however, give them little attention compared to the importance placed on the soil microflora (Schaller, 1968). Some of the soil animals have a well developed capacity for burrowing. These include wireworms, earthworms, some insect larvae, termites, ants, and millipedes. This activity may favor soil aeration and water penetration. Many of the nonburrowing animals such as protozoa, nematodes, and flatworms live in water-filled soil capillaries or water films around soil particles. When the soils become too dry, they survive by forming cysts. The remainder of the soil animals can live apart from the soil water and can withstand the desiccating action of the atmosphere. These include isopods, mites, collembola, beetles, earwigs, millipedes, and centipedes.

Soil animals feed on the soil organic fraction or on other soil organisms. Only a few, such as isopods and some millipedes, collembola and mites are able to utilize the fresh organic residues. Most must have material which has undergone or is undergoing decomposition by the soil flora. Their feeding habits further the overall process of organic residue decomposition and soil humus formation.

1. PROTOZOA

The protozoa are the simplest and most abundant of the soil animals. Their numbers in top soil usually range from 10,000 to 100,000/g but occasionally may reach 300,000 or even 500,000. Although total numbers are much smaller than for bacteria their total biomass may be greater. The soil forms consist primarily of ciliates (Fig. 5), flagellates, amoeba, and testacea (Kevan, 1955). Many are similar to water species but the soil types are much smaller and better adapted to live in the thin water films around soil particles. They most generally reproduce asexually by division but some also reproduce sexually.

The soil protozoa are largely heterotrophic. They feed on soluble and sometimes insoluble organic substances but the most common food source is bacteria. Species belonging to the genera *Aerobacter, Agrobacterium, Escherichia, Micrococcus, Pseudomonas,* and *Bacillus* are favored food

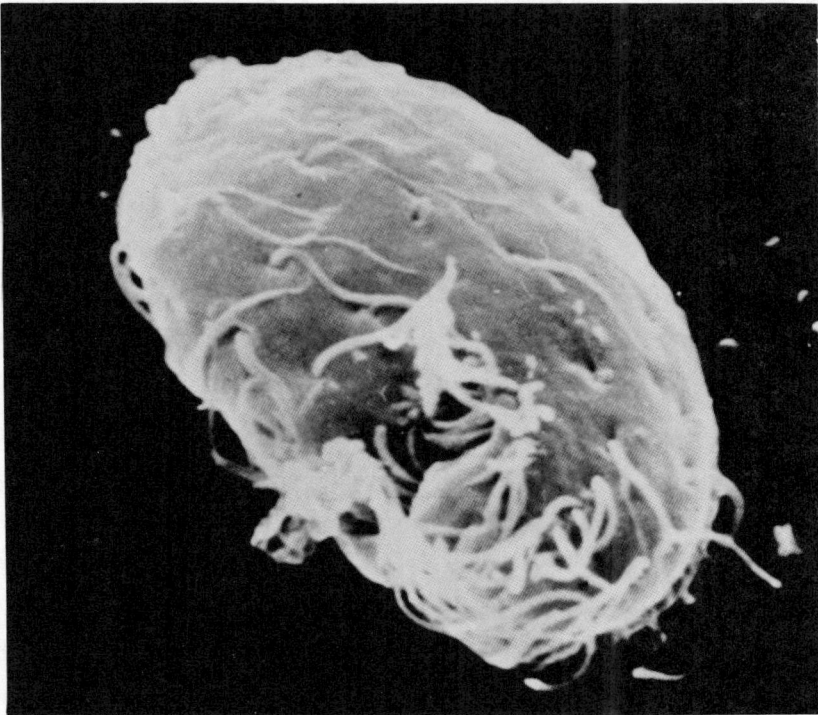

Fig. 5—A soil protozoan belonging to the ciliate group. (Courtesy of J. Darbyshire and D. Webley, Aberdeen, Scotland.)

sources. Yeasts, actinomycetes, fungi, and algae are not prefered food sources although some protozoan forms may occasionally ingest them (Stout & Neal, 1967). Soil management practices which increase microbial activities such as manure and fertilizer applications and cover cropping will increase the protozoan population. They are found at pH ranges from 3 to 10 but 6 to 8 is optimum. They grow best at soil temperatures of 18 to 32C. When adverse conditions develop such as exhausted food supply, lack of O_2, and low moisture most species form a resting body or cyst. Their activity in soil contributes to the process of organic residue decomposition with the release of plant nutrient elements.

2. NEMATODES

Soil nematodes (Fig. 6) are small, round worms which generally vary from 0.5 to 4 mm in length and from 50 to 250 μm in width (Nielsen, 1967; Thorne, 1961). Over 2,000 species inhabit the soil but about half of these are plant root parasites. The true soil nematodes live in the films of water around soil particles and in water-filled capillaries. They feed on soluble organic materials, bacteria, fungi, algae, protozoa, and other nematodes. Some forms pierce the cells of their prey with a stylet and suck out the cell contents while others ingest the whole microbial cell. The part of the food source utilized is largely protoplasm. Their numbers vary greatly from one soil microhabitat to another but usually there are 5 to 10 million individuals in 1 m^2, but counts as high as 30 million m^2 or more have been reported.

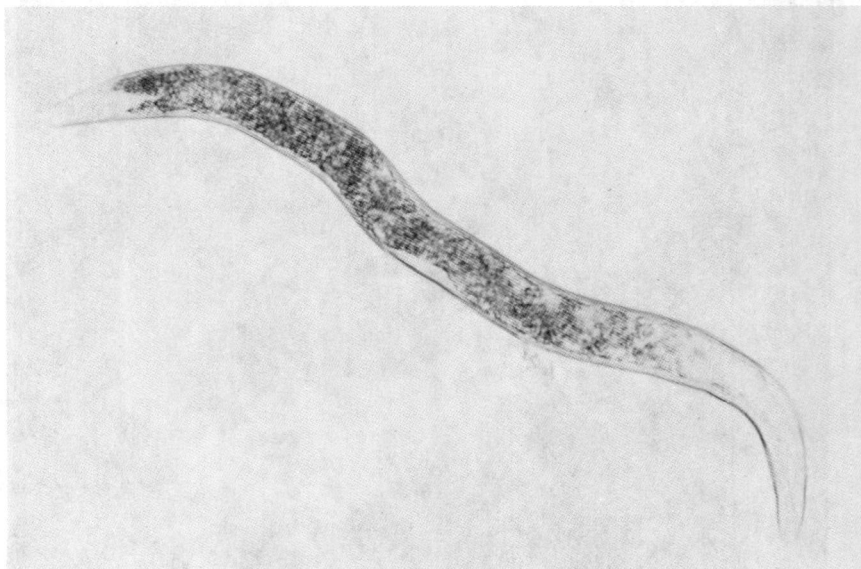

Fig. 6—*Caenorhabditis* sp., a soil nematode which feeds on bacteria. (Courtesy of S. VanGundy, Riverside, Calif.)

The nematodes probably do not participate in the direct decomposition of dead organic residues and do not significantly affect the physical properties of the soil. Their ecological importance, therefore, is concerned with their effect on total microbial activity through destruction of the species consumed. The significance of this activity is not known although it may well be of importance in certain microhabitats.

3. EARTHWORMS

There are about 1,800 known species of the common earthworm. Their numbers and activities depend primarily upon the availability of water and food (Satchell, 1968). They contain from 50 to 90% water and if this is decreased by 18% they cannot burrow or move. In temperate climates most species are inactive at acidity levels below pH 4.0 but, in some tropical soils, acid-tolerant species may be found (Kevan, 1955). Some species ingest largely organic debris on soil surfaces while others ingest the normal soil-organic

Table 4—Decomposition of organic compounds and residues in Greenfield sandy loam

| | Percentage of applied C evolved as CO_2 | | | | |
| | Weeks | | | | |
Compound or residue	1	2	4	8	12
Glycine	74	78	83	89	89
Algal amino acids	66	75	82	88	89
Glucose	75	82	84	86	87
Glucosamine	76	82	84	86	87
Peptone	66	71	78	82	85
Arthrobacter sp. cells	60	72	79	83	85
Starch	48	65	72	78	81
Microcoleus sp. protein	48	62	69	79	81
Acetic acid	60	66	74	77	80
Protein	49	62	70	76	78
Penicillium vinaceum cells	56	64	72	75	75
Azotobacter chroococcum polysaccharide	61	67	69	72	74
Vanillic acid	50	61	65	66	69
Hansenula holstii cells	30	49	57	64	68
Chromobacterium violaceum cells	45	52	56	61	66
Oat straw	31	39	47	56	62
Corn stalks	31	40	46	49	57
Stachybotrys atra mycelium	7	13	27	41	47
Walnut wood (*Juglans* L.)	7	20	28	40	46
Almond shells (*Prunus amygdalus*)	12	18	24	34	37
Avocado leaves (*Persea americana*)	14	19	23	32	36
Chromobacterium violaceum polysaccharide	12	24	29	32	35
Azotobacter indicus polysaccharide (strain 1)	10	15	20	26	30
Redwood tree wood (*Sequoia sempervirens*)	<1	1	3	6	12
Peat moss	<1	2	6	7	8
Hendersonula toruloidea phenolic polymer	3	3	4	5	5
Leonardite humic acid	<1	1	2	4	5
Linné loam humic acid	<1	<1	<1	1	1

complexes of deeper soil horizons. Estimates of their numbers in soil range from $<$ 1 to over 400/m^2 with a biomass of $<$ 1 g to over 250 g/m^2. As with other soil organisms any soil management practice which increases the available organic material in the soil will increase the earthworm population. Adequate N in the organic food material enhances growth and reproduction.

The passage of organic debris and mineral soil through the worms aids the process of organic matter decomposition, helps to solubilize plant nutrient elements from the insoluble soil minerals, and increases the structural stability of the mineral soil ingested. Most workers believe the stabilizing action results from increased microbial activity in the worm gut or subsequent fungal activity in the casts. Extensive burrowing of earthworms may improve soil aeration. In some high clay and poorly aerated soils this action is of sufficient magnitude that plant growth may be increased (Hopp & Slater, 1949). In such soils the earthworm burrows may account for two-thirds of the air capacity of the soil. The burrows may also increase water penetration and drainage.

4. ARTHROPODS

The soil arthropods include wood lice, scorpions, spiders, mites, millipedes, centipedes, springtails, silverfish, earwigs, the larvae of many insects, and numerous other forms (Raw, 1968). The full significance of these soil animals in organic matter decomposition and soil humus formation is not known but they are undoubtedly of great importance in many soils. Their feeding habits accelerate the degradation of surface litter in noncultivated soils. Microscopic examination of the humus layers of natural soils reveal large quantities of arthropod fecal pellets (Schaller, 1968).

III. IMPORTANT ACTIVITIES OF SOIL ORGANISMS

A. Decomposition of Organic Residues

Probably the most important function of the soil organisms is the decomposition of organic residues with the release of plant nutrient elements such as C, N, and S so that they may be utilized by new generations of living things (Allison, 1973). It has been estimated that 1/25 of the CO_2 in the atmosphere is utilized each year for organic matter synthesis primarily through photosynthesis (Alexander, 1961) and that up to 70% of the C from organic substances returned to the atmosphere each year as CO_2 is metabolized by microorganisms (Clark & Paul, 1970).

Organic residues consist primarily of polysaccharides, lignins, proteins, and fats with smaller amounts of simple sugars, aliphatic acids, phenols, and numerous other substances (Bernfield, 1963). As soon as residues are returned to the soil and environmental conditions are favorable the organisms begin the degradative processes. Different residues and different constituents of the same residues decompose at different rates (Table 4). Simple sugars, amino acids, aliphatic acids, some proteins, and some polysaccharides decompose very quickly and may be completely utilized in a few hours or days. More resistant materials such as plant lignins, other phenolic substances, and waxes decay more slowly (Haider et al., 1974). Dark-colored humic polymers decompose very slowly, namely 2 to 5% per year depending upon environmental conditions. Relatively higher temperatures increase the decomposition rate while lower temperatures slow degradation. A favorable moisture supply, a soil pH range of 5 to 9, and aerobic conditions also hasten decomposition. The organisms involved in decomposition processes require nutrient elements just as do higher plants and animals. If amounts in the organic residues are low, nutrients must be obtained from the soil or decomposition is delayed until the nutrients first used are released and made available to other organisms. The N content of the residue is especially important. If below about 1.5%, decomposition will be delayed and the process will be accelerated by supplemental N (Alexander, 1961).

All of the C utilized by the soil organisms is not evolved as CO_2. The organisms use a portion for the synthesis of cells and products (Behera & Wagner, 1974; Shields et al., 1973). Up to 50% or more of the substrate C may be initially incorporated into the microbial tissues.

The microbial products and cells are subject to microbial degradation upon death of the cells (Reisinger & Kilbertus, 1974: Webley & Jones, 1971) and, as with higher plant and animal residues, the rate of decay varies with the nature of the compounds synthesized (Table 4). Most microbial proteins, amino acids, and other simple compounds decompose quickly while fungus cells and microbial melanins or phenolic polymers are more resistant (Haider et al., 1974; Hurst & Wagner, 1969; Martin & Haider, 1971).

The large polymeric molecules which make up the bulk of the plant, animal, and microbial residues returning to the soil must first be broken into smaller units before they can enter the microbial cells and be further utilized for energy and cell synthesis. To accomplish this, the organisms must first excrete enzymes which break the large molecules into smaller units and into constituent units. Polysaccharides such as cellulose are cleaved to form sugars. Proteins are broken down to form peptides and amino acids. The large fatty acid molecules are oxidized by β-oxidation with the release of acetic acid units. When lignin is decomposed by white rot fungi (Flaig, 1966; Shubert, 1965) many simple phenolic compounds are released (Table 5). These are further degraded (Dagley, 1967) to aliphatic compounds, CO_2, and water (Fig. 7).

Table 5—Some phenols released during decomposition of lignin by soil fungi

p-Hydroxybenzaldehyde	Vanillin
p-Hydroxybenzoic acid	Vanillic acid
Coniferaldehyde	Guaiacylglycerol
Ferulic acid	Protocatechuic acid
p-Hydroxycinnamylaldehyde	Syringic acid
p-Hydroxycinnamic acid	3-Methoxy-4-hydroxyphenylpyruvic acid

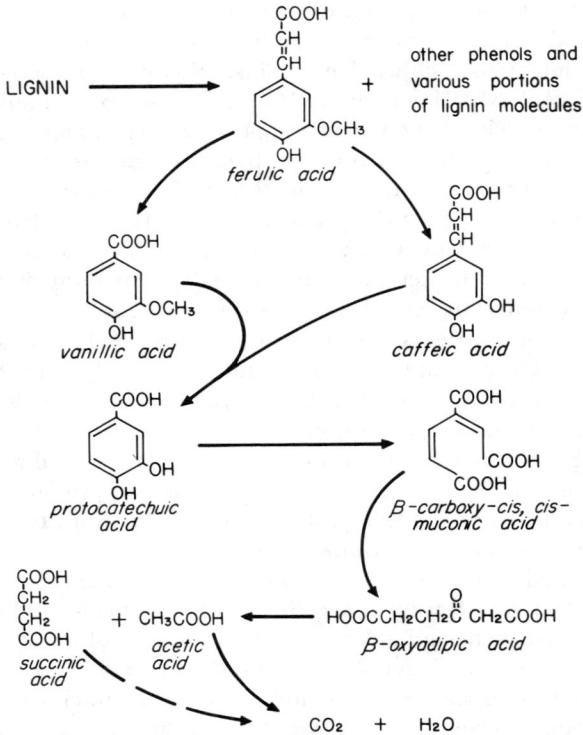

Fig. 7—Steps in the microbial degradation of lignin.

Generally from 60 to 85% of the applied C in most fresh residues is re-leased as CO_2 with in a few weeks to 2 or 3 months (Jenkinson, 1971). The C in more resistant residues such as highly lignified leaves and woods and fungus melanins is released more slowly. If woody material is finely ground it will decompose faster than if it is left in larger pieces (Fig. 8). The amounts of added residues will have little effect on decomposition rate up to about 1.5 to 2.0% of the dry weight of the soil. Also when two types of residues such as oat straw (*Avena sativa* L.) and microbial polysaccharide or peach wood (*Prunus persica* [L.] Batsch [*Amygdalus persica* L.]) and phenolic compounds are applied together, both types of residues will decompose at essentially the same rate as when applied alone (Haider & Martin, 1975; Martin et al., 1974). Soil humus generally will decompose at the rate of 2 to 5% a year depending upon the climatic conditions. Clays will often increase microbial growth rate (Table 6) but will reduce the total loss of carbon through increasing the efficiency of C utilization, and by complexing with decomposition products and humic substances which will reduce the C avail-able to microbes (Bondietti et al., 1972; Filip, 1973; Lynch et al., 1956; Stotzky, 1966). The addition of easily decomposable organic substances to the soil may temporarily increase the decomposition rate of the more re-sistant humus (Sørenson, 1974).

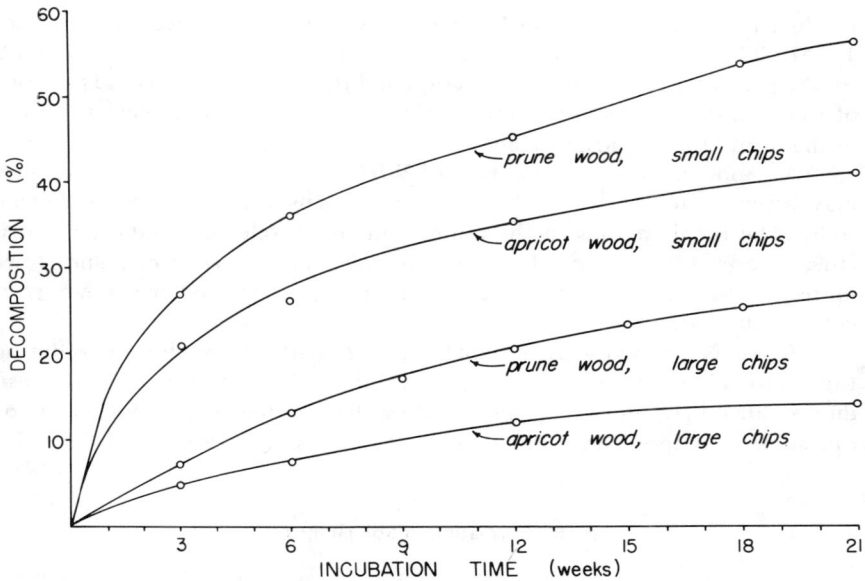

Fig. 8—Effect of particle size on decomposition of prune (*Prunus domestica*) and apricot (*Prunus armeniaca*) wood in soil.

The most common method of estimating overall microbial activity in soils is to determine CO_2 evolution. Best results are obtained when the soil is constantly aerated with a stream of humidified air. Other methods include measurement of O_2 utilization and enzyme activity (Skujins, 1967; Stotzky, 1965).

Nitrogen is released as ammonia during the decomposition of organic nitrogenous compounds. If the N content of the residue is about 1.5 to 2.0% or above, N is released quickly (Allison, 1973). If it is below this amount the organisms utilize all the N present for biomass synthesis and N will not be available for plant use until it is concentrated through loss of C as CO_2.

From applications of good quality steer manure about 25 to 50% of the N would be released the first year after application. About 10 to 15% of the

Table 6—Influence of montmorillonite on growth of and CO_2 evolution by *Candida utilis* in a glucose-NaNO$_3$ medium (Haider et al., 1970)

Incubation time	Biomass		CO_2 evolution[†]	
	Control	With clay	Control	With clay
days	——— g/liter ———			
1	2.2	7.2	14	35
2	4.9	10.4	24	40
3	5.6	10.0	33	42
5	6.4	9.1	48	47
10	6.4	8.7	58	52

† Percentage of applied carbon.

residual nitrogen would probably be released the second year (Pratt et al., 1973). The N in the more resistant humus is released at the rate of about 2 to 5% per year depending on climatic conditions (Jenkinson, 1971). More of the N in fresh chicken manure is released during the first year as most is in the form of urea and uric acid.

A good example of the release of N from decomposing organic residues may often be noted where the fairy ring fungus is growing in N-deficient sods. The fungi growing in the thatch and humus release N and other nutrients. The grass utilizes the elements, especially the N, turns green, and grows more luxuriously. After the N is utilized the grass again assumes the normal color of the sod.

Sulfur is released as H_2S or SO_4^{2-} and P as the orthophosphate during the microbial decomposition of organic residues. If the S content of the residue is above 0.15 and the P above 0.20% the supplies will be sufficient for optimum decomposition of the residues (Stewart et al., 1966).

B. Formation of Soil Humus

Humus is a very important constituent of the soil and is a natural product of the soil environment. It is formed during the microbial decomposition or alteration of plant and animal residues and of cellular constituents and products synthesized by soil organisms (Felbeck, 1965; Kononova, 1972; Martin & Haider, 1971). Generally, about 15 to 30% of the original residue C remains in the soil after 1 year. Some is present in the more resistant microbial cells or resting bodies and the remainder in new humus (Jenkinson, 1971; Shields et al., 1973). Some of the important and beneficial properties of humus are listed in Table 7.

1. HUMIC ACID

The soil humus is a mixture of numerous organic compounds; however, two types of substances, humic acids, and polysaccharides, may constitute up to 80% or more of the total (Gascho & Stevenson, 1968; Swinzer et al.,

Table 7—Beneficial properties of soil humus

1) Slow release plant nutrient element fertilizer
2) Improves soil physical properties
3) Aids in micronutrient element nutrition of plants through chelation reactions
4) Aids in solubilization of plant nutrient elements from insoluble minerals
5) Has high adsorptive or exchange capacity for plant nutrient elements
6) Increases soil buffer capacity
7) Dark color favors heat absorption and earlier spring planting
8) Certain components may exert plant growth promoting effects
9) Supports a greater and more varied microbial population which favors biological control
10) Reduces toxicity of toxic substances both natural and man made
11) Increases soil water-holding capacity

Table 8—Some phenols synthesized by soil fungi

Acids	Toluenes	Others
Orsellinic	Orcinol	Resorcinol
Cresorsellinic	2,4-Dihydroxy	5-Methylpyrogallol
6-Methylsalicylic	2,6-Dihydroxy	Phloroglucinol
3,5-Dihydroxybenzoic	2,3,5-Trihydroxy	Pyrogallol
2,4-Dihydroxybenzoic	2,4,6-Trihydroxy	
2,5-Dihydroxybenzoic	2,4,5-Trihydroxy	
Salicylic	4,Methyl-2,6-dihydroxy	
Protocatechuic	p-Cresol	
p-Hydroxybenzoic	m-Cresol	
m-Hydroxybenzoic		
2,6-Dihydroxybenzoic		
3,5-Dihydroxy-4-methylbenzoic		
2,4,6-Trihydroxybenzoic		
2,3,4-Trihydroxybenzoic		
p-Hydroxycinnamic		
Caffeic		
Gallic		

1968). Humic acid denotes that fraction of the soil humus which is soluble in alkali and precipitated by acid (pH 1 to 2.0). It is dark brown to black in color. The material which does not precipitate is often called fulvic acid. It consists primarily of humic acid-type molecules possibly of relatively smaller molecular weights, and of polysaccharides with a very large molecular weight range (Forsyth, 1950).

The humic acid-type molecules appear to be complex polymers of phenolic units with linked amino acids, peptides, amino sugars, and other organic constituents (Bondietti et al., 1972; Flaig, 1966; Kononova, 1961; Stevenson & Butler, 1969). Phenolic compounds and polymers synthesized by plants, especially lignin and microorganisms, are important sources of these phenolic units. Portions of lignin molecules and other phenolic poly-mers, and simple phenolic compounds released during microbial degradation or synthesized by soil microbes (Tables 5 and 8) are transformed by β-oxida-tion of side chains, introduction of additional hydroxyl groups, oxidation of methyl groups, and decarboxylation to form numerous phenolic compounds (Fig. 9). Some of these, especially the trihydroxyphenols, are highly reactive and may undergo autoxidative polymerization reactions (Haider & Martin, 1967). Others undergo polymerization reactions through the action of phenolase or peroxidase enzymes synthesized by soil organisms (Fig. 10 and 11). Less reactive phenols and other aromatic compounds and substances with free amino groups such as peptides, amino acids, and amino sugars, free or in polysaccharide chains may be linked into the developing molecules through nucleophylic addition to the quinones formed through oxidation. After linkage, susceptible phenolic molecules could be reoxidized and under-go further linkage. The polymerization reactions may also involve radical formation. Phenolic radicals formed through the action of phenolases or peroxidases could become stabilized through linkage (Fig. 11). The humic

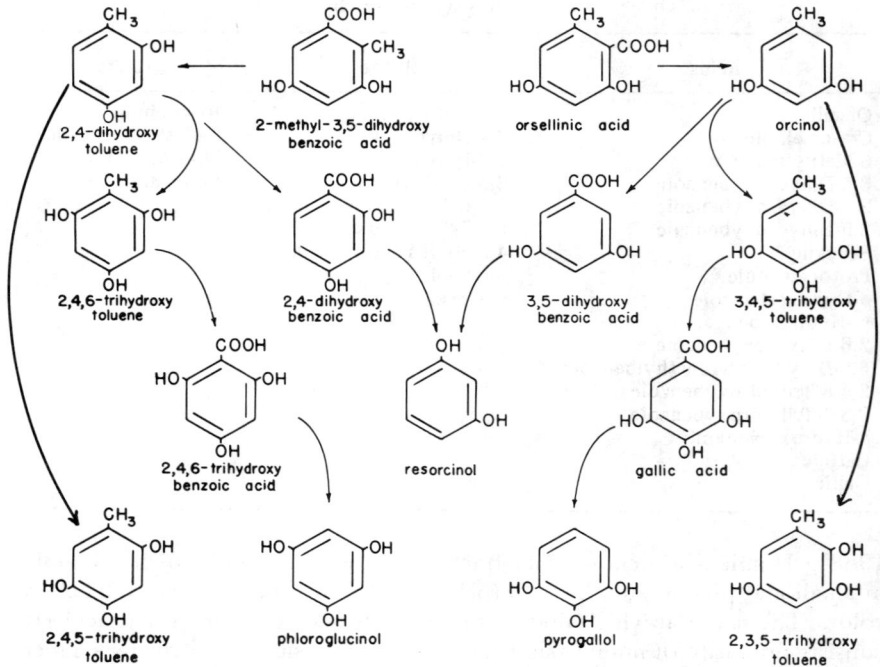

Fig. 9—Some microbial transformations of orsellinic and cresorsellinic acids, two phenolic substances synthesized by certain soil fungi (Haider & Martin, 1967).

acid molecules could vary greatly in composition related to constituent units available in a given microenvironment during formation. All would have similar properties, however, which would be related to the numerous functional groups, primarily the COOH and phenolic OH groups.

Certain microbes, especially some fungi and streptomycetes, synthesize dark-colored pigments which are similar to humic acids (Haider et al., 1964). They may be formed in the culture medium, in the cells, or both. They are phenolic polymers (Martin et al., 1974) with associated peptides and other substances including anthraquinones and possibly naphthalenic compounds. The phenolic or peptide content may vary greatly depending upon the amount and kind of N source and other cultural conditions (Bondietti et al., 1972). These fungal polymers are similar to soil humic acids with respect to exchange capacity, total acidity, carbon content, phenolic OH groups, types of phenols recovered upon Na-amalgam degradation (Table 9), resistance to decomposition in soil, appearance under the electron microscope (Fig. 12), and amino acids released upon 6N HCl hydrolysis (Burges et al., 1964; Martin et al., 1974; Piper & Posner, 1972). A summary scheme for the formation of humic acid is presented in Fig. 13.

Simple phenolic substances in the soil do not decompose as completely as many other readily available organic substances such as glucose, acetic acid, amino acids, proteins, and polysaccharides (Table 4). Also the more

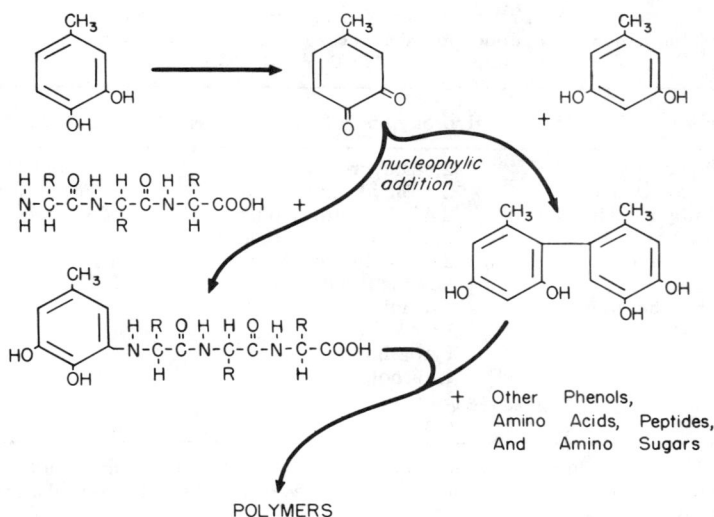

Fig. 10—Some possible enzymatic oxidative polymerization reactions involved in the formation of humic polymers.

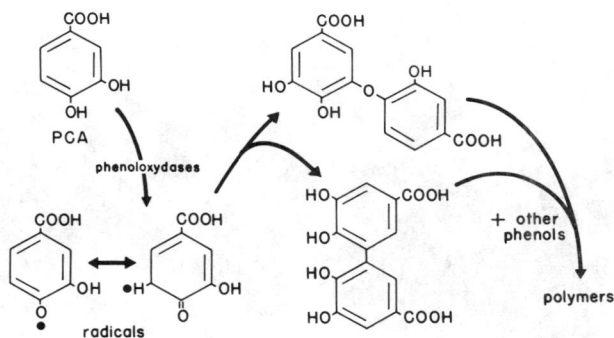

Fig. 11—Some possible enzymatic oxidative polymerization reactions of phenols involving radical formation.

chemically reactive phenols are decomposed to a lesser extent than the less reactive (Haider & Martin, 1975). If the aromatic ring were broken, the loss of phenol carbon would be similar to that lost from the more available compounds (Fig. 4). This would indicate that a portion of the intact phenolic ring is incorporated into the soil humus, is protected by the existing humus or clay, or is incorporated into the phenolic polymers formed by certain soil microbes.

Studies with ^{14}C-labeled organic compounds have shown that the new humic substances, although relatively resistant to microbial degradation, still decompose at a faster rate than the older humus (Jenkinson, 1971). This would be expected as portions of new molecules would be more susceptible to microbial cleavage than other parts. Fungus humic acid-type polymers

Table 9—Some phenols released upon Na-amalgam reductive degradation of fungal
phenolic polymers, model polymers incorporating the same phenols and soil
humic acids (Martin et al., 1974)

Hydroxybenzoic acids	Hydroxy phenols and toluenes	Lignin-type phenols[†]
2,4-Dihydroxybenzoic	2,6-Dihydroxytoluene	Vanillic acid
3,5-Dihydroxybenzoic	2,4-Dihydroxytoluene	Ferulic acid
2,3,4-Trihydroxybenzoic	2,4,5-Trihydroxytoluene	Syringic acid
p-Hydroxybenzoic	2,3,5-Trihydroxytoluene	
Protocatechuic	2,4,6-Trihydroxytoluene	
Gallic	Phloroglucinol	
m-Hydroxybenzoic	Orcinol	
p-Hydroxycinnamic	Resorcinol	
	Pyrogallol	
	Catechol	
	p-cresol	
	m-cresol	

† The lignin-type phenols were present in the fungal polymers when the cultures were
grown on plant residues or when the phenols were added to culture media or placed
under fungal pads.

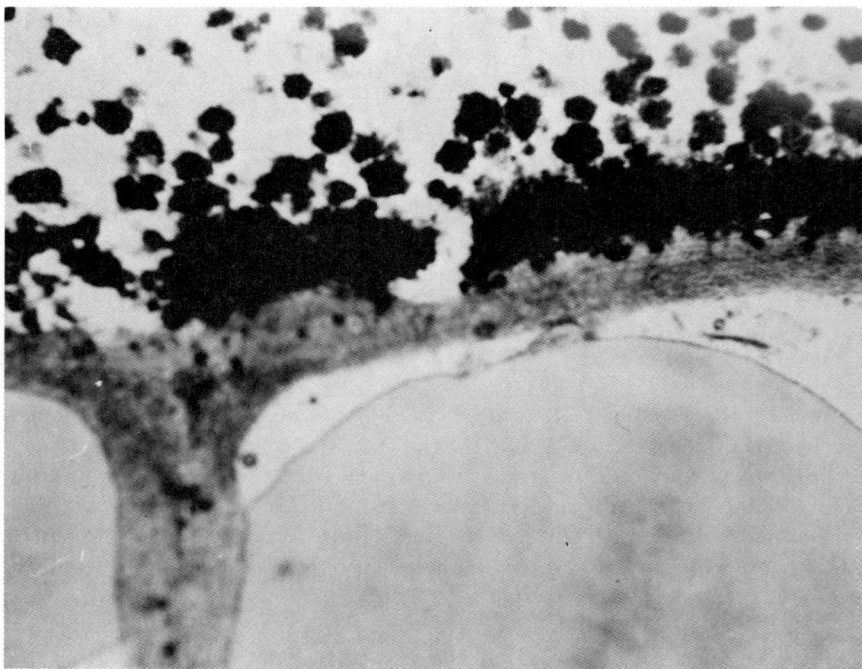

Fig. 12—Phenolic polymer (melanin) granules formed in the outer layers of the cell walls
of the fungus, *Aureobasidium pullulans*. (Courtesy of O. Reisinger and G. Kilbertus,
Nancy, France.)

may undergo 5 to 30% decomposition in 3 to 6 months. Model phenolase polymers made with specific ^{14}C-labeled components show that amino acids, peptides, and amino sugar units linked into the polymers decompose at a faster rate than the ring carbons of the phenolic units (Bondietti et al., 1972; Verma et al., 1975), and carboxyl, side chains, and OCH_3 carbons of phenolic units are more readily cleaved than the ring phenolic carbons (Haider & Martin, 1975). An increase in the number of carboxyl groups in the phenolic polymers may reduce the availability of side chains, amino sugars, and amino acid carbons of the polymers to the microorganisms. With time, the more readily available carbons are utilized and the residue becomes more resistant to further degradation.

2. SOIL POLYSACCHARIDES

Polysaccharides are common constituents or metabolic products of essentially all living organisms. Most plant, animal, and microbial polysac-

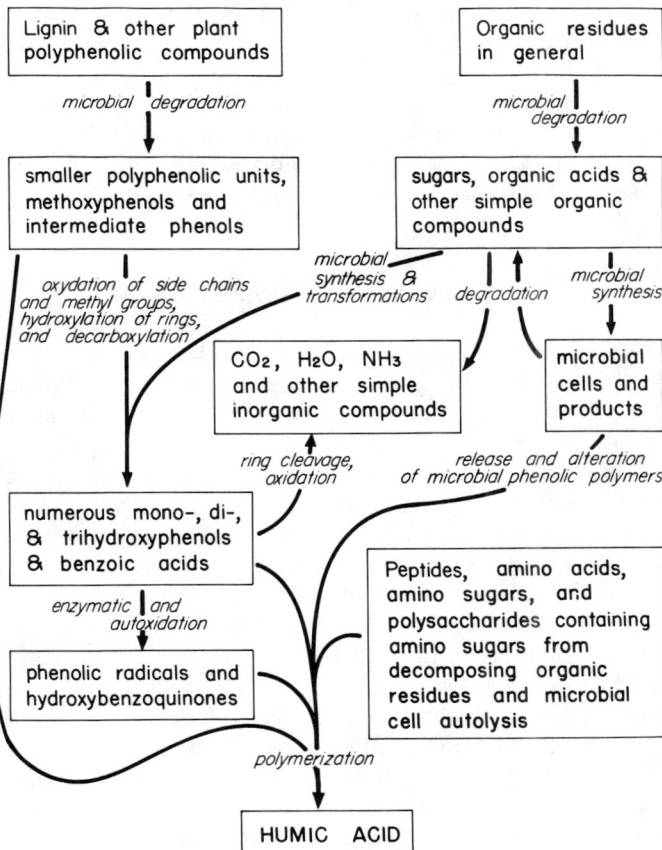

Fig. 13—A scheme for the formation of humic acid.

charides are readily susceptible to microbial decomposition (see Table 4) but some are more resistant, and from 10 to 30% of the soil humus consists of a polysaccharide fraction which is relatively stable to microbial degradation (Martin, 1971b; Swinzer et al., 1968). Most plant and microbial polysaccharides have 1 to 3 constituent units and occasionally 5 or 6, but the soil polysaccharide fraction has 12 or more common structural units (Cheshire et al., 1973; Swinzer et al., 1969). All the modern methods of separating mixtures of organic molecules have been applied to soil polysaccharides but all fractions obtained contain 10 or more structural units. It has been concluded that the procedures have failed to separate a complex mixture of polysaccharides, but it is also possible that the soil polysaccharide fraction is built up in a manner similar to the humic acid. Plant and microbial polysaccharide units in all stages of degradation could serve as constituent units of new polymers peculiar to the soil environment. Microbial enzymes could be involved in the new linkages. Combinations which are relatively resistant to decomposition, or which become resistant through salt or complex formation with metal ions or clays would constitute the resistant polysaccharide fraction of the soil humus. Polysaccharides with amino sugar units could be stabilized by linkage through free amino groups to the humic acid molecules. Polysaccharide-type constituents of microbial cell walls consist of numerous structural units including amino sugars and amino acids. Some of these or parts of the complex molecules could undergo linkage to soil phenolic polymers through free amino groups and thus become stabilized.

C. Improvement of Soil Physical Properties

If the sand, silt, and clay particles of a soil are so highly dispersed that water cannot penetrate into the soil and plant roots cannot break through a soil crust, crop yields will be reduced even though plant nutrient elements are present in adequate amounts (Harris et al., 1966; Martin et al., 1955). From the physical point of view a good soil is one in which the small soil particles are bound into water-stable aggregates or granules (Fig. 14). Such a soil does not crust as readily, rain and irrigation water penetrate the soil more readily, erosion is reduced, the soil can be cultivated with greater ease, aeration is more favorable, and root respiration and microbial activity are enhanced. In the best agricultural soils of the world the binding substances are largely organic in nature, and they are produced during the biological decomposition of organic residues in the soil (Russell, 1961).

In general organic residues containing relatively large percentages of readily available constituents exert the quickest and greatest binding action on the soil but the aggregating effect is of short duration (Table 10). More resistant materials require a longer time for maximum aggregation to occur, but the binding action continues to ʋe effective over a longer period of time. Large applications are more effective than small applications and aggregation is increased more in soils which already are poorly aggregated. Growth of

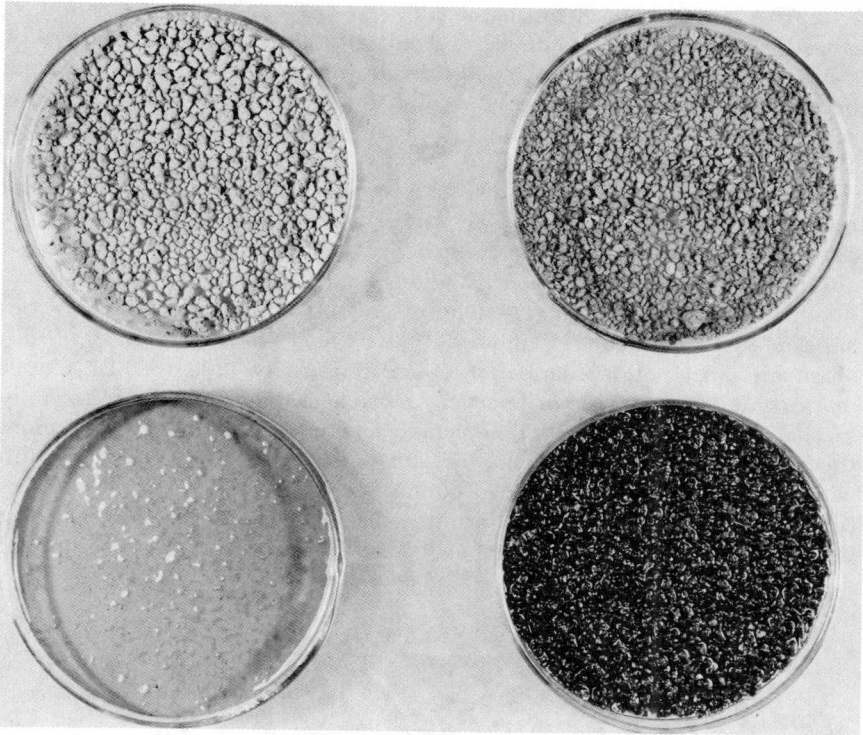

Fig. 14—Relation of soil humus to stability of soil aggregates. The dry aggregates on the top left are from a cropped silt loam containing < 1% organic matter. Those on the top right are from a silt loam containing about 4.0% organic matter. After addition of water (bottom) the aggregates on the left quickly dispersed while those on the right were stable. The soil on the right developed under grassland.

crops, especially sod crops, may improve aggregation (Johnston et al., 1942). This is associated with the abundance of root residue left in the soil which can be utilized by the soil organisms and the good distribution of the root residues throughout the soil mass.

Table 10—Influence of some organic residues on aggregation of a loam soil

| | Aggregation of silt and clay particles‡ | | | | | |
| | Days | | | | | |
Residue†	5	10	20	50	100	200
Sucrose	75	70	62	52	47	36
Alfalfa (*Medicago sativa* L.)	64	69	66	65	58	55
Cow manure	34	46	53	56	54	54
Peat	0	0	0	0	4	12

† 0.5%
‡ Percentage increase over control.

After application of organic residues to soil, or plowing under sod, aggregation declines. To maintain good soil structure, periodic addition of organic residues or rotation with soil building crops is necessary. Lower temperatures favor a more prolonged aggregation while higher temperatures accelerate the destruction of the soil binding substances (Martin & Craggs, 1946).

During the intense microbial activity following application of organic residues to soil, the microbial filaments and cells may mechanically bind soil particles together but substances synthesized by the soil organisms are generally regarded as more important. Various fractions of the soil humus are undoubtedly involved in aggregate stabilization, but the polysaccharide fraction appears to be particularly important (Martin, 1946; Harris et al., 1966). Most microbial polysaccharides in concentrations of 0.02 to 0.2% exert a marked binding action on soil particles. The high binding action of the polysaccharides is related to (i) their length and linear structure which allows them contact with more than one soil particle, (ii) their flexible nature which allows a greater number of contact points with soil particle surfaces, (iii) to the large number of OH groups which may be involved in hydrogen bonding, and (iv) to COOH groups which may allow ionic bonding through di- and trivalent metal ions (Greenland et al., 1962).

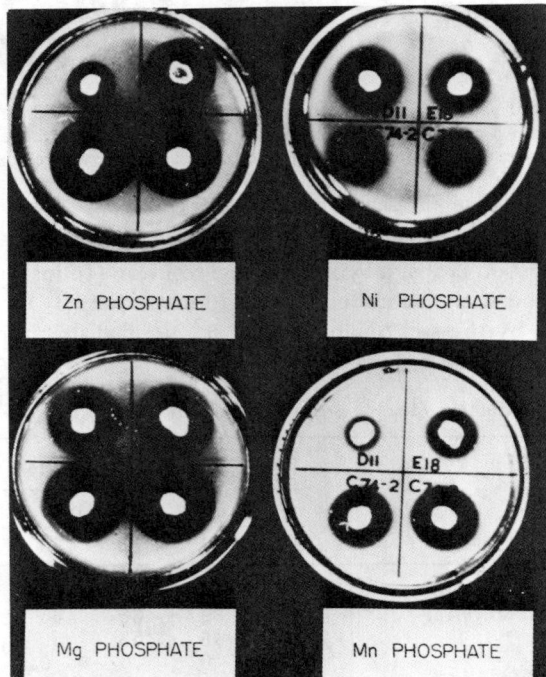

Fig. 15—Solubilization of insoluble phosphates by soil bacteria. (Courtesy of D. M. Webley and associates, Aberdeen, Scotland.)

Pure culture studies have demonstrated that microbial species vary widely in their ability to bind soil particles. Bacterial species which synthesize appreciably amounts of polysaccharides are highly effective. In general, fungi are more effective than most bacterial species (McCalla, 1946).

D. Release of Plant Nutrient Elements from Insoluble Inorganic Minerals

Soil organisms aid in weathering processes that break down rocks into soils and solubilize insoluble soil minerals, which increases the availability of phosphorus, potassium, and other plant nutrient elements (Fig. 15). Lichens and algae are among the first organisms to colonize barren rock and new rock or volcanic surfaces (Shields & Durrell, 1964). These organisms aid in the weathering of rocks and provide a source of organic residues for other organisms, especially bacteria and fungi (Webley et al., 1963). A high proportion of the bacteria actinomycetes and fungi associated with rock surfaces and crevasses are able to dissolve many mineral silicates and phosphates (Henderson & Duff, 1963). The fungi with high solubilizing action synthesize organic acids such as citric and oxalic. The most active bacteria synthesize 2-ketogluconic acid. A high percentage of the bacteria isolated from rock surfaces and crevasses synthesize this acid. Compounds of soil humus especially phenolic polymers of the fulvic acid fraction are active in solubilizing mineral elements in soil (Davies, 1971).

E. Mycorrhizal Relationships

Plant root surfaces and soil in contact with roots constitute a region of intense microbial activity. This region is commonly referred to as the *rhizosphere* (Rovira & McDougal, 1967). Most of the organisms present are saprophytes which are there because organic matter is available. The available C material is in the form of numerous excreted sugars, organic acids, amino acids, and of sloughed off old and injured root surface cells. The numbers of microorganisms in the rhizosphere are 5 to 100 times more numerous than in the soil away from the roots. The full significance with respect to plant growth, of this increased activity of saprophytic organisms in the root area is not known.

More highly specialized relationships exist between most plant species and microorganisms. An important N-fixing symbiosis will be discussed in the following subsection. Most plant species have a root-fungus symbiotic relationship called *Mycorrhizae* (Gerdeman, 1968; Marx, 1972). There are two main types of mycorrhizae. In one, ectotrophic or ectomycorrhizae, the fungi form a mantle of fungus cells around the feeder roots (Fig. 16). Fungus hyphae penetrate between the epidermis and cortical cells and eventually surround many or most of the cortical cells. Single hyphae or strands of hyphae grow from the mantle into the surrounding soil. In some

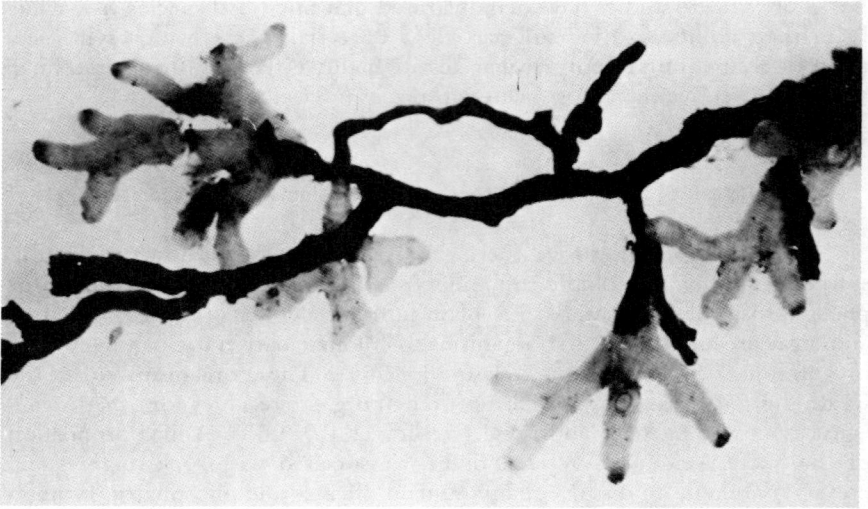

Fig. 16—Mycorrhizal shortleaf pine roots (*Pinus echinata* Mill.). (Courtesy of D. H. Marx, Athens, Ga.)

plants, the only contact the root has with the soil is through the fungus. The fungus obtains carbohydrates and other organic materials from the plants and the plants obtain nutrient elements and water through the fungus.

The ectotrophic mycorrhizae occur primarily on pine (*Pinus* L.), fir (*Abies* Mill.), spruce (*Picea* Dietr.), larch (*Larix* Mill.), eucalyptus (*Eucalyptus* sp.), beech (*Fagus grandifolia* Ehrh.), birch (*Betula* L.), oak (*Quercus* L.), hickory (*Carya* Nutt.), and other trees. The fungi involved belong to the basidiomycetes, fungi which form mushrooms. Most of these fungi can form a mycorrhizal relationship with more than one tree species.

A second mycorrhizal relationship is referred to as endotrophic or endomycorrhizae. The endomycorrhizae occur on numerous cultivated and noncultivated plants including corn (*Zea mays* L.), wheat (*Triticum aestivum* L.), cotton (*Gossypium hirsutum* L.), grasses, legumes, maple (*Acer* L.), dogwood (*Cornus* L.), citrus (*Citrus* sp.), roses (*Rosa* sp.), holly (*Ilex* sp.), and cottonwood (*Populus* L.). The most common forms produce very large, thick-walled spores (Fig. 17). Hyphae of the fungi grow between the epidermal and cortical cells of the feeder roots and then penetrate cortical cells. Within the cells it forms highly branched structures called arbuscules which are digested by the plant and, thereby, plant nutrients obtained from the soil by the fungus are made available to the plants. The fungus gets energy material from the plant.

The most common forms of endotrophic mycorrhizae belong to the genus *Glomus* (formerly *Endogone*) and are probably *Phycomycetes* (Gerdeman & Trappe, 1974). They do not grow on the common media used for isolation of soil fungi and, therefore, have not been extensively studied. Actually they are one of the most common fungus types in the soil.

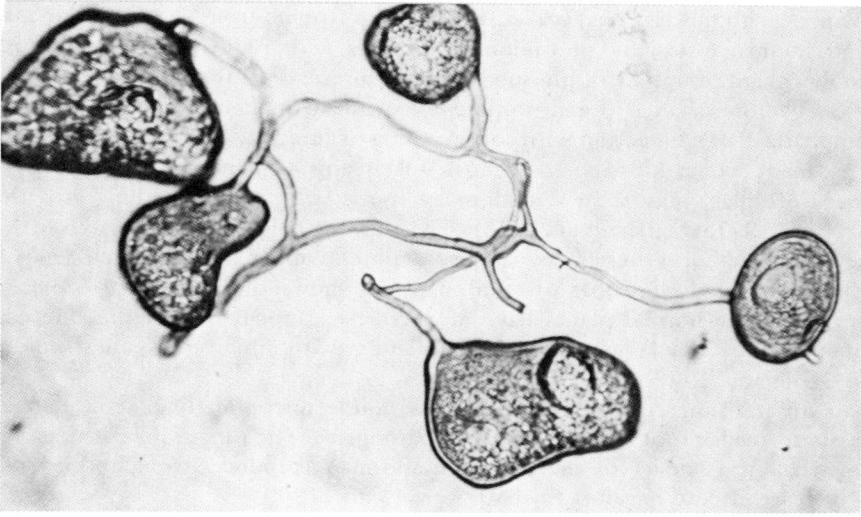

Fig. 17—An endomycorrhizal fungus (*Glomus* sp.) obtained from plant roots from virgin grass-shrub land. (Courtesy of T. DeWolf, Riverside, Calif.)

The mycorrhizal association appears to be of great importance in plant nutrition for both natural and cultivated plants, especially in soils of low fertility (Gerdeman, 1968; Marx, 1972). The prevalence of mycorrhizae on plant roots is also much greater in infertile soils. The mycorrhizae enable the

Fig. 18—Stunted and normal citrus seedlings growing in fumigated soil. The stunting is associated with P, Cu, and Zn deficiency and absence of endotrophic mycorrhizae.

plants to obtain less readily available plant nutrients from the soil minerals and humus, especially phosphorus. Studies with radioactive phosphorus show enhanced uptake of phosphorus in the presence of the fungus.

The importance of endotrophic mycorrhizae in plant nutrition has been emphasized in connection with soil steaming or fumigation practices. After a treatment which kills many or most soil organisms many plants, especially tree seedlings, may show a temporary spotty stunting of seedling growth (Martin, 1971a; Martin et al., 1973). The stunting (Fig. 18) is associated with severe phosphorus deficiency and sometimes copper and zinc deficiency. Examination of the roots of stunted plants shows the absence of *Glomus* sp. while the normal plants have mycorrhizae. Inoculation of the stunted plants corrects or largely corrects the plant growth inhibition (Klineschmidt & Gerdeman, 1972).

In addition to improving plant nutrition, ectotrophic fungi act as deterrents to feeder root infection by root pathogens. The fungus mantle acts as a mechanical barrier to the pathogen and many produce strong antibiotics which are effective against the pathogens (Marx, 1969).

F. Nitrogen Fixation

Under favorable environmental conditions and with a suitable carbon and energy source, many microorganisms are able to utilize elemental nitrogen for the synthesis of proteins and other nitrogenous cellular components (Alexander, 1961). Biological nitrogen fixation is usually divided into two types (i) nonsymbiotic nitrogen fixation brought about by certain free-living bacteria and blue-green algae, and (ii) symbiotic nitrogen fixation which occurs only in an association of a plant root or leaf and a microorganism.

1. NONSYMBIOTIC NITROGEN FIXATION

The most important or classical nonsymbiotic, nitrogen-fixing organisms are: (i) *Azotobacter* and related genera, *Azomonas, Beijerinckia,* and *Derxia*; (ii) *Clostridium butyricium* and related species; and (iii) many blue-green algae (*Cyanophyceae*) belonging primarily to the genera *Anabaena, Anabaenopsis, Aulosira, Calothrix, Cylindrospermum, Mastiglocladus, Nostoc,* and *Tolypothrix* (Jensen, 1965; Fogg, 1973). Some of the blue-green algae occur as components of lichens.

Environmental conditions which favor nitrogen fixation by the nonphotosynthetic, nitrogen-fixing bacteria are (i) a temperature range of 15 to 35C, (ii) a pH range of about 5 to 9 although *Beijerinckia* thrive at pH 3 to 9, (iii) adequate supply of available nutrients espécially P and Mo, (iv) adequate moisture, (v) absence of readily available N compounds particularly NH_3, and most important (vi) the presence of available C and energy sources.

There has been a considerable difference of opinion as to the actual importance of the nonsymbiotic, nitrogen-fixing bacteria under natural and cropping conditions. Some investigators estimate a fixation of 5 to 35 kg N ha^{-1} year^{-1} (Allison, 1973). Others believe these estimates are high and point out that the organisms utilize fixed forms of N if they are available, that they have to compete with numerous other bacterial species for the available C and energy, and that their numbers in the soil in comparison with other types are small.

Nitrogen fixation by blue-green algae, however, is generally considered to be of great importance in flooded soils such as rice crops (*Oryza sativa* L.), on barren or new rock surfaces, and on both virgin and cultivated arid soils (Cameron & Fuller, 1960; De & Mandal, 1956). Pot culture studies have indicated that gains equivalent to about 15 to 79 kg/ha could be expected through fixation by effective strains, and in southeast Asia the inoculation of rice paddies with effective N fixing blue-green algal strains has been recommended (Watanabe, 1959). Nitrogen fixation by desert blue-green algal or blue-green algal-lichen crusts is believed to constitute an important source of N for desert soils (MacGregor & Johnson, 1971; Rychert & Skujins, 1974). Nitrogen fixation by the photosynthetic blue-green algae is of particular importance because it is not dependent upon an available supply of carbohydrate or other organic energy sources.

2. SYMBIOTIC NITROGEN FIXATION

Fixation of N in nodules on the roots or leaves of plants is probably of much greater importance than nonsymbiotic N fixation. Studies on the fixation in the nodules on the roots or leguminous plants have received the greatest attention (Nutman, 1965). Leguminosae is one of the world's largest families of flowering plants. Almost all species are subject to nodulation. The importance of legumes in building and conserving soil fertility has been recognized since the beginning of agriculture.

The organisms which form nodules on legume roots belong to the genus, *Rhizobium*. They are aerobic, gram negative, nonsporulating rods with simple nutritional requirements (Alexander, 1961). They can grow in the soil without the host plant and are common rhizosphere organisms. They are classified according to their host range, that is, the number of legume species each can infect. The legume species nodulated by a particular *Rhizobium* sp. is referred to as the cross inoculation group.

The amount of N fixed by a nodulated legume crop will vary from about 60 to 600 kg/ha (Allen & Allen, 1958) and will vary depending upon the *Rhizobium* strain, plant nutrient availability, soil pH, soil moisture, plant species, and other factors. If inorganic N is present, the plants will utilize the soil supply and the amount of gaseous nitrogen fixed will be reduced.

Although the legume-*Rhizobium* symbiotic relationship has received the most attention, numerous nonleguminous plants fix N through symbiosis with various organisms (Fig. 19). The genera of nonleguminous plants include: *Alnus* (Alder), *Cassuarina, Coriaria, Elaeagnus* (Silverberry); *Shep-*

Fig. 19—Nodules on the roots of the sage, *Artemisia ludoviciana*. (Courtesy of R. B. Farnsworth and M. A. Clawson, Provo, Utah.)

herdia (Buffaloberry); *Myrica* (Wax Myrtle); *Ceanothus* (Snowberry, Buck-brush, California Lilac); *Dryas, Purshia* (Bitterbush); *Cercocarpus* (Mountain Mahogany); *Opuntia* (Prickly Pear); *Psychotria, Chrysothamnus* (Rabbit-brush); and *Artemisia* (Sagebrush) (Allen & Allen, 1965; Clawson, 1973[1]; Youngberg & Wollum, 1968). In the future additional plant genera will un-doubtedly be added to this list.

Another type of symbiotic N-fixation not involving nodule formation has recently been reported for a number of tropical grasses (Dobereiner et al., 1972). These include *Paspalum notatum, Digitaria decumbens,* and *Pen-nisetum purpureum.* The symbiotic association supplies a major portion of the plant's nitrogen requirements. The active organism from the roots of *Digitaria decumbens* was identified as *Spirillum lipoferum* which is presum-ably a widespread soil organism. In pure culture it fixed 7.2 mg N/g of malate compared to 7.9 mg N/g for *Azotobacter paspali.* The organism enters the root and develops within the inner cortical cells. The plant pro-vides the energy source, primarily malate, and the organism fixes N_2. The N

[1]M. A. Clawson. 1973. Nitrogen fixation by *Artemisia ludoviciana* (gray sagewort), characterization of the endophyte and factors influencing nodulation. M.S. Thesis. Brigham Young University, Provo, Utah.

is immediately available to the plant and does not depend on digestion or decomposition of the bacterial cells. These observations suggest that through breeding, selection, and genetic manipulation it may be possible to extend N_2-fixation to other grasses and even to grains.

A symbiotic relationship of N_2-fixing blue-green algae with several plant species has also been reported (Shields & Durrell, 1964). Although large amounts of chemically fixed N are added to soils each year, biologically fixed N still accounts for a greater amount and probably 80% or more of the total (Dalton & Mortenson, 1972). For a recent summary of the mechanism of N fixation the reader is referred to the review of Hardy and Havelka (1975).

G. Antagonistic Relationships

Competition among the soil microbes is intense. They compete primarily for the available C and energy material and possibly for other essential nutrients (Clark, 1967). Some produce toxic substances which inhibit or kill other organisms (Fig. 20). Many of these toxic substances have been isolated and now constitute the highly important antibiotic drugs (Waksman & Lechevalier, 1962). One organism may parasitize another or trap it and then consume it. Protozoa feed on bacteria, but certain fungi and other organisms

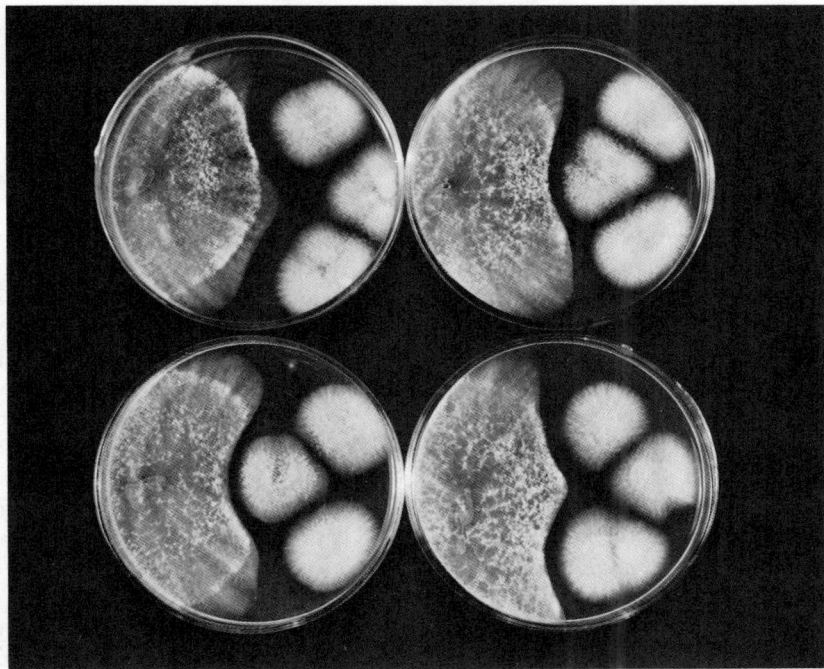

Fig. 20—Antibiotic action of four *Fusarium solani* isolates against the soil fungus, *Trichoderma viride*.

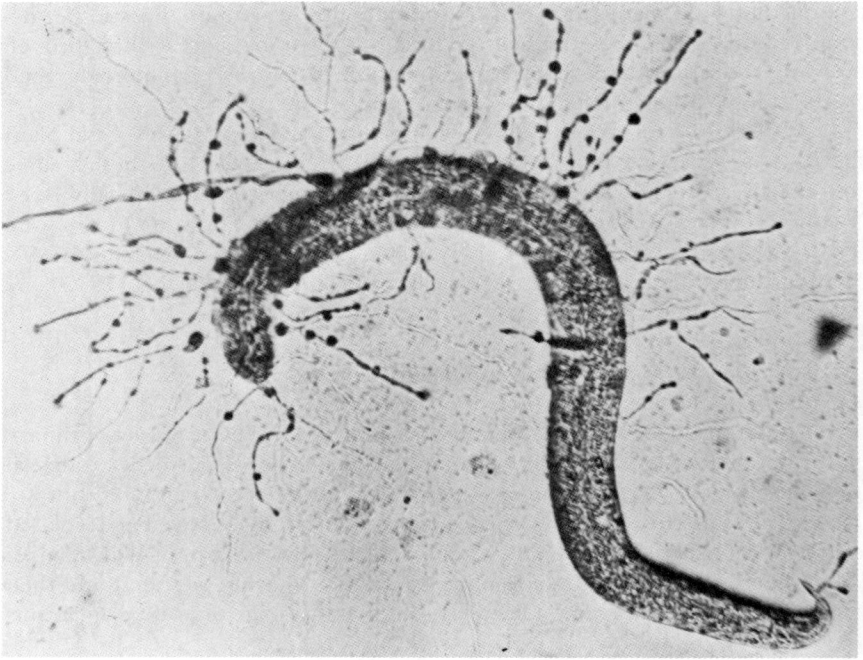

Fig. 21—A nematode being destroyed by a fungus. (Courtesy of R. C. Baines, Riverside, Calif.)

parasitize and kill protozoa. Some nematodes feed on fungi, but other fungi and certain actinomycetes attack nematodes (Fig. 21). *Antagonism* refers to all these species interactions in which at least one of the organisms involved is harmed.

The antagonistic activities of the soil saprophytic populations are important in helping to control pathogenic or disease organisms. Most human pathogenic microbes are quickly destroyed when introduced into the soil (Waksman & Lechevalier, 1962). *Bdellovibrio bacteriovorus* is a small endoparasitic bacterium that multiples within the host bacterial cell and eventually causes its lysis. Klein and Casida (1967) showed that a single strain of *Escherichia coli* OX9 was parasitized by *Bdellovibrio* isolated from 23 different soils obtained throughout the United States. They also demonstrated that strains of *Bdellovibrio* isolated as parasites of two different *E. coli* serotypes parasitized all of the other 25 *E. coli* strains tested. They concluded that *Bdellovibrio* was widespread in soils and active against many *E. coli* serotypes, and suggested that the failure of *E. coli* to survive in soil may be due to the ubiquitous occurrence of *Bdellovibrio*.

Generally the damage to plant roots caused by root parasites is reduced by the presence of the saprophytic population (Fig. 22). If the soil is sterilized and inoculated with the pathogen the damage is usually more severe (Kreutzer, 1965). Simultaneous inoculation with a good antagonist such as

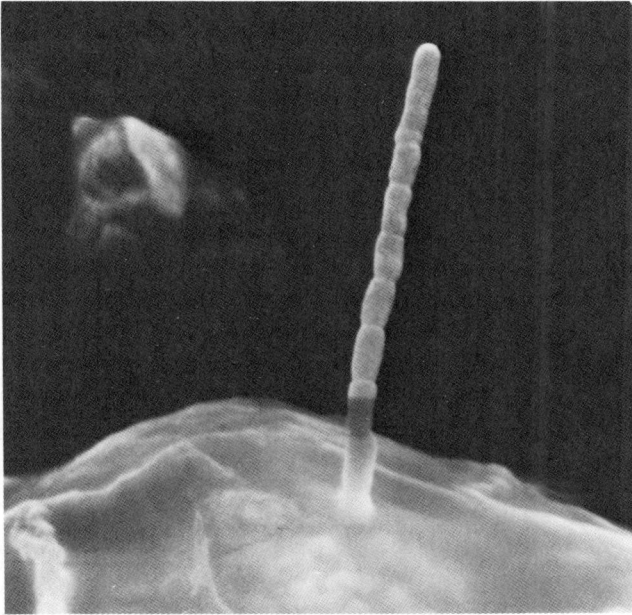

Fig. 22—The fruiting structure of a soil streptomycete piercing the wall of the fruiting structure of the plant root parasite, *Phytophthora parasitica*, a fungus. The *Streptomyces* sp. has destroyed the fungus oospore inside the oogonium. (Courtesy of R. Honour and P. Tsao, Riverside, Calif.)

Fig. 23—Citrus seedlings growing in *(1)* old citrus soil infested with citrus root parasites, *(2)* old citrus soil previously fumigated to kill root parasites, *(3)* fumigated soil inoculated with the root parasite, *Thielaviopsis basicola*, and *(4)* fumigated soil inoculated with both *Thielaviopsis basicola* and the saprophytic fungus, *Penicillium nigricans*. Note how the *Penicillium nigricans* has reduced the injury caused by the root pathogen.

Myrothecium verrucaria or *Penicillium nigricans* will often reduce the injury (Fig. 23).

The addition of organic amendments including organic wastes to the soil will often control or reduce the severity of certain plant root diseases (Patrick & Toussoun, 1965). Plowing under a green manure crop may control potato scab (*Solanum tuberosum* L.), *Phymatotrichum* root rot of cotton, foot-rot of snap beans (*Phaseolus vulgaris* L.), and other plant root diseases. It is believed that the addition of the correct organic residue stimulates the growth of organisms antagonistic to the pathogen. Crop rotation has long been a recognized method for control of many plant root parasites. One factor involved may be that the rotation crop enhances the growth of antagonistic species. Applications of animal manure to soil will enhance microbial antagonism by increasing the microbial numbers and activity. Growth of *Trichoderma viride*, a well recognized antagonist, may be enhanced by soil acidification (Weindling & Fawcett, 1936) and by application of specific chemicals (Martin, 1971a).

IV. TRANSFORMATIONS OF INORGANIC ELEMENTS

A. Nitrification and Denitrification

1. NITRIFICATION

Nitrification is a two-step chemoautotrophic process effected sequentially by oxidation of ammonium to nitrite by *Nitrosomonas* and of NO_2^- to NO_3^- by *Nitrobacter* (Meiklejohn, 1954). The energy obtained from oxidation of NH_4^+ or NO_2^- is coupled with the reduction of carbon dioxide or bicarbonate to carbohydrate for cellular synthesis. Chemoautotrophic nitrification is thus proportional to cellular yields. The process is aerobic involving the use of O_2 mainly as a terminal electron acceptor although one atom of O_2 in the NO_2^- ion is derived from molecular O_2 (Rees & Nason, 1965). The initial step of ammonium oxidation appears to involve the participation of molecular O_2 into the substrate to yield hydroxylamine (Verstraete & Alexander, 1972). Hydroxylamine condenses with NO_2^- to form nitrohydroxylamine, which is oxidized to two moles of NO_2^- (Aleem & Lees, 1963). Oxidation of NO_2^- to NO_3^- involves the incorporation of water, while O_2 functions solely as an electron acceptor. Not surprisingly, reduced O_2 tensions diminish nitrification rates.

The optimum pH for nitrification is slightly on the alkaline side. Pure cultures of *Nitrosomonas* and *Nitrobacter* have an optimal pH range of 7.8 to 8.8 (Hofman & Lees, 1953; Meyerhof, 1916). The extreme limits for growth of *Nitrosomonas* and *Nitrobacter* are between 5.0 and 9.4 and 5.5 to 10.2, respectively. In nature, it is unlikely that *Nitrobacter* grows at pH values as high as in pure culture because free ammonia is toxic to the bacterium and inhibits its growth. Under such conditions nitrite would tend to accumulate. Thus pH has the effect of regulating the equilibrium between ammonium and ammonia ($pK_a = 4.2$).

The optimal temperature for growth of *Nitrosomonas* and *Nitrobacter* is between 34–37C in culture, although higher temperature optima of 42C have been reported with maximal oxygen consumption rates at 49C. The generation times of the nitrifying bacteria are much longer than those for most heterotrophic bacteria. The shortest generation reported by Skinner and Walker (1961) for *Nitrosomonas* is 8 hours. However, most studies show that both *Nitrosomonas* and *Nitrobacter* have generation times ranging from about 18–32 hours.

The nitrifying bacteria are difficult to isolate and to maintain in pure culture. Recent studies with fluorescent antibody techniques have made it possible to observe and quantify these and other bacteria in their natural habitats (Schmidt, 1974).

Heterotrohpic nitrification is a process having wide species diversity among fungi, bacteria, and actinomycetes. Energy is apparently not generated from this type of nitrification since no relationship has been established with cell yields. Speculation has been raised that heterotrophic nitrification may be significant in alkaline (Verstraete & Alexander, 1973) or acid (Becker & Schmidt, 1964) soils, where autotrophic nitrification would not occur. Though inhibition of the latter process with N-serve (2-chloro-6-[trichloromethyl] -pyridine) has failed to reveal the significance of the former process. Since an organic pathway, as well as an inorganic pathway is recognized, it is speculated that heterotrophic nitrification may represent a means of detoxifying oximes, C-nitroso, and nitro compounds that would be toxic to the organisms.

The accumulation of nitrite and amines in soils receiving heavy animal waste applications has generated some environmental discussion concerning the potential formation of hazardous nitrosamines, which are potent carcinogens. Nitrosamines can be produced chemically by condensation of nitrous acid with a secondary amine, or microbiologically at neutral or alkaline pH with nitrite ion. Laboratory studies with lake water, sewage, and soil, each of which were spiked with dimethylamine and nitrite at concentrations in excess of those found in nature, showed the bacterial formation of trace quantities of dimethyl nitrosamine (Ayanaba & Alexander, 1974). Voets et al. (1975) did not detect nitrosamines in waste waters containing high nitrite concentrations, and concluded that they presented no problems in waste disposal. Presently, there is no evidence that nitrosamines are formed in soils receiving heavy applications of animal wastes.

2. DENITRIFICATION

Denitrification is the reduction of NO_3^- or NO_2^- to gaseous forms of nitrogen, usually N_2O (nitrous oxide) or N_2. It is extremely rapid since it is a respiratory process that involves the transfer of electrons to NO_3^- in lieu of O_2. The branch point in the electron transport chain occurs at cytochrome c and involves a molybdenum-iron complex (Payne, 1973). Denitrifying bacteria grown anaerobically in the presence of nitrate produce more cytochrome c (Daniel & Appleby, 1972; Sapshead and Winpenny, 1972), while the levels of cytochrome a, the terminal cytochrome oxidase, remain unchanged or are reduced. The free energy change resulting from the transfer of an electron pair to NO_3^- is lower than for O_2, yet a mole of NO_3^- accepts 20% more electrons than O_2. Consequently, the energetics for denitrification and respiration are very similar on a molar basis. This has led some investigators to suggest that the same kinetics for O_2 consumption would apply for denitrification. Limited experiments in sewage suggest that this is true (Mulbarger, 1971), but it has not been tested in soil.

Denitrification is inherent only to the bacteria, most of which are aerobes comprising the genera of *Pseudomonas, Bacillus, Micrococcus,* and *Achromobacter.* The term *dissimilatory nitrate reduction* is frequently used to designate the participation of the respiratory cytochrome system in the reduction of NO_3^- to NO_2^-. Thus, the first step in denitrification may occasionally be referred to in this way. However, some bacteria, usually facultative anaerobes (e.g., *Escherichia, Staphylococcus,* anaerobes [*Clostridium*], and actinomycetes [*Mycobacterium, Nocardia, Streptomyces*]) do not carry out reduction of NO_2^- to gaseous N_2. This group of microorganisms is usually referred to as "NO_3^- respiring" or "NO_3^- dissimilating." "Assimilatory NO_3^- reduction" involves the reduction of NO_3^- to NH_4^+. and is associated strictly with biosynthesis processes (i.e., amino acid synthesis). This process is not only common to microorganisms, but to higher green plants as well. Unlike dissimilatory NO_3^- reduction, NO_3^- assimilation is unaffected by the presence or absence of oxygen.

The pathway of NO_3^- assimilation is believed to follow the reverse of nitrification. The simplest and most logical sequence is proposed by Payne (1973) from $NO_3^- \rightarrow NO_2^- \rightarrow NO \rightarrow N_2O \rightarrow N_2$ with compelling enzymatic evidence and positive product identification existing for each step in the pathway of *Pseudomonas perfectomarinus.* Some reports claim that nitrous oxide is not an intermediate (Sacks & Barker, 1952) because the compound is reduced only after a lag period and that cells poisoned with azide or cyanide to prevent reduction of N_2O still liberate N_2 from nitrite. The interpretation of these results have been questioned on grounds that the lag period reflects the time required for equilibrium of gaseous-solute diffusion (Focht, 1974) and that this lag period is eliminated when cells are incubated in a N_2O atmosphere. Compelling evidence for the obligatory role of N_2O is also presented in light of data which show that only NO_3^- and N_2O reduction are mediated by particulate-type cytochromes, which involve phosphorylation (Payne, 1973). Cell yields were identical regardless of whether nitrite, nitric oxide, or nitrous oxide are utilized as the terminal electron acceptor, which proves that energy is transferred only with the latter compound.

Nitrite is occasionally found as a transient intermediate (Bremner & Shaw, 1958). This is largely dependent on the available C and the dissolved O_2 content (or redox potential, E_h). Where C is limiting, nitrite tends to accumulate. Nitrite is also reduced at a lower E_h (200 mV) than NO_3^- (Patrick, 1960) so that NO_3^- tends to be reduced first under low oxygen tensions. Surprisingly NO_2^- is reduced faster than NO_3^- in pure culture and in soils (Nommik, 1956; Sacks & Barker, 1949). This apparent anomaly stems from the observation that high nitrite concentrations repress NO_3^- reduction severely and N_2O reduction somewhat less (Payne, 1973). Thus the observation that NO is generated from neutral to alkaline soils when NO_2^- is added as substrate cannot unequivocally be attributed to chemodenitrification, a reaction that is prevalent primarily in acid soils (Bollag et al., 1973).

Denitrification has an optimal temperature range of 65–75C, which is presumably affected by thermophillic *Bacillus* species. The effect of temperature upon rate adheres to Arrenhius kinetics except for temperatures below 20C (Focht, 1973). The rate is most drastically affected at lower temperatures, with reduction of nitrate being more affected than reduction of nitrite (Bailey & Beauchamp, 1973). There is also evidence that the proportion of N_2O/N_2 is greater at reduced temperatures ($<$ 5C). The greater retardation of NO_3^- and N_2O reduction in comparison to NO_2^- reduction at reduced temperatures may be due to the nature of the enzymes. The NO_3^- and N_2O reductases are membrane-bound while NO_2^- reductase is soluble (Nommik, 1956; Payne, 1973).

The optimal pH range for denitrification is slightly alkaline with most reports showing about 7.0–8.0 being highest. The proportions of N_2O and NO diminish as the pH increases, and molecular N becomes the predominant gas evolved (Bremner & Shaw, 1958; Focht, 1973).

Denitrification occurs frequently in what appears to be "well aerated" soils which has raised questions as to whether or not the unaccounted-for N deficit in N fertility studies is truly due to denitrification rather than experimental error. Certain misconceptions frequently arise concerning the oxygen status of anaerobic microsites in soil. Greenwood (1961) has shown that such zones do in fact exist, and Wuhrman (1964) has shown that anoxic conditions may be present within a 100 μm diameter floc in an aerobic activated sludge tank. Studies with pure cultures (Collins, 1955) showed that denitrification occurs in flasks that are aerated on a shaking platform, and that the geometry of these flasks is an important factor governing the rate of oxygen diffusion. Because respiration rates may exceed the diffusion rate of oxygen into liquid, the soil atmosphere may bear little or no reality to what is occurring in the liquid. Meek et al. (1969) showed that reduction of NO_3^- correlated well with redox potential but not with O_2 concentration. Similarly, Focht et al. (1975) showed that atmospheric O_2 concentrations never fell below 14% and were usually around 17% despite the occurrence of denitrification in field plots receiving manure applications.

Wesseling and Van Wijk (1957) showed that diffusion of O_2 becomes critical between 85–90% saturation of the pore space. This appears to be the point where highest concentrations of N_2O are observed in laboratory and field studies (Focht et al., 1975). Starr et al. (1974) observed the concomitant occurrence of nitrification and denitrification in soil columns maintained at an overall average pore-filled porosity of about 85%. Denitrification ceases at about 25–30 centibars suction in laboratory (Pilot & Patrick, 1972) and field studies.

Since denitrification requires a reductant, it is not surprising that increased available C accelerates the process. Generally less nitrous oxide is produced as the available C content of the soil is increased (Nommik, 1956). Because the respiratory quotient (CO_2 produced/O_2 consumed) should be stoicheometrically equivalent to the denitrification quotient, the C/N ratio for denitrification is very low compared to that required for assimilation (i.e., immobilization). The optimal C/N ratio for denitrification is between 2–3 and would depend on the reducing power of the carbonaceous substrates (Bowman & Focht, 1974). It is not surprising that assimilation of NO_3^- under anaerobic conditions accounts for only a small amount by comparison to that reduced for respiratory purposes (Bowman & Focht, 1974), though at low nitrate concentrations, where N is limiting, up to 1/3 of the NO_3^- may be assimilated (Keeney et al., 1971). High C/N ratios and/or high sugar concentrations appear to inhibit denitrification.

Denitrification was thought to be independent of NO_3^- concentration (Broadbent & Clark, 1965), but these studies were all conducted at high nitrate concentrations when calculated on a solution basis as opposed to the expressed dry weight basis. Thus, Stanford et al. (1975) found that the rate was dependent in all cases when solution concentrations below 32 ppm N were used. Bowman and Focht (1974) resolved this anomaly by showing that denitrification followed Michaelis-Menten kinetics, providing the reductant or oxidant was not limiting when the concentration was varied. Frequently, failure to detect an increase in the rate by addition of more NO_3^- may be due to available C limitations.

Though some denitrifying bacteria can use molecular H (*Micrococcus denitrificans*) or S (*Thiobacillus denitrificans*) as an energy source it is difficult to assess their significance in soil. However, Martin and Ervin (1953) observed that amendments of S to citrus groves resulted in N deficiency, which they attributed to autotrophic denitrification. Similarly, Mann et al. (1972) showed that denitrification rates were greatly accelerated when S was added to soil columns receiving continuous perfusion of a NO_3^- solution.

B. Transformations of Sulfur

Sulfur, like nitrogen, is found in many different valence states in nature, largely as a result of the oxidation-reduction potential or its incorporation into biomass. All forms of life are capable of taking up SO_4^{2-}, the most oxidized form of S, for fixation into protoplasmic constituents of organic sulfates (e.g., adenosine phosphosulfate), sulfonates (e.g., taurine), and reduced sulfhydryl, and thiol forms such as the sulfur amino acids of cysteine, cystine, and methionine. The reverse reaction from reduced S organic compounds to SO_4^{2-} is also common to all forms of life (Alexander, 1961).

Sulfide is toxic to most higher forms of life and is generally liberated from organic sulfur compounds by bacterial action, usually under anaerobic conditions. The production of hydrogen sulfide (H_2S) is a test used in diagnostic bacteriology to distinguish between genera and species of the *Enterobacteriacea*. Only a few soil bacteria, primarily *Aeromonas* and *Xanthomonas*, carry out this conversion. Sulfide can be utilized at low concentrations by plants, certain species of yeasts, and bacteria, specifically *Salmonella*.

The most significant ecological production of sulfide in soil and aquatic systems is brought about by reduction of SO_4^{2-} under anaerobic conditions. The best characterized microorganism affecting this conversion is the bacterium, *Desulfovibrio desulfuricans*. The organism is an obligate anaerobe containing a partial cytochrome system that couples the oxidation of organic matter with the reduction of SO_4^{2-} to sulfite and finally to H_2S. Estuaries are common habitats of this bacterium, owing to the relative abundance of SO_4^{2-} and organic matter. At high temperatures, SO_4^{2-} is converted to H_2S by the thermophilic anaerobe *Clostridium nigrificans* (Alexander, 1961).

Sulfate reduction is not common to agricultural soils, with the possible exception of rice paddies because of the low redox potential (−200 mV) required for SO_4^{2-} and sulfite reduction (Latimer, 1952). The precipitation of ferrous sulfide in iron pipes buried under the soil or in marine sediments is due to SO_4^{2-} reduction. The process can be of serious economic consequences to the petroleum industry, though frequent washing of the pipes with dilute acid inhibits SO_4^{2-} reduction, which does not occur below pH 5.5. Though the species diversity is very restricted for SO_4^{2-} reduction, many diverse genera of fungi, bacteria, and actinomycetes can reduce thiosulfite, sulfite, and tetrathionate to H_2S.

Oxidation of sulfide to S is normally brought about by photosynthetic bacteria belonging to the *Chlorobacteriaceae* and *Thiorhodaceae* families, which are commonly referred to as the green bacteria and the purple-sulfur bacteria, respectively. Bacterial photosynthesis is similar to that of green plants only with respect to CO_2 fixation. Hydrogen sulfide is used instead of water as the initial reductant and S is formed instead of O_2. *Chlorobium*, one of the green bacteria, can also use S or thiosulfate as well, liberating SO_4^{2-} (Stanier et al., 1970).

Bacterial photosynthesis is an anaerobic process that generates molecular sulfur, if H_2S is used, and never oxygen. This process is thus restricted to habitats where sunlight is present and conditions are anaerobic, such as waste lagoons, and would not be of significance to arable soils. Nonphotosynthetic oxidation of H_2S to S has been demonstrated with *Beggiatoa, Thiothrix,* and *Thioplaca,* which deposit S granules within the cell.

The oxidation of S or sulfide is perhaps the most familiar and best understood S transformation in soil. The reaction is effected by the bacterium *Thiobacillus*. The pH optimum varies for different species and may run as high as 6.8 for *T. denitrificans* or 2.0 for *T. thioxidans*. Oxidation of S is an aerobic, chemoautotrophic process involving the production of acid. The S-oxidizing bacteria are ubiquitous in soils and are responsible for lowering the pH upon addition of S. This is significant not only for the reclamation of highly alkaline soils but for control of potato-scab and rot by species of *Streptomyces*. These diseases are not severe below pH 5.

The pathway of S oxidation is relatively complex, and varies between different species though all of them oxidize molecular S and thiosulfate (Doelle, 1969). Thus the role of thiosulfate is clearly defined. It appears as though the general over-all pathway involves reduction of S^0 to S^{2-} to transport the sulfur into the cell, whereupon it is reoxidized to S^0. Sulfite, dithionate, and tetrathionate have also been identified as intermediates.

The S-oxidizing bacteria represent growth and survival at extreme environments ranging from pH 1 to pH 10.2. Many of the sulfur hot springs also abound with thermophilic species of *Thiobacillus* with growth temperature optima of 65C. *Sulfolobus,* which may be a morphological variant of *Thiobacillus,* has a reported temperature optimum of 85C, and not only survives but supposedly *grows* at boiling temperatures (Brock et al., 1972).

C. Transformations of Metal Elements

1. IRON

Iron transformations in soil may be grouped into the following categories: (i) oxidation of Fe^{2+} to Fe^{3+} ion, (ii) reduction of Fe^{3+} to Fe^{2+}, (iii) degradation of organic Fe compounds, and (iv) production of organic Fe chelants. The first group of microorganisms are usually referred to as the Fe bacteria, and consist of the chemoautotrophs, which derive energy from the process, and the heterotrophs which deposit Fe^{3+} hydroxide on the surface of the cells or as slime capsules. The two recognized species of autotrophs, *Thiobacillus ferrooxidans* and *Ferrobacillus ferrooxidans,* have been characterized. Both bacteria are simple rod-shaped organisms that grow optimally between pH 2.0 to 4.5 (Alexander, 1961). *Thiobacillus ferrooxidans* also oxidizes S, while *F. ferrooxidans* does not. This process has drastic environmental effects where drainage water of bituminous coal mines mixes with aquatic life. Vast quantities of H_2SO_4 and dissolved Fe are produced microbiologically from the high content of Fe sulfides in the coal (Dugan, 1973).

The heterotrophic Fe-oxidizing bacteria deposit Fe from solution at concentrations as low as 1.0 ppm. All of them form stalks, sheaths, capsules, or some sessile structure that enables attachment to a surface, which is a highly ecological selection factor in nutrient-poor environments. Though some earlier studies have suggested that *Gallinonella* (stalk-forming bacteria) and *Leptothrix* (sheathed bacteria) may be facultative autotrophs, it has been difficult to substantiate whether the oxidation of Fe that occurs at neutral to alkaline pH, which is their optimal growth range, is chemical or biological. The Fe bacteria cause considerable problems in clogging of Fe pipes and tile drains where dissolved O_2 and organic matter are present to support their growth (Meek et al., 1973).

Reduction of Fe^{3+} to Fe^{2+} ion is coupled with the oxidation of organic matter brought about by common heterotrophic bacteria of the genera

Bacillus, Pseudomonas, and *Clostridium* in soil. Several genera of the enteric bacteria have also been shown to reduce Fe. Iron reduction occurs at an E_h of about 200 mV (Patrick, 1960) and is somewhat analogous to dissimilatory nitrate reduction. Poorly drained soils (gleys) usually exhibit a grayish coloration due to the accumulation of ferrous sulfide, which is believed to be due to the bacterial reduction of Fe and SO_4^{2-}. The bacterial reduction of Fe, but not SO_4^{2-}, in soils of the Imperial Valley of California facilitates leaching of the more soluble Fe^{2+} ion to tile drains. The tiles contain sufficient O_2, presumably due to periods of drying, which enables growth of Fe bacteria responsible for clogging the tiles (Meek et al., 1973).

Degradation of organic Fe compounds is a widespread biological occurrence which tends to precipitate the Fe, which was formerly bound as a chelant or as a soluble salt of an acid. The reverse procedure, the production of organic acids, is widespread among microorganisms and may render the Fe soluble by reducing the pH or by chelation. The hydroxamic acids are particularly good chelating agents.

2. MANGANESE

Manganese exists in soil primarily in two valence states as manganous (Mn^{2+}) and manganic (Mn^{4+}) ions. Above pH 8.0, Mn may be oxidized chemically to MnO_2 by the addition of citric and butyric acids (Zajic, 1969) though autoxidation normally takes place at pH 10. Below pH 5.5, chemical equilibria tend to favor the Mn^{2+} ion.

Though oxidation of Mn is an exothermic reaction, it has not been established whether or not autotrophic manganese oxidizing bacteria exist. Oxidation is affected by the simple unicellular bacteria, *Aerobacter, Flavobacterium, Bacillus, Corynebacterium,* and *Pseudomonas,* and by the iron bacteria *Leptothrix, Crenothrix, Clonothrix,* and *Cladothrix.* Several fungal genera, *Cladosporium, Curvularia, Cephalosporium,* and *Helminthosporium* also oxidize Mn. Oxidation of Mn brings about a dark brown to black precipitate in the form of MnO_2. Many of the filamentous and sheathed bacteria which cause problems in clogging of pipes through deposition of iron and biomass also bring about the same problem in tile drains containing abundant amounts of the soluble manganous sulfate and soluble organic matter.

Manganese reduction occurs under anaerobic conditions and is effected by many of the common aerobic soil bacteria. Since the redox potential at which NO_3 is reduced is very close to that for reduction of Mn^{4+} ion (300 mV), Grass et al. (1973) have shown that concentration of soluble Mn^{2+} can be used as a qualitative index for ascertaining the presence of NO_3^- in soil. They, thus, observed an inverse correlation of Mn^{2+} and NO_3^- and concluded that Mn^{4+} functioned as a terminal electron acceptor, and was preferentially reduced after all the O_2 and NO_3^- was depleted.

Manganic ion is solubilized to an even greater extent by the generation of H_2SO_4 from S oxidation by *Thiobacillus.* The bacterium presumably couples the reduction of Mn^{4+} by utilizing it as a terminal electron acceptor during the oxidation of S.

The microbial reduction of trace metals, with respect to their liberation from ores, is an aspect that has been exploited by mining industries for recovery of certain metals. A book written by Zajic (1969) covers most of this material.

D. Methylation of Metals

In the presence of arsenical compounds, the gas trimethylarsine is produced by fungi of the genera *Aspergillus, Mucor, Fusarium, Trichophyton, Candida, Gliocladium, Penicillium, Paecilomyces,* and *Scopulariopsis,* and represents the major pathway by which arsenicals are removed from soil. The reaction from arsenious acid (As[OH]$_3$) involves a successive sequence of methylations to form the respective methylarsenic acid, cacodylic acid, and trimethyl arsine. Each reaction, therefore, involves an apparent displacement of an OH with a CH$_3$ group. The specified mechanism was proposed by Challenger et al. (1954). Cacodylic acid, the dimethylarsenic acid, is frequently used as a broad scale, top-killing herbicide. Though most studies have focused on fungi, McBride and Wolfe (1971) showed that *Methanobacterium* could convert arsenite to dimethylarsine when incubated anaerobically.

The environmental hazards of mercury in marine sediments came to the forefront in 1956 with the outbreak of chronic mercury poisoning and deaths in Japan. The accelerated biomagnification of the element into the food chain was attributed to bacterial methylation of mercurous mercury

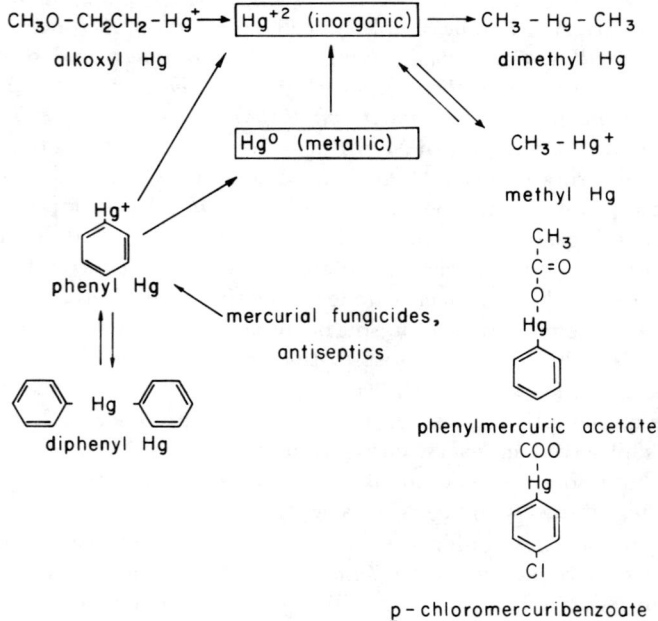

Fig. 24—Microbial transformations of mercurial compounds.

and/or dimethylation of mercuric ion (Jernelov, 1969). Mercury, aside from silver, is the most common toxic metal to bacteria. Methylation represents a detoxification step, which also makes the compound more volatile. The process is anaerobic and appears to be found among many diverse bacterial genera.

Mercurial compounds enter the environment from industrial waste discharge, antiseptics and disinfectants, pulp mills, and fungicides. Many of the phenylmercuric compounds have been used in controlling wheat rust. At relatively low concentrations in the soil, however, these compounds can be broken down (Fig. 24) and converted eventually to phenyl, methylated, or ionic forms. With well-aerated soils, it is probably unlikely that methylation of mercury would occur. Thus phenyl and diphenyl mercury would be incorporated into humic polymers, or the aromatic ring could be degraded and the toxic mercurous or mercuric ion that would be generated could be tied up in the soil by cation exchange, chelation, or precipitation.

Methylation of selenium has also been demonstrated (Challenger et al., 1954), but it does not appear to be a significant detoxification mechanism. Dimethyl selenide was found as the end product of the reduction of selenate and selenide by *Penicillium* (Fleming & Alexander, 1972). This same fungus also reduced tellurite and tellurate to dimethyl telluride.

V. PESTICIDES AND SOIL ORGANISMS

Organic pesticides may be applied directly to soils or may get into the soil through plant pest control or organic waste disposal operations. When used properly, most of these chemicals will exert little or no harmful effect on soil properties and are soon lost by microbial degradation or other mechanisms. Some of the pesticides, however, may have a profound temporary influence on soil properties which may be beneficial or harmful to plant growth. A few may be resistant to degradation and may accumulate if used in excess. Due to limited space this important area cannot be discussed. The interested reader may obtain information in this area from the following (Domsch, 1963; Guenzi, 1974; Kearney & Kaufman, 1969; Martin, 1971a).

VI. SOIL INOCULATION

Because the soil organisms exert many beneficial effects on soil properties which favor plant growth, many individuals have concluded that if one could inoculate the soil with the right organism or organisms or establish favorable species in the plant rhizosphere, crop yields could be greatly increased. There are specific conditions where soil or plant root or seed inoculation may prove beneficial. Inoculating legume seed with effective *Rhizobia* species especially if the legume is to be planted on a soil for the first time, may prove highly effective in supplying the plant with nitrogen. When lands not previously forested are planted with forest trees, inoculation of the seedlings with proper mycorrhizal fungi may be highly effective in improving stands and increasing tree growth. As pointed out in previous sections, inoculation of fumigated soils with antagonistic species or mycorrhizal organisms may prove successful in improving growth, nutrition, and health of the plants.

Most of the soil inoculants or preparations which are commonly sold on the markets are general soil inoculants or "enzyme activators." It is claimed that a very small amount is needed and that they contain all the bacteria or enzymes essential for maintenance of high soil productivity. The preparations are dripped or placed in the irrigation water or spread or sprinkled on the soil where they allegedly go to work and release plant nutrient elements from soil minerals and organic matter or tie up salts in saline soils. Actually the soil is already teaming with organisms which produce enzymes and which are present because they are best adapted to the environmental conditions which exist. Introduced species usually die off quickly and are used as food sources by the indigenous population. In addition, the species introduced in the "natural" organic products or inoculants are already present in most soils.

Many of the general soil inoculants and other similar products have been tested at Riverside, Calif. and at other Agricultural Experiment Stations (Golueke et al., 1954; Martin and Ervin, 1952; Rauschkolb et al., 1970; Weaver et al., 1974). Significant effects on numbers of soil organisms, on plant growth, and on soil physical properties have not been observed. Actually the numbers of organisms added in the inoculants are infinitesimal compared to the numbers present in normal soils. The addition of a few more would not be expected to exert measurable effects on soil properties or plant growth.

The only way to put the general soil organisms to work is to feed them. This means application of bulk organic residues and manures, or proper soil management practices which increase the amounts of organic residues returning to the soil.

VII. CONCLUSIONS

A good soil is a living soil. It is the home of vast numbers of microbes and other organisms. The important forms include bacteria, actinomycetes, fungi, algae, lichens, protozoa, nematodes, arthropods, and earthworms. Although these organisms live according to the rule of survival of the fittest, they carry out many important and beneficial functions. Most of their activities are associated with or occur during the decomposition of organic residues and soil humus. They release inorganic plant nutrient elements from the organic substances and synthesize soil humus. Their activity in these processes forms substances which improve soil structure or tilth and solubilize nutrient elements from inorganic soil minerals. They decompose toxic organic substances synthesized by plants or microbes or added to the soil as pesticides. Many exert antagonistic actions against plant root and other pathogenic organisms. Specific species form a symbiotic relationship with plants which results in fixation of N_2 for use by the plant or in improved mineral nutrition of the plant.

Some bacteria oxidize ammonia-N to nitrates, while in the absence of O_2 other forms reduce the nitrate to N_2 and other reduced forms of nitrogen. This function, when properly manipulated, may be useful in reducing contamination of ground waters with nitrates. Other inorganic elements are transformed in a similar manner by soil organisms.

To enhance the growth and activities of the soil organisms it is necessary to feed them. This is accomplished by applying organic residues including animal manures to the soil, growing cover crops to be disked or plowed under, crop rotation with plant species such as grasses which provide greater amounts of root residues to the soil, and by proper fertilization and soil management procedures. Any practice which increases plant growth will normally increase microbial activity which will be related to the greater quantities of root and top residues returned to the soil. Applications of trace amounts of materials in solution or general soil inoculants will have little or no value.

Soil is the proper and natural place for the disposal of both solid and liquid organic wastes. By serving as a carbon and energy source the added residues will increase the numbers and activities of most of the beneficial soil organisms including bacteria, fungi, soil insects, and earthworms. The increased activity will enhance microbial antagonisms which will reduce the activity of many plant root parasites and destroy most animal pathogens that may be present or may enter the soil with the organic waste materials. The organic matter additions will normally accelerate the decomposition of natural and man-made toxic organic chemicals which may get into the soil.

By supplying nitrogen, which is released as ammonia and oxidized to nitrate, the organic wastes will stimulate the activity of the soil nitrifying bacterial populations. Transformations of sulfur and other elements will be influenced in a similar manner and the solubilization of inorganic minerals will be increased. Sewage sludges and animal wastes represent good source material for increasing the quantities of the beneficial soil humus and providing organic binding material for soil aggregation. One must apply the organic wastes intelligently or excesses of certain plant nutrients and salts may reduce growth of some plants even though microbial activity may be enhanced.

There are a few microbial activities which could be adversely affected by large scale applications of organic wastes to soil. Increases in nitrogen from manures and sludges would tend to reduce nitrogen fixation both in cultivated and natural soils. Increases in available phosphorus and other essential plant nutrient elements would reduce mycorrhizal activity as the plants generally become resistant to infection if the available supplies of the nutrients are adequate.

Disposal of organic wastes on specific soils such as acid forest soils, as well as greatly affecting the widespread mycorrhizal relationships, could exert a profound influence on microbial and soil animal ecology. The added cations such as calcium and sodium could increase the soil pH which would make conditions more favorable for bacteria and actinomycetes. The increased competition may reduce the activity of the basidiomycete fungi in the forest soils and other soil insects may be better adapted to compete in the changed environment than the ones normally present. Reduced ectotrophic mycorrhizal activity may possibly affect the resistance of the plant to certain root parasites. The significance of these and other effects should be followed.

VIII. ADDENDUM

Since the completion of our manuscript, several new taxonomic concepts have been set forth. In the latest edition of *Bergey's Manual of Determinative Bacteriology* (Buchanan & Gibbins, 1974), the *Actinomycetes* group has been expanded to include bacteria which do not form mycelia; the genus *Arthrobacter* is now included in this group.

The distinct difference of the procaryotic blue-green algae from the eucaryotic algae is noted by their classification as blue-green bacteria (Cyanobacteria) under Division I of the Kingdom, *Procaryotae.* The portions of the text on N fixation by "algae" refer only to the procaryotic blue-green group, as N fixation by eucaryotic algae has not been established.

Gerdeman and Trappe (1974) have divided the endomycorrhizal fungus, *Endogone,* into four genera. The fungus in Fig. 17 is now *Glomus* sp.

LITERATURE CITED

Ahmadjian, V., and M. E. Hale, ed. 1973. The lichens. Academic Press, Inc., New York. p. 1–630.

Ainsworth, G. C., F. K. Sparrow, and A. S. Sussman, ed. 1965. The fungi, Vol. I. The fungal cell. Academic Press, New York. p. 1–748

Ainsworth, G. C., F. K. Sparrow, and A. S. Sussman, ed. 1973. The fungi. Vol. IV A, A taxonomic review with keys; ascomycetes and fungi imperfecti. p. 1–621 and Vol. IV B, Basidiomycetes and lower fungi. p. 1–504. Academic Press, Inc., New York.

Aleem, M. I. H., and H. Lees. 1963. Autotrophic enzyme systems. I. Electron transport systems concerned with hydroxylamine oxidation in *Nitrosomonas.* Can. J. Biochem. Physiol. 41:763–778.

Alexander, M. 1961. Introduction to soil microbiology. John Wiley and Sons, Inc., New York. p. 1–472.

Allen, E. K., and O. N. Allen. 1958. Biological aspects of symbiotic nitrogen fixation. p. 48–118. *In* W. Ruhland (ed.) Encyclopedia of plant physiology. Vol. 8. Springer-Verlag, Berlin.

Allen, E. K., and O. N. Allen. 1965. Non-leguminous plant symbiosis. p. 77–106. *In* C. M. Gilmour and O. N. Allen (ed.) Microbiology and soil fertility. Oregon State University Press, Corvallis.

Allison, F. E. 1973. Soil organic matter and its role in crop production. Elsevier Scientific Publishing Co., Amsterdam, London, and New York. p. 1–637.

Ayanaba, A., and M. Alexander. 1974. Transformation of methyl amines and formation of a hazardous product, dimethylnitrosamine in samples of treated sewage and lake water. J. Environ. Qual. 3:83–89.

Bailey, L. D., and E. G. Beauchamp. 1973. Effects of temperature on NO_3^- and NO_2^- reduction, nitrogenous gas production, and redox potential in a saturated soil. Can. J. Soil Sci. 53:213-218.

Barron, G. L. 1968. The genera of hyphomycetes from soil. Williams and Wilkins Co., Baltimore. p. 1-364.

Becker, G. F., and E. L. Schmidt. 1964. β-Nitropropionic acid and nitrite in relation to nitrate formation by *Asperigillus flavus.* Arch. Microbiol. 49:167-175.

Behera, B., and G. H. Wagner. 1974. Microbial growth rate in glucose-amended soil. Soil Sci. Soc. Am. Proc. 38:591-597.

Bernfield, P. 1963. Biogenesis of natural compounds. The MacMillan Co., New York. p. 1-930.

Bollag, J. M., S. Drzymala, and L. T. Kardos. 1973. Biological vs. chemical nitrite decomposition in soil. Soil Sci. 116:44-50.

Bondietti, E., J. P. Martin, and K. Haider. 1972. Influence of nitrogen source and clay on growth and phenolic polymer production by *Stachybotrys* spp., *Hendersonula toruloidea* and *Aspergillus sydowi.* Soil Sci. Soc. Am. Proc. 35:917-922.

Bowman, R. A., and D. D. Focht. 1974. The influence of glucose and nitrate concentrations upon denitrification rates in sandy soils. Soil Biol. Biochem. 6:297-301.

Bremner, J. M., and K. Shaw. 1958. Denitrification in soil. II. Factors affecting denitrification. J. Agric. Sci. 51:40-52.

Broadbent, F. E., and F. Clark. 1965. Denitrification. *In* W. V. Bartholomew and F. E. Clark (ed.) Soil nitrogen. Agronomy 10:344-359. Am. Soc. of Agron., Madison, Wis.

Brock, T. D., K. M. Brock, R. T. Belley, and R. L. Weiss. 1972. *Sulfolobus*: A new genus of sulfur-oxidizing bacteria living at low pH and high temperature. Arch. Microbiol. 84:54-68.

Buchanan, R. E., and N. E. Gibbins, ed. 1974. Bergey's manual of determinative bacteriology. Williams and Williams Co., Baltimore, Md. 1,246 p.

Burges, N. A., H. M. Hurst, and S. B. Walkden. 1964. The phenolic constituents of humic acid and their relation to lignin of the plant cover. Geochim. Cosmochim. Acta 28:1547-1564.

Cadman, C. H. 1965. Pathogenesis of soil-borne viruses. p. 302-313. *In* K. F. Baker and W. C. Snyder (ed.) Ecology of soil-borne plant pathogens. University of California Press, Berkeley.

Cameron, R. E., and W. H. Fuller. 1960. Nitrogen fixation by some algae in Arizona soils. Soil Sci. Soc. Am. Proc. 24:353-356.

Challenger, F., D. B. Lisle, and P. B. Dransfield. 1954. Studies on biological methylation. Part XIV. The formation of trimethylarsine and dimethyl selenide in mould cultures from methyl sources containing ^{14}C. J. Chem. Soc. 1954:1760-1771.

Cheshire, M. V., C. M. Mundie, and H. Shepherd. 1973. The origin of soil polysaccharides: Transformation of sugars during the decomposition in soil of plant material labeled with ^{14}C. J. Soil Sci. 24:54-68.

Clark, F. E. 1967. Bacteria. p. 1-49. *In* A. Burges and F. Raw (ed.) Soil biology. Academic Press, Inc., New York.

Clark, F. E., and E. A. Paul. 1970. The microflora of grassland. Adv. Agron. 22:375-435.

Collins, F. M. 1955. Effect of aeration on the formation of nitrate-reducing enzymes by *P. aeruginosa.* Nature, London 175:173-174.

Dagley, S. 1967. The microbial metabolism of phenolics. p. 290-317. *In* A. D. McLaren and G. H. Peterson (ed.) Soil biochemistry. Marcel Dekker, Inc., New York.

Dalton, H., and L. E. Mortenson. 1972. Dinitrogen (N_2) fixation (with a biochemical emphasis). Bacteriol. Rev. 36:231-260.

Daniel, R. M., and C. A. Appleby. 1972. Anaerobic-nitrate, symbiotic and aerobic growth of *Rhizobium japonicum.* Effects of cytochrome p 450, other hemoproteins, nitrate, and nitrite reductases. Biochim. Biophys. Acta 275:347-354.

Darland, G., T. D. Brock, W. Samsonoff, and S. F. Conti. 1970. A thermophilic acidophilic mycoplasma isolated from a coal refuse pile. Science 170:1416-1418.

Davies, R. F. 1971. Relation of polyphenols to decomposition of organic matter and to pedogenic processes. Soil Sci. 111:80-85.

De, P. K., and L. M. Mandal. 1956. Fixation of nitrogen by algae in rice soils. Soil Sci. 81:453-458.

Dobereiner, J., J. M. Day, and P. J. Dart. 1972. Nitrogenase activity and oxygen sensitivity of the *Paspalum notatum—Azotobacter paspali* association. J. Gen. Microbiol. 71:103–116.

Doelle, H. W. 1969. Bacterial metabolism. Academic Press, New York. 486 p.

Domsch, K. 1963. Einflüsse von Pflanzenschutzmitteln auf die Bodenmikroflora. Mitteilungen aus der Biologischen Bundesanstalt für Land-und Forstwirtschaft, Berlin-Dahlem. Heft 107. p. 1–52.

Domsch, K. H., and W. Gams. 1972. Fungi in agricultural soils. John Wiley and Sons, Inc., New York. 289 p.

Dugan, P. R. 1973. Biochemical ecology of water pollution. Plenum Publishing Corp., New York.

Felbeck, G. T. 1965. Structural chemistry of soil humic acids. Adv. Agron. 17:317–368.

Filip, Z. 1973. Clay minerals as factors influencing biochemical activity of soil microorganisms. Folia Microbiologica 18:56–74.

Flaig, W. 1966. Humusstoffe. p. 328–358. *In* H. Linser (ed.) Hanbuch der Pflanzenernahrung und Düngung. II. Boden und Dungemittel. Springer-Verlag, Wien and New York. p. 328–358.

Fleming, R. W., and M. Alexander. 1972. Dimethylselenide and dimethyltelluride formation by a strain of *Penicillium*. Appl. Microbiol. 24:424–429.

Focht, D. D. 1974. The effect of temperature, pH, and aeration on the production of nitrous oxide and gaseous nitrogen: A zero-order kinetic model. Soil Sci. 118:173–179.

Focht, D. D., N. R. Fetter, W. Lonkerd, and L. H. Stolzy. 1975. Effects of moisture and manure upon gaseous concentrations of nitrous oxide in soils. Proc. 2nd Annu. NSF-RANN Trace Contam. Conf., Asilomar, Calif., 1974. p. 232–235.

Fogg, G. E., W. D. P. Stewart, P. Fay, and A. E. Walsby. 1973. The blue-green algae. Academic Press, New York. p. 1–459.

Forsyth, W. G. C. 1950. Studies on the more soluble complexes of soil organic matter. Biochem. J. 46:141–146.

Fuller, W. H., and R. N. Rogers. 1952. Utilization of phosphorus of algal cells as measured by the Neubauer technique. Soil Sci. 74:417–430.

Gascho, G. F., and F. J. Stevenson. 1968. An improved method for extracting organic matter from soil. Soil Sci. Soc. Am. Proc. 32:117–118.

Gerdeman, J. W. 1968. Vesicular-arbuscular mycorrhiza and plant growth. Annu. Rev. Phytopathol. 6:397–418.

Gerdeman, J. W., and J. M. Trappe. 1974. The *Endogenaceae* in the Pacific Northwest. Mycologia Memoir No. 5. New York Botanical Gardens, Mycol. Soc. of Am., New York. 76 p.

Golueke, C. G., J. C. Bradley, and P. H. McGaukey. 1954. A critical evaluation of inoculums in composting. Appl. Microbiol. 2:45–53.

Grass, L. B., A. J. MacKenzie, B. D. Meek, and W. F. Spencer. 1973. Manganese and iron solubility changes as a factor in tile drain clogging: I. Observations during flooding and drying; and II. Observations during the growth of cotton. Soil Sci. Soc. Am. Proc. 37:14–17 and 17–21.

Greenland, D. J., G. R. Lindstrom, and J. P. Quick. 1962. Organic materials which stabilize natural soil aggregates. Soil Sci. Soc. Am. Proc. 26:366–371.

Greenwood, D. J. 1961. The effect of oxygen concentration on the decomposition of organic materials in soil. Plant Soil 14:360–375.

Guenzi, W. D., ed. 1974. Pesticides in soil and water. Soil Sci. Soc. Am., Inc., Madison, Wis. 562 p.

Hagendorn, C., and J. G. Holt. 1975. Ecology of soil arthrobacters in Clarion-Webster toposequences of Iowa. Appl. Microbiol. 29:211–218.

Haider, K., Z. Filip, and J. P. Martin. 1970. Einfluss von Montmorillonit auf die Bildung von Biomasse und Stoffwechselzwischenprodukten durch einige Mikroorganismen. Arch. Microbiol. 73:201–215.

Haider, K., and J. P. Martin. 1967. Synthesis and transformation of phenolic compounds by *Epicoccum nigrum* in relation to humic acid formation. Soil Sci. Soc. Am. Proc. 31:776–772.

Haider, K., and J. P. Martin. 1975. Decomposition of specifically ^{14}C-labeled benzoic and cinnamic acid derivatives in soil. Soil Sci. Soc. Am. Proc. 39:657–662.

Haider, K., J. P. Martin, and Z. Filip. 1974. Humus biochemistry. p. 195–244. *In* E. A. Paul and A. D. McLaren (ed.) Soil biochemistry. Vol. 4. Marcel Dekker, Inc., New York.

Hardy, R. W. F., and U. D. Havelka. 1975. Nitrogen-fixation research: A key to world food. Science 188:633–643.

Harris, R. F., G. Chesters, and O. N. Allen. 1966. Dynamics of soil aggregation. Adv. Agron. 18:107–169.

Henderson, M. E. K., and R. B. Duff. 1963. The release of metalic and silicate ions from minerals, rocks and soils by fungal activity. J. Soil Sci. 14:236–246.

Hofman, T., and H. Lees. 1953. The biochemistry of the nitrifying organisms. 4. The respiration and intermediary metabolism of *Nitrosomonas*. Biochem. J. 54:579–583.

Hopp, H., and C. S. Slater. 1949. The effect of earthworms on the productivity of agricultural soil. J. Agric. Res. 78:325–339.

Hurst, H. M., and G. H. Wagner. 1969. Decomposition of ^{14}C-labeled cell wall and cytoplasmic fractions from hyalin and melanic fungi. Soil Sci. Soc. Am. Proc. 33:707–711.

Jenkinson, D. S. 1971. Studies on decomposition of ^{14}C-labeled organic matter in soil. Soil Sci. 111:64–70.

Jensen, H. L. 1965. Nonsymbiotic nitrogen fixation. *In* W. V. Bartholomew and F. E. Clark (ed.) Soil nitrogen. Agronomy 10:436–480. Am. Soc. of Agron., Madison, Wis.

Jernelov, A. 1969. Conversion of mercury compounds. p. 68–74. *In* M. W. Miller and G. Berg (ed.) Chemical fall-out. Charles C. Thomas, Springfield, Ill.

Johnson, D. D., and W. D. Guenzi. 1963. Influence of salts on ammonium oxidation and carbon dioxide evolution from soil. Soil Sci. Soc. Am. Proc. 27:663–666.

Johnston, J. R., G. M. Browning, and M. B. Russell. 1942. The effect of cropping practices on aggregation, organic matter content, and loss of soil and water in Marshall silt loam. Soil Sci. Soc. Am. Proc. 7:105–110.

Kearney, P. C., and C. S. Kaufman. 1969. Degradation of herbicides. Marcel Dekker, Inc., New York. p. 1–394.

Keeney, D. R., R. L. Chen, and D. A. Graetz. 1971. Importance of denitrification and nitrate reduction in sediments to the nitrogen budgets of lakes. Nature, London 233:66–67.

Kevan, D. K. M., ed. 1955. Soil zoology. Butterworths, London. p. 1–512.

Klein, D. A., and L. E. Casida, Jr. 1967. Occurrence and enumeration of *Bdellovibrio bacteriovorus* in soil capable of parasitizing *Escherichia coli* and indigenous soil bacteria. Can. J. Microbiol. 13:1235–1241.

Kleinschmidt, G. D., and J. W. Gerdeman. 1971. Stunting of citrus seedlings in fumigated nursery soils related to the absence of endomycorrhizae. Phytopathology 62:1447–1453.

Kononova, M. M. 1961. Soil organic matter. Pergammon Press, Oxford. p. 1–450.

Kononova, M. M. 1972. Current problems in the study of soil organic matter. Pochvovedenie 7:27–36.

Kreutzer, W. A. 1965. The reinfestation of treated soil. p. 495–508. *In* K. F. Baker and W. C. Snyder (ed.) Ecology of soil-borne pathogens. University of California Press, Berkeley.

Küster, E. 1967. The actinomycetes. p. 111–127. *In* A. Burges and F. Raw (ed.) Soil biology. Academic Press, New York.

Lang, W. 1971. Enhancement of algal growth in *Cyanophyta*-bacteria systems by carbonaceous compounds. Can. J. Microbiol. 17:303–314.

Latimer, W. M. 1952. The oxidation states of the elements and their potentials in aqueous solutions. 2nd ed. Prentice-Hall, Inc., Englewood Cliffs, N. J. 312 p.

Lewin, J. C. 1962. Physiology and biochemistry of algae. Academic Press, New York. p. 1–929.

Lynch, D. L., L. M. Wright, and L. J. Cotnoir. 1956. The adsorption of carbohydrates and related compounds on clay minerals. Soil Sci. Soc. Am. Proc. 20:6–9.

McBride, B. C., and R. S. Wolfe. 1971. Biosynthesis of dimethylarsine by methanobacterium dimethylarsine. Biochemistry 10:4312–4317.

McCalla, T. M. 1946. Influence of some microbial groups on stabilizing soil structure against falling water drops. Soil Sci. Soc. Am. Proc. 11:260–263.

MacGregor, A. N., and D. E. Johnson. 1971. Capacity of desert algal crusts to fix nitrogen. Soil Sci. Soc. Am. Proc. 35:843–844.

MacLeod, N. H., E. H. Chappelle, and A. M. Crawford. 1969. ATP assay of terrestrial soils: A test of an exobiological experiment. Nature 223:267–268.

Mann, L. D., D. D. Focht, H. A. Joseph, and L. H. Stolzy. 1972. Increased denitrification in soils by additions of sulfur as an energy source. J. Environ. Qual. 1:329–332.

Martin, J. P. 1946. Microorganisms and soil aggregation: II. Influence of bacterial polysaccharides on soil structure. Soil Sci. 61:157–166.

Martin, J. P. 1950. Use of acid, rose bengal and streptomycin in the plate method for estimating soil fungi. Soil Sci. 69:215–232.

Martin, J. P. 1971a. Side effects of pesticides on soil properties and plant growth. p. 733–792. In C. A. I. Goring and J. W. Hamaker (ed.) Organic chemicals in the soil environment. Vol. 2. Marcel Dekker, Inc., New York.

Martin, J. P. 1971b. Decomposition and binding action of polysaccharides in soil. Soil Biol. Biochem. 3:33–41.

Martin, J. P., and B. A. Craags. 1946. Influence of temperature and moisture on the soil aggregating effect of organic residues. J. Am. Soc. Agron. 38:332–339.

Martin, J. P., and J. O. Ervin. 1952. Soil organisms: Fact and fiction. West. Grower Shipper 23:30–39.

Martin, J. P., and J. O. Ervin. 1953. Nitrogen losses during oxidation of sulfur in soils. Calif. Citrogr. 39:54–56.

Martin, J. P., and J. O. Ervin. 1958. Changes in the fungus population of old citrus orchard soils when cropped to orange seedlings in the greenhouse. Soil Sci. 86:152–155.

Martin, J. P., W. J. Farmer, and J. O. Ervin. 1973. Influence of steam treatment and fumigation of soil on growth and elemental composition of avocado seedlings. Soil Sci. Soc. Am. Proc. 37:56–60.

Martin, J. P., and K. Haider. 1971. Microbial activity in relation to soil humus formation. Soil Sci. 111:54–63.

Martin, J. P., K. Haider, W. J. Farmer, and E. Fustec-Mathon. 1974. Decomposition and distribution of residual activity of some ^{14}C-microbial polysaccharides and cells, glucose and wheat straw in soil. Soil Biol. Biochem. 6:221–230.

Martin, J. P., K. Haider, and C. Saiz-Jimenez. 1974. Sodium amalgam reductive degradation of fungal and model phenolic polymers, soil humic acids and simple phenolic compounds. Soil Sci. Soc. Am. Proc. 38:760–764.

Martin, J. P., W. P. Martin, J. B. Page, W. A. Raney, and J. D. DeMent. 1955. Soil aggregation. Adv. Agron. 7:1–38.

Martin, T. L., D. A. Anderson, and R. Goates. 1942. Influence of chemical composition of organic matter on development of mold flora in soil. Soil Sci. 54:297–302.

Marx, D. H. 1969. The influence of ectotrophic mycorrhizal fungi on the resistance of pine roots to pathogenic infections. II. Production, identification and biological activity of antibiotics produced by Leuropaxillus cerealis var. piceina. Phytopathology 59:411–417.

Marx, D. H. 1972. Mycorrhizae. p. 13–16. In Agrichemical age. January 1972.

Meek, B. D., L. B. Grass, and A. J. MacKenzie. 1969. Applied nitrogen losses in relation to oxygen status of soils. Soil Sci. Soc. Am. Proc. 33:575–578.

Meek, B. D., A. L. Page, and J. P. Martin. 1973. The oxidation of divalent manganese under conditions present in the tile lines as related to temperature, solid surfaces, microorganisms, and solution chemical composition. Soil Sci. Soc. Am. Proc. 37:542–548.

Meiklejohn, J. 1954. Some aspects of the physiology of the nitrifying bacteria. p. 68–83. In B. A. Fry and J. L. Peel (ed.) Autotrophic micro-organisms. Cambridge Univ. Press, London.

Meyerhof, O. 1916. The respiration of the nitrate forming bacteria. Arch. Ges. Physiol. 164:353–427.

Mulbarger, M. C. 1971. Nitrification and denitrification in activated sludge systems. Water Pollut. Conf. Fed. J. 43:2059–2070.

Nielsen, C. O. 1967. Nematodes. p. 197–211. In A. Burges and F. Raw (ed.) Soil biology. Academic Press, New York.

Nommik, H. 1956. Investigations on denitrification in soil. Acta Agric. Scand. 6:195–228.

Nutman, P. S. 1965. Symbiotic nitrogen fixation. *In* W. V. Bartholomew and F. E. Clark (ed.) Soil nitrogen. Agronomy 10:360–383. Am. Soc. of Agron., Madison, Wis.

Patrick, W. H. 1960. Nitrate reduction rates in submerged soil as affected by redox potential. Int. Congr. Soil Sci., Trans. 7th (Madison, Wis.) p. 494–500.

Patrick, Z. A., and T. A. Toussoun. 1965. Plant residues and organic amendments in relation to biological control. p. 440–459. *In* K. F. Baker and W. C. Snyder (ed.) Ecology of soil-borne plant pathogens. University of California Press, Berkeley.

Payne, W. J. 1973. Reduction of nitrogenous oxides by microorganisms. Bacteriol. Rev. 37:409–452.

Pilot, L., and W. H. Patrick, Jr. 1972. Nitrate reduction in soils: Effect of soil moisture tension. Soil Sci. 114:312–316.

Piper, T. J., and A. M. Posner. 1972. Sodium amalgam reduction of humic acid. II. Application of the method. Soil Biol. Biochem. 4:525–531.

Pratt, P., F. E. Broadbent, and J. P. Martin. 1973. Using organic wastes as nitrogen fertilizers. Calif. Agric. 27:10–13.

Rauschkolb, R., A. D. Halderman, and L. True. 1970. Evaluation of biochemical soil additive vineyard study. Univ. of Ariz. Agric. Exp. Sta. Rep. 259. p. 1–10.

Raw, F. 1968. Arthropods. p. 323–362. *In* A. Burges and F. Raw (ed.) Soil biology. Academic Press, London and New York.

Rees, M., and S. Nason. 1965. Incorporation of atmospheric oxygen into nitrite formed during ammonia oxidation by *Nitrosomonas europaea*. Biochim. Biophys. Acta 113:398–401.

Reisinger, O., and G. Kilbertus. 1974. Biodegradation et humification. IV. Microorganismen intevenant dans la decomposition des cellules d *Aureobasidium pullulans* (De Bary) Arnaud. Can. J. Microbiol. 20:299–306.

Rose, A. H., and J. S. Harrison (ed.). 1969. The yeasts. Vol. 1, Biology of yeasts. Academic Press, New York. p. 1–508.

Rovira, A. D., and B. M. McDougal. 1967. Microbiological and biochemical aspects of the rhizosphere. p. 417–463. *In* A. D. McLaren and G. H. Peterson (ed.) Soil biochemistry. Marcel Dekker, Inc., New York.

Russell, E. W. 1961. Soil conditions and plant growth. Longmans, New York. 668 p.

Rychert, R. C., and J. Skujins. 1974. Nitrogen fixation by blue-green algae-lichen crusts in the Great Basin Desert. Soil Sci. Soc. Am. Proc. 38:768–771.

Sacks, L. E., and H. A. Barker. 1949. The influence of oxygen on nitrate and nitrite reduction. J. Bacteriol. 58:11–22.

Sacks, L. E., and H. A. Barker. 1952. Substrate oxidation and nitrous oxide utilization in denitrification. J. Bacteriol. 64:247–252.

Safferman, R. S., and M. Morris. 1964. Growth characteristics of the blue-green algal virus LPP-1. J. Bacteriol. 88:771–775.

Sapshead, L. M., and J. M. T. Winpenny. 1972. The influence of oxygen and nitrate on the formation of the cytochrome pigments of the aerobic and anaerobic respiratory chain of *Micrococcus denitrificans*. Biochim. Biophys. Acta 267:388–397.

Satchell, J. F. 1968. Lumbricidae. p. 332. *In* A. Burges and F. Raw (ed.) Soil biology. Academic Press, London and New York.

Schaller, F. 1968. Soil animals. University of Michigan Press, Ann Arbor. p. 1–144.

Schisler, L. C., J. W. Sinden, and E. M. Sigel. 1967. Etiology, symtomology and epidemiology of a virus disease of cultivated mushrooms. Phytopathology 57:519–526.

Schmidt, E. L. 1974. Quantitative autecological study of microorganisms in soil by immunofluorescence. Soil Sci. 118:141–149.

Shields, J. A., E. A. Paul, and W. E. Lowe. 1973. Turnover of microbial tissue in soil under field conditions. Soil Biol. Biochem. 5:743–764.

Shields, L. M., and L. W. Durrell. 1964. Algae in relation to soil fertility. Bot. Rev. 30: 92–128.

Shubert, W. J. 1965. Lignin biochemistry. Academic Press, New York. p. 1–131.

Skinner, F. A., and N. Walker. 1952. Growth of *Nitrosomonas europaea* in batch and continuous culture. Arch. Microbiol. 38:339–349.

Skujins, J. J. 1967. Enzymes in soil. p. 371–417. *In* A. D. McLaren and G. H. Peterson (ed.) Soil biochemistry. Marcel Dekker, Inc., New York.

Smith, P. F. 1971. The biology of mycoplasmas. Academic Press, New York and London. p. 1–257.

Sørensen, L. H. 1974. Rate of decomposition of organic matter in soil as influenced by repeated air drying, rewetting and repeated additions of organic material. Soil Biol. Biochem. 6:287-292.

Stanford, G., R. A. Vander Pol, and S. Dzienzia. 1975. Potential denitrification rates in relation to total and extractable soil carbon. Soil Sci. Soc. Am. Proc. 39:284-289.

Stanier, R. Y., M. Doudoroff, and E. A. Adelberg. 1970. The microbial world. Prentice Hall, Inc., Englewood Cliffs, N. J. p. 1-873.

Starr, J. L., F. E. Broadbent, and D. R. Nielsen. 1974. Nitrogen transformations during continuous leaching. Soil Sci. Soc. Am. Proc. 38:283-289.

Stevenson, F. J., and J. H. A. Butler. 1969. Chemistry of humic acids and related pigments. p. 534-557. In G. Eglinton and M. T. J. Murphy (ed.) Organic geochemistry. Springer-Verlag, New York.

Stewart, B. A., L. K. Porter, and F. G. Viets. 1966. Effect of sulfur content of straws on rates of decomposition and plant growth. Soil Sci. Soc. Am. Proc. 30:355-358.

Stotzky, G. 1965. Microbial respiration. In C. A. Black (ed.) Methods of soil analysis. Part 2. Agronomy 9:1550-1572. Am. Soc. of Agron., Madison, Wis.

Stotzky, G. 1966. Influence of clay minerals on microorganisms II. Effect of various clay species, homionic clays and other particles on bacteria. 12:831-848.

Stout, J. P., and O. W. Neal. 1967. Protozoa. p. 149-193. In A. Burges and F. Raw (ed.) Soil biology. Academic Press, London and New York.

Swinzer, G. D., J. M. Oades, and D. J. Greenland. 1968. Studies on soil polysaccharides. I. The isolation of polysaccharides from soil. Aust. J. Soil Res. 6:211-224.

Swinzer, G. D., J. M. Oades, and D. J. Greenland. 1969. The extraction, characterization and significance of soil polysaccharides. Adv. Agron. 21:195-235.

Thorne, G. 1961. Principles of nematology. McGraw-Hill, New York. p. 1-553.

Timonin, M. I. 1965. Interaction of higher plants and soil microorganisms. p. 135-158. In C. M. Gilmour and O. N. Allen (ed.) Microbiology and soil fertility. Oregon State University Press, Corvallis.

Tsao, P. H. 1964. Effect of certain fungal isolation agar media on *Thielaviopsis basicola* and on its recovery on soil dilution plates. Phytopathology 54:548-555.

Verma, L., J. P. Martin, and K. Haider. 1975. Decomposition of ^{14}C-labeled proteins, peptides and amino acids; free and complexed with humic polymers. Soil Sci. Soc. Am. Proc. 39:279-284.

Verstraete, W., and M. Alexander. 1972. Formation of hydroxylamine from ammonium by N-oxygenation. Biochim. Biophys. Acta 261-59-62.

Verstraete, W., and M. Alexander. 1973. Heterotrophic nitrification in samples of natural ecosystems. Environ. Sci. Technol. 7:39-42.

Voets, J. P., H. Vanstaer, and W. Verstraete. 1975. Removal of nitrogen from highly nitrogenous wastewaters. J. Water Pollut. Contr. Fed. 47:394-398.

Waksman, S. A. 1961. The actinomycetes. Vol. II. Classification, identification and descriptions of genera and species. The Williams and Wilkins Co., Baltimore. p. 1-363.

Waksman, S. A., and H. A. Lechevalier. 1962. The actinomycetes. Vol. III. Antibiotics of actinomycetes. The Williams and Wilkins Co., Baltimore. 430 p.

Warcup, J. H. 1967. Fungi in soil. p. 51-110. In A. Burges and F. Raw (ed.) Soil biology. Academic Press, New York.

Watanabe, A. 1959. On the mass culture of the nitrogen-fixing blue-green alga, *Tolypothrix tenuis*. J. Gen. Appl. Microbiol. Tokyo 5:85-91.

Weaver, R. W., E. P. Dunigan, J. F. Parr, and A. E. Hiltbolt, ed. 1974. Effect of two soil activators on crop yields and activities of soil microorganisms in the southern United States. South. Coop. Ser. Bull. 189:1-24.

Webley, D. M., M. E. K. Henderson, and I. F. Taylor. 1963. The microbiology of rocks and weathered stones. J. Soil Sci. 14:102-112.

Webley, D. M., and D. Jones. 1971. Biological transformations of microbial residues in soil. p. 448-484. In A. D. McLaren and J. Skujins (ed.) Soil biology and biochemistry. Vol. 2. Marcel Dekker, Inc., New York.

Weindling, R., and H. S. Fawcett. 1936. Experiments in the control of Rhizoctonia damping-off of citrus seedlings. Hilgardia 10:1-16.

Wesseling, J., and W. R. van Wijk. 1957. Land drainage in relation to soils and crops: I. Soil physical conditions in relation to drain depth. In L. N. Luthin (ed.) Drainage of agricultural lands. Agronomy 7:461-504. Am. Soc. Agron., Madison, Wis.

Wolk, C. P. 1973. Physiology and cytological chemistry of blue-green algae. Bact. Rev. 37:32-101.

Wong, P. T. W., and D. M. Griffin. 1974. Effect of osmotic potential on streptomycete growth, antibiotic production and antagonism to fungi. Soil Biol. Biochem. 6:319-325.

Wuhrman, K. 1964. Microbial aspects of water pollution control. Adv. Appl. Microbiol. 6:119-151.

Youngberg, C. T., and A. G. Wollum. 1968. Nonleguminous symbiotic nitrogen fixation. p. 383-395. In C. T. Youngberg and C. B. Davey (ed.) Tree growth and forest soils. Oregon State University Press, Corvallis.

Zajic, J. E. 1969. Microbial biogeochemistry. Academic Press, New York. 345 p.

chapter 7

Effluent from food processing plant is filtered; the solid waste is used as feed for livestock and the water is used to irrigate grazing pastures (Photo courtesy of USDA).

Recycling of Carbon and Nitrogen through Land Disposal of Various Wastes[1]

C. M. GILMOUR, University of Idaho, Moscow, Idaho

F. E. BROADBENT, University of California, Davis, California

S. M. BECK, University of Idaho, Moscow, Idaho

I. INTRODUCTION

Land areas used for the utilization and disposal of solid and liquid waste are unique ecosystems. Energy flow patterns measured as net gains (inputs) and losses (outputs) of carbon (C) and nitrogen (N) receive special emphasis (Olson, 1963). The incorporation of C and N into the soil biomass and concomitant formation of humus are key C and N conservation pathways. On the other hand, CO_2, CH_4, N_2 and oxides of N, and NO_3^- leaching losses are primary examples of loss or output routes. Kinetically, the retention and release of C and N must be considered from both short- and long-term viewpoints. Thus, the study of cyclical events is requisite for a better understanding of metabolic interactions involving C and N or indeed for the attainment of an acceptable C and N balance sheet.

II. CYCLING OF CARBON

A. Waste Carbon in Soil

Organic waste contains a broad spectrum of C species. Chemical analysis reveals varying states of degradation for each species, depending on the degree of stabilization of the waste. Other chapters will provide a more detailed coverage of the elemental composition of various waste materials. Suffice to say that in the overall cycling of C, the following substrates occur: (i) readily oxidizable, soluble organic C; (ii) proteins; (iii) hemicellulose; (iv) cellulose; (v) lipids; and (vi) lignin, given in the approximate order of biodegradability. Relationships among these species are shown schematically in Fig. 1.

Certain salient features emerge in terms of the sequential breakdown of C species. These are: (i) original waste C is largely insoluble and this insolubility is reaction-rate limiting; (ii) overall decomposition of manure C species is largely determined by the individual reaction rates (k values) of the different waste fractions, and in consequence may be defined in terms of a stepwise series of steady state reactions; (iii) during C mineralization and immobilization, a dynamic kinetic condition prevails—true equilibrium does not exist until waste stabilization occurs; and (iv) resynthesis of cellular constituents by the soil biomass shows that the breakdown sequence can be reversed, attesting to the dynamic nature of the C flux.

As with most "black boxes," the flow chart oversimplifies the biological energy flow between input C and reservoir-humus C, yet the concept provides a workable model to aid in understanding the waste degradation system.

---

[1] Section II entitled "Cycling of Carbon" was prepared by C. M. Gilmour and S. M. Beck; F. E. Broadbent is the author of section III entitled "Cycling of Nitrogen."

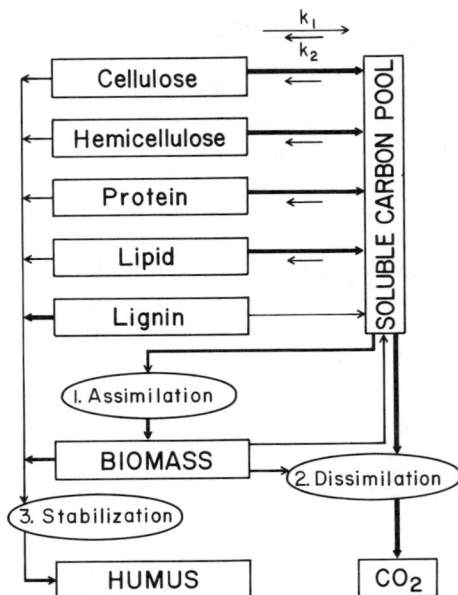

Fig. 1—Energy flow via mineralization of various C species.

B. The Biosphere

The soil environment for organic waste biodegradation has distinct eco-logical features. The solid, liquid, and gaseous components of the soil com-plex provide a continuum of microniches for the biochemical and physical transformation of organic matter. Norman (1961) states "the microenviron-ment of most soil organisms is a water film" and that organisms exist "in communities linked together by common nutritional capabilities." In such environments, bacteria, actinomycetes, fungi, algae, protozoa, and macro-animals play key roles in the initial cleavage and terminal oxidation of di-verse C substrates. In this regard, total soil biomass has been estimated to range from 3,000 to 3,500 kg/ha; it is the combined activity of this entire soil population which carries decomposition to completion.

Variation in environmental conditions may exert a specific population pressure, thereby shifting the dominance of a particular microbial group. Oxygen and moisture tensions, pH, temperature, and substrate specificity are important population determinants. At low O_2 tensions, intermediate C oxi-dation products tend to accumulate and substrate oxidation does not go to completion. For example, Elliott and McCalla (1972) noted that at very low redox potentials, high CH_4 and decreased O_2 levels were found in the soil profile below animal feedlots. Cropped soil profiles showed no evidence of CH_4. In addition, the optimum temperature and moisture tensions for microbial activity markedly influence the nature and rate of C turnover. The latter factors will be discussed shortly in respect to the kinetics of organic C biodegradation.

C. Mineralization and Immobilization

Mineralization is a general term for the oxidation of organic C to the inorganic state. From the biochemical point of view it involves a number of pathways of carbohydrate catabolism mediated by specific enzyme systems. In the case of waste organic C, the more complex C species undergo initial degradation and eventual solubilization via reactions mediated by extracellular enzymes—amylase, cellulase, lipase, protease, etc. The soluble C substrates may then enter the microbial cell and be oxidized via intracellular enzymes to CO_2 or temporarily immobilized as cellular material.

Jenkinson (1965) found that after 1 year of decomposition, 31% of ^{14}C-labelled ryegrass straw (*Lolium annuum* var. *westerwoldicum*) remained in the soil. After approximately 3 years, some stabilization had occurred, and about 20% of the input organic C had been immobilized or conserved as soil organic matter. The total loss thus approximated 80% for the 3-year period. Early studies of composting of animal manure reported losses of organic matter in excess of 50% during 3- to 4-month decomposition periods (Salter & Schollenberger, 1939). Recently, McCalla et al. (1969) reported that about 50% of the volatile solids of beef cattle manure supplemented with urine decomposed over a 4-month period. Assuming first-order kinetics, Broadbent (1973) calculated that the activated sludge process could achieve a 10%/day reduction in C. On the other hand, ARS-USDA investigators (1972) showed that soil amended with raw sewage lost only 27% of its C over a 54-day period. Miller (1973a, b) called attention to the slow biodegradation of anaerobically digested sewage sludges in soil. Approximately 17–20% of the waste C was recovered as CO_2 during a 6-month period. It appears that digested wastes added to soils may be largely immobilized with a concomitant buildup of the soil organic matter.

Soil also may be used for the disposal of various domestic and industrial waste waters. Ordinarily, C inputs are expressed in terms of biochemical oxygen demand (BOD) or chemical oxygen demand (COD). Some caution has to be exercised, since the 5-day BOD may account for only about 60% of the total BOD. In addition, if toxic compounds are present in effluents, the BOD may not always provide a sufficiently accurate basis for the determination of C inputs. This was discussed by Gilmour et al. (1961) and an alternative procedure proposed. Land disposal of domestic and industrial waste effluents has the potential to provide the equivalent of aerobic tertiary sewage treatment and thereby provide at least a 90% reduction in the oxidizable C components of such wastes.

Studies at the University of Idaho have centered on land disposal of feedlot waste. It was recognized that laboratory and field methodologies do not always complement one another, particularly from the viewpoint of the variable stresses imposed by fluctuations in temperature and moisture. Consequently, a field respirometer was designed for use in conjunction with conventional laboratory cabinet respirometers. Of some significance was the finding that the field and laboratory procedures gave compatible results.

D. Pattern of Carbon Loss

The interval and cumulative plots in Fig. 2 are based on the decrease in C load by measurement of CO_2 evolved from feedlot manure in the respiration cabinet. Similar trends were obtained with the field respirometer. The initial rapid mineralization of organic C and the progressive decreases in C losses are evident, and were observed also by Sikora (1974). When these data are plotted as reaction velocities (Fig. 3), the sequential decrease in reaction velocity reflects the initial rapid utilization of existing soluble C substrates and the progressively slower turnover of the more complex C moieties. A relatively steady state is attained between day 25 and 42. The rate constants or k values were derived by use of the conventional decay equation:

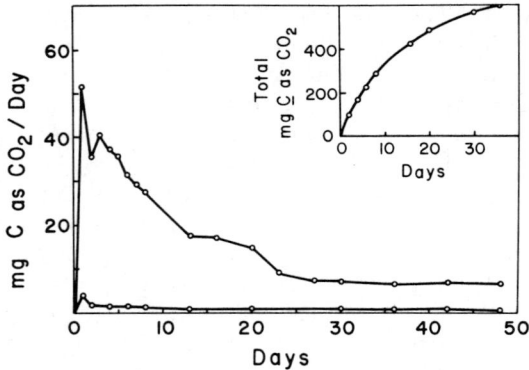

Fig. 2—Interval and cumulative CO_2 production observed during decomposition of feedlot manure.

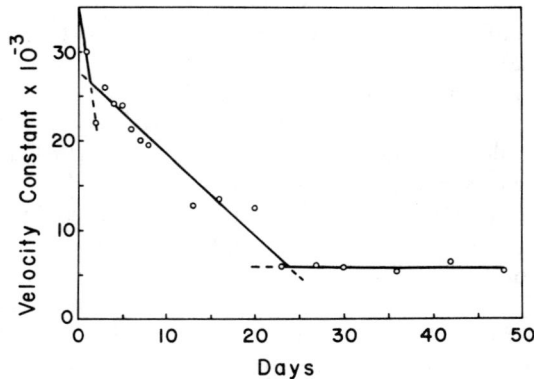

Fig. 3—Observed progressive decrease in the decomposition or decay rate of feedlot manure.

$$k = [2.303/(t_2 - t_1)] \log C_1/C_2 \qquad [1]$$

where k = reaction velocity constant, t_1 = initial time, t_2 = final time, C_1 = input carbon at t_1, and C_2 = residual carbon at t_2.

This equation implies that the rate of C mineralization can be represented by first-order kinetics. However, it is not reasonable to apply a single rate constant to the entire decomposition sequence.

Figure 4 indicates that over the 42-day decomposition period, three rather distinct C loss phases characterize the overall degradation of applied waste manure. A relatively steady state prevailed from day 20 to day 42. This yielded a k value which was used to calculate the half-life[2] of the remaining organic C according to:

$$t_{1/2} = (0.693/k). \qquad [2]$$

The half-lives at different loading rates are presented in Table 1. In agreement with other investigators, the loading rate did not alter the percentage mineralization and immobilization of the added C. Approximately 60% of the organic C remained after 42 days incubation at 27C and at 60% of the soil water holding capacity (WHC). With all treatments, an average of 112 additional days were required for a 50% decrease in the remaining C. Thus in 5 months, approximately 70% of the manure C was completely mineralized. The residual 30% was probably in the form of semistabilized humus-like material with a very slow turnover rate similar to that of native soil organic matter. The half-life of this resistant fraction was estimated to be approximately 4.3 years. Under low moisture and adverse temperature condi-

---

[2] It should be recognized that use of the term *half-life* with mixed substrates and implied first-order kinetics leaves something to be desired. In consequence, perhaps the term *average age*, or the time required for 50% disappearance of the input C is more meaningful. With these reservations in mind, the terms *half-life* and *average age* become synonymous.

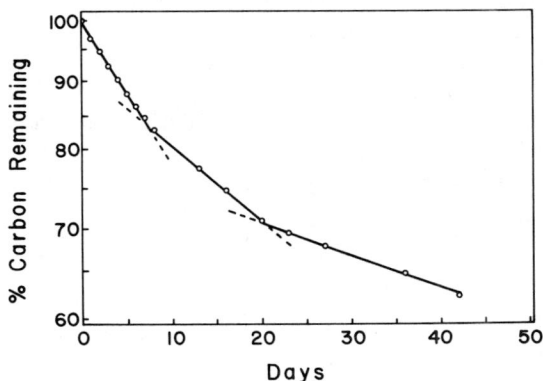

Fig. 4—Carbon loss phases observed during mineralization of feedlot manure.

Table 1—Observed mineralization and immobilization of organic C

Loading rate	Carbon		Predicted half-life of immobilized fraction‡
	Mineralized to CO_2†	Immobilized	
metric tons/ha	— % —		days
0	2.4	97.6	1,530
45	38.4	61.6	118
90	36.9	63.1	123
135	37.7	62.3	119
180	39.3	60.7	102
225	38.3	61.7	99

† After 42 days incubation.
‡ Calculated on rate of decomposition between 30 and 42 days at 27C and 60% WHC.

tions, a 10-year half-life would not be unrealistic. We should point out that these half-life estimates are not intended to indicate the specific age of soil C or "mean residence time" as defined by Campbell et al. (1967). Undoubtedly, the highly resistant refractile humus fractions will contain C dating back many years and no attempt will be made to relate such long term events to the turnover of waste C.

E. Temperature Effects

The various microbial species involved in waste decomposition may have optimum and maximum growth temperatures running from the mesophilic to the higher thermophilic range. Nonetheless, the Van't Hoff rule appears to hold for the rate of manure mineralization if due attention is given to conditions during decomposition. The key determinant is the reference temperature where maximum mineralization occurs. The reaction velocity at this temperature is decreased to 93.3% of its value for each 1C decrement: thus a 50% decrease occurs with each 10C decrement ($[0.933]^{10}$ = 0.50). In situations where halving the decomposition rate occurs within $n°C$ instead of 10C, the velocity constant (k) decreases not by a factor of 0.933 for each $°C$, but by a factor b such that $b^n = 0.5$. Also, k values at temperatures somewhat above the reference temperature may be approximated by increasing k by a factor of $1/b$ for each $°C$ increase. For any temperature expressed in heat units such as Celsius degree days, Eq. [2] may be modified to give the temperature effect on decomposition rate:

$$t_{1/2} = \frac{0.693}{k \times 0.933 \,^{-[(\text{Annual Heat Units}/365) - (T_2 - T_1)]}} \qquad [3]$$

where
k = reaction velocity constant at temperature T_2,
T_2 = experimental reference temperature, $°C$,
T_1 = 5C = temperature at which annual heat units = 0,

Annual Heat Units = (MAT − 5) × 365, and
MAT = mean annual temperature, °C.

$$\text{MAT} = \sum_{\text{Day 1}}^{\text{Day 365}} \left(\frac{\text{Maximum Daily Temperature, (°C)} + \text{Minimum Daily Temperature, (°C)}}{2} \right) \times \frac{1}{365} . \quad [4]$$

In the mesophilic range very few degrees may separate optimal microbial activity from the point where enzymic activity ceases. Thus, at elevated soil temperatures it becomes difficult to predict the net decrease or increase in the rate of decomposition using the Van't Hoff rule.

The laboratory manure decomposition studies in Idaho were carried out at a reference temperature of 27C (81F), near the optimum. No attempt was made to apply the Van't Hoff rule at temperatures below 5C (41F) because of the extremely low C turnover rates. The theoretical data to be presented relate only to temperature effects below the reference temperature. Reaction velocities were established in part from field respirometry. Heat units or Celsius degree days[3] are included to give a partial demarcation of climatic zonal effects.

The data in Table 2 call attention to the step-wise decrease in reaction velocity and related increase in decomposition time as a function of temperature. The 10C temperature reduction brought about a 50% decrease in reaction rate and doubling of the half-life for the mineralization of manure C. A similar pattern was observed for the second drop to 7C (45F). It should be stressed that the velocity trends given in Table 2 are based on calculated values. The intent is not to regard such trends as absolute but to provide a workable basis upon which predictive decomposition rates can be obtained.

[3]One day's Celsius heat units = that day's temperature (°C) −5, because a temperature of 5C (41F) is considered to contribute "0" heat units.

Table 2—Calculated effect of temperature on biodegradation of organic C

Mean annual temperature	Annual heat units based on °C	Specific reaction velocity ($k \times 10^{-3}$)	Carbon half-life ($t_{1/2}$)
°C			days
27†	8,030	5.78	120
26	7,665	5.39	128
25	7,300	5.03	138
22	6,205	4.09	170
20	5,475	3.56	195
17	4,380	2.89	240
15	3,650	2.51	276
12	2,555	2.04	339
10	1,825	1.78	390
7	730	1.44	480
5	0	--	--

† Reference temperature for determination of k = 27C (81F).

Table 3—Calculated effect of moisture on biodegradation of organic C

Soil moisture (M_2)	Reaction velocity $(k \times 10^{-3})$	Carbon half-life $(t\frac{1}{2})$
% of WHC		days
60	5.80	120†
50	4.83	143
40	3.87	179
30	2.90	239†
20	1.93	358
15	1.45	478

† Determined by field respirometry.

F. Moisture Effects

As with temperature, it is often difficult to gauge the exact influence of moisture on decomposition. Pal and Broadbent (1975) observed that biological activity decreased markedly at 30% of the soil WHC as compared to that at 60% WHC, resulting in long C turnover times. Field respirometry studies at 60% and 30% of the soil WHC indicated an approximately linear relationship between reaction velocity and moisture content of soil amended with feedlot manure with the reaction velocity decreasing by 50% and the half-life doubling. In consequence, we used the following equation to estimate the magnitude of the moisture effect:

$$t\frac{1}{2} = 0.693/[k \times (M_2/M_1)]$$ [5]

where k = reaction velocity constant at optimum temperature (27C), M_1 = soil moisture for optimum mineralization (% of WHC), and M_2 = actual soil moisture (% of WHC).

The effects of moisture are similar to those already observed for temperature (Table 3). As the moisture decreases below 30% to 15% of the soil WHC, the half-life for mineralization again doubles. It should be stressed that with the exception of the 60% and 30% WHC data, the remaining half-life and k values were calculated from Eq. [5].

G. Summary

The compartments shown in Fig. 1 represent the primary components of the energy flow associated with the mineralization and immobilization of waste organic C. Linkages occur via specific reaction rates characteristic of the individual waste fractions (Olson, 1963; Minderman, 1968). Most investigators agree that rates of mineralization can be expressed as an exponential function, Eq. [1]. If corrections are made for effects of temperature and moisture, Eq. [3] and [5], C loss values may be predicted. Thus, a mathematical model based on experimental reaction velocities can be derived Eq. [6]. Not all parameters influencing predicted C loss values are included. However, testing of this equation in the field will aid in our understanding of the degradation of organic C.

$$t_{1/2} = \frac{0.693}{k \times (M_2/M_1) \times 0.933 \ ^{-[(\text{Annual Heat Units}/365) - (T_2 - T_1)]}}. \qquad [6]$$

H. Model Application

It was considered appropriate to apply the carbon turnover model Eq. [3] to other published data, and thereby provide additional information as to the general applicability of the model. For this purpose, the sludge decomposition data of Miller (1973a) were used. Points were selected and read as accurately as possible from the published graphs.

It is necessary to have a "reference" determination on the graph to represent what, in actual practice, would be an accurate, carefully carried out, and well-monitored decomposition experiment (field or laboratory). The reference point was chosen at 11,000 "monthly degree days" (equivalent to a mean daily temperature of 16.2C), at which there was 19.5% "sludge C evolved as CO_2" (or 80.5% C remaining). A reference k value for this point was calculated (0.0012053) using Eq. [1], and a reference half-life (575 days) using Eq. [2].

Using these calculated "reference" values, any additional point may then be chosen (on any of the three curves) to test the model. Decomposition, expressed as "percent C remaining" at "TC", may be used in Eq. [3] to calculate a new k value, which then is used in Eq. [1] to calculate the new "percent C remaining", and in Eq. [2] to give the new "half-life".

Five sample points then were chosen to test the accuracy of the model in predicting amounts of sludge carbon remaining after decomposition under what might represent other "field" temperatures and times as compared to the laboratory or "reference" conditions. Results are presented in Table 4. The calculated values for "percent C remaining", are seen to be in excellent agreement with the values obtained from Miller's (1973a) graph, attesting to the accuracy of the model, and also attesting to the applicability of modified first order kinetics to the decomposition process.

III. CYCLING OF NITROGEN

Transformations of N in wastes applied to soil encompass the whole spectrum of the N cycle. Often much of the N in the waste is organic, and the rate of mineralization becomes the rate-limiting step for all the changes that follow. Since concern over pollutional aspects usually focuses on NO_3^-, the rate of production of NO_3^- from various kinds of wastes assumes con-

Table 4—Tests of decomposition model using published sludge decomposition data of Miller (1973a)

Selected test points†			Mean daily temperature (calculated from monthly degree days), °C	Calculated k value ($\times 10^{-5}$)	Calculated half-life, days	Percent C remaining	
Decomposition curve, days	Monthly degree days	Percent sludge C evolved as CO_2				Calculated from graph	Calculated from model
180‡	11,000‡	19.5‡	16.2‡	120.5‡	575‡	80.5‡	--
180	8,000	15.6	6.9	89.5	775	84.4	85.1
90	4,000	11.0	6.9	89.5	775	89.0	92.3
90§	6,500	19.8	22.3	260.2	266	80.2	79.1
30	1,000	3.3	0.7	58.2	1,191	96.7	98.3
30§	2,000	12.0	19.3	211.4	328	88.0	93.8

† From Miller (1973a).
‡ Reference determination.
§ Temperature is above reference temperature. Modify Eq. [3] to increase k value 1.0718 times for each °C increase.

siderable importance, as does denitrification. Frequently the quantities of wastes which can be applied to land are limited by the level of NO_3^- found in the subsoil after decomposition of the waste (Webber & Lane, 1960; Hinesly et al., 1971).

A. Agricultural Wastes

1. CROP RESIDUES

The effects of returning crop residues to soil on N transformations are reasonably well understood. The most dramatic effects are on the relative rates of N immobilization and N mineralization. In general, mature plant residues contain insufficient N to meet the needs of the large microbial population which arises in response to addition of decomposable organic material, and net immobilization of inorganic N occurs. This condition is temporary, and gives way to net mineralization after available C is depleted in a few days or weeks. A comprehensive review of this subject has been written by Bartholomew (1965).

2. ANIMAL MANURES

The application of animal manures to soil is as old as agriculture. In recent years this practice has received renewed attention because of the need to dispose of and utilize large quantities of manures concentrated in small areas. Another factor is an increased awareness of the potential for pollution from applications of manures on land. One of the major concerns is the higher NO_3^- levels usually encountered in soils receiving substantial amounts of animal manures, which may contribute to pollution of surface waters or ground water.

Mielke et al. (1974) described the soil profile under cattle feedlots as having three layers. At the surface is a layer of manure, overlying an interface layer of mixed organic and mineral material. Below this is the top of the mineral soil profile. The two upper layers are quite impermeable when moist and afford little opportunity for downward movement of NO_3^-. These authors reported higher NO_3^- concentrations in one feedlot profile than in comparable cropland down to 1.7 m, but little difference in another location in the same feedlot. Clearly, however, some mineralization of organic N and subsequent nitrification occurs in these feedlots. Adriano et al. (1973) found NO_3^- concentrations under dairy corrals to be higher than under adjacent pastures of cropland at 3-6 m depths.

a. **Nitrogen-Mineralization Rates**—Estimates of rates of mineralization of organic N in animal manures, although seldom attempted by direct experimentation, can be obtained from field trials comparing manures with various levels of inorganic fertilizers as sources of N. For example, such a comparison based on the date of Tyler et al. (1964) indicates that about 41%

Table 5—Nitrogen mineralization rates for cattle feedlot manures calculated from
data of Herron and Erhart (1965)

Application rate	Year			
	1	2	3	4
metric tons/ha	—— kg N mineralized/metric ton manure ——			
11.2	5.22	2.00	2.00	0.91
22.4	5.22	2.49	1.77	0.95
44.8	5.49	2.99	2.09	0.68
Average	5.31	2.49	1.95	0.85

of the N in dry steer manure was mineralized the first year and 25% the second year. Mineralization rates are highly variable, depending on the nature of the manure, conditions of storage prior to application, climatic conditions, and other variables such as the amount of bedding mixed with the manure. Nitrogen in poultry manure is more readily mineralized than that from large animals (MacMillan et al., 1972). Pratt et al. (1973) suggested the determination of N mineralization rates based on "decay series" for manures which assume progressively decreasing decomposition rates. Some decay series are suggested for various manures together with N mineralization curves for various levels of manure application. As these authors point out, well designed, long term field trials are needed to obtain data on N mineralization under a variety of soil and climatic conditions.

Herron and Erhart (1965) compared the N supplied to sorghum (*Sorghum bicolor* Moench.) by manure at three levels with that furnished by NH_4NO_3 in field trials. Table 5, based on their data, gives N-equivalent values over a 4-year period. Total N mineralized, expressed as the percent of total N in the manure, as a function of time is described by the equation:

$$N = 11.15 + 25.7t - 2.9t^2.$$

Thus, 34% of the total N in the manure was mineralized after 1 year, an additional 17% the second year, plus 11.2% in the third year, and 5.4% in the fourth year for a cumulative total of 67.6% in 4 years, reaching a maximum of 68.2% in 4.4 years. Although this function fits the data well, it has little value for extrapolation because N mineralization would obviously continue for some time at a low rate.

b. **Nitrification**—Although the oxidation of NH_4^+ to NO_3^- is inhibited by high salt concentrations (Johnson & Guenzi, 1963) and by high levels of NH_3 (Broadbent et al., 1957), there is little evidence in the literature to indicate that even very high rates of applications of manure retard nitrification. Usually, manure applications increase soil NO_3^- levels as a result of increased input of organic and/or NH_4^+-N (Weeks et al., 1972). Meek et al. (1974) applied manure at rates up to 360 metric tons/ha and observed increased NO_3^- concentrations in the soil solution down to the 140-cm depth in an irrigated clay soil. Similar results were obtained by Mathers and Stewart (1974) who found that manure applications ranging from 45 to 224 metric tons/ha increased NO_3^- concentrations in the soil profile. At levels of 112 and 224 metric tons/ha, NO_3^- increased down to a depth of 360 cm. Some buildup of NH_4^+-N in a soil receiving 70 metric tons/ha manure annually for 40 years was reported by Sommerfeldt et al. (1973) in Alberta. Apparently, the low mean annual temperature slowed the nitrification process, since NH_4^+-N in unmanured fields was about 24 ppm near the soil surface, compared with NO_3^--N of about 4 ppm. By comparison manured fields had about 36 ppm NH_4^+-N and 55 ppm NO_3^--N near the surface.

c. **Denitrification**—Application of manures to soil favors the activities of denitrifying bacteria since it increases the supply of available C, some of which is soluble, and increases the O_2 demand within the soil. Olsen et al. (1970) added manure to Plainfield sand at rates varying from 23 to 621 metric tons/ha. Under aerobic conditions nitrates accumulated to 25–180 ppm N but when the soil was maintained in a saturated condition virtually no NO_3^- was found at any application level. Average recovery of N after 37 weeks was 77% for aerobic conditions but only 24% in the flooded soil, indicating that considerable denitrification had occurred. Meek et al. (1974) observed that redox potentials in calcareous Holtville clay receiving 180 metric tons manure in each of 2 successive years usually did not fall below 400 mv with the normal irrigation schedule. However, by doubling the number of irrigations the redox potential dropped to zero and NO_3^- reduction occurred. These authors suggest that it is possible to adjust manure application rates and irrigation schedules for fine-textured soils to promote denitrification and thereby minimize leaching of NO_3^-. This requires maintenance of aerobic conditions at the surface and anaerobic conditions in the subsoil, which is dependent to a considerable extent on the downward movement of soluble C from the manure. In this connection it is interesting that Murphy et al. (1972) observed that concentrations of NO_3^- in the subsoil were higher in plots which had received only one heavy application of beef feedlot manure as compared to plots which had received continued heavy applications. Sukovaty et al. (1974) reported that after 2 years of application of beef feedlot effluent to a Colo silty clay loam soil cropped to sorghum, soil solution NO_3^- was higher beneath plots receiving 2.5 cm effluent/week than beneath plots given 5 cm/week.

B. Processing Wastes

Wastes from processing agricultural and forestry products include those from fruit and vegetable canneries, packing houses, dairies, sawmills, pulp and paper mills, etc. These materials often have a high O_2 demand owing to the presence of soluble and suspended organic matter. Gambrell and Peele (1973) reported that additions of NH_4NO_3 to peach (*Prunus persica*) cannery waste decreased COD reduction as the waste percolated through soil columns.

They attributed this decrease to greater microbial activity resulting from a more favorable C/N ratio. Law et al. (1970) reported 90% removal of N from cannery waste water with a N loading of 577 kg ha^{-1} year^{-1}. Denitrification was considered to be primarily responsible.

C. Urban and Industrial Wastes

1. MUNICIPAL COMPOSTS

The effects of composted municipal refuse on soil N transformations are related primarily to the initial N content of the material applied and usually have been evaluated in terms of crop yield. Bunting (1963) found that addition of sewage sludges increased the N content of municipal compost. Garner (1962) reported that town refuse increased yields of several crops, but the increases were smaller than those achieved with animal manures. Tietjen and Hart (1969) found that cured compost increased potato yields whereas fresh compost decreased them. This depression may be associated with the low available N of such composts. Terman et al. (1973) observed in greenhouse pot experiments that highly carbonaceous compost induced N deficiency in corn (*Zea mays* L.) grown in a soil already very deficient in N. Nitrogen deficiency was not seen in compost-treated soils with more available N. Mays et al. (1973) measured yield responses to compost applications in field trials with bermuda grass (*Cynodon dactylon* L.), sorghum, and corn, but equal or better yields of bermuda grass and sorghum were achieved with mineral fertilizers.

Bengtson and Cornette (1973) reported that the N concentration in young slash pine foliage (*Pinus elliottii* Engelm) decreased following incorporation of municipal waste compost at 44 metric tons/ha, but recovery to pretreatment levels was rapid, indicating a short period of net immobilization.

2. LIQUID WASTES

a. **Sewage Sludge**—Peterson et al. (1973) cite analyses of sewage sludges from various waste water treatment plants in the U. S. which varied in total N content from 3.5 to 6.4%, and from a trace to 3.6% NH_4^+-N. Net immobilization would not be expected to occur with incorporation of sludge having such low C/N ratios, estimated to be in the range below 15:1. However, N mineralization is usually far less rapid than in some other biological materials of comparable N content since the digestion process removes most of the so-called "volatile solids," or readily decomposed substances, leaving the more refractory material. King (1973) reported that NO_3^- accumulations equivalent to 22% of the N in surface-applied sludge and 38% of the N in incorporated sludge were produced in 18 weeks. Since substantial gaseous losses of N were noted, total N mineralization including losses was 59% from the incorporated sludge. Ryan et al. (1973) determined that from 4 to 48%

of the organic N in an anaerobic digested sludge was converted to NO_2^- and NO_3^- over a 16-week period. At high rates of sludge application some of the NH_3 was not nitrified during this incubation period. These investigators noted substantial losses of N from the system, which were attributed to denitrification. At 1,880 ppm sludge N only 62% of the added N could be accounted for.

Placement of sludge has a significant effect on the rate at which N is subsequently mineralized. King and Morris (1974) reported from 52 to 64% of the N in surface-applied sludge remained in the sludge crust. Total recoveries in their experiments, which includes volatilized NH_3, varied from 73 to 82%, indicating that a significant amount of denitrification also occurred.

Where sludge has been used for reclamation of acid strip mine spoil banks (Lejcher & Kunkle, 1973; Hunt et al., 1971; Dalton et al., 1968), nitrification may be initially slow due to the low pH and probably a lack of nitrifying bacteria until the environment is made more hospitable by sludge application.

The recent upsurge in research involving land application of sludge should provide considerable new information on soil N transformations. Although it has been suggested that the heavy metals in some sewage sludges may inhibit microorganisms involved in the N-cycle, experimental evidence in support of this suggestion is very meager.

b. **Septic Tank Effluent**—Septic tank seepage beds receive effluent which has been maintained in a predominantly anaerobic condition, and on that account, most of the N is present as NH_4^+ rather than NO_3^-. Walker et al. (1973) observed a "crust zone" at the boundary between the gravel bed and adjacent soil. They found nitrification occurred within about 2 cm of the crust and was essentially complete within about 15 cm of the crust. These authors consider that denitrification is the only feasible means of reducing the NO_3^- content of the percolating effluent, but regarded a significant level of denitrification as unlikely where seepage beds are located in deep sandy soils with little organic C.

c. **Municipal Waste Water**—Effluents from sewage treatment plants have most of the readily decomposable organic matter removed. Of the remaining organic matter, a portion is particulate, sometimes colloidal, and consists of a rather complex mixture including amino acids, amino sugars, and proteins (Hunter & Kotalik, 1973). Usually more than half the total N is soluble, occurring primarily as NH_4^+, but with variable amounts of NO_3^-, depending on the nature of the waste water treatment. Effluent from anaerobic treatment processes will contain little or no NO_3^-, with NH_4^+-N usually < 35 mg/liter. Even where an oxidative process has followed an anaerobic digester, as for example in a trickling filter, nitrification may be very slow due to lack of soluble iron in the waste water. Formation of highly insoluble iron sulfides during anaerobic digestion may reduce the concentration below the level required by nitrifying bacteria.

Table 6—Changes of inorganic N in four municipal sewage samples
during aerobic oxidation

	Days					
	0	4	10	17	24	31
	ppm					
Riverside						
NH_4^+-N	8.1	9.6	0.3	0.4	0.0	0.0
NO_2^--N	0.0	0.0	8.2	3.6	0.0	0.0
NO_3^--N	0.0	0.48	0.72	3.03	5.06	5.06
Sum	8.1	10.5	9.3	7.0	5.1	5.1
Ontario						
NH_4^+-N	10.8	10.8	9.8	0.4	0.0	0.0
NO_2^--N	0.0	0.0	1.7	4.5	0.7	0.5
NO_3^--N	0.0	0.3	0.3	0.0	12.5	10.6
Sum	10.8	11.1	11.8	4.9	13.2	11.1
Redlands						
NH_4^+-N	18.2	18.4	0.3	0.3	0.0	0.0
NO_2^--N	0.0	0.0	15.5	6.9	0.0	0.0
NO_3^--N	0.0	0.8	1.3	5.1	11.9	14.3
Sum	18.2	19.2	17.0	12.3	11.9	14.3
Sun City						
NH_4^+-N	19.3	18.4	0.4	0.3	0.0	0.0
NO_2^--N	trace	0.0	19.3	9.2	0.0	0.0
NO_3^--N	0.0	0.6	1.5	4.2	14.0	22.2
Sum	19.3	19.0	21.1	13.8	14.0	22.2

Temporary accumulation of NO_2^- is a common feature in nitrification of waste waters, as shown in Table 6 giving data on secondary effluents from four municipal sewage treatment plants. There is also evidence of denitrification during aerobic treatment, as indicated by the decreasing values for total inorganic N up to 24 days. Nitrite in waste water would be expected to be readily oxidized to NO_3^- after application to soil unless some inhibitory factor such as high NH_3 level is present. This inhibition may account for the persistence of NO_2^- in a neutral soil treated with stored oxidation ditch poultry manure which apparently contained considerable NO_2^- and NH_3 when applied (MacMillan et al., 1972).

Application of waste waters to soil normally results in rapid conversion of organic N in the water to NH_4^+ and hence to NO_3^-. A major concern is the potential for pollution of ground water by NO_3^-. Kardos and Sopper (1973) reported that application of sewage effluent to forest and cropland at the rate of 2.5 cm per week over a period of 6 years did not increase the NO_3^--N concentration of soil solution samples above the 10 ppm Public Health Service standard, but this limit was exceeded when waste water was applied at 5 cm per week. In their cropland experiments, there was considerable removal of N from the system in harvested crops.

In addition to removal by crops, a critical consideration with respect to the possible pollution of ground or surface waters by excess NO_3^- is the degree to which denitrification takes place in the soil profile. Bouwer (1973) reported consistent N removal by denitrification in a system with very high inputs (11–15 metric tons N ha^{-1} $year^{-1}$) by controlling the periods of flooding and drying. With short, frequent flooding (2 days wet, 2 days dry) NO_3^--N levels were close to zero except for short periods following a new inundation, when a peak of nitrate which formed during the drying period moved downward. These observations were confirmed by Lance and Whisler (1972) in N balance experiments with soil columns. They found no net removal of N with cycles of 2 days flooded and 5 days dry, but with 9 to 23 days flooded followed by 5 days dry, net N removal was 30%, and half the remaining N in the water was concentrated in a wave of high NO_3^- water immediately after the dry period. Somewhat greater denitrification would likely occur in finer textured soils. Obviously, the organic content of the input sewage as well as the organic matter status of the soil play an important role in determining the quantity of denitrification which will occur. Figure 5 shows a comparison of NO_3^- concentrations in the effluents of a column of surface soil and a subsoil of much lower organic matter content receiving an input of 4.5 cm/week secondary sewage. The period of several weeks before NO_3^- appeared in the effluent was due to nearly complete removal by denitrification. Nitrogen recovery was somewhat higher in the subsoil because of its low organic matter content. The effect of adding organic substrate during a flooding cycle of a subsoil otherwise inactive in denitrification is illustrated in Fig. 6. The subsoil was made comparable to the surface soil in denitrification activity under flooded conditions by addition of rice straw (*Oryza sativa* L.) or sucrose.

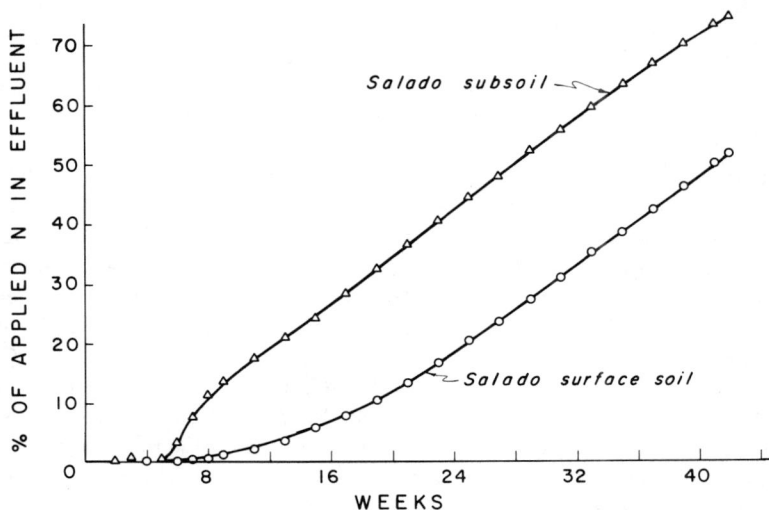

Fig. 5—Nitrate in effluents from columns of Salado fine sandy loam expressed as percent of N in applied sewage.

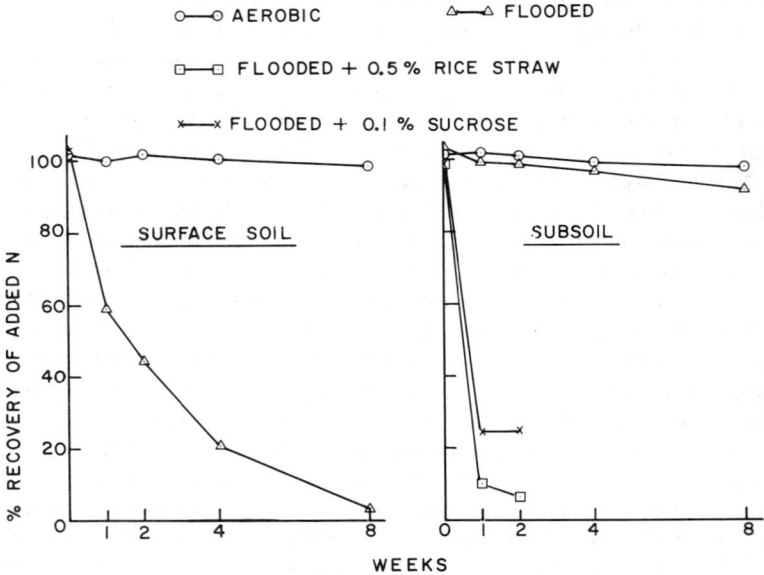

Fig. 6—Effect of added substrate on N loss from a flooded subsoil otherwise inactive in denitrification.

D. Acid Production in Nitrification

A side effect of the application of sewage effluents to soil is the production of acid in the nitrification reactions:

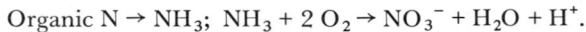

$$NH_4^+ + 2\,O_2 \rightarrow NO_3^- + H_2O + 2\,H^+$$

$$\text{Organic N} \rightarrow NH_3; \quad NH_3 + 2\,O_2 \rightarrow NO_3^- + H_2O + H^+.$$

This may have a secondary effect upon ground-water quality by dissolution of soil minerals, or by displacement of soil cations, in particular calcium and magnesium, which contribute to the hardness of ground water. Figure 7 shows concentrations of Ca in the effluents from columns of a calcareous soil treated with nitrified and nonnitrified sewage at the rate of 4.5 cm/week for 42 weeks. These data suggest that although the application of NH_4^+-containing waste water to soils may cause a small increase in the leaching of Ca, the waste water itself is sufficiently well buffered that the increase in leached bases is not equivalent to the quantity of NH_4^+ nitrified. Kardos and Sopper (1973) found no significant changes in exchangeable Ca in soil samples from a corn rotation area which received 0, 2.5, and 5 cm waste water per week over a 6-year period. Values were averaged over five 30-cm depth intervals. In fact, in their experiments, the pH of the surface 30 cm of soil was significantly increased as a result of waste water application.

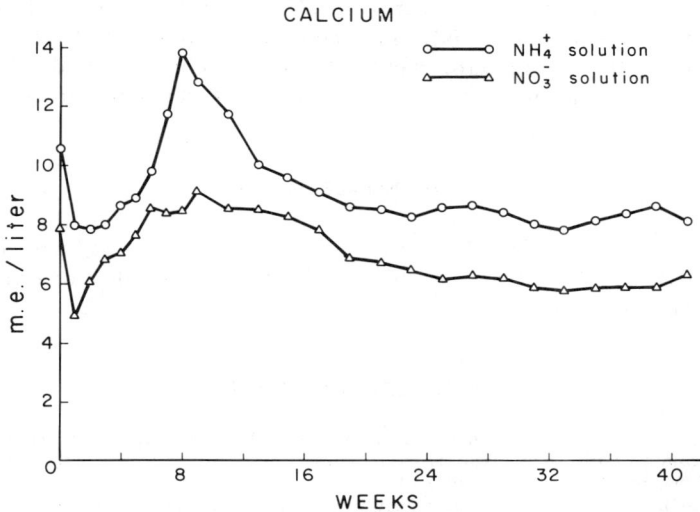

Fig. 7—Calcium concentration in effluents from soil columns treated with sewage containing N as NH_4^+ or NO_3^-.

IV. SUMMARY

The disposal of animal, domestic, and industrial wastes on land, and subsequent recycling of plant nutrients is an ancient agronomic practice, since farmers soon observed that soil tilth and crop yields improved by the addition of these wastes. Yet, over the years several problems required further study. These relate primarily to proper application rates for particular geographic regions and to the potentially harmful effects of overloading the soil system. Loading rates cannot be viewed in a simplistic manner. Due attention must be given to all factors which include: (i) chemical nature of the waste, (ii) geographic and climatic region, (iii) soil type, (iv) irrigation practices, and (v) underground-surface water features.

From the biological viewpoint, the mineralization and immobilization of C and N become inseparable entities. One index that has proven useful for estimating the rate of organic waste breakdown is the waste "half-life" or the time required for a particular amount of waste carbon to undergo 50% decomposition.

Waste loading rates also may be established on the basis of crop N needs. Unfortunately this index is subject to considerable error because of the wide variation in the N content of the waste. For example, feedlot manure is normally stored for a variable time period, thereby allowing for variable losses of N. Thus, a total N analysis for both soil and manure is required before accurate soil amendments can be made. It should be stressed that the N fertilizer value of any waste product is, to a large extent, dependent on the C/N ratio of the individual waste. At C/N ratios > 20:1, less and less N is released for plant uptake. In other words, N is often conserved with respect to the C turnover process. Under such conditions, additional N fertilizer may be required for optimal crop yields. Thus the cycling of C and N becomes a matter of the balance between assimilation and dissimilation of both elements.

ACKNOWLEDGMENT

Appreciation is expressed to the U. S. Army Corps of Engineers, Walla Walla District, Walla Walla, Washington for financial support received for the C recycling studies.

LITERATURE CITED

Adriano, D. C., P. F. Pratt, and S. E. Bishop. 1973. Fate of inorganic forms of N and salt from land-disposed manures from dairies. Proc. Int. Symp. on Livestock Wastes. Am. Soc. Agric. Eng. p. 243–246.

Bartholomew, W. V. 1965. Mineralization and immobilization of nitrogen in the decomposition of plant and animal residues. In W. V. Bartholomew and F. E. Clark (ed.) Soil nitrogen. Agronomy 10:287–306. Am. Soc. of Agron., Madison, Wis.

Bengtson, G. W., and J. J. Cornette. 1973. Disposal of composted municipal waste in a plantation of young slash pine: Effects on soil and trees. J. Environ. Qual. 2: 441–444.

Bouwer, H. 1973. Renovating secondary effluent by groundwater recharge with infiltration basins. p. 164–175. In Recycling treated municipal wastewater and sludge through forest and cropland. Penn. State Univ. Press, University Park, Pa.

Broadbent, F. E. 1973. Organics. p. 97–101. In Proc. of Joint Conf. on Recycling Municipal Sludges and Effluents on Land. Champaign, Ill. Nat. Assoc. of State Univ. and Land-Grant Coll., Washington, D. C.

Broadbent, F. E., K. B. Tyler, and G. N. Hill. 1957. Nitrification of ammonical fertilizers in some California soils. Hilgardia 27:247–267.

Bunting, A. H. 1963. Experiments on organic manures, 1942–49. J. Agric. Sci. 60:121–140.

Campbell, C. A., E. A. Paul, D. A. Rennie, and K. J. McCallum. 1967. Applicability of the carbon-dating method of analysis to soil humus studies. Soil Sci. 104:217–224.

Dalton, F. E., J. E. Stein, and B. T. Lynam. 1968. Land reclamation—A complete solution of the sludge and solids disposal problem. J. Water Pollut. Control Fed. 40(5): 789–800.

Elliott, L. F., and T. M. McCalla. 1972. The composition of the soil atmosphere beneath a beef cattle feedlot and a cropped field. Soil Sci. Soc. Am. Proc. 36:68–70.

Gambrell, H. V., and T. C. Peele. 1973. Disposal of peach cannery waste by application to soil. J. Environ. Qual. 2:100–104.

Garner, H. V. 1962. Experiments with farmyard manure, sewage sludges, and two refuses on microplots at schools, 1940–49. Emp. J. Exp. Agric. 30(120):295–304.

Gilmour, C. M., F. Merryfield, F. Burgess, L. Punkerson, and K. Carswell. 1961. Persulfate-oxidizable carbon and B.O.D. as a measure of organic pollution in water. Proc. 15th Ind. Waste Conf., Purdue Univ. Eng. Bull. (Eng. Ext. Ser. No. 106), 45:143–149.

Herron, G. M., and A. B. Erhart. 1965. Value of manure on an irrigated calcareous soil. Soil Sci. Soc. Am. Proc. 29:278–281.

Hinesly, T. D., C. C. Braids, and J. E. Molina. 1971. Agricultural benefits and environmental changes resulting from the use of digested sewage sludge on field crops. U. S. Environ. Protect. Agency Interim Rep. SW-30d.

Hunt, C. F., W. E. Sopper, and L. T. Kardos. 1971. Renovation of treated municipal sewage effluent and digested liquid sludge through irrigation of bituminous coal strip mine spoil. The Penn. State Univ., Tech. Pap., Inst. for Research on Land and Water Resources. 118 p.

Hunter, J. V., and T. A. Kotalik. 1973. Chemical and biological quality of treated sewage effluents. p. 6-25. *In* Recycling treated municipal wastewater and sludge through forest and cropland. Penn. State Univ. Press, University Park, Pa.

Jenkinson, D. S. 1965. Studies of the decomposition of plant material in soil. I. Losses of carbon from ^{14}C-labelled ryegrass incubated with soil in the field. J. Soil Sci. 16:104-115.

Johnson, D. D., and W. E. Guenzi. 1963. Influence of salts on ammonium oxidation and carbon dioxide evolution from soil. Soil Sci. Soc. Am. Proc. 27:663-666.

Kardos, L. T., and W. E. Sopper. 1973. Effects of land disposal of wastewater on exchangeable cations and other chemical elements in the soil. p. 220-231. *In* Recycling treated municipal wastewater and sludge through forest and cropland. Penn. State Univ. Press, University Park, Pa.

King, L. D. 1973. Mineralization and gaseous loss of nitrogen in soil-applied liquid sewage sludge. J. Environ. Qual. 2:356-358.

King, L. D., and H. D. Morris. 1974. Nitrogen movement resulting from surface application of liquid sewage sludge. J. Environ. Qual. 3:238-242.

Lance, J. C., and F. E. Whisler. 1972. Nitrogen balance in soil columns intermittently flooded with secondary sewage effluent. J. Environ. Qual. 1:180-186.

Law, J. P., Jr., R. E. Thomas, and L. H. Meyers. 1970. Cannery wastewater treatment by high-rate spray on grassland. J. Water Pollut. Control Fed. 42:1621-1631.

Lejcher, T. R., and S. H. Kunkle. 1973. Restoration of acid spoil banks with treated sewage sludge. p. 184-199. *In* Recycling treated municipal wastewater and sludge through forest and cropland. Penn. State Univ. Press, University Park, Pa.

McCalla, T. M., J. R. Ellis, and W. R. Woods. 1969. Changes in the chemical and biological properties of beef cattle manure during decomposition. Proc. Am. Soc. Microbiol. p. 4-5.

MacMillan, K., T. W. Scott, and T. W. Bateman. 1972. A study of corn response and soil nitrogen transformations upon application of different rates and sources of chicken manure. p. 481-494. *In* Waste management research. Proc. 1972 Cornell Agric. Waste Manage. Conf., Ithaca, N. Y.

Mathers, A. C., and B. A. Stewart. 1974. Corn silage yield and soil chemical properties affected by cattle feedlot manure. J. Environ. Qual. 3:143-147.

Mays, D. A., G. L. Terman, and J. C. Duggan. 1973. Municipal compost: Effects on crop yields and soil properties. J. Environ. Qual. 2:89-92.

Meek, B. D., A. J. MacKenzie, T. J. Donovan, and W. F. Spencer. 1974. The effect of large applications of manure on movement of nitrate and carbon in an irrigated desert soil. J. Environ. Qual. 2:89-92.

Mielke, L. N., N. P. Swanson, and T. M. McCalla. 1974. Soil profile conditions of cattle feedlots. J. Environ. Qual. 3:14-17.

Miller, R. H. 1973a. Soil microbiological aspects of recycling sewage sludges and waste effluents on land. p. 79-90. *In* Proc. of Joint Conf. on Recycling Municipal Sludges and Effluents on Land. Champaign, Ill. Nat. Assoc. of State Univ. and Land-Grant Coll., Washington, D. C.

Miller, R. H. 1973b. Soil as a biological filter. Proc. of Symp. on Recycling Treated Municipal Wastewater and Sludge Through Forest and Cropland. Penn. State Univ. Press, University Park, Pa.

Minderman, C. 1968. Addition, decomposition and accumulation of organic matter in forests. J. Ecol. 56:355-362.

Murphy, L. S., G. W. Wallingford, W. L. Powers, and H. L. Manges. 1972. Effects of solid beef feedlot wastes on soil conditions and plant growth. p. 449-464. *In* Waste management research. Proc. 1972 Cornell Agric. Waste Manage. Conf., Ithaca, N. Y.

Norman, A. G. 1961. The biological environment of roots. p. 653-664. *In* M. X. Zarrow (ed.) Growth in living systems. Proc. Int. Symp. on Growth. Purdue Univ., Lafayette, Ind.

Olsen, R. J., R. F. Hensler, and O. J. Attoe. 1970. Effect of manure application, aeration, and soil pH on soil nitrogen transformations and on certain soil test values. Soil Sci. Soc. Am. Proc. 34:222-225.

Olson, J. 1963. Energy storage and the balance of producers and decomposers in ecological systems. Ecology 44:322-331.

Pal, D., and F. E. Broadbent. 1975. Influence of moisture on rice straw decomposition in soils. Soil Sci. Soc. Am. Proc. 39:59-63.

Peterson, J. R., C. Lue-Hing, and D. R. Zenz. 1973. Chemical and biological quality of municipal sludge. p. 26–37. *In* Recycling Treated Municipal Wastewater and Sludge Through Forest and Cropland. Penn. State Univ. Press, University Park, Pa.

Pratt, P. F., F. E. Broadbent, and J. P. Martin. 1973. Using organic wastes as nitrogen fertilizers. Calif. Agric. 27(6):10–13.

Ryan, J. A., D. R. Keeney, and L. M. Walsh. 1973. Nitrogen transformations and availability of an anaerobically digested sewage sludge in soil. J. Environ. Qual. 2:489–492.

Salter, P. J., and C. J. Schollenberger. 1939. Farm manure. Ohio Agric. Exp. Sta. Bull. 605:69.

Sikora, L. J. 1974. Soil as an animal wastes disposal system. Dissertation Abstracts International. B. The Sciences and Engineering 35(2):654-B. Xerox University Microfilms, Ann Arbor, Mich.

Sommerfeldt, T. G., U. J. Pittman, and R. A. Milne. 1973. Effect of feedlot manure on soil and water quality. J. Environ. Qual. 2:423–427.

Sukovaty, J. E., L. F. Elliott, and N. P. Swanson. 1974. Some effects of beef feedlot effluent applied to forage sorghum grown on a Yolo silty clay loam soil. J. Environ. Qual. 3:381–388.

Terman, G. L., J. M. Soileau, and S. E. Allen. 1973. Municipal waste compost: Effects of crop yields and nutrient content in greenhouse pot experiments. J. Environ. Qual. 2:84–89.

Tietjen, C., and S. A. Hart. 1969. Compost for agricultural land? J. Sanit. Eng. Div., Proc. Am. Soc. Civil Eng. 95:269–287.

Tyler, K. B., A. F. van Maren, O. A. Lorenz, and F. H. Takatori. 1964. Sweet corn fertility experiments in the Coachella Valley. Calif. Agric. Exp. Sta. Bull. 808.

U. S. Department of Agriculture, Agricultural Research Service. 1972. Progress report. Incorporation of sewage and sludge in soil to maximize benefits and minimize hazards to the environment. Beltsville, Md.

Walker, W. G., J. Bouma, D. R. Keeney, and P. G. Olcott. 1973. Nitrogen transformations during subsurface disposal of septic tank effluent in sands: II. Ground-water quality. J. Environ. Qual. 2:521–525.

Webber, L. R., and T. H. Lane. 1969. The nitrogen problem in the land disposal of liquid manure. p. 124–130. *In* Animal waste management. Cornell Univ. Conf. on Agric. Waste Manage., Syracuse, N. Y.

Weeks, M. E., M. E. Hill, S. Karczmarzyk, and A. Blackmer. 1972. Heavy manure applications: Benefit or waste? p. 441–447. *In* Waste management research. Proc. 1972 Cornell Agric. Waste Manage. Conf., Ithaca, N. Y.

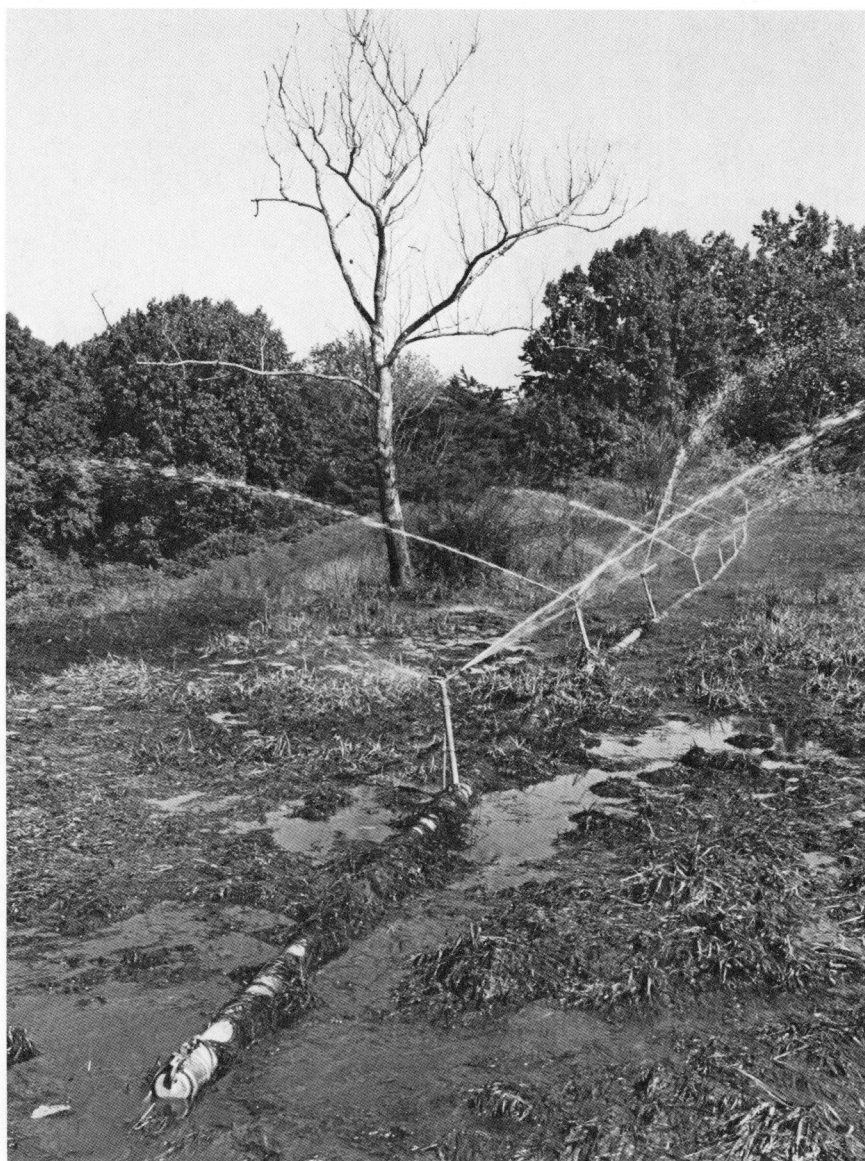

chapter 8

This field has been oversprayed with filtered effluent from a food processing plant (Photo courtesy of USDA).

Effect of Waste Application on Soil Phosphorus and Potassium[1]

S. R. OLSEN, USDA-ARS, Fort Collins, Colorado

S. A. BARBER, Purdue University, Lafayette, Indiana

I. INTRODUCTION

Animal manure has always been valuable in soil management practices for its plant nutrient content and organic matter. However, its economic value fluctuates in response to changing costs of commercial fertilizers. Because commercial fertilizer costs have risen since 1973, animal manures have become more attractive as nutrient sources. A need to conserve plant nutrients as resources, especially P, requires a wise policy of utilizing manure on farmlands to produce crops.

Other organic wastes can also be utilized as nutrient sources. If their long-term use adds elements that eventually become toxic to plants, or to animals and people that consume the plants, then this potential hazard must be taken into account.

Azevedo and Stout (1974) clearly indicated that if animal manures are to maintain a firm position as marketable commodities in developed countries, better recognition of their functions as fertilizers, soil amendments, and plant nutrients hold the only immediate promise for their general acceptance.

Plant nutrients, especially N, and organic matter in animal manure can be lost or drastically reduced because of its perishable nature. Very few farmers realize the full benefits of animal manure, unless they follow recommended disposal and storage practices. The best guideline is to incorporate manure into the soil as soon as possible (daily) after spreading to prevent N losses. Proper storage practices to retain the liquid portion of animal manure will preserve half of the N and two-thirds of the K. As a fertilizer, the value of manure depends mainly on its content of N, P, and K which varies with animal species, age, and feeding program. Therefore, measuring the amount of N and other nutrients applied in manure will be difficult. Soil tests provide a more dependable guide of available nutrients resulting from long term manure use.

If a long-term practice of adding manure is followed, the limiting factor in determining the application rate is mainly the N added and its subsequent mineralization in the soil. Addition of N from manure or commercial fertilizer must be controlled at a level which supplies the optimum amount of N required by a given crop. Variations in yield levels due to environmental factors must be considered for long term uses of manure. For example, a constant annual manure (25% water) rate of 27 metric tons/ha will release N at 112 kg/ha the first year and 233 kg/ha the 20th year based on 1.5% N, dry weight basis (Pratt et al., 1973). The N status of a soil for an expected crop and yield must be assessed to assure an adequate but not an excessive supply of N. Thus, rates of manure on farmland must be regulated according to N requirements of crops to preclude NO_3^- leaching and ground-water con-

[1]Contribution from Agricultural Research Service, USDA, and the Department of Agronomy, Purdue University, in cooperation with Colorado Agricultural Experiment Station. Journal Paper No. 5977, Agricultural Experiment Station, Purdue University, West Lafayette, IN 47907. Cosponsored by Tennessee Valley Authority and Soil Science Society of America Symposium.

tamination. Consequently, the amount of P added from manure will necessarily depend on the rate required for N needs. Such practices will gradually increase the supply of available P in soils.

Annual recovery of P by crops from manure is less than recovery of N. Thus, long term use of manure builds the supply of available P in soils, except in cases of low maintenance rates. For short term use on soils low in available P, a manure application rate adequate for N may be inadequate for P, but for land that receives annual applications of manure the P content of the manure will be adequate to ample after a few years.

The main question about P from manure in soil fertility management concerns trends in available P levels when manure rates are adjusted to meet the N requirements of the crop. If this guideline is followed for recommending application rates, what changes in available P and K levels may be predicted? The main purpose of this review is to examine this question using experimental data from soils which have received manure for long periods. Other related objectives concern the benefits, problems, potential problems, and their solutions connected with the long term use of manure and other organic wastes on farmlands.

II. LONG TERM EFFECTS ON PHOSPHORUS FROM MANURE APPLICATIONS

A. Effect on Available Phosphorus

The effect on available P from annual manure applications up to 117 years has been measured. Changes in extractable P (Olsen et al., 1974a) and P concentration in $0.01M$ $CaCl_2$ (Aslyng, 1954) were measured in Hoosfield and Broadbalk plots at the Rothamsted Experiment Station in England. The results are shown in Table 1. Long term applications of manure or superphosphate have greatly increased soluble and available P to higher levels than

Table 1—Long term effects of manure and superphosphate on soluble P in soil at Rothamsted, England

Field and plot	Years	Annual rate		pH	Soluble P	
		Manure†	P		$0.01M$ $CaCl_2$	$0.5M$ $NaNO_3$
		metric tons/ha	kg/ha		moles/ liter $\times 10^6$	ppm
Hoosfield 1/0	1852–1966	0	0	7.5	0.2	5
Hoosfield 7/2	1852–1966	31.4	0	7.4	25.4	102
Hoosfield 4/0	1852–1966	0	33	7.4	21.2	126
Broadbalk 10	1843–1952	0	0	7.4	0.5	--
Broadbalk 2	1843–1952	31.4	0	7.4	24.7	--
Broadbalk 7	1843–1952	0	33	7.5	10.9	--

† Manure contained about 28 kg P/ha.

required for an adequate supply of P to crops (Aslyng, 1954; Warren & Johnston, 1967; Olsen & Watanabe, 1970).

The solubility of P in these two soils (Table 1) was calculated and the results are shown in Fig. 1. In the Hoosfield plots both P treatments produced a solid phase phosphate in equilibrium with octocalcium phosphate (OCP), $Ca_4H(PO_4)_3$. In the Broadbalk plots the manure treatment apparently caused OCP to form, but the superphosphate treatment produced a solid phase phosphate less soluble than OCP.

A similar 117-year experiment at Barnfield, Rothamsted Experiment Station, was conducted with manure and superphosphate treatments as shown in Table 2. Large increases in soluble P are evident. Data for plot 2/N showed a solubility of solid phase P near the OCP line in Fig. 1; however, the manure plot 1/0 was undersaturated with OCP. The superphosphate treatment on this soil was much less effective than manure in raising soluble P levels. Evidently, the Barnfield soil caused greater reversion of added P to less soluble forms than soil at the Hoosfield or Broadbalk plots, although all soils contained $CaCO_3$.

Yields and P uptake by mangels and sugarbeet (*Beta vulgaris* L.) at

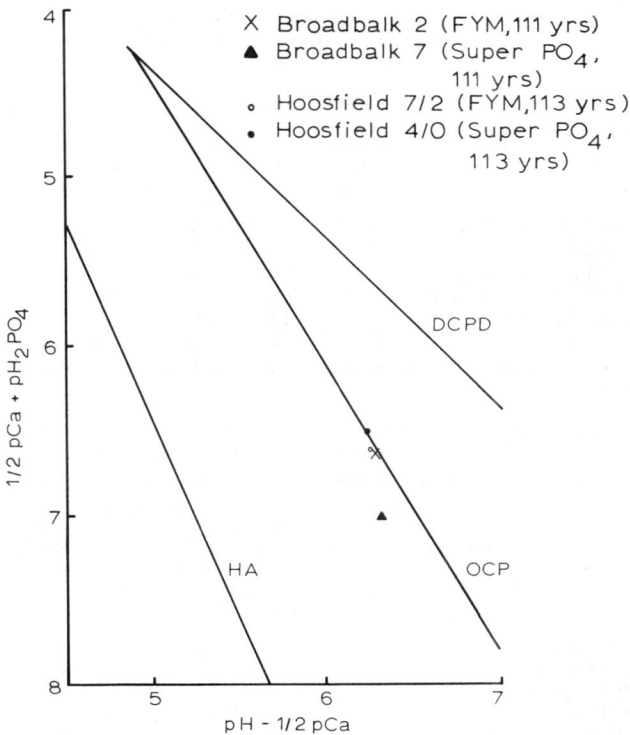

Fig. 1—Lime and phosphate potentials after long term applications of manure and superphosphate (HA, hydroxyapatite; OCP, octocalcium phosphate; DCPD, dicalcium phosphate dihydrate; FYM, farmyard manure).

Barnfield showed that manure provided enough P for crop needs. Adding superphosphate with manure did not increase yields or P uptake. However, this rate of manure (Table 2) did not supply sufficient N to meet the needs of these crops (Warren & Johnston, 1962). When additional inorganic N was applied the yield increased and caused a proportional increase in P uptake, so that the P supply in the soil from these P treatments was ample for maximum yields.

Total P in the surface soil (0 to 23 cm) on Barnfield plots increased over twofold as a result of some of the long term treatments shown in Table 2. Manure and superphosphate increased total P about equally (compare plot 8/0 with 1/0 and 6/0). However, manure produced a larger increase in available P.

The long term effects on available P from additions of rape cake and castor meal for 116 years are shown in Table 3. This source of P increased available or soluble P more than superphosphate but less than manure on Barnfield plots. Total P increased an average of 270 ppm from rap cake or castor meal, or 59% of the amount of P supplied by manure. The combina-

Table 2—Long term effects of manure and superphosphate on soluble P levels at Barnfield, Rothamsted

Field and plot	Years	Annual rate Manure†	P	pH	0.01M CaCl$_2$	0.5M NaHCO$_3$	Total P
		metric tons/ha	kg/ha		moles/ liter $\times 10^6$	——— ppm ———	
Barnfield 8/0	1843-1959	0	0	7.5	0.6	23	780
Barnfield 1/0	1843-1959	31.4	0	7.5	13.2	83	1,240
Barnfield 2/0	1843-1959	31.4	33	7.5	14.9	140	1,950
Barnfield 2/N	1843-1959	31.4	33	7.5	20.0	132	1,840
Barnfield 6/0	1843-1959	0	33	7.5	1.9	66	1,220

† Manure contained about 40 kg P/ha (Warren & Johnston, 1962).

Table 3—Long term effects on P of manure, castor meal, and superphosphate on soluble P and total P levels at Barnfield, Rothamsted, 1843 to 1958

Plot	Annual rate Manure	P	0.01M CaCl$_2$	0.5M NaHCO$_3$	Total P (0-23 cm)
	metric tons/ha	kg/ha	moles/ liter $\times 10^6$	——— ppm ———	
Barnfield 8/0	0	0	0.6	23	780
Barnfield 8/C†	0	0	4.1	56	940
Barnfield 1/0	31.4	0	13.2	83	1,240
Barnfield 1/C†	31.4	0	19.9	128	1,550
Barnfield 6/0	0	33	1.9	66	1,220
Barnfield 6/C†	0	33	4.5	93	1,560

† C indicates addition of rape cake or castor meal at nearly 2 metric tons/ha.

tion of manure and rape cake produced a solid phase phosphate in near equilibrium with OCP.

At Saxmundham, Suffolk, manure was applied annually for 56 years (1901 to 1956) to a calcareous clay loam at 13.4 metric tons/ha. The P addition was estimated as 29.6 kg/ha annually or 1,714 kg/ha for the observed period. Cooke et al. (1958) measured changes in soluble P as a result of the manure treatment as compared with superphosphate applied annually at 17.9 kg P/ha. Manure increased the $NaHCO_3$-soluble P from 2 to 33 ppm while superphosphate increased this value to 23 ppm. Soluble P in $0.01M$ $CaCl_2$ was increased from 0.1 to 6.3 μmoles/liter by manure and to 1.6 μmoles/liter by superphosphate. The solid-phase P was undersaturated with OCP for all treatments. Fresh superphosphate applications of 43.7 kg P/ha indicated that the long term P treatments contained adequate available P for near maximum yields (Williams & Cooke, 1971).

Another experiment with lower manure rates on Rotation II at Saxmundham included an application of 22.4 metric tons/ha for 1 year in a 4-year rotation. Total P applied between 1899 to 1964 was estimated at 790 kg/ha. This treatment increased $NaHCO_3$-soluble P from 6 ppm on the control to 12 ppm. Manure treatments at the same rate plus superphosphate at 67.2 kg P/ha every 4 years increased $NaHCO_3$-soluble P to 41 ppm (Mattingly et al., 1970). This combination treatment produced better yields than manure alone so this manure rate was insufficient for maintaining an adequate supply of P for crops.

In Rotations I and II a linear relationship was found between labile P (isotope exchangeable, P_e), $NaHCO_3$-soluble P, and 1/2 p Ca + pH_2PO_4 and increases in total P resulting from manure or superphosphate. The increase in total P (P_t) ranged from 71 to 290 ppm (Mattingly et al., 1970). The regression lines gave the following changes:

$$\Delta P_e/\Delta P_t = 0.37 \pm 0.038,$$

$$\Delta NaHCO_3\text{-soluble P}/\Delta P_t = 0.11 \pm 0.007, \text{ and}$$

$$\Delta (1/2 pCa + pH_2PO_4)/\Delta P_t = -57 \pm 5.3 \times 10^{-4}.$$

These regression equations apply to "old" residues accumulated during 65 years. With "new" residues applied during a 4-year period the $NaHCO_3$-soluble P increased 25% and labile P increased 60% of the total P.

Pratt et al. (1956) reported data on changes in P as a result of a 28-year period of fertilization with manure, other organic materials, and concentrated superphosphate (CSP). Changes in $NaHCO_3$-soluble P and total P are shown in Table 4 for a Ramona sandy loam soil derived from granitic alluvium. The surface soil has a pH of 7.6 at 1:1 dilution with water and contains a trace of free carbonates. All of the source materials containing P increased the $NaHCO_3$-soluble P to levels more than adequate to supply the crop needs. The $NaHCO_3$-soluble P in the soil increased as total P increased in Table 4,

Table 4—Long term effects of organic materials and concentrated superphosphate on NaHCO$_3$-soluble P and total P in surface Ramona sandy loam soil (pH 7.6)

Source of P	NaHCO$_3$-soluble P	Total P
	ppm	
Manure†	69.2	841
Cottonseed meal	61.8	659
Dried blood	19.2	403
CSP‡	44.2	902
Check (none)	6.3	357

† Manure applied at 19.4 metric tons ha^{-1} year^{-1} (1972 to 1938) and 48.2 metric tons ha^{-1} year^{-1} (1939 to 1955).
‡ Concentrated superphosphate (CSP) applied at 145 kg P ha^{-1} year^{-1}.

but the NaHCO$_3$-soluble P increased more for the organic materials per unit of total P than for CSP.

Pratt et al. (1956) showed results from a 16-year period (1939 to 1955) on the same soil when manure was applied at 14.3 metric tons/ha (6.4 tons/ acre) annually. This treatment increased NaHCO$_3$-soluble P from 6.3 ppm in the check to 24.5 ppm. This rate of application supplied the needs of citrus crop (*Citrus* sp.) and also raised available P to a level adequate for most other crops.

Olsen et al. (1954b) measured NaHCO$_3$-soluble P in old rotation plots which had received manure for a 39-year period. Treatments and soluble P are shown in Table 5. The NaHCO$_3$-soluble P increased as the frequency of manure application increased. The increase in NaHCO$_3$-soluble P was greater for the fine sandy loam soil than for the silty clay. Manure rates were sufficient to maintain an adequate P supply for crops on the Tripp very fine sandy loam, except for the lower rate. The Pryor silty clay has a higher capacity to adsorb soluble P so the manure rates were only adequate or marginal for maintaining an adequate P supply for crops and the lowest rate was inadequate.

Data for NaHCO$_3$-soluble P (Table 5) and for CSP treatments on a Fort Collins loam are plotted in Fig. 2 against the increases in total P resulting from long term treatments. All soils are calcareous. Relations between in-

Table 5—Extractable P in rotation plots receiving manure from 1912 to 1950

			NaHCO$_3$-soluble P	
Manure	Frequency	Years	Pryor silty clay	Tripp very fine sandy loam
metric tons/ha	years		ppm	
0	--	--	5.56	6.22
26.9	annual	--	--	36.1
26.9	alternate	191? to 1950	19.4	27.7
26.9	1 to 3	1912 to 1950	14.2	26.0
26.9	1 to 6	1912 to 1950	6.1	7.2

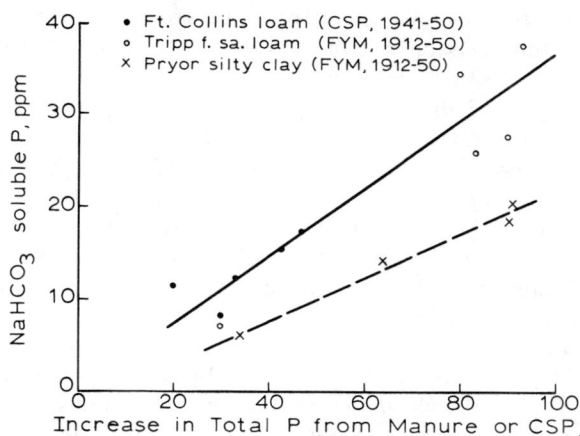

Fig. 2—Relation between NaHCO₃-soluble P and the increase in total P resulting from long term applications of farmyard manure (FYM) and concentrated superphosphate (CSP).

creases in total P and NaHCO$_3$-soluble P were linear, but the slope was less with Pryor silty clay. Fort Collins loam received CSP at 41 and 53 ppm P during 1941 to 1950.

Vitosh et al. (1973) evaluated long term effects of annual manure applications over 6, 7, and 9 years on two soils. On Conover-Hodunk loam (pH 6.5), available P (Bray P$_2$) levels were 38, 59, and 93 ppm for 0, 44.8, and 67.2 metric tons/ha, respectively, after seven annual applications. Similar treatments on Matea sandy loam (pH 6.5) resulted in available P levels of 81, 110, 133, and 136 ppm for 0, 22.4, 44.8, and 67.2 metric tons/ha respectively, after nine annual applications. For all treatments, there was a blanket application of 20 kg P/ha annually of commercial fertilizer. Sampling by depth revealed that nearly all the soluble P remained in the surface 30 cm of soil, but the soil test measured only 13% of the applied P on these soils.

B. Effect of Organic Matter on Phosphorus Concentration in Soil Solution

The organic matter from manure or rape cake increased soluble P in water or 0.01M CaCl$_2$ from soil by some chemical or physical mechanism. More P was added by superphosphate than by either the manure or rape cake treatments, but soluble P was higher with the extra organic matter application, especially manure. These effects of organic matter can be examined from the data in Table 2 and 3. Although increases in total P in plots 1/0 and 6/0 were similar, the manure-treated plot had more soluble P than the superphosphate-treated plot. This effect of organic matter on P concentration in solution is important because this level of concentration in soil directly affects the rate of P uptake by roots (Olsen & Watanabe, 1970).

Several mechanisms may cause a higher soluble P concentration with manure since it contains a mixture of organic P forms, including inositol

penta- and hexaphosphates (IPP and IHP). Anderson et al. (1974) showed
that IHP adsorbed on the same sites in acid soils as orthophosphate. The organic P depressed inorganic P sorption, particularly at high organic P application rates but inorganic P did not reduce sorption of organic P. This action
of organic P in soils would promote a higher degree of saturation of total
sorption sites with inorganic P and thus favor conditions required to form
OCP.

    If organic P reacts significantly through this mechanism, an increase in
organic P and IHP may be expected in soils that have received manure over
long periods. Oniani et al. (1973) measured changes in organic P and IPP
and IHP forms of the Barnfield plots shown in Table 2. Organic P changed
somewhat with an increase of 31 ppm and the inositol P was essentially unchanged. This data and other results from a 50-year field experiment cited
by Peperzak et al. (1959) indicated that organic P does not accumulate appreciably in soil receiving manure. However, this lack of evidence for organic
P increases does not necessarily nullify the suggested mechanism of action of
organic P and soluble P. Obviously, there are inputs of organic P to the soil
followed by mineralization to inorganic P and the suggested mechanism of
organic P could function during the temporary periods of an increase in organic P. In addition, sorption of organic anions may release P to solution.

    Manure nearly tripled the C and N contents of the Barnfield soil (Table
2) but the C/organic P ratios were 190 with manure, 100 with superphosphate, and 72 without phosphate (Oniani et al., 1973). In this soil, organic
P did not increase proportionally to organic matter. Sadler and Stewart
(1974) made similar observations for organic P.

    Another possible mechanism for soluble P increases with manure seems
to be the organic part of manure retarding P fixation by mechanically
separating soluble P from the mineral part of the soil (Peperzak et al., 1959).

    Changes in soil pH due to treatments can modify soluble P. Pratt et al.
(1956) observed a high correlation between pH and soluble P in field plots.
The $NaHCO_3$-soluble P significantly increased as acidity increased (initial pH
of 7.6). However, long term manure applications (Table 1, 2, and 3) have not
caused significant changes in pH (Aslyng, 1954; Warren & Johnston, 1962).
Vitosh et al. (1973) observed no change in pH for two soils after nine annual
manure applications of 67.2 metric tons/ha. Olsen et al. (1970) found pH
increased from 6.2 to 6.5 in a Plainfield sand and from 4.1 to 4.7 in an Ella
loamy sand after incubation with manure at rates of 23 to 621 metric tons/
ha for 37 and 21 weeks, respectively. Thus, changes in soil pH from manure
seem unlikely to be a cause of changes in soluble P.

C. Movement of Phosphorus from Applied Manure

    Penetration of P below the plow depth but within the root zone of
plants is beneficial for crops, especially in acid soils where root growth is
better. Cooke (1967) stated that P leached into subsoils of the classical ex-

periments at Rothamsted only when manure was applied annually because P solubility was increased by organic matter. Phosphorus penetrated to 45 cm in the Barnfield experiment after manure was applied annually at 35 metric tons/ha since 1845. Cooke and Williams (1973) and Pratt et al. (1956) observed, however, that P moved into the 61 to 91 cm layer of Ramona sandy loam of (pH 7.6) after 28 years of fertilization with CSP (Table 4).

III. SHORT TERM EFFECTS FROM MANURE AND OTHER ORGANIC WASTE APPLICATIONS

A. High Disposal Rates

The objective of many manure experiments during the last 10 years has been to determine the maximum application rate for land disposal of waste. To reduce handling costs and to avoid storage, rates over 700 metric tons/ha have been tested experimentally (Manges et al., 1971). Toxic effects on yields or inferior crop quality must be avoided or considered in a management program. Environmental pollution hazards to streams or ground waters must be eliminated or reduced to acceptable levels.

B. Potential Problems

Excessively high manure rates for disposal caused problems. These problems centered around salt injury and NH_3 toxicity to crops, high NO_3- in forage, adverse $K/(Ca + Mg)$ ratios in forage, and trace element accumulations to possible toxic levels (Azevedo & Stout, 1974).

Although high rates of manure application are limited because of distribution costs (usually, use of manure is limited to within 16.09 km of the source), a more realistic limiting factor is the amount of N the crop can assimilate. This management concept is less binding for short term than for long term effects, but for over 2- or 3-year periods, long term aspects should control application rates. Land for final disposal of animal wastes is ample because more fertilizer nutrients are required for food and fiber production than are available in animal manure. If manure from all livestock were spread equally on approximately 405 million ha (1 billion acres) of cropland, the annual N rate would be 14.8 kg/ha with the P level about one-fifth of this level (National Academy of Science, 1972). However, a farmer is likely to apply manure at high annual rates to supply or replace fertilizer N needed by a crop. This practice will also add more than adequate amounts of P and K for most crops, except in very deficient soils.

If a farmer's goal is to produce 7,840 kg/ha (125 bushels/acre) of corn (*Zea mays* L.), according to Stanford (1973), the fertilizer N required is 336 kg/ha (300 pounds/acre) assuming 50% efficiency of fertilizer N. One-half of the total N in manure may be considered as available to the crop (Azevedo

Table 6—Nutrient content of animal manures†

Animal	H₂O	N	P	K	S	Ca	Mg
	%				kg/metric ton		
Dairy cattle	79	5.6	1.0	5.0	0.5	2.8	1.1
Fattening cattle	80	7.0	2.0	4.5	0.85	1.2	1.0
Hogs	75	5.0	1.4	3.8	1.35	5.7	0.8
Horse	60	6.9	1.0	6.0	0.7	7.85	1.4
Sheep	65	14.0	2.1	10.0	0.9	5.85	1.85
Broiler	25	17.0	8.1	12.5	--	--	--
Hen	37	13.0	12.0	11.4	--	--	--

† From Loehr (1968).

& Stout, 1974; Pratt et al., 1973; McCalla, 1974), so 48 metric tons/ha (21.4 tons/acre) of manure would be needed to supply this N for the composition of manure (Table 6) from fattening cattle. This application rate would supply 96 kg P/ha (85.6 pounds P/acre) and 216 kg K/ha (193 pounds K/acre), which should be adequate for most crops.

In contrast, disposal rates of manure over 700 metric tons/ha have been tested (Manges et al., 1971), but crop yields were depressed when rates exceeded 204 metric tons/ha. Grain sorghum (*Sorghum bicolor* Moench.) yields under irrigation were reduced when the rate exceeded 134 metric tons/ha, but 22 metric tons/ha was adequate for maximum yield (Mathers & Stewart, 1971). Nitrate pollution hazards were minimal only when the crop utilized most of the applied N.

Although high land disposal rates of manure are not recommended if the objective is to supply the crop's N needs, there is still an interest in knowing other possible problems in terms of resulting soil P and K levels. High levels of available P reportedly causes Zn deficiencies in crops, especially when Zn supply was below adequate levels in coarse-textured or calcareous soils. Such interactions of P from manure with Zn have been observed in California orchards (Azevedo & Stout, 1974). Chandler (1937) could not induce Zn deficiency in a fine-textured soil, however, with four annual manure applications, at 134 metric tons/ha. Other P times Zn interactions have been re-

Table 7—Effect to field beans of P level on availability of Zn†

Treatment		Yield of pods and seeds		Zn content of leaves	
P	Zn	Low residual P	High residual P	Low residual P	High residual P
kg/ha		g/pot		ppm	
48.9	0	3.6	5.1	19.3	14.7
489.0	0	2.6	1.2	12.3	11.0
48.9	11.2	4.1	7.4	19.7	17.7
489.0	11.2	8.0	7.1	16.3	16.3

† Kawkawlin-Wisner loam (calcareous). From Ellis et al. (1964).

ported for high P levels either from manure or CSP (Olsen, 1972). Table 7 shows an example for field beans (*Phaseolus vulgaris* L.) where the high residual P could represent P levels accumulated from CSP or manure. Field bean yields were reduced 50% after sugar beets which had heavy P applications (Ellis et al., 1964). Boawn and Brown (1968) found evidence of a P-Zn imbalance in beans and potatoes (*Solanum tuberosum* L.). Increasing soil Zn level at a given P level resulted in plants with normal growth. A specific effect of sugar beets on beans is another possible explanation in the experiment by Ellis et al. (1964). Boawn et al. (1954) observed no effect of P applications on uptake of applied or native soil Zn by field beans, but P fertilizer applications caused severe growth disorder of Russet Burbank potatoes caused by Zn deficiency (Boawn & Leggett, 1963).

High levels of soluble P gradually appear from long term applications of manure at relatively low rates (Table 1), but high annual rates also produce high soluble P levels. Olsen et al. (1970) observed from 36 to 94 ppm soluble P (Bray No. 1) in a Plainfield sand and from 34 to 159 ppm in an Ella loamy sand when manure rates were increased from 0 to 621 metric tons/ha.

Herron and Erhart (1965) added 11.2, 22.4, and 44.8 metric tons/ha of feedlot manure containing 4.5 kg P/metric ton to a Colby silt loam (pH 7.8). Figure 3 shows the soluble P (Bray no. 1) 3 years after application. Manure increased soluble P linearly with rate. The increase in soluble P was 0.46 ppm/metric ton. Soluble P was 23% of the total P added. Mattingly et al. (1970) found that $NaHCO_3$-soluble P was 25% of the total P added from manure over a 4-year period on a calcareous soil at Saxmundham, England.

Table 8 shows P recovered as $NaHCO_3$-soluble P in a study with three calcareous soils to which three rates of feedlot manure was added (11.2, 22.4, and 44.8 metric tons/ha) followed by five wetting and drying cycles. These recoveries increased somewhat as clay content decreased but the values are similar to recoveries observed in field experiments. These data allow

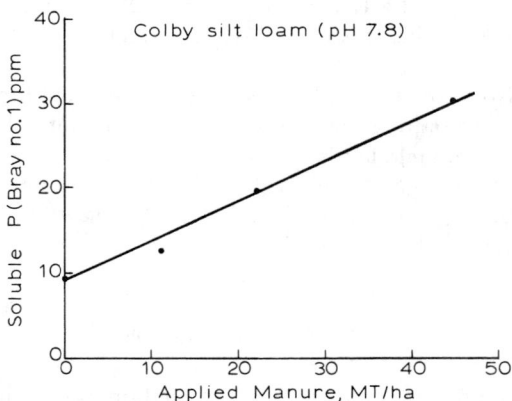

Fig. 3—Relation between soluble P and applied manure after 3 years following a single application.

Table 8—Recovery of phosphorus as NaHCO$_3$-soluble P from feedlot manure added
to three calcareous soils

Soil	pH	Clay	CaCO$_3$	Recovery as NaHCO$_3$-soluble P[†]
			%	
Pierre clay	7.53	51.0	2.9	27.2
Apishapa silty clay loam	7.53	36.6	6.2	29.8
Tripp very fine sandy loam	7.20	15.0	0.2	36.0

† Average recovery from three rates: 11.2, 22.4, and 44.8 metric tons/ha (dry weight).
Manure contained 0.77% P.

changes in soluble P to be estimated from variable rates of manure for short
periods (S. R. Olsen, unpublished data).

IV. SOURCES OF POTASSIUM FROM MANURE AND WASTE MATERIALS

Levels of K in most waste materials are not high compared to crop's K
requirements. Animal manures contain from 1 to 2% K (dry weight basis), a
level similar to that of the feed eaten by the animals. The K/N ratio is about
1. Thus, when animal wastes are applied at a rate adequate for crop N requirements, they also meet K requirements (Powers et al., 1974).

Beef-feedlot-lagoon waters contain 240 to 1,850 μg/g K (Wallingford
et al., 1974), while municipal waste waters from sewage disposal plants contain 5 to 20 μg/g K (Blakeslee, 1973), or similar to the K content of most soil
solutions (Barber et al., 1963). Thus K deficiencies rather than excessive K
are more likely when municipal waste waters are used to supply nutrients for
crops.

Municipal waste compost containing up to 20% sewage sludge mixed
with garbage waste had a K content of 0.8 to 0.97%. When this was applied
to the soil the K in this waste was 64% as effective as K fertilizer (Terman
et al., 1973).

When wastes contain 1% K (dry weight basis), applications of 40 to 50
metric tons/ha are required annually to supply the normal K requirement of
crops grown at high yield levels.

When liquid sludge was applied to K deficient bermudagrass (*Cynodon
dactylon* L.) at rates to give 178 and 328 kg K/ha the bermudagrass completely recovered the K (King & Morris, 1973).

A. Reactions with Soils

Since much of the K added to soils is tied up, it does not move in either
runoff or percolating water. There is movement of K ions between ions in
solution, ions held as exchangeable ions, and ions held as nonexchangeable
ions.

1. EXCHANGEABLE K

The soil's clay minerals in the temperate region have a net negative charge from isomorphous substitution of a lower valence cation, like Al, for a higher valence cation, like Si, in the crystal structure of the clay mineral. These net negative charges are balanced by exchangeable cations that are external to the crystal lattice of the clay mineral. Also K may be held by charges due to broken bonds on the edges of clay crystals and by carboxyl and phenolic hydroxyl groups on organic matter which bond cations exchangeably.

The relative strength that exchange sites hold K depends on their nature and other cations present. If complimentary cations are held very tightly, they tend to push K to less tightly held sites, and, hence, K is more subject to loss by leaching. The magnitude of the cation exchange capacity varies from 1 meq/100 g in a sandy soil to 30 or 40 meq/100 g in a clay soil with montmorillonite types of clay or a soil high in organic matter, since organic matter has a high cation exchange capacity.

2. NONEXCHANGEABLE K

Some of the K added to soil may react with clay minerals and be held so that it cannot be readily displaced with another cation, like NH_4^+, H^+, or Ca^{2+}. This K is held between the clay layers of illite-type clays, which have a high degree of isomorphous substitution and pores on their surfaces nearly the same size as the K ion. When K goes between the clay layers to balance these charges, it fits snugly into the pore in the lattice and the clay sheets come together and trap the K so that it cannot be readily exchanged. Some clays have a very high capacity to tie-up K so that a considerable part of added K goes into this position and is relatively unavailable to the plant and not subject to leaching.

With exhaustive cropping, K "fixed" between the clay layers gradually becomes available. However, when large quantities of K are added, K flows into these "fixed" positions rather than out of them.

B. Historical Rates and Potential Benefits

Crop removal of K varies from 20 kg/ha for grain crops to 600 kg/ha for a forage crop cut several times per year. The amount of K applied to K-deficient soils is usually larger than the amount removed, especially when only the grain is harvested. For example, in corn production about 35 kg/ha K is removed in the grain for a crop yield of 9,000 kg/ha, yet annual K application rates (100 kg/ha) are required for K-deficient soils to provide plants with the amount they require since part of the soil K will be tied up in an unavailable form. When wastes are applied to soil, they will react similarly to

Table 9—Long term effects of manure and potash fertilizers on soluble K at Rothamsted

Field and plot	Years	Annual rate		Soluble K†	
		Manure	K	1N NH$_4$ acetate	0.01M CaCl$_2$
		metric tons/ha	kg/ha	——— mg/100 g ———	
Hoosfield 1/0	1852–1966	0	0	8.7	1.1
Hoosfield 7/2	1852–1966	31.4	0	75.8	28.8
Hoosfield 3/0	1852–1966	--	89.6	49.6	16.1
Barnfield 8	1843–1959	0	0	18	--
Barnfield 6	1843–1959	--	224	65	--
Barnfield 1	1843–1959	31.4	--	54	--
Barnfield 2	1903–1959	31.4	224	89	--

† Potassium added as K$_2$SO$_4$.

fertilizer K, since most of the K in wastes will be in a soluble inorganic form. The effect on the level of available K in the soil will be determined by the balance between the amounts applied, the amount absorbed in an unavailable form, and the amount that may be leached from the soil.

Table 9 shows long term effects on exchangeable and soluble K in soil from additions of manure and potash fertilizer. Exchangeable K increased greatly at 31.4 metric tons/ha manure application. Removal of K was much less than input of K, so large available K reserves have accumulated (Warren & Johnston, 1962, 1967). An exchangeable K level of 15 mg/100 g of soil is generally adequate for most crops.

C. Modern Rates and Requirements

The soil has a large capacity to adsorb K; however, amounts absorbed and not deleterious to crop growth depend on the soil. Principle factors are soil cation exchange capacity and degree of K fixation by the clay minerals. For example, in a Chalmers silt loam with 25 meq/100 g cation exchange capacity, applying 800 kg K/ha annually for 4 years had no adverse effect on corn yield and K fixation was large enough so that at the end of 4 years exchangeable K had only increased from 250 to 1,250 kg/ha (Barber, unpublished data). On soils with below 5 meq/100 g exchange capacities, large K applications may increase available K relative to Ca and Mg levels so that the crop uptake of Ca and Mg are depressed, causing crop deficiencies. Since many soils have a relatively high capacity to adsorb K, as exchangeable and nonexchangeable cations, large manure applications may be made once so it should not be necessary to spread the application over a year.

Besides the soil's capactiy to adsorb K, uptake of K by plants increases as available soil K concentration increases, so that at high K applications, crops grown in the soil adsorb larger amounts. Thus, if the total aboveground portion of the crop is removed, the amount of available soil K is soon reduced.

While wastes may be applied as either solids or liquids, most of the K is water soluble. Thus, it readily reacts with the soil so the form of waste application has little, other than indirect effects, on the K balance of soil.

D. Losses from Runoff

1. SURFACE WATERS

With heavy precipitation on sloping lands, part of the precipitation may run off over the surface and enter lakes and rivers. Since K is held predominately as an exchangeable cation, K equilibration between that in solution in surface runoff and that in the soil reduces the K level in the runoff. When organic wastes containing soluble K are applied to saturated soils, K in runoff waters will be higher because it has not had an opportunity to leach out of the waste into the soil before runoff begins. The K content of river water (3 to 5 μg/g) is probably a good estimate of possible K loss. This amounts to 400 g K ha^{-1} cm^{-1} runoff.

2. SEDIMENT

Losses of K are much larger with sediment in runoff because available K is held as exchangeable K on the colloid removed. Also, the total K content of the sediment will be 1 to 2%. While this total K loss can be quite large, it is tied up so tightly by the soil that it has little plant availability. An average solid content of runoff water is approximately 5%. Assuming this sediment is higher in clay and organic matter than the average soil and has 200 μg/g exchangeable K content, the exchangeable K loss would be 1,000 g K ha^{-1} cm^{-1} runoff. If the sediment had a 2% total K content, the loss would be 100 kg K ha^{-1} cm^{-1} runoff.

E. Losses from Deep Percolation

1. DRAINAGE WATERS

Water percolating through the soil equilibrates with soil exchangeable K. The anion content of this percolating water and other complimentary cations will influence the K level. Since K in wastes would be applied to the soil's upper 15 cm, at this depth solution K should be higher than at lower depths. Hence, when this water percolates through the subsurface horizons (usually low in exchangeable K), solution K will decrease because of the equilibration between percolate and soil. In acid soils where many of the exchangeable cations are Al, solution K is higher because Al is held much more tightly than other cations present and the K level is higher. However, the amount of K in these solutions is usually between 2 to 40 $\mu g/g$. Hence, K removal would be between 200 and 4,000 g K ha^{-1} cm^{-1} percolating water.

2. GROUND-WATER AQUIFERS

Loss of K in water moving to ground water is probably less than surface drainage water. In drainage water the water flows through about 1 m of soil and then frequently is removed by a tile line. When the water moves to the ground-water level, it passes through more soil which would remove K from the water as it passes.

F. Crop Removals

The amount of K removed by a crop is large, particularly when the total plant is harvested, but much less when only the seed is removed. Table 10 shows some typical K-removal values by various crops.

When soil K levels are very high and the total plant is harvested, K removals at the same yield level will be greater because K concentration in the plant is high, due to "luxury consumption." On soils where K is not high, K removed will be lower because the K level in the soil has a large effect on the K content of the plant even when yield is not affected greatly.

G. Microbial Uptake and Storage Time

The K level of soil organic matter is low; although microflora contain K, they do not tie-up enough K during decomposition of organic materials to appreciably affect available soil K.

Table 10—Approximate K removals for several crops[†]

Crop	Yield	K removed
	metric tons/ha	kg/ha
Alfalfa (*Medicago sativa* L.)	8	480
Coastal bermudagrass	20	420
Maize		
Grain	11.2	54
Stover	9.0	215
Wheat (*Triticum aestivum* L.)		
Grain	5.4	30
Stover	6.7	150
Soybeans (*Glycine max* [L.] Merr.)		
Grain	4.0	97
Stover	2.5	60

† From Tisdale and Nelson (1974).

V. OUTLOOK

Rising commercial fertilizer costs and increasing demand to produce more food will continue to increase the value of manure and waste organic materials as nutrient sources for crops. Conservation of P is especially needed because of limited, known, long term sources of this essential element. Nitrogen in manure contributes most to its value but it can be lost readily through improper handling, storage, and application. The cost of NH_3 has increased over 2.5-fold from 1973 to 1975 (Greek, 1975). Such trends emphasize a strong need to use and manage manure wisely. Good methods of handling have been described (Hinish, 1974; Humenik et al., 1974). If crop N requirements are supplied by manure, the P and K requirements will be met on a long term basis.

When manure or organic waste materials are applied for over 2 or 3 years to soils to supply nutrients for crops, the N, P, K levels, micronutrients, soluble salts, and hazardous trace elements should be monitored initially and at regular intervals to assess soil changes. The plants should be observed for phytotoxic effects, nutrient imbalances, or excesses, when such concentrations in the plants may be harmful to animals or man. With these guidelines, manure and similar materials can be used wisely as a valuable resource of plant nutrients. A farmer may receive valuable advice from knowledgeable technical experts available to the agricultural community.

Removing and hauling manure from feedlots or collection areas to the land is a major cost as compared with its value as a source of nutrients for crops and organic matter for the soil. Salter and Schollenberger (1938) assessed the average value of manure as fertilizer at \$2.75/metric ton and the cost was less for commercial fertilizer equivalent to manure nutrients. Hinish (1974) estimated \$2.93/metric ton as the fertilizer value of manure on a typical dairy farm, but he indicated a greater value of manure if the fertilizer is not available. Alternative disposal systems have been tested and considered, but the traditional land disposal method still offers the best answer to utilize a valuable resource (Azevedo & Stout, 1974; Hinish, 1974; Humenik et al., 1974; McCalla, 1974; Salter & Schollenberger, 1938; Warren & Johnston, 1962). The actual value of manure to the animal producer will depend mainly on local manure supply and demand.

LITERATURE CITED

Anderson, G., E. G. Williams, and J. O. Moir. 1974. A comparison of the sorption of inorganic orthophosphate and inositol hexaphosphate by six acid soils. J. Soil Sci. 25:51–62.

Aslyng, H. C. 1954. The lime and phosphate potentials of soils; the solubility and availability of phosphates. Roy. Vet. and Agric. Coll. Yearbook. Copenhagen, Denmark. p. 1–50.

Azevedo, J., and P. R. Stout. 1974. Farm animal manures: an overview of their role in the agricultural environment. Calif. Agric. Exp. Sta. Ext. Serv. Manual 44.

Barber, S. A., J. M. Walker, and E. H. Vasey. 1963. Mechanisms for the movement of plant nutrients from the soil and fertilizer to the plant root. J. Agric. Food Chem.

11:204-207.

Blakeslee, P. A. 1973. Monitoring considerations for municipal waste water effluent and sludge application to land. p. 183-198. *In* Proc. of Joint Conf. on Recycling Municipal Sludges and Effluents on Land. Champaign, Ill. Nat. Assoc. of State Univ. and Land-Grant Coll., washington, D. C.

Boawn, L. C., and J. C. Brown. 1968. Further evidence for a P-Zn imbalance in plants. Soil Sci. Soc. Am. Proc. 32:94-97.

Boawn, L. C., and G. E. Leggett. 1963. Zinc deficiency of Russet Burbank potato. Soil Sci. 95:137-141.

Boawn, L. C., F. G. Viets, Jr., and C. L. Crawford. 1954. Effect of phosphate fertilizers on zinc nutrition of field beans. Soil Sci. 78:1-7.

Chandler, W. H. 1937. Zinc as a nutrient for plants. Bot. Gaz. 98:625-646.

Cooke, G. W. 1967. The availability of plant nutrients in soils and their uptake by crops. p. 48-69. *In* Annu. Rep. of the East Malling Res. Stn. for 1966. Rothamsted Exp. Stn. Rep., Harpenden, Herts., England.

Cooke, G. W., G. E. G. Mattingly, and R. J. B. Williams. 1958. Changes in the soil of a long-continued field experiment at Saxmundham, Suffolk. J. Soil Sci. 9:298-305.

Cooke, G. W., and R. J. G. Williams. 1973. Significance of man-made sources of phosphorus: fertilizers and farming. p. 19-33. *In* Water research. Pergamon Press, Great Britain.

Ellis, R., J. F. Davis, and D. L. Thurlow. 1964. Zinc availability in calcareous Michigan soils as influenced by phosphorus level and temperature. Soil Sci. Soc. Am. Proc. 28:81-86.

Greek, B. F. 1975. Ammonia squeeze loosens, outlook stays good. Chem. Eng. News 53(17):10-13.

Herron, G. M., and A. B. Erhart. 1965. Value of manure on an irrigated calcareous soil. Soil Sci. Soc. Am. Proc. 29:278-281.

Hinish, W. W. 1974. Manure doesn't smell so bad anymore. Crops Soils 27(3):12-15.

Humenik, F. J., M. R. Overcash, L. B. Driggen, and G. J. Kriz. 1974. Cleaning the animal farm environment. Environ. Sci. Technol. 8:984-989.

King, L. D., and H. D. Morris. 1973. Land disposal of liquid sludge: IV. Effect of soil phosphorus, potassium, calcium, and magnesium. J. Environ. Qual. 2:411-414.

Loehr, R. C. 1968. Pollution implications of animal wastes—a forward-oriented review. Robert S. Kerr Water Res. Center, U. S. Department of the Interior, Ada, Okla.

McCalla, T. M. 1974. Use of animal wastes as a soil amendment. J. Soil Water Conserv. 29:213-216.

Manges, H. L., L. A. Schmid, and L. S. Murphy. 1971. Land disposal of cattle feedlot wastes. p. 62-65. *In* Livestock waste management and pollution abatement. Proc. Int. Symp. on Livestock Wastes. Am. Soc. Agric. Eng. Publ. Proc. 271.

Mathers, A. C., and B. A. Stewart. 1971. Crop production and soil analyses as affected by applications of cattle feedlot waste. p. 229-231. *In* Livestock waste management and pollution abatement. Proc. Int. Symp. on Livestock Wastes. Am. Soc. Agric. Eng. Publ. Proc. 271.

Mattingly, G. E. G., A. E. Johnston, and M. Chater. 1970. The residual value of farmyard manure and superphosphate in the Saxmundham Rotation II experiment, 1899-1968. p. 91-112. *In* Rep. Rothamsted Exp. Stn. for 1969. Part 2. Harpenden, Herts., England.

National Academy of Science, Committee on Nitrate Accumulation. M. Alexander, chairman. 1972. Accumulation of nitrate. Nat. Acad. Sci., Washington, D. C.

Olsen, S. R. 1972. Micronutrient interactions. p. 243-264. *In* J. J. Mortvedt, P. M. Giordano, and W. L. Lindsay (ed.) Micronutrients in agriculture. Soil Sci. Soc. of Am., Madison, Wis.

Olsen, S. R., C. V. Cole, F. S. Watanabe, and L. A. Dean. 1954a. Estimation of available phosphorus in soils by extraction with sodium bicarbonate. USDA Circ. 939, Washington, D. C.

Olsen, R. J., R. F. Hensler, and O. J. Attoe. 1970. Effect of manure application, aeration, and soil pH on soil nitrogen transformations and on certain soil test values. Soil Sci. Soc. Am. Proc. 34:222-225.

Olsen, S. R., and F. S. Watanabe. 1970. Diffusive supply of phosphorus in relation to soil textural variations. Soil Sci. 110:318-327.

Olsen, S. R., F. S. Watanabe, H. R. Casper, W. E. Larson, and L. B. Nelson. 1954b.

Residual phosphorus availability in long-time rotations on calcareous soils. Soil Sci. 78:141-151.

Oniani, O. G., M. Chater, and G. E. G. Mattingly. 1973. Some effects of fertilizers and farmyard manure on the organic phosphorus in soils. J. Soil Sci. 24:1-9.

Peperzak, P., A. G. Caldwell, R. R. Hunziker, and C. A. Black. 1959. Phosphorus fractions in manures. Soil Sci. 87:293-302.

Powers, G. W., G. W. Wallingford, L. S. Murphy, D. A. Whitney, H. L. Manges, and H. E. Jones. 1974. Guidelines for applying beef feedlot manure to fields. Kansas State Univ. Ext. Serv. C-502. Manhattan, Kans.

Pratt, P. F., F. E. Broadbent, and J. P. Martin. 1973. Using organic wastes as nitrogen fertilizers. Calif. Agric. 27(6):10-13.

Pratt, P. F., W. W. Jones, and H. D. Chapman. 1956. Changes in phosphorus in an irrigated soil during 28 years of differential fertilization. Soil Sci. 82:295-306.

Sadler, J. M., and J. W. B. Stewart. 1974. Residual fertilizer phosphorus in Western Canadian soils: a review. Saskatchewan Inst. of Pedology. Publ. No. R136.

Salter, R. M., and C. J. Schollenberger. 1938. Farm manure. p. 445-461. In Soils and men, yearbook of agriculture. USDA, Washington, D. C.

Stanford, G. 1973. Rationale for optimum nitrogen fertilization in corn production. J. Environ. Qual. 2:159-166.

Terman, G. L., J. Soileau, and S. E. Allen. 1973. Municipal waste compost: Effects on crop yield and nutrient content of greenhouse pot experiments. J. Environ. Qual. 2:84-89.

Tisdale, S. L., and W. L. Nelson. 1974. Soil fertility and fertilizers. 3rd ed. The Macmillan Co., New York, N. Y.

Vitosh, M. L., J. F. Davis, and B. D. Knezek. 1973. Long-term effects of manure, fertilizer, and plow depth on chemical properties of soils and nutrient movement in a monoculture corn system. J. Environ. Qual. 2:296-299.

Wallingford, G. W., L. S. Murphy, W. L. Powers, and H. L. Manges. 1974. Effect of beef feedlot lagoon water on soil chemical properties and growth and composition of corn forage. J. Environ. Qual. 3:74-78.

Warren, R. G., and A. E. Johnston. 1962. Barnfield. p. 227-247. In Rep. Rothamsted Exp. Stn. for 1961. Harpenden, Herts., England.

Warren, R. G., and A. E. Johnston. 1967. Hoosfield continuous barley. p. 320-338. In Rep. Rothamsted Exp. Stn. for 1966. Harpenden, Herts., England.

Williams, R. J. B., and G. W. Cooke. 1971. Results of the Rotation I experiment at Saxmundham, 1964-69. p. 68-98. In Rep. Rothamsted Exp. Stn. for 1970. Part 2. Harpenden, Herts., England.

chapter 9

ARS soil scientist checks the salt content of brackish water—a necessary first step before using the water for irrigation (Photo courtesy of USDA).

Soluble Salt Considerations with Waste Application

B. A. STEWART, USDA-ARS, Southwestern Great Plains Research Center,
Bushland, Texas

B. D. MEEK, USDA-ARS, Imperial Valley Conservation Research Center,
Brawley, California

I. INTRODUCTION

Wastes vary greatly in salt content as does their application rate to cropland. Also, wastes accumulate under climatic conditions ranging from arid to humid, are applied to soils ranging from sands to clays, and are applied to both nonirrigated and irrigated cropland. Consequently, there is no simple way of indicating waste handling practices to avoid salinity problems. The basic principles of salinity management apply, however, and all factors must be considered to alleviate salinity hazards associated with applying wastes to cropland.

The two hazards most often encountered are excess total salts and high Na levels. Excess total salts can reduce germination and growth. High Na levels and K, to a lesser degree, cause dispersion of soil particles, poor soil structure, and reduced infiltration rates. Therefore, the nature of the salts, as well as application rates, should be determined.

II. SALINITY CHARACTERISTICS OF WASTES

A. Variability in Soluble Salt Content

1. ANALYTICAL METHODS

The soluble salt content of wastes can be determined by analyzing each of the constituents using the decantate of an organic waste water mixture. A simpler method is to measure the electrical conductivity (EC) of the organic waste water mixture and then to use a factor to calculate total soluble salt content. Soluble salt content will be influenced by dilution of the organic waste with water because some salts have low solubility, like gypsum. Therefore, the dilution ratio must be standardized to compare analyses.

Estimating soluble salt content by measuring EC generally is accurate enough to evaluate the effects of organic waste application rates. A satisfactory procedure is to add 10 g of organic waste to a 1-liter volumetric flask and to fill with distilled water. The EC of the decantate is measured and then the soluble salt content of the organic waste is estimated by multiplying the EC value by an appropriate factor. For NaCl, the factor is 640 ppm for 1 mmho/cm of EC, but it varies for water samples containing different amounts of Ca. A satisfactory estimate, however, usually will result if a factor of about 700 ppm is assumed.

Sometimes it may be necessary to measure the concentration of each salt in the decantate of the organic waste water mixture. This information is valuable for estimating salt effects in soils, such as changes in exchangeable Na or K. The amounts of exchangeable Na or K and Ca and Mg are used to calculate sodium adsorption ratios (SAR) and potassium adsorption ratios (PAR). These ratios,

$$SAR = Na^+/[(Ca^{2+} + Mg^{2+}/2)]^{\frac{1}{2}} \text{ and } PAR = K^+/[(Ca^{2+} + Mg^{2+}/2)]^{\frac{1}{2}}$$

for soil extracts and irrigation waters are used to express relative activities of Na or K in exchange reactions with soil. Ionic concentrations are expressed in meq/liter. Also, a complete analysis aids in identifying toxic ions present in the soil solution. However, a complete analysis requires substantial time and is often not necessary for evaluating the effects of soluble salts in organic wastes.

2. ORGANIC WASTE COMPOSITION

Table 1 gives the Na, K, Ca, and Mg contents of representative organic waste materials. The sum of these four cations will be approximately one-half that of total salt content. Generally there is a good balance between Ca plus Mg, and Na and K in solid wastes. The primary salinity hazard of utilizing these wastes on soils is total salts, rather than the Na hazard. However, the Na content should be carefully watched, and management practices should be followed that will keep the Na content of the wastes as low as possible. Beef cattle wastes often have higher salt contents in arid areas because of less leaching. Sewage sludge is usually low in soluble salts because many of the salts have been removed with the sewage effluent. The principles developed to reduce the effects of soluble salts contained in animal waste should apply equally well for sewage sludge.

Table 1—Amounts of Na, K, Ca, and Mg in organic waste materials

Organic waste	Na	K	Ca	Mg
		% dry weight		
Beef cattle manure				
Nebraska[†]	0.10	0.46	0.26	0.17
Kansas[‡]	0.23	1.09	0.78	0.39
Texas Panhandle[§]	1.13	2.29	1.98	0.76
Arid Southwest[¶]	1.12	2.30	2.80	1.53
Dairy cattle manure				
California[#]	0.40	1.72	1.93	0.86
Poultry manure				
Georgia[††]	0.40	1.51	1.55	0.35
Sewage sludge[‡‡]				
	0.39	0.74	3.43	0.64

[†] McCalla et al., 1972.
[‡] Murphy et al., 1972.
[§] Mathers et al., 1972.
[¶] Meek et al., 1974.
[#] B. D. Meek, unpublished data.
[††] Wilkinson et al., 1971.
[‡‡] Vlamis and Williams, 1971.

III. WASTE WATER COMPOSITION

Table 2 gives the cation composition values for selected waste waters. The Na adsorption ratios (SAR) and K adsorption ratios (PAR) are also shown. Both Na and K, to a lesser degree, can cause soil structure, infiltration, and permeability rate problems. A high percentage of exchangeable Na in soils containing swelling-type clays results in dispersed conditions that restrict plant growth and water movement. The amount of K adsorbed (exchangeable) by a soil from irrigation water is a function of the proportion of Na to Ca and Mg in the water. Therefore, the higher the SAR, the higher the soil exchangeable Na will become. For sensitive fruits, the tolerance limit for the SAR of irrigation water is about 4. For field crops, a limit of 8 to 18 is usually considered within a usable range, although this depends somewhat on the type of clay mineral, electrolyte concentration in the water, and other variables.

The general relationships presented above were derived for neutral and alkaline soils of arid areas and do not apply equally as well to acid soils found

Table 2—Sodium, K, Ca, and Mg concentrations and sodium (SAR) and potassium (PAR) adsorption ratios for some waste waters

	Na	K	Ca	Mg	SAR	PAR
	meq/liter					
Sewage effluents						
University Park, Pa.[†]	2.0	0.4	1.6	1.5	1.7	0.3
St. Paul, Minn.[‡]	12.2	0.4	3.5	2	7.4	0.2
Abilene, Tex.[§]	8.3	0.9	0.9	3.7	5.5	0.6
Pomona, Calif.[§]	4.4	0.3	2.1	2.2	3.0	0.2
Beef cattle feedyard runoff						
Bushland, Tex.[¶]	26	34	22	17	5.9	7.8
Mead, Nebr.[#]	37	65	40	41	5.6	10.2
Pratt, Kans.[††]	13	18	13	8	4.1	5.3
Waste waters from plants processing selected food products in the Mid-Atlantic States[‡‡]						
Green beans	4.3	0.6	1.0	1.2	4.1	0.6
Lima beans	34.4	1.1	1.6	1.3	28.7	0.9
Tomatoes	5.3	1.5	4.0	1.3	3.3	0.9
Corn	3.2	1.5	0.9	1.5	2.9	1.4
Sweet potatoes, steam-peeled	2.2	1.1	0.8	1.2	2.2	1.1
Sweet potatoes, lye-peeled	163.0	9.7	3.0	3.3	91.6	5.4
Poultry	1.6	0.6	1.6	0.7	1.5	0.6

[†] Parizek et al., 1967.
[‡] W. E. Larson (private communication).
[§] Pound and Crites, 1973.
[¶] Clark et al., 1975.
[#] McCalla et al., 1972.
[††] Wallingford et al., 1974.
[‡‡] Pearson et al., 1972.

in humid regions. Lunin and Batchelder (1960) found that the amount of Na adsorbed increased as the level of Ca saturation in the soil increased. For this reason, soil characteristics, which differ greatly between arid and humid areas, must be considered.

The waste waters presented in Table 2 vary greatly in amounts of salts, SAR, and PAR. Sewage effluents usually are satisfactory for irrigation water, but vary considerably, so a careful analysis of the effluent is desirable. Waste waters from processing plants can vary extremely. Based on samples obtained from food processors in the Mid-Atlantic area, Pearson et al. (1972) concluded that waste water from plants processing green beans (*Phaseolus vulgaris* L.), squash (*Cucurbita maxima*), tomatoes (*Lycopersicon esculentum* Mill), corn (*Zea mays* L.), steam-peeled sweet potatoes (*Ipomoea batatas* Lam.), and poultry was suitable for irrigation. Waste water from pea (*Pisum sativum* L.) and lima bean (*P. limensis*) processing may be suitable, but special attention is often required because of high concentrations of Na and Cl in the water from some processing plants. Waste water from lye-peeled sweet potato processing plants is not suitable because of extremely high Na concentrations.

Beef cattle feedyard runoff usually has a relatively high SAR and accompanying high PAR. A high amount of exchangeable K also can have a negative effect on the physical properties of soils, but it is generally regarded as less damaging than Na.

A. Leaching of Soluble Salts

1. EFFECT OF PARTICLE SIZE

The leaching of soluble salts from organic wastes is influenced by particle size. Soluble salts leach more rapidly from wastes which have a small particle size. Amoozegar-Fard et al. (1975) found that reducing the size of manure aggregates, from 4.8 by 2.6 cm, so they were small enough to pass through a screen, with 0.42-mm openings, increased conductivity of the leachate from 10.9 to 20.6 mmhos/cm during the first leaching period, but decreased it from 3.8 to 2.7 during the fifth leaching period. Also, hazardous salts (K and Na) in the manure were highly soluble and were readily leached. However, divalent cations, Ca and Mg, the nutrient materials (N and P), and trace elements were less soluble and migrated more slowly.

2. RELEASE OF SALTS DURING DECOMPOSITION

Part of the salts in organic wastes is not soluble initially but is released during decomposition of the waste. Nitrogen is a good example because much of the total N in wastes is in the organic form. During decomposition, organic N is converted to inorganic forms with a subsequent rise in salinity.

Table 3—Percent soluble† Na, K, Ca, Mg, and SO_4 in manure (from four locations)

Location	Na	K	Ca	Mg	SO_4
			%		
Feedlot manure					
Brawley, Calif.	77	71	4	10	55
Blythe, Calif.	82	89	8	23	56
Bushland, Tex.	85	70	12	38	57
Dairy manure					
Hemet, Calif.	92	64	4	9	80

† Soluble equals amount in solution when 10 g of manure are diluted in 1 liter of water. Total was determined on a triple-acid digest.

Pratt et al. (1973) published tables giving representative data on decomposition rates of organic wastes.

In addition to N, significant amounts of Ca and Mg may be released during decomposition (Table 3).

IV. FACTORS THAT INFLUENCE RATES OF WASTE APPLICATIONS IN RELATION TO CROP GROWTH

A. General

Wastes are applied to cropland both for recycling nutrients and for disposal. Even when emphasis is on disposal, it is desirable to obtain good plant growth. Several factors are important and they will be considered individually, although they are interrelated and the complete waste disposal system must be evaluated.

B. Soil Characteristics

1. SOIL TEXTURE

The range of conductivity values found in soils and the effect of texture on conductance are shown in Fig. 1. The line for a saturation percentage (SP) of 50% (water percentage on a dry weight basis) represents a fine-textured soil. The 25% SP line typifies a coarse-textured soil. A fine-textured soil containing 0.1% salt (equivalent to about 2,200 kg/ha in 15 cm of soil) would have a saturation extract EC value of 3 mmhos/cm. This amount of salt would be added with 30 metric tons/ha of waste containing 7.5% salt. The same amount of salt in a sandy soil would result in an EC value of over 6 mmhos/cm. This illustrates the importance of soil texture in salinity management. However, even though salt additions influence the EC of sandy soils more than clay soils, salts are more easily leached from sandy

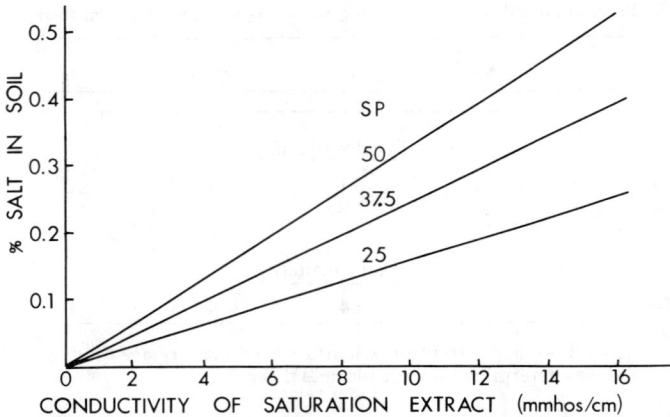

Fig. 1—Range of conductance values found in saturation extracts of soils as influenced by texture and salt content. Saturation percentages (SP) of 50 to 25 represent fine-textured and coarse-textured soils, respectively (U. S. Salinity Laboratory Staff, 1954).

soils. This results in a greater buildup of salinity on fine-textured soils from repeated annual applications of organic waste materials.

2. INFILTRATION RATE

Infiltration rate of a soil is the maximum rate that water can enter under specified conditions. The rate is influenced by factors like condition of the soil surface, nature and physical status of the soil profile, and distribution of water in the soil profile. Some soils have very low infiltration rates and generally contain high initial salt contents. For these soils, it may not be possible to maintain a favorable salt balance when extra salt is added by organic waste applications, unless these applications increase infiltration rate.

C. Water Management

1. GENERAL

Water management is the most important factor in preventing harmful effects of salts added in organic wastes. It is also the factor most readily controlled by management practices, both under irrigated and nonirrigated conditions.

2. AMOUNT OF LEACHING

To maintain the EC in the soil solution at a suitable level, the amount of water moving through the soil profile must be increased as salt added with organic waste applications is increased. This is illustrated in Fig. 2 and 3.

Figure 2 shows the amounts of leaching water required to maintain a

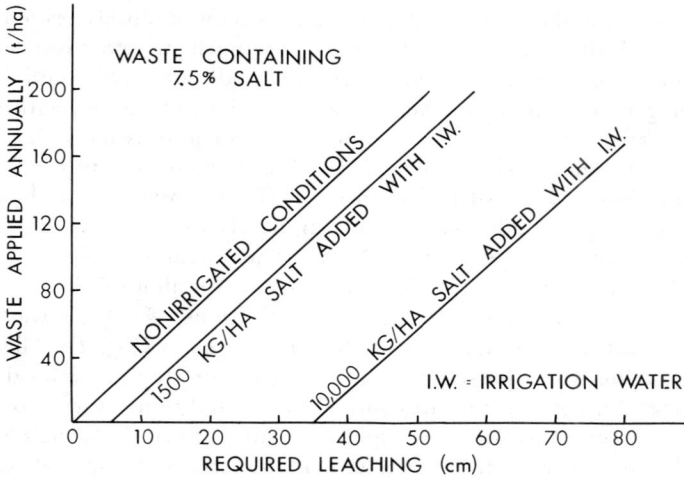

Fig. 2—Water (cm) required to move through the root zone so as not to exceed 4 mmhos/ cm EC in the soil solution as a function of dry waste applied.

conductivity of 4 mmhos/cm or less in the root zone when varying rates of organic waste, containing 7.5% soluble salts, are applied. For example, under nonirrigated conditions, approximately 40 metric tons/ha of wastes (dry weight basis) could be added annually without buildup of salinity when 10 cm of water is moved through the profile annually. However, much of the eastern United States has >10 cm of leaching water annually. Therefore, salinity is not likely to be a problem unless manure is applied at very high rates for disposal. In arid areas there is little, if any, leaching and waste ap-

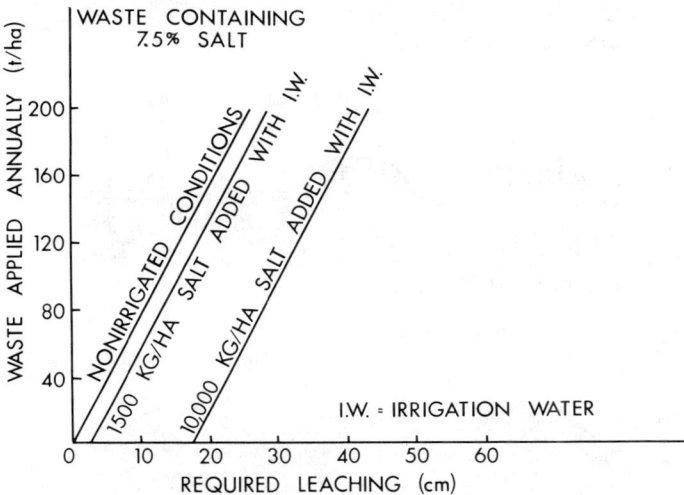

Fig. 3—Water (cm) required to move through the root zone so as not to exceed 8 mmhos/ cm EC in the soil solution as a function of dry waste applied.

plication rates should be low and soil salinity followed closely, except under irrigated conditions. Even here, however, care must be taken to insure that salts do not accumulate. Leaching is required under irrigated conditions and the leaching requirement will be increased as the salts added with organic wastes increase. An annual application of 40 metric tons/ha of dry organic waste (Fig. 2) requires 10 cm of additional leaching water. In the Southern High Plains about 45 cm of irrigation water (EC of about 0.5 mmhos/cm) is applied annually which adds about 1,500 kg/ha of salt. This has an estimated leaching requirement of only about 5 cm. The actual leaching requirement may be even less because Ca and Mg salts account for about 60% of the total. Consequently, 20 metric tons/ha annual applications of organic wastes containing 7.5% salt would double the leaching requirement (Fig. 2). This would not be desirable from the standpoint of cost or supply in a water-deficient area. Therefore, application rates should be limited under these conditions to amounts required to supply adequate fertility. Even then, the salt status of the soil should be monitored. Under many desert conditions, where irrigation water quality is low and evapotranspiration is high, 10,000 kg/ha or more salt may be added annually in irrigation water. Under these conditions, salts from organic wastes will be a smaller percentage of the total salt but their impact may be great because many soils are presently being leached at or near their maximum rate. These examples (Fig. 2) are based on a maxi-

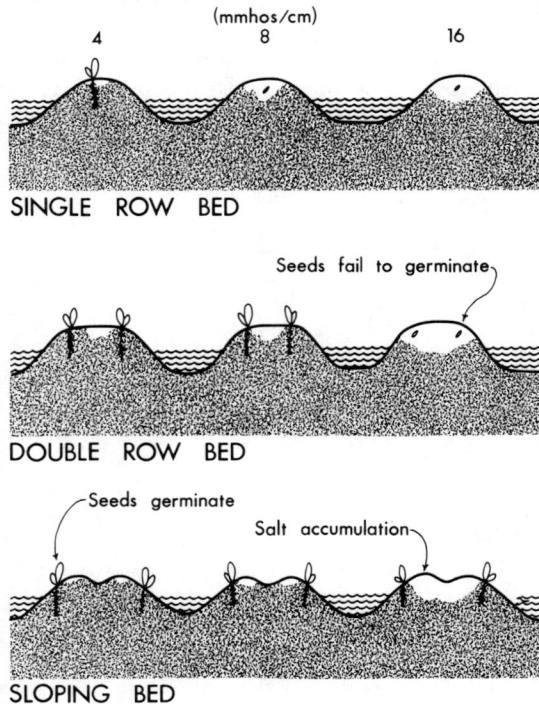

Fig. 4—Soil salinity at planting time under single and double row beds and sloping beds.

mum EC of 4 mmhos/cm in the root zone, which would allow essentially all crops to be grown without salt management practices.

The leaching requirement can be reduced by 50% if the EC value is allowed to increase to 8 mmhos/cm (Fig. 3). This requires some crop selection and salt management practices; however, many crops can be grown satisfactorily at these levels.

3. IRRIGATION METHODS

The irrigation method is often controlled by the cropping system, soil type, topography, water supply, and economic considerations. Different irrigation methods can vary considerably in their leaching effectiveness. In general, several small applications are more effective than a few large applications because more time is allowed for salts to equilibrate with the leaching water. Sprinkler systems are more efficient than border or furrow irrigation methods because they allow irrigations at shorter intervals and provide uniform leaching over the irrigated area. Furrow irrigation is least efficient because salts accumulate in the ridges. Figure 4 illustrates this accumulation of salt at three salinity levels (4, 8, 16 mmhos) for single row beds. Two other bed shapes (double row and sloping) are also shown and these allow germination in saline soils.

4. TIMING OF IRRIGATION

Organic waste applications should be timed so that at least one irrigation is applied before planting to disperse the added salt into a larger part of the soil profile. Deeper incorporation will distribute salt into a larger volume of soil and allow higher manure rates to be applied.

5. EFFECT OF WASTE APPLICATION ON INFILTRATION RATE

Organic wastes will usually improve the infiltration rate of soil if the surface layer is restricting. This improvement in infiltration rate helps to counteract salts added in organic wastes. B. D. Meek (unpublished data) found manure applications resulted in an increase in the initial infiltration rate (1.25 hours), but did not affect the basic infiltration rate (after 12 hours) (Table 4). Also, the infiltration rate was much higher during the growing season for plots which received manure. For example, during the third irrigation, the infiltration rate was 0.47 cm/hour for no manure treatment and 4.60 cm/hour for the treatment receiving 144 metric tons/ha of manure. The effect of manure application on the infiltration rate may last for years—the plots that received 144 or 288 metric tons/ha (dry weight) manure rates in 1971 still had higher initial infiltration rates in 1973 than plots that did not receive manure. The fiber content of animal manures may be the most important parameter controlling soil hydraulic conductivity after waste applications according to Azevedo and Stout (1974). Solid organic waste materials

Table 4—Infiltration rates for manure treatments in 1973 during the growing season and after post-harvest tillage

Manure treatment†	Third irrigation‡	After harvest	
		1.25-hour flooding	9-hour flooding
metric tons/ha		cm/hour	
0-0-0	0.47 e§	0.86 a	0.25 a
36-36-36	1.14 c	1.65 b	0.30 a
72-72-72	2.48 ab	1.42 b	0.46 a
144-144-144	4.60 a	2.23 c	0.51 a
288-0-0	0.61 d	1.65 b	0.46 a
144-0-0	0.71 d	1.65 b	0.38 a
144-288-0	1.51 b	1.58 b	0.28 a

† Dry weight of manure applied in 1971, 1972, and 1973, respectively.
‡ Applied 9.67 cm of water to each plot and the infiltration rate is for the total time necessary for the water to infiltrate.
§ Values not followed by the same letter are significantly different at the 0.05% level.

usually have a good balance between monovalent and divalent ions and usually do not cause deterioration of the physical properties of the soil.

Beef feedlot lagoon water and sewage effluent may contain an excess of monovalent ions and cause deterioration of the physical properties of soils. Travis et al. (1971) found that feedlot lagoon water reduced infiltration rate in soil columns. Powers et al. (1973) and Clark et al. (1974) recommended dilutions of at least four parts water to one part feedyard runoff to prevent salinizing disposal areas. In areas of high rainfall, natural dilution is considerable and salinity hazard is reduced.

D. Crop Effects

1. SELECT CROP FOR SOLUBLE SALT LEVEL

Crops can be selected so that there will be adequate growth at the salinity level present in the root zone. Table 5 is a generalized index of the effect of soil salinity on plant growth. Plants that are salt sensitive or only moderately tolerant decrease in growth and yield as salinity levels increase. Figure

Table 5—Response of crops grown in soils of varying electrical conductivity (EC) values at 25C

Salinity EC	Crop responses
mmhos/cm at 25C	
0-2	Salinity effects mostly negligible
2-4	Yields of very sensitive crops may be restricted
4-8	Yields of many crops restricted
8-16	Only tolerant crops yield satisfactorily
Above 16	Only a few very tolerant crops yield satisfactorily

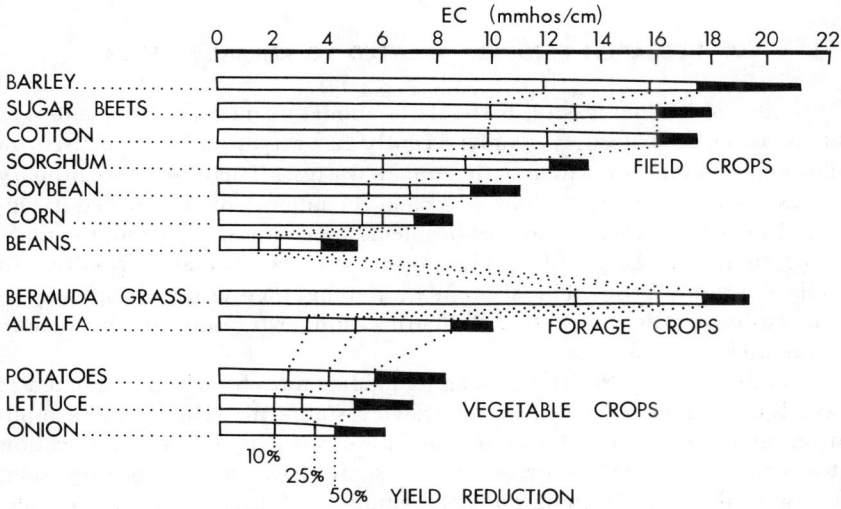

Fig. 5—Effect of soil salinity on crop growth (adapted from Bernstein, 1964).

5 shows salt tolerance of some important crop plants. Indicated tolerances apply to the period of rapid plant growth and maturation, from the late seedling stage onward. These data show that barley (*Hordeum vulgare* L.) and bermudagrass (*Cynodon dactylon* L.) can tolerate very high salinity, while beans, lettuce (*Lactuca sativa* L.), and onions (*Allium cepa* L.) are affected by small amounts of soluble salts.

2. EFFECT OF STAGE OF GROWTH

In selecting crops, particular attention should be given to the effect of salinity on germination. Poor crops often result because of poor stands. Yet, some plant species are sensitive to salinity during germination while during later stages of growth they are very tolerant. An example is sugarbeets (*Beta vulgaris* L.). Modifying seeding practices (mentioned under the section on irrigation methods) can minimize salt accumulation near the seed.

3. TILLAGE CONSIDERATIONS

Tillage plays an important part in the rate of organic waste that can be applied. A uniform mixture of organic waste and soil should be obtained by tillage operations. Tillage can mix the organic waste with a greater volume of soil and, thus, reduce salt content. Reddell (1973) incorporated 200 metric tons/ha of manure (51% moisture, wet weight basis) to a 90-cm depth and still was able to increase yield over that of the check. This study was conducted near El Paso, Texas, on a Vinton fine sandy loam using good quality irrigation water.

V. RESEARCH FINDINGS RELATED TO ANIMAL WASTES

Utilizing organic wastes, particularly animal manures, on cropland is as old as agriculture itself, yet only recently has attention been focused on salinity problems associated with animal wastes. There are two primary reasons for this increased attention. First, the number of animals fed has increased in arid areas where salinity problems are common. Second, the number of animals in large-scale feeding operations has increased, resulting in localized accumulation. This accumulation sometimes results in applying organic wastes for disposal, rather than utilization, or transporting the manure several miles for utilization.

Soluble salt aspects of high manure loading rates have been evaluated at Pratt, Kansas; Bushland, Texas; and Brawley, California. High manure application rates decreased yields at all three locations (Fig. 6). The application rates required for yield reductions varied somewhat, which is not surprising, because of different leaching rates and amounts of leaching water and soils. Yields were reduced at a lower manure application rate at Bushland, Texas, which was due to less leaching because of lower water application (Fig. 6).

The harmful salinity effects of these high rates of manure application generally do not persist. Figure 7 shows yield of grain sorghum (*Sorghum bicolor* Moench.) at Bushland, Texas, on plots treated with 268 metric tons/ ha of manure for 1 or 3 years, as compared with 11 metric tons/ha for 3 years. There was only a small reduction 1 year after manure applications and no reduction the second year. Yield recoveries at Brawley, California, have been even more rapid. An application of 288 metric tons/ha of manure in January reduced grain sorghum yields in August by 50%. The area was double-cropped with lettuce, a salt sensitive crop, which showed no yield reduction.

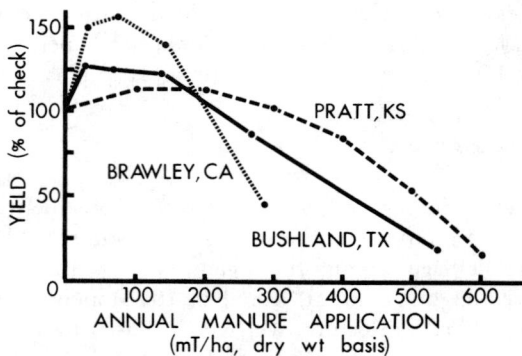

Fig. 6—Effect of manure on plant yields (grain sorghum at Bushland, Tex. [Mathers et al., 1972] and Brawley, Calif. [Meek et al., 1974] and corn silage at Pratt, Kans. [Wallingford et al., 1974]).

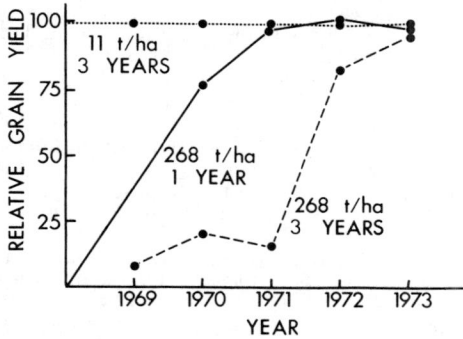

Fig. 7—Recovery of grain sorghum yields from plots at Bushland, Tex., treated with 268 metric tons/ha of manure (dry weight) in 1969 only, and in 1969, 1970, and 1971 as compared with plots treated with 11 metric tons/ha of manure each year (A. C. Mathers and B. A. Stewart, unpublished data).

VI. GUIDELINES FOR PREVENTING SALINITY PROBLEMS FROM ORGANIC WASTE APPLICATION

1) Determine soluble salt content of wastes before applying wastes to the field.
2) Determine soil characteristics.
3) Time applications so that added salts will leach somewhat before seeding.
4) Use Fig. 2 and 3 as a guide for estimating the maximum rate that can be applied.
5) When feasible, apply wastes at rates to maximize nutrient utilization. Salinity will not usually be a problem under these conditions.

LITERATURE CITED

Amoozegar-Fard, A., W. H. Fuller, and A. W. Warrick. 1975. Migration of salt from feedlot waste affected by moisture regime and aggregate size. J. Environ. Qual. 4:468-·472.

Azevedo, J., and P. R. Stout. 1974. Farm animal manures: an overview of their role in the agricultural environment. California Agric. Exp. Sta. Ext. Serv. Manual No. 44.

Bernstein, L. 1964. Salt tolerance of plants. USDA Agric. Inform. Bull. No. 283. 24 p.

Clark, R. N., A. D. Schneider, and B. A. Stewart. 1975. Analysis of runoff from Southern Great Plains feedlots. Trans. Am. Soc. Agric. Eng. 18:319-322.

Lunin, J., and A. R. Batchelder. 1960. Cation exchange in acid soils upon treatment with saline conditions. Int. Congr. of Soil Sci. Trans. 7th (Madison, Wis.) 1:507-515.

McCalla, T. M., J. R. Ellis, C. B. Gilbertson, and W. R. Woods. 1972. Chemical studies of solids, runoff, soil profile, and groundwater from beef cattle feedlots at Mead, Nebraska. Proc. Cornell Agric. Waste Manage. Conf. p. 211-223.

Mathers, A. C., B. A. Stewart, J. D. Thomas, and B. J. Blair. 1972. Effects of cattle feedlot manure on crop yields and soil conditions. USDA Southwestern Great Plains Res. Center Tech. Rep. No. 11.

Meek, B. D., A. M. MacKenzie, T. J. Donovan, and W. F. Spencer. 1974. The effect of large applications of manure on movement of nitrate and carbon in an irrigated desert region. J. Environ. Qual. 3:253-258.

Murphy, L. S., G. W. Wallingford, W. L. Powers, and H. L. Manges. 1972. Effects of solid beef feedlot wastes on soil conditions and plant growth. Proc. Cornell Agric. Waste Manage. Conf. p. 449–464.

Parizek, R. R., L. T. Kardos, W. E. Sopper, E. A. Myers, D. E. Davis, M. A. Farrell, and J. B. Nesbitt. 1967. Waste water renovation and conservation. The Pennsylvania State Univ. Studies No. 23. University Park, Pa.

Pearson, G. A., W. G. J. Knibbe, and H. L. Worley. 1972. Composition and variation of waste water from food-processing plants. USDA-ARS 41–186. 10 p.

Pound, C. E., and R. W. Crites. 1973. Characteristics of municipal effluents. Proc. of Joint Conf. on Recycling Municipal Sludges and Effluents on Land. Champaign, Ill. p. 49–62.

Powers, W. L., R. L. Herpich, L. S. Murphy, D. A. Whitney, H. L. Manges, and G. W. Wallingford. 1973. Guidelines for land disposal of feedlot lagoon water. Coop. Ext. Serv. Circ. No. C-485. Kansas State Univ., Manhattan, Kans.

Pratt, P. F., F. E. Broadbent, and J. P. Martin. 1973. Using organic wastes as nitrogen fertilizers. Calif. Agric. 27:10–13.

Reddell, D. L. 1973. Crop yields from land receiving large manure applications. Symp. on Animal Waste Management, USDA Southwestern Great Plains Research Center, Bushland, Tex. p. 14–26.

Travis, D. O., W. L. Powers, L. S. Murphy, and R. I. Lipper. 1971. Effect of feedlot lagoon water on some physical and chemical properties of soils. Soil Sci. Soc. Am. Proc. 35:122–126.

U. S. Salinity Laboratory Staff. 1954. Diagnosis of saline and alkali soils. U. S. Dep. of Agric. Handb. No. 60.

Vlamis, J., and D. E. Williams. 1971. Utilization of municipal organic wastes as agricultural fertilizers. Calif. Agric. 25:7–9.

Wallingford, G. W., L. S. Murphy, W. L. Powers, and H. L. Manges. 1974. Effect of beef-feedlot-lagoon water on soil chemical properties and growth and composition of corn forage. J. Environ. Qual. 3:74–78.

Wilkinson, R. R., J. A. Studemann, D. J. Williams, J. B. Jones, Jr., R. N. Dawson, and W. A. Jackson. 1971. Recycling broiler house litter on tall fescue pastures at disposal rates and evidence of beef cow health problems. p. 321–324. In Proc. Int. Symp. on Livestock Wastes. Columbus, Ohio. Am. Soc. Agric. Eng., St. Joseph, Mich.

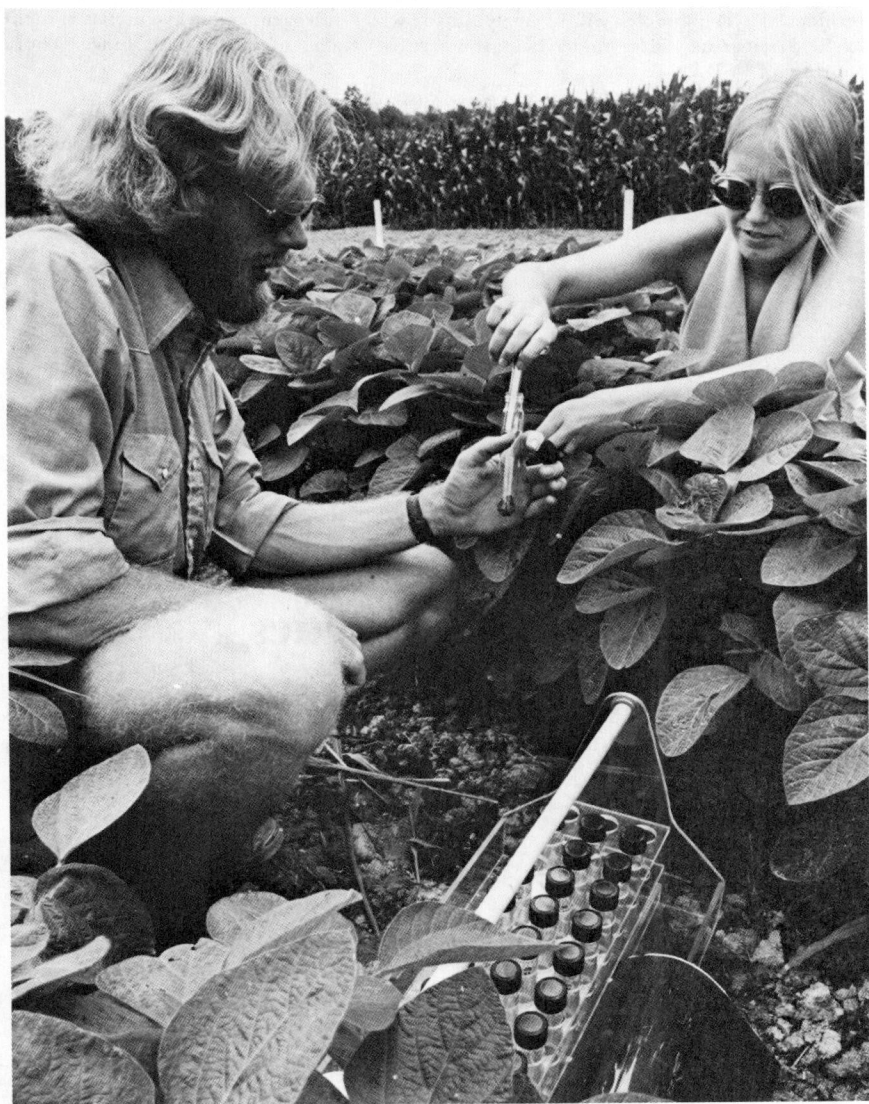

chapter 10

Leaf samples from soybeans and corn fertilized with composted sludge are taken by technicians. The samples will be analyzed for water, nitrogen, and heavy metals content to help scientists better understand the fertilizer value of composted sludge (Photo courtesy of USDA).

Microelements as Related to Plant Deficiencies and Toxicities

RUFUS L. CHANEY, USDA-ARS, Beltsville, Maryland

PAUL M. GIORDANO, Tennessee Valley Authority, Muscle Shoals, Alabama

I. INTRODUCTION

Utilization or disposal of organic wastes and waste waters on land adds micro-elements to the land. Because nearly all sewage wastes and refuse contain levels of microelements higher than those present in soils, agronomists are studying the long term effects of waste-applied microelements. Some micro-nutrient elements (Zn, Cu, B, Fe) in wastes may serve to correct plant de-ficiencies. Alternatively, repeated or unwise use of wastes might lead to phytotoxicities of Zn, Cu, Ni, or B. Further, certain nonnutrient micro-elements (Cd, Pb, Hg) present in some wastes are of concern to agronomists, nutritionists, and environmentalists because of the hazard to animal health.

Although the correction of micronutrient deficiencies has become routine agricultural practice, the study of potentially phytotoxic additions of microelements is a young science. Research on the micronutrients has provided much information on the waste-soil-plant-animal relationships of some of these elements; however, little is known about many of the factors controlling these relationships at high levels. The focus of this research is the definition of the maximum application rates of waste-borne microelements which are safe for crops and the food chain in the long term. Agronomists want to be sure that the apparent economic saving for cities and farmers gained by land application of wastes does not become an environmental or fertility problem of the future. Most of the identified potential problems from waste-borne microelements can be avoided by good management—proper selection of wastes, soils, and crops, and proper reliance on soil and plant testing. This review summarizes the knowledge available about micro-elements in organic wastes, focusing on the management tools available to avoid long term problems, and on research needed to improve these manage-ment tools.

II. CORRECTION OF PLANT MICRONUTRIENT DEFICIENCIES BY UTILIZING ORGANIC WASTES AND WASTE WATERS ON AGRICULTURAL LAND

Plant deficiencies of Zn, Fe, and B are commonly observed in the United States, and deficiencies of Mn, Cu, Mo, and Co are occasionally found. Kubota and Allaway (1972) described the geographic distribution and soil properties related to these deficiencies. Many aspects of the use of micro-nutrients in agriculture were reviewed recently (see Mortvedt et al., 1972), including climatic and soil conditions which influence deficiencies (Lucas & Knezek, 1972), crop differences in micronutrient requirements (Brown et al., 1972), and the use of micronutrient fertilizers to correct these deficiencies (Murphy & Walsh, 1972). Most micronutrient fertilizers are soluble inorganic compounds or soluble metal chelates, although some insoluble inorganic com-pounds (e.g., $ZnCO_3$, ZnO, CuO) are good fertilizers. Micronutrient fertiliza-tion has been increasing nationwide. Because most elements needed as micronutrient fertilizers are present in one or more of the wastes considered herein, it is appropriate to review the use of these wastes as micronutrient fertilizers.

The concept of using organic wastes to correct micronutrient deficiencies is linked to the potential of these same wastes to cause toxicities. Any waste which contains enough of an element to cause toxicity if used in excess, could be considered a fertilizer if applied at lower rates to correct a deficiency of that element. Thus, wastes may be used to correct Zn, Cu, and possibly B, Mn, and Fe deficiencies.

Animal manures were used to correct Zn deficiency on eroded and leveled land (Grunes et al., 1961b) long before the specific need for Zn on these areas was established. Carlson et al. (1961), Grunes et al. (1961a), and Follet et al. (1974) studied the response of corn (Zea mays L.) to amendments of N, P, manure, Zn, etc., on Gardena surface soil and exposed subsoil in North Dakota. The exposed subsoil was Zn deficient. A 44 metric ton/ha manure application supplied enough Zn to give near maximum yields of corn on the exposed subsoil, although the manure was not as effective as $ZnSO_4$ in increasing the Zn content of the corn leaves. Viets (1966) suggested that the increasing problem of Zn deficiency may be related to the failure to return manure to the land which grew the crops; ordinarily the added Zn should be considerably more available to plants than the native soil Zn.

A study of crop response to chemical and organic N fertilizers in Florida showed that Cu in activated sludge corrected an unrecognized Cu deficiency of watermelon (Citrullus vulgaris) and cucumber (Cucumis sativus L.) (Fiskell et al., 1964). Field plots on virgin soils showed marked variation in plant response to the organic N sources with very large responses in some years being due apparently to Cu. Fiskell et al. (1964) concluded that sludge could be used to correct Cu deficiency.

Subsequently, Miller et al. (1969) and Parsa (1970)[1] reported detailed studies of the utility of poultry manure, sewage sludge, and composted refuse to correct Zn and Fe deficiencies of corn on a calcareous soil. Miller et al. (1969) added only 3.5 mg Zn (in poultry manure)/kg soil, yet this corrected severe Zn deficiency. Parsa (1970)[1] compared several methods of Zn fertilization of a severely Zn deficient, slightly acid, sandy clay loam. Treatments included $ZnSO_4$, a lime plus $FeCl_3$-treated activated sludge and its ash, and a refuse compost and its ash. At equal Zn rates the organic wastes were as effective as $ZnSO_4$ in correcting Zn deficiency of corn, but the ashes of the wastes were ineffective. At higher application rates, the activated sludge produced some phytotoxicity which complicated interpretation of the yield results; this type of phytotoxicity commonly occurs when raw activated sludge is applied at rates exceeding the N requirement.

The potential use of organic wastes as Fe fertilizers is especially attractive because inorganic Fe is seldom effective except at high rates (Hodgson et al., 1972; Mortvedt & Giordano, 1973), and synthetic chelates of Fe are

---

[1]A. A. Parsa. 1970. Solid wastes as Zn fertilizers. Ph.D. Thesis. Colorado State University, Fort Collins, Colo. 44 p.

expensive. Mortvedt and Giordano (unpublished) compared sewage sludge with $FeSO_4$ and FeEDDHA as sources of Fe for sorghum (*Sorghum bicolor* Moench.) in a greenhouse trial. Forage yields and Fe content in two successive crops of sorghum were highest with FeEDDHA. However, the 175 mg of Fe in the 15 metric tons/ha of sewage sludge (0.35% Fe) were more effective than 24 mg of Fe as $FeSO_4$, especially for the second crop. Lindsay and Park (1970) compared the ability of refuse compost (4.75% Fe), lime plus $FeCl_3$-treated activated sludge (1.98% Fe), and FeEDDHA to correct Fe deficiency of sorghum grown on a calcareous clay loam. They also compared $FeSO_4$ additions in the presence and absence of organic wastes. Addition of $FeSO_4$ to the refuse compost increased yields well beyond levels obtained with $FeSO_4$ alone, at least for the first crop. Activated sludge was as effective as FeEDDHA and more effective than the refuse compost; inclusion of $FeSO_4$ was of no added benefit; yields were as high as with FeEDDHA. Miller et al. (1969) observed a partial correction of Fe deficiency of corn with poultry manure.

Whether organic wastes can be safely used to correct Fe deficiency remains uncertain. The wastes vary widely in Fe content and chemical form. Many sludges will be relatively high in Fe when $FeCl_3$ is used to precipitate the P and suspended solids during improved secondary treatment. On the other hand, other heavy metals (Zn, Cu, etc.) in some organic wastes may eventually accentuate Fe deficiency stress (Brown & Jones, 1975), even though the deficiency is temporarily corrected.

Thus, it has been experimentally demonstrated that manure, sewage sludge, and composted refuse can be used to correct Zn and Cu deficiencies, and they may alleviate Fe deficiency. Because application of high rates of unleached refuse and refuse compost can cause B toxicity, if applied at lower rates, these materials should correct B deficiency. A similar situation exists for fly ash (Mulford & Martens, 1971). However, normal applications (5 metric tons/ha) of liquid digested sewage sludge did not correct B deficiency in alfalfa (*Medicago sativa* L.) on a limed Hagerstown silt loam (Chaney et al. [1974]).[2]

No reports were found concerning correction of Mn, Mo, or other microelement deficiencies by use of organic wastes on land. Perhaps some wastes high in these elements could be used as fertilizers, although "domestic" sources would be ineffective.

Use of a waste to add both macronutrients and needed microelements depends on its composition. It seems reasonable that wastes *which are otherwise safe for use on cropland,* could be applied to satisfy N or P requirements, and the waste could simultaneously correct any Zn, Cu, etc., deficiencies. A N fertilizer application (10 metric tons/ha) of a "domestic" sewage sludge (Table 1) could add 200 kg Fe, 25 kg Zn, 10 kg Cu, 5 kg Mn, 0.5 kg B, and 0.05 kg Mo/ha. Although excessive applications of the same

---

[2]R. L. Chaney, J. C. Baxter, D. S. Fanning, and P. W. Simon. 1974. Heavy metal relationships on a sewage sludge utilization farm at Hagerstown, Md. Agron. Abstr. p. 24–25.

Table 1—Metal contents of digested sewage sludges

Metal	Reachable levels[†]	Maximum domestic[‡]	Observed maximum
Zn, ppm	750	2,500	50,000
Cu, ppm	250	1,000	17,000
Ni, ppm	25	200	8,000
Cd, ppm	5.0	25	3,400
Cd/Zn, %	0.8	1.0	117
Pb, ppm	--	1,000	10,000
Hg, ppm	2.0	10	100
Cr, ppm	50	1,000	30,000

† Observed in sludges generated from waste water of newer suburban communities.
‡ Typical of sludges from communities without excessive industrial waste sources or with adequate source abatement.

organic wastes could lead to microelement-induced phytotoxicity, use at low rates could supplant use of inorganic micronutrient fertilizers.

Zinc levels in forages are often marginal or deficient for domestic animals, especially high producing dairy cows (Reid & Jung, 1974; Pringle et al., 1973). Recently Hambridge et al. (1972) reported clinical Zn deficiency in U. S. teenagers because of the low Zn content of their diet. Vegetarians may similarly have diets marginal in Zn (Sanstead et al., 1974). Use of organic wastes such as sewage sludge and composted refuse on cropland would increase the Zn content of leafy crops from 2- to 10-fold, and could increase the Zn content of grains 2-fold. Because many people do not, or will not supplement their diet with Zn, use of wastes containing Zn on cropland could be a device to increase dietary Zn for humans and domestic animals. Because the Zn content of skeletal muscle meats are relatively unaffected by increased feed Zn, the benefit for human diets of crops enriched in Zn would be largely removed when these are fed to animals. It remains unclear whether higher Zn in plant-based food would have enough influence on human dietary Zn to make it worth the effort to increase plant Zn above levels needed to produce maximum crop yields.

The mechanism whereby organic wastes correct Zn deficiency is unclear, although the simple addition of the waste-borne Zn could account for this correction. On the other hand, Miller and Ohlrogge (1958a, 1958b); Tan et al. (1971); and Tan, King, and Morris (1971) have characterized Zn chelating organic materials present in manure and sludge. Some have suggested that these increase the plant availability of waste and soil Zn. From the work of Parsa (1970)[1] it was clear that ash of an organic waste is less effective as Zn fertilizer; this work indicated that organic waste Zn was more available to corn than soluble inorganic Zn. Studies on acid soils have usually shown that soluble inorganic Zn is more available to plants than Zn in digested sewage sludge or composted refuse (Giordano et al., 1975; Mortvedt & Giordano, 1975; Giordano & Mays, 1976a; Cunningham et al., 1975b). Wastes relatively high in Zn, such as sewage sludge or composted refuse, are better Zn fertilizers than most animal manures.

III. PLANT MICROELEMENT DEFICIENCIES INDUCED BY LAND UTILIZATION OF ORGANIC WASTES AND WASTE WATERS

In several cases, land utilization of wastes has led to plant deficiencies of a microelement. The deficiencies were generally caused by excessive applications of the wastes, or by changes in pH, P, or redox caused by waste application.

A. Induced Mn Deficiency

Trocmé et al. (1950) suspected Mn deficiency at the Paris sewage farm on soil that had increased in organic matter and Zn and Cu contents, and was near neutral pH. The chlorosis was diagnosed as a Mn deficiency because (i) foliar Mn in chlorotic plants was as low as 9 ppm, considerably lower than unaffected plants, and (ii) the chlorosis was corrected by foliar sprays of Mn. Two mechanisms have been proposed to explain the cause of this Mn deficiency: (i) reducing conditions caused by frequent, heavy applications of raw waste water led to reduction of the active soil MnO_2; Mn^{2+} was subsequently leached down the profile; and (ii) sewage-borne Zn accumulated in the surface soil causing Zn-induced Mn deficiency. Zinc can inhibit Mn uptake (E. V. Mass. 1967. Manganese absorption by barley roots. Ph.D. Thesis. Oregon State University, Corvallis, Oreg. 153 p.; Singh & Steenberg, 1974). G. W. Leeper (personal communication) attempted to correct "grayspeck" (Mn deficiency) in oat (*Avena sativa* L.) with different Mn compounds and protected his plots with galvanized "bird-cages." He observed a Zn-induced Mn deficiency under the cages, while nearby the same Mn compounds corrected the Mn deficiency in the absence of excessive Zn. It is likely that both leaching of Mn and inhibition of Mn uptake by Zn contributed to the deficiency at Paris.

In some cases, organic waste applications may raise the soil pH enough to reduce plant Mn uptake (Page, 1966; Atkinson et al., 1958). This could lead to waste-induced Mn deficiency, especially where Zn-rich wastes are applied to acidic soils marginal or deficient in Mn. In Paris continued foliar sprays of Mn are required to correct the chlorosis.

B. Induced Fe Deficiencies

Many microelements cause an interveinal chlorosis, similar to that of simple Fe deficiency when they are present in phytotoxic concentrations. Excessive Zn, Cu, Ni, Co, Cd, and Mn in nutrient solutions and sand culture will produce this symptom (Hewitt, 1948, 1953; Chapman et al., 1939). With soil-grown plants, however, this chlorosis symptom is quite complex.

Simple Fe deficiency (interveinal chlorosis) usually occurs whenever the

Fe content of a newly matured leaf is < 40 ppm. When phytotoxic levels of Cu or Ni are added to nutrient solutions, the Fe content of young leaves falls to < 40 ppm and chlorosis ensues, hence, the name Cu- or Ni-induced Fe deficiency. Copper or Ni inhibition of translocation of Fe from roots to tops has been demonstrated. The chlorosis caused by Cu or Ni can be alleviated by Fe sprays. However, when excessive Zn is added, the chlorosis may occur in leaves containing apparently normal Fe levels. The chlorosis can still be corrected by Fe sprays (Chapman et al., 1939; Guest & Chapman, 1944; Hewitt, 1948).

Nicholas et al. (1957) showed that Ni and Co additions to soil induced low Fe chlorosis. Spencer (1966) found that Cu-induced chlorosis of citrus (*Citrus* sp.) occurred with Fe in leaves < 40 ppm; excess P caused more severe chlorosis. Depending on the level of added Zn, Cd, Cu, Co, or Ni, and soil properties, metal-induced Fe chlorosis may occur at soil pH values as high as 6.5, although usually no chlorosis from excess Zn occurs above pH 6.1 (Chaney et al., 1975; Lott, 1939; Walsh & Clark, 1945) (Fig. 1a). Nicholas (1950) found that chlorosis was more severe with added Co and Ni than with Cu or Zn. Nickel and Cu caused chlorosis in oat and soybean (*Glycine max* [L.] Merr.) in an organic soil at pH 6.5 (Roth et al., 1971).

Although many authors have described Zn-induced chlorosis, Boawn and Rasmussen (1971) failed to observe chlorosis in any crop experiencing Zn-toxicity in a neutral soil. Except for Zn additions to Fe-deficient soils or potting mixtures, or studies using Fe-inefficient crops (Brown & Jones, 1975; Milbocker, 1974), Zn-induced chlorosis is not usually observed at pH 7. One would have expected that the increased difficulty a plant has in obtaining Fe in a neutral soil would tend to lead to a more severe chlorosis. However, the greater reaction of the heavy metals with the soil as the pH is raised generally alleviates phytotoxicity and chlorosis (Chaney et al., 1975).

The observations on Cu-induced chlorosis in Florida provide much information on induced Fe-chlorosis in the field. The first symptoms were chlorosis of annual cover crops and weeds. With greater Cu additions, the citrus crops became unproductive and some trees became severely chlorotic. Until very high levels of Cu had been added, the chlorosis could be corrected by liming. The chlorosis was generally observed on light-textured, acidic soils high in Cu and P (Reuther & Smith, 1954). Although increased P tended to reduce Cu phytotoxicity (growth reduction), it was related to increased potential for chlorosis. The chlorosis could be corrected by soil applications of FeEDTA (Leonard & Stewart, 1952), or by liming to > pH 6. The chlorosis was also related to available Fe in the subsoil. For soil types with low subsoil Fe, high Cu led to severe chlorosis while on soil types with high subsoil Fe, chlorosis was infrequent (Reuther & Smith, 1954; Ford, 1953).

Related work in Florida was done by Westgate (1952) who observed severe chlorosis of many crops growing on an old celery field which had been sprayed repeatedly with Cu. The celery (*Apium graveolens* L.) tolerated Cu, while other crops did not.

Because some of the organic wastes contain high levels of available Fe (e.g., FeCl₃-conditioned sludge vacuum filter cake or any sludge where the Fe is added to precipitate P), the wastes may tend to temporarily counteract their own potential to induce Fe chlorosis. The high P content of many wastes should tend to alleviate metal-induced stunting, yet increase the potential for chlorosis. Each of these (heavy metals, Fe, P) tend to react slowly with the soil and become less available to plants; thus, the long term pattern is unclear.

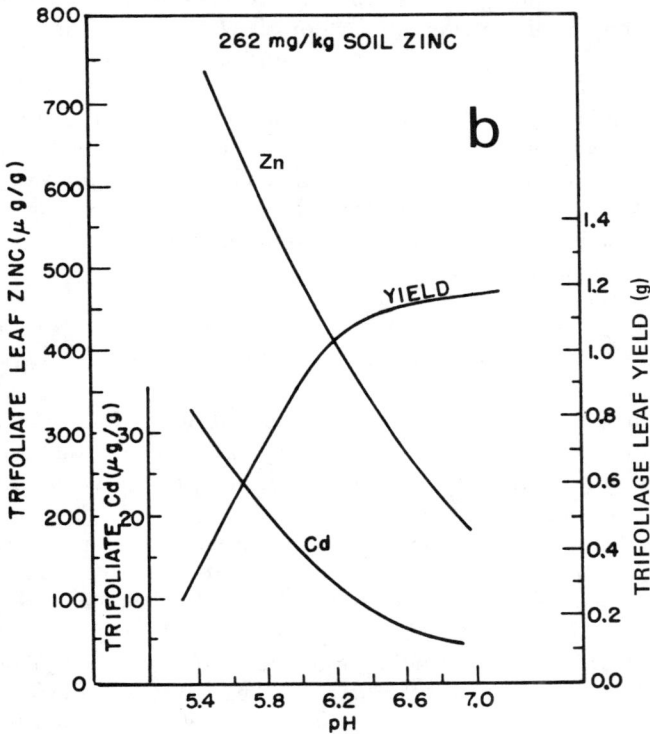

Fig. 1—Effect of soil pH on growth and metal uptake by soybean on Sassafras sandy loam amended with 262 ppm Zn plus 7.9 ppm Cd: *(a)* Photograph showing chlorosis correction above pH 6.2, and *(b)* Yield and Zn and Cd contents of trifoliate leaves; (Chaney et al., 1975).

Metal-induced chlorosis has been observed in a number of situations where excessive additions of heavy metals occurred, such as noted above for Cu toxicity in Florida. A similar Zn-induced chlorosis was observed in cotton (*Gossypium hirsutum* L.) following peach orchards (*Prunus persica*) on acid, light-textured soils in South Carolina (Lee & Page, 1967). Sewage sludge caused Fe-chlorosis of beets (*Beta vulgaris* L.), spinach (*Spinacia oleracea*), and bean (*Phaseolus vulgaris* L.) in experiments by Lunt (1953, 1959). In pot experiments at Beltsville, Md., Chaney (unpublished) observed chlorosis on chard (*Beta vulgaris* Cicla) and snapbean at pH 5.5 and chard at pH 6.5; the sludge (5,400 mg Zn, 2,100 mg Cu, 400 mg Ni/kg) had been leached and allowed to become peat-like over a 10-month period before incorporation into Evesboro loamy sand (ls) and Christiana silt loam (sil). Rohde (1961, 1962) described chlorosis of metal-sensitive vegetable crops at the Berlin sewage farm. High levels of Cu and Zn had accumulated, and local pH variations and metal variations led to the distribution of chlorosis. Doring (1960) found that lime and especially $CaCN_2$ alleviated growth inhibition and chlorosis. Rinno (1964) confirmed the work of both Rohde (1961, 1962) and Doring (1960). Rinno (1964) found higher Cu and Zn where chlorosis occurred and that liming or liming plus $CaCN_2$ alleviated the "sewage-exhaustion." Examples of metal-induced chlorosis were reported in Great Britain (Patterson, 1971). In most cases, the chlorosis was corrected by liming the soil to pH 6.5–7.0.

Thus, metal-induced Fe-deficiency chlorosis is related to crop, kind and amount of metal, pH, P, Fe, and other soil properties. In the field, chlorotic seedings often outgrow chlorosis and little yield reduction occurs provided the subsoil contains adequate available Fe. Low soil temperatures, high soil H_2O content, and other factors known to enhance simple Fe-deficiency should heighten metal-induced Fe chlorosis.

C. Induced Zn Deficiency

Zinc deficiency was observed on old corral sites in California (Chandler et al., 1946) and later confirmed in other areas. The explanation is incomplete, but the Zn deficiency appears to be related to high available P content in corral soils, coupled with high soil pH.

Land disposal of sugarbeet tops, and growth of sugarbeets in general, may lead to Zn deficiency if the following crop is susceptible to Zn deficiency (Boawn, 1965). DeRemer and Smith (1964) summarized a number of these organic matter-induced Zn deficiencies, and suggested that they were caused by microbial immobilization or by inactivation of the available soil Zn by strong binding to sites on the new soil organic matter. Further study by Smith et al. (1965) and Smith and Shoukry (1968) suggested that Zn deficiency was caused by Zn becoming inactivated by binding to the organic matter in calcareous low Zn soils.

IV. PHYTOTOXICITY FROM MICROELEMENTS IN ORGANIC WASTES
AND WASTE WATERS

The accumulation of phytotoxic levels of microelements in the soil from repeated applications of metal-rich wastes is a serious concern of ultimate disposal of urban wastes and waste waters on agricultural land. Application of "domestic" wastes at very high cumulative rates, or application of wastes enriched in metal by mixing with industrial waste metals would lead to large increases in microelements in surface soils. Because removal of most of these microelements by cropping would be slow, their accumulation presents a long term potential for phytotoxicity. Phytotoxicity is dependent on management factors such as soil pH and crop species, and high accumulations of some waste-borne microelements could require more intensive management of these factors than is ordinarily observed.

A. Historic Phytotoxicity

Reduction in crop yields due to an excess of accumulated or naturally occurring microelements have been observed. Phytotoxicity has resulted in many countries from repeated use of unneeded micronutrient fertilizers, from excessive use of fungicides, from soil contamination by mine wastes, smelter emissions, etc. This body of evidence has defined the present concern about specific microelements. Some of these examples involve organic wastes and waste waters; however, the bulk of the examples are from practices that occurred over many years, while sludge, refuse compost, etc., applications have occurred over only a few years in most countries. Much of our knowledge about microelement-induced phytotoxicity has developed in relation to these problems rather than additions from organic wastes and waste waters. Most of these phytotoxicities occurred under poor management conditions, e.g., Cu and Zn accumulation in strongly acid, coarse-textured soils where Cu-, Zn-, and Ni-sensitive vegetable crops were commonly grown.

1. COPPER TOXICITY FROM EXCESS FUNGICIDE AND FERTILIZER COPPER

The most widely studied problem has been accumulations of Cu from repeated applications of Bordeaux mixture which began before 1900. Delas (1963) and Reuther and Smith (1954) have reviewed different aspects of Cu toxicity in vineyards and orchards. Instances of Cu phytotoxicity on fungicide-contaminated soils are well known in Florida and in France. The early observations were chlorosis or stunting of weeds and annual crops used as ground cover. Subsequently production of the perennial orchard crop was reduced. Attempts to replant the orchards produced chlorotic and/or

stunted citrus seedlings because the roots of the young plants are more restricted to the topsoil than are those of mature trees.

A similar situation developed on an old celery field (Westgate, 1952).
Celery is relatively tolerant to excess Cu, but not to excess Zn or Ni (Webber,
1972). Thus, when other crops (even corn) not as tolerant of excess Cu were
grown, chlorosis and crop failure resulted.

Smith et al. (1950) found that the unusual chlorosis of citrus on acid
soils was an Fe-deficiency chlorosis. Then Reuther and Smith (1952a,
1952b) reported that Cu and P had accumulated in the topsoil from repeated
spraying with Cu fungicides and repeated application of unneeded Cu fertilizers. High Cu, low pH, high P, and low Fe content of the soils were significantly correlated with the observation of chlorosis. Reuther et al. (1953)
described experiments to define limits on safe Cu additions to Florida citrus
soils. Citrus seedlings were grown on virgin and old grove soils (Lakeland
sand and Orlando fine sand) with varied pH and varied Cu addition. The old
grove soil contained high Cu and P levels. Other tests included the effect of
Cu vs. that of Cu + Zn + Mn and Cu + P. They concluded that (i) Zn + Mn
levels found in the field were not a problem, (ii) the presence of high levels of
P enhance the Fe-chlorosis symptom from excess Cu, and (iii) restricted root
growth could lead to an apparent N deficiency. The ability of roots to escape
Cu-toxic soil by growing into a nontoxic subsoil can alleviate the chlorosis.
They suggest that damage to trees from toxic levels of Cu would be much
greater on shallow, poorly drained soils that do not provide favorable conditions for rooting in the subsoil. They felt that Cu levels at 10% of the cation
exchange capacity (CEC) (32 ppm Cu/meq per 100 g) in these acid sandy
soils would be unfavorable for the normal growth of citrus roots. Although
increased P reduced injury due to excess Cu, liming to higher pH was much
more effective. Spencer's (1966) work largely confirmed the early work of
Ruether and Smith (1954).

Walsh, Erhardt, and Seibel (1972) recently examined the potential effects of long term Cu fungicide applications on growth of snapbeans. Copper
was added up to 486 kg/ha (equal to 27 years of application) to a Plainsfield
loamy sand (pH 6.7; 0.7% organic matter) as either $CuSO_4$ or $Cu(OH)_2$. Slight
yield reduction occurred with addition of 130 kg Cu/ha, while severe yield
reduction occurred when 405 kg Cu/ha (as Cu[OH]$_2$) or 486 kg Cu/ha (as
$CuSO_4$) was applied. Phytotoxicity was not reduced during a second year of
the trial. No chlorosis was observed.

2. HISTORIC ZINC TOXICITY FROM EXCESS FUNGICIDE

Zinc from Zn oxysulfate antibacterial sprays accumulated in the soils of
peach orchards in South Carolina (Lee & Page, 1967); sulfur sprays simultaneously caused low soil pH. Studies were conducted with cotton (Lee &
Page, 1967) and soybean (Lee & Craddock, 1969) to find the cause and some
corrective treatments. The actual amount of Zn that accumulated in the soil
was much lower than would be expected to cause phytotoxicity (45 mg acid

extractable Zn/kg Faceville sand, pH 5.3; 13 mg acid extractable Zn/kg Wagram loamy sand, pH 5.3, CEC 1.04 meq/100 g). However, extractable Mn also was increased by these treatments.

In their plant trials, Lee and Craddock (1969) observed Mn toxicity symptoms (crinkle leaf; Adams & Wear, 1957), with added Zn and added Zn increased plant Mn. Lime lowered plant Zn and Mn, and alleviated the toxicity. Although they observed slight chlorosis, FeEDDHA had little effect on yield. Other workers have noted that soil additions of zinc (at constant soil pH) increase plant uptake of Mn; the increase in Zn and Mn are highly correlated (White et al., 1974).[3] Thus, at low soil pH, Zn can cause phytotoxicity both directly and because it increases translocation of Mn to the plant leaves where Zn induced Mn-toxicity results. Similar results were reported by Singh and Steenberg (1975).

In a study simulating many years of current Zn fungicidal treatments, Walsh et al. (1972) applied up to 363 kg Zn/ha as $ZnSO_4$ (equal to 81 years of application) to a Plainsfield loamy sand (pH 6.7; 0.7% organic matter). No phytotoxicity was observed for snapbean, cucumber, or corn. Foliar Zn reached 350 mg/kg; actually higher than levels found in the studies by Lee and Craddock (1969). The soil pH was high, which reduced the Zn phytotoxicity and avoided the Zn-induced Mn toxicity.

3. HISTORIC NICKEL TOXICITY

Nickel toxicity in nature has been observed chiefly on serpentine soils or Ni ore outcrops. Nickel has been used to combat blister blight of tea (*Camellia sinensis*), and Venkata Ram (1964) reported that soil applications could be made and the uptake of Ni might combat this disease at levels below those phytotoxic to tea. Hardison (1963) described use of nickel to control rust in *Poa pratensis*, but warned against accumulation of Ni in the soil to excessive levels. The role of Ni in plant growth and metabolism was recently reviewed by Mishra and Kar (1974).

The effects of Ni pollution from stack emissions was reported by Ashton (1972), Hutchinson and Whitby (1974), and Whitby and Hutchinson (1974). In each case high levels of Ni had accumulated in surrounding soils.

Studies of serpentine soils provide much of our available knowledge about the effect of Ni on field-grown plants. Several classes of soils are reported, each supporting locally unusual vegetation (Wild, 1970). Although all are high in exchangeable Mg, some have adequate Ca; some have low available Ni, while others have very high available Ni, usually controlled by the field soil pH. In some cases plant injury was due simply to high Mg and low Ca (Proctor, 1971). This can be corrected with $CaSO_4$ (Martin et al., 1953). Others required both $CaSO_4$ and Mo (Walker, 1948; Johnson et al., 1952). However, in other soils the available Ni was high, the pH was low, and Ca and

---

[3]M. C. White, R. L. Chaney, A. M. Decker. 1974. Differential varietal tolerance in soybean to toxic levels of zinc in Sassafras sandy loam. Agron. Abstr. p. 144–145.

Mo were not growth limiting (Robinson et al., 1935; Hunter & Vergnano, 1952; Halstead, 1968; Anderson et al., 1973). All the studies showed that liming to raise the soil pH corrected the toxicity (e.g., Crooke, 1956). Application of Na_2CO_3 was equal to that of $CaCO_3$, indicating a pH effect rather than a Ca effect. Although other plant species are more sensitive to Ni than oat, this crop has been grown in nearly every test because it displays a characteristic symptom of Ni toxicity (Hunter & Vergnano, 1952).

Anderson et al. (1973) clarified the roles of Ni and Cr in plant injury on an acid serpentine soil (pH about 5.3). Chromium was present in the soil solution at only 1% of that needed to cause injury, while Ni reached 3.25 mg/liter. Cobalt was much lower, and these results clearly showed that the toxicity was due to excess available Ni.

4. HISTORIC BORON TOXICITY

Bradford (1966) summarized B toxicity experiences. Soils naturally high in B or soils made toxic from irrigation water with high B were usually involved. One incidence of B toxicity to sensitive citrus crops occurred when waste waters from a citrus packing house were used for irrigation (Kelly & Brown, 1928). The B was present because it was used as a fungicide.

5. PHYTOTOXICITY FROM NATURAL OCCURRENCES OF MICROELEMENTS, OR IN MINE WASTES, STACK EMISSIONS, AND INDUSTRIAL WASTE WATERS

Microelement ores occur naturally, and nearby or overlying soils are often high in the respective metal. Plants tolerant of individual metals have led prospectors to the ores; the calamine violet (Zn indicator), copper flower, and nickel flower are examples of plants that have indicated particular metal ore outcrops (Cannon, 1960). Subspecies or "ecotypes" have evolved that are tolerant to high levels of a particular metal (Antonovics et al., 1971).

Dykeman and deSousa (1966) described the flora on a copper-enriched peat (up to 7% Cu). The soil was neutral to alkaline, and the peat remained anaerobic most of the year. The vegetation bore no toxic symptoms, but was high in Cu (analyses unclear as the tissues were not washed). Fraser (1961) reported that this site was "found" when a nearby farmer hauled some of the peat to fertilize his garden and it killed his plants. The chemical species of Cu involved are unclear, but it seems that CuS should have predominated. Under the reducing, neutral conditions prevailing in the peat, plants thrived at very high Cu levels.

Wallace and Hewitt (1946) describe chlorosis of cereal crops on a calcareous soil high in Zn. Chlorosis was corrected by $FeSO_4$, but the chlorotic leaves contained 500 ppm Fe along with 700 ppm Zn.

Staker and Cummings (1941) reported a Zn toxicity on a moderately acid peat soil area in Orleans County, N. Y. Crops tolerated the high Zn if the peat remained poorly drained, but crops suffered severe chlorosis after drainage. Sphalerite (ZnS) was subsequently proven to exist in the undrained peat (Cannon, 1955). Staker and Cummings (1941) and Staker (1942) found that liming usually corrected the Zn phytotoxicity; liming to pH 7, plus additional Fe, Mn, and B was slightly better. The Zn-induced chlorosis patterns in crops and native vegetation was a good biogeochemical prospecting tool (Cannon, 1955). On many areas, only potatoes (*Solanum tuberosum* L.) could be grown. Other farmed areas were abandoned. Some of the plants contained as much as 10,000 mg Zn/kg.

Robinson et al. (1947) analyzed plants and "soil" at a Zn ore tailings deposit at Friedensville, Pa. The soil was calcareous, yet few plants survived; the soil contained up to 12.5% Zn. Plant Zn was as high as 5,400 ppm, while corn at the edge of the area had foliar Zn of 790 ppm.

Knowles (1945) described phytotoxicity on a farm where foundry wastes containing Cu and Zn were spread during 1914–1918. During World War II the farm was cropped. Oat, wheat (*Triticum aestivum* L.), sugarbeet, and potato all failed; they were stunted and/or chlorotic. Similar problems develop on land where rubber tires are burned, or where the insulation is burned off copper wire.

Pollution of soils with mine wastes led to "itai-itai" disease caused by increased dietary Cd (Kobayashi, 1970) and levels of metals that would be phytotoxic for sensitive crops under "upland" conditions (Takijima & Katsumi, 1973a). Buchauer (1973) described accumulation of Zn, Cd, Cu, and Pb in soils downwind of a Zn smelter. Hutchinson and Whitby (1974) measured Ni and other metals in soils downwind of a Ni smelter. Lagerwerff et al. (1972) described accumulation of Zn, Cd, Cu, and Pb downwind of a Pb and Zn smelter. In these and similar cases, the simultaneous emission of SO_2 caused the soils to have low pH; SO_2 caused most of the observed injury to vegetation. Phytotoxicity from metals resulted even when the soils were cropped in a greenhouse free of SO_2 problems. Lagerwerff (1975) summarized other environmental hazards associated with smelter emissions.

6. HISTORIC PHYTOTOXICITY FROM ORGANIC WASTES AND WASTE WATERS

Phytotoxicity and induced Mn deficiency (Trocmé et al., 1950), and induced Fe deficiency (Rohde, 1961, 1962) on the Paris and Berlin sewage irrigation farms were described earlier. Metal-sensitive vegetable crops were severely injured. Blood (1963) described a sewage irrigation farm in Great Britain where 500 ppm Zn had accumulated in the surface 25 cm of a sandy soil; sugarbeet yields were reduced in patches where the soil pH had fallen below 6.4. Liming provided partial recovery. Spotswood and Raymer (1973) describe a sewage irrigation farm converted to a sludge disposal farm where very high levels of total Zn (400–1,300 ppm in soil) and total Cu (50–300 ppm) were found. The average soil pH was 6.2, but some samples were as low as 5.0. Of the 875 ppm total Zn, 270 ppm were extractable with 0.5N acetic acid. Attempts to grow sugarbeet failed. Cereal crops must be grown for 1 to 2 years after addition of 30 cm of liquid sludge before potato can be grown. However, even then yields may be reduced. The current sludge contains only 2,000 ppm Zn and 350 ppm Cu on a dry matter basis. Spotswood and Raymer (1973) believed that the continued farming was possible at these high levels of Zn and Cu, and low soil pH because of the high soil organic matter content and their choice of metal-tolerant crops.

Because sludge disposal on land has been used so widely in the United Kingdom, and the soils chosen were often course-textured and acid, and metal-sensitive crops (e.g. vegetables or sugarbeets) are "normal" crops, there have been numerous observations of crop failure caused by sludge-borne metals. The metal toxicity research program at Bristol University was begun in response to questions about phytotoxicity from an industrial sewage sludge (Wallace & Hewitt, 1946). Patterson (1971) wrote "Instances of crop damage caused by the build up of toxic levels of one or more metals in the soil, usually after dressings of sewage sludge, repeated over a number of years, are known to every agriculture advisor." Repeated heavy applications of domestic sludges or use of sludges high in metals were generally involved. Although most metal phytotoxicity problems were corrected with liming, some required growing only metal-tolerant crops. On a "market garden in Somerset" levels of 5,000 ppm Zn and 500 ppm Cu in soil were reached precluding further use as a market garden.

Descriptions of crop failure requiring liming or change to tolerant crops can be found in the discussion section of many papers appearing in *Water Pollution Control* (London) (formerly J. Inst. Sewage Purif.). In the discussion following Webber's (1972) paper, V. H. Lewin (of Oxford) described failure of potatoes but satisfactory yield of small grains even though they limed regularly.

In field experiments with sludges high in metals (Webber, 1972, 1974), lettuce (*Lactuca sativa* L.) and redbeet failed where 112 metric tons/ha of a sludge containing 8,000 or 16,000 ppm Zn was applied to a pH 6.4 gravelly silt loam. Damage was more severe the second and third years than the first. In a field experiment with a low metal sludge in Georgia, rye (*Secale cereale* L.) was injured but bermudagrass (*Cynodon dactylon* L.) was not; the soil pH fell to 4.2, and Zn and Cu were implicated in the phytotoxicity (King & Morris, 1972).

In other situations, particularly where low metal sludges or waste waters were applied to fine-textured soils at neutral pH, and grains or grasses were grown, satisfactory crops resulted. Instead of the problems encountered in Berlin and Paris, a successful sewage irrigation farm has been run in Melbourne since 1897 (Johnson et al., 1974). Sewage sludge has been used successfully for revegetation of barren ore processing wastes already high in metals; success was better with Cu- and Fe-ore wastes than Zn-ore wastes (Street & Goodman, 1967; Weston et al., 1964; Gadgil, 1969). Sludge or refuse application and seeding to standard grasses was more successful in revegetation than only introduction of ecotypes of grasses tolerant to the metals (Gadgil, 1969). Sludge alleviated phototoxicity from Zn-contaminated soil in galvanized steel lysimeters (Hinesly et al., 1971).

B. Potential Phytotoxic Elements in Wastes

A major difficulty in predicting potential problems from an organic waste is the inherent variability in the composition of wastes. Nearly every city has industries that dump metals into the sewer in "slugs" (Oliver & Cosgrove, 1974). A grab sample of primary or activated sludge represents only a few hours composition, while digested sludge samples represent a week or more due to mixing. Refuse is extremely heterogenous. Manures from unknown origin can vary widely. These statements apply equally to macronutrients and microelements.

1. MANURES

Most manures are relatively low in microelements (Cd, Mo, Cu, Se) because feeds (plant tissue) must be safe for animals. However, some swine and poultry are fed diets containing up to 250 ppm Cu and 100–200 ppm Zn to increase feed efficiency in the absence of antibiotic compounds. The manures from these animals are thus 10–40 times higher in Cu and 4–10 times higher in Zn than normal. Application of these high-metal manures at N fertilizer rates would apply 3–6 kg Cu ha^{-1} year^{-1} (Baker, 1974; Davis, 1974). Some of these high Cu manures could be sprayed on pastures, or the effluent from anaerobic lagoons could be sprayed on grass-covered land. Batey et al. (1972) and Humenik et al. (1972) have described contamination of forages with Cu-enriched wastes. Davis (1974) argues that the Cu would be a problem only where the soils and plants are simultaneously low in Mo. Much of the Cu is present as CuS, at least initially (Robel & Ross, 1975).

2. SEWAGE SLUDGE

Sewage sludges can contain any element used in modern industry (McCalla et al., 1976 [Chapter 2, this book]; Berrow & Webber, 1972) (Table 1). Comparison of the metal content of sludges with safe levels in soils (Page 1974; Berrow & Webber, 1972) has led to a list of potential problem microelements. Of primary hazard to plants are Zn, Cu, and Ni and to animals are Cd, Pb, Hg, Se, As, and Mo. Researchers are also interested in B, Co, Cr, and even Be, Ba, Sr, Sb, Ge, Ga, and Ag (Chaney, 1973; Leeper, 1972; Lisk, 1972; Allaway, 1968; Baker & Chesnin, 1976; Braude et al., 1975; Larson et al., 1974). Of these, only Zn, Cu, Ni, and Cd will commonly be potential agronomic problems. The chemical species of the individual metals present are not known, even for Zn and Cu. Jenkins and Cooper (1964) were unable to demonstrate whether hydroxides were dominant. Bloomfield and Pruden (1975) found that the extractability (by EDTA or 0.5N acetic acid) of Zn, Cu, Ni, Cd, and Pb was increased by aerobic incubation and decreased by anaerobic incubation. Perhaps this explains the wide variation in percent extractability of metals in sludges shown by Berrow and Webber (1972) and

Baxter and Chaney (1973).[4] These results suggest that metal sulfides might predominate. Tan, King, and Morris (1971) described organic macromolecules extracted from sludge which may bind or chelate heavy metals, but whether organic matter in sludges is important in binding metals is unclear. It is known, however, that very insoluble compounds formed during anaerobic incubation are readily transformed to extractable forms during aerobic incubation of sludge or of sludge-soil mixtures. Some sludges may contain metal compounds which will remain unextractable and unavailable to plants.

3. WASTE WATER

Potential hazards from microelements in waste waters applied to land have been reviewed (Murrmann & Koutz, 1972; Bouwer & Chaney, 1975; Leeper, 1972; Knezek, 1972; National Academy of Sciences-National Academy of Engineering, 1974). Metals in waste water must accumulate in the surface soil over many years before problems occur. Generally, the same elements listed under sewage sludge (Zn, Cu, Ni, Cd) are important potential hazards if present in waste water. In addition, B is recognized as potentially hazardous in any irrigation water if present at greater than about 0.75 mg/liter (for sensitive crops) or 2-4 mg/liter (for tolerant crops) (see Bradford, 1966; Richards, 1954; National Academy of Science-National Academy of Engineering, 1974). Pound and Crites (1973) reported that waste waters presently irrigated on cropland contain 0.3-1.5 mg B/liter. Boron in waste water may increase as B compounds are increasingly substituted for phosphates in household detergents. Neary et al. (1975) recently described B toxicity in red pine (*Pinus resinosa* Ait.) irrigated with municipal waste water containing only 0.90 mg B/liter. After two seasons, B had accumulated in the surface acid humus layer and, because many roots were present in this layer due to the frequent irrigation, B toxicity resulted. This situation has not been observed in row crops on tilled land. Red pine does appear to be unusually sensitive to B; little information is available on the relative B tolerance of fast-growing tree crops.

4. REFUSE AND REFUSE COMPOSTS

Microelements in refuse composts are generally present at their levels in low metal content sewage sludges. Unleached composts may also be high in available B (Purves & Mackenzie, 1973, 1974). Boron caused severe toxicity when shredded refuse was incorporated; B toxicity was more severe where NH_4NO_3 was added to correct N deficiency and it lowered the soil pH. Much of the B was present in the glue used on labels and in corrugated paperboard (N. M. Cottrell. 1975. Disposal of municipal wastes on sandy soil: Effect on plant nutrient uptake. M.S. Thesis. Oregon State University, Corvallis, Oreg. 201 p.). Thus, potential phytotoxicity exists for B, Zn, Cu,

---

[4]J. C. Baxter and R. L. Chaney. 1973. Content and extractability of toxic heavy metals in sewage sludge. Agron. Abstr. p. 185.

and Ni. Not only is refuse heterogeneous in total content of an element, but the chemical forms of the elements also vary. Zinc may be present as sulfide, hydroxide, carbonate, and chromate, or Zn metal or galvanized steel; it can be in rubber or within colored glass; it can be an insoluble pigment used in paint or on paper as ink (Eaton et al., 1975). Each of the elements can be present in different forms.

C. Reactions of Waste-borne Microelements with Soils

A number of reviews of microelement reactions with soils have appeared (Hodgson, 1963; Jenne, 1968; Ellis & Knezek, 1972; Ellis, 1973; Lindsay, 1972, 1973). Reactions can generally be characterized by Langmuir or Freundlich isotherms; thus, soils bind small additions of metals considerably more strongly than large additions. Metals are bound more strongly with increasing soil pH, and adsorption capacities are increased. Many components of soils have been shown to bind microelements; clays, organic matter (Stevenson & Ardakani, 1972; Schnitzer & Khan, 1972) hydrous Fe oxides, hydrous Mn oxides, carbonates, and inorganic compounds of the individual elements (Lindsay, 1972, 1973). The relative importance of each of these soil components in binding metals in forms no longer phytotoxic is unclear. The kinetics of these reactions have received little study. Interactions among several metals added simultaneously (e.g., Cd in the presence of 100 times more Zn) have not been reported.

Andersson and Nilsson (1974) described the binding of Cd by several soil components vs. pH. Organic matter bound Cd strongly across a wide pH range while Fe oxide binding increased from none to a maximum when the pH increased from 4.5 to 5.8. Each soil component had a slightly different Cd binding vs. pH curve (Fig. 2). The importance of Fe and Mn oxides is now being seriously appreciated (Gadde & Laitinen, 1974; Jenne, 1968).

An important implication of basic studies on metal binding by soils is that as the proportion of its maximum adsorption capacity filled by a metal increases, the relative strength of binding decreases and, hence, the relative plant availability of the metal should increase. The apparent stability constant of Cu and Cd with soil organic matter falls 2 to 3 logs as the percent saturation with Cu or Cd rises from about 5% to 50% (E. A. Bondietti & F. H. Sweeton. 1973. Investigations on the metal binding ability of soil and microbial humic acids using cadmium, copper, and lead ion electrodes. Agron. Abstr. p. 89.). Ennis and Brogan (1961) found that up to 26 ppm Cu in a peat soil was bound so strongly that it was unavailable to oats. The metal binding sites of organic matter are quite heterogeneous, and the strength of binding decreases as the more specific binding sites are filled. In studies of Zn-adsorption by soils, Shuman (1975) found that Zn adsorption at low and high Zn concentrations had markedly different strengths; perhaps plant uptake will increase sharply when one type of binding site becomes saturated.

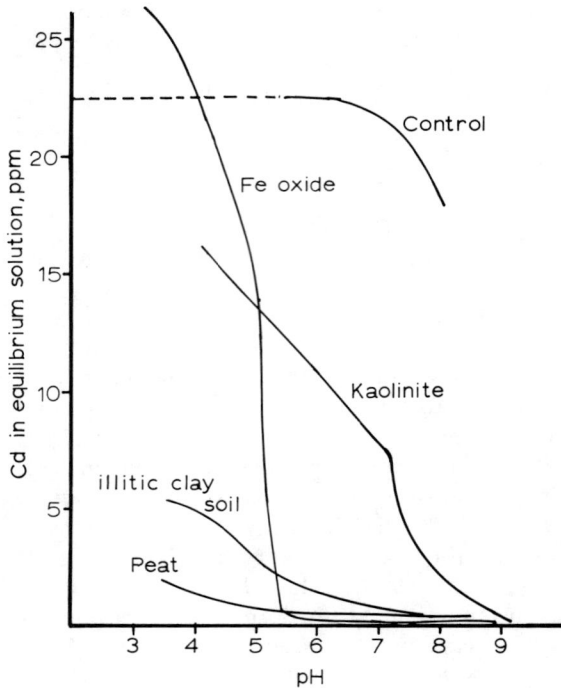

Fig. 2—Distribution of Cd between soil material and equilibrium solution as a function of pH. Solid/liquid ratio was 1:50, and 2 μeq of Cd was added per g soil material.

Recent work on uptake of Pb by corn and soybean has shown a high positive correlation between Pb uptake and ratio of added Pb to the maximum Pb adsorption capacity (Miller et al., 1975a, 1975b; Hassett, 1974). Further, higher soil pH and CEC significantly reduced Pb uptake by both soybean and corn, and increased the soil Pb adsorption capacity. The studies utilized additions of soluble Pb salts; waste-borne Pb usually doesn't increase plant uptake of Pb. Cadmium uptake by radish (*Raphanus sativus* L.) and lettuce from 30 Cd-amended soils was also highly correlated with a "Cd-adsorption-capacity" measurement on these soils (John et al., 1972).

An important factor in reactions of metals with soils is a slow process of conversion of bound metals into forms no longer plant available referred to as *reversion* by Leeper (1972). Brown et al. (1964) described reversion in regard to decreasing plant availability of fertilizer Zn. Boawn (1974) found that DTPA extractability of Zn added as $ZnSO_4$ (22 kg Zn/ha) decreased for 4 years, and appeared to reach an equilibrium availability thereafter (Fig. 3). Follett and Lindsay (1971) found reversion of Zn, Cu, Fe, and Mn added as soluble salts or chelates in soils of all textures and pH; the extractability fell more than 25% in 14 weeks.

What soil constituents are involved in reversion? Lindsay (1973) believes that the inorganic compounds of the metals are involved. On the other

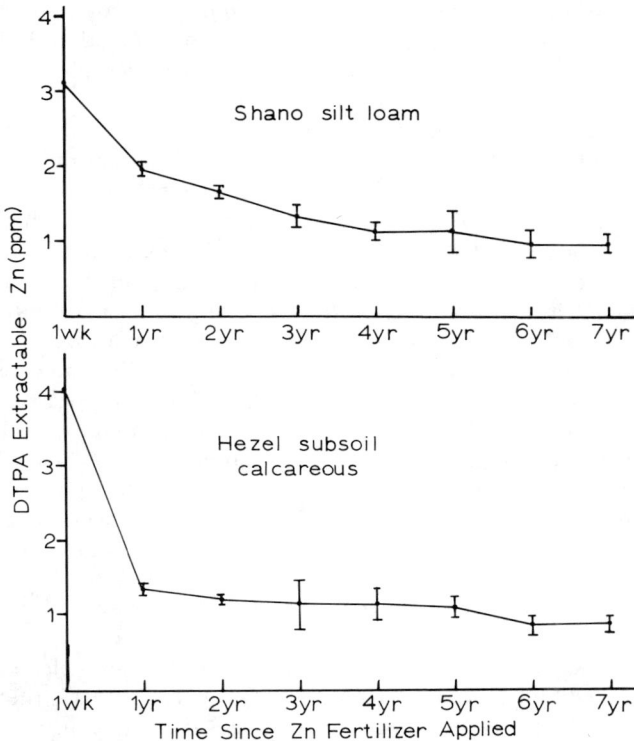

Fig. 3—Residual DTPA-extractability of Zn during 7 years following addition of 22.4 kg Zn (as $ZnSO_4$)/ha to a noncalcareous Shano silt loam or a calcareous Hezel subsoil (Boawn, 1974 and unpublished).

hand, Mn oxides are clearly the soil constituent which causes reversion of Co (McKenzie, 1972). Cobalt availability from fertilizers is so controlled by the Mn oxide content of soils that it is uneconomic to add Co salts to soils containing > 1,000 ppm Mn (Adams et al., 1969). The other potentially toxic metals revert much more slowly than Co and their reversion is not known to be related to a particular soil constituent. The effect of soil pH on the rate or extent of microelement reversion is unclear, as is the effect of the amount added. Boawn (1974) found that the shape of the curves of declining Zn availability was similar for Zn rates, and that an apparent equilibrium value was reached after a few years (Fig. 3). After 5 years, about 16% of the added Zn remained DTPA extractable. Soil $CaCO_3$ content and addition of wheat straw had no effect on this equilibrium value.

What about reversion of metals added at phytotoxic levels? Although many descriptions of phytotoxicity correction by liming contain observations of decreasing injury over several years, the basis of the decreasing injury is unclear because agricultural grade limestone reacts slowly. Giordano et al. (1975) were unable to find rapid reversion of Zn added as $ZnSO_4$ to a pH 4.9

Sango sil; snapbean yield reduction and Zn content were not appreciably corrected during the second year after application. Walsh et al. (1972) did not find reversion of Zn added as $ZnSO_4$ at high rates; however, phytotoxic levels were not studied.

In nearly every study on the effect of incubation on the extractability of microelements added as sludge or refuse compost, a continuing decline in extractability is observed. An initial increase is often observed, apparently due to conversion of insoluble metal sulfides to more extractable forms (Bloomfield & Pruden, 1975). Levels of $0.5N$ HCl extractable Cd, Cu, Zn, and Pb increased during the second year after sludge addition to a silt loam soil (pH 6.4) (Giordano & Mays, 1976b); accordingly, concentrations of these metals were higher in several vegetable species. On long term sewage farms (Johnson et al., 1974; Chaney et al., 1974[2]) Zn, Cu, and Cd remained readily extractable by $0.1N$ HCl or DTPA-TEA after many years. However, Andersson and Nilsson (1972) found no increase in uptake of Cd after 15 years of sludge use although extractable Cd had increased. Thus, whether or not we can depend on reversion to decrease the long term potential for microelement phytotoxicity and uptake by plants remains uncertain.

D. Plant Uptake and Translocation of Microelements

The effect of land disposal of wastes on levels of microelements in the leaves and the edible parts of plants is important because this is the first step in the food chain. However, most organic wastes simultaneously add several metals (Zn, Cu, Cd, Fe, etc.) and P, and influence soil pH. This, coupled with plant differences, clouds any ordered picture of plant uptake of waste-borne microelements. Generally, certain elements can readily move to plant tops when applied to soil (Mn, Zn, Cd, B, Mo, Se), others with intermediate ease (Ni, Co, Cu), and still others hardly reach the tops (Cr, Pb, Hg). These differences, along with other soil and plant factors strongly influence the movements of microelements to edible plant parts.

Plant absorption of microelements depends on many soil factors, most of which affect microelement soil solution levels and rates of diffusion to plant roots. Movement from the bulk soil to the root has been described by Barber (1974) and Wilkinson (1972), and summarized by Chaney (1975). Hodgson (1969) and Lindsay (1974) summarized the role of soluble chelates of metals both in convection (mass-flow) and diffusion of the metals (e.g., Hodgson et al., 1967, for Zn). Soil pH is very important in Zn diffusion (Clarke & Graham, 1968; Melton et al., 1973). Other factors affecting Zn diffusion include Zn concentration, soil texture, clay and organic matter contents, other nutrients, soil bulk density, and soil moisture content (Ellis et al., 1970; Warncke & Barber, 1971a, 1971b, 1973). Growing roots can cause local pH and bulk density changes which influence microelement uptake (Barber, 1974). Roots can change the rhizocylinder pH several units (Smiley, 1974), and the pH drop with ammonium fertilizers does increase Zn uptake

(Boawn et al., 1960). Graham (1973) has considered the role of both labile pools (available) and selective distribution ([metal bound to soil/metal in solution] relative to that for Ca) on plant accumulation of micronutrients. Graham's (1973) and Barber's (1971a, 1973, 1974) work are compelling evidence for these models of microelement movement to roots. Tiller et al. (1972) also characterized labile-pool Zn in relation to plant uptake. With the use of these models, chemical characterization of soils may eventually define the maximum safe metal loadings.

The relationship of metal phytotoxicity to the models of microelement movement from the soil to root absorption sites remains unclear. At some level of added metal, the rate of metal convection to the root may exceed the rate of uptake; its concentration will then build up at the root, leading to greater uptake. Because all the factors known to influence metal toxicity are considered and participate in these models, they should apply to phytotoxicity. This research approach appears fruitful.

Soil pH is predominant among soil factors affecting microelement uptake because it controls the extent of reaction of these elements with the soil. Zinc uptake vs. soil pH (Fig. 1b) has been studied at phytotoxic levels (Chaney et al., 1975; Lott, 1939; Walsh & Clarke, 1945) and at nonphytotoxic levels (Wear, 1956). In each case, Zn uptake decreased linearly with increased pH. Wear (1956) showed that Zn uptake by sorghum fit the same linear relationship with pH with both $CaSO_4$ (lowered pH) and $CaCO_3$ (raised pH) additions; raised pH, not added Ca^{2+}, reduced Zn uptake. The level of soluble salts in soils can also influence microelement uptake (Maas et al., 1972).

Addition of chelators to a soil may allow formation of soluble microelement chelates. Wallace and Mueller (1973) have shown that addition of strong, synthetic chelators (e.g., EDTA, DTPA) increases plant uptake of Cu and Co to phytotoxic levels. Similar results were found for NTA (Wallace, Mueller, and Alexander, 1974) and for the other crops and metals (Wallace et al., 1974; Wallace & Mueller, 1969). Whether natural chelating compounds formed during decomposition of wastes react similarly remains unclear.

Competition among metals in uptake has received some study at low levels of microelements (Moore, 1972; Olsen, 1972). However, competition at phytotoxic levels has received little study. Singh and Steenberg (1975) and White et al. (1974)[2] have shown that Zn and Mn interact strongly, but this may be more in translocation than uptake. Chaney (1973) showed possible Cu-Zn interactions on Zn uptake and translocation which differed considerably among crops. The metal interactions may be either in uptake by roots or translocation to plant tops.

Interaction occurs between P and phytotoxic levels of microelements. Previously, we noted that P reduced Cu injury and concentration in citrus leaves (Spencer, 1966), but simultaneously enhanced Fe chlorosis. Smilde et al. (1974) studied P and lime amendments to soil polluted with Zn by smelter emissions. For some crops, P alleviated Zn toxicity and reduced Zn uptake, while for others, P had no effect unless deficient for growth.

Tiffin (1972, 1976) reviewed many aspects of micronutrient translocation in plants. Translocation is controlled by chemical properties of natural
chelates of the microelements found in the root cytoplasm, and on charged
surfaces in the root cell walls (Antonovics et al., 1971), cytoplasm, and
xylem walls. For Zn, Mn, Cd, etc., the strength of the chelates, types of
chelates, and their rates of exchange favor transport to the tops. For Cu and
Ni, most chelates in the root are only slowly released to the tops, while for
Cr, Pb, and Hg the types of chelates in the roots and strength of adsorption
in the soil keep plant top levels very low (Tiffin, 1976; Chaney, 1975).

E. Microelements in Plants from Land-applied Wastes

Based on the previous sections, we should expect that wastes would increase uptake of elements such as Zn, while others should increase little or
perhaps decrease depending on changes in pH, organic matter, P, etc. A
general summary of most observations on plant uptake of metals from sludge-
or compost-amended soil is: Zn, large increases; Mn, usually increased; Cd,
nearly always increased somewhat; Cu, increased a little with small additions
then reaches a plateau; Ni, increase somewhat; Pb, often decreased a little;
Hg, only small changes; and Se, Mo, As, etc., too little data to make judge-
ments at this time. Increases in Cd, Hg, and Pb are most important because
of their potential impact on the food chain (Lagerwerff, 1972; Braude et al.,
1975). Thus, these elements will be considered at greater length.

1. INCREASED PLANT MERCURY

Wastes generally have not increased plant Hg. Hinesly et al. (1971,
1972), and Andersson and Nilsson (1972) found insignificant changes in Hg
content of corn and rape (*Brassica napus* L.) following sewage sludge applications. Van Loon (1974) found similar results for carrot (*Daucus carota* L.),
lettuce, and grasses, but found much higher Hg in tomato fruit (*Lycoperiscon
esculentum* Mill.) and possibly snapbean pods where sludge was applied. D.
J. Lisk (Ithaca, N. Y., personal communication) failed to observe increased
Hg in tomato or snapbean in studies with two sludges. Work with more
crops and sludges must be done to see if a few crops (e.g., tomato) are unusual accumulators of Hg.

2. INCREASED LEAD IN PLANTS

Application of wastes in the field has nearly always decreased the Pb
content of plants. Hinesly et al. (1971, 1972), Andersson and Nilsson
(1972), Sabey and Hart (1975), Dowdy and Larson (1975a, 1975b), and
Chaney et al. (1975) generally found that low levels of sludge slightly increased plant Pb, while higher sludge levels decreased plant Pb, especially in
grain, fruit, or edible roots. Liebhart and Koske (1974), Giordano et al.

(1975), and King et al. (1974) found lower plant Pb with additions of refuse compost.

Other information on Pb uptake from soils treated with soluble Pb salts suggests that large Pb additions (in the absence of the huge P additions of sludge and compost) should cause some increase in plant Pb. Jones and Hatch (1945) found that many plants grown on soils contaminated with Pb arsenate increased only slightly in Pb; however, levels in eggplant fruit (*Solanum melongena*) rose to 35 ppm. Jones, Jarvis, and Cowling (1973) found that S deficiency allowed more Pb to reach plant tops. Miller et al. (1975a, 1975b) found that Pb uptake by corn and soybeans was negatively correlated with soil pH and CEC. Lead in corn, but not soybean, was negatively correlated with soil-available P. Zimdahl and Foster (1976) found that applications of P, manure, or lime reduced Pb uptake by corn. MacLean et al. (1969) found that increasing soil pH, CEC, and available P decreased Pb levels in oats and alfalfa. Baumhardt and Welch (1972) added up to 3,200 kg Pb/ha to Drummer sil (pH 5.9); although foliar Pb increased, grain was unaffected. An unusual increase in Pb content in late fall to early winter has been observed for orchardgrass (*Dactylis glomerata* L.), ryegrass, and clover (*Trifolium* sp.) (Mitchell & Reith, 1966) and confirmed by O. L. Bennett (Morgantown, W. Va., personal communication). The potential effects of this increase of Pb in winter pastures is unclear.

3. INCREASED CADMIUM IN PLANTS

a. **Why Limit Levels of Cd in Foods?**—Increases in plant Cd from sewage wastes and refuse applications on cropland is a major issue in the safety of land disposal. When people absorb Cd it is accumulated in the kidney and liver. Recent work has found that kidney Cd builds up until about age 50. Absorbed Cd is excreted very slowly. Two major exposures to Cd have been found: food and smoking. Normal water and air Cd levels are low enough that these are not appreciable Cd sources in the absence of pollution or industrial exposure. There is little available information on the Cd content of foods or raw agricultural produce at this time (Page & Bingham, 1973).

Tobacco (*Nicotiana tabacum* L.) contains 1–6 ppm Cd; because about 30% of inhaled Cd is adsorbed, smoking one pack of cigarettes per day approximately doubled the usual level of absorbed Cd (Lewis et al., 1972; Menden et al., 1972; Elia et al., 1973). For the nonsmoker, absorbed Cd comes mostly from food, and that mostly from grain products and fruits (because they constitute the bulk of our diet) (Fulkerson & Goeller, 1973; Friberg et al., 1971; Flick et al., 1971; Braude et al., 1975). Approximately 3 to 6% of dietary Cd is absorbed. The Food and Drug Administration calculates that we now ingest 72 to 90% of the desired maximum daily dietary levels and that we should not elect to increase food Cd levels (Braude et al., 1975). Estimates on the current protection factor (kidney ppm Cd which causes failure at age 50 ÷ current kidney ppm Cd at age 50) varies from 4 for the general population to 12.5 for nonsmokers without industrial exposure. People have suffered kidney failure and bone malformation due to increased Cd in food and water in the Cd-polluted area in Japan (Kobayashi, 1970; Fulkerson & Goeller, 1973; Friberg et al., 1973). Most researchers feel that there is sufficient information about Cd that we should limit Cd exposure to humans where possible. Thus, the effect of land utilization of organic wastes and waste water on levels of Cd in food chain crops and tobacco is an important factor which may limit this practice.

b. **Cadmium Levels in Crops Grown on Long Term Sludge and Waste Water Application Sites**—There have been several studies on the microelement content of crops growing on fields where sewage wastes have been applied for many years as compared to nearby control fields. Andersson and Nilsson (1972) found no increase in Cd in rape fodder after 15 years of sludge utilization. Chaney et al. (1974)[2] found no increase in Cd in the existing pasture grasses where sludge was applied for 24 years. Johnson et al. (1974) found no increase in the Cd content of perennial ryegrass (*Lolium perenne* L.) growing on the Werribee sewage irrigation farm (Melbourne, Australia) after over 70 years of irrigation. These studies involved use of low Cd sewage or waste water.

Kirkham (1975) analyzed corn plants growing on a sludge-disposal farm at Dayton, Ohio. The soil had increased from 2 to 50–70 ppm Cd. The corn leaves increased from 2.1 to 14 and 24 ppm Cd, but the corn grain was increased only from 0.8 to 0.9 and 1.0 ppm Cd on control, sludge-treated, and supernatent-treated areas. Although these Cd levels in the control soil and plants are higher than others have generally reported, the exclusion of Cd from corn grain is remarkable. R. L. Chaney and P. W. Simon (unpublished) have grown Swiss chard, soybean, and orchardgrass on fields where sludge was applied for many years. They found little increase in foliar or grain Cd where domestic (Table 1) sludges were applied. However, where higher Cd and Cd/Zn sludges were applied, foliar Cd and grain Cd levels were higher than controls (Table 2). R. L. Chaney and P. W. Simon (unpublshed) have also found substantially higher Cd in corn grain on several farms in the Northeast where high Cd and Cd/Zn sludges were applied over many years.

c. **Cadmium Levels in Crops Grown on Recently Established Research Plots**—Recent field experiments have generally shown that plant Cd levels increase with sludge, refuse compost, or waste water application. Hinesly et al. (1972) and Jones et al. (1975) applied high Cd sludge and observed large increases in corn foliar Cd (0.5 to 20 ppm Cd) and corn grain (< 0.1 to 1.0 ppm Cd). Soybean grain Cd levels have exceeded 3 ppm on some of their plots. Dowdy and Larson (1975b) found 2- to 4-fold higher levels of Cd in nearly every crop tissue from large applications of a domestic sludge; however, pea fruit (*Pisum sativum* L.) and corn grain levels remained low. Chaney et al. (1975 and unpublished) found corn silage and grain increased from

Table 2—Cadmium content of Swiss chard and soybean tissues grown on long term sludge utilization sites in 1975 (R. L. Chaney and P. W. Simon, unpublished)†

City	Treatment	pH	Soil DTPA‡ Cd	Soil DTPA‡ Zn	Chard	Soybean Leaf	Soybean Grain
						ppm Cd	
4	Control	5.66	0.13	6.8	0.62	0.27	0.17
4	Control	6.68	0.14	6.0	0.52	0.26	0.15
4	Sludged	5.16	0.53	40.8	1.91	--	--
4	Sludged	6.16	0.57	43.8	0.63	--	--
9	Control	5.28	0.13	3.0	3.64	1.04	0.36
9	Control	6.71	0.10	3.3	1.15	0.55	0.28
9	Sludged	4.84	1.13	14.7	73.0	10.7	3.70
9	Sludged	6.58	1.59	19.9	5.46	1.87	1.51
13	Control	5.30	0.93	4.2	0.89	0.24	0.16
13	Control	6.38	0.96	4.1	0.49	0.17	0.13
13	Sludged	5.61	7.15	53.2	7.04	5.70	2.64
13	Sludged	6.63	5.45	37.7	1.77	2.38	0.65

† Sludge was applied from 1962–1975 at city 4, from 1961–1973 at city 9, and from 1967–1974 at city 13.
‡ The DTPA extract was a 1:2, soil/extractant ratio.

0.73 and 0.09 ppm Cd (control) to 4.6 and 0.70 ppm Cd following application of 224 metric tons/ha of a domestic sludge; soybean leaves and grain were increased from 0.21 and 0.07 (control) to 0.44 and 0.24 ppm Cd (224 metric tons/ha). Giordano et al. (1975) found that Cd in sweet corn grain had increased from about 0.3 to as high as 1.2 ppm Cd. They found that repeated application of waste-borne Cd did not lead to a proportional increase in foliar or grain Cd levels (Giordano & Mays, 1976a) (Table 3). Sabey and Hart (1975) reported a 3-fold increase in wheat grain Cd when 100 metric tons/ha of Denver sludge was applied. Refuse and refuse-sludge mixtures increased Cd in rye and corn leaves (King et al., 1974). Levels in corn grain were unchanged. Refuse-sludge compost application raised corn stover and grain Cd (Giordano et al., 1975).

Comparison of corn grain Cd levels among these reports reveals that Cd levels in control grain was about 0.03 ppm and only slightly increased by waste applications in north-central U. S. and Canada, while studies in Maryland and Alabama find much higher Cd increases in corn grain at similar stover Cd levels. Whether this result is due to varietal difference or to soil, sludge, or climatic differences is unclear. No regional differences in Cd or Zn/Cd were found in many states east of the Rocky Mountains (Huffman & Hodgson, 1973).

Microelement distribution in effluent, soil, and crops on the Pennsylvania State University effluent irrigation plots (stated in 1964) was reported by Sidle et al. (1976). Digested sludge was injected into the effluent from 1971 through 1973. Only about 7% of the added metals were removed in

Table 3—Effect of single and repeated organic waste applications on Cd content of corn grain and leaves and snapbean pods and leaves grown in 1975 (Giordano & Mays, 1976b)

Treatment†	Application rate		Cd concentration							
			Corn				Snapbean			
			A‡		B‡		A‡		B‡	
	A‡	B‡	Grain	Leaves	Grain	Leaves	Pods	Leaves	Pods	Leaves
	metric tons/ha		ppm							
Check	0	0	0.18	0.72	0.21	0.73	0.07	0.15	0.07	0.16
Compost	56	224	0.38	1.91	0.98	2.04	0.14	0.36	0.13	0.24
	112	448	0.42	1.93	0.65	2.53	0.10	0.24	0.12	0.26
	224	996	0.43	2.06	0.64	3.51	0.10	0.32	0.13	0.23
Sludge	50	200	0.75	3.42	0.87	4.94	0.17	0.75	0.19	1.14
	100	400	0.90	4.72	1.06	6.61	0.17	0.98	0.30	1.35
	200	800	1.00	5.91	1.17	7.04	0.15	0.82	0.27	1.29

† Cadmium content of compost and sludge on dry weight basis was 10 and 40 ppm, respectively.
‡ A applied for 1972; B (same rate) applied for 1972, 1973, 1974, and 1975.

harvested reed canarygrass (*Phalaris arundinacea*) or corn silage. Zinc, Cu, and Cd were increased in reed canarygrass; part of the increase could be adhering particulates from the sludge or waste water. The effluent and sludge were relatively low in microelements, and only plant levels of Cu were above the normal range.

d. **Greenhouse Studies on Cd Uptake by Plants**—A number of greenhouse studies on the effect of sludge Cd additions on Cd uptake by crops have been reported. Bingham et al. (1975, 1976) tested the phytotoxicity of Cd to many crops by adding different levels of Cd to a sludge, and adding 1% Cd-amended sludge to a calcareous soil. It was a "diagnostic criteria" test (Cd additions up to 640 ppm). Their results show that crops differ greatly in both Cd uptake and in soil Cd levels which cause phytotoxicity. Soil levels related with 25% yield reduction varied from 4 ppm for spinach to 170 ppm for cabbage (*Brassica oleracea* var. *capitata*); Cd content of diagnostic leaves at 25% yield reduction varied from 15 for field bean to 160 for cabbage. Table 4 shows the Cd content of diagnostic leaves and edible tissues of many crops grown in their trials (Bingham et al., 1975, 1976). Both relative crop uptake and tolerance of Cd could be different at lower soil pH and other sludge and soil composition. These results show that we can not rely on visual phytotoxicity to indicate unsafe levels of Cd in plant-based foods.

In two pot experiments studying the effect of sludge and lime rates on uptake of Cd by wheat and rape, Linnmann et al. (1973) and Andersson and Nilsson (1974) found that Cd was increased significantly in both wheat grain and rape fodder. Both sludge and lime rates affected soil pH, and their Cd uptake results are confounded by these pH changes; without CaO, sludge raised soil pH, while at the high CaO additions, sludge lowered soil pH. Soil pH had more effect on Cd in rape than sludge additions. Other Swedish

Table 4—Effect of addition of Cd-amended sewage sludge on yield and Cd content of diagnostic and edible tissues of plants (Bingham et al., 1975, 1976, unpublished)

	Control soil†		Cd added, 10 ppm‡		Soil Cd at 50% YD¶
	Diagnostic§	Edible§	Diagnostic§	Edible§	
			ppm Cd		
Paddy rice	<0.1	<0.1	<0.1	0.2	>640
Upland rice	0.4	<0.1	0.9	0.4	36
Sudangrass (*Sorghum vulgare sudanense*)	0.2	0.2	5.7	5.7	58
White clover (*T. repens*)	0.2	0.2	6.0	6.0	120
Alfalfa	0.3	0.3	8.2	8.3	145
Bermudagrass	0.3	0.3	9.4	9.4	400
Field bean	0.6	<0.1	10.3	0.7	65
Wheat	<0.1	<0.1	11.6	5.8	80
Zuchinni squash (*Cucurbita* sp.)	0.6	<0.1	12.5	0.7	200
Soybean	0.4	0.7	15.6	10.7	11
Tall fescue	1.4	1.4	17.3	17.3	320
Corn	3.9	<0.1	27.0	1.4	35
Carrot	1.4	0.9	38.0	16.0	65
Cabbage	0.7	0.2	39.0	1.8	350
Radish	4.2	0.3	40.0	4.0	160
Swiss chard	1.4	1.4	42.0	42.0	320
Table beet	0.8	0.2	47.0	4.5	133
Romaine lettuce	0.8	0.8	62.0	62.0	35
Tomato	2.6	<0.1	71.0	2.4	415
Curly cress	2.4	2.4	89.0	89.0	15
Spinach	3.6	3.6	161.0	161.0	10
Turnip (*Brassica rapa*)	1.8	<0.1	162.0	9.2	100

† Domino silt loam; contained 0.1 ppm Cd.
‡ Soil amended with 1% of a digested sludge containing 1,000 ppm Cd.
§ Leaves normally used for diagnostic foliar analysis, and plant tissue normally eaten.
¶ Yield decrement.

studies revealed that Cd levels in fall wheat have slowly increased during the 20th Century (Kjellstrom et al., 1975) and that monocot grains differ in Cd content (Kjellstrom et al., 1974). Small plot studies with ^{109}Cd, ^{115m}Cd, and sludge by Stenstrom and Lonsjo (1974) led them to conclude that sludge will increase food Cd.

Greenhouse pot studies on factors influencing Cd uptake have been reported. John et al. (1972) found that Cd adsorption capacity, exchangeable Cd levels, soil pH, and soil organic matter content were significantly related to Cd uptake by oats from 30 Cd-amended soils. Lime application reduced Cd uptake from Cd-amended soils by radish and lettuce (John, 1972) and other vegetables (John, 1973); the crops differed widely (at 40 mg Cd/kg soil, spinach Cd was 208 ppm, while pea pods were only 9.5 ppm).

Haghiri (1973, 1974) studied Cd uptake from several soils at pH 6.5 to 6.7. Added Cd reduced yields, and caused chlorosis. Crops differed in Cd level required to cause phytotoxicity and in Cd uptake at equal soil Cd levels. Haghiri (1974) added a muck soil (as an organic matter source) to a soil low in organic matter to test the relationship of soil CEC and soil organic matter to Cd uptake by oat. He concluded that except for its CEC effect, organic matter did not influence oat Cd levels. In another trial, he found that increasing soil temperature increased uptake of Cd by soybean. The importance of soil pH in plant uptake of Cd has been shown in many reports of both the pot and field studies. Figure 1b shows that Cd in soybean leaves was reduced from 33 ppm at pH 5.3 to 4 ppm at pH 6.9 (Chaney et al., 1975).

Interactions between microelements and Cd have been observed in nutrient solution and pot experiments. Francis and Rush (1974) found that increased Se reduced Cd transport to snapbean tops. Cunningham et al. (1975) found that increased Cu led to greater translocation of Cd to tops of corn and rye. Several researchers have investigated the effects of Zn on Cd uptake by plants. Lagerwerff and Biersdorf (1972) found that increased Zn lowered Cd levels in radish leaves and tubers grown in nutrient solutions. Francis and Rush (1974) found that increased Zn reduced Cd uptake and, to a lesser extent, translocation by snapbean grown in nutrient solutions. Haghiri (1974) grew soybean in a soil amended with 10 ppm Cd and with 0 to 400 ppm Zn; low Zn additions increased Cd in the tops, while higher Zn additions markedly reduced Cd in the soybean tops (Fig. 4). These results with dicots indicate that high Zn/Cd ratios in organic wastes would tend to restrict increases in plant Cd. In studies on corn in nutrient solution (Iwai et al., 1975), corn and rye grown in sludge and metal amended (Cunningham et al., 1975b), and tall fescue grown on Cd- and Zn-amended Lakeland sand (Giordano & Mays, 1976b) (Table 5), Zn did not decrease Cd uptake or translocation. F. Haghiri (personal communication) found that excessive K reduced Cd transport to plant tops. Perhaps these crop differences in the ef-

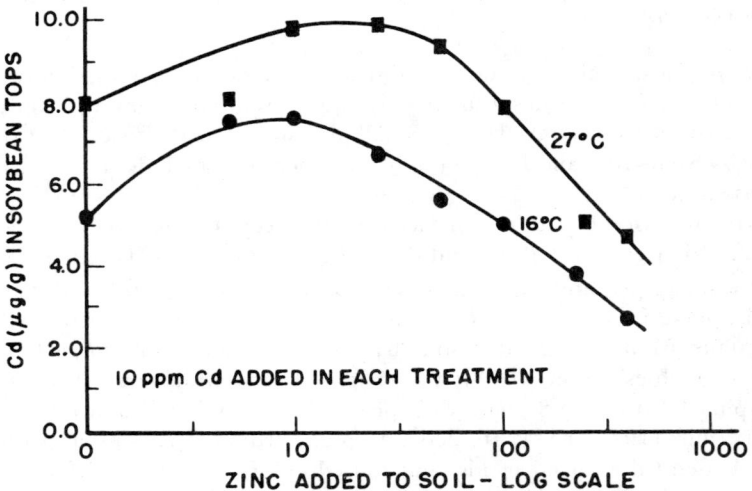

Fig. 4—The effects of Zn addition and soil temperature on Cd content of soybean tops grown on soil amended with 10 ppm Cd (Haghiri, 1974).

Table 5—Effect of additions of Zn and Cd to lakeland sand on yield and Cd content of four cuttings of tall fescue (Giordano & Mays, 1976a)

Cd added, ppm	Fescue yield (Zn added, ppm)			Cd content (Zn added, ppm)		
	0	50	250	0	50	250
	——— g/pot ———			——— ppm ———		
0	6.4	4.4	1.2	0.4	3.4	3.5
2	6.6	3.7	1.2	25.0	26.0	25.0
10	6.1	3.7	1.0	81.0	74.0	99.0
100	1.4	1.8	0.6	229.0	423.0	412.0

fects of other added metals and nutrients and soil pH on Cd uptake and translocation will help to explain the differences in Cd uptake due to additions of different sludges or composts (Cunningham et al., 1975a).

In Japan, rice (*Oryza sativa* L.) grain Cd levels were implicated in "itai-itai" disease (Yamagata & Shigematsu, 1970). Considerable research effort has focused on amendments or management techniques which might reduce Cd uptake by rice, although rice is actually much lower in Cd than other crops grown on these soils. Takijima, Katsumi, and Koizuma (1973); Takijima, Katsumi, and Takezawa (1973); and Takijima and Katsumi (1973a, 1973b) found that soil pH and Eh were important controls on Cd uptake; low soil pH favored uptake, while reducing conditions hindered uptake. Lime amendments and maintaining anaerobic soil conditions during grain filling by irrigation could reduce grain Cd to acceptable levels. Of course, other crops will not have the benefit of reducing available soil Cd by formation of CdS in submerged rice fields.

Phosphorus fertilizers also add Cd to cropland, although yearly Cd additions are much lower than with sludge. Williams and David (1973) found that Cd in Australian fertilizers was higher than in U. S. fertilizers; fertilizer-applied Cd remained extractable and crop Cd was increased on fertilized fields. They noted that Australian wheat remained generally low in Cd, although the only long term test of changes in grain Cd showed increases (Kjellstrom et al., 1975). Mortvedt and Giordano (1976) also found that Cd in fertilizers could increase Cd in plants.

Urban wastes generally increase crop Cd levels somewhat. Fassett (1975), Fulkerson and Goeller (1973), Friberg et al. (1971), Braude et al. (1975), Baker and Chesnin (1976), and others have discussed the movement of Cd into the food chain. It appears that increases in crop Cd can be kept low by excluding Cd wastes from sewers (see Chaney et al., 1975; Fulkerson & Goeller, 1973).

4. DIRECT CONTAMINATION OF CROPS WITH WASTE-BORNE MICROELEMENTS

When liquid sludge or manure, filter cake sludge, or waste waters are applied to fields of established forage crops, there is a possibility of increas-

ing plant levels of microelements by adhering particulates from the wastes.
Batey et al. (1972) found high levels of Cu and Zn in or on crops sprayed
with swine manure slurries. Boswell (1975) found extremely high foliar contamination with Cd, Pb, Cr, etc., when he applied sludge filter cake. Chaney
et al. (1974)[2] found that recent liquid sludge applications on orchardgrass
constituted the only significant increase in Cd, Pb, and Cu at Hagerstown,
Md.

The increased level of microelements in contaminated forages may exceed that from plant uptake from sludge-treated soil for elements such as Hg,
Pb, Cd, Cu, and Fe. R. L. Chaney, C. A. Lloyd, and P. W. Simon (unpublished) measured foliar contamination of tall fescue (*Festuca eliator* L.) by
applied liquid sludge, and the subsequent decline in microelements with
weathering and plant growth. "Apparent sludge content" of harvested fescue
fell from as much as 32% to as low as 2.5% (Fig. 5). Cattle have been observed grazing in pastures contaminated with adhering sludge.

Another potential route of waste-applied microelement movement into
the food chain is ingestion of soil by grazing animals. Healy et al. (1974)
have found high levels of soil in feces of grazing animals. Soil usually comprises 2 to 14% of the dietary dry matter of grazing cows and sheep. Whether ingested soil-borne microelements are absorbed by animals has not been
reported.

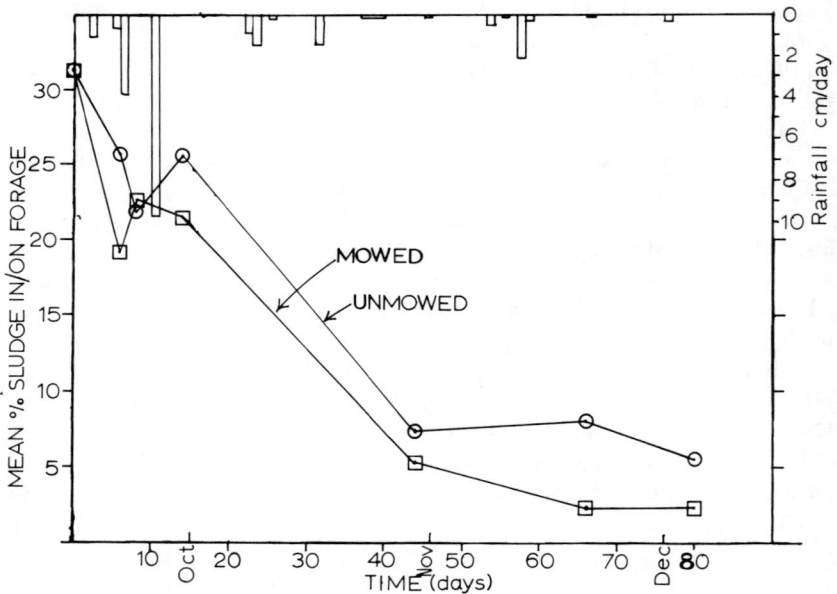

Fig. 5—The effect of mowing and rainfall on sludge adherence to tall fescue; 1.03 cm of
liquid digested sludge was applied on 16 Sept. 1975 (R. L. Chaney, C. A. Lloyd, &
P. W. Simon, unpublished).

F. Differential Crop Response to Microelements

Brown et al. (1972) summarized differences in crop response to deficient supplies of micronutrients. However, little work has been done to characterize differential crop response to phytotoxic levels of microelements. Differential crop response to Cd toxicity in a calcareous soil was reported by Bingham et al. (1975, 1976). Eaton (1944) and Richards (1954) summarized the response of many crops to phytotoxic B levels. Crop responses to phytotoxic Zn in a calcareous soil were reported by Boawn and Rasmussen (1971) and in a slightly acid (pH 6.5) soil by Boawn (1971). In nutrient solutions, crops differed in response to phytotoxic Zn (Caroll & Loneragen, 1969); the relative Zn tolerance of crops differed in these studies. Crop response to Ni-toxic serpentine soil was reported by Hunter and Vergnano (1952). Generally, monocots were more tolerant than dicots, and vegetable plants were most sensitive to metals. The results of Boawn and Rasmussen (1971) showed that monocots were less tolerant to toxic Zn in a calcareous soil than were dicots; this pattern was opposite that of nearly all the field experience with phytotoxicity (e.g., Smilde et al., 1974).

Plant species differ in their uptake of microelements when growing on the same soil. Beeson et al. (1947) showed differences in foliar Cu; Boawn (1971) and Boawn and Rasmussen (1971), in foliar Zn; Hunter and Vergnano (1952), in foliar Ni; and Bingham et al. (1975, 1976), in foliar Cd. A few plant families appear to be accumulators of one or more elements: clovers are high in Cu, and the beet family high in Zn and Cd. Cannon (1960), Antonovics et al. (1971), and Brown et al. (1972) have described more of these differences. Soybean varieties differ in Zn uptake at high levels (White et al. [1974][3]), as do other crops at low levels (see Brown et al., 1972).

Chumbley (1971) introduced the Zn-equivalent calculation (Zn[eq] = 1 × ppm Zn + 2 × ppm Cu + 8 × ppm Ni) to sum the potential phytotoxicity of the different metals in a waste. Chaney (1974), Chaney et al. (1975), and Stewart and Chaney (1975) have discussed the problems of the Zn-equivalent equation, but suggest that some measure of the relative phytotoxicity of Zn, Cu, and Ni is needed in setting maximum loading rates (Walker, 1975; Shipp & Baker, 1975).

Plant breeding may be capable of introducing tolerance to microelements or exclusion of microelements (Antonovics et al., 1971). However, not all crops display "ecotype" tolerance (Gartside & McNeilly, 1974). Walley et al. (1974) found some species which had Zn-tolerant ecotypes but not Cu-tolerant ecotypes.

The tolerance of soil organisms to excess microelements is unknown (Huisingh, 1974). Gish and Christenson (1973) and Van Hook (1974) found that earthworms concentrated Cd and Zn up to 200-fold above soil total levels; Zn/Cd fell from 137 in soil to 56 in earthworms. Lu et al. (1975) found that several organisms on their model ecosystem were very sensitive to sludge-borne metal. It would be appropriate to learn the tolerance of organisms other than higher plants before potentially toxic levels are added in organic wastes.

V. DISCUSSION AND SUMMARY

A. Factors Reducing the Phytotoxicity of Microelements in Organic Wastes

Greenhouse pot studies of the phytotoxicity of metals have usually utilized soluble salts of the metals. These studies revealed characteristic symptoms and "diagnostic criteria" for tolerable levels of metals in soils for specific crops and crop tissue concentrations at particular yield decrements (e.g., Boawn & Rasmussen, 1971; Bingham et al., 1975, 1976). Cunningham et al. (1975b) found that the phytotoxicity to corn and rye of metals added in digested sewage sludge was considerably less than when added as soluble salts. Based on these pot studies, one would predict metal phytotoxicity when many sludges are land applied. However, sludge applications have seldom caused yield reductions in field studies even when the predicted phytotoxic levels were exceeded in the surface soil (Jones et al., 1975). These observations have led many to extrapolate from short term studies and to conclude that no phytotoxicity would ever result; e.g., Rohde (1974) suggested that aerobically composted municipal wastes can not cause phytotoxicity, even though he demonstrated metal phytotoxic effects earlier on a sewage irrigation farm (Rohde, 1962).

We feel that a number of factors present in organic wastes tend to reduce or prevent toxicity of the waste-borne heavy metals: (i) wastes often cause a temporary rise in pH from mineralization of organic-N to ammonia, or because they may have enough alkalinity to "lime" acid soils; (ii) compounds of metals present initially may be extremely insoluble, e.g., ZnS is slow release Zn fertilizer (Wear et al., 1968; Qureshi & Gammon, 1971; Gammon, 1975), however, in sludges the sulfides, if present, appear to be rapidly converted (Bloomfield & Pruden, 1975); (iii) organic wastes increase the soil organic matter content and also increase the cation exchange capacity and metal adsorption capacity which could temporarily prevent phytotoxicity; (iv) wastes are initially poorly mixed in the soil and metals stay in the plow layer while pot studies expose all roots to the metals; (v) wastes generally add large amounts of P which prevents phytotoxicity directly or indirectly by reactions in the soil or plant; (vi) organic wastes increase available soil Fe which should reduce the ability of waste-borne metals to cause phytotoxicity by inducing Fe chlorosis; (vii) usually the potentially phytotoxic microelements in wastes are balanced; when single elements are added they can cause phytotoxicity by interference with another microelement (as Ni–Zn, Cu–Zn, Zn–Mn interactions); and (viii) the higher fertility of waste-amended soils (N, P, organic matter) could cause a greater yield increase if the heavy metals were absent; yields equal to the control may already indicate some phytotoxicity.

Making direct comparison between waste-borne metals and inorganic salts is complicated (Cunningham et al., 1975b; Jones, Hinesly, and Ziegler, 1973). Salts, unidentified phytotoxic organics and gases, ammonia, etc., formed during the initial rapid decomposition of the wastes may temporarily limit plant growth in waste-amended soils. These problems, plus the slow conversion of metals to available forms and possible reversion to unavailable forms, and changes in pH going on in waste-amended soils, prevent satisfactory estimation of the long term result. However, it is clear that phytotoxicity to vegetable crops from waste-applied metals at pH 6.5 has occurred in the field as summarized above; usually this requires higher metal concentrations than are present in the inorganic residue of domestic sludges and refuse composts.

B. Problems in Interpreting Published Reports

Microelement phytotoxicity, and metal accumulation in edible plant parts are very complicated. Further, some published reports fail to provide enough information to allow valid interpretation, or are done in such a way that no reasonable interpretation may be made. Terman (1974) described the unique requirements of pot experiments; frequently, inadequate macronutrients are provided to obtain maximum yields and the results of many experiments are uninterpretable.

When field or pot experiments are run, the reports should contain pertinent information about the chemical and physical nature of the waste and the soil so that results can be properly interpreted. A metal toxicity study with a sand from Florida would be very different than one with sand from Minnesota. Soil properties which should be reported include: pH, cation exchange capacity, exchangeable cations, organic matter, P and metal contents, 1/3-bar moisture percentage, kind and amounts of clays, and Mn and Fe oxide levels. Crop species and varieties should be listed. When an infertile sand at pH 4.5 is fertilized with a sludge compost and the mixture is pH 7, one should not be surprised when the plants on the compost-amended plots have lower metal content than the controls. This need for soil and plant characterization may be particularly true for studies involving old sludged sites (Kirkham, 1975; LeRiche, 1968).

Another problem is the accuracy of analyses for difficult-to-analyze elements like Cd, Pb, Ni, Hg, As, Se, etc. (Lisk, 1974). Koirtyohann and Pickett (1965, 1966) showed that "background correction" was necessary for analysis of low levels of many elements by atomic absorption, yet no statement is made in many papers regarding background correction. Some papers significant to heavy metal science have reported uncorrected values (e.g., Schroeder et al., 1967; Hinesly et al., 1972; Lagerwerff & Specht, 1970). Fulkerson and Goeller (1973) and Friberg et al. (1971) have discussed the problems these uncorrected results create in attempting to understand the environmental flow and tolerances of Cd.

C. Microelements in Organic Wastes: What Should be Done?

Potential phytotoxicity from microelements is an important long term problem in land application of some wastes, while other wastes may be safely applied indefinately. We have summarized information in this review showing that wastes, soils, and plants differ in their content of and reactions with microelements. These differences complicate the management of utilization or disposal of wastes on cropland. However, it has become clear that most problems with waste disposal on land can be managed judiciously by (i) excluding wastes contaminated with excessive amounts of microelements, (ii) limiting maximum total applications to levels safe for crops and the food chain, and (iii) managing waste-amended soils according to plant and soil tests.

Many agronomists are concerned that highly favorable yield responses of corn to applied wastes will lead farmers to add excessive amounts of waste to their land, or to accept polluted wastes because they are "free." Some wastes do have significant potential to cause phytotoxicity, and some will increase human dietary intake of Cd. Regulatory controls and, where necessary, monitoring programs must be established to protect against the unwise use of polluted wastes on land.

The trend toward land disposal of urban wastes can be attributed to the Environmental Protection Agency (EPA) which has a Congressional mandate to fund only cost-effective practices. Since billions of dollars are involved, the low energy requirements, low capital costs, and potential benefits lead cities to seriously consider land disposal. For some time the EPA has needed answers to questions about maximum safe soil loading and maximum safe waste compositions so that it can decide whether the benefits exceed the risks. Putting high levels of metals on noncalcareous soils is bound to lead to at least occasional crop failure because a desirable soil pH is not always maintained by private land owners. The suggestions of Chaney (1974) and Stewart and Chaney (1975) would specify a maximum safe loading for any crop at pH 6.5. Even here the land use commitment is appreciable, for yields of some vegetable crops would be reduced at pH 5.5 or below. It is doubtful whether higher loading should be considered since they may further limit potential land use. Several states have already published guidelines for sludge use on land which take these factors into account (Ohio [Thomas, 1975], Wisconsin [Keeney et al., 1975], Maryland [Walker, 1975], and [Illinois Environmental Protection Agency, 1975[5]]).

The research needs in this field are immense. Only sketchy knowledge is available about any of the identified aspects of microelements in waste-soil-plant-animal systems. For example, there are no established criteria for setting the maximum loading of one waste for one soil, to allow growth of one variety of one crop, to be fed to one animal. Limiting loadings of un-

---

[5] Illinois Environmental Protection Agency. 1975. Design criteria for municipal sludge utilization on agricultural land. Tech. Policy WPC-3, draft copy. p. 24.

polluted wastes can be tolerated now. Hopefully, with proper management, urban wastes can join manure and crop residues as wastes which may be utilized on farmland without negative impact on soil fertility, food quality, and human health.

LITERATURE CITED

Adams, F., and J. I. Wear. 1957. Manganese toxicity and soil acidity in relation to crinkle leaf of cotton. Soil Sci. Soc. Am. Proc. 21:305–308.

Adams, S. N., J. L. Honeysett, K. G. Tiller, and K. Norrish. 1969. Factors controlling the increase of cobalt in plants following the addition of a cobalt fertilizer. Aust. J. Soil Res. 7:29–42.

Allaway, W. H. 1968. Agronomic controls over the environmental cycling of trace elements. Adv. Agron. 20:235–274.

Anderson, A. J., D. R. Meyer, and F. K. Mayer. 1973. Heavy metal toxicities: levels of nickel, cobalt, and chromium in the soil and plants associated with visual symptoms and variation in growth of an oat crop. Aust. J. Agric. Res. 24:557–571.

Andersson, A., and K. O. Nilsson. 1972. Enrichment of trace elements from sewage sludge fertilizer in soils and plants. Ambio 1:176–179.

Andersson, A., and K. O. Nilsson. 1974. Influence of lime and soil pH on Cd availability to plants. Ambio 3:198–200.

Antonovics, J., A. D. Bradshaw, and R. G. Turner. 1971. Heavy metal tolerance in plants. Adv. Ecol. Res. 7:1–85.

Ashton, W. M. 1972. Nickel pollution. Nature 237:46–47.

Atkinson, H. J., G. R. Giles, and J. G. Desjardins. 1958. Effect of farmyard manure on the trace element content of soils and of plants grown thereon. Plant Soil 10:32–36.

Baker, D. E. 1974. Copper: Soil, water, plant relationships. Proc. Fed. Am. Soc. Exp. Biol. 33:1188–1193.

Baker, D. E., and L. Chesnin. 1976. Chemical monitoring of soils for environmental quality and animal and human health. Adv. Agron. 27:305–374.

Barber, S. A. 1974. Influence of the plant root on ion movement in soil. p. 525–564. In E. W. Carson (ed.) The plant root and its environment. University Press of Virginia, Charlottesville, Va.

Batey, T., C. Berryman, and C. Line. 1972. The disposal of copper-enriched pig-manure slurry on grassland. J. Br. Grassl. Soc. 27:139–143.

Baumhardt, G. R., and L. F. Welch. 1972. Lead uptake and corn growth with soil-applied lead. J. Environ. Qual. 1:92–94.

Beeson, K. C., L. Gray, and M. B. Adams. 1947. The absorption of mineral elements by forage plants: 1. The phosphorus, cobalt, manganese, and copper content of some common grasses. J. Am. Soc. Agron. 39:356–362.

Berrow, M. L., and J. Webber. 1972. Trace elements in sewage sludges. J. Sci. Food Agric. 23:93–100.

Bingham, F. T., A. L. Page, R. J. Mahler, and T. J. Ganje. 1975. Growth and cadmium accumulation of plants grown on a soil treated with a cadmium-enriched sewage sludge. J. Environ. Qual. 4:207–211.

Bingham, F. T., A. L. Page, R. J. Mahler, and T. J. Ganje. 1976. Yield and cadmium accumulation of forage species in relation to cadmium content of sludge-amended soil. J. Environ. Qual. 5:57–60.

Blood, J. W. 1963. Problems of toxic elements in crop husbandry. NAAS Q. Rev. 14(59):97–100.

Bloomfield, C., and G. Pruden. 1975. The effects of aerobic and anaerobic incubation on the extractabilities of heavy metals in digested sewage sludge. Environ. Pollut. 8:217–232.

Boawn, L. C. 1965. Sugarbeet induced zinc deficiency. Agron. J. 57:509.

Boawn, L. C. 1971. Zinc accumulation characteristics of some leafy vegetables. Commun. Soil Sci. Plant Anal. 2:31–36.

Boawn, L. C. 1974. Residual availability of fertilizer zinc. Soil Sci. Soc. Am. Proc. 38:800–803.

Boawn, L. C., and P. E. Rasmussen. 1971. Crop response to excessive zinc fertilization of alkaline soil. Agron. J. 63:874–876.

Boawn, L. C., F. G. Viets, Jr., C. L. Crawford, and J. L. Nelson. 1960. Effect of nitrogen carrier, nitrogen rate, zinc rate, and soil pH on zinc uptake by sorghum, potatoes, and sugarbeets. Soil Sci. 90:329–337.

Boswell, F. C. 1975. Municipal sewage sludge and selected element applications to soil: Effect on soil and fescue. J. Environ. Qual. 4:267–273.

Bouwer, H., and R. L. Chaney. 1975. Land disposal of wastewater. Adv. Agron. 26:133–176.

Bradford, G. R. 1966. Boron. p. 33–61. *In* H. D. Chapman (ed.) Diagnostic criteria for plants and soils. Div. of Agric. Sci., University of California, Berkely, Calif.

Braude, G. L., C. F. Jelinek, and P. Corneliussen. 1975. FDA's overview of the potential health hazards associated with land application of municipal wastewater sludge. Proc. 2nd Nat. Conf. Municipal Sludge Management, Information Transfer, Inc., Rockville, Md. p. 214–217.

Brown, A. L., B. A. Krantz, and P. E. Martin. 1964. The residual effect of zinc applied to soils. Soil Sci. Soc. Am. Proc. 28:236–238.

Brown, J. C., J. E. Ambler, R. L. Chaney, and C. D. Foy. 1972. Differential responses of plant genotypes to micronutrients. p. 389–418. *In* J. J. Mortvedt, P. M. Giordano, and W. L. Lindsay (ed.) Micronutrients in agriculture. Soil Sci. Soc. Am., Madison, Wis.

Brown, J. C., and W. E. Jones. 1975. Heavy metal toxicity in plants: 1. A crisis in embryo. Commun. Soil Sci. Plant Anal. 6:421–438.

Buchauer, M. J. 1973. Contamination of soil and vegetation near a zinc smelter by zinc, cadmium, copper, and lead. Environ. Sci. Technol. 7:131–135.

Cannon, H. L. 1955. Geochemical relations of zinc-bearing peat to the Lockport Dolomite, Orleans County, New York. Geol. Surv. Bull. 1000-D:119–185.

Cannon, H. L. 1960. Botanical prospecting for ore deposits. Science 132:591–598.

Carlson, C. W., D. L. Grunes, J. Alessi, and G. A. Reichman. 1961. Corn growth on Gardena surface and subsoil as affected by applications of fertilizer and manure. Soil Sci. Soc. Am. Proc. 25:44–47.

Carroll, M. D., and J. F. Loneragen. 1969. Response of plant species to concentrations of zinc in solution. II. Rates of zinc absorption and their relation to growth. Aust. J. Agric. Res. 20:457–463.

Chandler, W. H., D. R. Hoagland, and J. C. Martin. 1946. Littleleaf or rosette of fruit trees: VIII. Zinc and copper deficiency in corral soils. Proc. Am. Soc. Hortic. Sci. 47:15–19.

Chaney, R. L. 1973. Crop and food chain effects of toxic elements in sludges and effluents. p. 129–141. *In* Proc. Joint Conf. on Recycling municipal sludges and effluents on land. Nat. Assoc. State Univ. and Land-Grant Coll., Washington, D. C.

Chaney, R. L. 1974. Recommendations for management of potentially toxic elements in agricultural and municipal wastes. p. 97–120. *In* Factors involved in land application of agricultural and municipal wastes. USDA Nat. Program Staff, Soil, Water, and Air Sciences, Beltsville, Md.

Chaney, R. L. 1975. Metals in plants—absorption mechanisms, accumulation, and tolerance. Proc. Symp. Metals in the Biosphere. Dep. Land Resour. Sci., University of Guelph, Guelph, Ontario, Canada. p. 79–99.

Chaney, R. L., M. C. White, and P. W. Simon. 1975. Plant uptake of heavy metals from sewage sludge applied to land. p. 169–178. *In* Proc. 2nd Nat. Conf. Municipal Sludge Management. Information Transfer, Inc., Rockville, Md.

Chapman, H. D., G. F. Liebig, Jr., and A. P. Vanselow. 1939. Some nutritional relationships as revealed by a study of mineral deficiency and excess symptoms on citrus. Soil Sci. Soc. Am. Proc. 4:196–200.

Chumbley, C. G. 1971. Permissible levels of toxic metals in sewage used on agricultural land. Agric. Dev. Advis. Serv., Advis. Pap. No. 10. 12 p.

Clarke, A. L., and E. R. Graham. 1968. Zinc diffusion and distribution coefficients in soil as affected by soil texture, zinc concentrations, and pH. Soil Sci. 105:409–417.

Crooke, W. M. 1956. Effect of soil reaction on uptake of nickel from a serpentine soil. Soil Sci. 81:269–276.

Cunningham, J. D., D. R. Keeney, and J. A. Ryan. 1975a. Yield and metal composition of corn and rye grown on sewage sludge-amended soil. J. Environ. Qual. 4:449–454.

Cunningham, J. D., D. R. Keeney, and J. A. Ryan. 1975b. Phytotoxicity and uptake of metals added to soils as inorganic salts or in sewage sludge. J. Environ. Qual. 4: 460–462.

Cunningham, J. D., J. A. Ryan, and D. R. Keeney. 1975. Phytotoxicity in and metal uptake from soil treated with metal-amended sewage sludge. J. Environ. Qual. 4: 455–460.

Davis, G. K. 1974. High-level copper feeding of swine and poultry and the ecology. Proc. Fed. Am. Soc. Exp. Biol. 33:1194–1196.

Delas, J. 1963. The toxicity of copper accumulating in soils. Agrochimica 7:258–288.

DeRemer, E. D., and R. L. Smith. 1964. A preliminary study on the nature of a zinc deficiency in field beans as determined by radioactive zinc. Agron. J. 56:67–70.

Doring, H. 1960. Chemical reasons for the "fatigue" of Berlin sewage farm soils and possibilities for correcting it. Dtsch. Landwirtsch. 11:342–345.

Dowdy, R. H., and W. E. Larson. 1975a. Metal uptake by barley seedlings grown on soils amended with sewage sludge. J. Environ. Qual. 4:229–233.

Dowdy, R. H., and W. E. Larson. 1975b. The availability of sludge-borne metals to various vegetable crops. J. Environ. Qual. 4:278–282.

Dykeman, W. R., and A. S. deSousa. 1966. Natural mechanisms of copper tolerance in a copper swamp forest. Can. J. Bot. 44:871–878.

Eaton, D. F., G. W. A. Fowles, M. W. Thomas, and G. B. Turnbull. 1975. Chromium and lead in colored printing inks used for children's magazines. Environ. Sci. Technol. 9:768–770.

Eaton, F. M. 1944. Deficiency, toxicity, and accumulation of boron in plants. J. Agric. Res. 69:237–277.

Elia, V. J., E. E. Menden, and H. G. Petering. 1973. Cadmium and nickel—common characteristics of lettuce leaf and tobacco cigarette smoke. Environ. Lett. 4:317–324.

Ellis, B. G. 1973. The soil as a chemical filter. p. 46–70. In W. E. Sopper and L. T. Kardos (ed.) Recycling treated municipal wastewater and sludge through forest and cropland. Pennsylvania State University, University Park, Pa.

Ellis, B. G., and B. D. Knezek. 1972. Adsorption reactions of micronutrients in soils. p. 59–78. In J. J. Mortvedt, P. M. Giordano, and W. L. Lindsay (ed.) Micronutrients in agriculture. Soil Sci. Soc. Am., Madison, Wis.

Ellis, J. H., R. I. Barnhisel, and R. E. Phillips. 1970. The diffusion of copper, manganese, and zinc as affected by concentration, clay mineralogy, and associated anions. Soil Sci. Soc. Am. Proc. 34:866–870.

Ennis, M. T., and J. C. Brogan. 1961. The availability of copper from copper humic acid complexes. Ir. J. Agric. Res. 1:35–42.

Fassett, D. W. 1975. Cadmium: Biological effects and occurrences in the environment. Annu. Rev. Pharmacol. 15:425–435.

Fiskell, F. G. A., P. H. Everett, and S. J. Locascio. 1964. Minor element release from organo-N fertilizer materials in laboratory and field studies. J. Agric. Food Chem. 12:363–367.

Flick, D. F., H. F. Kraybill, and J. J. Dimitroff. 1971. Toxic effects of cadmium. A review. Environ. Res. 4:71–85.

Follett, R. H., D. L. Grunes, G. A. Reichman, and C. W. Carlson. 1974. Recovery of corn yield on Gardena subsoil as related to leaf and soil analysis. Soil Sci. Soc. Am. Proc. 38:327–331.

Follett, R. H., and W. L. Lindsay. 1971. Changes in DTPA-extractable zinc, iron, manganese, and copper in soils following fertilization. Soil Sci. Soc. Am. Proc. 35: 600–602.

Ford, H. W. 1953. Root distribution of chlorotic and iron-chelate-treated citrus trees. Proc. Fla. State Hortic. Soc. 66:22–26.

Francis, C. W., and S. G. Rush. 1974. Factors affecting uptake and distribution of cadmium in plants. p. 75–81. In D. D. Hemphill (ed.) Trace substances in environmental health VII. University of Missouri, Columbia, Mo.

Fraser, D. C. 1961. A syngenetic copper deposit of recent age. Econ. Geol. 56:951–962.

Friberg, L., M. Piscator, and G. Nordberg. 1971. Cadmium in the environment. CRC Press, Cleveland, Ohio. 166 p.

Friberg, L., M. Piscator, G. Nordberg, and T. Kjellstrom. 1973. Cadmium in the environment, II. Environ. Prot. Agency Rep. No. EPA-R2-73-190. NTIS No. PB 221 198. 147 p.

Fulkerson, W., and H. E. Goeller, ed. 1973. Cadmium, the dissipated element. ORNL Rep. No. ORNL-NSF-EP-21. Oak Ridge National Laboratory, Oak Ridge, Tenn. 473 p.

Gadde, R. R., and H. A. Laitinen. 1974. Studies of heavy metal adsorption by hydrous iron and manganese oxides. Anal. Chem. 46:2022-2026.

Gadgil, R. L. 1969. Tolerance of heavy metals and the reclamation of industrial waste. J. Appl. Ecol. 6:247-259.

Gammon, N., Jr. 1975. Persistence of ZnS in Leon fine sand and its relative availability to plants. Soil Crop Sci. Soc. Fla. Proc. 34:52-54.

Gartside, D. W., and T. McNeilly. 1974. The potential for evolution of heavy metal tolerance in plants. II. Copper tolerance in normal populations of different plant species. Heredity 32:335-348.

Giordano, P. M., and D. A. Mays. 1976a. Yield and heavy metal content of several vegetable species grown in soil amended with sewage sludge. *In* Biological implications of metals in the environment. 15th Annu. Hanford Life Sci. Symp., Richland, Wash. (In press.)

Giordano, P. M., and D. A. Mays. 1976b. Implications of land disposal applications of municipal wastes. Environ. Prot. Technol. Ser. EPA-000-000. (In press.)

Giordano, P. M., J. J. Mortvedt, and D. A. Mays. 1975. Effect of municipal wastes on crop yields and uptake of heavy metals. J. Environ. Qual. 4:394-399.

Gish, C. D., and R. E. Christensen. 1973. Cadmium, nickel, lead, and zinc in earthworms from roadside soil. Environ. Sci. Technol. 7:1060-1062.

Graham, E. R. 1973. Selective distribution and labile pools of micronutrient elements as factors affecting plant uptake. Soil Sci. Soc. Am. Proc. 37:70-74.

Grunes, D. L., L. C. Boawn, C. W. Carlson, and F. G. Viets, Jr. 1961a. Zinc deficiency of corn and potatoes as related to soil and plant analyses. Agron. J. 53:68-71.

Grunes, D. L., L. C. Boawn, C. W. Carlson, and F. G. Viets, Jr. 1961b. Land l-e-v-e-l-i-n-g: it may cause zinc deficiency. N. D. Farm Res. 21(11):4-7.

Guest, P. L., and H. D. Chapman. 1944. Some effects of pH on the growth of citrus in sand and solution cultures. Soil Sci. 58:455-465.

Haghiri, F. 1973. Cadmium uptake by plants. J. Environ. Qual. 2:93-96.

Haghiri, F. 1974. Plant uptake of cadmium as influenced by cation exchange capacity, organic matter, zinc, and soil temperature. J. Environ. Qual. 3:180-183.

Halstead, R. L. 1968. Effect of different amendments on yield and composition of oats grown on a soil derived from serpentine materials. Can. J. Soil Sci. 48:301-305.

Hambridge, K. M., C. Hambridge, M. Jacobs, and J. D. Baum. 1972. Low levels of zinc in hair, anorexia, poor growth, and hypogeusia in children. Pediatr. Res. 6:868-874.

Hardison, J. R. 1963. Commercial control of *Puccinia striiformis* and other rusts in seed crops of *Poa pratensis* by nickel fungicides. Phytopathology 53:209-216.

Hassett, J. J. 1974. Capacity of selected Illinois soils to remove lead from aqueous solutions. Commun. Soil Sci. Plant Anal. 5:499-505.

Healy, W. B., P. C. Rankin, and H. M. Watts. 1974. Elemental composition of grazed and ungrazed herbage. N. Z. J. Agric. Res. 17:59-61.

Hewitt, E. J. 1948. Relation of manganese and some other metals to the iron status of plants. Nature 161:489-490.

Hewitt, E. J. 1953. Metal interrelationships in plant nutrition. 1. Effects of some metal toxicities on sugar beet, tomato, oat, potato, and marrowstem kale grown in sand culture. J. Exp. Bot. 4:59-64.

Hinesly, T. D., O. C. Braids, and J. E. Molina. 1971. Agricultural benefits and environmental changes resulting from the use of digested sewage sludge on field crops. Environ. Prot. Agency Solid Waste Manage. Ser. Rep. SW-30d. 62 p.

Hinesly, T. D., R. L. Jones, and E. L. Ziegler. 1972. Effects on corn by applications of heated anaerobically digested sludge. Compost Sci. 13:26-30.

Hodgson, J. F. 1963. Chemistry of the micronutrient elements in soil. Adv. Agron. 15:119-159.

Hodgson, J. F. 1969. Contribution of metal-organic complexing agents to the transport of metals to roots. Soil Sci. Soc. Am. Proc. 33:68-75.

Hodgson, J. F., W. L. Lindsay, and W. D. Kemper. 1967. Contribution of fixed charge and mobile complexing agents to the diffusion of Zn. Soil Sci. Soc. Am. Proc. 31:410-413.

Hodgson, J. F., K. L. Neeley, and J. C. Pushee. 1972. Iron fertilization of calcareous soils in the greenhouse and laboratory. Soil Sci. Soc. Am. Proc. 36:320-323.

Huffman, E. W. D., Jr., and J. F. Hodgson. 1973. Distribution of cadmium and zinc/ cadmium ratios in crops from 19 states east of the Rocky Mountains. J. Environ. Qual. 2:289-291.

Huisingh, D. 1974. Heavy metals—implications for agriculture. Annu. Rev. Phytopathol. 12:375-388.

Humenik, F. J., R. W. Skaggs, C. R. Willey, and D. Huisingh. 1972. Evaluation of swine waste treatment alternatives. Proc. Cornell Agric. Waste Manage. Conf. p. 341-352.

Hunter, J. G., and O. Vergnano. 1952. Nickel toxicity in plants. Ann. Appl. Biol. 39: 279-284.

Hutchinson, T. C., and L. M. Whitby. 1974. Heavy-metal pollution in the Sudbury mining and smelting region of Canada. I. Soil and vegetation contamination by nickel, copper, and other metals. Environ. Conserv. 1:123-132.

Iwai, I., T. Hara, and Y. Sonoda. 1975. Factors affecting cadmium uptake by the corn plant. Soil Sci. Plant Nutr. 21:37-46.

Jenkins, S. H., and J. S. Cooper. 1964. The solubility of heavy metal hydroxides in water, sewage and sewage sludge. III. The solubility of heavy metals present in digested sewage sludge. Int. J. Air Water Pollut. 8:695-703.

Jenne, E. A. 1968. Controls on Mn, Fe, Co, Ni, Cu, and Zn concentrations in soils and water: the significant role of hydrous Mn and Fe oxides. Adv. Chem. 73:337-387.

John, M. K. 1972. Effect of lime on soil extraction and on availability of soil applied cadmium to radish and leaf lettuce plants. Sci. Total Environ. 1:303-308.

John, M. K. 1973. Cadmium uptake by eight food crops as influenced by various soil levels of cadmium. Environ. Pollut. 4:7-15.

John, M. K., C. J. VanLaerhoven, and H. H. Chuah. 1972. Factors affecting plant uptake and phytotoxicity of cadmium added to soils. Environ. Sci. Technol. 6:1005-1009.

Johnson, C. M., G. A. Pearson, and P. R. Stout. 1952. Molybdenum nutrition of crop plants: II. Plant and soil factors concerned with molybdenum deficiencies in crop plants. Plant Soil 4:178-196.

Johnson, R. D., R. L. Jones, T. D. Hinesly, and D. J. David. 1974. Selected chemical characteristics of soils, forages, and drainage water from the sewage farm serving Melbourne, Australia. Department of the Army, Corps of Engineers, Washington, D. C. 54 p.

Jones, J. S., and M. B. Hatch. 1945. Spray residues and crop assimilation of arsenic and lead. Soil Sci. 60:277-288.

Jones, L. H. P., S. C. Jarvis, and D. W. Cowling. 1973. Lead uptake from soils by perennial ryegrass and its relation to the supply of an essential element (sulphur). Plant Soil 38:605-619.

Jones, R. L., T. D. Hinesly, and E. L. Ziegler. 1973. Cadmium content of soybeans grown in sewage sludge-amended soil. J. Environ. Qual. 2:351-353.

Jones, R. L., T. D. Hinesly, E. L. Ziegler, and J. J. Tyler. 1975. Cadmium and zinc contents of corn leaf and grain produced by sludge-amended soil. J. Environ. Qual. 4: 509-514.

Keeney, D. R., K. W. Lee, and L. M. Walsh. 1975. Guidelines for the application of wastewater sludge to agricultural land in Wisconsin. Wisconsin Dep. Nat. Resour. Tech. Bull. 88. 36 p.

Kelley, W. P., and S. M. Brown. 1928. Boron in the soils and irrigation waters of southern California, and its relation to citrus and walnut culture. Hilgardia 3(16):445-458.

King, L. D., and H. D. Morris. 1972. Land disposal of liquid sewage sludge: II. The effect on soil pH, manganese, zinc, and growth and chemical composition of rye (Secale cereale L.). J. Environ. Qual. 1:425-429.

King, L. D., L. A. Rudgers, and L. R. Webber. 1974. Application of municipal refuse and liquid sewage sludge to agricultural land: 1. Field study. J. Environ. Qual. 3: 361-366.

Kirkham, M. B. 1975. Trace elements in corn grown on long-term sludge disposal site. Environ. Sci. Technol. 9:765-768.

Kjellstrom, T., B. Lind, L. Linnman, and C. G. Elinder. 1975. Variations of cadmium concentrations in Swedish wheat and barley. Arch. Environ. Health 30:321-328.

Kjellstrom, T., B. Lind, L. Linnman, and G. Nordberg. 1974. A comparative study on methods for cadmium analysis of grain with an application to pollution evaluation. Environ. Res. 8:92-106.

Knezek, B. D. 1972. Heavy metal reactions in the soil. Mich. State Univ. Inst. Water Res. Tech. Rep. 30:27-43.

Knowles, F. 1945. The poisoning of plants by zinc. Agric. Prog. 20:16-19.

Kobayashi, J. 1970. Relation between the "itai-itai" disease and the pollution of river water by cadmium from a mine. Proc. 5th Int. Water Pollut. Res. Conf., San Francisco, Calif. I-25:1-7.

Koirtyohann, S. R., and E. E. Pickett. 1965. Background corrections in long path atomic absorption spectrometry. Anal. Chem. 37:601-603.

Koirtyohann, S. R., and E. E. Pickett. 1966. Light scattering by particles in atomic absorption spectrometry. Anal. Chem. 38:1087-1088.

Kubota, J., and W. H. Allaway. 1972. Geographic distribution of trace element problems. p. 525-554. In J. J. Mortvedt, P. M. Giordano, and W. L. Lindsay (ed.) Micronutrients in agriculture. Soil Sci. Soc. Am., Madison, Wis.

Lagerwerff, J. V. 1972. Lead, mercury, and cadmium as environmental contaminants. p. 593-636. In J. J. Mortvedt, P. M. Giordano, and W. L. Lindsay (ed.) Micronutrients in agriculture. Soil Sci. Soc. Am., Madison, Wis.

Lagerwerff, J. V. 1975. Current research in heavy metals in soil, sediments, and water: January 1973, October 1974. Proc. 2nd Annu. Nat. Sci. Found. Trace Contam. Conf., Asilomar, Calif. p. 16-47.

Lagerwerff, J. V., and G. T. Biersdorf. 1972. Interactions of zinc with uptake and translocation of cadmium in radish. p. 515-522. In D. D. Hemphill (ed.) Trace substances in environmental health-V. University of Missouri, Columbia, Mo.

Lagerwerff, J. V., D. L. Brower, and G. T. Biersdorf. 1972. Accumulation of cadmium, copper, lead, and zinc in soil and vegetation in the proximity of a smelter. p. 71-78. In D. D. Hemphill (ed.) Trace substances in environmental health-VI. University of Missouri, Columbia, Mo.

Lagerwerff, J. V., and A. W. Specht. 1970. Contamination of roadside soil and vegetation with cadmium, nickel, lead, and zinc. Environ. Sci. Technol. 4:583-586.

Larson, W. E., R. H. Susag, R. H. Dowdy, C. E. Clapp, and R. E. Larson. 1974. Use of sewage sludge in agriculture with adequate environmental safeguards. Proc. Sludge Handling and Disposal Seminar, Toronto, 18-19 September. Environment Canada and Ontario Ministry of the Environment, Ottawa, Ontario, Canada. (In press.)

Lee, C. R., and G. R. Craddock. 1969. Factors affecting plant growth in high-zinc medium: II. Influence of soil treatments on growth of soybeans on strongly acid soil containing zinc from peach sprays. Agron. J. 61:565-567.

Lee, C. R., and N. R. Page. 1967. Soil factors influencing the growth of cotton following peach orchards. Agron. J. 59:237-240.

Leeper, G. W. 1972. Reactions of heavy metals with soil with special regard to their application of sewage wastes. Department of Army, Corps of Engineers, Washington, D. C. Contract No. DACW73-73-C-0026. 70 p.

Leonard, C. D., and I. Stewart. 1952. Correction of iron chlorosis in citrus with chelated iron. Proc. Fla. State Hortic. Soc. 65:20-24.

LeRiche, H. H. 1968. Metal contamination of soil in the Woburn Market-Garden experiment resulting from the application of sewage sludge. J. Agric. Sci. 71:205-208.

Lewis, G. P., W. J. Jusko, L. L. Coughlin, and S. Hartz. 1972. Cadmium accumulation in man: Influence of smoking, occupation, alcoholic habit, and disease. J. Chronic Dis. 25:717-726.

Liebhardt, W. C., and T. J. Koske. 1974. The lead content of various plant species as affected by Cycle-lite humus. Commun. Soil Sci. Plant Anal. 5:85-92.

Lindsay, W. L. 1972. Inorganic phase equilibria of micronutrients in soils. p. 41-57. In J. J. Mortvedt, P. M. Giordano, and W. L. Lindsay (ed.) Micronutrients in agriculture. Soil Sci. Soc. Am., Madison, Wis.

Lindsay, W. L. 1973. Inorganic reactions of sewage wastes with soils. p. 91-96. In Proc. Joint Conf. on Recycling Municipal Sludges and Effluents on Land. Nat. Assoc. State Univ. and Land-Grant Coll., Washington, D. C.

Lindsay, W. L. 1974. Role of chelation in micronutrient availability. p. 508-524. In E. W. Carson (ed.) The plant root and its environment. University Press of Virginia, Charlottesville, Va.

Lindsay, W. L., and N. J. Park. 1970. Waste composts as chelating agents in plant nutrition. Progr. Rep. No. 2 on Solid Waste Res. Grant No. 8 R01 EC 00273-02. p. 24.

Linnman, L., A. Andersson, K. O. Nilsson, B. Lind, T. Kjellstrom, and L. Friberg. 1973. Cadmium uptake by wheat from sewage sludge used as a plant nutrient source. Arch. Environ. Health 27:45-47.

Lisk, D. J. 1972. Trace metals in soils, plants, and animals. Adv. Agron. 24:267-325.

Lisk, D. J. 1974. Recent developments in the analysis of toxic elements. Science 184: 1137-1141.

Lott, W. L. 1939. The relation of hydrogen ion concentrations to the availability of zinc in the soil. Soil Sci. Soc. Am. Proc. 3:115-121.

Lu, P. Y., R. L. Metcalf, R. Furman, R. Vogel, and J. Hassett. 1975. Model ecosystem studies of lead and cadmium and of urban sewage sludge containing these elements. J. Environ. Qual. 4:505-509.

Lucus, R. E., and B. D. Knezek. 1972. Climatic and soil conditions promoting micronutrient deficiencies in plants. p. 265-288. In J. J. Mortvedt, P. M. Giordano, and W. L. Lindsay (ed.) Micronutrients in agriculture. Soil Sci. Soc. Am., Madison, Wis.

Lunt, H. A. 1953. The case for sludge as a soil improver with emphasis on value of pH control and toxicity of minor elements. Water & Sewage Works 100:295-301.

Lunt, H. A. 1959. Digested sewage sludge for soil improvement. Conn. Agric. Exp. Stn. Bull. 622:1-30.

Maas, E. V., G. Ogata, and M. J. Garber. 1972. Influence of salinity on Fe, Mn, and Zn uptake by plants. Agron. J. 64:793-795.

McKenzie, R. M. 1972. The manganese oxides in soils—a review. Z. Pflanzenernaehr. Bodenkd. 131:221-242.

Maclean, A. J., R. L. Halstead, and B. J. Finn. 1969. Extractability of added lead in soils and its concentration in plants. Can. J. Soil Sci. 49:327-334.

Martin, W. E., J. Vlamis, and N. W. Stice. 1953. Field correction of calcium deficiency on a serpentine soil. Agron. J. 45:204-208.

Melton, J. R., S. K. Mahtab, and A. R. Swoboda. 1973. Diffusion of zinc in soils as a function of applied zinc, phosphorus, and soil pH. Soil Sci. Soc. Am. Proc. 37: 379-381.

Menden, E. E., V. J. Elia, L. W. Michael, and H. G. Petering. 1972. Distribution of cadmium and nickel of tobacco during cigarette smoking. Environ. Sci. Technol. 6: 830-832.

Milbocker, D. C. 1974. Zinc toxicity to plants grown in media containing polyrubber. HortScience 9:545-546.

Miller, B. F., W. L. Lindsay, and A. A. Parsa. 1969. Use of poultry manure for correction of Zn and Fe deficiencies in plants. Proc. Conf. Animal Waste Manage., Cornell University, Ithaca, N. Y. p. 120-123.

Miller, J. E., J. J. Hassett, and D. E. Koeppe. 1975a. The effect of soil lead sorption capacity on the uptake of lead by corn. Commun. Soil Sci. Plant Anal. 6:349-358.

Miller, J. E., J. J. Hassett, and D. E. Koeppe. 1975b. The effect of soil properties and extractable lead levels on lead uptake by soybeans. Commun. Soil Sci. Plant Anal. 6:339-347.

Miller, M. H., and A. J. Ohlrogge. 1958a. Water-soluble chelating agents in organic materials. I. Characterization of chelating agents and their reaction with trace metals in soils. Soil Sci. Soc. Am. Proc. 22:225-228.

Miller, M. H., and A. J. Ohlrogge. 1958b. Water-soluble chelating agents in organic materials. II. Influence of chelate containing materials on the availability of trace metals to plants. Soil Sci. Soc. Am. Proc. 22:228-231.

Mishra, D., and M. Kar. 1974. Nickel in plant growth and metabolism. Bot. Rev. 40: 395-452.

Mitchell, R. L., and J. W. S. Reith. 1966. The lead content of pasture herbage. J. Sci. Food Agric. 17:437-440.

Moore, D. P. 1972. Mechanisms of micronutrient uptake by plants. p. 171-198. In J. J. Mortvedt, P. M. Giordano, and W. L. Lindsay (ed.) Micronutrients in agriculture. Soil Sci. Soc. Am., Madison, Wis.

Mortvedt, J. J., and P. M. Giordano. 1973. Grain sorghum response to iron in a ferrous sulfate-ammonium thiosulfate-ammonium polyphosphate suspension. Soil Sci. Soc. Am. Proc. 37:951-955.

Mortvedt, J. J., and P. M. Giordano. 1975. Response of corn to zinc and chromium in municipal wastes applied to soil. J. Environ. Qual. 4:170–174.

Mortvedt, J. J., and P. M. Giordano. 1976. Crop uptake of heavy metal contaminants in fertilizers. p. 000–000. *In* Biological implications of metals in the environment. Proc. 15th Annu. Hanford Life Sci. Symp., Richland, Wash. (In press.)

Mortvedt, J. J., P. M. Giordano, and W. L. Lindsay, ed. 1972. Micronutrients in agriculture. Soil Sci. Soc. of Am., Madison, Wis. 666 p.

Mulford, F. R., and D. C. Martens. 1971. Response of alflafa to boron in flyash. Soil Sci. Soc. Am. Proc. 35:296–300.

Murphy, L. S., and L. M. Walsh. 1972. Correction of micronutrient deficiencies with fertilizers. p. 347–388. *In* J. J. Mortvedt, P. M. Giordano, and W. L. Lindsay (ed.) Micronutrients in agriculture. Soil Sci. Soc. Am., Madison, Wis.

Murrmann, R. P., and F. R. Koutz. 1972. Role of soil chemical processes in reclamation of wastewater applied to land. *In* S. C. Reed (ed.) Wastewater management by disposal on the land. Cold Reg. Res. & Eng. Lab. (CRREL) Spec. Rep. 171:48–76. Hanover, N. H.

National Academy of Sciences-National Academy of Engineering. 1974. Water quality criteria, 1972. Nat. Acad. of Sci., Washington, D. C.

Neary, D. G., G. Schneider, and D. P. White. 1975. Boron toxicity in red pine following municipal wastewater irrigation. Soil Sci. Soc. Am. Proc. 39:981–982.

Nicholas, D. J. D., C. P. Lloyd-Jones, and D. J. Fisher. 1957. Some problems associated with determining iron in plants. Plant Soil 8:367–377.

Nicholas, D. J. D. 1950. Some effects of metals in excess on crop plants grown in soil culture: 1. Effects of copper, zinc, lead, cobalt, nickel, and manganese on tomato grown in an acid soil. Bristol Agric. Hortic. Res. Stn. Annu. Rep. 1950:96–108.

Oliver, B. G., and E. G. Cosgrove. 1974. The efficiency of heavy metal removal by a conventional activated sludge treatment plant. Water Res. 8:869–874.

Olsen, S. R. 1972. Micronutrient interactions. p. 243–264. *In* J. J. Mortvedt, P. M. Giordano, and W. L. Lindsay (ed.) Micronutrients in agriculture. Soil Sci. Soc. Am., Madison, Wis.

Page, A. L. 1974. Fate and effects of trace elements in sewage sludge when applied to agricultural lands. A literature review study. U. S. Environ. Prot. Agency, Rep. No. EPA-670/2-274-005. 108 p.

Page, A. L., and F. T. Bingham. 1973. Cadmium residues in the environment. Residue Rev. 48:1–44.

Page, E. R. 1966. The micronutrient content of young plants as affected by farmyard manure. J. Hortic. Sci. 41:257–261.

Patterson, J. B. E. 1971. Metal toxicities arising from industry. *In* Trace elements in soils and crops. Min. Agric. Fish. Food Tech. Bull. 21:193–207.

Pound, C. E., and R. W. Crites. 1973. Characteristics of municipal effluents. p. 49–61. *In* Proc. Joint Conf. on Recycling Municipal Sludges and Effluents on Land. Nat. Assoc. State Univ. and Land-Grant Coll., Washington, D. C.

Pringle, W. L., W. K. Dawley, and J. E. Miltimore. 1973. Sufficiency of Cu and Zn in barley, forage, and corn silage rations as measured by response to supplements by beef cattle. Can. J. Anim. Sci. 53:497–502.

Proctor, J. 1971. The plant ecology of serpentine. II. Plant response to serpentine soils. J. Ecol. 59:397–410.

Purves, D., and E. J. Mackenzie. 1973. Effects of applications of municipal compost on uptake of copper, zinc, and boron by garden vegetables. Plant Soil 39:361–371.

Purves, D., and E. J. Mackenzie. 1974. Phytotoxicity due to boron in municipal compost. Plant Soil 40:231–235.

Qureshi, J. N., and N. Gammon, Jr. 1971. Special characteristics of zinc sulfide, a potential source of fertilizer zinc. Soil Crop Sci. Soc. Fla. Proc. 31:125–127.

Reid, R. L., and G. A. Jung. 1974. Effects of elements other than nitrogen on the nutritive value of forage. p. 395–435. *In* D. A. Mays (ed.) Forage fertilization. Am. Soc. of Agron., Madison, Wis.

Reuther, W., and P. F. Smith. 1952a. Iron chlorosis in Florida citrus groves in relation to certain soil constituents. Fla. State Hortic. Soc. Proc. 65:62–69.

Reuther, W., and P. F. Smith. 1952b. Toxic effect of copper on growth of citrus seedlings and its possible relation to acid-soil chlorosis in Florida citrus groves. Citrus Mag. 14(11):25–27.

Reuther, W., and P. F. Smith. 1954. Toxic effects of accumulated copper in Florida soils. Soil Sci. Soc. Fla. Proc. 14:17–23.

Reuther, W., P. F. Smith, and G. K. Scudder, Jr. 1953. Relation of pH and soil type to toxicity of copper to citrus seedlings. Fla. State Hortic. Soc. Proc. 66:73–80.

Richards, L. A., ed. 1954. Diagnosis and improvement of saline and alkali soils. USDA Agric. Handb. No. 60. 160 p.

Rinno, G. 1964. A contribution on the cause of sewage-exhaustion of soil. Albrecht Thaer. Arch. 8:699–710.

Robel, E. J., and W. C. Ross. 1975. Simplified spectrophotometric analysis of copper from cupric sulfide synthesized in porcine fecal matter. J. Agric. Food Chem. 23: 973–974

Robinson, W. O., G. Edgington, and H. G. Byers. 1935. Chemical studies of infertile soils derived from rocks high in magnesium and generally high in chromium and nickel. U. S. Dep. Agric. Tech. Bull. 471:1–28.

Robinson, W. O., H. W. Lakin, and L. E. Reichen. 1947. The zinc content of plants on the Friedensville zinc slime ponds in relation to biogeochemical prospecting. Econ. Geol. 42:572–582.

Rohde, G. 1961. Trace element-enrichment causes sewage-exhaustion. Wasserwirtsch. Wassertech. 11:542–550.

Rohde, G. 1962. The effects of trace elements on the exhaustion of sewage-irrigated land. J. Inst. Sewage Purif. 1962:581–585.

Rohde, G. 1974. Heavy metals in plant and soil. Soil Assoc. 2(7):1–5.

Roth, J. A., E. F. Wallihan, and R. G. Sharpless. 1971. Uptake by oats and soybeans of copper and nickel added to a peat soil. Soil Sci. 112:338–342.

Sabey, B. R., and W. E. Hart. 1975. Land application of sewage sludge: 1. Effect on growth and chemical composition of plants. J. Environ. Qual. 4:252–256.

Sanstead, H. H., W. H. Allaway, R. G. Burau, W. Fulkerson, H. A. Laitinen, P. M. Newberne, J. O. Pierce, and B. G. Wixson. 1974. Cadmium, zinc, and lead, p. 43–56. In Geochemistry and the environment. Vol. 1. The relation of selected trace elements to health and disease. Nat. Acad. of Sci., Washington, D. C.

Schnitzer, M., and S. V. Khan. 1972. Reaction of humic substances with metal ions and hydrous oxides. p. 203–251. In Humic substances in the environment. Marcel Dekker, Inc., New York.

Schroeder, H. A., A. P. Nason, I. H. Tipton, and J. J. Balassa. 1967. Essential trace metals in man: zinc. Relation to environmental cadmium. J. Chronic Dis. 20: 179–210.

Shipp, R. F., and D. E. Baker. 1975. Pennsylvania's sewage sludge research and extension program. Compost Sci. 16:6–8.

Shuman, L. M. 1975. The effect of soil properties on zinc adsorption by soils. Soil Sci. Soc. Am. Proc. 39:454–458.

Sidle, R. C., J. E. Hook, and L. T. Kardos. 1976. Heavy metals application and plant uptake in a land disposal system for waste water. J. Environ. Qual. 5:97–102.

Singh, B. R., and K. Steenberg. 1974. Plant response to micronutrients. III. Interaction between manganese and zinc in maize and barley plants. Plant Soil 40:655–667.

Singh, B. R., and K. Steenberg. 1975. Interactions of micronutrients in barley grown on zinc-polluted soil. Soil Sci. Soc. Am. Proc. 39:674–679.

Smilde, K. W., P. Koukoulakis, and B. Van Luit. 1974. Crop response to phosphate and lime on acid sandy soils high in zinc. Plant Soil 41:455–457.

Smiley, R. W. 1974. Rhizosphere pH as influenced by plants, soils, and nitrogen fertilizers. Soil Sci. Soc. Am. Proc. 38:795–799.

Smith, P. F., W. Reuther, and A. W. Specht. 1950. Mineral composition of chlorotic orange leaves and some observations on the relation of sample preparation technique to the interpretation of results. Plant Physiol. 25:496–506.

Smith, R. L., A. S. Henry, and K. S. M. Shoukry. 1965. Inactivation of radioactive and soil zinc by soil organic matter. p. 21–34. In Proc. Symp. Use of Isotopes and Radiation in Soil-Plant Nutrition Studies. IAEA, Ankara, Turkey.

Smith, R. L., and K. S. M. Shoukry. 1968. Changes in the zinc distribution within three soils and zinc uptake by field beans caused by decomposing organic matter. p. 397–410. In Isotopes and radiation in soil organic matter studies. IAEA, Vienna, Austria.

Spencer, W. F. 1966. Effect of copper on yield and uptake of phosphorus and iron by citrus seedlings grown at various phosphorus levels. Soil Sci. 102:296-299.

Spotswood, A., and M. Raymer. 1973. Some aspects of sludge disposal on agricultural land. Water Pollut. Control 72:71-77.

Staker, E. V. 1942. Progress report on the control of zinc toxicity in peat soils. Soil Sci. Soc. Am. Proc. 7:387-392.

Staker, E. V., and R. W. Cummings. 1941. The influence of zinc on the productivity of certain New York peat soils. Soil Sci. Soc. Am. Proc. 6:207-214.

Stenstrom, T., and H. Lonsjo. 1974. Cadmium availability to wheat: A study with radioactive tracers under field conditions. Ambio 3:87-90.

Stevenson, F. J., and M. S. Ardakani. 1972. Organic matter reactions involving micronutrients in soils. p. 79-114. In J. J. Mortvedt, P. M. Giordano, and W. L. Lindsay (ed.) Micronutrients in agriculture. Soil Sci. Soc. Am., Madison, Wis.

Stewart, B. A., and R. L. Chaney. 1975. Wastes: Use or discard. Proc. Soil Conserv. Soc. Am. 30:160-166.

Street, H. E., and G. T. Goodman. 1967. Revegetation techniques in the Lower Swansea Valley. p. 71-110. In K. J. Hilton (ed.) The Lower Swansea Valley project. Longmans, London.

Takijima, Y., and F. Katsumi. 1973a. Cadmium contamination of soils and rice plants caused by zinc mining. I. Production of high-cadmium rice on the paddy fields in lower reaches of the mine station. Soil Sci. Plant Nutr. 19:29-38.

Takijima, Y., and F. Katsumi. 1973b. Cadmium contamination of soils and rice plants caused by zinc mining. IV. Use of soil amendment materials for the control of Cd uptake of plants. Soil Sci. Plant Nutr. 19:235-244.

Takijima, Y., F. Katsumi, and S. Koizumi. 1973. Cadmium contamination of soils and rice plants caused by zinc mining. III. Effects of water management and applied organic manures on the control of Cd uptake by plants. Soil Sci. Plant Nutr. 19: 183-193.

Takijima, Y., F. Katsumi, and K. Takezawa. 1973. Cadmium contamination of soils and rice plants caused by zinc mining. II. Soil conditions of contaminated paddy fields which influence heavy metal contents in rice. Soil Sci. Plant Nutr. 19:173-182.

Tan, K. H., L. D. King, and H. D. Morris. 1971. Complex reactions of zinc with organic matter extracted from sewage sludge. Soil Sci. Soc. Am. Proc. 35:748-752.

Tan, K. H., R. A. Leonard, A. R. Bertrand, and S. R. Wilkinson. 1971. The metal complexing capacity and the nature of the chelating ligands of water extract of poultry litter. Soil Sci. Soc. Am. Proc. 35:265-269.

Terman, G. L. 1974. Amounts of nutrients supplied for crops grown in pot experiments. Commun. Soil Sci. Plant Anal. 5:115-121.

Thomas, P. R., ed. 1975. Ohio guide for land application of sewage sludge; Report of the task force on land application of sewage sludge. Ohio Coop. Ext. Bull. 598:1-12.

Tiffin, L. O. 1972. Translocation of micronutrients in plants. p. 199-229. In J. J. Mortvedt, P. M. Giordano, and W. L. Lindsay (ed.) Micronutrients in agriculture. Soil Sci. Soc. Am., Madison, Wis.

Tiffin, L. O. 1976. The form and distribution of metals in plants: an overview. In Biological implications of metals in the environment. Proc. 15th Annu. Hanford Life Sci. Symp., Richland, Wash. (In press.)

Tiller, K. G., J. L. Honeysett, and M. P. C. DeVries. 1972. Soil zinc and its uptake by plants. II. Soil chemistry in relation to prediction of availability. Aust. J. Soil. Res. 10:165-182.

Trocmé, S., G. Barbier, and J. Chabannes. 1950. Chlorosis, caused by lack of manganese, of crops irrigated with filtered water from Paris sewers. Ann. Agron. 1:663-685.

Van Hook, R. I. 1974. Cadmium, lead, and zinc distributions between earthworms and soils: Potential for biological accumulation. Bull. Environ. Contam. Toxicol. 12: 509-512.

Van Loon, J. C. 1974. Mercury contamination of vegetation due to the application of sewage sludge as a fertilizer. Environ. Lett. 6:211-218.

Venkata Ram, C. S. 1964. Resistance of tea plants to blister blight in soils augmented with nickel. Nature 204:1227.

Viets, F. G., Jr. 1966. Zinc deficiency in the soil-plant system. p. 90-128. In A. S. Prasad (ed.) Zinc metabolism. C. C. Thomas, Springfield, Ill.

Walker, J. M. 1975. Sewage sludges. Management aspects for land application. Compost Sci. 16:12-21.

Walker, R. B. 1948. Molybdenum deficiency in serpentine barren soils. Science 108: 473-475.

Wallace, A., and R. T. Mueller. 1969. Effects of chelating agents on the availability to plants of carrier-free ^{59}Fe and ^{65}Zn added to soils to simulate contamination from fallout. Soil Sci. Soc. Am. Proc. 33:912-914.

Wallace, A., and R. T. Mueller. 1973. Effects of chelated and nonchelated cobalt and copper on yields and microelement composition of bush beans grown on calcareous soil in a glasshouse. Soil Sci. Soc. Am. Proc. 37:907-908.

Wallace, A., R. T. Mueller, and G. V. Alexander. 1974. Effect of high levels of nitrilotriacetate on metal uptake by plants grown in soil. Agron. J. 66:707-708.

Wallace, A., R. T. Mueller, J. W. Cha, and G. V. Alexander. 1974. Soil pH, excess lime, and chelating agent on micronutrients in soybeans and bush beans. Agron. J. 66: 698-700.

Wallace, T., and E. J. Hewitt. 1946. Studies on iron deficiency of crops. I. Problems of iron deficiency and the interrelationships of mineral elements in iron nutrition. J. Pomol. Hortic. Sci. 22:153-161.

Walley, K. A., M. S. I. Khan, and A. D. Bradshaw. 1974. The potential for evolution of heavy metal tolerance in plants. I. Copper and zinc tolerance in *Agrostis tenuis*. Heredity 32:309-319.

Walsh, L. M., W. H. Erhardt, and H. D. Seibel. 1972. Copper toxicity in snapbeans (*Phaseolus vulgaris* L.). J. Environ. Qual 1:197-200.

Walsh, L. M., D. R. Stevens, H. D. Seibel, and G. G. Weis. 1972. Effects of high rates of zinc on several crops grown on an irrigated Plainfield sand. Commun. Soil Sci. Plant Anal. 3:187-195.

Walsh, T., and E. J. Clarke. 1945. Iron deficiency in tomato plants grown in an acid peat medium. Proc. R. Ir. Acad. 50-B:359-372.

Warncke, D. D., and S. A. Barber. 1971a. Diffusion of Zn in soils: I. The influence of soil moisture. Soil Sci. Soc. Am. Proc. 36:39-42.

Warncke, D. D., and S. A. Barber. 1971b. Diffusion of Zn in soils: II. The influence of soil bulk density and its interaction with soil moisture. Soil Sci. Soc. Am. Proc. 36: 42-46.

Warncke, D. D., and S. A. Barber. 1973. Diffusion of zinc in soils: III. Relation to zinc adsorption isotherms. Soil Sci. Soc. Am. Proc. 37:355-358.

Wear, J. I. 1956. Effect of soil pH and calcium on uptake of zinc. Soil Sci. 81:311-315.

Wear, J. I., D. L. Hartzog, and E. M. Evans. 1968. Sources of zinc for plants. Highlights Agric. Res., Auburn Univ. Agric. Exp. Sta. 15(1):8.

Webber, J. 1972. Effects of toxic metals in sewage on crops. Water Pollut. Control. 71: 404-413.

Webber, J. 1974. Sludge handling and disposal practices in England. Proc. Sludge Handling and Disposal Seminar, Toronto, 18-19 September. Environment Canada and Ontario Ministry of the Environment, Ottawa, Ontario, Canada. p. B1-B17.

Westgate, P. J. 1952. Preliminary report on copper toxicity and iron chlorosis in old vegetable fields. Proc. Fla. State Hortic. Soc. 65:143-146.

Weston, R. L., P. D. Gadgil, B. R. Salter, and G. T. Goodman. 1964. Problems of revegetation in the Lower Swansea Valley, an area of extensive industrial dereliction. Br. Ecol. Soc. Symp. 5:297-325.

Whitby, L. M., and T. C. Hutchinson. 1974. Heavy-metal pollution in the Sudbury mining and smelting region of Canada. II. Soil toxicity tests. Environ. Conserv. 1: 191-200.

Wild, H. 1970. Geobotanical anomalies in Rhodesia. 3. The vegetation of nickel-bearing soils. Kirkia 7 (Supplement):1-62.

Wilkinson, H. F. 1972. Movement of micronutrients to plant roots. p. 139-169. *In* J. J. Mortvedt, P. M. Giordano, and W. L. Lindsay (ed.) Micronutrients in agriculture. Soil Sci. Soc. Am., Madison, Wis.

Williams, C. H., and D. J. David. 1973. The effect of superphosphate on the cadmium content of soils and plants. Aust. J. Agric. Res. 11:43-56.

Yamagata, N., and I. Shigematsu. 1970. Cadmium pollution in perspective. Bull. Inst. Publ. Health 19:1-27.

Zimdahl, R. L., and J. M. Foster. 1976. The influence of applied phosphorus, manure, or lime on uptake of lead from soil. J. Environ. Qual. 5:31-34.

chapter 11

Food Chain Aspects of the Use of Organic Residues

W. H. ALLAWAY, USDA-ARS, U. S. Plant, Soil, and Nutrition Laboratory, Ithaca, New York

I. INTRODUCTION

Opinions concerning the effect of organic residue use on agricultural land upon the nutritional quality of the crop are highly varied. Proponents of organic fertilizers presume that crops of enhanced nutritional quality will result from the use of organic material on farm fields. On the other hand, there is concern over the potential for accumulation of toxic elements in soils and plants as a result of sewage sludge disposal. Examples of both enhanced or damaged animal nutritional status that can be traced to application of organic residues on agricultural land have been documented. Comparable examples involving human health and nutrition are rare.

In any survey of the results of field experiments using animal manures, sewage sludge, etc. the most common finding is increased yields of plants on the treated areas in comparison with nearby untreated areas. Thus, an increase in total food production is the most common effect of use of organic materials on agricultural land, as far as food chains are concerned.

This chapter will primarily consider possibilities that food and feed crops produced on land treated with organic residues may contain materials from these residues that will be detrimental to the nutritional quality of these crops. Secondary attention will be directed toward possibilities that crops of enhanced nutritional quality will result from the transfer of essential nutrient elements from the residues to the edible portion of the crop.

This discussion will be concerned with animal manures and the products of municipal sewage disposal systems. Factors affecting the transfer of materials that usually enter plant roots as ions will be considered in this chapter, with pesticides and living pathogens being considered in other chapters.

II. ORGANIC RESIDUE PRODUCTION AND UTILIZATION PROCESSES

Animal manures may be deposited directly upon farm fields and then incorporated into the soil prior to the production of the next crop. Organic residue production and application processes range from this simple and direct transfer to complicated schemes involving several different types of transport, physical treatments such as filtration, one or more biological decomposition processes, and chemical treatments. These complicated processes are used most commonly in connection with municipal sewage treatment plants. Some of the potential processes that may be involved in the transfer of organic residues to farm fields and transfer of elements contained in these residues into human and animal tissue are diagramed in Fig. 1.

III. SELF-LIMITING OR SELF-DESTRUCT POINTS

Points marked by "X" in Fig. 1 represent built-in limitations to the movement of potentially toxic elements from organic residues to human and

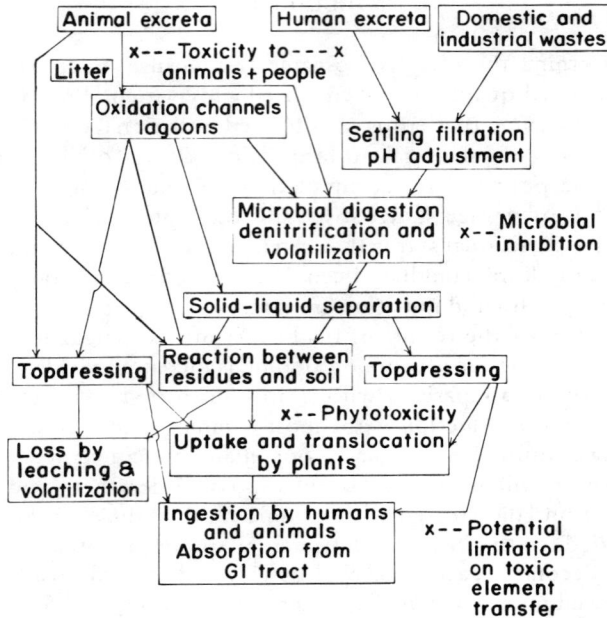

Fig. 1—Some of the possible pathways and limiting points in the movement of organic residues.

animal tissues. The first of these self-limiting points controls the concentration of toxic materials in human and animal excreta, in that any concentration of an element or compound in diets that would be acutely toxic to animals or people would soon stop the entire process. Animals usually digest about one-half of their diet, so the concentration of toxic elements in the excreta is limited to about double the acutely toxic concentration in the diet itself.

The next self-limiting point in organic waste disposal is applicable to wastes that go through a biological digestion process. Concentrations of toxic elements that inhibit microbial growth will lead to stoppage of the process until the problem is corrected. Failures of digestors in sewage treatment plants that appear to be due to excessive levels of metals have been described by Regan and Peters (1970). Establishment of precise upper limits of potentially toxic metals that are consistent with satisfactory operation of digestors is difficult due to the wide variation in properties of raw sewage from different sources, and to variation in the toxicity of different compounds of the same element.

Where the organic material is incorporated into the soil, and potentially toxic elements must be taken up by plant roots and translocated to the edible part of the plant, toxicity of these elements to plants provides a major barrier against toxicity to people and animals. This barrier or self-limiting point is very selective with some elements being substantially more toxic to plants than to animals, while other toxic elements may be taken up and incorporated into edible parts of plants up to concentrations that may be acutely or chronically toxic to the person or animal that eats the plant. Many of the common food and feed crops, if they show fairly normal growth and have not been contaminated by airborne pollutants, are sufficiently low in Pb, As, F, Zn, B, and Cr to represent no acute hazard from these elements when the crop is eaten by people or animals. Copper is more toxic to plants than to monogastric animals, but ruminants may be injured by Cu in certain plant species. On the other hand, plants of normal appearance and yield may contain concentrations of NO_3^-, Mo, Se, and possibly Cd and other elements that are damaging in human and animal diets. Phytotoxicity is therefore an important but imperfect barrier to movement of potentially toxic materials from organic residues to human and animal diets.

One of the objectives of organic residue utilization programs is to increase crop production and to substitute plant nutrients contained in the residues for those that would otherwise be applied in commercial fertilizers. Reliance on phytotoxicity to prevent damage to food chains is contrary to this objective and may represent critical losses to total food and feed crop production.

IV. CONTROL OR MANAGEMENT POINTS

In Fig. 1 essentially all of the arrows constitute a potential opportunity to control food chain aspects of organic residue utilization. In most cases this opportunity is circumscribed by both economic and biological considerations.

The soil and plant specialist who is called upon to advise on organic residue utilization is frequently in the position of having little or no control over the chemical composition of the organic material. Under these circumstances his opportunities are primarily in identifying those soils most suitable for organic residue application, advising on rates and methods of application for different soils, the selecting of crops to be grown on the area, and recommending the use of the crop. It may also be possible to suggest soil applications of fertilizers or lime that will minimize the solubility or plant availability of potentially toxic elements contained in organic residue.

Animal scientists concerned with organic residue disposal may be able to adjust the diet of the animals so as to change the amount and composition of animal manures. They may also be able to suggest dietary supplements to inhibit toxic effects of certain elements in feed crops produced on treated soils.

In situations when agricultural scientists can choose from a number of alternative disposal sites or where extremely heavy rates of waste application are not dictated by economic factors, it is generally possible to select waste disposal procedures that will minimize danger to food chains.

Measures that may be useful in preventing food chain damage from toxic metals have been described by Chaney (1973).

V. FOOD CHAIN TRANSFER OF SOME SPECIFIC ELEMENTS

In this section some of the features of transfer of specific elements from organic residues to the edible portion of food and feed crops are described. Special attention is directed toward potentially toxic elements that have been shown to occur in either animal manures or in municipal sewage effluents and sludges.

A. Arsenic

The use of organic arsenicals in poultry feeds is the principal source of As in organic wastes. Concentrations of 15 to 30 ppm As have been reported in poultry house litter (Morrison, 1969). Where poultry litter of this type had been applied to soils for as much as 20 years, no increase in the As concentration in alfalfa (*Medicago sativa* L.) and clover (*Trifolium* sp.) was detected.

Although there are fewer data on As concentration in municipal sewage sludges than for many of the other trace elements, Page (1974) gives a range of 1 to 18 ppm As in some sludges from Michigan.

Predictions of the hazard from As in organic materials can be derived from experiments on As residues from pesticides. The fate of As from pesticide residue has been discussed by Woolson et al. (1971). Substantial evidence points to phytotoxicity as the first effect of accumulation of As in soils. Crop failure, or sharply diminished yields, usually precede the production of crops that have hazardous levels of As in the edible portion.

Soil tests for predicting the hazard of phytotoxicity of As residues have been developed (Woolson et al., 1971). Use of these tests to monitor As accumulations and prevent development of phytotoxic levels offers very substantial protection against food chain damage from As in organic materials.

B. Boron

Problems with B are likely to be confined to soluble B compounds in sewage effluents. Boron is not required by animals and is of low toxicity to them. The first impact of application of high B effluents to agricultural soils is reduced crop growth due to B toxicity. Boron injury to plants caused by irrigation with high B sewage effluents was noted in the United States as early as 1928. Techniques for evaluating the B status of soils and plants are described by Bradford (1966). Alkaline soils used for production of B-tolerant crops represent the better opportunities to minimize hazard of phytotoxicity from high B effluents.

Use of high B effluents as sources of B to meet plant requirements on low B soils is of uncertain value due to the narrow margin between B requirements and B toxicity for plant species. Only effluents containing very constant levels of soluble B could be used in this way.

C. Cadmium

Concern over Cd in foods stems primarily from the experiments and writings of Schroeder and his associates (1964), plus the occurrence of itai-itai disease in the Jintsu Valley of Japan (Kobayashi, 1971). Schroeder's work carried the implication that dietary Cd might accumulate in the kidneys and result in hypertension. Itai-itai disease is characterized by a very painful disintegration of bone.

Recent reviews of the Cd problem (Fleischer et al., 1974; National Academy of Sciences, 1974) indicate that adverse effects of dietary Cd on human health are not clearly established. Although workers in Cd industries are subject to respiratory diseases, excessive incidence of hypertension among these workers has not always been found. Itai-itai disease appeared to have affected only that part of the population most likely to be deficient in Ca or vitamin D.

More research effort has probably been directed toward Cd than toward any other trace element in municipal sewage sludge. The following points are relevant to this problem:

1) Although many municipal sludges are low in Cd, some from industrialized cities contain more than 500 ppm Cd on a dry basis (Page, 1974). Sludges containing more than 500 ppm Cd are probably rare, in that levels in this range may be associated with sterilization of digestors (Regan & Peters, 1970). Chaney (1973) reports an observed range of 5 to 2,000 ppm Cd in sewage sludge.

2) Current background levels of Cd are probably about 0.05 to 0.40 ppm in human diets and the average daily human intake of Cd is probably 40 to 150 µg/day. Cadmium intake of people affected by itai-itai disease may have been 600 to 1,000 µg/day. Vegetables containing over 40 ppm Cd have been collected in places where pollution by airborne Cd has occurred (National Academy of Sciences, 1974).

3) Levels of Cd found in plants grown on field soils treated with sewage sludge, including some sludges containing over 200 ppm Cd, have been substantially lower than in plants exposed to airborne Cd, or in plants grown in soils or culture solutions containing added inorganic Cd (Chaney, 1973; John et al., 1973; King et al., 1974; Jones et al., 1973; Page et al., 1972). Similarly, Cd in foods of plant origin is very likely less digestible, and therefore less hazardous, than equal amounts of Cd in soluble inorganic forms contained in water or inhaled as dusts.

4) The ratio of Zn to Cd in plant material grown in unpolluted areas, and in human diets is usually in the range of 100 to 1. Under some

circumstances Zn competes with Cd in uptake and translocation processes in plants. Dietary Zn may also decrease the toxicity of Cd in
animals. Similar protection against Cd toxicity may be afforded by
dietary selenite (National Academy of Sciences, 1974).

5) Plant species vary widely in their tendency to accumulate Cd added
as soluble inorganic Cd compounds to soils or culture solutions. The
concentration of Cd in edible seeds or fruits generally is lower than
in leafy vegetables. The concentration of Cd in the kernels of corn
(*Zea mays* L.) has been especially low, even when the corn is grown
in soils treated with heavy rates of high Cd sludge (Chaney, 1973).

The most important feature of any program for minimizing the hazard
of Cd from sewage sludge is probably the need for continuous monitoring
of Cd concentrations in sludges and in plants growing on sludge-treated areas.
Although safe upper limits of Cd in food and feed crops have not been established, uncontrolled increases in the concentration of Cd in these crops are
undesirable. Procedures for minimizing hazards from Cd have been described
by Chaney (1973). Adjustment of Cd/Zn ratios in wastes to keep the level of
Cd < 0.5% of the Zn, selection of finer textured neutral or alkaline soils as
disposal sites, regulation of annual and total application rates, and cropping
the disposal site to corn used for grain are measures that help to minimize
the hazard to food chains from Cd contained in sludges. In the absence of
well-defined values for the permissible limit of Cd concentrations in human
and animal diets, it is uncertain whether or not phytotoxicity of Cd will act
as a safeguard against detrimental effects of Cd in food and feed crops.

D. Copper

Most of the present concern over Cu in organic residues stems from
proposals to use relatively high levels of Cu (> 200 ppm) in rations for hogs
for growth stimulation or antibiotic effects (Underwood, 1971). Batey et al.
(1972) concluded that there was little danger from Cu toxicity in animals
from use of manure from Cu-supplemented pigs unless the slurry was topdressed on a growing sward of grass. Incorporation of manure from Cu-
supplemented pigs into the soil did not consistently increase the Cu concentration in forage grasses.

The Cu concentration in municipal sewage sludges is reported to range
between 250 and 17,000 ppm (Chaney, 1973; Page, 1974). Median or average values are frequently about 1,000 ppm. Thus the Cu concentration in
municipal sludges approximates maximum values expected in manure from
Cu-supplemented pigs. Copper concentrations in crops produced on lands
treated with municipal sludges have not been elevated to levels indicative of
a hazard from Cu toxicity to consumers of the plants (Peterson et al., 1971).

Monogastric animals including man are not highly susceptible to Cu
toxicity and the primary ill effect of diets containing 200 ppm Cu or more is
ordinarily a correctable interference with absorption of Zn or Fe (Under-

wood, 1971). Phytotoxicity of Cu will ordinarily inhibit plant growth before Cu concentrations in this range would be found in plants (National Academy of Sciences, 1974). Ruminants, especially sheep, are more susceptible to Cu toxicity and may be adversely affected by dietary Cu concentrations found in plants growing on Cu-polluted areas. This Cu toxicity can be controlled by careful use of dietary supplements of Mo.

Application of Cu in organic wastes may help to prevent Cu deficiency in plants, and possibly in animals, but this is a crude and inefficient method of correcting these deficiencies.

E. Chromium

Chromium is one of the most recent additions to the list of elements essential in the nutrition of man and animals. The nutritionally effective form of Cr appears to be an organic complex of currently unknown structure. The Cr in leafy plants has not been found to be nutritionally effective and seed crops, especially wheat (*Triticum aestivum* L.), contain very low levels of Cr (Welch & Cary, 1975). The current status of research on Cr in nutrition has recently been reviewed by Mertz et al. (1974).

Although some instances of plant response to soil applications of Cr have been reported, tests of the essentiality of Cr to plants growing in highly purified culture solutions have been negative (Huffman & Allaway, 1973). Hexavalent Cr is toxic to plants when applied in solution to soils (Turner & Rust, 1971). Trivalent Cr is insoluble in soils, and additions of chromic hydroxide up to 0.5% of the soil have little effect on growth or Cr concentration in plants (E. E. Cary & W. H. Allaway, unpublished).

In view of the limited possibility for hexavalent Cr to be present in organic materials, it appears that Cr in these sludges presents no hazard to food chains, and it will not be effective in improving the nutritional quality of plants by increasing their content of nutritionally effective forms of Cr.

F. Fluorine

Factors involved in the environmental cycling of F have been reviewed by a committee sponsored by the National Academy of Sciences (1971a). Fluorine is probably essential to animals. Improvement in the structure of teeth and bones has been attributed to fluoride supplementation of potable waters. High levels of fluorides are toxic to both plants and animals, and fluoride toxicity traceable to the deposition of airborne fluorides on plant foliage has been an important practical problem.

When insoluble fluorides have been added to moderately acid, neutral, or alkaline soils, no damage to plants and no increase in F concentration in the plant have been observed. The hazard of F in organic residues is apparently minor and confined to residues containing soluble fluorides when these residues are applied to strongly acid soils. Any damage that occurs from application of fluorides to soils can be substantially corrected by liming.

G. Lead

Where the organic residue is incorporated into the soil, Pb contained in the residue appears to offer very little hazard as far as toxicity to humans and animals is concerned, and only modest toxicity to plants (Baumhardt & Welch, 1972). On soils that have been treated with as much as 3,200 kg/ha of Pb in the form of soluble salts, the concentration of Pb in corn leaves was increased but the Pb concentration in the grain was unchanged. Therefore, production of crops where only the edible seed is harvested may be an effective way of preventing Pb toxicity to humans and animals. Uptake of Pb by plants may also be reduced by liming acid soils. In general, potential hazards of Pb toxicity are reduced whenever the Pb is incorporated in soil, in contrast to the potential hazards from atmospheric Pb.

H. Mercury

Concern over hazard to food chains from Hg in sewage sludge has stemmed from evidence of methylation of Hg in aquatic systems, and from recent episodes of methylmercury poisoning from inadvertent human consumption of treated seeds. More recent research has indicated that the Hg concentration in aboveground parts of plants is generally very low, even though Hg seed treatments or other additions of Hg to soils had been made. Chaney (1973) indicates that hazards from Hg in terrestrial food chains are much less than those in aquatic food chains.

I. Molybdenum

The potential hazard from Mo in organic residues is closely related to the nature of the soils available for treatment with organic residues, to the crops produced, and to the use of these crops. Molybdenum is required by plants but the requirement is low in terms of normal nutrient application rates. Where levels of available Mo in soils are very high, plants may accumulate concentrations of Mo that are detrimental to ruminants that eat these plants. The effect of excessive Mo on animals is primarily a disturbance of Cu metabolism and Mo toxicity can be corrected by supplementation of the animal's diet with Cu, or by use of injectable forms of Cu. Ruminants are much more susceptible to Mo toxicity than are monogastric animals. Molybdenum-Cu relationships in animal nutrition are discussed by Underwood (1971).

Molybdenum toxicity has been a serious practical problem to livestock producers in the intermountain areas of the western United States. All areas producing forages containing toxic concentrations of Mo are occupied by wet, alkaline, or neutral soils containing relatively high levels of organic matter. Nearby well-drained soils, even though formed from high Mo parent materials do not produce high Mo forages (Kubota & Allaway, 1972).

On the basis of the naturally occurring instances of Mo toxicity, it seems safe to assume that Mo in organic residues presents no hazard to the food chain provided that these residues are used on well-drained soils with pH values of 6 or less. Where it is necessary to use high Mo residues on poorly drained alkaline soils, the use of these areas as pastures for ruminants should be discouraged, or else careful attention should be directed to Cu supplementation of the animals.

J. Nickel

The role of Ni in animal nutrition has been reviewed by Underwood (1971). Nickel may soon be established as an essential element for animals, but required dietary levels are likely to be very low. Dietary Ni levels of 500 ppm or more have been fed to experimental animals without acutely toxic effects, but chronic effects of lower levels of Ni ingestion are possible. Vanselow (1966) has summarized data on Ni in plants and indicates that Ni concentrations of 50 to 100 ppm dry basis are usually indicative of Ni toxicity to plants. Most of the information on Ni toxicity to plants has been developed from studies of plants growing on soil derived from serpentine, and other factors in addition to high Ni concentrations may have been responsible for plant symptoms observed. Phytotoxicity appears to provide a substantial barrier against Ni toxicity to humans and animals. Monitoring of Ni levels in sewage sludge and in crops grown on sludge-treated areas is highly desirable, however, since Page (1974) reports Ni concentrations in sludges of 3,000 ppm or more and cites evidence that plants grown on sludge-treated soils are substantially higher in Ni than controls.

K. Nitrogen

Nitrogen is perhaps the most valuable, and at the same time, the most troublesome element in organic residues. Nitrogen concentrations in animal manures and sewage sludges, and crop responses attributable to this N, are documented by Peterson et al. (1971). A more recent striking example is provided by Garcia et al. (1974). In this example, a moderately heavy application of sewage sludge applied to strip mine soil resulted in a fourfold increase in the yield of corn, and an increase of 2.5% in the protein concentration of the corn grain. Concentrations of heavy metals in the corn grain were similar to those in corn grown with normal farm practices.

Problems associated with N in organic wastes include the hazard of high concentrations of nitrate in crops or drainage water, including the possibility of nitrosamine formation. These problems have been reviewed by a committee of the National Academy of Sciences (1972). There are also reports (Wilkinson et al., 1972) of increased incidence of grass tetany in cattle where high rates of broiler litter have been applied to pastures, but in these instances K in addition to the N may have contributed to the problem.

There is no evidence that N in organic wastes presents either more or less hazard than equivalent amounts of nitrifiable or plant available N from inorganic sources. The rate at which N from organic wastes is nitrified must, however, be considered in appraising the hazard of producing high nitrate crops.

The management of N in organic residues has been discussed by a number of people including Larson et al. (1974). The objectives of management of N in organic wastes are to provide needed amounts of N to the crop, to avoid production of crops high in NO_3^-, and to minimize NO_3^- accretion to waters. Where costs of transportation and application of the organic wastes favor application rates that will exceed these objectives, sludge digestion processes or manure management systems that promote denitrification may be useful. The hazard from high NO_3^- crops can be minimized by growing corn or other cereals for grain since cereal seeds have not been found to contain high levels of NO_3^- even when the leaves and stems contain hazardous levels. Where appropriate crops and crop uses are selected, potential for NO_3^- accumulation in waters constitutes the major criterion for safe application of N in organic residues to farm land.

L. Selenium

Selenium is essential for animals and probably for people but not for plants. Concentrations of about 0.04 ppm Se or more in diets are required in order to prevent Se deficiency in young animals. Higher concentrations of dietary Se are toxic to animals, and dietary concentrations of < 4 to 5 ppm are often suggested as maximum amounts that will avoid Se toxicity. Both Se deficiency and Se toxicity are problems of practical importance in livestock production in the United States and in other countries. There are a number of recent reviews describing nutritional effects and environmental cycling of Se (Underwood, 1971; National Academy of Sciences, 1971b; Allaway, 1973).

The potential value and potential hazard of Se in organic residues are influenced by the stability and biological activity of different forms of Se. The Se in manures and sewage sludge is very likely to be present as either elemental Se, as selenities strongly bound to hydrous iron oxides, as heavy metal selenides, or as trimethylselenonium salts. All of these compounds are of very low availability to plants or animals, and when incorporated into soils they are rarely taken up by plants in potentially toxic concentrations. There appears to be a minimal hazard to food chains from Se in organic residues, and minimal opportunity to improve the nutritional quality of very low Se crops through organic residue utilization practices.

M. Zinc

Zinc is the most abundant, and most valuable in terms of nutritional quality of crops, of the trace elements in organic residues. Zinc deficiency in crops has been noted with increasing frequency and it has been suggested that some of this increase is due to feed crop removal without return of animal manures to the soil (Viets, 1966). Application of animal manure to the cut areas in fields leveled for irrigation in order to correct Zn and Fe deficiency is a common practice.

Zinc deficiency in livestock, especially hogs, is a problem of long standing and there is evidence of Zn deficiency in people in many countries, including the United States (National Academy of Sciences, 1974). Zinc deficiency of people and animals is frequently associated with low availability or digestibility of dietary Zn rather than with low total Zn intake. Zinc from plant sources, and especially Zn in seed plants seems to be of lower digestibility than Zn in foods of animal origin. However, the concentration of digestible Zn in seed crops can be increased through increase in the supply of available Zn to the plant (Welch et al., 1974).

Zinc is more toxic to plants than to animals or humans. One of the recent examples of phytotoxic effects of Zn contained in liquid sewage sludge has been described by King and Morris (1972). In this instance, the yield of rye (*Secale cereale* L.) declined whenever rye foliage contained concentrations of Zn of about 500 ppm or more. Levels of dietary Zn of this order and higher have been fed to animals with very little evidence of ill effects (Underwood, 1971). Therefore it appears safe, as pointed out by Chaney (1973), to assume that phytotoxicity of Zn provides a safeguard against extensive production of food or feed crops that may contain Zn concentrations hazardous to consumers.

Heavy applications of organic residues that are high in P and relatively low in Zn may lead to Zn deficiency in crops from phosphate-induced Zn deficiency (Viets, 1966).

N. Other Elements Posing Potential Hazards

In addition to the elements discussed above, there are many others that can be present in municipal sewage sludge or effluents. These elements include Sr, Li, Be, Sb, V, Bi, Br, I, and Sn. Some of them undoubtedly pose potential hazards to food chains, but the extent and nature of these hazards and possible procedures for minimizing them have not been worked out. Some of these elements, especially I, may have beneficial effects on the nutritional quality of crops unless application rates are such that uptake by plants reach excessive levels.

An effective monitoring system that measures concentration of a very large number of elements in sewage sludge is essential to their use in food crop production.

VI. PLANNING ORGANIC RESIDUE UTILIZATION PROGRAMS

Economic and engineering factors involved in planning residue utilization programs are discussed in other chapters. This section will consider factors that should be considered in planning residue utilization programs in order to minimize damage to food production potentials and to the nutritional quality of crops produced.

A. Analysis of the Residues

The first step in planning for the safe utilization of organic residues is information on the composition of the residue. This is especially important for municipal sewage sludges and effluents. Animal manures, and especially those from smaller operations involving direct application of manure to farm fields, have less stringent requirements for this type of information. Analysis of sludges and effluents must include, as mentioned previously, a large number of different elements.

B. Evaluation of Site for Residue Application

Alternative sites for residue application can best be evaluated after the composition of the residue is known, and on the basis of judgment concerning the relative hazard of the different elements that are present in the residue. Medium to fine-textured neutral or alkaline soils are usually preferred sites if the sludge is high in Cd or Pb. However, in alkaline soils, inert forms of Se are more likely to be oxidized to selenates and become available to plants, so the preferred sites for minimizing hazard from Cd or Pb may increase the potential hazard from Se. Similarly, the selection of a site that is poorly aerated for part of the year may enhance denitrification and minimize the possibility of N overloading, but Mo contained in residues applied to poorly drained soils will be more available to plants.

The crops most likely to be grown must be considered in evaluation of alternative sites. For residues containing fairly high levels of Pb, Cd, or NO_3^-, use of the site for production of annual seed crops such as corn harvested for grain with the crop residues left in the field tends to minimize potential hazards. Use of the site for production of leafy vegetables provides much less protection from potential hazard.

C. Reaction of the Organic Residue with the Soil

In the process of planning residue utilization programs, the reactions between the specific residue and the soils on the site to which the residue is applied must be studied. Of particular importance are measurements of nitrification rates, effects on soluble salt levels, and effects on pH. These measurements must normally be made under laboratory conditions, but the rates of residue application and conditions of moisture and temperature should approximate those expected in the field.

D. Selection of Residue Application Methods and Rates

Where a potential hazard to nutritional quality of crops is of primary concern, organic residue application procedures that involve incorporation into the soil, such as plowing, are usually indicated. Incorporation into the soil permits fixation of potentially toxic elements by the solid phase, thereby lowering their concentrations in the crop. Similarly, the tendency of roots of many crops to exclude elements such as Pb or As is utilized to a maximum.

Topdressing permanent pasture and perennial forage crop fields with animal manure has been used for many years without serious problems. Sewage sludges or effluents containing high concentrations of potentially toxic elements can rarely be safely topdressed on feed and food crops.

E. Monitoring of Trends in Soil and Plant Composition

Many residue utilization programs involve repeated applications of organic materials from the same source to a limited number of farms and fields. These operations pose a special need for monitoring, and offer special opportunities to monitor trends in soil and plant composition. Appropriate analysis of soils and crops can provide an early warning system against hazards to crop production and crop quality.

Methods used for this monitoring need to be fitted to individual elements of concern. Rising concentrations of soluble As can probably be detected by appropriate analysis of soil or soil extracts. Similarly, levels of soluble phosphorus adequate to meet crop requirements can be monitored by soil extractions, and accumulations detrimental to crop growth can be detected. For certain other elements, regularly spaced analyses of the crop produced will probably provide the most effective monitoring system.

In situations where possible development of phytotoxic effects is of primary concern, use of strips of sensitive plants may be a valuable aid in monitoring. Occasional strips of a B-sensitive plant in a field used for production of B-tolerant crops can provide early warning of potential buildup of B toxicity. Analysis of strips of leafy vegetable crops for Cd or Ni may provide early warning against detrimental effects of these metals on less sensitive plants. Also, analysis of leaves of crops such as corn, soybeans (*Glycine max* [L.] merr.), or wheat may be used to warn of buildups of potentially toxic concentrations of plant available Cd before any damage to the nutritional quality or yield of the harvested seed takes place.

F. Use of Information in Planning and Controlling Residue Application Programs

Properly designed advance studies and monitoring can be used to divert some sludges that are high in potentially toxic elements to sites that are used for ornamental plants, fiber crops, or other uses that do not involve a potential hazard to food chains. The problem of land application of sewage sludge will be most severe for large industrialized cities with a limited area of potential application sites. In selected cases, some method such as incineration may be more feasible than land application. It should be recognized, however, that the hazard to food chains from certain elements in ashes may be even more severe than from these same elements in the unburned organic material.

Monitoring of an ongoing residue application program can be used to regulate application rates or to direct the operation to new alternative sites before adverse effects occur. Investigations of many older residue disposal sites indicate reductions in plant uptake of many elements with time following the last application of sludges. Additional research is needed on long term effects on the availability of potentially toxic elements applied in organic residues.

VII. LONG TERM CONSIDERATIONS

From the previous sections there is evidence of many opportunities to solve problems of organic residue utilization without immediate damage to food chains. But it must be recognized that these solutions usually involve increases in the total concentration of potentially toxic elements in soils used for food production. For example, in the work of Garcia et al. (1974), the total Cd in strip-mined soil was increased by a factor of about 50, and the total Pb and Hg increased by a factor of about 10, by a sludge application that had very little effect on the concentration of these metals in the corn grain produced.

Some people may elect a viewpoint that any increase in the concentration of a potentially toxic element in agricultural soils represents undesirable pollution, even though the residues of this element in soil are very inert and not taken up by plants. This viewpoint must be reconciled with the fact that many of the best agricultural soils of the world have been polluted (by this definition of pollution) with F from phosphate fertilizers for over a century, without evidence of damage to food chains.

At the opposite end of the scale, some may elect the viewpoint that accumulation of inert forms of toxic elements in agricultural soils can be permitted up to the point of evidence of depressed crop growth or damage to human and animal health. This viewpoint must be reconciled with experience in the use of As in pesticides for fruit trees, where phytotoxicity which developed over a long period was not recognized until perennial orchards were replaced by other crops. One must also consider the possibility of delayed recognition of chronic toxicity in people, and delayed identification of the causative factors.

Attempts to resolve these differing points of view need to be based on considerations of each element individually, and the chemical changes that element is most likely to undergo over long periods of time in weathering soils. Even for single elements, opportunities for controlled experiments upon which to base objective decisions will be rare.

LITERATURE CITED

Allaway, W. H. 1973. Selenium in the food chain. Cornell Vet. 63:151–170.

Batey, T., C. Berryman, and C. Line. 1972. The disposal of copper-enriched pig-manure slurry on grassland. J. Br. Grassl. Soc. 27:139–143.

Baumhardt, G. R., and L. F. Welch. 1972. Lead uptake and corn growth with soil-applied lead. J. Environ. Qual. 1:92–94.

Bradford, G. R. 1966. Boron. p. 33–61. In H. D. Chapman (ed.) Diagnostic criteria for plants and soils. Div. of Agric. Sci., Univ. of Calif., Riverside.

Chaney, R. L. 1973. Crop and food chain effects of toxic elements in sludges and effluents. p. 129–141. In Recycling municipal sludges and effluents on land. Proc. Joint Conf. 9–13 July 1973. Champaign, Ill. Nat. Assoc. State Univ. and Land Grant Coll., Washington, D. C.

Fleischer, M., A. F. Sarofim, D. W. Fassett, P. Hammond, H. T. Shacklette, I. C. T. Nisbet, and S. Epstein. 1974. Environmental impact of cadmium: A review by the Panel on Hazardous Substances. Environ. Health Perspect. Exp. Issue No. 7. p. 253–323.

Garcia, W. J., C. W. Blessin, G. E. Inglett, and R. O. Carlson. 1974. Physical-chemical characteristics and heavy metal content of corn grown on sludge-treated strip-mine soil. J. Agric. Food Chem. 22:810–815.

Huffman, E. W. D., Jr., and W. H. Allaway. 1973. Growth of plants in solution culture containing low levels of chromium. Plant Physiol. 52:72–75.

John, M. K., C. J. Van Laerhoven, and H. H. Chuah. 1972. Factors affecting plant uptake and phyto-toxicity of cadmium added to soils. Environ. Sci. Technol. 6:1005–1009.

Jones, R. L., T. D. Hinesly, and E. L. Ziegler. 1973. Cadmium content of soybeans grown on sewage-sludge amended soil. J. Environ. Qual. 2:351–353.

King, L. D., and H. D. Morris. 1972. Land disposal of liquid sewage sludge. II. The effect of soil pH, manganese, zinc and growth and chemical composition of rye (Secale cereale L.). J. Environ. Qual. 1:425–429.

King, L. D., L. A. Rudgers, and L. R. Webber. 1974. Application of municipal refuse and liquid sewage sludge to agricultural land: I. Field study. J. Environ. Qual. 3:361–366.

Kobayashi, J. 1971. Relation between the "itai-itai" disease and the pollution of river water by cadmium from a mine. Proc. 5th Int. Water Pollut. Res. Conf., July–August 1970. San Francisco, Calif. p. 1–7.

Kubota, J., and W. H. Allaway. 1972. Geographic distribution of trace element problems. p. 525–554. In J. J. Mortvedt, P. M. Giordano, and W. L. Lindsay (ed.) Micronutrients in agriculture. Soil Sci. Soc. of Am., Madison, Wis.

Larson, W. E., R. H. Susag, R. H. Dowdy, C. E. Clapp, and R. E. Larson. 1974. Use of sewage sludge in agriculture with adequate environmental safeguards. Sludge Handling and Disposal Seminar. 18–19 Sept., 1974. Toronto, Ontario, Canada.

Mertz, W., E. W. Toepfer, E. E. Roginsky, and M. M. Polansky. 1974. Present knowledge of the role of chromium. Fed. Proc. 33:2275-2280.

Morrison, J. L. 1969. Distribution of arsenic from poultry litter in broiler chickens, soil and crops. J. Agric. Food Chem. 17:1288-1290.

National Academy of Sciences. 1971a. Fluorides. Nat. Acad. of Sci., Washington, D. C.

National Academy of Sciences. 1971b. Selenium in nutrition. Nat. Acad. of Sci., Washington, D. C.

National Academy of Sciences. 1972. Accumulation of nitrate. Nat. Acad. of Sci., Washington, D. C.

National Academy of Sciences. 1974. Geochemistry and the environment. I. The relation of selected trace elements to health and disease. Nat. Acad. of Sci., Washington, D. C.

Page, A. L. 1974. Fate and effects of trace elements in sewage sludge when applied to agricultural lands. A literature review study. Environ. Protect. Tech. Ser. EPA-670/2-74-005. Environ. Protect. Agency, Cincinnati, Ohio.

Page, A. L., F. T. Bingham, and C. Nelson. 1972. Cadmium absorption and growth of various plant species as influenced by solution cadmium concentration. J. Environ. Qual. 1:288-291.

Peterson, J. R., T. M. McCalla, and G. E. Smith. 1971. Human and animal wastes as fertilizers. p. 557-596. In R. A. Olson, T. J. Army, J. J. Hanway, and V. J. Kilmer (ed.) Fertilizer technology and use. Soil Sci. Soc. of Am., Madison, Wis.

Regan, T. M., and M. M. Peters. 1970. Heavy metals in digesters: Failure and cure. J. Water Pollut. Control. Fed. 42:1832-1839.

Schroeder, H. A. 1964. Cadmium hypertension in rats. Am. J. Physiol. 207:62-66.

Turner, M. A., and R. H. Rust. 1971. Effects of chromium on growth and mineral nutrition of soybeans. Soil Sci. Soc. Am. Proc. 35:755-758.

Underwood, E. J. 1971. Trace elements in human and animal nutrition. 3rd ed. Academic Press, New York.

Vanselow, A. P. 1966. Nickel. p. 302-309. In H. D. Chapman (ed.) Diagnostic criteria for plants and soils. Div. of Agric. Sci., Univ. of Calif., Riverside.

Viets, F. G., Jr. 1966. Zinc deficiency in the soil-plant system. p. 90-128. In A. S. Prasad (ed.) Zinc metabolism. C. C. Thomas Publisher, Springfield, Ill.

Welch, R. M., and E. E. Cary. 1975. Concentration of chromium, nickel and vanadium in plant materials. J. Agric. Food Chem. 23:479-482.

Welch, R. M., W. A. House, and W. H. Allaway. 1974. Availability of zinc from pea seeds to rats. J. Nutr. 104:733-740.

Wilkinson, S. R., J. A. Stuedemann, J. B. Jones, Jr., W. A. Jackson, and J. W. Dobson. 1972. Environmental factors affecting magnesium concentrations and tetanigenicity of pastures. p. 153-175. In J. B. Jones, Jr., M. C. Blount, and S. R. Wilkinson (ed.) Magnesium in the environment. Soils, crops, animals and man. Div. of Agric., Fort Valley, Ga.

Woolson, E. A., J. H. Axley, and P. C. Kearney. 1971. The chemistry and phytotoxicity of arsenic in soils: I. Contaminated field soils. Soil Sci. Soc. Am. Proc. 35:938-943.

chapter 12

Tank trailer with knife injectors applies waste slurries to field (Photo courtesy of the Metropolitan Sanitary District of Greater Chicago).

Recycling Elements in Wastes through Soil-Plant Systems

L. T. KARDOS, Pennsylvania State University, University Park, Pennsylvania

C. E. SCARSBROOK, Auburn University, Auburn, Alabama

V. V. VOLK, Oregon State University, Corvallis, Oregon

I. INTRODUCTION

Prior to commercial manufacture of synthetic fertilizers, agricultural waste products, especially animal manures, were highly valued for use as fertilizer materials. Animal and plant residues were returned to the soil to enhance production of another crop, which in turn could be returned to the food chain. Thus, the concept of elemental recycling is not new, but has merely resurfaced with the current economic and environmental questions.

Recently, several factors have contributed to a larger concern for the application of organic waste, not only from agriculture, but from many other sources to land surfaces. The supply of available energy has become critical, the human population has burgeoned, and concern for the preservation of land, water, and air resources and quality for future mankind has developed. Standards for air and water quality have been established and must be met prior to approved disposal of waste products. Costs for producing a marketable product must include the cost of waste management to meet the quality standards. Thus, the economics of waste management have emerged, with land application of the waste product becoming a significant consideration.

Accumulation of waste product constituents in a productive agricultural soil should not be detrimental to the ability of the soil to produce food and fiber. With this concept, one must think of the soil as a dynamic entity through which elements pass during recycling from the waste to plants, water, or air.

The reaction of waste products applied to the soil depends upon soil type, topography, climate, nature of and constituents in the waste product, vegetation, and, very importantly, management. Management practices, such as irrigation, drainage, crop selection, and fertilization, will markedly affect sorption, precipitation, microbiological degradation, and elemental uptake by plants.

Chemical and physical processes of dissolution, oxidation, reduction, precipitation, volatilization, movement, and complexation affect the rate and extent of breakdown of waste products applied to soils. Each of these processes is affected by environmental and man imposed conditions in the soil. For example, the rate at which organic N is mineralized and nitrified is directly related to the soil microbial population, moisture content, and partial pressure of oxygen. Installation of tile drains may extend the time during which oxidative reactions can occur. On the other hand, denitrification of NO_3^- may be enhanced by closing tile drains and over-irrigating to reduce the partial pressure of oxygen and create anaerobic conditions, assuming the presence of an energy supply. Thus, the valuable plant nutrient (N) or the potential environmental problem (N) would be transformed to either N_2 or nitrogen oxide gases and lost to the atmosphere.

Emphasis in this chapter will be given to use of the plant and soil as vehicles whereby chemical elements in municipal waste waters, sewage sludges, municipal solid wastes, food processing wastes, industrial wastes, and agricultural wastes may be recycled in agricultural production. Elemental losses from the soil by water transport, mass movement, and volatilization are discussed in other chapters.

A simplistic approach to evaluate the recycling potential of a waste added to soil is to examine the relationship between waste composition and the harvested portion of the crop grown on the waste treated site. Elements removed from the soil by erosion, leaching, or volatilization may also be recycled through the plant but the recycling time may be extended significantly.

II. NUTRIENTS AND TRACE ELEMENTS IN WASTE PRODUCTS

A. Municipal Waste and Waste Composts

The N, P, and K contents of municipal waste vary considerably depending on such factors as the amount and type of materials recycled, season of the year, economic level of the area, usage of garbage disposal units, and residential living style. Since 50 to 75% of most municipal waste consists of paper products with a low NPK content, the waste is usually characterized by a total NPK content of < 2% (Table 1 and 2). The NPK level of municipal waste generally increases with reduction in paper content. However, regardless of the input variations, the waste is always characterized by a high C/N ratio resulting in an initially low availability of N. Kaiser (1966) assayed numerous refuse components and observed C contents from 40 to 60%. The

Table 1—Partial chemical composition of municipal refuse, municipal compost, paper mill sludge, and cannery waste

Material	N	P	K	Ca	Mg	Reference
			%			
Municipal refuse	0.83	0.25	0.31	--	--	Garner (1966)
Municipal refuse	0.57	0.08	0.31	0.85	0.21	King et al. (1974)
Municipal compost	0.57	0.26	0.22	1.80	1.20	Hortenstine and Rothwell (1972)
Municipal compost	1.66	1.36	1.14	3.75	--	Toth (1968)
Municipal compost	0.75	1.52	0.25	1.30	--	Toth (1968)
Municipal compost	0.79	0.81	0.63	1.30	--	Toth (1968)
Municipal compost (plus some sewage sludge)	2.27	0.45	0.20	0.01	--	Hortenstine and Rothwell (1973)
Municipal compost (plus some sewage sludge)	1.3	0.26	0.97	4.6	0.67	Terman et al. (1973)
Municipal compost (plus some sewage sludge)	1.2	0.24	0.80	3.6	0.49	Mays et al. (1973)
Paper mill sludge	2.33	0.50	0.74	1.53	0.15	Dolar et al. (1972)
Paper mill sludge	0.15	0.29	0.85	0.10	0.09	Dolar et al. (1972)
Paper mill sludge	0.27	0.16	0.44	1.59	0.10	Dolar et al. (1972)
Paper mill sludge	0.99	0.42	0.62	0.37	0.12	Dolar et al. (1972)
Paper mill sludge	0.62	0.29	0.52	0.85	0.09	Dolar et al. (1972)
Cannery waste	0.97	0.14	0.13	--	--	Reed et al. (1973)

Table 2—Elemental analysis of solid waste from Vancouver, Washington†

Fraction	Percent of total	N	P	S	Ca	Mg	K	Na
				% dry weight				
Solid waste > 2 mm‡ (organic material)	76	0.64	0.16	1.6	1.97	0.11	0.25	0.26
Solid waste < 2 mm	12	0.45	0.14	5.3	4.24	0.20	0.22	0.16
Solid waste < 1 mm	6	0.32	0.12	6.1	4.61	0.24	0.24	0.13

Fraction	Fe	Mn	Cu	Zn	B	Mo	Co	Cr
				ppm dry weight				
Solid waste > 2 mm	4,575	242	6	15	7	2.9	3.7	34
Solid waste < 2 mm	9,250	286	334	615	50	2.1	2.9	17
Solid waste < 1 mm	15,000	360	270	685	ND§	ND§	ND§	17

† Cottrell (1975).[1]
‡ Pieces of metal, rubber, glass etc. > 2 mm (12% of total) were not analyzed.
§ Not determined.

C/N ratio of food wastes may be 15:1 while the C/N ratio of paper may exceed 100:1 (Galler & Partridge, 1969). After removal of metal, rubber, and glass from a sample of the Vancouver, Washington municipal waste, Cottrell (1975)[1] observed S and Ca as the only plant nutrients to exceed 1% (Table 2).

The range in elemental composition of municipal waste composts is usually greater than for noncomposted waste (Table 1). The percentage of most elements in the compost increases with composting duration. Although some N may be volatilized during composting, the relative N content increases during the composting period. With increased curing time, the C/N ratio is reduced, resulting in more rapid availability of N and P when the compost is added to cropped soils. If sewage sludge is added to the waste before composting, the C/N ratio may decrease to < 30 and the total NPK content may exceed 4%.

B. Industrial Waste

1. PAPER MILL SLUDGE

Paper mill sludges have a wide range in elemental composition (Table 1). Residual sludges from primary treatment processes have C/N ratios of about 150 whereas those from secondary processes may have C/N ratios ranging from 12 to 50. Since N is released when the C/N ratio of organic matter is lower than approximately 20 and immobilized at higher ratios, there is a large range in N availability from these products. However, in general, the total

---

[1]N. M. Cottrell. 1975. Disposal of municipal wastes on sandy soils: Effect on plant nutrient uptake. M.S. Thesis. Oregon State Univ., Corvallis, Oreg.

NPK content of paper mill sludges is low and similar to municipal waste composts.

2. CANNERY WASTE

Cannery wastes have N contents similar to municipal waste composts but may have lower P and K contents (Table 1). When an alkali is used in the peeling process, the waste has a high Na content and a high pH. Some cannery wastes require considerable decomposition before significant quantities of N and P become available for plant use.

C. Animal Waste

The elemental content of animal waste varies with the type of animal, diet, age and degree of decomposition, and inclusions of nonwastes such as dirt or bedding (Table 3). Nitrogen in animal waste is found in both the urine and solid fractions with the N in the urine generally more available than that in the solid fraction. Potassium and Na occur in the urine fraction as their highly soluble chloride salts. Calcium and Mg are mainly excreted with the feces while P occurs mostly in the feces for herbivores, in the urine for carnivores, and in both the urine and feces for omnivores (Azevedo & Stout, 1974).

With the appearance of certain elements in different waste fractions, the amount of a given element applied to the land will depend strongly upon waste collection facilities and management of the stored waste. Elements, especially N, occurring in the organic fraction undergo a series of transformations dependent upon management procedures. Organically bound N may be mineralized to NH_4^+-N. Ammonia gas may then be evolved, especially under warm, high pH conditions. The NH_3 loss may be greatly enhanced by air movement (Heck, 1931a). The C/N ratio may be reduced considerably during animal waste composting and storage operations (Salter & Schollenberger, 1939). McCalla and Elliot (1971) have observed up to a 50% loss in the organic matter of wastes in some feedlot operations.

D. Sewage Effluent and Sludge

The composition of treated municipal sewage effluents is relatively constant and approaches a 2.5:1:2.7 N/P/K ratio for the major fertilizer elements. The proportions of organic-, NH_4^+- and NO_3^--N may vary more widely depending on degree and mode of treatment. Effluents which have received only primary treatment will have relatively more organic-N, 5 to 20% of the total N. With secondary treatment, the NH_4^+-N will generally dominate but in trickling filter or activated sludge or extended aeration sewage treatment plants, the NO_3^--N content of the effluent may equal or exceed

Table 3—Chemical composition of selected animal wastes†

Type of animal waste	Chemical composition												Reference
	N	P	K	Ca	Mg	S	Mn	Fe	B	Cu	Zn	Mo	
	% dry weight						ppm						
Broiler litter	2.3	1.1	1.7	2.0	0.4	0.4	272	1,244	33	29	128	13	Perkins and Parker (1971)
Hen litter	2.0	1.9	1.9	3.4	0.5	0.5	333	1,347	28	31	120	14	Perkins and Parker (1971)
Dairy cow	2.7	0.5	2.4	1.6	0.6	0.3	56	222	83	28	83	6	Benne et al. (1961)
Fattening cattle	3.5	1.0	2.3	0.6	0.5	0.4	23	182	91	23	68	2	Benne et al. (1961)
Hog	2.0	0.6	1.5	2.0	0.3	0.5	72	1,002	143	18	215	4	Benne et al. (1961)
Horse	1.7	0.3	1.5	2.9	0.5	0.3	37	500	56	19	56	4	Benne et al. (1961)
Sheep	4.0	0.6	2.9	1.9	0.6	0.3	32	518	32	16	81	3	Benne et al. (1961)

† Cited in Azevedo and Stout (1974).

Table 4—Average composition of chlorinated secondary treated sewage effluent from the Pennsylvania State University sewage treatment plant, 1968–1970

Substance	Concentration
	mg/liter
Total solids	360†
Suspended solids	25†
Biochemical oxygen demand (BOD_5)	15†
Detergent (MBAS)	0.5
Organic N	4.9
NH_4^+-N	11.0
NO_3^--N	5.1
Total-N	21.0
Orthophosphate-P	5.3
Potassium	15.0
Calcium	27.0
Magnesium	13.5
Sodium	41.6
Chloride	48.3
Boron	0.37
Copper	0.060
Zinc	0.203
pH	7.8

† Average from August 1960 to August 1970.

the NH_4^+-N level.   Effluents from single cell or multicell lagoon systems usually contain dominantly NH_4^+-N and organic-N with relatively little NO_3^--N.

Other elements commonly found in sewage treatment plant effluents at concentrations exceeding 1 ppm include: orthophosphate P, K, Ca, Mg, Na, and Cl (Table 4).

The elemental content of sewage sludges will vary, depending upon sample collection and processing procedure, and the sewage treatment plant process. The N content of sludge may range from 1 to 15% (Table 5); however, the more common N levels range from 3 to 6%. Sewage sludge contains virtually no NO_3-N, with from 40 to 75% of the total N occurring in organic

Table 5—Composition of digested sewage sludges

	Content			
	Minimum	Maximum	Minimum†	Maximum†
	%			
Total N	4.69	7.27	1	15
NH_4^+-N	1.33	3.63	--	--
P	2.20	3.90	1	6
K	0.24	0.68	0.05	1
N/P/K	1.2:1:0.17‡	3.3:1:0.11‡		

† Page and Chang (1975).
‡ Computed from Peterson et al. (1973), dry weight basis.

Table 6—Trace element concentrations in sewage sludges from the USA†

Element	Concentration		
	Minimum	Maximum	Median
		µg/g	
B	4	680	20
Cd	1	1,100	10
Cr	20	33,000	400
Cu	100	11,700	700
Ni	10	4,500	50
Mo	2	1,000	5
Hg	0.1	50	3
Pb	10	26,000	500
Zn	100	28,500	2,200
As	0.5	30	4
Se	<0.1	81	3

† Page and Chang (1975).

combinations. Sludges may lose from 30 to 90% of their NH_4^+-N when dried. The P and K content of sewage sludges will generally range from 1 to 4% and 0.2 to 1%, respectively (Table 5, Sommers et al., 1973).

Page and Chang (1975) have reported that the median concentration of trace elements (Cd, Cu, Pb, Zn, and Hg) in sewage sludge is greater than the same elements found in most soils (Table 6).

III. ELEMENTAL CONCENTRATIONS IN AND REMOVAL BY CROPS

The element concentration in a plant depends upon plant species, climate, soil, and management practices. With respect to the major nutrients, N, P, and K, the elemental concentrations in the harvested tissue fall in a relatively narrow range for most crops. For example, Miller (1973) indicates that the grains (barley [*Hordeum vulgare* L.], corn [*Zea mays* L.], rice [*Oryza sativa* L.], rye [*Secale cereale* L.], and wheat [*Triticum aestivum* L.]) have a range of N from 1.6% for rice to 2.3% for wheat; of P from 0.3% for rice to 0.5% for barley; and of K from 0.2% for rice to 0.6% for barley. For forage crops, including alfalfa (*Medicago sativa* L.), bromegrass (*Bromus inermis* Leyss.), red clover (*Trifolium pratense* L.), orchardgrass (*Dactylis glomerata* L.), and timothy (*Phleum pratense* L.), the % N ranges from 1.2 for timothy to 2.8 for alfalfa; the % P ranges from 0.1 for timothy to 0.4 for orchardgrass; and the % K ranges from 1.7 for timothy to 2.4 for bromegrass.

The total quantities of nutrient elements removed in the harvested crop may range from as little as 60 kg N ha^{-1} year^{-1} for small grains to as much as 390 kg N ha^{-1} year^{-1} for three cuttings of reed canarygrass (*Phalaris arundinacea*); from as little as 9 kg P ha^{-1} year^{-1} for small grains to as much as 62 kg P ha^{-1} year^{-1} with three cuttings of reed canarygrass; and from as little as 16 kg K ha^{-1} year^{-1} for grains to as much as 300 to 380 kg K ha^{-1} year^{-1} for forages such as alfalfa and reed canarygrass (Fried & Broeshart, 1967; Kardos et al., 1974).

Calcium removal from soil by small grain crops is usually < 20 kg ha^{-1} year^{-1}. Forage legumes generally remove larger quantities of Ca (100 to 170 kg ha^{-1} year^{-1}) than the monocotyledonous grasses (10 to 20 kg ha^{-1} year^{-1}) (Fried & Broeshart, 1967).

Magnesium removal from soils by corn grain plus stalk approximated 26 kg ha^{-1} year^{-1} whereas alfalfa and red clover remove 17 and 25 kg ha^{-1} year^{-1}, respectively (Fried & Broeshart, 1967). Barley and wheat grain remove Mg from the soil at a rate of 8 to 10 kg ha^{-1} year^{-1}.

Sulfur and P removal by crops is similar, ranging from as little as 2.5 kg ha^{-1} year^{-1} for potatoes (*Solanum tuberosum* L.) to 18 kg ha^{-1} year^{-1} for alfalfa and 20 kg ha^{-1} year^{-1} for tomatoes (*Lycopersicon esculentum* Mill.) (Fried & Broeshart, 1967).

Sodium removal by cropping is relatively small, usually < 5 kg ha^{-1} year^{-1}. Fried and Broeshart (1967) cited Na removals by wheat, barley, alfalfa, and red clover at < 8 kg ha^{-1} year^{-1}.

Chloride is taken up by plants much more readily than Na but is not removed to any great extent unless the stalk and leafy portions of the plant are removed. Kardos et al. (1974) found that corn grain removed only 1 to 3 kg ha^{-1} year^{-1} whereas the same crop harvested for silage removed 35 to 40 kg ha^{-1} year^{-1}.

The trace elements, B, Cu, Zn, Cd, Pb, Ni, Ca, Cr, Al, Fe, Mn, As, and Hg, are usually present in the ppm concentration range in plant tissue. Crop removals may fall in the range of 40 mg Cu ha^{-1} year^{-1} to 0.4 kg Mn ha^{-1} year^{-1} for a 4,000 kg/ha harvest (Allaway, 1968).

IV. CROP REMOVAL OF ELEMENTS IN RELATION TO WASTE APPLICATION

A. Municipal Waste and Waste Composts

In an extensive 8-year study in England where pulverized municipal refuse was applied to soils at rates of 13.2 or 36.5 metric tons/ha, average yields of most root crops, vegetables, and cereals increased (Garner, 1966). However, the refuse applications produced only 72% of the yield increase observed for the same rates of farm manure. Crop uptake of elements from the refuse was not reported.

Application of shredded metropolitan waste to irrigated calcareous sandy soils in eastern Oregon at rates up to 448 metric tons/ha (approximately 298 metric tons of dry matter/ha) did not decrease alfalfa or fescue (*Festuca* sp.) yields, but neither did it increase yields or improve plant forage quality (Cottrell, 1975).[1] Nitrogen was added as ammonium sulfate at a rate of 448 kg/ha. Additional N fertilizer did not increase plant yields. Some beneficial fertilizer effect was obtained from P, Zn, and Cu in the waste materials; but quantities of B which could be detrimental to sensitive plants were also added at all rates of solid waste application. With addition of 896 metric tons of solid waste/ha, good initial alfalfa or fescue stands were not obtained because of inadequate seedbed preparation. Large quantities of the micronutrients, Zn, Cu, and B, as well as Na, were added to the soil with the 896 metric tons/ha waste application such that Zn and B uptake by wheat and fescue approached or exceeded toxic levels during the first growing season. Sufficient Mo was added to increase uptake by alfalfa to levels potentially hazardous to livestock if only one feed source was used (Cottrell, 1975).[1]

Trace element toxicities diminished after the first growing season, with the exception of B, which remained in available form at levels detrimental to sensitive crops even on the soil which received the lower waste treatments. Elements such as Zn, Cu, Mn, and Mo were apparently immobilized in the soil, while excess Na was leached from the surface and distributed throughout the soil profile. Zinc and Mn uptake by the fescue remained high only where soil pH had decreased to < 6.0 following heavy applications of ammonium sulfate fertilizer. The Zn and Mn content of the waste-treated soil could present toxicity problems in acid soils.

The NO_3^--N produced in the upper portion of a waste-treated sandy soil did move with the water front; however, careful irrigation management could eliminate ground-water contamination (Halvorson, 1975).[2] Other cations (Ca, Mg, Na, K) also moved slightly down the soil profile while the Cu, Zn, and P remained essentially within the depth of waste incorporation.

Metropolitan waste may be applied to agricultural soils but this practice has not been followed because of health hazards to cattle from ingestion of foreign objects, equipment unavailability, high hauling costs, potential pathogenic problems, possible B toxicities to sensitive plants, and, very importantly, sociological acceptance. In past years sanitary landfills have been used while the mix of a waste product and agricultural soils was not seriously considered.

Rye plants grown for 3 months on Guelph loam soil treated with municipal compost at a rate of 188 metric tons/ha had N, P, K, Ca, Mg, and Na contents similar to plants grown on untreated soils (King et al., 1974). Nine months after planting, the rye contained about 70% more Zn and 33% less NO_3^--N than control plants. Corn, planted after rye without further additions of compost, produced yields about the same for the control and refuse treatments. Ear leaves of corn grown in refuse-treated soil contained the same concentrations of N and P as control plants. Corn grown on the untreated soil contained more Ca and Mg but less K than corn grown on refuse-treated plots. Both corn grain and stover contained 50 to 100% more Zn when refuse was applied. It appears that K and Zn were the only elements from the refuse which were recycled to any extent in the corn crop. No

---

[2]G. Halvorson. 1975. Movement of elemental constituents in Sagehill loamy sand treated with municipal waste. M.S. Thesis. Oregon State Univ., Corvallis, Oreg.

yield decrease was observed, probably because the soil, Guelph loam, had an organic matter content of 4.1% and had been in a grass sod for at least 20 years. Nitrogen release from the soil organic matter may have offset N immobilization by the refuse, which had a C/N ratio of 65.

Application of municipal compost (0, 23, 46, 82, 246, and 327 metric tons ha^{-1} 2 years^{-1}), containing up to 20% sewage sludge, to forage sorghum (*Sorghum bicolor* Moench.), common bermudagrass (*Cynodon dactylon* L.), and corn increased sorghum yields over a 3-year period for each increment of compost (Mays et al., 1973). Nitrogen added with the compost produced additional yield increases. The average bermudagrass forage yield increased with addition of compost up to a rate of 36 metric tons/ha. Higher rates of compost addition produced no additional bermudagrass yield increases. Corn yields over two seasons increased with compost additions up to 112 metric tons/ha but higher rates did not increase yield further. Over a 3-year period from 2 to 4% of the N added as compost was recycled in the sorghum. The amount of municipal compost applied had little effect on the percentage N recycled.

Tietjen and Hart (1968) reported on extensive experiments in Europe with municipal composts. No elemental analyses of the composts were reported but the importance of the time of composting was well illustrated. Fresh compost reduced potato yields the first year after application whereas ripe compost increased yields. One or 2 years after fresh compost applications, oat (*Avena sativa* L.) and rye yields increased markedly. A significant factor, not often observed, was that compost increased yields at about a constant rate irrespective of the N added. They observed that the principal use of municipal compost in Europe was for erosion and runoff control for which compost was most effective.

Initial nutrient levels of the soil relate strongly to crop response to first year municipal waste compost applications (Hortenstine & Rothwell, 1972). In their experiments, compost at rates of 0, 25, and 70 metric tons/ha, with and without added NPK, was added to phosphate mine sand tailings almost devoid of available plant nutrients. Plant growth was poor with all treatments, and essentially no yield of sorghum and oats was obtained the first year when no fertilizer source of NPK was added. While compost alone increased yields the second year after application, much higher yields were produced when both NPK and compost were applied.

Compost applications of both 8.96 and 17.92 metric tons/ha increased corn yields the first year after application with some residual effect on yield at high rates with a low fertility soil but none on soils medium or high in fertility (King & Morris, 1969).

Hortenstine and Rothwell (1973) reported that compost added at rates equivalent to 8 to 64 metric tons/ha increased yields of both the first and second crops of sorghum. However, the compost had a N content of 2.27%, which is more than twice that found in most municipal composts. With higher rates of compost application, the percent NPK recovered by the crop decreased. Recoveries ranged from 13 to 19% for N, 7 to 26% for P, and 71 to 98% for K. The compost application did not affect the pH of the Arredondo sand but did increase P and K contents.

Terman et al. (1973) found that corn grown on a N deficient Mountview silt loam soil required N additions to compost to prevent yield reduction. However, on a soil containing more available N, compost additions did not induce N deficiency. They found that N, P, and K in a compost was 16, 71, and 64% as effective, respectively, for corn as the same amount of these elements in soluble fertilizers.

1. SUMMARY

When fresh garbage or green compost is added to soils, N may limit plant growth. No availability problem occurs with K, while P occupies an intermediate position. When green compost is added to soils, and crops are planted without delay, the severity of the N deficiency depends on the N content of the compost, and the availability of soil N. Approximately 1 year after compost addition, decomposition has progressed to minimize N and P availability problems. If the composting process is from 6 months to 1 year in length, the cured compost may have a sufficiently low C/N ratio for net release of N to the crop immediately after application.

Crop production on irrigated sandy soils after incorporation of up to 448 metric tons solid waste/ha is feasible with proper crop and water management. Yields cannot be expected to exceed those produced on normally fertilized soil, however; economic benefit could come from successful disposal of the waste product.

B. Industrial Waste

1. PAPER MILL SLUDGE

When primary and secondary paper mill sludges were added at rates of 2.5% and 10%, by weight, to five Wisconsin soils, oat yields increased at the lower rate of secondary sludge additions but decreased at all rates for the high C/N ratio primary sludges and at high rates for the secondary sludges (Dolar et al., 1972). The primary sludges had C/N ratios of about 150:1 and secondary sludges, depending on the mill process, had C/N ratios ranging from 12:1 to 50:1. When sufficient NPK was added with the material having a C/N ratio of 12 and when applied at the 2.5% rate, 25%, 12%, and 46% of the N, P, and K, respectively, was recovered in the oats. No significant quantities of NPK were recovered at the 10% application rate for the low C/N material. A small amount of P but no N and K was recovered from the materials with C/N ratios higher than 12:1. Incubation studies confirmed that the high C/N ratio sludges released no N over a 90-day period.

The problem of N immobilization in primary paper mill sludges was shown by Aspitarte et al. (1973). They applied sludges with a C/N ratio of

150 at rates of 448, 896, and 1,344 metric tons/ha. Sufficient N was added to the soil to produce C/N ratios of 10, 50, and 100 (assuming 30% C, 16,072 kg of N was required to decrease the C/N ratio of 1,344 metric tons of sludge from 150:1 to 10:1). All sludge treatments to the Hesson clay soil reduced yields of green beans (*Phaseolus vulgaris* L.) and corn for the first year after application. One or more years after application, all sludge applications, with or without N additions, increased bean and corn yields. Apparently after 1 year of decomposition, even a material with an initial C/N ratio of 150:1 released sufficient N for the crop. No nutrient recovery data were reported.

2. CANNERY WASTES

Application of peach (*Prunus persica*) cannery wastes to Lakeland sand and Cecil sandy clay loam increased coastal bermudagrass yields with respect to a control which received rainfall only; but, bermudagrass growth on soils irrigated with distilled water exceeded that from all waste treatments (Hunt & Peele, 1968). Apparently, water supplied was the only beneficial effect from the cannery waste. Wheat yields increased when wastes from tomato and fruit canneries were added to soils (Reed et al., 1973).

C. Animal Waste

Nitrogen added in animal waste products occurs mainly as organically combined N or as NH_4^+-N. Upon application to the soil the organic N may be mineralized to NH_4^+-N and subsequently be oxidized to NO_3^--N. The NO_3^--N may be utilized by crops grown on the waste-treated soil, may accumulate, or may leach through the soil profile to ground water (Mathers & Stewart, 1971). The problem of NO_3^--N movement into ground water becomes more severe if the ground-water table in the application area is very shallow, if high N applications are made, and if the soils have a high hydraulic conductivity. The NO_3^--N may accumulate in the soil profile under conditions where no net water drainage occurs.

Nitrogen in animal waste is found in both urine and solid fractions with the N in urine generally more available than that in the solid fraction. Heck (1931b) reported that N recovery by barley grown in manure-treated soil varied from 17% for fresh dung to approximately 40% for fermented dung. Approximately 50% of the N was recovered when a complete manure was applied to a light sandy soil. The residual effect from N after application of the solid and liquid animal waste fractions to crops was quite different. With the liquid fraction, the N became available immediately and was largely recovered in the first crop; while in the solid fraction the N tended to become available over a much longer period of time.

Pratt et al. (1973) applied a decay series concept to the mineralization of organic N. They suggested that N from chicken manure may be 90% available the first year of application since it exists largely as urea or uric acid. On the other hand, N in dry corral manure may only be 20% available. Using this approach the mineral N available for plant use from a given waste application can be calculated. Thus one can adjust the waste application rate in any given year to maintain a constant N mineralization rate or a constant pool of plant available N. The decay series concept can also be used as an important input into an assessment of potential NO_3^--N accumulation in the soil profile or movement to ground water.

Nutrient recovery from animal waste may be depressed by addition of highly carbonaceous materials to soil. Heck (1931b) observed that the addition of 6% straw to fresh cow manure reduced N recovery by barley by 17%; N recovery by subsequent crops was not affected by straw additions. Immobilization of N by addition of straw or other similar highly carbonaceous material has been attributed to the enhanced microbial activity and resultant N tie-up by microorganisms. Because of the slow release of organically combined N, 0.91 metric tons (1 ton) of manure has been considered equivalent to 45.4 kg (100 pounds) of 3-5-10 or 4-5-10 fertilizer, rather than a 10-5-10 which would reflect more accurately the total N content (Salter & Schollenberger, 1939).

Schneidewind (1931) observed that an amount of N equivalent to 41% of the N present in stable manure was recovered by sugarbeets (*Beta vulgaris* L.) followed by barley. With the same application of stable manure, a potato crop followed by wheat recovered 36% of the N applied.

In Ohio, application of 4.5 metric tons (5 tons) of animal manure annually for 32 years resulted in decreases in the N found in the surface soil when cropped with continuous corn, oats, or wheat (Salter & Schollenberger, 1939). A corn, wheat, clover rotation in conjunction with the manure application maintained soil N levels. In Rothamsted, England, when manure was applied at a rate of 31.4 metric tons/ha (14 tons/acre) and continuous wheat was grown for 71 years, approximately 21% of the applied N was recovered in the soil (Russell, 1937). At the Sanborne Field plots in Missouri, application of only 3.4 metric tons ha^{-1} year^{-1} (1.5 tons acre^{-1} year^{-1}) of manure for 40 years in a 4-year rotation which included clover produced an apparent N recovery of 166% (Jenny, 1933). Nitrogen fixed by clover and by nonsymbiotic microorganisms must have contributed largely to the increased soil N content.

Mathers and Stewart (1971) found that grain sorghum yields were decreased upon application of animal manure at rates of 269 and 538 metric tons/ha to a Pullman clay loam in west Texas. The decreased yield was attributed to salt accumulations in the soil. With irrigation, forage sorghum yields again increased. Nitrate-N accumulated in the profiles of soils treated with high rates of animal manure. The NO_3^--N accumulations may cause concern for pollution of ground water or for high NO_3^--N concentrations in crops. Mathers and Stewart (1971) recommended that waste application to land be based upon N utilization by the crop to eliminate pollution hazards.

Turner and Proctor (1971) observed forage yield increases with manure slurry applications in western Washington. Nitrate-N did not accumulate in the forage to toxic levels, probably because of leaching and denitrification during the heavy winter rainfall period. They recommended that manure application to grass-legume forages should not exceed 2.5 cm of a slurry with 10% suspended solids and should be applied to stubble just after harvest. As much as 12.5 cm of manure slurry could be applied prior to seeding silage corn.

Webber et al. (1968) reported that continuous corn or grass could accept up to 336 kg N ha^{-1} year^{-1} from manure. Higher waste application rates led to crop yield depression and water contamination. To effectively use N and to assist in pollution control, they recommended that, in Ontario, 41 ha be required to utilize the manure produced by a 100-dairy cow, 365-day operation. Only 20 ha of land would be required if manure disposal was the primary objective. If N were added at a rate of 353 kg/ha as manure and in precipitation, a corn crop could recycle 150 kg/ha; the remainder of the applied N would be lost by leaching, denitrification, volatilization, or would accumulate in the soil profile.

Adriano et al. (1971) studied the distribution of NH_4^+, NO_2^-, and NO_3^--N in soils of the Chino-Corona dairy area of southern California which had been treated with dairy animal waste. The NO_3^--N concentration in ground waters increased as a function of manure application. The authors indicated that a cow/disposal ha ratio of 10:1 might be possible in southern California without ground-water contamination if two corn crops or three forage cuttings were removed each year.

Overman et al. (1971) observed better response by forage sorghum than oats from application of high rates of dairy manure slurries. The N, P, and K removed by the oats treated with dairy manure slurry at a rate of 0.63 cm/week exceeded, 153, 118, and 153%, respectively, that added with the animal waste. The Ca and Mg removed by the forage sorghum and oats did not exceed, 38% and 25%, respectively, that added in the lowest dairy manure treatment.

The extent of N recycling from animal wastes into the food chain through plants depends strongly upon animal waste handling. In addition to losses by leaching or surface runoff, N may be lost as volatile NH_3 or other nitrogenous gases after NO_3^--N denitrification. Ammonia volatilization from feedlots (Hutchinson & Viets, 1970), dairy areas (Luebs et al., 1973), and from grazed pastures (Denmead et al., 1974) may account for sizable N losses. Nitrogen which is lost through NH_3 volatilization would be readily nitrified and available to plants if incorporated into the soil. Usually, over 50% of the total N excreted by animals or poultry is lost through handling. Because of the importance of handling or pretreatment technique, data on N recovery or recycling by plants varies widely. The determination or recommendation of a waste application plan for a given operation must consider both handling and pretreatment of the waste as well as the soil and crop management program.

1. SUMMARY

Nitrogen in animal waste can be recycled through the soil-plant system. Nitrogen in the liquid waste becomes available to plants readily whereas the organically combined N in the solid animal waste becomes available more slowly and may remain as a N reserve in the soil. Nitrogen can serve as an effective parameter to determine waste application to soils, assuming that salt concentration, odor, insect, solids accumulation, and sociological effects of the waste application on neighboring communities do not present problems. Application of animal wastes to supply N will often eliminate micronutrient deficiencies of Zn or Fe and add considerable P and K to the soil.

D. Sewage Effluent and Sludge

Probably the most comprehensive and detailed study of the relation of inputs of chemical elements from sewage effluent and removal in harvested crops is that being carried out at the "living filter" project at Pennsylvania State University, State College, Pa. (Parizek et al., 1967; Kardos, 1967).

In the study, 2.5 cm of sewage effluent applied at weekly intervals during the growing season was not sufficient to balance crop removals of N by corn grain or corn silage (Tables 7 and 8). For example, corn grain (Table 7) in 4 of the 5 crop years removed N equivalent to 102% to 174% of the amount applied in the 2.5 cm/week crop season effluent application. Corn silage (Table 8) removed N equivalent to 115% to 286% of the amount applied in the 2.5 cm/week treatment. With reed canarygrass, even the 5 cm/week year-round effluent application did not supply sufficient N to equal the amount removed in 3 of the 5 years (Table 9).

Alfalfa and reed canarygrass removed more K from the soil than added by the 2.5 cm/week effluent treatment (Tables 9 and 10). Only with the corn grain and in 2 of the 4 years with corn silage, did the effluent supply as much K as was removed (Tables 7 and 8). The 5.0 cm/week crop season effluent application was sufficient to replenish K removal by corn grain and silage every year but in only 2 of 4 years by alfalfa. The 5.0 cm/week year-round application added K in amounts more than that removed by the reed canarygrass in 4 of the 5 years.

The 2.5 cm/week and 5.0 cm/week crop-season, and 5.0 cm/week year-round applications of sewage effluent supplied more P than was removed in the harvested crops except in 1969 with corn silage (Tables 7, 8, 9, 10, and 11). The one exception was due in part to the lower than normal P content of the effluent because P was being removed at the sewage treatment plant during part of the year by aluminum precipitation as part of a sanitary engineering research project.

Calcium, Mg, B, Na, and Cl were added in excess of crop removals with all three effluent treatments (Tables 7, 8, 9, and 10). The maximum utiliza-

Table 7—Quantities of chemical elements removed in corn grain and removal efficiency at two levels of sewage effluent irrigation†

	1963 Removal				1964 Removal				1965 Removal			
	kg/ha		% of added		kg/ha		% of added		kg/ha		% of added	
Element	2.5	5.0	2.5	5.0	2.5	5.0	2.5	5.0	2.5	5.0	2.5	5.0
N	119	120	174	89	114	155	76	52	107	107	166	88
P	26	32	52	31	15	15	21	10	37	38	78	40
K	23	33	24	18	33	32	25	12	36	37	26	13
Ca	0.3	0.3	0.2	0.1	4.7	4.5	19	0.9	1.3	3.5	0.6	0.9
Mg	7.1	7.6	7.6	4.6	6.4	5.9	4.9	2.3	12	13	11	6.1
Cl	ND‡	ND‡	--	--	4.2	3.7	1.2	0.5	2.9	2.1	0.9	0.3
B	ND‡	ND‡	--	--	0.04	0.03	1.3	1.0	0.02	0.02	1.0	0.5
Na	ND‡	ND‡	--	--	ND‡	ND‡	--	--	ND‡	ND‡	--	--

	1966 Removal				1967 Removal			
	kg/ha		% of added		kg/ha		% of added	
Element	2.5	5.0	2.5	5.0	2.5	5.0	2.5	5.0
N	119	139	102	69	97	91	162	77
P	31	44	64	51	26	29	86	34
K	26	32	16	10	25	24	22	10
Ca	ND‡	ND‡	--	--	ND‡	ND‡	--	--
Mg	10	12	6.5	4.1	7.8	7.8	11	5.0
Cl	3.0	4.9	0.7	0.6	3.2	1.1	1.0	0.2
B	ND‡	ND‡	--	--	0.01	0.01	0.4	0.2
Na	0.03	0.05	0.1	0.1	0.10	0.05	0.04	0.02

† Adapted from Kardos et al. (1974).
‡ Not determined.

Table 8—Quantities of chemical elements removed in corn silage and removal efficiency at two levels of sewage effluent irrigation†

Element	Effluent level, cm/week							
	2.5	5.0	2.5	5.0	2.5	5.0	2.5	5.0
	Removal				Removal			
	kg/ha		% of added		kg/ha		% of added	
	1965				1966			
N	118	121	184	100	134	156	115	78
P	19	21	39	22	30	42	65	51
K	86	111	62	40	104	145	66	49
Ca	35	35	17	8	30	34	13	8
Mg	19	18	17	8	20	24	13	9
Cl	32	36	10	6	40	49	9.5	6.2
B	0.04	0.04	1.9	1.0	0.05	0.07	2.0	1.0
Na	ND‡	ND‡	--	--	2.4	2.7	0.8	0.4
	1967				1969			
N	104	123	175	105	285	218	286	120
P	24	33	81	39	33	65	178	123
K	121	140	106	61	159	107	264	97
Ca	23	22	16	8	31	18	33	7
Mg	18	20	24	13	24	20	35	15
Cl	40	41	13	7	53	37	22	10
B	0.07	0.08	2.5	1.5	0.09	0.05	1.0	0.5
Na	1.5	1.9	0.6	0.4	2.2	2.0	6.5	1.8

† Adapted from Kardos et al. (1974).
‡ Not determined.

tions for these elements at the 5.0 cm/week, crop-season effluent application level were 19% with alfalfa, 0.9% with corn grain, 8% with corn silage, and 13% with reed canarygrass for Ca; and 22% with alfalfa, 6.1% with corn grain, 15% with corn silage, and 15% with reed canarygrass for Mg.

Boron which was added in amounts equivalent to 1.2 to 8.9 kg/ha in the various effluent treatment systems was only removed to the extent of 0.2% to 8% of the amount applied, with alfalfa removing the most and corn grain the least.

Two interesting sidelights of the crop removal data are the large amounts of Cl removed by the forage crops as contrasted with the grain, and the apparently strong exclusion of Na by all crops. Reed canarygrass removed 134 to 223 kg Cl/ha whereas corn grain removed only 1.1 to 4.9 kg Cl/ha. Sodium removals ranged from as little as 0.03 kg/ha for corn grain to a high of 5.3 kg/ha for reed canarygrass.

If one examines the situation for the trace elements, Cu and Zn, in silage corn and reed canarygrass (Table 12), it is apparent that reed canarygrass has a higher concentration of both Cu and Zn. However, removal of Cu and Zn, when expressed as percent of the amounts added in the effluent, was approximately the same with both crops because larger quantities of Cu and Zn had been added to the reed canarygrass area. Removals of Cu and Zn were

Table 9—Chemical elements removed in reed canarygrass (three cuttings) and removal efficiency when irrigated with 5.0 cm of sewage effluent per week

Element	1965		1966		1967		1968‡		1969	
	kg/ha	% of added	kg/ha	% of added	kg/ha	% of added	kg/ha	% of added	kg/ha	% of added
N	388	226	299	103	362	150	392	68	390	76
P	44	32	36	25	62	34	52	25	52	49
K	354	86	275	70	377	103	280	69	235	97
Ca	68	13	32	6.1	71	13	45	7	54	9
Mg	33	12	21	6.9	45	15	31	10	40	14
Cl	207	24	134	12	223	18	187	16	190	20
B	0.11	1.0	0.05	0.9	0.11	1.3	0.05	0.6	0.12	1.9
Na	5.3	0.7	1.3	0.2	5.2	0.6	3.4	0.3	3.0	0.6

† Adapted from Kardos et al. (1974)
‡ Only two cuttings harvested in 1968.

Table 10—Quantities of chemical elements removed by two cuttings of alfalfa hay and removal efficiency at two levels of sewage effluent irrigation†

Element	Effluent level, cm/week							
	2.5	5.0	2.5	5.0	2.5	5.0	2.5	5.0
	Removal				Removal			
	kg/ha		% of added		kg/ha		% of added	
	1963				1965			
N	157	211	NC‡	NC	273	312	NC‡	NC‡
P	25	35	44	30	31	37	64	39
K	185	257	177	125	262	328	188	119
Ca	55	49	29	13	66	79	32	19
Mg	12	15	11	7	22	26	18	11
Cl	ND§	ND§	--	--	119	146	38	23
B	ND§	ND§	--	--	0.18	0.23	8	5
Na	ND§	ND§	--	--	ND§	ND§	--	--
	1966				1967			
N	198	203	NC‡	NC‡	126	146	NC‡	NC‡
P	27	32	60	38	23	26	76	32
K	241	266	152	90	199	230	174	99
Ca	57	52	24	11	35	35	25	12
Mg	16	18	11	6	12	13	17	9
Cl	91	115	22	9	86	104	28	19
B	0.11	0.09	4	2	0.08	0.09	3	2
Na	2.0	2.1	0.6	0.3	1.3	1.4	0.5	0.3

† Adapted from Kardos et al. (1974).
‡ Not computed because of ambiguity of N supplied by legume bacteria.
§ Not determined.

Table 11—Average amounts of N, P, and K removed in the harvested crops (1963–1969) at the 5.0-cm/week sewage effluent level

Crop‡	Harvested years	Annual removal		
		N	P	K
		kg/ha		
Alfalfa hay	3	243	34.5	270
Red clover hay	2	250	29.7	291
Corn grain	5	122	31.6	32
Corn silage	4	155	40.1	126
Oats	2	51	13.7	18
Reed canarygrass	5	366	49.1	305

† Adapted from Kardos et al. (1974).
‡ Reed canarygrass was irrigated year-round; other crops only during the growing season, April–May through September–October.

equivalent to approximately 10% of the annual application. The other 90% of these added trace metals either accumulated in the soil or moved into the ground water and the preponderant evidence favors accumulation. Average annual applications of Cu and Zn on the year-round irrigated reed canary-

Table 12—Copper and Zn relationships in applied sewage effluent and in corn silage
and reed canarygrass (1965–1970), Penn State University†

	Sewage effluent‡				Crop§					
	Average concentration		Total added		Average concentration		Removed			
	Cu	Zn	Cu	Zn	Cu	Zn	Cu	Zn	Cu	Zn
	— μg/liter —		—kg/ha—		— μg/g —		–kg/ha—		% of added	
Corn silage	64	202	4.2	14.3	7.5	30.8	0.45	1.70	10.7	11.9
Reed canarygrass	57	204	8.4	28.2	10.5	38.9	0.76	2.85	9.0	10.1

† From unpublished file data, L. T. Kardos, Dep. of Agronomy, The Pennsylvania State University, University Park, Pa.
‡ Five centimeters of effluent applied at weekly intervals; May–September on silage corn; January–December on reed canarygrass.
§ For reed canarygrass, values represent weighted average of three cuttings per year.

grass area have been 1.4 and 4.7 kg/ha, respectively. These quantities should not pose any hazard to crop growth or to the animal food chain.

Information on elemental recycling through the soil-plant system after sewage sludge applications is limited. King and Morris (1972a) reported that in 1969, five cuttings of coastal bermudagrass removed approximately 430 kg N/ha on plots which had received 954 kg N/ha as a liquid digested sludge. Rye seeded on the bermudagrass sod removed an additional 84 kg N/ha (King & Morris, 1972b). They also reported total crop removals of Cu and Zn over a 2-year period by coastal bermudagrass (nine cuttings) and forage rye (one cutting each year) ranging from 0.9% to 0.3% of the applied Cu and from 1.3% to 1.0% of the applied Zn as the application of liquid digested sludge ranged from 11.3 cm to 60.0 cm over the 2-year period. In the case of the 60-cm application, 81% of the Cu and 56% of the Zn was found at the end of the experiment in the sludge crust on the sod surface.

Hinesly et al. (1971) grew corn, soybeans (*Glycine max* [L.] Merr.), reed canarygrass, and sorghum on field lysimeter plots treated with sewage sludge. They concluded that digested liquid sludge applied at a rate of 5 to 8 cm on the soil surface immediately after withdrawal from the digester would satisfy annual N needs of nonleguminous crops without producing excessive NO_3^--N levels in percolation water. The liquid sludge supplied 224 to 392 kg NH_4^+-N/ha, about the same amount of organic-N/ha, 280 to 540 kg P/ha, and 45 to 90 kg K/ha.

After the sludge application the Zn, Cd, and Fe content of plant tissue increased to the greatest extent; however, phytotoxicity was not observed and concentration levels of the trace elements in plant tissues did not reach proportions that constituted a health hazard to animals consuming all or any part of the plants. Sludge applied in amounts necessary to supply N and/or P was regarded as inadequate to supply K to a plant growing on a K-deficient soil.

Boswell (1975) found no phytotoxicity on fescue sod topdressed with a total of 16.8 metric tons/ha, dry sludge equivalent of digested, vacuum filter cake sludge applied in three 5.6 metric ton/ha increments over a 15-month period. Concentrations of Cu, Zn, Cd, Cr, Pb, and Mn in the harvested fescue increased significantly with sludge application. Using total yield and average metal concentration for the eight cuttings, crop removal of metals ranged from 3% of the applied Cr to 13% of the applied Mn.

It must be noted that large sewage sludge applications may cause Zn, Ni, and Cu toxicities to selected crops, dependent upon their sensitivity to trace elements (Weber, 1972; King & Morris, 1972b; Bingham et al., 1975). For example, growth of spinach (*Spinacia oleracea*) and swiss chard (*Beta vulgaris cicla*) was reduced 25% at 5 ppm Cd concentration whereas a comparable reduction in cabbage (*Brassica oleracea* var. *capitata*) growth was not observed until the Cd concentration was 200 ppm. Chaney (1973) recommended that, for a good benefit/risk ratio, the following elemental concentrations for sewage sludge slated for land application should not be exceeded: Zn, 2,000 ppm; Cu, 800 ppm; Ni, 100 ppm; Cd, 0.5% of Zn; B, 100 ppm; Pb, 1,000 ppm; and Hg, 15 ppm.

Nitrogen has been suggested as an appropriate element to use as a guide to recommend sewage sludge applications to soils. To apply 224 kg/ha (200 pounds) of total N to a soil would require about 2.3 metric tons of dry sludge or 1.25 cm of a sludge slurry containing 5% solids. An application of this magnitude would supply P in excess of most crop needs (Table 13). Phosphorus accumulations could be accentuated because additional N beyond crop removal may be required to account for the low initial N availability in organic-N compounds (Miller, 1973). If one desired to use plant nutrients in sewage sludge most efficiently, P would serve as a better monitor or guideline element. Phosphorus required by plants and P fertilizer applications to soils

Table 13—Relative elemental content of sewage sludge†

Element	Concentration, dry weight	Added to soil‡
		kg/ha
Total N	4.0%	224
NH_4-N	1.0%	56
P	1.5%	84
K	1.0%	56
Zn	2,000 ppm	11.2
Cu	800 ppm	4.5
Ni	100 ppm	0.56
Cd	1 ppm	0.0056
B	100 ppm	0.056
Pb	1,000 ppm	5.6
Hg	15 ppm	0.084

† Selected values to typify sewage sludge compositions.
‡ Equivalent to 2.3 metric tons dry sludge/ha or 1.25 cm of sludge containing 5% solids.

are generally lower than N; thus less sludge would be applied. In fact, supplemental N and K fertilization could likely be required on many soils.

1. SUMMARY

Substantial quantities of N, P, and K in treated sewage effluents and sludges can be recycled through harvested crops with benefits in both yield and quality. In most cropping situations, if sufficient effluent or sewage sludge is applied to meet N needs, P will be adequate but K may not be, particularly with sewage sludges. The trace elements are supplied by both effluents and especially sludges in excess of crop removals. If equal quantities of N were supplied by an effluent and a sludge, approximately 10 to 20 times as much and in some cases 100 times as much trace element, as required by plants, would be applied in the sludge. Much research needs to be done to evaluate the significance of the long term accumulation of trace elements in the food chain. If the effluents and sludges are considered in the context of a recycling concept rather than a disposal concept, the trace element problem becomes much less critical.

V. GENERAL SUMMARY

The extent of elemental recycling through the soil-plant system after application of organic wastes to soils varies widely. After application of a waste product to the soil, chemical and physical reactions begin to change elemental availability. These reactions may result in the accumulation of elements in the soil, volatilization, translocation across the soil surface or down the soil profile, or removal by plants and animals. Soil and crop management practices will largely control the nature of the chemical and physical processes which occur.

Prior to waste application to the land, one should know the chemical, physical, and microbiological properties of the material; topographical and climatological features of the area; soil chemical and physical properties; long range land use; cropping plans; crop nutrient requirements and elemental sensitivities; and public reaction or sentiment. With this information, management systems can be developed to optimize elemental recycling.

LITERATURE CITED

Adriano, D. C., P. F. Pratt, and S. E. Bishop. 1971. Nitrate and salt in soils and ground water from land disposal of dairy manure. Soil Sci. Soc. Am. Proc. 35:759–762.

Allaway, W. H. 1968. Agronomic controls over environmental cycling of trace elements. Adv. Agron. 20:235–274.

Aspitarte, T. R., A. S. Rosenfeld, B. C. Samle, and H. R. Amberg. 1973. Pulp and paper mill sludge disposal and crop production. Tappi (Tech. Assoc. Pulp Pap. Ind.) 56: 140–144.

Azevedo, J., and P. R. Stout. 1974. Farm animal manures: An overview of their role in the agricultural environment. Calif. Agric. Exp. Sta. Ext. Serv. Manual 44. 109 p.

Benne, E. J., C. R. Hoglund, E. D. Longnecker, and R. L. Cook. 1961. Animal manures—What are they worth today? Mich. State Univ. Agric. Exp. Sta. Circ. Bull. 231. 15 p.

Bingham, F. T., A. L. Page, R. J. Mahler, and T. J. Ganje. 1975. Growth and cadmium accumulation of plants grown on a soil treated with a cadmium-enriched sewage sludge. J. Environ. Qual. 4:207-211.

Boswell, F. C. 1975. Municipal sewage sludge and selected element applications to soil: Effect on soil and fescue. J. Environ. Qual. 4:267-273.

Chaney, R. L. 1973. Crop and food chain effects of toxic elements in sludge and effluents. Proc. Joint Conf. on Recycling Municipal Sludges and Effluents on Land. Champaign, Ill. p. 129-141.

Denmead, O. T., J. R. Simpson, and J. R. Freney. 1974. Ammonia flux into the atmosphere from a grazed pasture. Science 185:609-610.

Dolar, S. G., J. R. Boyle, and D. R. Keeney. 1972. Paper mill sludge disposal on soils: Effects on the yield and mineral composition of oats (Avena sativa L.). J. Environ. Qual. 1:405-409.

Fried, M., and H. Broeshart. 1967. The soil-plant system in relation to inorganic nutrition. Academic Press Inc., New York, N. Y. 358 p.

Galler, W. S., and L. J. Partridge. 1969. Physical and chemical analysis of domestic municipal refuse from Raleigh, North Carolina. Compost Sci. 10(3):12-15.

Garner, H. V. 1966. Experiments on the direct, cumulative, and residual effects of town refuse, manures, and sewage sludges at Rothamsted and other centres, 1940-47. J. Agric. Sci. 67:223-233.

Heck, A. F. 1931a. Conservation and availability of the nitrogen in farm manure. Soil Sci. 31:335-359.

Heck, A. F. 1931b. The availability of nitrogen in farm manure under field conditions. Soil Sci. 31:467-480.

Hinesly, T. D., O. C. Braids, R. I. Dick, R. L. Jones, and J. A. E. Molina. 1971. Agricultural benefits and environmental changes resulting from the use of digested sludge on field crops. Report to Metropolitan Sanitary District of Greater Chicago and to U. S. Environ. Protect. Agency, Office of Solid Waste Management. April 1967 to April 1971. University of Illinois, Urbana, Ill.

Hortenstine, C. C., and D. F. Rothwell. 1972. Use of municipal compost in reclamation of phosphate-mining sand tailings. J. Environ. Qual. 1:415-418.

Hortenstine, C. C., and D. F. Rothwell. 1973. Pelletized municipal refuse compost as a soil amendment and nutrient source for sorghum. J. Environ. Qual. 2:343-345.

Hunt, P. G., and T. C. Peele. 1968. Organic matter removed from peach wastes by percolation through soil and interrelations with plant growth and soil properties. Agron. J. 60:321-323.

Hutchinson, G. L., and F. G. Viets, Jr. 1970. Nitrogen enrichment of surface water by absorption of ammonia volatilized from cattle feedlots. Science 166:514-515.

Jenny, H. 1933. Soil fertility losses under Missouri conditions. Missouri Agric. Exp. Sta. Bull. 324. Columbia, Mo.

Kaiser, E. R. 1966. Chemical analysis of refuse components. Proc. Nat. Incinerator Conf., Am. Soc. Mech. Eng., New York. p. 84-88.

Kardos, L. T. 1967. Waste water renovation by the land-A living filter. p. 241-250. In N. C. Brady (ed.) Agriculture and the quality of our environment. Am. Assoc. Adv. Sci. Publ. No. 85.

Kardos, L. T., W. E. Sopper, E. A. Myers, R. R. Parizek, and J. B. Nesbitt. 1974. Renovation of secondary effluent for reuse as a water resource. Environ. Protect. Technol. Ser. EPA-660/2-74-016. 495 p.

King, L. D., and H. D. Morris. 1969. Municipal compost for crop production. Ga. Agric. Res. 10(3):10-12.

King, L. D., and H. D. Morris. 1972a. Land disposal of liquid sewage sludge: I. The effect on yield, in vivo digestibility, and chemical composition of coastal bermudagrass (Cynodon dactylon L. Pers.). J. Environ. Qual. 1:325-328.

King, L. D., and H. D. Morris. 1972b. Land disposal of liquid sewage sludge: II. The effect on soil pH, manganese, zinc, and growth and chemical composition of rye (Secale cereale L.). J. Environ. Qual. 1:425-429.

King, L. D., L. A. Rudgers, and L. R. Webber. 1974. Application of municipal refuse and liquid sewage sludge to agricultural land: I. Field study. J. Environ. Qual. 3:361-366.

Luebs, R. E., A. E. Laag, and K. R. Davis. 1973. Ammonia and related gases emanating from a large dairy area. Calif. Agric. 27(2):11-12.

McCalla, T. M., and L. F. Elliott. 1971. The role of microorganisms in the management of animal wastes on beef cattle feedlots. p. 132-134. In Livestock waste management and pollution abatement. Proc. Int. Symp. on Livestock Wastes. Am. Soc. Agric. Eng. Publ. PROC-271.

Mathers, A. C., and B. A. Stewart. 1971. Crop production and soil analysis as affected by applications of cattle feedlot waste. p. 229-31, 34. In Livestock waste management and pollution abatement. Proc. Int. Symp. on Livestock Wastes. 19-22 April. Columbus, Ohio. Am. Soc. Agric. Eng. PROC-271.

Mays, D. A., G. L. Terman, and J. C. Duggan. 1973. Municipal compost: Effects on crop yields and soil properties. J. Environ. Qual. 2:89-92.

Miller, D. F. 1973. Composition of cereal grains and forages. Comm. on Feed Composition, Nat. Agric. Board, NAS-NRC, Washington, D. C. 663 p.

Miller, H. R. 1973. Soil microbiological aspects of recycling sewage sludges and waste effluents on land. Proc. Joint Conf. on Recycling Municipal Sludges and Effluents on Land. Champaign, Ill. p. 79-90.

Overman, A. R., C. C. Hortenstine, and J. M. Wing. 1971. Growth response of plants under sprinkler irrigation with dairy waste. p. 334-337. In Livestock waste management and pollution abatement. Proc. Int. Symp. on Livestock Wastes. Am. Soc. Agric. Eng. Publ. PROC-271.

Page, A. L., and A. C. Chang. 1975. Trace element and plant nutrient constraints of recycling sewage sludges on agricultural land. Proc. Second Nat. Conf. on Water Reuse: Water's Interface with Energy, Air, and Solids. Sponsored by Am. Inst. Chem. Eng. and U. S. Environ. Protect. Agency. 4-8 May. Chicago, Ill.

Parizek, R. R., L. T. Kardos, W. E. Sopper, E. A. Myers, D. E. Davis, M. A. Farrell, and J. B. Nesbitt. 1967. Waste water renovation and conservation. Penn State Stud. No. 23. The Pennsylvania State Univ., University Park, Pa. 71 p.

Perkins, H. F., and M. B. Parker. 1971. Chemical composition of broiler and hen manures. Univ. Ga. Agric. Exp. Sta. Res. Bull. 90. 17 p.

Peterson, J. R., C. Lue-Hing, and D. R. Zenz. 1973. Chemical and biological quality of municipal sludge. p. 26-37. In W. E. Sopper and L. T. Kardos (ed.) Recycling treated municipal wastewater and sludge through forest and cropland. The Pennsylvania State University Press, University Park, Pa.

Pratt, P. F., F. E. Broadbent, and J. P. Martin. 1973. Using organic wastes as nitrogen fertilizers. Calif. Agric. 27(6):10-13.

Reed, A. D., W. E. Wildman, W. S. Seyman, R. S. Ayers, J. D. Prato, and R. S. Rauschkolb. 1973. Soil recycling of cannery wastes. Calif. Agric. 27(3):6-9.

Russell, E. J. 1937. Soil conditions and plant growth. Sixth ed. Longmans Green and Co., London. p. 277.

Salter, R. M., and C. J. Schollenberger. 1939. Farm manure. Bull. 605, Ohio Agric. Exp. Sta., Wooster, Ohio.

Schneidewind, W. 1931. Die Ernährung der landwirtschaftlichen Kulturpflanzen. Paul Parey, Berlin.

Sommers, L. E., D. W. Nelson, J. E. Yahner, and J. V. Mannering. 1973. Chemical composition of sewage sludge from selected Indiana cities. Ind. Acad. Sci. 82:424-432.

Terman, G. L., J. M. Soileau, and S. E. Allen. 1973. Municipal waste compost: Effects on crop yields and nutrient content in greenhouse pot experiments. J. Environ. Qual. 2:84-89.

Tietjen, C., and S. A. Hart. 1968. Compost for agricultural land? J. Sanit. Eng. Div. Proc. Am. Soc. of Civil Eng. Vol. 95, No. Sa2:269-287.

Toth, S. J. 1968. Chemical composition of seven garbage composts produced in the United States. Compost Sci. 9(3):27-28.

Turner, D. O., and D. E. Proctor. 1971. A farm scale dairy waste disposal system. Sci. Pap. 3360. Wash. Agric. Exp. Sta., Pullman, Wash.

Webber, L. R., T. H. Lane, and J. H. Nodwell. 1968. Guidelines to land requirements for disposal of liquid manure. Proc. 8th Ind. Water Wastewater Conf. Lubbock, Tex. p. 20-34.

Weber, J. 1972. Effects of toxic metals in sewage on crops. Water Pollut. Control. 71:404-413.

chapter 13

Site Selection as Related to Utilization and Disposal of Organic Wastes

J. E. WITTY, USDA-SCS, Broomall, Pennsylvania

K. W. FLACH, USDA-SCS, Washington, D. C.

I. INTRODUCTION

The purpose of this paper is to discuss site selection criteria for management and utilization of organic wastes and waste waters with emphasis on land and soil properties. Criteria or properties considered are those that will lead to the utilization or disposal of wastes without causing environmental problems outside the site perimeter and any buffer zones. The basic objective, therefore, is to utilize or dispose of the wastes in such a way that they are either rendered harmless or prevented from moving onto adjacent land, into surface waters, into the ground water, or into air.

In discussing site selection criteria, one can give only general principles that apply to wastes from many sources and to waste management systems that are in common use. Soil chemical, physical, and biological properties related to waste interactions with soils are discussed in earlier chapters. The list of soil properties, their limits, and intraactions is almost infinite. However, some properties may be crucial for a specific waste disposal problem at a given location but may be unimportant elsewhere. Also one set of properties of a given soil may maximize its ability to renovate wastes, another set may minimize its ability to accept significant amounts of wastes, and a third set may even influence management of the disposal site. Final decision as to whether a site should or should not be used for a specific system almost always represents a compromise. The properties of many soils are known and can be used to make initial selection of disposal sites. Additional studies may be needed, however, to determine soil properties that may be critical for a specific use.

Three general sets of criteria can be considered. First, there are those criteria that are important if the soil is to act primarily as a container for highly concentrated wastes and where the wastes do not interact with the soil to a significant degree, such as in sanitary landfills or in feedlots. Secondly, there are those criteria that are important if the soil is to react with important components of the wastes so as to immobilize or destroy them, and where utilization is not or cannot be a primary consideration. Examples are sewage effluent disposal sites or sludge disposal sites. The third set of criteria are important if waste utilization is the primary consideration.

Each of these three sets of criteria is discussed with emphasis on soil properties, followed by a discussion on the use of soil surveys as an aid for locating potential sites and some hydrological and geological considerations in selecting potential sites. Regional limitations such as soil temperature, length of growing season, or amount and distribution of precipitation are not discussed. The above items are important considerations for regional planning, however, because they do affect decisions on the feasibility of soil-based systems or on costs if winter storage facilities are necessary.

The following presentation is centered around guidelines (Tables 1-5) that have been developed and are now being used by the Soil Conservation Service, USDA (1973). However, these guidelines are under continual review and subject to change from time to time.

In these guides, individual critical soil properties are rated as to how severely they limit the usefulness of soils as treatment media for certain wastes. No attempt is made to evaluate the ease with which limitations can be overcome through appropriate design of the system or through modification of the soil.

The approach is simple and can serve as an initial guide in rating kinds of soils on the basis of criteria that have been published (Soil Conservation Service, 1971) or are available in computer storage for the 11,000 or so soil series of the United States.

II. SITE SELECTION CRITERIA FOR WASTES DISPOSED ON LAND AT HIGH RATES

Examples where wastes are concentrated or applied at high rates include sanitary landfills, sewage lagoons, feedlots, and areas of stockpiled organic material. The wastes, when disposed on land, are generally highly concentrated in small areas and have a high potential for causing environmental problems. Of prime importance is the design of facilities and proper management of the wastes because the soil will not normally have the capacity to dissipate them adequately. The basic function of the soil is to act as a container. Proper site selection can greatly reduce the problems of design and management.

A. Sanitary Landfills

The process of sanitary landfilling is to bury wastes in soil. Loughry (1974) described four functions that soil has in relation to landfills as follows:
1) Soil serves as container and support.
2) Soil serves as the most commonly used cover material.
3) Soil retains intermediate products, providing time and a favorable medium for change and recycling of some of the wastes.
4) Soil, if used as the final cover material, supports vegetation and can be used for farming, forestry, or recreation.

The Soil Conservation Service (1971) has published guides for assessing the suitability of different kinds of soil for sanitary landfills. Two guides are provided, one for the trench-type sanitary landfill and the other for the area-type sanitary landfill.

1. TRENCH-TYPE SANITARY LANDFILL

The trench-type sanitary landfill consists of trenches in which refuse is covered at least daily with a layer of soil material at least 15 cm thick. Soil excavated in digging the trench is used as the covering material. When the

Table 1—Soil limitation ratings for trench-type sanitary landfills† ‡

	Degree of soil limitation		
Item affecting use	Slight	Moderate	Severe
Depth to seasonal high water table	Not class determining if > 180 cm		< 180 cm
Soil drainage class	Excessively drained, somewhat excessively drained, well drained, and some § moderately well drained	Somewhat poorly drained and some § moderately well drained	Poorly drained and very poorly drained
Flooding	None	Rare	Occasional or frequent
Permeability¶	<5 cm/hour	<5 cm/hour	>5 cm/hour
Slope	0–15%	15–25%	>25%
Soil texture# (dominant to a depth of 150 cm)	Sandy loam, loam, silt loam, sandy clay loam	Silty clay loam††, clay loam, sandy clay, loamy sand	Silty clay, clay, muck, peat, gravel, sand
Depth to bedrock			
Hard	>180 cm	>180 cm	<180 cm
Rippable	>150 cm	<150 cm	<150 cm
Stoniness class‡‡	0 and 1	2	3, 4, and 5
Rockiness class‡‡	0	0	1, 2, 3, 4, and 5

† From *Guide for Interpreting Engineering Uses of Soils* (Soil Conservation Service, 1971).
‡ Based on soil depth (1.5–2 m) commonly investigated in making soil surveys.
§ Soil drainage classes do not correlate exactly with depth to seasonal water table. The overlap of moderately well-drained soils into two limitation classes allows some of the wetter moderately well-drained soils (mostly in the Northeast) to be given a limitation rating of *moderate*.
¶ Reflects ability of soil to retard movement of leachate from the landfills; may not reflect a limitation in arid and semiarid areas.
# Reflects ease of digging and moving (workability) and trafficability in the immediate area of the trench where there may not be surfaced roads.
†† Soils high in expansive clays may need to be given a limitation rating of *severe*.
‡‡For class definitions see *Soil Survey Manual*, p. 216–223 (Soil Survey Staff, 1951).

trench is full, the landfill is covered with a layer of soil material at least 60 cm thick.

Table 1 lists the soil limitation ratings for the trench-type sanitary landfill. Soil properties considered are depth to seasonal high water table, soil drainage class, flooding, permeability, slope, soil texture, depth to bedrock, stoniness class, and rockiness class. The degree and duration of soil wetness as related to seasonal water table, soil drainage class, and flooding are considered because they affect earth moving operations and the likelihood of contaminating the ground water. As degree of soil wetness increases, the site becomes increasingly less suitable as a sanitary landfill site.

Soil permeability is important because it affects vertical or lateral movement of leachate. Soils with low permeability are most desirable because seepage is minimized.

Soil slope is an important consideration since it may affect runoff and ease of constructing trenches and roads. On moderately steep soils, leachate may concentrate in downslope trenches (Apgar & Langmuir, 1971), thus increasing the potential for ground-water pollution.

Soil texture affects the workability and trafficability of the soil, both wet and dry. Soils with textures that are workable over a wide range of moisture content are most desirable. Many coarse-textured soils have a low degree of workability and trafficability when dry, while many fine-textured soils have low workability qualities when either wet or dry. The final cover should be soil material that is favorable for plant growth.

Bedrock, stoniness, and rockiness affect the ease of excavating trenches to suitable depths. Fractured bedrock immediately underlying the trench also creates a potential for the pollution of ground water.

2. AREA-TYPE SANITARY LANDFILL

In this type of landfill, waste is placed on the soil surface and covered with soil. The waste is covered daily with at least 15 cm of soil and is covered with soil at least 60 cm thick when the landfill is completed.

Table 2 lists soil limitation ratings for the area-type sanitary landfill. Soil properties considered are: depth to seasonal water table, soil drainage class, flooding, permeability, and slope. The importance of these properties

Table 2—Soil limitation ratings for area-type sanitary landfills†

| Item affecting use | Degree of soil limitation | | |
	Slight	Moderate	Severe
Depth to seasonal‡ water table	>150 cm	100–150 cm	<100 cm
Soil drainage ‡ class	Excessively drained, somewhat excessively drained, well drained, and moderately well drained	Somewhat poorly drained	Poorly drained and very poorly drained
Flooding	None	Rare	Occasional or frequent
Permeability §	Not class determining if <5 cm/hour		>5 cm/hour
Slope	0–8%	8–15%	>15%

† From *Guide for Interpreting Engineering Uses of Soils* (Soil Conservation Service, 1971).
‡ Reflects influence of wetness on operation of equipment.
§ Reflects ability of the soil to retard movement of leachate from landfills; may not reflect a limitation in arid and semiarid areas.

for workability or potential pollution of ground water is the same as discussed above for the trench-type sanitary landfill. Stoniness, rockiness, or bedrock are not important considerations because no excavating is done in the area-type sanitary landfill.

The daily cover material and final cover material for the area-type sanitary landfill generally must be imported from other soil areas. A table giving the "suitability ratings of soils as sources of cover material for area-type sanitary landfills" has been prepared by the Soil Conservation Service (1971). This table is not included in this paper. Soil properties listed for the cover material are moist consistence, texture, thickness of material, coarse fragments, stoniness, slope, and drainage class. Soils with very friable or friable consistence are good sources of cover material, those with loose or firm consistence are fair sources, and those with very firm or extremely firm consistence are poor sources. Soils with good textures for cover material include sandy loam, loam, silt loam, and sandy clay loam; those with fair textures are silty clay loam, clay loam, sandy clay, and loamy sand; and those with poor textures are silty clay, clay, muck, peat, and sand. Thick, well-drained soils with gentle slopes and without coarse fragments are better sources of cover material than shallow, gravelly, or stony soils or soils in wet areas.

B. Sewage Lagoons

A sewage lagoon or stabilization pond is a flat-bottomed pond used to hold sewage for the time required for its bacterial decomposition (Soil Conservation Service, 1971; Clark et al., 1971). In sewage lagoons the soil serves the following two functions: (i) as a container for the impounded sewage, and (ii) as material for the enclosing embankment. The lagoon must be capable of holding water with minimum seepage. Material for the enclosing embankment does not have to come from the sewage lagoon site.

Table 3 gives soil limitation ratings for sewage lagoons. Criteria considered are depth to water table, permeability, depth to bedrock, slope, coarse fragments < 25 cm in diameter, percentage of surface area covered by coarse fragments > 25 cm in diameter, and Unified Soil Classification Groups.

Depth to water table is important in that water should never rise high enough to enter the lagoon. If, however, the floor of the lagoon consists of at least 60 cm of essentially impermeable material, depth to water table can be disregarded. If the floor of the lagoon consists of slowly permeable material, at least 120 cm of material is needed between the bottom of the lagoon and the seasonal water table or any cracked and creviced bedrock.

Limitation classes for slope are determined by the requirement that, for the lagoon to function properly, the liquid depth should range from 60 to 150 cm. The slope must be sufficiently gentle and the soil material sufficiently thick over the bedrock to make land smoothing practical so as to obtain a uniform depth in the lagoon.

Table 3—Soil limitation ratings for sewage lagoons†

Item affecting use	Degree of soil limitation		
	Slight	Moderate	Severe
Depth to water table (seasonal or year-round)	>150 cm	100–150 cm‡	<100 cm‡
Permeability	<1.5 cm/hour	1.5–5 cm/hour	>5 cm/hour
Depth to bedrock	>150 cm	100–150 cm	<100 cm
Slope	<2%	2–7%	>7%
Coarse fragments, <25 cm in diameter; percent, by volume	<20%	20–50%	>50%
Percent of surface area covered by coarse fragments >25 cm	<3%	3–15%	>15%
Flooding §	None	None	Soils subject to flooding
Soil groups (Unified)¶ (rated for use mainly as floor of sewage lagoon)	GC, SC, CL, and CH	GM, ML, SM, and MH	GP, GW, SW, SP, OL, OH, and PT

† From *Guide for Interpreting Engineering Uses of Soils* (Soil Conservation Service, 1971).
‡ If the floor of the lagoon is nearly impermeable material at least 60 cm thick, disregard depth to water table.
§ Disregard flooding if it is not likely to enter or damage the lagoon (low velocity and the depth less than about 1.5 m).
¶ Disregard if permeability is < 1.5 cm/hour and it does not increase as a result of building the lagoon.

A high percentage of coarse fragments interferes with the manipulation and compaction needed to prepare the lagoon properly; hence, limitation classes for coarse fragments should be considered.

Soils subject to flooding are normally unsuited as sites for sewage lagoons because of the potential of floodwaters to mix with and carry away polluting sewage before sufficient decomposition has occurred. If, however, floodwaters do not damage the lagoon embankment or do not overflow the lagoon, this limitation does not apply.

Soil materials placed in the Unified Soil Classification Groups (U. S. Army Corps of Engineers, 1968) of GC, SC, CL, and CH (defined below) can be compacted to a satisfactory low permeability for a lagoon bottom. The coarse groups with few fines and soil materials high in organic matter have severe limitations and are poorly suited. Soil materials in the Unified Soil Classification Groups GM, ML, SM, and MH are suitable if properly compacted or used in combination with soils classified as GC, SC, CL, and CH.

The Soil Conservation Service (1971) has published a guide showing the general relationships between the Unified Soil Classification Groups and U. S. Department of Agriculture texture classes. The relationship is not perfect but it can be used for predicting the likely group or groups for each textural class. The following shows a simplified relationship between the Unified Soil Classification Groups listed above and the USDA texture classes:

GC — very gravelly silty clay loam, gravelly silty clay loam, and very gravelly silty clay.

SC — heavy sandy loam, sandy clay loam, and sandy clay.

CL — heavy silt loam, clay loam, and silty clay loam.

CH — heavy clay loam, heavy silty clay loam, silty clay, and clay.

GM — very gravelly sandy loam, very gravelly loam, very gravelly silt loam, and gravelly silt loam.

ML — fine sandy loam, very fine sandy loam, loam, silt loam, and silt.

SM — fine sand, very fine sand, loamy sand, sandy loam, and fine sandy loam.

MH — silty clay loam and clay loam.

The Soil Conservation Service (1971) has rated separately soils that are suitable for lagoon embankments and those that are suitable for lagoon floors. Properties considered in rating soil materials for their suitability as lagoon embankments are sheer strength, compressibility, permeability of compacted soil, susceptibility to piping, and compaction characteristics. They are evaluated for each Unified Soil Classification Group. Basically, soils in the Unified Soil Classification Groups listed as having slight limitations for the floor of a sewage lagoon are also suitable for the embankment.

C. Feedlots

Under this subheading, major emphasis is placed on site selection for animal pen areas. Criteria for selecting sites for lagoons or catch basins associated with pen areas are virtually the same as those discussed in the previous section on sewage lagoons. If the manure is stored outside the pen areas, then criteria discussed under the subheading "Areas for Stockpiled Organic Materials" apply.

General guidelines for evaluating soils for feedlots have been published by the U. S. Environmental Protection Agency (Kreis & Shuyler, 1972). The guidelines specify that soils with slopes of 2 to 6% are suitable and that highly permeable loose soils, shallow soils over fractured bedrock, and soils with a shallow water table should be avoided. Sloppy pen conditions may develop if the slope is < 2%, and uncontrollable runoff may occur if the slope is > 6%. Loose, shallow, or wet soils may lead to contamination of ground water.

If a feedlot is managed properly and continuously stocked, and a manure mulch left after cleaning, an impermeable layer forms at the manure-soil interface that effectively seals the floor of the feedlot against downward movement of pollutants (Elliott et al., 1973; Mielke et al., 1974). This seal apparently forms in any soil regardless of texture or permeability. Therefore, texture and permeability are not considered in rating except for very rapidly permeable soils ($>$ 50 cm/hour). These may have moderate limitations because of the potential instability and time lag before a seal forms.

Soil drainage is important because of its effect on trafficability. Well-drained, somewhat excessively drained, and excessively drained soils as well as sloping, moderately well-drained soils have slight limitations; poorly and very poorly drained soils have severe limitations. If slopes are $<$ 2 or 3%, however, moderately well-drained soils have moderate limitations and somewhat poorly drained soils have severe limitations.

Soil slope is important. Erosion is a hazard on steep slopes but sloppy pen conditions may result if the soils are level or nearly level. Gilbertson et al. (1970) reported no significant difference in runoff volume or solids removal from feedlots near Mead, Nebr. having slopes of 3, 6, and 9%. Swanson et al. (1971), however, found that in eastern Nebraska a feedlot with 13% slope lost more solids than one with an 8.5% slope. This indicates that possibly slopes as high as 10% have slight limitations while steeper slopes might have moderate or severe limitations. The slopes in these studies, however, were relatively short, about 30 m and less, and may not represent solids removal for longer slopes under similar precipitation characteristics. E. J. Monke (Dep. of Agricultural Engineering, Purdue University, personal communication) suggest general slope limitation classes to 2 to 6% as slight, 0 to 2% and 6 to 10% as moderate, and $>$ 10% as severe. Swanson also suggests that if snowmelt or rainfall is not a problem, soils with slopes of 15 to 20% should be useful for feedlots. Slopes steeper than about 20% present a safety hazard for machinery operations. Hence the general slope guidelines, as given above, should be adjusted according to snowmelt or precipitation characteristics.

Depth to bedrock should be considered because it affects feedlot construction if terracing or, on level soils, mounding is required. The soil must be deep enough so that the feedlot can be cleaned properly and revegetated when it is abandoned. This is important to remove nitrogen compounds that might otherwise pollute the ground water (McCalla, 1972). Depth to bedrock should probably be $>$ 1 m.

Stones affect feedlot construction and cleaning. Stoniness classes of 0 and 1 present slight limitations, 2 and 3 present moderate limitations, and 4 and 5 present severe limitations (Soil Survey Staff, 1951).

D. Areas for Stockpiled Organic Materials

The stockpiled materials considered here are organic materials handled as solids rather than as liquids. Materials are stockpiled in open piles and are not covered with soil material as in sanitary landfills. The materials may be stockpiled for either a short or long time but the site is used continuously. Of primary concern here are animal wastes, but included are organic materials such as logs in the lumbering or pulp industry, sewage sludge, leaves, or other kinds of organic materials that are composted in large quantities. It is assumed that the stockpiled materials are managed to minimize odor and vector problems. The primary function of the soil is that it serves as container and support.

Specific guidelines have not been published. The same soil properties and limitation ratings used for making soil limitation ratings for the area-type sanitary landfill (Table 2), however, can be considered.

III. SITE SELECTION CRITERIA FOR WASTES DISPOSED ON LAND AT LOW RATES

Under this heading are discussed criteria for selecting sites on which wastes can be applied at a rate that is in equilibrium with rate of decomposition. Hence, the site should be usable on a continuous basis. Side benefits may be realized, such as harvestable crops or recharge of ground water, but the primary objective is to dispose of wastes.

Kinds of wastes disposed of on land at low rates may be sewage sludge, sewage effluent, animal wastes, and cannery wastes. The major function of the soil is to dissipate the wastes, to recycle them through crops, or to purify them through filtering and adsorption.

The Soil Conservation Service, USDA (1973) has prepared an interim guide that is being tested (Tables 4 and 5). Soil properties used to rate kinds of soils by this guide are permeability, soil drainage class, runoff, flooding, and available water capacity.

Soil permeability influences length of time liquid wastes remain in the soil and potential loading rates. If permeability is very high, liquid wastes or soluble components of solid wastes may pass through a soil so fast that any potential pollutants are not adequately dissipated, especially during periods of high rainfall. On the other hand, if permeability is too low permissible application rates would be too low to be practical, or anaerobic conditions would be induced. Moderate and severe limitations do not apply for moderately slow, slow, or very slow permeabilities if layers having these permeabilities are below the rooting depth and evapotranspiration exceeds water added by rainfall and waste, or if solid waste is not plowed or injected into these layers.

In humid areas (udic moisture regimes), excess water in a soil can be predicted according to its soil drainage class. Soil drainage classes are a measure of the length of time the soil is naturally at or near saturation during the growing season. They reflect both the ability of the soil to remain aerobic and to support traffic. Well-drained and moderately well-drained soils are considered to have slight limitations, while excessively drained, or poorly and very poorly drained soils have severe limitations.

It is important that the applied waste stay on the site, therefore, soils are also rated for surface runoff and flooding. Runoff is closely related to infiltration rate, soil slope, and cover. It has been argued that the infiltration

Table 4—Soil limitations for accepting nontoxic biodegradable liquid waste†

Item affecting use	Degree of soil limitation		
	Slight	Moderate	Severe
Permeability of the most restricting layer above 150 cm	Moderately rapid and moderate, 1.5–15 cm/hour	Rapid and moderately slow‡, 15–50 and 0.5–1.5 cm/hour	Very rapid, slow, and very slow‡, >50 and <0.5 cm/hour
Soil drainage class§	Well drained and moderately well drained	Somewhat excessively drained and somewhat poorly drained	Excessively drained, poorly drained, and very poorly drained
Runoff¶	None, very slow, and slow	Medium	Rapid and very rapid
Flooding	None	Soils flooded only during nongrowing season	Soils flooded during growing season
Available water capacity from 0 to 150 cm or a limiting layer			
Humid#	>15 cm	8–15 cm	<8 cm
Arid††	>8 cm		<8 cm

  † Modified from an interim guide for use in the Soil Conservation Service.
  ‡ Moderate and severe limitations do not apply for moderately slow, slow, and very slow permeability if layers having these permeabilities are below the rooting depth and if evapotranspiration exceeds water added by rainfall and waste.
  § For class definition see *Soil Survey Manual,* p. 169–172 (Soil Survey Staff, 1951).
  ¶ For class definition see *Soil Survey Manual,* p. 166–167 (amended to use "None" for "Ponded") (Soil Survey Staff, 1951).
  # Humid, as used here, includes soils that have aquic, udic, or ustic moisture regimes if utilized throughout the year. For definitions, see *Soil Taxonomy* (Soil Survey Staff, 1975).
  †† Arid, as used here, includes soils that have aridic or torric moisture regimes, or xeric moisture regime if utilized only during the dry season. For definitions, see *Soil Taxonomy* (Soil Survey Staff, 1975).

rate should be considered in rating soils for receiving liquid wastes. However, the actual infiltration rate depends so much on management practices that it is omitted from Table 4. If soil is managed to maximize infiltration, e.g., by maintaining plant cover, by keeping traffic to a minimum, or by interjecting drying cycles, then the effective infiltration rate is primarily dependent on soil permeability. The degree of soil limitation for runoff is given in terms of runoff classes as defined in the *Soil Survey Manual* (Soil Survey Staff, 1951). In general, soils that flood are considered to have severe limitations for disposal of wastes. If the soils flood only during the nongrowing season, however, they are considered as having only moderate limitations at some localities.

The available water capacity is primarily a measure of the capacity of a soil to supply moisture to plants. It is used here as a measure of the mini-

Table 5—Soil limitations for accepting nontoxic biodegradable solids†

Item affecting use	Degree of soil limitations		
	Slight	Moderate	Severe
Permeability of the most restricting layer above 150 cm	Moderately rapid and moderate, 1.5-15 cm/hour	Rapid and moderately slow‡, 15-50 and 0.5-1.5 cm/hour	Very rapid, slow, and very slow‡, >50 and <0.5 cm/hour
Soil drainage class§	Well drained and moderately well drained	Somewhat excessively drained and somewhat poorly drained	Excessively drained, poorly drained, and very poorly drained
Runoff¶	None, very slow, and slow	Medium	Rapid and very rapid
Flooding	None		Soils flooded
Available water capacity from 0 to 150 cm or to a limiting layer	>15 cm	8-15 cm	<8 cm

† Modified from an interim guide for use in the Soil Conservation Service.
‡ Moderate and severe limitations do not apply for moderately slow, slow, and very slow permeability unless the waste is plowed or injected into the layers having these permeabilities or if evapotranspiration is less than water added by rainfall or irrigation.
§ For class definition see *Soil Survey Manual*, p. 169-172 (Soil Survey Staff, 1951).
¶ For class definition see *Soil Survey Manual*, p. 166-167 (amended to use "None" for "Ponded") (Soil Survey Staff, 1951).

mum soil volume needed to dissipate the wastes through plant nutrient uptake, microbial decomposition, and soil adsorption. The depth considered is from the soil surface to 150 cm, or to a limiting layer < 150 cm deep. Soils with > 15 cm available water have slight limitations, those with 8 to 15 cm have moderate limitations, and those with < 8 cm have severe limitations. The moderate limitation, however, does not apply for liquid wastes in an arid climate.

IV. SITE SELECTION CRITERIA FOR WASTES UTILIZED FOR CROP PRODUCTION

Considered here are the organic wastes and waste waters that can be used as fertilizer, soil amendment, or irrigation water to supplement precipitation. For example, wastes may be used on golf courses, on parks, or for crops. Yield of vegetation or crop, rather than disposal of waste, is the primary objective. Although the site selection criteria concerning soil properties are practically the same as for waste treatment on land at low rates, they are discussed separately because of possible differences in the extent and distribution of suitable soils. Furthermore, arrangements for use of the wastes are usually made with individual landowners or governing bodies, such as an irrigation district. The parcels of land may be widely scattered and economic factors may influence the feasibility of the system.

The success of a project in which the primary objective is to utilize the waste ultimately depends on the value of the wastes compared with costs of alternative methods of satisfying the landowner's needs. Liquid wastes have much greater value in the arid western part of the United States than in the humid eastern part. As a rule, if arrangements for utilization of the wastes have to be made with many landowners, the total extent of soils with suitable properties must be much greater than if the municipality or industry purchases or leases land for its waste disposal. Under these circumstances the amount of land needed is likely to be inversely proportional to the value of the waste in a given farming system.

V. SELECTING A SITE

Soil surveys are probably the most useful single source of information for making initial judgments on the suitability of potential sites for disposal or management of wastes on land (Flach & Carlisle, 1974). Soil surveys are available for > 40% of the country (Flach, 1973) and are generally available where soils are used most intensively. They consist of detailed soil maps usually at a scale of 1:31,680 to 1:15,840 on photographic background, a general soil map, description of the soils by series and mapping unit, data on engineering and agronomic properties of soils (usually with some characterization data on major soil series), and interpretive tables. Soil surveys are prepared by the Soil Conservation Service in cooperation with agricultural experiment stations and units of local government.

A report prepared by Sopper and Kardos (1972) regarding the suitability of soils in the Tocks Island Region of the Delaware River Basin for potential use of treated municipal sewage effluent is an excellent example of the use of soil surveys for making an inventory of potential disposal sites.

Sopper and Kardos (1972) reviewed published soil surveys and supplementary information for the area to establish criteria for the selection of desirable kinds of soil. After development of the criteria, the soils were evaluated and those that did not measure up to the standards were rejected. Next, suitable soils were located on soil maps, color coded, and acreage of the various soil parcels measured. This provided information on the extent and distribution of soils in the area which were potentially suitable for spray irrigation.

The guidelines discussed in our paper are useful as a first approximation for making a general survey of soil resources suitable for waste treatment systems.

The guidelines do not consider interaction among soil properties, between treatment systems, or combinations of soil properties. The guidelines consider soils in a pedologic sense. Also, they do not take into account underlying unconsolidated regolith that may be an important part of treatment systems, and they do not allow one to pinpoint soils with the best potential for a particular treatment system if all soils in the area available as treatment sites have the same degree of limitations.

More sophisticated and complex guides could be developed, but because of the large number of waste materials, treatment systems, and soils, the utility of such an effort is questionable.

Hence, after an initial screening using these guides, further evaluation is still necessary in which all information on the properties of soils of a given area, and the requirements and alternatives of the treatment system are used.

Information on the properties of individual soils can be obtained from soil descriptions and tables of soil properties in published soil surveys (Table 6). A computerized inventory of properties of the 11,000 or so soil series in the United States is being prepared by the Soil Conservation Service, USDA (1975). An example of the kind of data in the inventory is shown in Table 7. In addition, a great many site data, representative of many kinds of soils, are contained in the Soil Conservation Service, USDA, Soil Survey Investigations Reports and in other technical publications. In fact, for a first approximation, many soil properties important for waste treatment systems can be deduced from the placement of soils in *Soil Taxonomy* (Soil Survey Staff, 1975), the system of soil classification adopted by the National Cooperative Soil Survey. Hence, a competent soil classifier working closely with other specialists in soil science and with engineers, geologists, and hydrologists can identify potential sites that meet as many requirements as is possible for a given area.

In any case, practical experience with a specific kind of soil should be an overriding consideration in judging the suitability of a particular kind of soil. If a system works well in one area with a specific soil, it can be expected to do equally well with the same or a similar soil elsewhere.

Soil surveys, however, are concerned primarily with the top 2 m of the regolith. For many disposal systems, particularly trench-type sanitary landfills and lagoons, the nature of the underlying unconsolidated material and the depth of the regolith to inert bedrock also must be determined. This is particularly important if the regolith is permeable and chemically active and if the rock is jointed, fractured, or contains other open channels such as tubes in basalt or solution channels in limestone. In addition, the hydrology of the site as it might be affected by the construction and the operation of the disposal site must be evaluated. For example, the site may have limited capacity to accept added waste water and the addition of waste water may cause the ground-water level to rise (Keeley, 1972; Parizek, 1973).

Some information on the geology and the hydrology of the site can be obtained from geologic maps and the geologic literature of the area, but careful onsite studies are usually necessary. Onsite studies are also necessary for a detailed evaluation of the soil resource. Soil mapping units of the published soil survey may include small areas of contrasting soils that could not be shown at the scale of a published survey, but that may influence the design of the system or render a site unsuitable. Small areas of shallow soil where

Table 6—Information available from published soil surveys

A. Physical and Chemical Properties of Soils

Soil name and map symbol	Depth	Permeability	Available water capacity	Soil reaction	Shrink-swell potential	Risk of corrosion		Erosion factors	
						Uncoated steel	Concrete	K	T
	Inches	Inches/hour	Inches/inch	pH					
Addicks									
Ad	0–11	0.6–2.0	0.15–0.24	6.1–8.4	Low	High	Low	0.32	5
	11–49	0.6–2.0	0.15–0.24	6.6–8.4	Low	High	Low	0.37	
	49–78	0.6–2.0	0.15–0.24	6.6–8.4	Moderate	High	Low	0.37	
Ak†									
Addicks part	0–11	0.6–2.0	0.15–0.24	6.1–8.4	Low	High	Low	0.32	5
	11–49	0.6–2.0	0.15–0.24	6.6–8.4	Low	High	Low	0.37	
	49–78	0.6–2.0	0.15–0.24	6.6–8.4	Moderate	High	Low	0.37	

B. Engineering Properties and Classifications

Soil name and map symbol	Depth	USDA texture	Classification		Percentage passing sieve number				Liquid limit	Plasticity index
			Unified	AASHTO	4	10	40	200		
	Inches								%	
Addicks										
Ad	0–11	Loam	CL, CL–ML	A–4, A–6	100	95–100	95–100	51–75	20–30	5–14
	11–49	Loam, silt loam	CL, CL–ML	A–4, A–6	95–100	90–100	75–95	60–75	20–40	5–20
	49–78	Laom, silt loam, silty clay loam	CL	A–6, A–7	95–100	90–100	90–100	60–80	25–45	11–27
Ak†										
Addicks part	0–11	Loam	CL, Cl–ML	A–4, A–6	100	95–100	95–100	51–75	20–30	5–14
	11–49	Loam, silt loam	CL, CL–ML	A–4, A–6	95–100	90–100	75–95	60–75	20–40	5–20
	49–78	Loam, silt loam, silty clay loam	CL	A–6, A–7	95–100	90–100	90–100	60–80	25–45	11–27

(continued on next page)

Table 6—Continued.

C. Soil and Water Features

Soil name and map symbol	Hydrologic group	Flooding			High water table		
		Frequency	Duration	Months	Depth	Kind	Months
					feet		
Addicks							
Ad	D	None	--	--	1.0–2.5	Apparent	Jan.–Feb.
Ak†							
Addicks part	D	None	--	--	1.0–2.5	Apparent	Jan.–Feb.

† This mapping unit is made up of two kinds of soil.

Table 7—Data included in computer records for soil survey interpretations†

PA0134   S O I L   S U R V E Y   I N T E R P R E T A T I O N S   STEINSBURG SERIES

MLRA(S): 125, 130, 148, 149
SCE, 8-74
TYPIC DYSTROCHREPTS, COARSE-LOAMY, MIXED, MESIC

THE STEINSBURG SERIES CONSISTS OF MODERATELY DEEP, WELL DRAINED SOILS ON UPLANDS. THEY FORMED IN MATERIAL WEATHERED FROM SANDSTONE AND CONGLOMERATE. TYPICALLY THESE SOILS HAVE A BROWN GRAVELLY LOAM SURFACE LAYER 8 INCHES THICK. THE SUBSOIL FROM 8 TO 15 INCHES IS BROWN SANDY LOAM. THE SUBSTRATUM FROM 15 TO 30 INCHES IS STRONG BROWN GRAVELLY SANDY LOAM. BEDROCK IS AT 30 INCHES. SLOPES RANGE FROM 3 TO 35 PERCENT.

ESTIMATED SOIL PROPERTIES (A)

DEPTH (IN.)	USDA TEXTURE	UNIFIED	AASHO	FRACT >3 IN (PCT)	PERCENT OF MATERIAL LESS THAN 3" PASSING SIEVE NO.				LIQUID LIMIT	PLAS-TICITY INDEX
					4	10	40	200		
0-8	L, SL, FSL	ML, SM	A-4	0-5	95-100	90-100	65-90	35-70		
0-8	GR-L, GR-SL, GR-FSL	SM, ML	A-4	0-15	80-95	65-85	35-60	35-55		
8-15	L, GR-SL, FSL	SM, SC	A-2, A-4	0-10	75-95	65-85	35-60	15-40	<25	NP-5
15-30	GR-SL, GRV-LS	SM, GM	A-2	10-40	45-85	40-80	35-60	15-35	<25	NP-3
30	UWB									

DEPTH (IN.)	PERMEABILITY (IN./HR)	AVAILABLE WATER CAPACITY (IN/IN)	SOIL REACTION (PH)	SALINITY (MMHOS/CM)	SHRINK-SWELL POTENTIAL	CORROSIVITY		EROSION FACTORS		WIND EROD. GROUP
						STEEL	CONCRETE	K	T	
0-8	2.0-6.0	0.10-0.14	4.5-6.5	-	LOW	LOW	HIGH	.28		2
0-8	2.0-6.0	0.10-0.14	4.5-6.5	-	LOW	LOW	HIGH	-		-
8-15	2.0-6.0	0.10-0.14	4.5-6.5	-	LOW	LOW	HIGH	-		-
15-30	2.0-6.0	0.04-0.08	4.5-6.5	-	LOW	LOW	HIGH	-		
30										

HIGH WATER TABLE			CEMENTED PAN		BEDROCK		SUBSIDENCE		HYD GRP	POTENT'L FROST ACTION
DEPTH (FT)	KIND	MONTHS	DEPTH (IN)	HARDNESS	DEPTH (IN)	HARDNESS	INIT. (IN)	TOTAL (IN)		
>6.0					24-40	RIPPABLE	-	-	C	-

FLOODING		
FREQUENCY	DURATION	MONTHS
NONE		

SANITARY FACILITIES (B)

SEPTIC TANK ABSORPTION FIELDS	0-15%: SEVERE-DEPTH TO ROCK 15+%: SEVERE-SLOPE, DEPTH TO ROCK
SEWAGE LAGOON AREAS	0-7%: SEVERE-DEPTH TO ROCK 7+%: SEVERE-SLOPE, DEPTH TO ROCK

SOURCE MATERIAL (B)

ROADFILL	0-25%: POOR-THIN LAYER 25+%: POOR-SLOPE, THIN LAYER
SAND	UNSUITED-EXCESS FINES

SANITARY LANDFILL (TRENCH)	25+%: SEVERE-SLOPE, DEPTH TO ROCK, SEEPAGE
SANITARY LANDFILL (AREA)	0-15%: SEVERE-SEEPAGE 15+%: SEVERE-SLOPE, SEEPAGE
DAILY COVER FOR LANDFILL	0-8%: FAIR-THIN LAYER, SMALL STONES 8-15%: FAIR-SLOPE, THIN LAYER, SMALL STONES 15+%: POOR-SLOPE

COMMUNITY DEVELOPMENT (B)

SHALLOW EXCAVATIONS	0-15%: SEVERE-DEPTH TO ROCK 15+%: SEVERE-SLOPE, DEPTH TO ROCK
DWELLINGS WITHOUT BASEMENTS	0-8%: MODERATE-DEPTH TO ROCK 8-15%: MODERAGE-SLOPE, DEPTH TO ROCK 15+%: SEVERE-SLOPE
DWELLINGS WITH BASEMENTS	0-8%: MODERATE-DEPTH TO ROCK 8-15%: MODERATE-SLOPE, DEPTH TO ROCK 15+%: SEVERE-SLOPE
SMALL COMMERCIAL BUILDING	0-4%: SLIGHT 4-8%: MODERATE-SLOPE 8+%: SEVERE-SLOPE
LOCAL ROADS AND STREETS	0-8%: SLIGHT 8-15%: MODERATE-SLOPE 15+%: SEVERE-SLOPE

REGIONAL INTERPRETATIONS (C)

LAWNS, LANDSCAPING, AND GOLF FAIRWAYS	0-8% NON-GR: MODERATE-DEPTH TO ROCK 0-8% GR: MODERATE-DEPTH TO ROCK, SMALL STONE 8-15%: MODERATE-SLOPE 15+%: SEVERE-SLOPE

GRAVEL	0-15%: POOR-SMALL STONES 15+%: POOR-SLOPE, SMALL STONES
TOPSOIL	

WATER MANAGEMENT (B)

POND RESERVOIR AREA	DEPTH TO ROCK, SLOPE, SEEPAGE
EMBANKMENTS DIKES AND LEVEES	PIPING, LOW STRENGTH
EXCAVATED PONDS AQUIFIER FED	DEEP TO WATER
DRAINAGE	NOT NEEDED
IRRIGATION	SLOPE, ROOTING DEPTH
TERRACES AND DIVERSIONS	SLOPE, DEPTH TO ROCK, ROOTING DEPTH
GRASSED WATERWAYS	DROUGHTY, SLOPE

1/A copy of part of the form (SOILS-5) used for entering data into computer storage.

deep soils were delineated, for example, may create difficulties for trench-type sanitary landfills or pollution hazards for liquid waste disposal systems.

In the design of treatment sites for liquid wastes, other points to consider are the probable loading from rain and snow and the periods when the soil is warm enough to be microbiologically active. In considering climate, the probability and magnitude of extremes, particularly in precipitation, must be carefully evaluated.

VI. CONCLUSIONS

Site selection requires the following steps:

1) Determine kind of waste and method of disposal or utilization.
2) Assess the soil properties and select criteria to determine the suitability of the soil for receiving the waste in question. Various guides are available for rating suitability of soils for receiving many kinds of wastes.
3) Using soil surveys, determine which soils in the area are suited for receiving wastes.
4) Locate the suitable soils on the soil map to determine extent of potential sites.
5) Provide onsite investigations by a soil scientist, hydrologist, and geologist to determine the actual suitability of the potential site for receiving wastes.

LITERATURE CITED

Apgar, M. A., and D. Langmuir. 1971. Ground-water pollution potential of a landfill above the water table. Ground Water 9:76-96.

Clark, J. W., W. Viessman, and M. J. Hammer. 1971. Water supply and pollution control. International Textbook Co., Scranton, Pa.

Elliott, L. F., T. M. McCalla, N. P. Swanson, L. N. Mielke, and T. A. Travis. 1973. Soil water nitrate beneath a broad-basin terraced feedlot. Soc. Agric. Eng. Trans. 16: 285-293.

Flach, K. W. 1973. Land resources. p. 113-119. In Proc. Joint Conf. on Recycling Municipal Sludges and Effluent on Land. Nat. Assoc. of State Univ. and Land Grant Coll., Washington, D. C.

Flach, K. W., and F. J. Carlisle. 1974. Soils and site selection. p. 1-17. In Factors involved in land application of agricultural and municipal wastes. ARS, USDA, National Program Staff, Soil, Water, and Air Sciences, Beltsville, Md.

Gilbertson, C. B., T. M. McCalla, J. R. Ellis, O. E. Cross, and W. R. Woods. 1970. The effect of animal density and surface slope on characteristics of runoff, solid wastes and nitrate movement on unpaved beef feedlots. Univ. of Nebraska, Coll. of Agric. and Home Econ. No. SB 508.

Keeley, J. W., sess. chairman. 1972. Bull session 3-solid waste-its ground water pollution potential. Ground Water 10:27-41.

Kreis, R. D., and L. R. Shuyler. 1972. Beef cattle feedlot site selection for environmental protection. Nat. Environ. Res. Center, Office of Res. and Monit., U. S. Environ. Protect. Agency, Corvallis, Oreg. EPA-R2-72-129.

Loughry, F. G. 1974. The use of soil science in sanitary landfill selection and management. p. 131-139. In R. W. Simonson (ed.) Non-agricultural applications of soil survey. Elsevier Scientific Publ. Co., New York.

McCalla, T. M. 1972. Beef cattle feedlot waste management research in the Great Plains. p. 49-69. *In* Seminar on control of agriculture-related pollution in the Great Plains. Univ. of Nebraska, Coll. of Agric., Lincoln, Nebr. Great Plains Agric. Counc. Publ. No. 60.

Mielke, L. N., N. P. Swanson, and T. M. McCalla. 1974. Soil profile conditions of cattle feedlots. J. Environ. Qual. 3:14-17.

Parizek, R. R. 1973. Site selection criteria for wastewater disposal—soils and hydrogeologic considerations. p. 95-147. *In* Proceedings on Recycling Treated Municipal Wastewater and Sludge Through Forest and Cropland. Univ. Press, The Penn. State Univ., University Park, Pa.

Soil Conservation Service. 1971. Guide for interpreting engineering uses of soils. U. S. Government Printing Office, Washington, D. C. Stock No. 0107-0332.

Soil Conservation Service, USDA. 1973. Guide for rating limitations of soils for disposal of waste. Interim Guide, Advisory Soils-14. Washington, D. C. 26 p.

Soil Conservation Service, USDA. 1975. National soils handbook-Part II. NSH Notice 3-5 May 1975. Washington, D. C.

Soil Survey Staff. 1951. Soil survey manual. USDA Handb. No. 18.

Soil Survey Staff. 1975. Soil taxonomy: A basic system of soil classification for making and interpreting soil surveys. Agric. Handb. No. 436. U. S. Government Printing Office, Washington, D. C.

Sopper, W. E., and L. T. Kardos. 1972. Potential use of spray irrigation in the Tocks Island Regions. Prepared for Delaware River Basin Comm.

Swanson, N. P., L. N. Mielke, J. C. Lorimor, T. M. McCalla, and J. R. Ellis. 1971. Transportation of pollutants from sloping cattle feedlots as affected by rainfall intensity, duration, and recurrence. p. 51-55. *In* Livestock waste management and pollution abatement. Proc. of the Int. Symp. on Livestock Wastes. Am. Soc. Agric. Eng., St. Joseph, Mich. (PROC-271).

U. S. Army Corps of Engineers. 1968. The Unified soil classification system. Military Standard 619B.

chapter 14

A dredge removes the sludge from the lower left basin of a liquid digested sewage sludge storage basin on mine spoil land at Fulton County, Ill. (Photo courtesy of the Metropolitan Sanitary District of Greater Chicago).

Site Design and Management for Utilization and Disposal of Organic Wastes[1]

FRED A. NORSTADT, USDA-ARS, Fort Collins, Colorado

NORRIS P. SWANSON, USDA-ARS, Lincoln, Nebraska

BURNS R. SABEY, Colorado State University, Fort Collins, Colorado

I. INTRODUCTION

Site design and management aspects of waste disposal in soils are dependent upon the nature of the organic waste, climate, and soil type. Objective is also important—whether to recycle plant and animal nutrients or to utilize the soil as a dumping and decomposition ground. Hopefully, we can resolve the problems inherent in the managing and disposing of wastes to our mutual benefit by employing education, understanding, adequate technology, and the proper economic and legal constraints (Wadleigh, 1968). Dollar profit as an operational guideline for decision making, which is a narrow viewpoint, is causing serious problems for well-established businesses and governments. Attitudes toward organic waste, its management and disposal, must be revised sufficiently, but at the same time with flexibility, so that our wastes are accommodated and integrated into our soils, agriculture, and all activities of our national life. To combine flexibility and reassignment of profit in regard to organic waste disposal in soils requires that we accept three rigid conditions and their various interpretations and ramifications: (i) organic wastes should be beneficial or at least innocuous when applied to soil, (ii) they should not be detrimental to public health, and (iii) they should not cause air or drainage-water pollution (Menzies & Chaney, 1974). The field and the associated soil profile must be regarded as an in situ soil column of great areal extent integrated with the earth and with our environment (Lyon et al., 1952).

In general terms, carbon per se will not be a problem since it will be returned to the atmosphere as CO_2 from which it originated. The same is true to some extent for N which can be returned to the atmosphere as N_2 through denitrification. Accumulation of P, K, Na, the heavy metals like Zn, Cu, and Cd, and toxic elements like As, B, Mo, and Se will pose problems severely limiting application of organic waste on soil within the constraints of the three conditions mentioned above (see also Chapters 2, 3, 9, 10, and 11).

Climate is a major factor influencing design and management of soil-associated systems utilizing organic wastes. Long term operations require frequency studies on a specific location to obtain data for determining:

1) The probabilities of favorable temperatures, moisture, and sunshine to permit application and decomposition of wastes.
2) The probability of coincident soil and weather conditions suitable for timeliness of required operations of cropped land.
3) The improbabilities of destructive or grossly unfavorable conditions for waste utilization. All climatic mean magnitudes must be interpreted against the natural daily weather, which is highly variable (Brooks, 1961).

Wang and Barger (1961) classified 11,000 articles on agricultural meteorology. Applicable local climatic data must be used in decision making.

[1] Contribution of USDA, Agricultural Research Service, Western Region, in cooperation with the Colorado and Nebraska Agricultural Experiment Stations.

Current weather data is available from the National Weather Service. Many valuable articles relating to climate were published in the USDA Yearbook, *Climate and Man* (U. S. Department of Agriculture, 1941).

The proper design and management of sites for either utilizing wastes or confinement animal feeding requires control of movement of possible pollutants. Wastes or their products can be transported by air or water. Materials may be moved by air transport as particulates and gaseous compounds; water transport may be by overland flow, percolation to the ground water, or return flows. Transport by overland flow moves chemicals that are in solution, as well as those which occur in the tissues of microorganisms, or in organic materials, in suspended soil sediments and bedloads. Percolating water transports some chemicals in solution, particularly NO_3^-, Cl^-, and associated ions of Na and K (Mathers & Stewart, 1974; Viets, 1974a, 1974b), and possibly some pathogens and microorganisms.

Dry climates and those with continuing wind movement with high velocity gusts require special precautions for animal confinement areas, solid waste storage, or surface spreading of solid wastes subject to movement by wind. Periods of continuing rainfall, particularly during warm weather, may result in anaerobic decomposition of solid wastes and the generation of malodors. Air transport considerations must not be neglected in design and management. Drifting snow is another air transport problem of concern in many regions. Snow drifts provide concentrations of water, producing effects different from those of uniformly distributed precipitation. Windbreaks can alleviate many air transport problems.

Transport by overland flow results from snowmelt or rainfall. Snowmelt produces maximum runoff if the soil surface is frozen. Application of either solid or liquid wastes on frozen soil requires special design and management considerations (Zwerman et al., 1974). Rain on frozen soils creates an equally or more serious condition for runoff. In most areas of the United States, the most serious transport of pollutants occurs from rainfall runoff. Soil erosion initiated by rainfall is a combination of detachment of solid particles by raindrop impact and transport of solids and solutes by overland flow. Water erosion of soil by rainfall on an average annual basis was quantitatively expressed by Zingg (1940) as a function of land slope and slope length, infiltration rate, and physical properties of the soil. Ayres (1936) stated that the amount, intensity, and duration of rainfall has a profound effect on the amount and rate of the resultant runoff, as does the elapsed time since the preceding rain. Frequency and intensity of precipitation are most important considerations in design and management of animal confinement and waste utilization areas. Rainfall frequency data for the U. S. was prepared by Hershfield (1961).

Soil is the other major factor governing design and management of systems for utilization of organic wastes. The criteria for site selections must be examined and integrated with all of the physical, chemical, and biological factors of the soil, the waste, and the degradation and reuse of the decomposition products. The objectives of design and management are maximum degradation and utilization of the products within the constraints specified above, but the needed perspective of the system is that we must live and work with this extensive, in situ soil column which cannot be isolated.

Fortunately, we have available not only a large bank of data on wastes, soils, weather, and climate, but also many agencies, bureaus, and businesses manned by people with the expertise to disseminate, interpret, and implement the information to utilize organic wastes and waste waters in our soils. Especially helpful is the technical help provided by the Soil Conservation Service (SCS) and the Agricultural Research Service, USDA, the State Experiment Stations and Extension Services, the U. S. Environmental Protection Agency (EPA), and the departments related to environmental protection within each state.

The National Referral Center for Science and Technology, Library of Congress, is responsible for identifying all significant information resources in the fields of science and technology, acquiring data describing the specialized capabilities of these resources, and providing guidance about their use. Direct assistance is provided to requests for referral service and through the publication of directories and guides in various subject fields. Presently, two directories are available—one in the physical sciences and engineering (Library of Congress, 1971) and one in the biological sciences (Library of Congress, 1972).

II. TYPES OF ORGANIC WASTES SITES AND PRODUCTS

Our complex society is characterized by numerous agricultural, processing, and manufacturing activities. The many production facilities produce large quantities of organic wastes, concentrated at particular sites. In addition, wastes are generated by ordinary living activities of man in the urban complexes.

Pollution problems relating to wastes may arise at the point of waste production, during temporary storage, or at the point of disposal and subsequent degradation in the soil. Ifeadi and Lawhon (1974) proposed a system of environmental analysis to cope with the complexities of feedlot systems. Their approach could be adapted into any organic waste context as follows. Parameters are structured to form an interaction matrix to identify potential areas of pollution problems, and simple models are developed to analyze waste disposal impacts and/or parts of systems of waste disposal. Part of their data analysis is presented in Fig. 1 for illustration. Feedlot systems and processes are listed in the vertical left column, while across the top are listed categories of environmental and parameter impact—all are in a hierarchy of order. Interactions between the two are marked with an X, representing potential for environmental pollution. Identifying potential problems enables isolation of a particular system or process whose design can then be analyzed and corrective measures taken.

In discussing application of mathematical models to waste problem

Level 1. General Ecological and Feedlot Categories

Level 2. Intermediate Environmental and Feedlot Components

Level 3. Specific Environmental and Feedlot Parameters

SURFACE WATER POLLUTION

PARAMETER IMPACT

FEEDLOT SYSTEM & PROCESSES

FEEDLOT SYSTEM & PROCESSES	Biochemical Oxygen Demand (BOD)	Fecal Coliforms	Nutrient as Nitrogen	Nutrient as Phosphorus	Suspended Solids & Turbidity	Toxic Substances	Volatile Solids	Dissolved Solids
FEEDLOT DESIGN	X	X	X	X	X	X	X	X
OPEN FEEDLOTS								
Location Selection	X	X	X	X	X	X	X	X
Residential Area Proximity	X				X			
Ground Water Table Height								
Water Courses	X	X	X	X	X	X	X	X
Lot Slop	X	X	X	X	X	X	X	X
Soil Characteristics	X	X	X	X	X	X	X	X
Feedlots Environmental Design								
Grading Drainage System	X	X	X	X	X	X	X	X
Shades and Windbreaks								
Feeding and Watering	X	X	X	X	X	X	X	X
Density, Area /Animal	X	X	X	X	X	X	X	X
Unpaved Floor	X	X	X	X	X	X	X	X
Paved Floor	X	X	X	X	X	X	X	X

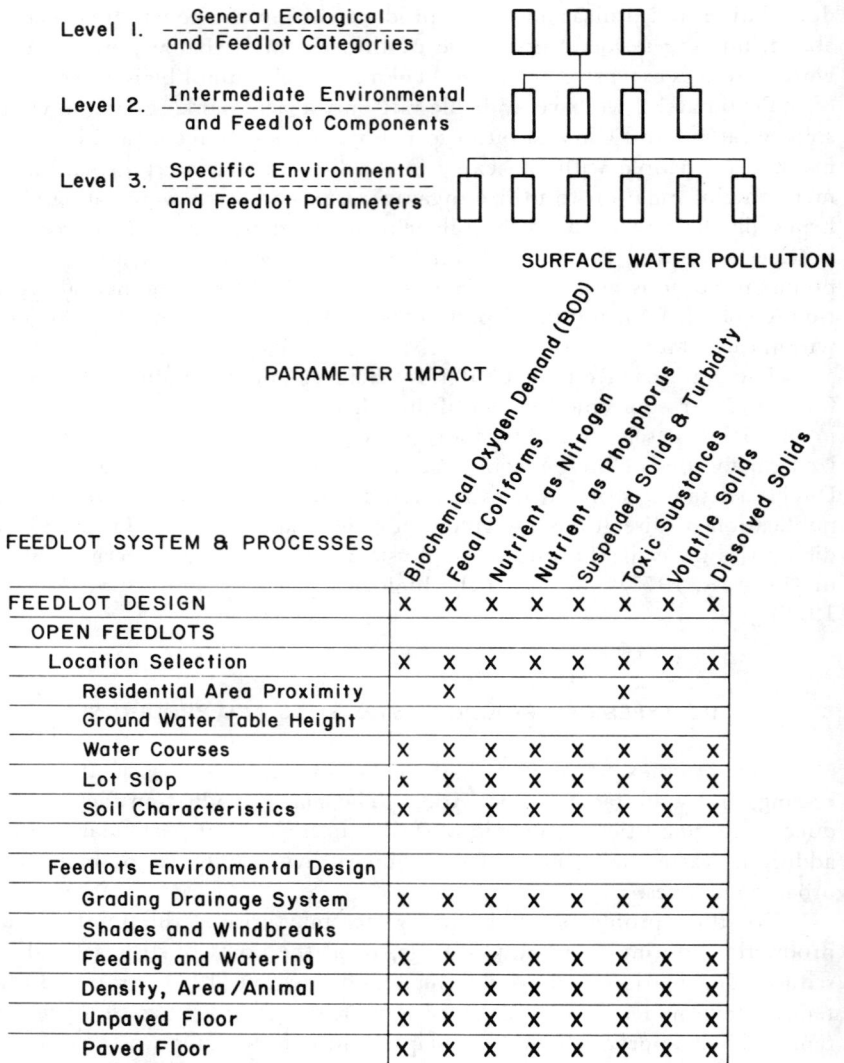

Fig. 1—Environmental impacts of feedlot system and processes (adapted from Ifeadi & Lawhon, 1974).

analysis, Ifeadi and Lawhon (1974) pointed out lack of data or insufficient accuracy for some specific parameters. The hierarchial matrix analysis system can be a valuable aid, however, since a manager of organic waste is interested in identifying what variables his process will affect. Pollution agencies with which the manager must deal are not "system" oriented, and his concern is meeting certain air quality or effluent standards, which are treated as independent, noninteracting variables.

Organic wastes are stored for such reasons as suitability of seasonal application to land because of unfavorable climate or cropping practices, poor weather conditions, economy in handling and transport, and economy in actual application practices. However, storage is at the time when decomposition is likely to be most rapid and anaerobic, leading to obnoxious odors and toxic gases.

At the disposal or utilization site, the processes of degradation and mineralization release elements and chemical compounds into the soil system by a complex series of steps mediated primarily by the soil microbes, but including also macroscopic animal and plant forms in the soil (Alexander, 1961; Waksman, 1966). At this juncture the cropping system is integrated into the plan. In addition, it is necessary to avoid losses of the organic waste by wind movement, overland water flow, and/or percolation. Finally, decomposition processes that generate obnoxious gases, and rates of waste application that cause percolation of NO_3-N or other salts at rates considered harmful to ground-water quality must be avoided.

The Ifeadi and Lawhon (1974) system of environmental analysis and the Bennett and McElroy (1974) analysis of balanced ecological systems for food production have the advantages including all foreseeable factors of organic waste disposal into an integrated system. Logically, any design and operational proposal must be meshed into the total environment insofar as knowledge and skills permit, and later modified as needed.

III. DESIGN ASPECTS OF CONTROLLING PRODUCTS OF POLLUTION POTENTIAL

Because of land costs, availability of suitable land, vagaries of climate and weather, and waste characteristics, the *ideal* combinations of factors for organic waste disposal and decomposition are seldom realized.

Given a suitable site, the design for controlling pollution products must include (i) control of wind movement, (ii) control of snow distribution, (iii) diversion of overland flows from off-site, (iv) control of runoff, transport, and soil erosion from the site, and (v) maintenance of adequate drainage if high water tables or possible surface inundation are problems.

A. Control of Wind Movement

Proper landscaping of a site will provide noise abatement, reduce wind movement and, where applicable, control snow distribution in addition to improving aesthetic considerations. When possible, trees and shrubs can be used for noise abatement (Cook & Van Haverbeke, 1974; Van Haverbeke & Cook, 1972). Woodruff et al. (1959) reported that a field windbreak can moderate summer wind movement and air temperature. Strong winds can move dry and even moist organic materials over bare soil before they can be incorporated. Such windblown materials will accumulate in depressions and

erosion rills and be subject to overland movement and leaching with succeeding rains or irrigation water. The design of natural, living windbreaks, and
recommendations for species and their management can be obtained from the
SCS. Live windbreaks of shrubs and trees would promote soil moisture removal from the disposal area by evapotranspiration. Manmade barriers can
serve in the early development of a disposal area until natural living windbreaks are established and effective. The character of some sites may make
them entirely dependent upon manmade barriers. For example, it is a common practice to use board windbreaks about 2.5 m high as a solid fence
about beef cattle feedlots. Evaporation in adjacent sheltered areas will be reduced by windbreaks and influenced by the porosity of the shelter (Skidmore & Hagen, 1970). A technical study revealing flow patterns, reductions
in velocity, streamlines, forces and velocities, and interpretations of the results to atmospheric conditions was made by Woodruff and Zingg (1952).

Shelterbelts can be utilized to control transport of unpleasant odors to
downwind locations. If this is a site design consideration of serious import, a
competent consultant should be contacted. Restriction of wind movement
can also result in the failure of odors or gases to dissipate under some conditions. Reduction of evaporation might be useful to promote degradation of
organic materials if lack of sufficient moisture or excessive drying were a
problem. Day and night air movement could promote dissipation of odors,
but cool, and therefore heavy, night air might move odor-laden air into a
critical area.

B. Control of Snow Distribution

Manmade barriers, usually snow fences, are effective for influencing
snow distribution (Martinelli, 1973; Price, 1961; Pugh, 1950). Barriers to
control snow distribution must be properly placed to avoid unwanted drifts
and excessive volumes of melt water concentrated in certain parts of the disposal area. Well-distributed snowmelt will minimize transport of organic materials and eroded soil by overland flow. Several aspects relating to the
spacing interval for supplemental shelterbelts were studied by Woodruff
(1956) in a wind tunnel and found to agree rather well with a prototype belt.
Configuration of slat fences and their design and spacing were studied by
Skidmore and Hagen (1970) and Woodruff (1954). Management of snowmelt with different kinds of barriers and level benches was reported by Willis
and Haas (1971) to minimize transport and promote infiltration.

C. Diversion of Overland Flows

Overland flows originating offsite create the same problems as runoff
originating onsite. Diversions can be utilized to protect an area or a structure
from runoff. They may also divert water out of active gullies, or shorten the
length of slope to control erosion. A diversion must have an adequate outlet

and be designed for safe velocities with the present or expected vegetation. Diversions subject to cultivation should be designed for velocities expected in bare channels. Normally, recently established, poor to fair vegetation will improve with time, except on extremely poor soil sites (Jacobson, 1961).

Proper site preparation and compaction of soil in construction are required in addition to proper design for diversions. Foster (1973) summarized approved practices in soil conservation. Failure of a diversion on some locations could create more damage than no structure. The SCS has engineering standards and specifications for diversions which can be obtained from local SCS offices. Some SCS handbooks may be purchased from the U. S. Government Printing Office.

D. Control of Runoff, Transport, and Soil Erosion

The control of onsite runoff and transport of chemicals in solution and solids in suspension or as bedloads is of major importance in the design of areas utilizing or accumulating organic wastes. The best topography, soil characteristics, and climatic and weather conditions, are seldom found in one place, and certain land modifications will improve the efficiency of an area.

Control of runoff, soil erosion, and water transport of organic materials on sloping fields can be achieved with terraces. Terraces function by restricting slope length and provide orderly disposal of runoff (Jacobson, 1961). Terraces can be designed to meet specific needs. An adaptation of broad-basin terraces for control of runoff and collection of transported solids has been widely accepted for use in cattle feedlots in Nebraska (Swanson et al., 1973). Level bench terraces are another design, especially useful to impound potential runoff in a large given area and allow infiltration (Black, 1968; Haas & Willis, 1968).

Raindrop impact and runoff flow detach and transport solids. The bed-loads and portions of the suspended solids are almost immediately deposited when flow velocity is decreased or runoff impounded. When runoff is temporarily stored, the accumulating solids reduce storage capacity of the impoundment. Problems arise with collections of organic materials. Depths, 30 cm or more, may dry too slowly to permit ready removal after drainage of the effluent.

Several researchers have reported organic materials in runoff from cattle feedlots. Gilbertson et al. (1972) reported the settling characteristics of materials transported with feedlot runoff. Swanson et al. (1971) studied the transport of solids by runoff on a cattle feedlot. While ultimate separation of transported solids from the effluent is desirable for management, it is not always feasible or necessarily desirable to design and manage a site to fully minimize transport. Installations can be provided to satisfactorily separate solids from effluent for temporary storage or for immediate disposal on soil. Basins associated with terraces and equipped with riser inlets can be used for onsite separation (Swanson et al., 1973; Linderman et al., 1976). Swanson and Mielke (1973) described a flow-through solids trap used for feedlot runoff.

The design of settling basins should be shallow and planned for removal of dry solids when accumulations do not exceed 20 to 75 cm. Deep settling basins, 150 to 200 cm or more, equipped with an overflow pipe permit deep accumulations of solids, but a dragline will be required for solids removal.

If an impoundment is required for storage of runoff before final disposal, the soil must be considered (Chapter 13). Specifications for construction of farm ponds are applicable (Matson, 1961). Consultation with SCS technicians may be desirable, particularly if both soil problems and a potential for ground-water pollution exist.

In many instances where pollution control is involved, the SCS can provide technical assistance in addition to consultation. Runoff control structures should not be constructed without a competent consultant. Many states have legal requirements for the diversion or impoundment of runoff. Publications are available in many states outlining both regulations for runoff control and suggested methods of compliance (Melvin, 1971).

The U. S. EPA's regulatory and permitting authority, Federal Water Pollution Control Administration, under the Amendments of 1972, applies to discharge of wastes to surface waters of the United States, and is not directly applicable to situations involving land disposal or percolation to the ground water (private communication from L. Edwin Coate, Deputy Regional Administrator, Region X, EPA). However, the EPA does have a vital concern that correction of a surface discharge problem does not result in either land or ground-water degradation. Moreover, in land disposal of organic wastes and waste waters, any resulting runoff that reaches navigable waters is still subject to control by the Non-Point Discharge Elimination System Permit Program, and the applicable effluent limitation guidelines are employed. For example, there are guidelines for disposal of wastes from canned and preserved fruits and vegetables (U. S. Environmental Protection Agency, 1973).

Where there is no runoff from a land disposal system, the states have primary responsibility for control (private communication from L. Edwin Coate, Deputy Regional Administrator, Region X, EPA). For example, Idaho has established basic waste water treatment requirements applicable to land treatment and/or disposal which are published by the Idaho Department of Environmental and Community Services.

E. Maintenance of Adequate Drainage

Ordinarily, high water table and drainage problems are avoided by judicious site selection. Sometimes this will not be possible. Fluctuating water tables resulting in inundation of the soil surface or capillary movement which keeps the soil surface too wet, or interferes with vegetative growth, will create serious management problems. Changed land use on the site, on adjacent land area, or the application of water containing organic wastes in continuing amounts exceeding evapotranspiration may create a high water table and salt problems.

Excess water may be removed from agricultural land by means of open or covered drains (Donnan & Schwab, 1974). Tile drains are used to lower the ground-water table and to provide drainage and aeration needed for proper plant growth (Sutton, 1961). Means for maintaining subsurface drains have been described by the SCS (Soil Conservation Service, USDA, 1971). Ham (1974) discussed the use of drainage wells for control of water tables.

Bouwer (1974) pointed out that the performance of either low-rate or high-rate systems for applying waste water to soil depends on local climate, topography, soil, and hydrology. Unless local experience exists, a small-scale experiment before any large-scale development is recommended. The best designed system will still be a failure without proper management.

IV. METHODS OF CONTROLLING POTENTIAL POLLUTION PRODUCTS

Once a suitable and practical land site has been selected and a design achieved that is compatible with the organic waste to be applied, there are several other management factors to consider. Beyond transport and application techniques, attention is needed to (i) control onsite runoff and transport of pollutants, (ii) control onsite percolation and movement of pollutants to ground water, and (iii) control onsite airborne pollutants. These three objectives must be integrated with maximum decomposition and degradation of the wastes, recycling of plant nutrients, and conversion of elements into non-available forms (e.g., phosphorus) to maximize the life of the disposal site.

A. Control of Onsite Runoff and Pollutant Transport

The Universal Soil Loss Equation (Smith & Wischmeier, 1962) indicates soil erosion is a function of rainfall, soil properties, slope length and steepness, cropping sequence, and supporting practices. Practices to minimize erosion will likewise control onsite runoff and pollutant transport.

Although essentially nothing can be done to change the amount, distribution, and intensity of natural rainfall, there are measures to reduce its erosiveness—i.e., decrease raindrop impact and splash energy, and decrease the amount and velocity of overland flow—to minimize sediment production (Amemiya, 1970). For waste water application by sprinklers or other means, methods should be chosen and managed to permit as low an impact as possible.

Soil properties affect both detachment and transport processes and can be modified somewhat to promote soil stability, size, shape, composition, and strength of soil aggregates and clods. These properties in turn affect transport, permeability to water, infiltration, crusting, and porosity.

Slope length and steepness affect transport which can be modified by cropping and supporting practices, including terraces. Runoff and transport are inversely related to water infiltration capacity of the soil, and any practice to increase infiltration and subsurface storage is worthwhile. Rough, cloddy surfaces enhance water intake (Moldenhauer & Kemper, 1969) and increase storage capacity.

Vegetative cover and surface mulch are effective means of controlling runoff and erosion (McCalla & Army, 1961; Smith & Wischmeier, 1962). Cover protects against raindrop impact energy, reduces detachment, and lessens surface sealing—all effects leading to high water intake. Mulch creates barriers and obstructions, reducing flow velocity and carrying capacity which reduces transport. Relatively modest reductions in flow velocity give large reductions in erosion rates, since the quantity of material moved is proportional to about the fourth power of velocity (Meyer & Mannering, 1968).

Water repellency, often developed as a result of fires, can curtail infiltration and promote runoff. Soil wettability can be increased by mechanical or chemical means (Osborn & Pelishek, 1964; DeBano, 1969).

Recently, the principles and practices of tillage methods to control erosion have been summarized in *A Handbook for Farmers Conservation Tillage* (Soil Conservation Society of America, 1974). Topics like no-till systems, till planting, chisel plow, strip tillage, and stubble mulching are discussed. A particular waste disposal system may develop runoff that must be controlled and kept out of a receiving stream. For example, initial runoff from cattle feedlots contain high NH_4^+-N, NO_3^--N, and biochemical oxygen demand (BOD) (Swanson et al., 1971). The NH_4^+-N and NO_3^--N contents of runoff decrease with continuing precipitation, indicating rapid leaching from the feedlot surface.

Infiltration rate on the site and the element of concentration for runoff discharges are important considerations. Early delivery of initial runoff to a stream adds serious dimensions to the pollution problem. If runoff has not developed from the adjacent watershed, the receiving water course may be at a low or normal flow resulting in minimum of immediate dilution and the concentrated runoff from the waste site may move downstream as a "slug." Early discharge of feedlot runoff into streams and the resulting pollution in Iowa was described by Dague (1968). Fish kills reported in Kansas by Smith (1965) were described as due to feedlot runoff with high concentrations of pollutants. Smith and Miner (1964) reported pollution of Kansas streams by runoff from feedlots.

Runoff from such events will have to be detained and disposed of on land capable of receiving the effluent. In the Texas high plains region, beef cattle feedlot runoff is commonly applied to the playa basins where the material is retained and does not enter existing water courses (Clark, 1975; Lehman et al., 1970). These unique geologic forms are underlain with essentially impervious materials, which prevent percolation of water and movement of pollutants. Runoff water is thereby disposed of by evaporation.

Feedlot effluent can be a valuable resource. For example, feedlot runoff effluent was applied over 3 years in quantities exceeding irrigation requirements on perennial forage crops in eastern Nebraska by Swanson et al. (1974). Yields were generally increased, and excellent quality forage was produced. Similar studies were conducted with corn and forage sorghums, and no deleterious effects on production or soils were found (Hinrichs et al., 1974; Sukovaty et al., 1974).

B. Control of Percolation and Ground-water Contamination

The environment below ground surface is complex and generally not easily determined. The part of the precipitation or irrigation water reaching the water table tends to be higher in dissolved salts than it was originally. Accumulation of salts in the soil from waste application will inevitably result in downward leaching of salts into the zone of saturation. Assuming no appreciable attenuation, potential pollutants would pass in sequence through: (i) the land surface; (ii) the zone of aeration (the zone between the land surface and the water table); (iii) the zone of saturation (the ground-water reservoir) to a stream; (iv) the stream course; and (v) the sea (LeGrand, 1970). Fortunately, the zone of aeration is active enough to consume most pollutants in normal agricultural practices. However, high rates of application of organic wastes and extended intervals of application tax the capacity of the soil to handle through decay, sorption, and inactivation. In addition, dilution contributes to attenuation of a given pollutant.

Permeability controls the rate of movement of water and associated pollutants. Soil textures ranging from sand to clay may have permeability rates varying by hundreds of times. Hydraulic conductivity ranged from about 1 cm/min for coarse sand (mean particle size, 0.5 mm) to about 0.05 cm/min for loamy sand (65% fine sand, 31% silt, and 4% clay) in a study reported by Holmes et al. (1967). Of course, much finer soil textures, those with a high percentage of clay, will have much lower conductivities. Water and included waste will take preferred paths, flowing readily through permeable zones and shunning or flowing with difficulty through relatively impermeable materials. In the case of a deep water table, percolation through overlying zones of sands, silts, and clays leads to marked attenuation of pollutants. In the more humid areas, frequent precipitation and sizeable water addition with organic wastes leads to mounding of water and continuous subsurface flow to nearby streams.

Nitrate is the ion of chief concern in ground-water contamination because of its mobility. Along with Cl^-, NO_3^- is a good index to study for monitoring pollution of ground water at a site. Phosphorus is strongly sorbed by most soils and is rarely a serious threat. The potential for polluting ground water is great where wastes are concentrated or where thin soils occur on cavernous limestone formations, or where thin, sandy soils occur on fractured rocks. If pollutants are not decayed, sorbed, or inactivated in the zone of aeration, subsequent attenuation is chiefly by dilution in ground water. The polluted zone is more pronounced at the water table than at greater depths, and the polluted zone tends to elongate in the direction of ground-water movement. Careful management and monitoring are desirable to prevent the occurrence of initial pollution, since "die-away" is likely to be very slow, particularly in Midwestern and Plains States and basins in the West (Adriano et al., 1973; Viets & Hageman, 1971).

C. Control of Airborne Pollutants

Fortunately, practices to control runoff and transport of pollutants and control percolation and ground water-pollution also minimize generation of airborne pollutants from a disposal site. If the area is in cultivated crops, injection or covering by disking or plowing might prevent wind dispersal of the organic waste and conservation of N. However, if odors were not a particular problem in the immediate area, leaving a waste, like manure, on the soil surface could promote aerial nitrogen loss by ammonia and amines and minimize the amount converted to nitrate (Salter & Schollenberger, 1939; Viets, 1974b). A judicious site location could promote mixing and dispersal of malodors away from inhabited areas by the prevailing winds to take advantage of air drainage patterns caused by day and night temperature changes.

Promotion of aerobic microbial processes in the upper soil profile will generate innocuous products of decay rather than the malodorous hydrogen sulfide and mercaptans. The upper soil profile can be maintained in an aerobic state by good soil drainage, cropping, cultivation, and regulation of rates of waste and water addition. Soil drainage is an inherent characteristic of a particular soil upon which any imposed artificial drainage techniques are dependent, while cropping, cultivation, and regulation of rates of waste and water additions are management techniques.

V. MANAGEMENT ASPECTS OF AN ORGANIC WASTE DISPOSAL SITE

We must assume the soil-based disposal system is capable of being managed, i.e., it will submit to control; it is governable and tractable. The manager of a land site for organic waste and waste water disposal or recycling must integrate: (i) rate of waste addition, (ii) rate of water addition, (iii) cultivation or tillage, (iv) cropping, (v) harvesting of crops, (vi) specialized amendments, (vii) runoff, (viii) use of settling basins, (ix) drainage, (x) weather, and (xi) monitoring.

A. Rate of Waste Addition

Table 1 summarizes some of the kinds of organic wastes and proposed application. From the standpoint of long-time maintenance, it is best to

Table 1—Rate of organic waste addition to soils

Waste description		Rates of amendments		Nutrients			Reference
Kind	State	Waste†	BOD	N	P	K	
			mg/liter	kg ha^{-1} year^{-1}			
Cattle feedlot manure	Moist solid‡	22	25,600§	300	70	180	Mathers and Stewart (1974)
Cattle feedlot manure	Moist solid¶	45	121,000¶	130	35	116	McCalla et al. (1970)
Cattle feedlot lagoon water	Liquid	30	1,400#	510	130	1,460	Powers et al. (1973)
Cattle feedlot lagoon water	Liquid	30	--	371	73	--	Linderman and Mielke (1975)
Poultry litter	Dry solid	11.2	--	566	188	253	Carreker et al. (1973)
Meatpacking waste water	Liquid	528	1,340	6,500	900	--	Tarquin et al. (1974)
Prepared dinners	Liquid	528	1,900	2,400	1,100	--	Schmidt et al. (1974)
Dressings, sauces, spreads	Liquid	528	2,600	740	580	--	Schmidt et al. (1974)
Meat specialties	Liquid	528	820	2,500	580	--	Schmidt et al. (1974)
Fish and vegetable	Liquid	395	387	1,820	560	--	Chawla (1973)
Potato processing	Liquid	234††	1,442	1,357	--	--	Smith (1976)
Fruits and vegetables	Liquid	300	582‡‡	510	105	--	Adriano et al. (1974)
Cheese and powdered milk	Liquid	150	1,062‡‡	570	510	--	Adriano et al. (1974)

† Rate of waste amendments—metric tons ha^{-1} year^{-1} for solids and cm/year for liquids. (Some values taken from articles abstracted in U. S. Environmental Protection Agency, 1974.)
‡ Cattle feedlot manure is variable in moisture content. This example had 45% water content on oven-dry weight basis.
§ BOD value from Dale and Day (1967).
¶ This example had 85% water content on an oven dry weight basis and analyzed for COD value rather than BOD.
# This example of COD from Kreis et al. (1972).
†† Average values for 2 years of data and COD value instead of BOD.
‡‡ COD value instead of BOD.

Table 2—Annual manure application rates needed to insure 200 kg of available N/ha which compensates for residual effects (adapted from Powers et al., 1974)

Nitrogen in manure	Year of application					
	1st	2nd	3rd	5th	10th	20th
%	metric tons/ha					
3.5	7.6	7.2	7.0	6.8	6.6	6.2
2.5	20.0	14.6	13.8	12.6	11.0	9.4
1.5	38.2	29.6	26.2	24.2	20.6	16.8
1.0	100.0	71.4	62.6	52.6	38.6	27.8

manage a disposal or recycling site as cropped agricultural land. When cattle feedlot wastes were returned at a rate to obtain maximum crop production with minimum nitrate leaching (Mathers & Stewart, 1974), the rate was 22 metric tons ha^{-1} $year^{-1}$, which supplied 300 kg/ha of nitrogen. Only about half of the nitrogen applied with manure was mineralized during the first season and lesser amounts in succeeding years. Table 2 illustrates the residual effect which is taken from a guide for applying beef feedlot manure to cropped fields. The data illustrate two points:

1) If manure is applied annually to the same field, lesser amounts should be used each succeeding year. By the 20th year, only about 28% of the original first-year rate should be applied, using a manure with a low nitrogen (1%) content.

2) A manure of high nitrogen content (3.5%) should be used in decreasing amounts each succeeding year, but by the 20th year, about 80% of the original first-year rate should be applied. Thus, an organic fertilizer of high nitrogen availability is very nearly comparable with an inorganic fertilizer in the relative amount required from year-to-year to achieve a given level of required fertility.

In short, the availability of the plant nutrients and residual effects must be considered when determining rates of organic waste application. Mathers and Stewart (1974) found a high annual rate of 224 metric tons/ha reduced yields, caused high accumulations of NO_3^--N in the soil, and lowered forage quality by NO_3^- accumulation. Their study showed soil receiving 112 metric tons/ha or more had saturated paste extract conductivities exceeding 3 mmhos/cm in the seed zone at seeding. Repeated heavy applications may lead to soil structure deterioration as well as serious plant growth problems. Supplying more N than a crop needs can result in leaching of NO_3^- below the root zone, with possible pollution of ground water. Data in Table 1 include several industrial sources of organic wastes carried in waste waters. At high application rates more nutrients are applied than can be removed by plants. For example, in the study reported by Chawla (1973), N and P were added at 1,820 and 560 kg ha^{-1} $year^{-1}$, respectively. Although nutrient removal efficiencies were 92% for the soluble P and 86% for the organic N with no cover crop, the test period was < 2 months which is much too short for evaluation of the system. Adriano et al. (1974) evaluated a much longer period of use. One site receiving waste water from processing fruits and vegetables had been used for 20 years. A cover crop of quackgrass (*Agropyron repens*) was not harvested, except during the last 2 years. High concentrations of NO_3^- in the soil solutions below the root zones and in underlying ground waters indicated this ion leached to below the root zone. Interpretations of the data suggested that the upper profile binding sites were saturated with P. The general conclusion was that soil infiltration capacities should not be used as the sole criterion for site selection and management. The capabilities of the plant-soil systems to intercept the nutrients should also be considered. For plants to effectively remove nutrients, the standing crop must be harvested and removed from the land.

B. Rate of Water Addition

Table 3 gives several examples of rates of waste water applications to soil used by dairy and food processing plants. Unless the land was managed to obtain a crop return, application rate was governed by infiltration rate and not by BOD or nutrient content of the waste water (Harding & Trebler, 1955). For a spray runoff system on a grassed slope, about 30 to 40% of the water applied ran off and was turned into a water course. In this spray runoff system, removal rates were 81 to 95% for chemical oxygen demand (COD), 73 to 93% for total N, and 65 to 84% for P after 10 years of operation (Bendixen et al., 1969).

Schraufnagel (1962) surveyed the ridge-and-furrow irrigation methods used by milk and dairy plants, meat and poultry processing plants, municipalities, tanneries, and chemical plants. The principal factor in loading was the ability of the soil to transmit water. He considered detailed soil maps and percolation tests to be useful in determining rates. Vegetation did not significantly improve infiltration with coarse-textured soils with high infiltra-

Table 3—Rates of water addition to soils with contained organic wastes

Rate of water addition	Irrigation system	Soil texture	Cover crop	Industrial process	Reference
cm/day					
12.5	Ridge and furrow	--	Yes	Dairy wastes	Harding and Trebler (1955)
15.2	Ridge and furrow	--	Yes	Citrus wastes	Anderson et al (1966)
1.3	Spray on slope	--	Yes	Food processing	Gilde (1970)
13.7	Spray	--	Yes	Food processing	Fisk (1964)
12.7	Ridge and furrow	Loam	Yes	Food processing	Sanborn (1953)
4.2	Spray	--	Woods	Vegetable cannery	Mather (1953)
1.3	Spray	Clay	Yes	Vegetable cannery	Drake and Bieri (1951)
5.5	Ridge and furrow	Sandy Loam	Yes	Vegetable cannery	Drake and Bieri (1951)

tion rate. The nonsignificance of vegetation effect on infiltration when surface stirring and mixing were used is in contrast with Bouwer's report (1974) in which vegetation improved infiltration on coarse-textured soils that were not cultivated.

Some unusual management techniques were used with pulp mill liquors. Voights (1955) reported spraying a total of 565 cm of liquid in 3 weeks at a rate of 27 cm/day and then letting the land rest for 3 years. Native vegetation was left in place. Lagooning had removed much of the semicolloidal lignin. In another example, a liquid waste with high sodium (16.4% of total dissolved solids) and high temperature (85 to 91C) with 11.3% solids was applied at 1.1 cm/week. Eventually the soil structure deteriorated and had to be rejuvenated by gypsum treatment and growing a cover crop (Guerri, 1971).

C. Cultivation and Tillage

Carbon compounds in soil are in continuous flux: organic wastes become substrates for the soil biota and undergo decomposition and synthesis transformations (Russell, 1961). Decomposition rate depends on factors of soil aeration, calcium, and other nutrients, water, and temperature. The more favorable a soil is for animal life (well aerated, adequately supplied with nutrients) the higher the decomposition rate. Poorly drained or acid soils generally have higher organic matter contents than those well aerated and near neutral in reaction. High temperatures within the range favorable for soil life promote organic matter decomposition. Tillage promotes organic matter decomposition by improved aeration, stirring, and mixing, which stimulates the activities of soil, plant, and animal life.

Thus, to promote decomposition of organic wastes at a disposal site, the management system should include soil cultivation alone or in conjunction with a cropping system. Using a ridge-and-furrow method of irrigation, one can have the land disked and refurrowed after sufficient drying following irrigation (Schraufnagel, 1962). Such stirring and mixing not only promotes organic matter decomposition, but would also promote infiltration during the next irrigation cycle. Decomposition of organic wastes requires more or less time, depending on the nature of the waste. In fact, the organic matter content of an arable soil tends to reach an equilibrium value (50 to 100 years under a given set of management conditions. Although regularly adding some organic amendment to a soil temporarily increases its organic matter content, the stimulated microbial and other life in the soil will consume most of the added organic substances, leaving only a small residue in the form of soil humus. The added soil humus improves many soil properties that promote plant growth and soil conservation.

A long term experiment described by Russell (1961) gives an indication of the proportion of added carbon and nitrogen that can remain in a cultivated, cropped soil. Over a period of 50 years, 224 kg/year of N and 4,252 kg ha^{-1} year^{-1} of C were added in the form of barnyard manure to continuous wheat. At the end of the period, one-fourth of the added N and one-eighth of the added C remained in the soil as slowly decomposable humus.

D. Cropping

A disposal site should be vegetated for control of erosion, improved water infiltration (if no cultivation is to be done), uptake of plant nutrients and water, and aesthetic consideration.

Growing and harvesting a crop for the uptake of the trace elements, e.g., Cu, Zn, Fe, and B, would have little effect on the balance of these elements in a disposal soil. However, a good forage crop yield can remove significant quantities of N, P, and K (National Plant Food Inst., 1963). For example, a corn crop of 130 hliters/ha requires 346 kg of N, 58 kg of P, 229 kg of K, and appreciable amounts of Ca, S, and Mg. In addition, the corn will utilize about 56 cm of water.

These nutrient removal values for corn are very modest as compared with the additions of N, P, and K added in 528 cm of waste water from an industry preparing package dinners (Table 1, Schmidt et al., 1974). However, if instead of applying 528 cm of waste water, only 52.8 cm were used to irrigate a corn crop, the N addition would be 240 kg and the P addition would be 110 kg, which more nearly equal crop needs. As mentioned previously, if wastes are applied at rates to supply the crop usage for the element of least content and deficiencies made up with fertilizer amendment, then a given disposal site should be useful indefinitely. Application of water and organic waste containing plant nutrients in excess of crop requirements constitutes land loading or dumping. The soil profile will by its sorbtive nature, retain many substances for a time, but eventually some materials will pass on and contaminate the ground water. Considerations concerning the use of perennials and annuals have been discussed by Sopper (1973), Day (1973), and Loehr (1974). Among the criteria are the following: (i) water requirements and tolerance, (ii) nutrient requirements and tolerance, (iii) optimum soil conditions for growth, (iv) season of growth and dormancy requirements, (v) sensitivity to toxic metals and salts, (vi) nutrient utilization and renovation efficiency, (vii) ecosystem stability, (viii) length of rotation, (ix) insect and disease problems, and (x) demand or market for the product.

Perennial grasses seem to be the most suitable for waste water disposal sites. They have fibrous root systems, are sod forming to aid in erosion control and promote high intake, are tolerant of a wide range of ecological conditions, have a long growing period, and have a high uptake of nutrients.

Table 4 illustrates salt tolerance levels of several common crops. (See also the article by Stewart [1974] regarding salinity problems associated with organic wastes.) Similar wide divergencies exist for most of the other criteria. A SCS plant specialist should be consulted in selecting a suitable crop for a given disposal site.

Table 4—Salt tolerance of selected crops (Powers et al., 1973)

Crop	Electrical conductivity of soil water extract	Salinity level
	mmhos/cm	
Bermudagrass (*Cynodon dactylon* L.)	13	
Barley (*Hordeum vulgare* L.)	12	
Tall wheatgrass (*Agropyron elongatum*)	11	Very high
Sugarbeet (*Beta vulgaris* L.)	10	
	8	
Wheat (*Triticum aestivum* L.)	7	
Tall fescue (*Festuca eliator* L.)	7	High
Sorghum (*Sorghum bicolor* Moench.)	6	
Soybean (*Glycine max* [L.] Merr.)	5.5	
Corn (*Zea mays* L.)	5	Medium
	4	
Alfalfa (*Medicago sativa* L.)	3	
Potato (*Solanum tuberosum* L.)	3	Low
Orchardgrass (*Dactylis glomerata* L.)	3	

Table 5—Quantities of nutrients removed by crops irrigated with 5 cm of sewage effluent weekly (adapted from Sopper, 1973)

Nutrient	Crop and nutrient removal		
	Reed canarygrass (*Phalaris arundinacea*)	Corn silage	Hardwood forest
	kg ha^{-1} year^{-1}		
Nitrogen	458	180	94
Phosphorus	63	47	9
Potassium	277	144	29
Calcium	50	30	25
Magnesium	45	26	6

E. Harvesting Crops

Obviously, if a standing crop is not cut and removed from the disposal site, most of the elements taken up from the soil will be returned as the vegetation decays. Of course, the crop will remove water from the soil profile, and perhaps more efficiently, if not harvested.

As the tonnage of crop removed increases, the quantity of mineral elements removed from the disposal site increases. Crops vary widely in nutrient uptake as shown in Table 5.

Forage species differ in their accumulation of NO_3^--N. For example, Long et al. (1975) found pearl millet, variety Gahi-1, (*Pennisetum americanum* [L.] K. Schum) contained six times as much NO_3^--N as Abruzzi rye (*Secale cereale* L.). The plots had received 45 metric tons/ha (dry weight basis) of dairy cattle manure each spring for 3 years, incorporated into the top 15 cm of soil. The manure was of medium N content (1.8%) and 807 kg/ha of N were applied each year. Coastal bermudagrass does not accumulate NO_3^--N like millet. Often, varieties within species exhibit differences as large as those between species (Crawford et al., 1973).

The NO_3^- content of forage is of concern because forage containing more than about 0.4% NO_3^--N may be toxic to ruminants (Bradley et al., 1939). The rye forage studied by Long et al. (1975) did not exceed about 0.1% NO_3^--N during the 3-year period, but the millet showed a gradual build-up of nitrate amounting to 0.2%/year NO_3^--N, and by the third year exceeded the limit if fed as the sole roughage. Forages fed fresh or as cured hay would need to be checked for NO_3^--N, but ensiling forages will lower NO_3^--N during fermentation (Crawford et al., 1973).

F. Specialized Soil Amendments

Soil assimilative capacity is related to microbial, chemical, and physical reactions in soil layers. A certain quantity of each waste can be applied per unit of soil depth or volume of soil infiltrated. Acceptable waste loading rates and undesirable reactions are known only in general terms; data are needed which can be used for engineered use of land disposal.

By extrapolating from results of experiments on the influence of inorganic fertilizers and other amendments on crop yields, one can envision using Ca to counter wastes containing Na from the lye peeling of fruits and vegetables and where salt has been used for food processing and pickling (Loehr, 1974; Wilcox, 1954). There is no convenient method for regenerating ion exchange capacity of soils in the field.

Toxic elements include B, Cd, Co, Cr, Cu, As, Hg, Ni, Pb, and Zn. There is some control of the availability of these to plants through soil conditions affected by amendments. Most of them are more available at pH's below 6.5 to 7.0 (Chaney, 1973) than at higher pH's. Liming can correct acidity in the plow layer. Organic matter can form insoluble precipitates and make the toxic metals less available. However, loss of soil organic matter could release toxic elements in sufficient quantity to become a problem in acid soils. Insufficient information is available regarding metal reversion to make useful management recommendations. Cobalt reversion is related to the quantity of Mn oxides in soil.

G. Runoff

The quantity of wastes that can be applied to soil largely depends upon infiltration rate, since a system may be expected to operate during periods when crop growth is dormant and evapotranspiration is small. If runoff is expected from a disposal site, the quality must be within acceptable limits for immediate discharge to a receiving stream or watercourse. Otherwise the runoff must be detained in some holding structure and reapplied to land or otherwise treated before disposal offsite.

H. Settling Basins

Using lagoons or settling ponds before land application of wastes is common. Removal of silt and suspended materials lessens clogging of the distribution system and soil pores. The BOD of a waste water may be reduced substantially. If appreciable organic material is present, then anaerobic conditions will generate malodors. Trapping suspended solids by a debris basin is feasible in some installations. These basins are often used in handling cattle feedlot runoff (Swanson & Mielke, 1973). An electron acceptor, like $Ca(NO_3)_2$, added to the lagoon can eliminate anaerobic conditions and control odors. Data are needed to quantify and regulate such a practice.

I. Drainage

An underdrain system to overcome adverse subsurface geologic or hydrologic conditions (impervious subsoil or high water table) is a common agricultural practice. Engineering and design assistance can be obtained from competent agricultural engineers and the SCS. Of course, drainage will result in an effluent that must be collected and discharged. Furthermore, removal efficiency of a disposal system depends partly on the detention time in the soil matrix, and water percolation at too high a rate could reduce effectiveness of the system.

J. Weather

Harvesting operations can be disrupted by rainfall if intense, and unusually intense precipitation could overload the hydraulic capacity of a disposal system. Therefore, an adequate safety factor, with weather conditions considered, must be part of every design and management schedule. Seasonal weather factors, like freezing and resultant loss of efficiency due to cold weather, must be considered. Storage is required when freezing temperatures do not permit winter operation or where regulations prohibit application on snow-covered ground. If the disposal site is in cultivated crops, application schedules of water must be adjusted to crop demands, as mitigated by weather, to avoid over-irrigation. Hence, management personnel must have a working knowledge of farming practices as well as principles of waste disposal.

K. Monitoring

Monitoring of the many interrelated systems of an organic wastes disposal system cannot be substituted for full understanding and adequate design beforehand (Blakeslee, 1973). Nevertheless, any disposal plan must have an ongoing monitoring schedule for observing, checking, and in general, monitoring the system. These would include observation of system performance, monitoring the quality of affected natural systems, like the underlying ground water, and evaluating environmental impacts with quality changes.

A detailed summary of monitoring problems and the parameters that need to be followed are discussed in articles by Blakeslee (1973) and Pound and Crites (1973). There are no overall formulas to guide a program since each waste and its disposal system is unique. Monitoring should begin with the quality of the waste to be utilized. Such information confirms on a day-to-day basis that the waste is acceptable and provides a record of land loading. In addition, there should be confirmation of ground-water quality. Suitability for crop utilization and soil compatibility must be ascertained. The health and welfare of workers within the site must be followed if there is body contact with the material or inhalation of aerosols. Public health control might demand adequate disinfection before application.

An effort has been made to develop a type of program to minimize the technical expertise necessary and control costs and still provide effective information. This approach is to use indicator parameters similar to indicator organisms used in water and waste water bacterial testing. Once research has established acceptable limitations with regard to quality of organic wastes, system loading rates, significance of airborne bacterial or vital movement, plant uptake of metals or toxic constituents, and all the other concerns, and a system developed to operate within these constraints, then monitorings of loadings, response, and performance can be readily employed at reasonable cost.

VI. NEEDED RESEARCH AND EDUCATION

The Federal Water Pollution Control Act Amendments of 1972 (U. S. House of Representatives, 1972) have been characterized as the most comprehensive and most complex legislation that has ever been enacted to clean up the Nation's waters. A sweeping Federal-State-local govenrment campaign is mandated, aimed toward reducing, preventing, and eliminating water pollution.

Sections of the act clearly indicate an intention to recycle and reclaim waste waters through production of agriculture, silviculture, or aquaculture products of combinations. But, organic wastes should not be disposed of in a manner causing environmental hazards. Further, facilities involving other industrial and municipal wastes (like solid wastes, waste heat, and thermal discharges) must be integrated with sewage treatment and recycling systems. Significantly, the legislation contains the provision that "The Administrator shall encourage waste treatment management which combines open space and recreational considerations"—acknowledgement of the importance of humane values, as well as environmental quality, energy, and shortages of raw materials.

Another important point arising from House Report No. 92–911 on the 1972 Amendments states [U. S. House of Representatives (1972)]: "There may be no net gain to the Nation if we adopt a technology to improve water quality without recognizing its possible adverse effect on our land and air resources." Obviously, we have three waste disposal alternatives: (i) treatment and discharge to receiving waters, (ii) treatment and reuse, and (iii) irrigation or some other land disposal method. No single treatment or technique can be considered a panacea or the final answer. In considering land disposal, the report pointed out that particular attention be given to treatment and disposal techniques which recycle organic matter and nutrients within the ecological cycle.

An inadequate knowledge of management practices and their effects—and/or a short-sighted unconcern—has contributed to the environmental quality problem. The challenge to us is to demonstrate that we can profit by mistakes.

Using soil to dispose of organic wastes likely cannot depart very far from sound agronomic techniques and practices of soil and water conservation without running into serious environmental problems. Some of the needed research and education can be classified into four categories as follows: (i) promotion of organic matter decomposition, (ii) educating the public and changing attitudes toward organic waste disposal, (iii) waste quality improvement, and (iv) site depletion of amendments.

A. Promotion of Organic Matter Decomposition

Assuming that a waste did not contain excessive amounts of metals or other undesirable substances, rate of addition to a site would be limited only by the capacity of the soil to degrade the organic matter. Conceivably there might be disposal systems which could effectively degrade organic matter at rates beyond those normally applied to arable soils. Techniques to promote decomposition would then be worthwhile.

Many kinds of organic substances degrade at different rates. However, not much is known about the degradation and release of nutrients from the wastes coming from the industrial processing of agricultural products, municipal wastes, and sludges.

Low grade waste heat, not useful for power production, could well promote biological reactions in organic waste decomposition. For example, sewage treatment plants, particularly larger ones, could profitably heat any of the biological treatment stages. Temperatures too low for waste decomposition, due to adverse climatic and weather conditions, might be improved by utilizing waste heat now vented to the atmosphere or streams.

Organic matter decomposing under sufficient aeration results in odorless products of CO_2, H_2O, residual soil organic matter (humus), and inorganic constituents. The general requirements to avoid formation of disagreeable odors and toxic substances are known from agricultural practices. However, the more intensive use of soil to degrade organic wastes will require more refined management and techniques than are currently known.

Modern technology injects into our wastes exotic substances like phenolic compounds, chlorinated hydrocarbons, chlorinated biphenyls, detergent residues, and petroleum products. The decomposition of these substances in the disposal systems, their fate in the soil, and their role in the cropping systems and the environmental must also be known.

B. Public Attitudes

Public attitudes toward organic wastes, their management, and disposition are part of the problem. We need to create new values as well as new technologies. People cannot be changed readily by telling them the facts or by coercion. The public must become involved by their interests or concerns. Detailed and carefully organized educational and information programs, using the assistance of experts, are needed to change the public attitudes toward wastes and the roles of wastes in our lives.

C. Waste Quality Improvement

Some wastes have properties that make them obviously unsuited for land application. Among these qualities are problems of marked acidity or basicity accompanied by a large buffering capacity, pathogens, toxic, and carcinogenic substances. In short, problems are not those developed from excessive amounts of wastes or organic waste itself. With time and experience, modifications of industrial processes and other techniques will be developed and employed which will bring about actual improvement of the quality of the waste that will be taken to the field for disposal.

Examples of waste quality improvement could include separation of storm, sanitary, and industrial sewer systems. Some toxic metals can be removed from industrial wastes at the factory before sewer disposal by ion-exchange techniques. Perhaps techniques to disinfect wastes can be improved to the point that we need not be concerned about health hazards.

If only relatively innocuous ions like Ca^{2+}, Mg^{2+}, Na^+, K^+, Cl^-, SO^{2-}, and HCO_3^- were left in the wastes and waters, possibly relative amounts of these could be adjusted and so diluted as to be harmless to the soils and ground waters for longtime application. Other substances like P, Cu, Zn, Co, Cd, Ni, Cr, Hg, Mn, As, Se, Mo, and Fe have been studied relatively little in relation to waste disposal and health. Recent soil chemistry work showed the importance of specific ion species in the soil solution and the involvement of many dissolution and precipitation reactions (Lindsay, 1973). Research must be done concerning the stability of metal complexes, metal chelates, and the solid phase precipitates.

D. Site Depletion of Amendments

Removal from the disposal site of part of the various components in an organic waste other than by evolution of gases to the atmosphere is by solution in the drainage waters and by harvesting of crops.

Research to improve the usefulness and efficiency of these avenues of loss is needed for organic waste disposal sites to prolong their useful lives. Conditions must be refined for cropping and harvesting of certain crops and management of a given soil and associated wastes.

More must be known about pH limits, tolerances for wooded areas, included solids, and effects of specific organic wastes on crops. Assuming rates of waste addition are adjusted to prevent soil structure and colloid degradation and to prevent ground-water deterioration, what should be the ratios of the many ions applied in the wastes?

We know much about vertical movement of ions and water in soils. However, comparatively little is known about horizontal movement in the water table of bacterial and chemical substances. Eventually, water in the soil moves to stream base flow or into other water bodies, and the substances in solution and suspension must be considered.

LITERATURE CITED

Adriano, D. C., R. S. Ayers, F. T. Bingham, R. L. Banson et al. 1973. Nitrates in the upper Santa Ana River basin in relation to groundwater pollution. Calif. Agric. Exp. Stn. Bull. 861. 59 p.

Adriano, D. C., A. E. Erickson, A. R. Wolcott, and B. G. Ellis. 1974. Certain environmental problems associated with long-term land disposal of food processing wastes. p. 222–233. *In* Processing and management of agricultural wastes. Proc. 1974 Cornell Agric. Waste Manage. Conf. Ithaca, N. Y.

Alexander, M. 1961. Organic matter decomposition. p. 139–162. *In* Introduction to soil microbiology. John Wiley and Sons, Inc., New York.

Amemiya, M. 1970. Land and water management for minimizing sediment. p. 35–45. *In* T. L. Willrich and G. E. Smith (ed.) Agricultural practices and water quality. Iowa State Univ. Press, Ames.

Anderson, D. R., W. O. Bishop, and H. F. Ludwig. 1966. Percolation of citrus wastes through soil. p. 892–901. *In* Proc. 21st Industrial Waste Conf., Purdue University, Lafayette, Ind.

Ayres, Q. C. 1936. Factors affecting rate of erosions. p. 21–22. *In* Soil erosion and its control. McGraw-Hill Book Co., Inc., New York.

Bendixen, T. W., R. D. Hill, F. T. DuByne, and G. G. Robeck. 1969. Cannery waste treatment by spray irrigation runoff. J. Water Pollut. Control Fed. 41:385–391.

Bennett, F. W., and A. D. McElroy. 1974. Analysis of balanced ecological systems for food production. p. 309–319. *In* Processing and management of agricultural wastes. Proc. 1974 Cornell Agric. Waste Manage. Conf. Ithaca, N. Y.

Black, A. A. 1968. Conservation bench terraces in Montana. Trans. ASAE 11:393–395.

Blakeslee, P. A. 1973. Monitoring considerations for municipal wastewater effluent and sludge application to the land. p. 183–198. *In* Proc. of Joint Conf. on Recycling Municipal Sludges and Effluents on Land. Champaign, Ill. U. S. Environ. Prot. Agency, U. S. Dep. of Agric., and Nat. Assoc. of State Univ. and Land-Grant Coll., Washington, D. C.

Bouwer, H. 1974. Hydraulic aspects of liquid-waste loadings. p. 161–177. *In* Factors involved in land application of agricultural and municipal wastes. A preliminary document of the U. S. Dep. of Agric., Agric. Res. Serv., Beltsville, Md.

Bradley, W. B., H. F. Eppson, and O. A. Beath. 1939. Nitrate as the cause of hay poisoning. J. Am. Vet. Med. Assoc. 94:541–542.

Brooks, F. A. 1961. Farm climates and solar energy. p. 817–866. In Agricultural engineers handbook. McGraw-Hill Book Co., Inc., New York.

Carreker, J. R., S. R. Wilkinson, J. E. Box, Jr., R. N. Dawson, E. R. Beaty, H. D. Morris, and J. B. Jones, Jr. et al. 1973. Using poultry litter, irrigation, and tall fescue for no-till corn production. J. Environ. Qual. 2:497–500.

Chaney, R. L. 1973. Crop and food chain effects of toxic elements in sludges and effluents. p. 129–141. In Proc. of Joint Conf. on Recycling Municipal Sludges and Effluents on Land. Champaign, Ill. U. S. Dep. of Agric., U. S. Environ. Prot. Agency, and Nat. Assoc. of State Univ. and Land-Grant Coll., Washington, D. C.

Chawla, U. K. 1973. Treatment of fish and vegetable processing waste-lagoon effluent by soil bio-filtration. p. 103–124. In Proc. Fourth Nat. Symp. on Food Processing Wastes. Syracuse, N. Y. U. S. Envion. Prot. Agency Rep. No. EPA-660/2-73-031.

Clark, R. N. 1975. Seepage beneath feedyard runoff catchments. p. 289–290. In Proc. 3rd Int. Symp. on Livestock Wastes. ASAE, University of Illinois, Urbana-Champaign. 21–24 Apr. 1975.

Cook, D. I., and D. F. Van Haverbeke. 1974. Trees and shrubs for noise abatement. Nebr. Agric. Exp. Stn. Res. Bull. 246. 77 p.

Crawford, R. F., W. K. Kennedy, and M. J. Wright. 1973. Nitrate in forage crops and silage: Benefits, hazards, and precautions. Cornell Univ. Misc. Bull. 37. 12 p.

Dague, R. R. 1968. Animal wastes—a major pollution problem. A paper published in the 50th Anniv. Brochure, Iowa Water Pollution Control Association.

Dale, A. C., and D. L. Day. 1967. Some aerobic decomposition properties of dairy-cattle manure. Trans. ASAE 10:546–548.

Day, A. D. 1973. Recycling urban effluents on urban land using annual crops. p. 155–160. In Proc. of the Joint Conf. on Recycling Municipal Sludges and Effluents on Land. Champaign, Ill. U. S. Environ. Prot. Agency, U. S. Dep. of Agric., and Nat. Assoc. of State Univ. and Land-Grant Coll., Washington, D. C.

DeBano, L. F. 1969. Water repellent soils. Agric. Sci. Rev. 7:11–18.

Donnan, W. E., and G. O. Schwab. 1974. Current drainage methods in the USA. In J. van Schilfgaarde (ed.) Drainage for agriculture. Agronomy 17:93–113. Am. Soc. of Agron., Madison, Wis.

Drake, J. A., and F. K. Bieri. 1951. Disposal of liquid wastes by the irrigation method at vegetable canning plants in Minnesota 1948–1950. p. 70–79. In Proc. 6th Ind. Wastes Conf. Purdue University, Lafayette, Ind.

Fisk, W. W. 1964. Food processing waste disposal. Water & Sewage Works 11:417–420.

Foster, A. B. 1973. Approved practices in soil conservation. Interstate Printers and Publishers, Inc., Danville, Ill. 497 p.

Gilbertson, C. B., J. A. Nienaber, T. M. McCalla, J. R. Ellis, and W. R. Woods. 1972. Beef cattle runoff, solids transport and settling characteristics. Trans. ASAE 15:1132–1134.

Gilde, L. 1970. Food processing waste treatment by surface infiltration. p. 311–326. In Proc. First Nat. Symp. Food Processing Wastes, Portland, Oreg. Fed. Water Qual. Admin. Rep. No. 12060-04/70.

Guerri, E. A. 1971. Sprayfield application handles spent pulping liquors efficiently. Pulp Pap. 45:93–95.

Haas, J. H., and W. O. Willis. 1968. Conservation bench terraces in North Dakota. Trans. ASAE 11:396–398, 402.

Ham, H. H. 1974. Use of drainage wells. Pap. No. 74-2518. December 1974. Am. Soc. of Agric. Eng., St. Joseph, Mich.

Harding, H. G., and H. A. Trebler. 1955. Fundamentals of the control and treatment of dairy wastes. Sewage Ind. Wastes 27:1369–1382.

Hershfield, D. M. 1961. Rainfall frequency atlas of the United States for durations from 30 minutes to 24 hours and return periods from 1 to 100 years. Tech. Pap. No. 40, U. S. Dep. of Commerce, Washington, D. C. 115 p.

Hinrichs, D. G., A. P. Mazurak, and N. P. Swanson. 1974. Effect of effluent from beef feedlots on the physical and chemical properties of soil. Soil Sci. Soc. Am. Proc. 38:661–663.

Holmes, J. W., S. A. Taylor, and S. J. Richards. 1967. Measurement of soil water. In R. M. Hagan, H. R. Haise, and T. W. Edminister (ed.) Irrigation of agricultural lands. Agronomy 11:295. Am. Soc. of Agron., Madison, Wis.

Ifeadi, C. N., and W. T. Lawhon. 1974. An environmental analysis of feedlot systems. p. 108-121. *In* Processing and management of agricultural waste. Proc. 1974 Cornell Agric. Waste Manage. Conf., Ithaca, N. Y.

Jacobson, P. 1961. Mechanics of water erosion. p. 401-413. *In* Agricultural engineers handbook. McGraw-Hill Book Co., Inc., New York.

Kreis, R. D., M. R. Scalf, and J. F. McNabb. 1972. Characteristics of rainfall runoff from a beef cattle feedlot. Robert S. Kerr Water Research Center, Ada, Okla. Rep. No. EPA R2-72-061.

LeGrand, H. E. 1970. Movement of agricultural pollutants with ground water. p. 303-313. *In* T. L. Willrich and G. E. Smith (ed.) Agricultural practices and water quality. The Iowa State University Press, Ames, Iowa.

Lehman, O. R., B. A. Stewart, and A. C. Mathers. 1970. Seepage of feedyard runoff water impounded in playas. Tex. Agric. Exp. Stn. MP-944. 7 p.

Library of Congress. 1971. A directory of information resources in the United States: Physical sciences and engineering. Suppl. Doc. U. S. Government Printing Office, Washington, D. C. 803 p.

Library of Congress. 1972. A directory of information resources in the United States: Biological sciences. Suppl. Doc. U. S. Government Printing Office, Washington, D. C. 577 p.

Linderman, C. L., and L. N. Mielke. 1975. Irrigation with feedlot runoff. p. 26-37. *In* Proc., Nebr. Irrigation Short Course. University of Nebraska, Lincoln. 20-21 January.

Linderman, C. L., N. P. Swanson, and L. N. Mielke. 1976. Riser intake designs for feedlot solids collection basins. Trans. ASAE 19:894-896.

Lindsay, W. L. 1973. Inorganic reactions of sewage wastes in soils. p. 91-96. *In* Proc. of the Joint Conf. on Recycling Municipal Sludges and Effluents on Land. Champaign, Ill. U. S. Environ. Prot. Agency, U. S. Dep. of Agric., and Nat. Assoc. of State Univ. and Land-Grant Coll., Washington, D. C.

Loehr, R. C. 1974. Land disposal of wastes. p. 353-390. *In* Agricultural waste management. Academic Press, New York.

Long, F. L., Z. F. Lund, and R. E. Hermanson. 1975. Effect of soil-incorporated dairy cattle manure on runoff water quality and soil properties. J. Environ. Qual. 4:163-166.

Lyon, T. L., H. O. Buckman, and N. C. Brady. 1952. The soil in perspective. p. 1-4. *In* Nature and properties of soils. The MacMillan Co., New York.

McCalla, T. M., and T. J. Army. 1961. Stubble mulch farming. Adv. Agron. 13:125-196.

McCalla, T. M., L. R. Frederick, and G. L. Palmer. 1970. Manure decomposition and fate of breakdown products in soil. p. 241-255. *In* T. L. Willrich and G. E. Smith (ed.) Agricultural practices and water quality. The Iowa State University Press, Ames, Iowa.

Martinelli, M., Jr. 1973. Snow fences for influencing snow accumulation. p. 1394-1398. *In* The role of snow and ice in hydrology. Symp. on Measurement and Forecasting Procedures. Vol. 2. Banff, Alberta, Canada. September 1972.

Mather, J. R. 1953. The disposal of industrial effluent by woods irrigation. p. 439-454. *In* Proc. 8th Ind. Waste Conf., Purdue University, Lafayette, Ind.

Mathers, A. C., and B. A. Stewart. 1974. Corn silage yield and soil chemical properties as affected by cattle feedlot manure. J. Environ. Qual. 3:143-147.

Matson, H. W. 1961. Water management, conservation use, and legal aspects. p. 492-503. *In* Agricultural engineering handbook. McGraw-Hill Book Co., Inc., New York.

Melvin, S. W. 1971. How to comply with Iowa's feedlot runoff control regulations. Iowa State Univ. Coop. Ext. Serv. Pm-511. 4 p.

Menzies, J. D., and R. L. Chaney. 1974. Waste characteristics. p. 18. *In* Factors involved in land application of agricultural and municipal wastes. A preliminary document of the U. S. Dep. of Agric., Agric. Res. Serv., Beltsville, Md.

Meyer, L. D., and J. V. Mannering. 1968. Tillage and land modification for water erosion control. p. 58-62. *In* Tillage for greater crop production. Am. Soc. of Agric. Eng., St. Joseph, Mich.

Moldenhauer, W. C., and W. D. Kemper. 1969. Interdependence of water drop energy and clod size in infiltration and clod stability. Soil Sci. Soc. Am. Proc. 33:297-301.

National Plant Food Institute. 1963. The fertilizer handbook. Nat. Plant Food Inst., Washington, D. C. 264 p.

Osborn, J. F., and R. E. Pelishek. 1964. Soil wettability as a factor in erodibility. Soil Sci. Soc. Am. Proc. 28:294-295.

Pound, C. E., and R. W. Crites. 1973. Wastewater treatment and reuse by land application. Vol. II. U. S. Environ. Prot. Agency Rep. No. EPA-660/2-73-006b.

Powers, W. L., R. L. Herpich, L. S. Murphy, D. A. Whitney, H. L. Manges, G. W. Wallingford et al. 1973. Guidelines for land disposal of feedlot lagoon water. Coop. Ext. Serv. Bull. C-485, Kansas State University, Manhattan. 7 p.

Powers, W. L., G. Wallingford, L. S. Murphy, D. A. Whitney, H. C. Manges, and H. E. Jones. 1974. Guidelines for applying beef feedlot manure to fields. Coop. Ext. Serv. Bull. C-502, Kansas State University, Manhattan. 11 p.

Price, W. I. J. 1961. The effects of characteristics of snow fences on the quantity and shape of deposited snow. p. 89-98. In Int. Assoc. of Sci. Hydrol. General Assembly Helsinki, IASH Publ. 54.

Pugh, H. 1950. Snow fences. Dep. of Sci. and Ind. Res., Road Res. Lab. Road Res. Tech. Pap. No. 19. His Majesty's Stationery Office, London.

Russell, E. W. 1961. The level of organic matter in soils. p. 290-295. In Soil conditions and plant growth. Longmans Green and Co., Ltd., London.

Salter, R. M., and C. J. Schollenberger. 1939. Farm manure. Ohio Agric. Exp. Stn. Bull. 605. 69 p.

Sanborn, N. H. 1953. Canning, freezing, and dehydration. p. 70. In W. Rudolphs (ed.) Industrial wastes. Reinhold Publ. Corp., New York.

Schmidt, C. J., E. V. Clements III, and J. Farquhar. 1974. Wastewater characterization for the specialty food industry. p. 218-245. In Proc. Fifth Nat. Symp. on Food Processing Wastes. Monterey, Calif. U. S. Environ. Prot. Agency Rep. No. EPA 660/2-74-058.

Schraufnagel, F. H. 1962. Ridge-and-furrow irrigation for industrial wastes disposal. J. Water Pollut. Control Fed. 34:1117-1132.

Skidmore, E. L., and L. J. Hagen. 1970. Evapotranspiration in sheltered areas as influenced by windbreak porosity. Agric. Meteorol. 7:363-374.

Smith, D. D., and W. H. Wischmeier. 1962. Rainfall erosion. Adv. Agron. 14:109-148.

Smith, J. H. 1976. Treatment of potato processing waste water on agricultural land. J. Environ. Qual. 5:113-116.

Smith, S. 1965. Pollution-caused fishkills in Kansas. Kans. State Water News, State Water Resour. Board 8(2):8.

Smith, S. M., and J. R. Miner. 1964. Stream pollution from feedlot runoff. p. 3. In Proc. 14th Sanit. Eng. Conf., University of Kansas, Lawrence.

Soil Conservation Service, USDA. 1971. Maintaining subsurface drains. Soil Conservation Service, U. S. Dep. of Agric. Leafl. No. 557. 8 p.

Soil Conservation Society of America. 1974. A handbook for farmers conservation tillage. Soil Conserv. Soc. of Am., Ankeny, Iowa. 52 p.

Sopper, W. E. 1973. Crop selection and management alternatives—perennials. p. 143-153. In Proc. of the Joint Conf. on Recycling Municipal Sludges and Effluents on Land. Champaign, Ill. U. S. Environ. Prot. Agency, U. S. Dep. of Agric. and Nat. Assoc. of State Univ. and Land-Grant Coll., Washington, D. C.

Stewart, B. A. 1974. Salinity problems associated with wastes. p. 140-160. In Factors involved in land application of agricultural and municipal wastes. A preliminary document of the U. S. Dep. of Agric., Agric. Res. Serv., Beltsville, Md.

Sukovaty, J. E., L. F. Elliott, and N. P. Swanson. 1974. Some effects of beef-feedlot effluent applied to forage sorghum grown on a Colo silty clay loam soil. J. Environ. Qual. 3:381-388.

Sutton, J. G. 1961. Agricultural drainage. p. 356-376. In Agricultural engineers handbook. McGraw-Hill Book Co., Inc., New York.

Swanson, N. P., C. L. Linderman, and J. R. Ellis. 1974. Irrigation of perennial forage crops with feedlot runoff. Trans. ASAE 17:114-147.

Swanson, N. P., J. C. Lorimor, and L. N. Mielke. 1973. Broad basin terraces for sloping cattle feedlots. Trans. ASAE 16:746-749.

Swanson, N. P., and L. N. Mielke. 1973. Solids trap for beef cattle feedlot runoff. Trans. ASAE 16:743-745.

Swanson, N. P., L. N. Mielke, J. C. Lorimor, T. M. McCalla, and J. R. Ellis. 1971. Transport of pollutants from sloping cattle feedlots as affected by rainfall intensity, duration and recurrence. p. 51-55. In Livestock waste management and pollution abatement. Am. Soc. of Agric. Eng., St. Joseph, Mich.

Tarquin, A., H. Applegate, F. Rizzo, and L. Jones. 1974. Design consideration for treatment of meatpacking plant wastewater by land application. p. 107–119. *In* Proc. Fifth Nat. Symp. on Food Processing Wastes. Monterey, Calif. U. S. Environ. Prot. Agency Dep. No. EPA-660/2-74-058.

U. S. Department of Agriculture. 1941. Climate and man. Yearbook of the U. S. Department of Agriculture. U. S. Government Printing Office, Washington, D. C. 1232 p.

U. S. Environmental Protection Agency. 1973. Canned and preserved fruits and vegetables processing industry category. Fed. Regist. 38(216):31076–31084.

U. S. Environmental Protection Agency. 1974. Land application of sewage effluents and sludges: Selected abstracts. U. S. Environ. Prot. Agency Technol. Ser. No. EPA-660/2-74-042.

U. S. House of Representatives. 1972. Federal Water Pollution Control Act Amendments. PL 92-500. 18 Oct. 1972. House Rep. No. 92-911. p. 87–88.

Van Haverbeke, D. F., and D. I. Cook. 1972. Green mufflers. Am. For. 78:28–31.

Viets, F. G., Jr. 1974a. Animal wastes and fertilizers as potential sources of nitrate pollution of water. p. 63–76. *In* Effects of agricultural production of nitrate in food and water with reference to isotope studies. Int. Atomic Energy Assoc., Vienna.

Viets, F. G., Jr. 1974b. Fate of nitrogen under intensive animal feeding. Fed. Proc. 33: 1178–1182.

Viets, F. G., Jr., and R. H. Hageman. 1971. Factors affecting the accumulation of nitrate in soil, water, and plants. USDA Agric. Handb. No. 413, U. S. Government Printing Office, Washington, D. C. 63 p.

Voights, D. 1955. Lagooning and spray disposal of neutral sulfite semi-chemical pulp mill liquors. p. 497–507. Proc. 10th Ind. Waste Conf. Purdue University, Lafayette, Ind.

Wadleigh, C. H. 1968. Introduction. p. 1. *In* Wastes in relation to agriculture and forestry. U. S. Dep. of Agric. Misc. Publ. 1065.

Waksman, S. A. 1966. Microbes and the survival of man on earth. Agric. Sci. Rev. 4: 1–14.

Wang, J. Y., and G. L. Barger. 1961. Bibliography of agricultural meteorology. University of Wisconsin, Madison. 673 p.

Wilcox, L. V. 1954. The quality of water for irrigation use. USDA Tech. Bull. 962. 40 p.

Willis, W. O., and H. J. Haas. 1971. Snow and snowmelt management with level benches, small grain stubble and windbreaks. p. 89–95. *In* Proc. Snow and Ice in Relation to Wildlife Recreation Symp. Iowa State University, Ames, Iowa.

Woodruff, N. P. 1954. Shelterbelt and surface barrier effects on wind velocities, evaporation, house heating, snowdrifting. Kans. State Coll. Tech. Bull. 77. 27 p.

Woodruff, N. P. 1956. The spacing interval for supplemental shelterbelts. J. For. 54: 115–122.

Woodruff, N. P., R. A. Read, and W. S. Chepil. 1959. Influence of a field windbreak on summer wind movement and air temperature. Kans. Agric. Exp. Stn. Tech. Bull. 100. 24 p.

Woodruff, N. P., and A. W. Zingg. 1952. Wind-tunnel studies of fundamental problems related to windbreaks. U. S. Soil Conserv. Serv. Tech. Publ. 112. 25 p.

Zingg, A. W. 1940. Degree and length of landslope as it affects soil loss in runoff. Agric. Eng. 21:59–64.

Zwerman, P. J., S. D. Klausner, and D. Ellis. 1974. Land disposal parameters for dairy manure. p. 211–221. *In* Processing and management of agricultural wastes. Proc. 1974 Cornell Agric. Waste Manage. Conf., Ithaca, N. Y.

chapter 15

Four ton capacity truck applies barnyard manure to replenish organic matter on cut areas following land leveling (Photo by F. P. Mika, USDA-SCS).

Transportation and Application of Organic Wastes to Land[1]

J. R. MINER, Oregon State University, Corvallis, Oregon

T. E. HAZEN, Iowa State University, Ames, Iowa

I. INTRODUCTION

The successful application of wastes to land is dependent upon an effective system for collection of waste at its source and final deposition, after suitable change, to the soil. Numerous alternatives exist to accomplish this task. Selection of the most suitable system is dependent upon waste source, location of the source, and location and nature of the soil to receive the waste. Systems vary in complexity from the pitchfork loading of manure from a stable onto a manure spreader and hauling to an adjacent pasture, to that of domestic sanitary sewage, which is pumped from the source in an underground pipeline to a sewage treatment plant, passed through a variety of physical, chemical, and biological treatment processes, disinfected, metered, stored, and finally applied to cropland through an electronically controlled sprinkler system dependent upon moisture within the soil for its activation.

Presently, management of wastes for land application is in the midst of rapidly changing technologies, affected by such variables as climate, soil type, topography, and production management, and influenced by public attitudes, economics, and local environmental restraints. Consequently, a wide variety of systems are being used and observed, particularly for handling livestock wastes (Loudon et al., 1975). Table 1 illustrates the more common combinations of components being recommended. This chapter, therefore, includes some discussion of selected components in the overall waste management system because they directly influence the selection of transportation and application methods. The end objective is to arrive at a workable system.

All systems are likely to be compromises, each component having its advantages and disadvantages, with the final selection being a series of balances between availability, convenience, cost, labor, etc. In the newer schemes of waste management, system components likely will be adaptations of items not specifically designed for the particular application until research, testing, and manufacturing processes have time to better meet the needs.

Publications that directly relate to land application of wastes, either as a comprehensive reporting on research or as a guide to systems planning, are now beginning to appear (Miner & Smith, 1975; Midwest Plan Service, 1975). Agriculturally oriented professional societies are preparing uniform terminology, standards, and recommendations for systems and components used in the transport and application of wastes to land (American Society of Agricultural Engineers, 1974).

The specific method of waste transport has a considerable influence on the treatment system configuration and the method of land application. Outstanding examples are municipal sewer treatment plant complexes and, more recently, flushing of livestock wastes where these systems are designed around hydraulic transport. On the other hand, whether the wastes are applied to

---

[1]Technical Paper Number 3936. Oregon Agricultural Experiment Station, Corvallis, Oreg.

Table 1—Partial list of livestock waste management systems being used. Lines connect the more common alternatives, but more combinations are used and many more are possible. (Midwest Plan Service, 1975)

A. Covered Confinement Systems

Building type	In-building storage	Transport (conveyor)	Treatment	Disposal or utilization
1) Solid floor	On floor None Deep narrow gutter	On floor Scraper Shovel Flushing gutter Gravity drain	Manure stack Compost aeration Anaerobic lagoon Aerated lagoon	Reuse as flush water Sale as soil additive Irrigation onto cropland
2) Slotted floor Partial Complete	Under-floor tank Oxidation ditch	Slurry pump Tank wagon Sewer line	Anaerobic digester Oxidation ditch	Soil injection Haul to cropland Barriered landscape
3) Suspended cages	On floor, litter no litter Droppings tray Deep pit, liquid dry	Mechanical scrapers Tank wagon Manure spreader	Dehydration	Incineration Refeeding Haul to cropland

B. Nonroofed Confinement Systems

From pen surface	Transport	Treatment	Storage	Disposal or utilization
1) Liquid	Open channels Pipes Culverts	Solids removal Aeration None	Retention pond aerated nonaerated Storage tank	Pump to cropland Haul to cropland Reuse as flush water
2) Solid	Scrapers Front-end loader	None Composting Drying	None Stockpile Trench Silo	Haul to cropland Refeed Incineration Landfill

land as solids or liquids is usually determined by prior selection of storage or treatment method.

Wastes for application to land arise from a variety of man's activities. Municipal refuse, sewage treatment plant effluent, sewage treatment sludge, food processing wastes, and animal manures are among the most common. The chemical, biological, and physical properties of these various wastes are covered elsewhere, but must be considered in selection of transport and application components.

A. Waste Classification

The single most important characteristic of a waste which determines methods of transport and application is moisture content. For the purposes of this chapter, wastes will be classified into three groups according to moisture content.

Solid wastes are defined as those which do not flow hydraulically, and would be typified by municipal refuse, animal manures containing a large quantity of bedding, and food processing refuse such as pea (*Pisum sativum* L.) vines and pods or other vegetable trimmings. When these wastes are managed in such a way as to eliminate the addition of water, their volume is maintained as low as possible and they become more adaptable to different handling techniques than the more fluid materials.

Liquid wastes are defined as those that behave hydraulically in the same manner as water. Typically these include solutions, free of large solids, containing < 5% total dry matter. Among wastes of this type are sewage treatment plant effluent, runoff from cattle feedlots, and the liquid wastes from most processing industries. These materials are managed and handled using normal hydraulic equipment. They are adaptable to use of centrifugal pumps, pipeline transport, and application to land with standard irrigation equipment.

The third classification of wastes is the intermediate moisture range called *slurry*. These wastes typically contain 5 to 15% solids and may be handled hydraulically; however, specialized equipment is required. Characteristic wastes in this classification are sewage sludge, fresh animal manures, and certain food processing wastes. These materials are subject to conversion to either solid or liquid wastes or handling with specialized equipment designed for the material.

B. Climatic Consideration

In addition to the characteristics of the waste material involved, climatic conditions of the area influence the choice of waste transport and application systems. Climatic conditions may dictate that wastes be applied to land only

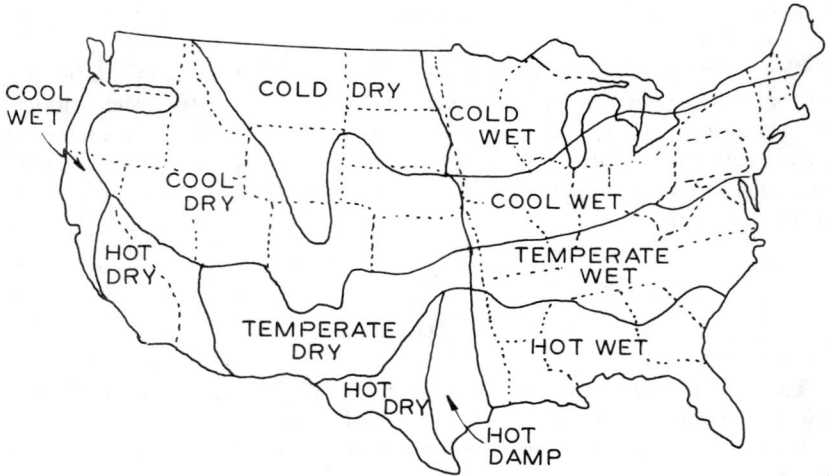

Fig. 1—Climatic zones for waste management selection.

during a restricted period of the year, or may eliminate the possibility of certain treatment systems. As an example, evaporation ponds may be used to dispose large quantities of liquid wastes in warm, dry regions of the country while they would be totally ineffective in areas where rainfall exceeds evaporation. In other locations, application of wastes to land during the winter period is impossible because of freezing conditions, snow cover, or the hazard of runoff when soil is impermeable. Figure 1 is a map of the United States divided into sections according to average temperature and rainfall evaporation data. Examination of this figure demonstrates that waste handling facilities must differ according to location.

With the variety of waste materials requiring application to land, the diverse climatological conditions under which the operation will be practiced, and the diversity of equipment available for use in designing a system, the alternatives available to the planner become innumerable. Table 2 is presented to serve as a guideline in system identification. Not every system will contain a component from each of the columns. In these cases, however, Table 2 will serve as an aid in system review to be certain that no important aspect has been overlooked. In beginning the system design process, assurance must first be obtained that the waste is chemically and biologically compatible with the available soil system. Identifying the waste material in terms of quantity and moisture content is of major importance. Thus, it is possible to classify the material into solid, slurry, or liquid and then to evaluate its availability in terms of being seasonal, sporadic, or continuous. Once these two decisions have been made, it is possible to move through the design process looking at various alternatives and evaluating their applicability.

Table 2—Alternatives for the transport and application of wastes to land

Waste generation	Onsite treatment	Transportation to treatment site	Treatment	Transportation to soil site	Application to land
Physical form	Storage	Truck	Solid-liquid separation	Truck	Irrigation
Solid					Surface
Slurry	Volume reduction	Trailer	Biological stabilization	Wagon	Sprinkler
Liquid		Pipeline	Dehydration	Railroad	Mechanical spreader
Timeliness	Byproduct recovery	Flush	Storage	Pipeline	
Continuous			Sorting		Soil injection
Intermittent			Size reduction		
Sporadic	Constituent removal	Scrape	Dilution	Open channel	
Seasonal					
Hazardous					

II. SOLID WASTES

A. Sources of Solid Wastes

Solid wastes suitable for application to cropland are those materials generated by industrial, agricultural, and municipal activities which yield an organic material that is sufficiently dry that it behaves as a solid, and which is not toxic to soils and plants under the intended conditions. Thus, solid wastes are different than other wastes applied to cropland primarily in their physical form. In many instances, the system planner has the choice of whether to handle a material as a solid waste or to add liquid and handle as a liquid or slurry. By maintaining the material in solid form, the quantity requiring disposal is minimized. Where storage is necessary before application, this has considerable advantage. If particle size is large enough to preclude handling diluted waste material as a liquid, any size reduction should be performed before combining waste flows. Having solid wastes in that form also frequently contributes to byproduct recovery. In this form, many wastes can be classified by hand-sorting or other techniques to recover those solids that have value in other operations.

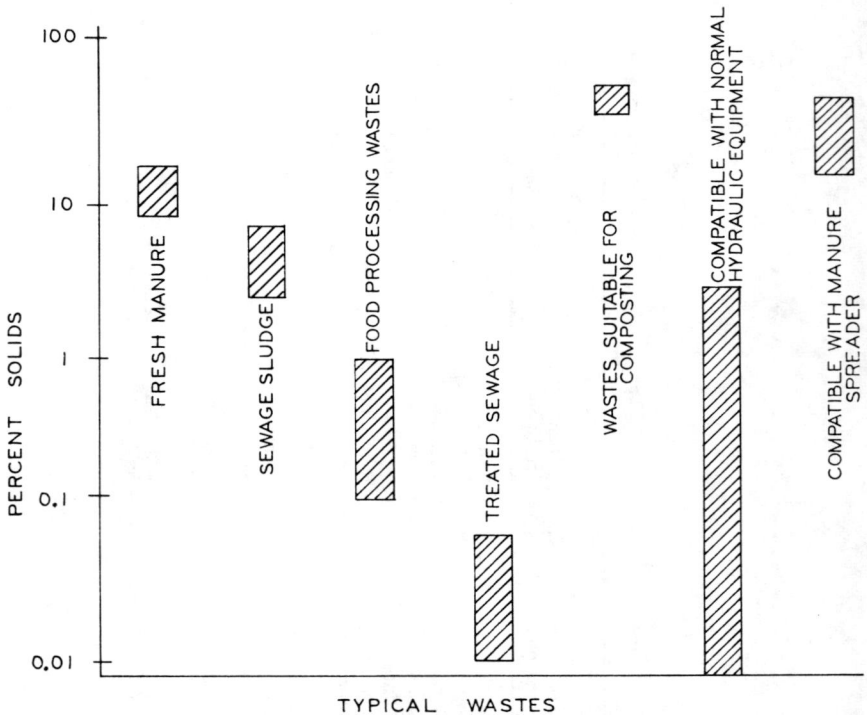

Fig. 2—Range of moisture content typical of wastes from municipal, agricultural, and commercial sources.

1. MUNICIPAL REFUSE

Municipal refuse is an abundant, bulky material produced by a variety of activities within communities and municipalities. It includes a large proportion of household refuse as well as miscellaneous commercial debris and, in some locations, demolition debris and assorted industrial wastes. Typically, it is collected by mobile-packer trucks serving homes, businesses, and industries on a predetermined schedule. Among the important considerations with respect to municipal refuse and its land application are its potential alternate uses. Currently, most municipal refuse is hauled to a sanitary landfill, compacted, and covered. Potential exists for recovery of many resources from municipal refuse. Sorting and separate storage are necessary if the full value is to be recovered. In the past, economics has not favored this approach. A second alternative that is receiving favorable consideration is the utilization of municipal refuse as a heat source suitable for combustion and production of power. Demonstration projects are under way to evaluate the technology of incinerator design and heat recovery.

When being handled for land application, municipal refuse may be pretreated in a variety of ways. Frequently, storage will be necessary to accommodate climatic interferences with regular land disposal. Composting is a second alternative for pretreatment of municipal refuse prior to land application. Refuse frequently is used as the dry component in a mixture of wastes for composting. Municipal sludge and animal wastes have both been mixed with refuse prior to composting.

Another frequent pretreatment process for municipal refuse is shredding and grinding. Large blowing debris can be eliminated and the material becomes more compatible for land application equipment. All these alternatives require that some means of sorting be applied to municipal refuse prior to land application.

2. SEWAGE SLUDGE

Sludge from the treatment of municipal sewage may be dried at the treatment plant to yield a material for land disposal that is solid in nature. Typically, this material is the result of both primary and secondary sewage treatment processes in which the sewage is first settled and then subjected to aerobic biological treatment. Solids from these two operations are then combined and subjected to anaerobic digestion for 30 to 60 days. After digestion, the sludge is frequently discharged to sand drying beds for dewatering. After drying, the sludge is suitable for land application. An alternative is filtering sludge prior to digestion. The latter material is raw sludge and, under some circumstances, has been applied to land.

The disposal or application of raw sewage sludge offers peculiar health and aesthetic problems. This material is odorous, highly biodegradable, and bears the array of pathogenic microorganisms typical of human excreta. Special precautions are required for land application, and for the management of that land following application. Immediate incorporation into the soil should be a requirement for application of this material.

The application of digested sludge is much more common and desirable. Digested sludge may come directly from sand drying beds or may be the filter cake from a filtering operation applied to digested sludge. In either case, moisture content is variable. This sludge may also contain large numbers of weed seeds and undesirable amounts of heavy metals. Even though the anaerobic digestion system is successful in eliminating a large number of pathogenic microorganisms, sludge must still be regarded as a potential mode for the transmission of communicable diseases and must be handled accordingly. Although digested sludge is much less odorous than raw sludge, many people object to the odor of wet digested sludge. Sludges are produced by sewage treatment plants on a nearly continuous basis. Availability of the sludge for application to cropland, however, may depend upon weather conditions, especially if sand drying beds are used. Thus, the disposal of sludge may not be fully compatible with crop needs. Sludge can be stored for application at a more convenient time or it may be applied in less than fully optimal periods.

3. FOOD PROCESSING WASTES

The food processing industry creates a great variety of waste materials, many of them suitable for land application. The material may be inedible plant parts, trimmings of edible materials, or spoiled or unacceptable food. Whatever the source, food processing wastes are intermittent in their generation and highly seasonal in nature. To devise a system compatible with the food processing industry and land management, each application must be regarded separately.

Frequently, food processing wastes are subject to alternate uses which may compete with land application. Many plant materials are being processed for use as livestock feed, industrial raw material, or more recently, potential fuel supply. A prime example has been the ryegrass straw (*Lolium multiflorum* Lam.) of the Oregon Willamette Valley. In the past, ryegrass straw was a solid waste generated in the production of ryegrass seed. Its disposal was achieved by open field burning both to remove the straw and to provide heat treatment to the soil surface. With recent shortages in animal roughages and fuel, ryegrass straw has increased sufficiently in value so that current concern is not for disposal on the field but to minimize the quantity required to achieve satisfactory heat treatment. This has given rise to the need for methods to handle and process the straw to make it useful in areas of higher costs.

4. ANIMAL MANURES

Animal manures are also handled as solid wastes. Fresh animal excreta is typically 10 to 15% solids, which makes it somewhat between the slurry and solid waste classifications. By adding bedding or by natural drying, animal manures frequently assume the physical properties of solid wastes.

As such, animal manures are bulky, subject to odor release when wet, and of variable nutrient quality depending upon handling technique and length of storage prior to land application. Of all the solid waste materials mentioned, animal manures have the longest history of utilization on soils. Their soil enrichment properties are well documented and techniques are well established for the handling of manures from small herds onto adjacent cropland. More challenging problems are arising as systems are devised to produce livestock in larger units or where manure is produced in excess of the land requirement near the livestock facility. Solid animal manures are frequently stored prior to application to land and a small quantity is dried and bagged for sale for home gardening. More importantly, large quantities of animal manures are stockpiled and available to land owners for use as an alternative to commerical fertilizers.

B. Pretreatment Possibilities

Several operations have proved beneficial before wastes are applied to land. These operations can be designed to transform wastes into more easily managed forms, to change physical quality to avoid nuisance problems, to store wastes so they become available at a time when application to land is more advantageous, to recover certain valuable materials from wastes for alternate uses, to mix wastes with other material, thereby making disposal of both more convenient and less expensive. Choice of pretreatment steps is a function of final disposal technique and climatic region in which the operation is to occur. Among the pretreatment operations commonly applied are drying, storage, composting, incineration, shredding and grinding, and sorting.

1. DRYING

Drying is desirable when wastes are to be handled as solids and particularly if they are to be transported long distances. By reducing moisture content from 85% (usual for digested municipal sludge) to 60%, the weight of material to be applied can be reduced by approximately 50%. Typically, this operation is done on sand drying beds following anaerobic digestion of sludges.

In addition to reducing weight, drying enhances other desirable features. Principally, the material loses its tendency to stick to and foul processing equipment and also becomes less odorous and less suitable for insect propagation. Where sludges are to be provided to individual farmers, drying is especially important.

Another technique used to achieve drying is to mix normally wet solid waste with a very dry material such as municipal refuse. This practice has been widely used prior to composting of materials as a moisture adjustment technique.

Mechanical drying has also been practiced. Only those wastes that have a more valuable use than application to cropland can justify the cost of mechanical drying. Drying of animal manures has been practiced on a rather limited scale to provide material suitable for refeeding to animals on an experimental basis and also to meet the limited demand for dried and processed animal manures for sale to home gardeners. In the past, this has produced a high cost material which could not compete with chemical sources of nutrients, and for which other attributes were not sufficient to encourage its purchase in large quantities.

2. STORAGE

Storage is perhaps the most commonly practiced pretreatment technique before applying wastes to cropland. Storage can be used to meet a number of needs, the principal one being to make the material available at a time when it normally is not produced or not produced in a sufficient quantity to justify handling. In much of the country, wastes cannot be applied to land economically during winter and early spring nor can they be applied without danger of runoff and ensuing water quality degradation. In addition, it is frequently desirable to avoid handling wastes during times when other more demanding operations must be done, such as during planting and harvesting.

For storage to be feasible, it must be devised as part of the overall waste handling scheme so that it does not materially add to the operative effort nor greatly increase the cost of waste handling. For solid wastes, provisions generally are made to scrape or dump the material into some type of holding reservoir, which, if climate demands, is roofed and designed so that it may be loaded with a front-end loader or some type of pneumatic loading equipment. Storage units for animal manures have been built with sloping walls and an entry ramp so front-end loaders can be used. An alternate technique is an above-ground manure storage structure into which the manure is scraped or elevated, with the floor sloping in such a direction that any drained liquid will flow into a liquid manure handling system.

3. COMPOSTING

Composting is a process in which volatile solids in garbage, fecal solids, or virtually any other organic waste are digested aerobically. The process differs from conventional aerobic waste treatment in that it takes place at a much larger ratio of solids to water. This low water content allows the development of a loose matrix of material which can be aerated with less mixing than would be required in a liquid system; consequently, biological activity in a viable compost has sufficient heat energy to raise the temperature into the thermophilic zone, 65–75C, without external heat supplies. Maintenance of aerobic, thermophilic conditions is inhibitory to most pathogenic organisms. Because the process is aerobic, it is free of offensive odors and fly breeding does not occur (Howes, 1966). Composting reduces the weight and volume of the original waste and produces an inoffensive, stable material suitable as a soil amendment.

Manures from domestic livestock and other solid organic wastes can be stabilized by composting for 20 days to a condition that will not permit putrefaction, flybreeding, rodent infestation, or pathogen survival (Toth & Gold, 1971; Willson, 1971; Martin et al., 1972). Such stabilization is dependent upon proper management of the composting process. Air must be supplied at such a rate that aerobic conditions are maintained, yet not at a rate that leads to excessive cooling. The structure of a compost must be porous; this requires dilution of raw livestock waste with straw, wood shavings, or dry compost. The best indicators of composting progress are temperature and oxygen level in the core of the compost. Aeration rates and mixing should be controlled to achieve temperatures between 50 and 70C. Mixing and aeration must be more frequent at the beginning of a batch process. Continuous flow composting would be a desirable goal because of ease of controlling environmental conditions, but this is not yet feasible with machines that are economically acceptable. High rate composting (10 to 20 days) will stabilize the organic fraction of manure, but extended storage (2 to 6 months) is required to degrade cellulose, or to reduce moisture content to 10 to 20%.

4. INCINERATION

Incineration is a controlled combustion process for burning solid wastes to gases and a residue containing little or no combustible material. In this regard, incineration is a disposal process because incinerated materials are converted to water and gases that may be released to the atmosphere. However, there are residues from incineration processes requiring further handling. Typically, these materials are applied to land as the final disposal step. Considerable work has been done on the incineration of municipal wastes. As such, incineration may be considered an alternative to land application of wastes; however, as indicated above, if incineration is practiced, there is generally a residue for land application. Generally the volume of municipal solid wastes can be reduced 80 to 90% by incineration. In the process, 98 to 99% by weight of the combustible materials is converted to carbon dioxide and water. Total weight reduction is commonly 75 to 80% based on the weight of the as-charged solid wastes, including moisture. Since the residue is more easily compacted than fresh wastes, further volume reduction results. Typically, if the incinerated material is then hauled to a sanitary landfill, the volume reduction is somewhere in the range of 90 to 96%.

5. SHREDDING AND GRINDING

The processing of solid waste by size reduction has received some application in Europe but only limited studies have been conducted in this country. The principal purpose has been to prepare solid waste for incorporation into sanitary landfills. Advantages are that a more dense fill can be obtained, flying debris problems are minimized, and rat infestations are reduced. Typically, the wastes are shredded in a hammer mill so that particles will pass through a 7.6- to 15.4-cm (3- to 6-inch) mesh screen. This results in a more compactable homogeneous material.

Two hammer mills were studied by Ham et al. (1972). In both cases, the mills were able to grind the municipal refuse satisfactorily to control blowing litter. Other than handling problems getting refuse into the mill, no major difficulties were encountered. Costs for milling and landfilling municipal refuse at that location ranged from $2.12 to $2.80 per ton, exclusive of land charges or administration. Presorting was not required. The milled refuse was used in landfills without cover for up to 4 years without complaints about flying debris, odors, or insects or rodents. Fully loaded trucks were able to drive on the landfill surfaces, which is a major advantage of the system.

Shredding and grinding, although practiced to a relatively small extent, have been indicated as necessary when municipal refuse or other large particulate material is to be applied to land. By so doing, distribution problems are greatly reduced and the obvious aesthetic problem of having identifiable refuse items visible is eliminated. Additional operating experience and cost data are needed, however.

6. SORTING

The sorting of solid wastes to either reclaim valuable byproducts or eliminate materials which would otherwise interfere with the final disposal operation has proven to be expensive and time-consuming. Economics of sorting have not proven attractive and operations involving sorting have been short-lived. Ideally, sorting should take place at the original waste generation site. Early data indicated that American homeowners were not willing to sort refuse by placing bottles and cans in containers separately from biodegradable materials; however, more recent interest in environmental quality and waste management costs are challenging that position. A second factor which may play heavily upon this operation is the ultimate value of reclaimed materials. As metals and wood fiber become more expensive, either from source costs or energy and processing, reclamation of used materials may become more attractive. Recovery processes are already established in certain fields—scrap metal, for example—and others are evolving, such as waste glass, paper, and so forth. Some interest has been exhibited in the use of dry solid waste as an alternative fuel source to replace conventional fossil fuels. Should these uses develop, they potentially could reduce the amount of solid wastes now applied to land.

C. Transportation

Transportation of solid wastes has received little unique attention. The principal goal in the selection of solid waste disposal sites has been to minimize the haul distance. Where distances are short, trucking equipment used for collection has been used for transport to the disposal site. Where more extensive hauls are involved, larger-bodied trucks have been utilized.

Some work has been done relative to transport of manure from generation point to field disposal. For manures having moisture contents < 80%, manure spreaders are the most typical method of transport. In these systems, which generally contain bedding or represent air-dried manure, these function relatively well. Manure spreaders may be either box type or of the open tank design. In addition to serving as the transport vehicle, they also provide a means of uniform distribution of the manure on cropland.

Generally open tank spreaders are built with a shaft mounted near the open top and parallel to the main axis of the tank. Chains on this shaft act as flails when the shaft turns and manure is thrown out of the side of the spreader. Box spreaders may be obtained as pull-type machines or they may be truck-mounted. Pull-type spreaders range in capacity from 2.5 to 18 m^3 (70 to 520 bushels). The spreading mechanism on all spreaders having capacities under 3.4 m^3 (100 bushels) is available with either ground-driven mechanisms or PTO (power takeoff) drives. PTO drives are used on spreaders having capacities of more than 3.4 m^3 (100 bushels). Two or four wheels are used depending on the capacity of the spreader and tire size. Truck-mounted box type spreaders range in size from 6 to 18 m^3 (175 to 524 bushels) capacity. Both steel and wood are used to fabricate the spreader boxes, and both are available in water-tight models. Paddles, flails, and augers are among the devices used for the spreader mechanism. It is common practice to provide variable speed on the apron. Some spreaders have moving front-end gates.

D. Application Technology

The application of solid wastes to land has received considerable attention with regard to disposal of animal manures to cropland. The systems may be divided into surface and soil incorporation techniques. For surface systems, the goal is to provide uniform distribution on the land surface. Manure spreaders and other similar devices which distribute the material across the surface of the soil are used. Application rate is usually restricted to that required for adequate drying so that severe odor problems and insect proliferation are avoided.

In the soil incorporation technique, an effort is made to get the manure mixed with soil immediately after application. Deep plowing equipment was used for application of cattle feedlot manure to cropland in Texas according to Reddell et al. (1972). The equipment consisted of: (i) 76 cm (30 inches) moldboard plowing 75 to 90 cm (30 to 36 inches) deep; (ii) a 45 cm (18 inch) moldboard plowing 53 cm (21 inches) deep; (iii) a trencher with a 68 cm (27 inch) digger wheel working 76 cm (30 inches) deep; (iv) a 127 cm

(50 inch) disk plowing 53 cm (21 inches) deep; and (v) a 30 cm (12 inch) moldboard plowing 35 cm (14 inches) deep. Application rates were from 67 to 200 kg/m^2 (300 to 900 tons/acre) of wet manure per season, equivalent to 35 to 140 kg/m^2 (150 to 600 tons/acre) of dry matter. Corn (*Zea mays* L.) and grain sorghum (*Sorghum bicolor* Moench.) yields were considered excellent for all plots at one site receiving up to 67 kg/m^2 (300 tons/acre) of manure. The 200 kg/m^2 (900 tons/acre) application reduced yields to approximately one-third of the peak yield, although yields on plots receiving 200 kg/m^2 (900 tons/acre) exceeded yields of the check plots. Yields did increase the second and third years after the 200 kg/m^2 (900 tons/acre) manure application.

Cost of manure handling for a 14,000-head cattle feedlot in Los Angeles county was discussed by van Dam and Perry (1968). They estimated mounding in the corral and removal to a compost stockpile cost $0.19/m^3 ($0.25/yard^3). The completed compost if marketed in bulk had a cost of approximately $0.49/m^3 ($0.65/yard^3). If packaged in paper bags of 0.6-m (2-foot) capacity, the processing cost was $1.31/m^3 ($1.88/yard^3).

III. WASTE SLURRIES

Large quantities of waste materials are produced in slurry form; these behave intermediately between solid and liquid forms. When regarded as a liquid, they exhibit high viscosity and do not have some of the desirable handling features of typical liquids. Handled as solids, they exhibit sufficient fluid behavior to be troublesome. Typically, slurries contain from 4 to 15% solids. Certain materials, however, which lie beyond these solid concentration ranges, must be classified as slurries in handling behavior.

Although any number of waste materials may be generated as slurries, the ones encountered for potential land application are domestic sewage sludge and animal manures. The most obvious characteristic of these two wastes is their high viscosity and associated tendency to defy efficient handling with either solid or liquid handling equipment. Sewage sludges, if well digested, are relatively stable with respect to further decomposition and odor production, while animal manures are prone to rapid biological decomposition and associated odor and fly problems.

Viscosity has two effects which influence handling decisions relative to waste slurries. First, viscosity determines how readily the material will flow in a pump suction, and second, it determines the head loss which will be encountered while forcing the material through a closed pipe. Various efforts have been made to classify materials according to solids content, viscosity, and source. Sobel (1966) tried to classify manures as semisolid, semiliquid, and liquid, but was unable to assign any numerical indices. Hart et al. (1966) examined various forms of pumps, and concluded that positive displacement pumps usually failed because of inability of the slurry to enter the suction

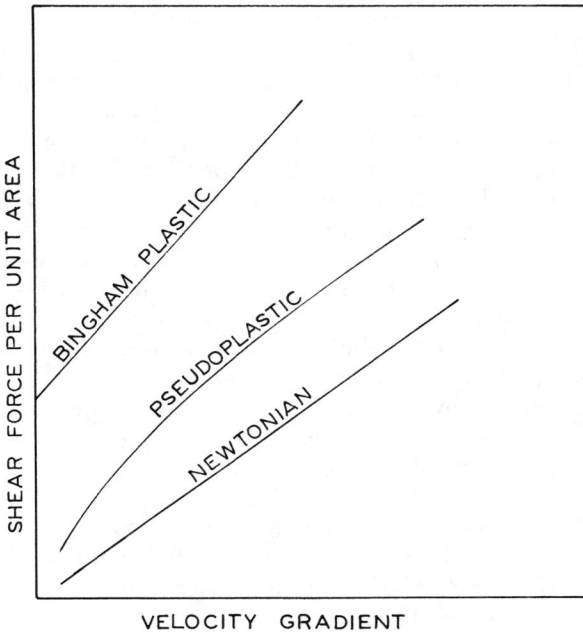

Fig. 3—Definition of flow properties based upon shear force as a function of velocity gradient.

line when the solids content exceeded 15%; centrifugal pumps failed to draw the material when solids content exceeded 7%.

The flow of a fluid into a pump suction is related to its rheological properties. Fluids may be characterized into three classes depending upon the shape of their shear force to velocity gradient curves (Fig. 3). A Bingham plastic has an initial stress to overcome before motion can occur. Babbitt and Caldwell (1939, 1940) have indicated that municipal sludges behave as Bingham plastics. The presence of an initial shear stress would not preclude pumping, but would entail designing adequate pump inlet conditions so that gravity would overcome this initial yield stress. Kumar et al. (1970) used a specifically designed coaxial cylinder viscometer on various samples of dairy manure, with and without sawdust. Their graphs indicated pseudoplastic behavior and did not show any shear stress at very low strain rates. More dilute slurries behave as typical Newtonian fluids and at this point may be regarded as liquids in their handling characteristics. They would, however, have higher viscosities, thus resulting in greater head requirements.

Slurries are frequently pumped from storage containers to tank wagons or other storage. They can also be moved through properly designed pipelines and nozzles for distribution on farmland. Solid-liquid separation occurs when livestock wastes are stored, so slurries must be agitated before pumping. Fibrous solids can plug pumps and nozzles of improper design.

A. Pretreatment Possibilities

The usual handling practices for manure slurries are to either handle them as such or to transform them to solid or liquid forms so more conventional, less costly handling techniques can be utilized. The most common treatment is natural drying, a typical example being open livestock feeding areas in which fresh manure is sun-dried to a solid which can be transported with conventional dirt-moving equipment. The design of cattle feedlots is predicated on the natural drying ability of that particular climate and soil. During wet periods, the cattle loading rate is frequently decreased to avoid an excessive accumulation of moisture on the feedlot's surface. In contrast, cattle density is frequently increased during summer months to increase the moisture content of the manure pack for dust control.

A common alternative for the pretreatment of waste slurries is to mix the waste material with a drier substance to produce a final product that can be handled more easily. Typical of this practice is the use of bedding in dairies, litter in poultry houses, and mixing of sewage sludges with municipal refuse. In all these instances, the dry material is used to absorb the excess water, yielding a material which is then subject to handling with solid waste equipment or manure spreaders. A second application of this technique is to utilize the mixture in a composting operation. With a judicious mixture of waste slurries and dry material, a substance can be produced which is amenable to composting both in terms of moisture content and C/N ratios.

Another common practice for management of waste slurries is to dilute the slurry with water of lower solids content. For example, in manure management systems, flushing is becoming increasingly popular. This is a system in which either clean water or treated waste water is used to dilute and transport the manure from its point of production to other points in the system. Once this has been done, waste can be treated as a liquid according to procedures outlined later.

Storage is a pretreatment concept that is widely applied to waste slurries. By maintaining the waste as a slurry, the additional volume of water required to produce a liquid waste is avoided and the additional volume contribution by adding bedding is also eliminated. Thus, a slurry storage system may be economical and technically feasible where waste must be hauled to cropland. Storage unit design will depend upon type of waste and, for manure, type of animal housing system.

Storage tanks are popular for livestock operations. Most tanks are located under the floor of the livestock building, but they may be located adjacent to or completely away from the barn. Manure is either scraped into the tank through openings in the floor or worked through a slotted floor. Construction design procedures have been developed for manure tanks and are available from the Midwest Plan Service (1971) and the American Society of Agricultural Engineers (1974).

Recommended storage time will vary but 180 days is recommended for northern states. In milder climates where manure may be applied to cropland during a greater portion of the year, shorter storage periods are accept-able. Recommendations from the Texas Agricultural Extension Service sug-gest a 30-day accumulation capacity (Stewart & Sweeten, 1974). The vol-ume of dilution water used in the system is a major factor in controlling the size of manure storage tank required. Although eliminating dilution water will tend to reduce the size of the storage tank, dilution water is often a slurry that is sufficiently liquid to allow pumping.

An additional consideration for selecting the size of manure storage tank has to do with operation of the unit as a livestock production system. For example, if swine are being produced over a slotted floor with manure storage beneath the floor, pumping of the tank should be planned for those periods in which the building is free of animals. Thus, in a farrowing building in which animals are farrowed four times per year, the storage capacity should be sufficient for 40 to 90 days to allow the pit to be emptied between cycles.

An alternative to beneath-the-building storage tanks or separate storage tanks is the use of a manure storage pit. The manure storage pit differs from an anaerobic lagoon in that the goal is to provide storage for a temporary period rather than to achieve a significant degree of biological decomposition. Provision must be made for precipitation that enters the pit during the wet period and thereby decreases the volume available for manure storage. Con-verse et al. (1974) described the use of a manure storage pit in Wisconsin which was 3 to 5 m (10 to 15 feet) deep. Manure was pumped into the pit from a collection sump and stored during the entire winter season. As part of the functional design of the pit, a low-head, high-capacity mixing pump was installed to bring the manure solids into suspension prior to removal. From this kind of operation, the manure could either be hauled to cropland using a tank wagon or, if sufficiently dilute, handled through irrigation equip-ment.

The potential of odor nuisance and safety hazards of manure storage tanks should be noted by anyone contemplating their construction. Upon agitation prior to removal of wastes, considerable quantities of volatile ma-terials may be evolved. Incidents have occurred in which animals were killed because toxic gases were released or sufficient quantities of inert gases re-placed the available oxygen. Large quantities of odorous gases usually are released from animal manure storage tanks and may be a nuisance factor to nearby residents.

B. Transportation

The transport of waste slurries may be accomplished by either mobile tank-bearing equipment or pipelines. Regardless of the transport device, some means of waste agitation is necessary before removing the material from a storage basin or pit. Some wastes can be agitated with ease; others require large energy inputs. For example, hog manure, which generally contains only fine particles, is easily agitated, but cattle manure is frequently difficult.

Agitation devices include paddle wheels with vertical shafts, inclined augers, blow back from vacuum-filled tanks, and submerged centrifugal recirculating pumps. All these devices are acceptable for agitating easily stirred wastes; however, pump recirculation is the only one that has proved dependable for agitating cow manure and other difficult slurries. Paddle wheels, augers, and blow back from tankers have proved effective when used on small tanks. The auger is probably the least effective. If pump recirculation is used, maximum tank dimensions should be 12 m (40 feet). Diaphragm pumps generally do not have enough capacity to be used for slurry agitation.

Slurries in round or square tanks generally are easier to agitate than those in rectangular tanks. In round tanks, pump discharge should be directed across the tank either from the edge or center. In rectangular basins, flow should be directed across the tank rather than along the length. Sewell (1971) made a rather detailed study of the mixing theory involved in slurry storage tanks. He concluded that the configuration of internal baffles was highly important. A rectangular slurry storage tank with a two-to-one length to width ratio with baffles showed appreciably fewer solids remaining on the bottom after agitation than the similar tank without internal baffling.

1. TANK WAGON OR TRUCK

Usually slurries with high solids contents are pumped from storage containers to tank wagons, and on occasion, to other more permanent storage devices. As indicated above, agitation is necessary before removing most slurries from storage tanks. Selection of pumping technique is highly important, otherwise fibrous solids can plug pumps and result in considerable operating difficulty (Table 3). Frequently centrifugal pumps are used to fill tank wagons as well as to provide the necessary agitation in the storage tank. Pumping capacities of up to 0.157 m^3/second (2,500 gallons/min) are used.

Other pumps suitable for transferring heavy slurries from storage tanks to transport vehicles include augers, diaphragm pumps, and vacuum suction hoses. Augers are generally inefficient as movers of slurry but are suitable for small quantities. They are commonly powered with PTO (power takeoff)

Table 3—Preferred type of pump for liquid handling of livestock wastes according to solids content and consistency

Pump	Consistency	Source of waste	% Solids
Diaphragm or helical screw	Thick malted	Poultry and swine Dairy and beef	>25 >12
Centrifugal trash handling	Catsup	Poultry and swine Dairy and beef	20-25 10-12
Open impeller (preferably with chopper)	Pancake batter	Poultry and swine Dairy and beef	10-20 7-10
Standard	Water to buttermilk	Poultry and swine Dairy and beef	<10 < 7

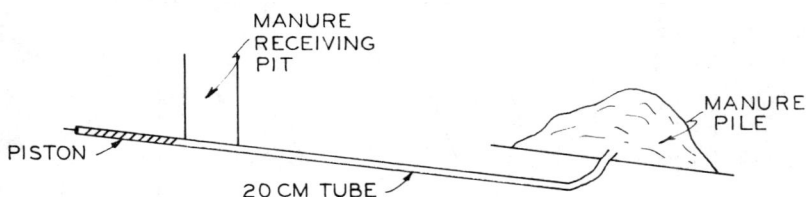

Fig. 4—"Mole hill" concept of manure handling.

driven hydraulic motors. One-to-five horsepower augers generally will pass 0.0025 to 0.0114 m³/second (40 to 180 gallons/min) through 10 to 15 cm (4 to 6 inch) augers.

Diaphragm pumps have low capacities but can handle solids. Taiganides et al. (1964) reported that a 7.6-cm (3-inch), two-horsepower pump would lift 1.6 liters/second (25 gallons/min) against a head of 3.2 m (10.5 feet). Vacuum pumps usually are mounted on the tank wagon itself. A vacuum is created in the tank wagon which pulls the manure slurry from the pit through a 7.6- or 10-cm hose into the tanks. When the tank wagon has reached the field, the air pump is reversed, causing pressure to develop in the tank to discharge the tank contents rapidly.

Another device being used for the movement of waste slurries is the "mole hill" manure handling system developed in Sweden. Within a barn, manure is scraped into a receiving hopper from which it is fed into a 23-cm (9-inch) steel tube, where a piston forces it horizontally through a 20-cm (8-inch) tube to an outside manure pile or lagoon (Fig. 4). If it is moved to a manure pile, the conveyor pipe is turned upward at the pile and the manure is forced up to form a "mole hill." Proponents claim that because the pile is built from below by newly excreted manure, odor and freezing problems are minimized.

Tank wagons or tank trucks are selected whenever it is desirable to transport small quantities of solid waste slurry. Since extra dilution water is not needed, the quantity of wastes to be handled is generally minimized. They represent one of the lower capital investment alternatives for the application of small waste quantities where long haul distance is required, and where control of odors and surface runoff is of high priority. The disadvantage is that tank wagons require more labor than pipeline transport of waste.

Tank wagons are commercially available in capacities from 2.8 to 13.2 m³ (750 to 3,500 gallons). Selections of size depend on storage capacity, speed of travel while emptying, flotation requirements, tractor horsepower available, and accessibility of land for disposal. The tractor horsepower required to pull loaded tanks of various sizes through plowed ground is reported in Fig. 5 (Stewart & Sweeten, 1974). Power requirements are lower for pastures than for plowed fields, and vary with tractor speed and soil moisture content. Distribution systems (pressure or mechanical type) require approximately 10 additional PTO horsepower.

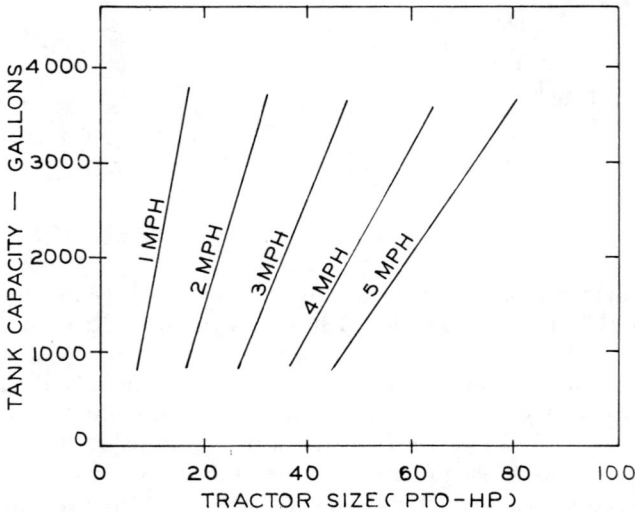

Fig. 5—Approximate tractor size required to pull a manure tank of given capacity through
a plowed field at speeds of 1–5 miles per hour (Stewart & Sweeten, 1974).

2. PIPELINE TRANSPORT

Long distance pumping of slurries is relatively uncommon because costs increase proportionately to distance and this is not generally true for most alternatives. Raw manure is usually handled by spreaders or tank wagons but pipelines are sometimes used for land application of liquid manure or processing wastes through a manure gun. Design data relating to pumping slurries through pipes are limited. For Newtonian fluids, head loss in a circular pipe flowing full can be estimated by the following equation:

$$h = (4fLV^2)/2gD$$

where h = head loss, f = Fanning friction factor, L = length of pipe, V = velocity of fluid, g = acceleration of gravity, and D = pipe diameter.

The value of f can be determined from a table of f values vs. Reynold's number. This approach is valid only for Newtonian fluids, a fluid of known viscosity, and for turbulent flow. Hart et al. (1966) showed that dairy, swine, and poultry manures could be considered equivalent to water with respect to pressure drop if the solids contents were < 2.5, 2.97, and 5.0%, respectively. In all instances, head losses for slurries with higher solids contents than those listed above exceeded those for water.

Staley et al. (1971) measured head losses when pumping dairy manure slurries of 2 to 9% solids through 5.1- and 7.6-cm (2- and 3-inch) aluminum irrigation pipe. They showed that a small quantity of solids, $< 2\%$, can actually reduce head loss to less than that measured for water at the same flow rate. No reasons were given for this phenomenon.

Staley et al. (1973) collected data on the viscosities of dairy manure slurries varying from 5 to 26.6% solids. Flow behavior was non-Newtonian, and the power law (pseudoplastic behavior) model adequately correlated their data. The power law as used in their publication is:

$$\tau = m\Upsilon^n$$

where τ = shear stress, Υ = shear rate, m = a parameter, the consistency coefficient, and n = a parameter, the flow behavior index.

These parameters were correlated, and from these data a generalized Reynold's number and a Fanning friction factor were calculated. Then, utilizing the Darcy formula, the anticipated pressure drop in the pipe is calculable.

C. Land Application

The same techniques are generally used for land application of municipal sludges, food processing slurries, and manure slurries. Application is either on the surface or by incorporation with soil. Where soil surface application is selected, most tank wagons are either fitted with splash plates or other distribution devices, or a large bore nozzle is used to distribute manure from a pressurized transport line. Whichever technique is used, prudent decisions concerning potential runoff and potential odor nuisance must be considered. Where these problems are considered too risky or if greater capture of the nutrient value of the manure is desired, subsurface application is preferred.

1. LARGE BORE NOZZLES (MANURE GUNS)

Large bore irrigation nozzles adapt to handling relatively heavy slurries as well as liquid wastes with lower solids contents. The large sprinklers generally have a capacity of 380 to 1,500 liters/min (100 to 400 gallons/min) and cover from 0.2 to 1.0 ha (0.5 to 2.0 acres) per setting. The large nozzles can often pass up to 2-cm (3/4-inch) diameter solids. Some models use rubber nozzles.

Manure guns can be adapted to hand carry, permanent set, or traveling irrigation systems. Polyvinylchloride (PVC) plastic pipe is suitable for pipeline installation. It has been reported that liquid will drain from portable aluminum irrigation pipe following shutdown, leaving a deposit of solids in the pipe. Later removal of these solids can present a difficult management problem. The key is to avoid this problem by immediate flushing with clear water. Another problem encountered when using the manure gun is the accumulation of solids on the leaves of growing crops. These residues coat the leaves, thereby reducing photosynthesis. When possible, it is desirable to operate the system with clean water for ten to fifteen minutes after each set. This flushes the pipe, washes solids from the foliage, and prevents loss of crops from the pool of concentrated liquid that can result when the sprinkler is detached from the pipe. Several techniques are used to provide the necessary rinse water for the system, e.g., an alternative water supply to provide the necessary water, water directly added to the manure storage pit at the end of the day's operation, or a second smaller centrifugal pump to supply clean water into the irrigation line.

2. SUBSURFACE APPLICATION

Among the techniques available for subsurface application of slurries are direct injection beneath the soil surface or injection ahead of plow blades. Another alternative is to spread manure on land and incorporate it into the soil by deep plowing. Tank wagons, when used for surface application of manure, may be of the reversing pressure design. This results in a rapid discharge of manure and uniform distribution onto the land. This same pump may be used to agitate the manure slurry during transport from the storage tank to the field. In small manure storage tanks, agitation may be effected by rapid discharge of the tank wagon contents under pressure as an alternative to more complex agitation equipment.

Most manufacturers of liquid manure wagons can supply soil injection attachments for injecting liquid wastes into soil. "Plow-furrow-cover" consists of depositing a manure slurry in a 15.2- to 30.5-cm (6- to 12-inch) deep furrow and covering the furrow with a trailing moldboard plow attachment. "Subsod injection" equipment is designed for deposition of liquid manure at high rates beneath a sod surface without appreciable sod disturbance. In either case, odors and flies are effectively controlled, pollution of surface waters through runoff is minimized, and losses of ammonia nitrogen through volatilization are curtailed. Equipment of this type can accommodate slurries of up to 25% solids. Tractor horsepower (hp) requirements for soil injection, as affected by tractor speed and chisel penetration, are shown in Fig. 6. This power requirement must be added to that required to pull the tank wagon. Where vacuum pumps are used, an additional 10 horsepower will be required. For example, if a 7.6 m^3 (2,000-gallon) tank is to be used with two chisel units for 15.2-cm (6-inch) plow down, and field speed is 80 m/min (3 miles per hour), the total tractor horsepower rating should be:

$$towing - 36 \text{ hp (from Fig. 5)}$$
$$plowing - 20 \text{ hp (2} \times 10 \text{ hp, from Fig. 6)}$$
$$pumping - 10 \text{ hp}$$
$$Total - 66 \text{ hp.}$$

An alternative which reduces tractor horsepower is to use three operations. A furrow can be prepared in one operation, the slurry pumped into the furrow in a second, and the furrow covered in the final operation. Normally, the first and last steps are combined so that as a new furrow is

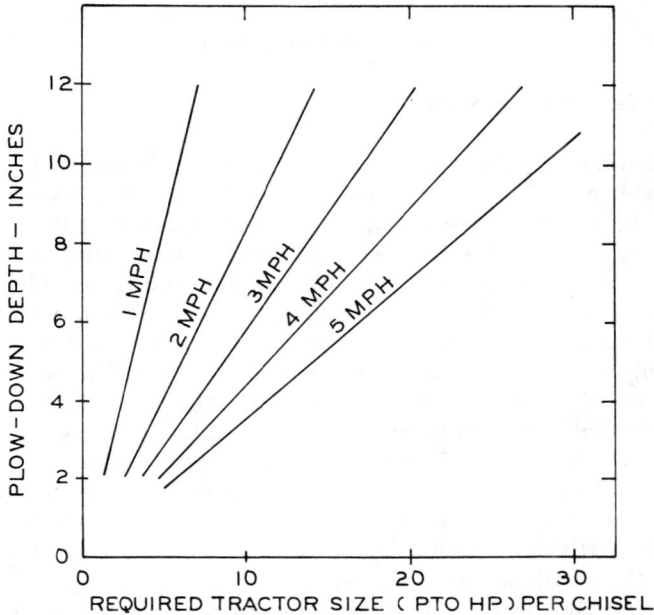

Fig. 6—Approximate tractor size required to plow down manure with chisels in heavy soils and sod at various speeds (Stewart & Sweeten, 1974).

opened, the slurried furrow is covered. Either two tractors are required, or enough open furrow provided for at least a full tank of slurry.

The application of dairy manure using a plow-furrow-cover method was described by Reed (1975). In his experiments, the disposal of 38 to 50 kg/ m^2 (170 to 225 tons) of wet manure per acre was accomplished without odor. The manure was distributed from a tank into 15.2- to 20.3-cm (6- to 8-inch) deep furrows and covered in the same or subsequent operation.

IV. LIQUID WASTES

Liquid wastes are perhaps the most common materials currently being considered for land application. Their hydraulic behavior is similar to water. Typically, handling is with traditional irrigation equipment; however, transportation in tanks also works well although it is more costly. Among the wastes applied in this way are domestic sewage treatment plant effluent, dilute sewage sludge, liquid animal wastes, and food processing waste waters. Each of these has its own unique characteristics and a successful system for application to land will accommodate those particular features.

A. Sources of Liquid Wastes

1. SEWAGE TREATMENT PLANT EFFLUENT

Treated municipal sewage is usually continuous in supply and available in predictable quantities. Except in communities with large industrial developments, the quantity of waste water effluent is well established before a land application system is initiated and changes in quantity are relatively long term in nature. When sewage treatment plant effluent is applied to land, it is relatively dilute. The specific quality of sewage effluent will be a function of the sewage-generating community, its economic activity, amount of water used, and treatment provided the sewage. Typical sewage treatment processes include primary sedimentation and either trickling filter or activated sludge biological treatment, waste stabilization ponds, and frequently, chlorination. The degree of organic matter removal, nutrient removal, and bacterial disinfection achieved by these processes will depend upon their specific design and the operating attention which the devices receive. Effluent characteristics for various types of sewage treatment plants are shown in Table 4 along with raw sewage data.

Of principal concern especially with respect to municipal sewage effluent is that pathogenic organisms are present. Except where adequate chlorination facilities are operated, and sufficient detention time is provided for control of pathogenic organisms, sewage effluent must be regarded as a source of disease-producing microorganisms. As such, operators must be

Table 4—Typical characteristics of untreated sewage and effluent from various types of sewage treatment plants (Pound & Crites, 1973)

Constituent	Untreated sewage	Primary plant	Trickling filter	Activated sludge	Waste stabilization ponds
			mg/liter		
Total dissolved solids		1,402	1,166	917	1,330
BOD$_5$	200	152	17	20	70
Nitrogen					
Total	40	37	16	23	23
Ammonia	25	23	6	17	8
Nitrate	0	0.3	6.3	3.9	0.7
Total phosphorus	10	11	13	13	7
Chlorides†	50‡	461	276	185	138
Potassium†		22	14	20	14
Alkalinity†	100‡	635	491	--	682
Boron†	--	1.2	0.7	0.7	1.2
Sodium†	--	329	267	192	257
Calcium†	--	96	80	52	92
Magnesium†	--	34	50	37	48

† Values dependent upon water supply quality.
‡ Values should be increased by amount in water supply.

given adequate protection and nearby residents must receive assurances that precautions are being taken which will insure their continued health and well-being.

2. SEWAGE TREATMENT PLANT SLUDGE

Sewage sludge is often applied to land as a liquid. It is produced in much smaller quantities than effluent, generally about 1%, but is a much more concentrated material. The composition of sewage sludge is highly variable, since it is based partially on industrial and commercial activities in the source city. Typically, sewage sludge has undergone anaerobic digestion which removes most of the readily biodegradable organic matter. Anaerobic digested sewage sludge is often stored prior to application, such as in a sludge-holding lagoon. Well-digested sludge has little odor and, when the storage lagoon is remote from housing areas, creates little difficulty. Under this system, the sludge is pumped from the storage basin to the land and generally applied with irrigation equipment. Domestic sewage sludge can also be handled as a slurry and applied to land with a tank wagon system similar to those described under slurries.

Particular problems that are encountered with sewage sludges are similar to those mentioned for sewage effluent. The material must be regarded as potentially infectious, although considerable die-off of pathogenic organisms occurs during the digestion process. Depending upon the degree of screening, large solids may be present in the sludge. Regarding application rates, the presence of heavy metals and salts becomes much more critical in sludge application than with most other liquid waste sources. The heavy metal problem is still a developing technology and will be reviewed in greater detail in other chapters.

3. LIQUID ANIMAL MANURES

Animal manures are frequently applied to land as liquids. These liquid wastes arise from either of two sources. Both involve the addition of water to manure to produce the liquid for disposal. When roofed or partially roofed livestock systems utilize liquid manure handling systems, manure is typically removed from the animal confinement area on a daily or more frequent basis and transported to the storage or treatment unit. Additional water is introduced to the manure either to hydraulically remove manure from the unit or from rain water which enters the system in partially roofed buildings. Leaking waters and miscellaneous cleanup water are also sufficient in many operations to render the manure suitable for handling as a liquid. In unroofed systems, the manure for application to land is washed from the feedlot surface by rainfall or snowmelt, carrying with it soluble and suspended materials. This liquid waste source is intermittent and highly variable in quality, depending upon the nature of the runoff-producing event and the condition of

the open lot prior to runoff. To design and operate an economically reasonable system, runoff waste requires some means of storage prior to land application.

4. LIQUID MANURE FROM ROOFED FACILITIES

With the advent of environmentally controlled livestock and poultry
confinement buildings, the quantity and quality of manure requiring disposal
has undergone rapid and dramatic change. Confinement buildings have grown
into major animal production facilities involving large numbers of animals on
very small land area operated with minimum labor input. Under these conditions, mechanized manure collection systems have evolved to a high degree
of refinement. Basically two types of manure collection systems are in use
for liquid manure. One is the hydraulic flushing system, in which fresh water
or treated waste water is used to hydraulically transport manure from the
livestock confinement area to the treatment or storage unit. The alternative
systems are those in which manure is transported from the animal to a storage or treatment device without addition of water, but in which sufficient
water is added or produced during treatment to yield the desired liquid characteristics. In this category are those systems which involve a manure storage
pit beneath slotted floors to which dilution water is added as an aid in removing the accumulated waste material. Another is the scraper system which
operates beneath slotted floor sections and scrapes manure on a frequent
basis into a storage area, flushing channel, or other further handling device.
In either case, the characteristics of the animal manure are changed considerably from those of the fresh excreta. Information presented in Table 5 concerning the characteristics of animal manures is based upon fresh feces and
urine and does not reflect changes in quality accomplished by treatment and
storage techniques.

A transport system must accomplish the following:

1) The manure must be separated from the animal in such a way as to
   yield a healthful environment and promote efficient animal husbandry.
2) The manure must be transported from the point of production to a
   treatment-storage device.
3) The wastes must be treated to make them amenable to final disposal,
   to reduce environmental pollution hazards, or to transform the material in time to a more economic and convenient application point.
4) Adequate application or disposal techniques must be utilized. The
   quantity of liquid manures generated is a function of the amount of
   water used to liquefy the waste as well as the number of animals
   being served by the system. Thus, the storage devices are frequently
   designed on the basis of a demand for hydraulic transport rather
   than animal manure production.

Table 5—Manure production and characteristics†

Parameter		Dairy cow	Beef feeder	Swine feeder	Sheep feeder	Poultry		Horse
						Layer	Broiler	
				kg/1,000 kg live weight§				
Raw manure (RM)‡	wt/day	82	60	65	40	53	71	45
	ratio	2.2	2.4	1.2	1.0	--	--	4.0
Total solids (TS)	wt/day	10.4	6.9	6.0	10.0	13.4	17.1	9.4
	% RM	12.7	11.6	9.2	25	25.2	25.2	20.9
Volatile solids (VS)	wt/day	8.6	5.9	4.8	8.5	9.4	12.0	7.5
	% TS	82.5	85	80	85	70	70	80
BOD_5	wt/day	1.7	1.6	2.0	0.9	3.5	--	--
	% TS	16.5	23	33	9.0	27	--	--
COD	wt/day	9.1	6.6	5.7	11.8	12.0	--	--
	% TS	88	95	95	118	90	--	--
Nitrogen (Total, as N)	wt/day	0.41	0.34	0.45	0.45	0.72	1.16	0.27
	% TS	3.9	4.9	7.5	4.5	5.4	6.8	2.9
Phosphorus (P)	wt/day	0.073	0.11	0.15	0.066	0.28	0.26	0.046
	% TS	0.7	1.6	2.5	0.66	2.1	1.5	0.49
Potassium (K)	wt/day	0.27	0.24	0.30	0.32	0.31	0.36	0.17
	% TS	2.6	3.6	4.9	3.2	2.3	2.1	1.8

† Source: American Society of Agricultural Engineers. Data adapted from Structures and Environment Comm. 412 Rep. AW-D.1. Revised 14 June 1973. Am. Soc. Agric. Eng., St. Joseph, Mich.
‡ Feces and urine with no bedding.
§ The numbers may be interpreted as either pounds/1,000 pounds or kilograms/1,000 kilograms; they are equivalent.

5. MANURE FLUSHING SYSTEMS

Moving livestock wastes with water is an old practice, but automation for confined livestock housing is relatively new. Two early systems of mechanized hydraulic manure removal were applied to dairy housing by Gribble and Bennett (1965) and to swine housing by Smith and Hazen (1967). A theoretical analysis of the process was presented by Jones, Willson, and Schwiesow (1971). After relating manure transport to bedload transport in rivers, they developed a manure channel with a stairstep cross sectional configuration. Their objective was to improve manure transport at low flows by increasing velocity. Hazen, reported by Smith et al. (1971), and Smith et al. (1973), developed a cross sectional channel which improved scouring at low flows. This concept has proven satisfactory under slats. The need for improved scouring in the case of the covered gutter has been suggested by Smith (1971), who observed that in spite of hydraulic deficiencies, the action of the pigs' feet in the open gutter was instrumental in keeping the manure moving.

George (1973) and Smith et al. (1973) published design recommendations for swine-flushing gutters. Both investigators assumed a quasi-steady state flow to allow application of Manning's equation for open channel flow. George (1973) attempted to correlate flushing volume requirements with manure deposition rates, but Smith et al. (1973) limited their discussion to hourly flushing. George (1973) suggested a variable slope and variable width channel, while Smith et al. (1973) specified only one width and slope. George's (1973) approach conserves flushing water, but the approach of Smith et al. (1973) is valid for recycling systems where flow quantities do not alter fresh water usage. Both George (1973) and Smith et al. (1973) follow the 56 m/min (3 feet/second) velocity recommendation of Jones, Willson, and Schwiesow (1971). Smith et al. (1973) specifies a 36 m/min (2 feet/second) velocity in open gutters and 56 m/min (3 feet/second) below slats where animal hoof action is not involved.

All flushing systems require initial discharge at the channel head at such a rate that a steady state velocity of 56 m/min (3 feet/second) or more is attained. Early work by Gribble and Bennett (1965) described the use of an over-center tilting tank and a falling dam of water discharge. Both of these methods are described in relation to dairy cows and milking operations. Jones, Willson, and Schwiesow (1971) and Taiganides and White (1971) used siphon flush tanks for flushing swine wastes. At present, the siphon, falling dam, and tilting tank would seem to be the methods of choice.

Fehr and Smith (1975) reported no flushing underneath poultry cages using renovated and recycled waste water. Satisfactory movement of waste through the flushing trays was obtained. However, the material had to be passed through a screening device and a specially constructed chopper to prevent feathers from clogging the system.

As an alternative to the flushing gutter concept, a system has been devised for flushing between slats. This process was described by Koelliker et al. (1972). Several Midwestern cattle feeders have expanded the concept into a flume system, in which manure enters by falling through widely spaced slats in the floor. Liquid from the waste treatment process is recirculated continuously or at frequent intervals within flumes, rapidly carrying the manure from the confinement building to the waste treatment site. A waste water recovery system to provide adequate flushing liquid is a necessary adjunct of the narrow-slat system, and probably is desirable for other flushing systems.

6. FEEDLOT RUNOFF

A common source of liquid animal wastes is runoff from unroofed livestock units—feedlots. Feedlot runoff is characterized by its intermittent nature and by its variable composition. The first step in feedlot runoff management is to minimize runoff. This is commonly done by minimizing lot surface area, by planning and constructing diversion facilities to prevent uncontaminated waters from crossing manure-covered surfaces, and by constructing runoff channels to convey the runoff to a treatment and storage facility. Prior to discharge into a storage basin, solids are commonly removed from the runoff water by small sedimentation basins ahead of the storage reservoir dams which slow the water and allow solids to settle. The design of feedlots is related to the effectiveness of runoff control facilities. It is desirable to have the lots sloped so that drainage of runoff water is rapid, to minimize wet spots within the feedlot, thereby producing a more desirable confinement area for animals. In addition, maintenance of a dry lot is effective in reducing odors.

Since runoff is produced by precipitation, runoff rates are related to rainfall intensity, and the total volume of runoff to be anticipated is a function of rainfall intensity, duration, and frequency. Maximum intensities are usually experienced for only short times, and data for a 10-year, 1-hour storm are frequently applied as design storms. Total quantity of runoff to be collected is frequently associated with extended rainfall periods when field application is not feasible. The most recent regulation by the Environmental Protection Agency (1975) indicates that by 1983 "runoff and process wastewater may be discharged to navigable streams only when rainfall events, either chronic or catastrophic, cause an overflow of process wastewater from a facility designed, constructed, and operated to contain all process generated wastewaters plus the runoff from a 25-year, 24-hour rainfall event."

Because dewatering of storage basins is not always practical immediately after a storm, Koelliker et al. (1974) considered designs based on a 10-day or chronic wet period. Many state regulatory agencies have made allowances for this condition by requiring storage capacity to contain 90–180 days of accumulated runoff. It is reasoned that the cold season in many cattle-producing states is unsuitable for land application of liquid from storage ponds; this season lasts for 120 to 180 days depending on location. Local precipitation patterns and probability data are the best techniques to use in the design of runoff control facilities.

Various techniques have been used to predict runoff volumes anticipated from particular storms. The "rational equation" used by hydrologists is used to predict peak runoff rates. This equation is:

$$Q = CIA$$

where Q = peak runoff volume, C = runoff coefficient (0.7 to 1.0), I = rainfall intensity for the design recurrent interval and for a duration equal to the time of concentration of the watershed, and A = area of the watershed.

Although discharge rates are important for designing transport channels, the primary concern for most designers is detention pond volume. This volume will have two components: the volume required for storage during periods when dewatering is not possible, and the volume required to contain the design storm as defined by the waste discharge permit. These two volumes are not necessarily additive; if the design storm normally occurs during a dry period, management could be defined so that the reservoir was kept at a low level during this period. The time period for cold season storage normally varies from < 90 days in Oklahoma and other southern states to 180 days in areas of extended freezing conditions.

In predicting the total quantity of runoff from a particular storm, Shuyler et al. (1973) recommend use of the standard SCS equation

$$R = (P - 0.352)^2/(P + 1.41)$$

where R = runoff measured as equivalent rainfall in the feedlot area (inches) and P = actual rainfall (inches). This formula specifies that 0.894 cm (0.352 inches) of rainfall will be stored on a lot surface before runoff occurs. Wise and Reddell (1973) used a regression analysis to show that in Texas

$$R = 0.863 P - 0.458.$$

Data from other states (Gilbertson et al., 1971) can be made to fit an equation of the same form,

$$R = aP - b.$$

Unless more precise data are available, values a = 0.8 and b = 0.5 generally give acceptable results.

Runoff from feedlot snowmelt is different in both quality and characteristics than runoff from rainfall. Sublimation is appreciable on southern exposures. The radiation-absorbing characteristics of snow are markedly influenced by foreign substances such as dust, twigs, and manure solids. Animal action rapidly packs snow on the feedlot to an ice-like density. Deposits

Table 6—Characteristics of feedlot runoff at Pratt, Kans.†

Parameter	Range	Mean	Standard deviation
		mg/liter	
	From rainfall‡		
COD	1,514-14,309	6,111	2,631
N	85-962	494	211
P	19-482	87	89
TS	2,971-17,669	7,528	2,622
VS	1,429-11,437	3,891	1,627
	From snowmelt‡		
COD	7,299-36,764	13,767	8,087
N	590-2,337	1,033	617
P	65-459	209	171
TS	9,282-36,684	19,308	11,425
VS	5,253-23,551	11,620	7,924

† Source: W. J. Fields. 1971. Hydrologic and water quality characteristics of beef feedlot runoff. M.S. Thesis. Kansas State University Library, Manhattan, Kans.
‡ Chemical oxygen demand (COD), kjeldahl nitrogen (N), phosphorus (P), total solids (TS), and volatile solids (VS).

of fresh manure on the surface of the snowpack are particularly vulnerable to transport by overland flow when runoff is produced by melting snow or rainfall. The dark color promotes rapid absorption of heat from the sun and on slopes in excess of 3%, the mass of manure flows readily over the frozen surface below. Gilbertson et al. (1970) reported lava-like snowmelt flows of manure and water around densely stocked, sloping lots. The suspended solids transported by snowmelt were much higher than those in rainfall runoff. Gilbertson et al. (1970) reported snowmelt total solids content ranging from 1.4 to 10.7%. It was estimated that the total solids in winter runoff ranged from 55 to 70 kg/m³ (6.2 to 7.9 tons/acre-inch) of runoff in lots stocked at 60 m² (200 feet²) per animal.

Feedlot runoff, as compared to domestic sewage, is a highly concentrated organic waste of variable quality. Table 6 emphasizes this variability by giving characteristics of feedlot runoff for both rainfall and snowmelt from the same feedlot.

7. FOOD PROCESSING WASTES

Various other industrial wastes are generated in liquid form which are suitable for land application. Among these are the waste waters from food processing plants. These wastes are typified by their seasonality, high flow rates, and variable composition. Data provided by Loehr (1974) on a number of food processing wastes serve as a general guideline to this material. As he indicates, however, because of variability in processing techniques and water consumption, specific data need to be collected before a disposal system is established.

B. Pretreatment Possibilities

Although many liquid wastes may be applied to land as generated, frequently there are advantages to providing some degree of pretreatment prior to application. The most common is storage. By providing storage, an otherwise undesirable or impossible land application system may become usable.

Frequently, liquid waste has physical properties that make it incompatible with low-cost application techniques. A high solids concentration tends to plug pipes, sprinklers, and other liquid-handling devices. By providing solid-liquid separation, the material becomes much more amenable to land application.

Some wastes as generated are not compatible with the soil to which they would be applied. For example, they may be excessively high in salts or toxic materials. Treatment may be designed to remove one or more of these constituents to either minimize the land required for disposal or to enhance the value of the waste as a soil additive. The impact of a waste material on the total environment may be altered by pretreatment. Odor control is a prime example. Where wastes are to be applied to cropland near housing areas, recreational sites, or commercial developments, consideration must be given to odor. An appropriate selection of pretreatment techniques may be beneficial prior to application in these areas.

1. STORAGE DEVICES

Storage devices for liquid wastes are of several types. Where the quantity of waste water is small, tanks similar to those used for waste slurries are used. Their design is covered elsewhere, and few unique considerations are required. A tank or reservoir sufficiently large to hold the accumulated waste water during the period between applications to cropland is necessary. Where solid-liquid separation is anticipated during the storage interval, some means of agitation is required.

a. **Detention Basins**—Liquid wastes are often generated in sufficiently large quantities to make construction of a metal or concrete tank economically unattractive. In these cases, a storage reservoir or detention basin is the most common means of storage. Typical of these applications are storage reservoirs for the accumulation of feedlot runoff. Where land areas are unsuitable for the application of liquid during a portion of the year because of frozen or snow-covered conditions, storage is generally provided for that interval.

In dealing with storage of liquid wastes generated by rainfall conditions, statistical consideration of rainfall data is required. Most often, storage capacity is defined in terms of a storm of specified duration and frequency of occurrence. The reservoir would be sized to hold the design storm plus allowance for some accumulation.

b. Anaerobic Lagoons—Anaerobic lagoons are often used for pretreatment of liquid wastes prior to land application. They may perform a dual function of storage and biological treatment. In appearance, they may closely resemble a retention basin for the storage of liquid wastes; however, in design and intended function, they are radically different. They are basically ponds that receive a mixture of waste and dilution water. The biological processes which take place in the lagoons are much the same as those in a heated, mixed anaerobic digester, except there is no heating or mixing equipment, and the processes are less rapid and more influenced by weather variations. An anaerobic lagoon is designed to provide at least the minimum volume necessary for anaerobic breakdown of organic matter. The lagoons have had their greatest application in the treatment of animal manures; however, they have also been used for the treatment of other process waste waters containing organic matter suitable for anaerobic breakdown.

An anaerobic lagoon can be expected to perform a significant reduction in the organic content of waste water and to cause solids to settle, removing them from flow. Most solids will tend to liquefy because of long term retention on the bottom of the lagoon. In some structures designed according to the above procedures, sludge removal has proved necessary. Because of anaerobic decomposition, organic N will be converted to free NH_3 with much of it lost from the lagoon's surface. It may be anticipated that in a lagoon of long term retention, 60 to 80% of the N originally in the waste will escape to the atmosphere. Organic matter removal will be a function of detention time; again, reductions of 80% or more are not unreasonable.

Effluent from anaerobic lagoons is suitable for application to cropland. The odor downwind of sprinkler irrigation equipment may be detectable; however, on most occasions, it has not proved highly objectionable. Lagoon effluent is not suitable for discharge to a receiving stream as it is generally dark-colored and contains sufficient organic matter and plant nutrients to contribute to water quality degradation. In some locations, pumping problems have been associated with anaerobic lagoon effluents because of precipitation of magnesium ammonium phosphate crystals on pump impellers and other metal surfaces. These crystals may cause excessive pump wear, plugging of valves, and general fouling of hydraulic equipment. This precipitate is subject to removal by a dilute acid treatment, and is most often countered in this way (Booram & Smith, 1974).

c. Aerobic Lagoons—Naturally aerobic lagoons are similar to municipal waste stabilization ponds or facultative lagoons. The waste stabilization pond concept is illustrated in Fig. 7. In this system, a symbiotic relationship is developed between anaerobic or facultative bacteria degrading solid materials in the bottom of the lagoon and oxygen production by algae near the surface. For this process to continue, the oxygen demand must not exceed the ability of the algae to supply the necessary oxygen. Otherwise, the pond will become anaerobic and objectionable odors will ensue.

The design of aerobic lagoons is generally based upon surface area.

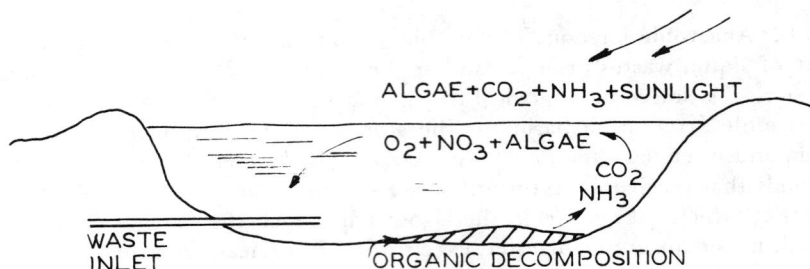

Fig. 7—Operation of a naturally aerobic lagoon (waste stabilization pond).

Daily loadings up to 0.11 kg/m²-day (45 pounds/acre-day) of ultimate bio-chemical oxygen demand (BOD) are generally accepted under moderate climatic conditions. When transformed into design criteria for livestock wastes, the surface areas shown in Table 7 are recommended. The large surface areas suggested by Table 7 illustrate why these devices have proved un-attractive in livestock waste management. The surface area becomes exces-sively expensive to provide and the water balance difficult to maintain.

To function satisfactorily, lagoons are typically operated with a mini-mum depth of 1.0 m (3 feet) and a maximum depth of 1.6 m (5 feet). Shal-lower lagoons are subject to infestation with bottom-growing plants, while deeper lagoons do not receive sufficient wind mixing to provide aerobic con-ditions throughout the water mass.

Aerobic lagoons overcome odor problems associated with anaerobic lagoons during normal operating conditions. Effluent from an aerobic lagoon is suitable for application to cropland, but is not acceptable for dis-charge into surface streams. It is high in algae content and undergoes rapid and extreme changes in dissolved O_2 during a typical day. Nitrogen losses due to aerobic lagoon treatment and land application are generally regarded to be 50 to 70%.

Mechanically aerated lagoons may be used as an alternative to naturally aerobic lagoons. In this system, oxygen is provided to the lagoon by a float-ing surface mechanism, submerged aeration tubing, or other mechanical de-vice to force O_2 into the liquid. A mechanically aerated lagoon may be used to provide a low odor means of liquid waste storage, but it is expensive.

Table 7—Suggested surface area of naturally aerobic lagoons, 1 to 1.6 m (3 to 5 feet) deep, used for the treatment of livestock manure (Miner & Smith, 1975)

Species	Surface area	
	m²/kg	ft²/lb
Poultry	0.70	3.4
Swine	0.39	1.9
Dairy cattle	0.33	1.6
Beef cattle	0.31	1.5

Since the mechanically aerated lagoon does not depend upon algae to provide the necessary O_2 or surface-diffused aeration, it may be constructed much deeper than naturally aerobic lagoons, thereby reducing the surface area required. Volumes of approximately 50 times the daily manure production have proved satisfactory for treatment of animal manures (Jones, Dale, and Day, 1971). However, where storage of the effluent within the lagoon is desired, suitable additional volumetric capacity should be provided.

d. **Oxidation Ditch**—The oxidation ditch is an aerobic storage and treatment device first applied to the treatment of municipal wastes and, more recently, to the treatment and storage of livestock wastes. As used in the treatment of municipal sewage, the oxidation ditch is a direct modification of the activated sludge process. Physically, the oxidation ditch consists of an oval-shaped tank which contains a center partition so that the water circulated within the tank travels a racetrack-shaped route. The rotor (a horizontally shafted aeration device) introduces O_2 into the liquid and provides sufficient velocity to the water to maintain the solids in suspension. When used in the treatment of municipal wastes, the rotor is stopped one or more times daily, solids allowed to settle, and the supernatant discharged as effluent.

To utilize the oxidation ditch in the management of more concentrated livestock wastes, certain modifications in both function and purpose have been necessary. The oxidation ditch is highly effective in maintaining odor-free manure or other waste storage conditions. In addition, the oxidation ditch results in significant reduction in the organic matter in the waste. It also allows the loss of a large quantity of N due to NH_3 volatilization, as well as to alternate nitrification and denitrification. The power consumption of an oxidation ditch is higher than most other treatment devices for the storage of animal manures; however, where a high level of odor control and working conditions is critical, it has received widespread application.

The usual design approach for oxidation ditches serving livestock wastes is to supply approximately 6 m^3 of liquid volume in the ditch per kg (30 feet^3/pound) of daily BOD added. With these loading rates, starting with water in the ditch, operations may be expected for an extended period if the solids in the ditch are managed at about 25,000 to 30,000 mg/liter by periodic or continuous sludge removal. The oxidation ditch rotor should have sufficient capacity to add O_2 to the waste at a rate of 2 kg/kg (2 pounds/pound) of BOD contributed to the animals daily, additionally. There should be sufficient propelling capacity to provide a minimum velocity of 18 m/min (1 foot/second) throughout the ditch contents.

2. SOLID-LIQUID SEPARATION

Where the liquid contains solids, it may be desirable to remove the solids prior to soil application. One purpose of solid-liquid separation is to reclaim the solids for some alternate use for which they have higher value than application to land. A second purpose is to protect the application system from the deleterious effects of the solids. For example, where small diameter nozzles are anticipated for an application system, solids, if not removed, contribute to plugging problems and increased operational cost. In the design of feedlot runoff retention basins, solids are frequently removed from runoff by sedimentation to prolong the life of the storage reservoir. Solids removal may also be an aid in further biological treatment. If the solids can be removed and handled as an inert material, the expense of providing biological treatment is greatly reduced. Often these solids can be reused as livestock food or otherwise elevated in economic value.

There are two fundamental methods of solid-liquid separation. One utilizes the difference in density between particulate matter and liquid; the second utilizes the size and shape of the particles to effect separation.

a.   Settling—Gravitational settling utilizes the different densities of particles and transporting liquid to effect separation. If transport velocity of fairly dilute waste waters is reduced temporarily almost to quiescence, the particles in the water will settle to the bottom or float to the surface. Sobel (1966) reported that dairy and poultry manures would have to be diluted with twice their volume of water before significant settling would occur. Greater dilution was accompanied by a rapid period of settling followed by a long period of compaction; the duration of the initial settling period decreased as the dilution ratio was increased.

Moore et al. (1973) measured gravitational settling of beef, dairy, horse, poultry, and swine manures. Wastes from each of these species were mixed to give slurries ranging from 0.01 to 1.0% solids. Chemical oxygen demand (COD), total solids, and total volatile solids were measured on all samples. All the manure slurries showed similar behavior, and the shapes of the settling curves were almost identical, for the dilutions tested, to those found by Sobel (1966). Moore et al. (1973) concluded that settling can provide useful removal of total solids and COD from a dilute slurry.

Witz and Pratt (1971) described a system in which wastes from beef cattle on a slotted floor fell through the slats onto a solid floor with a 2% slope. A chain-and-flight conveyor moved slowly up the slope; some solid-liquid separation took place as liquid ran down the slope and solids were conveyed to the top. The moisture content of the solid fraction was reported to be 80%.

Verley and Miner (1974) devised a solid-liquid separating device using gravity settling. The idea was to provide a series of settling basins with liquid effluent from one flowing into the next lower one; at the same time, these basins moved up the slope, discharging the dewatered solids at the top. This was achieved in the spaces formed between flights of a hollow auger and its cylindrical sheath.

Centrifugal separation has received little application in the waste management field because of the expensive equipment involved and the related skilled maintenance requirements. Glerum et al. (1971) described the use of two types of centrifuges on pig slurries. The first was a centrisieve that had a conical drum with a filter cloth liner. In this device, liquids were forced through the cloth and dewatered solids were discharged at the wider end of the drum. Between 30 and 50% of the solids could be removed from the influent, but the higher yield corresponded to a wetter cake.

Ross et al. (1971) examined the centrifuging characteristics of poultry manure using a solid bowl machine. Influent slurry concentrations were 5, 15, and 25% solids. At the lowest solids content, all slurries showed the same dewatering characteristics over the centrifuging range of 2,000–10,000 g. As the initial solids concentration increased, the amount of water that could be removed decreased, and the effect of centrifugal forces became important.

Holmes et al. (1971) reported on the centrifugation of mixed liquor from an oxidation ditch treating swine manure. The objective was recovery of microbial protein, so coarse particles and hair were removed on a 20-mesh screen before centrifugation. After some laboratory scale studies, they obtained a pilot scale solid bowl machine. Using an influent solids concentration of 12%, they were able to recover 90 to 95% of the influent liquid.

b. Screening—There are two methods of screening: the first uses a slow relative motion between slurry and screen, and the second uses a rapid vibratory motion. Equipment required is complex and only available through commercial sources.

Graves et al. (1971) described tests with a model hydrasieve (Fig. 8) with 0.05-cm (0.02-inch) bar spacing. Dilutions of fresh dairy manure were used to test the equipment. If dilution exceeded six volumes of water per volume of manure, good separation was achieved by the screen, with total solids reduced by about 60% in all instances. Solids content of a particulate material flowing from the bottom of the unit was about 62%.

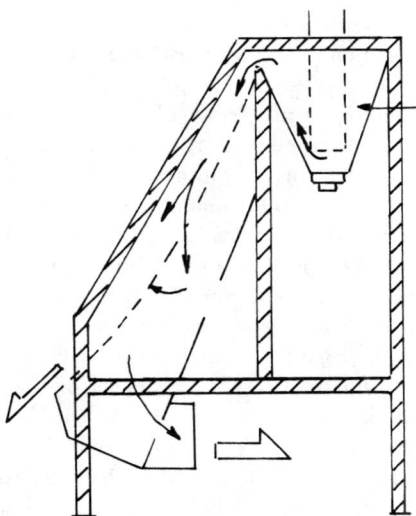

Fig. 8—Hydrasieve solid-liquid separator evaluated by Graves et al. (1971) and by Taiganides and White (1972).

Taiganides and White (1972) used a full-scale hydrasieve with 0.15-cm (0.06-inch) bar spacings to screen effluent from a hydraulically cleaned swine confinement building. No screening efficiency results were reported, although the authors mentioned some problems with slime growth on the screen. These growths, particularly troublesome in warm weather, were controlled with periodic chlorine disinfection.

Separating solids from liquid using a 20-mesh vibration screen was described by Fairbank and Bramhall (1968). Their work was related to the waste water from hydraulically cleaned holding pens of dairies in California. For influent slurry containing approximately 0.4% dry matter, the washed manure solids from the screen contained about 20% dry matter and had a bulk density of 430 kg/m^3 (27 pounds/foot^3). The washed solids were judged inoffensive and did not attract flies. In warm climates such as California, these solids may be heaped and composted without further attention. Fairbank and Bramhall (1968) indicated screening such dilute slurries causes a negligible reduction in liquid volume.

A detailed study of the engineering parameters associated with vibrating screens was reported by Ngoddy et al. (1971), using diluted beef and swine wastes. A 74-mesh screen gave the best separation. Virtually all test results showed that 55 to 60% of the total solids from swine manure could be removed; solids removal from beef manure was somewhat less.

c. **Filtration**—Screening and filtration have some features in common. The filter medium usually has a finer mesh size than a screen and is often made of organic fibers. The most significant difference is in the method of operation. In screening, the separation is achieved mainly by the screen mesh size because the layer of solids from slurry is kept minimal. In filtration, particles in the slurry form the filtering medium, and the filter cloth mesh only determines the characteristics of the initial filtrate.

A continuous belt machine with metal mesh belting which passed under press rolls was described by Bartlett et al. (1973). The machine was tested on cattle manure. The final product had a solids content of 25% and was judged inoffensive. The material was crumbly and easily handled.

Dale (1973) reported on a commercial device developed by Babson Brothers Co., Oak Brook, Ill. This machine used an inclined screen before a porous belt and a roller squeezer. Manure must be diluted with an equal volume of water to make the equipment function properly. Dale (1973) claimed that the solids discharged were 50 to 55% dry matter. Liquid from the separation contained 3 to 4% dry matter.

The vacuum filter (Fig. 9) has performed well with municipal and industrial wastes. Waste slurry enters the chamber of the bottom of the machine, and a submerged filter cake is formed. As the drum rotates, valving to the various segments of the drum operates so that the cake is dried by drawing air through it and in some cases, is lifted by a positive pressure to facilitate removal. The machines are costly to install and require competent management; thus, few commercial units have been installed.

Fig. 9—Vacuum filter as traditionally used in dewatering municipal sewage sludges.

d. Conclusions—Solid-liquid separation during the transport of dilute wastes is most easily achieved by sedimentation or screening. Poultry manure is among the more difficult materials to handle because of the presence of feathers and small particle size. When it is desirable to separate manure solids from transport liquid, both the vibrating screen and the stationary sloping screen produce satisfactory results. Because the vibrating screen has moving parts, it may require more maintenance than the stationary type; conversely, routine removal of biological growths on the stationary screen is important. Many areas require a screening device be housed in a heated building.

C. Application

Liquid wastes are usually applied to cropland by irrigation equipment. These wastes may be applied using tank wagons and trucks; however, only small quantities of wastes can be applied because of economic considerations. Where it is desired to use tank wagons, the considerations listed under waste slurries are applicable.

The type of irrigation system selected for the application of liquid wastes to land depends upon topography, soil type and cropping practices, area to be covered, labor requirements, and characteristics of the waste. The design of the system should include provision to control runoff and to prevent erosion within the irrigation site. The amount and timing of waste application is most often established by the nutrient needs and tolerances of the cover crop.

In most instances, dilute wastes are most economically applied to cropland by pipeline irrigation. For example, 1.03 ha-cm (1 acre-inch) of liquid waste is 103,160 liters (27,150 gallons), which represents 18 trips with a 5,700-liter (1,500-gallon) tank wagon. If an irrigation pump of 0.09 ha-cm/ hour (40 gallons/min) capacity were used, the wastes could be applied unattended, in 11.5 hours. Part of the pumping costs might be recovered from the fertilizer value of nutrients in the waste water.

Disadvantages of waste water irrigation are as follows: high initial investment is required, with heavy operating costs; good management is recessary to avoid runoff and water pollution; low-cost irrigation equipment may have a high labor demand; odor problem may be more widespread; untimely application may reduce crop yields; sprinkler irrigation causes NH_3 loss by volatilization; tight soils may only accept small quantities of liquid before runoff occurs; and in humid areas, tile drainage may be necessary. In spite of these potential disadvantages, irrigation is the most realistic means of applying sewage treatment plant effluent, feedlot runoff, and other liquid wastes which occur in large quantities.

Design of an irrigation system for liquid wastes involves selection of the distribution system (gated pipe, sprinklers, or others), sizing of pipes and laterals, and pump selection. The spreading system selected will influence pump capacity and will contribute to the pressure that the pump must work against as well as friction loss in the pipe. Total head in the system and pump capacity must be calculated to complete the design of the system.

Liquid wastes from pipeline systems can be gravity-spread by using open ditches, flat irrigation tubing, or gated pipe. Sprinkler systems available include hand-moved, towline, side-roll boom, center pivot self-propelled, traveling gun, and solid set.

1. SURFACE IRRIGATION

Four types of surface irrigation systems are suitable for disposal of liquid wastes: border irrigation, furrow irrigation, corrugations, and wild flooding. Waste water can be supplied to the disposal field by gated pipe, lay-flat irrigation tubing, open ditches with siphon tubes, or open ditches with turn-out gates. Gated pipe or open ditches with turn-out gates are recommended because of ease of cleaning. Because runoff must be controlled when disposing of liquid wastes, additional labor is required for surface irrigation, such as changing gates and checking distance of flow. A more expensive sprinkler system is frequently chosen to reduce labor.

Gated pipe is available in diameters from 10 to 30 cm (4 to 12 inches), of aluminum or plastic, with openings or gates every 75 to 200 cm (30 to 80 inches). This pipe is portable, giving flexibility to the system. The cost of gated pipe is greater than siphon tubes, but less than sprinkler systems. Also, less labor is required and the desired flow can be more easily managed. Open ditches with siphon tubes require a minimum of investment, but labor and maintenance may be expensive.

Waste water should not be applied to a wet disposal area. The border, furrow, or corrugated systems should be shut off before waste water reaches the low end of the field to eliminate runoff. This should be done when the water is two-thirds to three-fourths of the way to the end. Catching the runoff from the field in a basin and returning it to the irrigation system is another alternative receiving widespread application.

Irrigation with furrows provides relatively uniform distribution of water for row crops. Furrow irrigation is generally not recommended on slopes > 1%; steeper slopes, however, may be irrigated for short distances. Slopes up to 10% can be irrigated with contour furrows when furrow grade is kept $<$ 1%. Often some land grading is required for good distribution. The row crop may be planted on ridges. The furrows may be constructed with disk or middlebuster furrow openers. Furrow flow rates should not exceed the furrow-carrying capacity or cause erosion. Furrow irrigation runs generally do not exceed 400 m (0.25 mile) in length.

Corrugations provide a means of irrigating relatively steep, irregular land by surface irrigation. Adapting best to close-growing crops, the corrugations ("V" notches) are constructed by pulling a tool bar instrument over the ground after the seed bed is worked but not planted. It is especially useful for permanent grasses. The "V" notches are generally spaced closer than furrows and are shallower. Guidelines for furrow flow rates may be used for corrugations, unless this exceeds the corrugation-carrying capacity.

Low parallel soil berms constructed in the direction of the maximum slope of the field are used for border irrigation. The berms or borders are spaced 10 to 30 cm (30 to 100 feet) apart. Pasture and other close-growing crops can be irrigated with a border system. The length of run for border irrigation should generally not exceed 400 m (0.25 mile), and should be proportionately less for slopes exceeding 1% and for soils with moderate or high intake rates. Maximum slopes should not exceed 4%. Uniform water distribution depends on a sheet of water passing down the border at a depth of 8 to 12 cm (3 to 5 inches). Consequently, the fall between berms should not exceed 2 to 5 cm (1 to 2 inches). Because of this limitation, border irrigation requires an even, gently sloping field, either naturally or through land grading. Berms may be spaced closer together with a border maker, a road maintainer, or rear blade of a tractor.

Wild flooding refers to applying waste water to land with no control structures within the distribution pipe or ditch. This method has a low initial cost, adapts to a wide range of irrigation flows, and can be used on close-growing crops, rolling land, and shallow soil. However, wild flooding often subdivides the field, has a higher labor requirement, and an uneven water distribution.

2. SPRINKLER APPLICATION

Rolling and irregular land that would require extensive reshaping or surface irrigation can often be irrigated with sprinkler systems. Although initial and operating costs are generally higher for sprinklers, labor requirements are sometimes reduced, and some systems may be automated. Uniformity of application is also improved. For sprinkler systems, it is important to select sprinklers and spacings that will not cause runoff for the particular soil type, topography, crop, and time of application. Flushing the system with clean water after use is desirable to prolong equipment life and to avoid leaving a coating of solids on the crop.

Hand-carry sprinkler systems, unpopular for crop irrigation because of high labor requirements, have special merit where small acreages are managed as waste utilization sites. Adaptability to diverse topography, relatively low initial cost, and availability of used systems are advantages. Large manure guns can be handled in this way.

The towline system can be moved from one set to another by tractor. This system usually has lower initial and operating costs than traveling sprinkler systems. For corn, travel lanes and turn spaces are required, which amount to about 10% of the land area. Underground mainline pipe may be laid in a shallow ditch enabling the operator to cross the mainline pipe. The lateral is disconnected from the mainline for each move. A short flexible hose connects the lateral to the mainline. Towline systems are often designed for laterals spaced 20 m (60 feet) apart, with sprinklers spaced 15 m (40 feet) apart on the lateral. The towline has special merit for waste disposal over hand-carry systems because of the relative ease of moving the lateral line with a tractor at little increase in cost.

Side-roll lateral sprinkler systems can be used only with low-growing crops on rectangular fields because the aluminum pipe is the axle for movement of the lateral through the field. Crop clearance is slightly less than one-half the diameter of the wheels used at each lateral joint. A separate air-cooled engine is generally used to roll the lateral from one setting to the next.

The traveling sprinkler consists of a single large sprinkler mounted on a four-wheel carriage. Power to travel is provided by a small auxiliary engine, a water turbine, or a water cylinder. The auxiliary engine unit generally travels at more uniform speed but has a higher initial cost. These traveling units range in capacity from 0.2 to 5 m³/min (50 to 1,500 gallons/min). A cable, mounted on a winch on the machine, is extended 200 to 400 m (660 to 1,320 feet) across the field and anchored at the far end. As the power source drives the cable winch, the unit is pulled across the field, irrigating as it travels. A flexible hose connects the traveling sprinkler to a rigid pipe at the midpoint of its travel path. As the self-propelled unit passes adjacent to the waste water supply outlet (usually portable aluminum pipe), the hose forms an elongated "U" behind the unit. As the self-propelled unit proceeds along its travel path, the hose is extended full length in the opposite direction from its original layout. This allows for continuous movement of twice the length of the hose. Hoses can be purchased in different lengths. A popular size for waste disposal uses 100 m (330 feet) of flexible hose, and makes a 200-m (660-foot) pass. Travel distances of not more than 400 m (1,320 feet) should be selected. The supply line to the gun is flexible hose that is dragged on the ground as the gun moves; 10.2-cm (4-inch) hose is generally used. Speed of water turbines varies as load varies. Speed of movement can be varied from 0.15 to 2.5 m/min (0.5 to 8 feet/min) supplying 1 to 10 cm (0.3 to 3.5 inches) per application. The traveling guns are adapted to irregularly shaped fields and rough terrain, and cross a terrace best when moving perpendicular to it. When row crops are to be irrigated, most units will require that two or three rows be left out of production (or moved if rows are not straight) to provide lanes for traveling units.

Solid set sprinklers are the most labor-saving and flexible of all systems, and are particularly suited to small areas managed as utilization sites. They consist of lateral lines and sprinklers covering the entire area. If automation is desired, remote control valves can be programmed from an automatic timer to operate separate laterals or blocks in sequence (Koelliker et al., 1972).

Sprinkler head selection is of major concern for waste application systems. The uniformity of application is more important for a waste disposal system than for clean water irrigation. Nutrients in the effluent often limit the quantity applied (Booram & Smith, 1974) and uniformity becomes critical under conditions where excessive water or nutrients may be applied to those areas near the sprinkler. The application rate of a system should be less than the intake rate of the soil. This is controlled by the selection of nozzle size, pressure, and spacing. Small nozzles have lower pressure requirements, cover smaller areas, and have lower application rates. In contrast, large nozzles require higher pressure, cover larger areas, generally have poorer distribution patterns, and are more affected by wind. For waste disposal, single nozzle sprinklers will probably be more successful than two-nozzle sprinklers. The size of particles in the water may limit the minimum nozzle size. Generally nozzles < 0.5 cm (3/16 inch) are not practical. When larger particles or organic matter are present, nozzles of 1 cm (3/4 inch) may be required.

3. BARRIERED LANDSCAPE

The soil and its microbial population are capable of removing N by nitrification and denitrification; phosphorus may be removed by adsorption. Natural field situations are variable in N and P removal. Erickson et al. (1971) attempted to control N and P removal by tailoring the soil profile for this specific purpose. The result is the Barriered Landscape Water Renovation System (BLWRS). The BLWRS has three layers (Fig. 10). The top layer has slag or limestone and a cover crop; this layer is designed to act as a phosphorus trap. The second layer is a topsoil that serves to degrade organic matter and promote nitrification. Finally, there is a lower layer that can be kept flooded and anaerobic. The BLWRS is isolated from the soil under it by

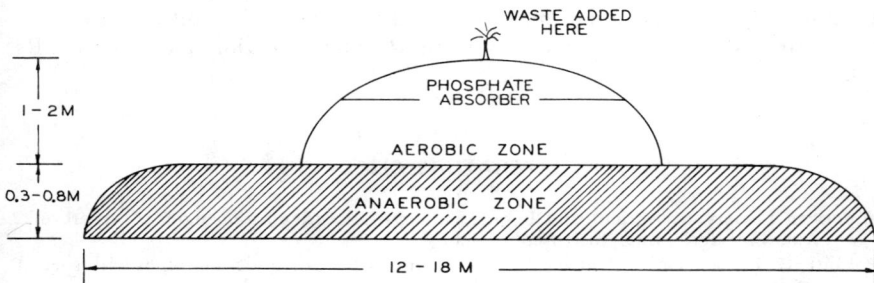

Fig. 10—Schematic cross section of a barriered landscape water renovation system (BLWRS) as described by Erickson et al. (1972).

Table 8—Average analysis of waste applied to and effluent from an experimental barriered landscape water renovation system (BLWRS) in Michigan (Erickson et al., 1972)

Constituent	Swine		Dairy	
	Waste	BLWRS effluent	Waste	BLWRS effluent
		mg/liter		
Total N	650	8	300	13
Nitrate N	10	6	10	10
Phosphate P	20	0.02	40	0.02
BOD	1,100	5	1,200	5
COD	2,000	40	3,000	70

an impermeable membrane. Water that has percolated through the system can either be collected by perimeter drains or allowed to seep horizontally into undisturbed soil. The lower anaerobic layer is fitted with a perforated pipe that allows a carbon source to be added. Erickson et al. (1971) used molasses as a carbon source; Koelliker and Miner (1970) have shown that raw swine wastes also may be used to promote denitrification. Their work used a liquid system in a digester only and was not extended to denitrification in the soil.

After the encouraging performance of the prototype BLWRS, Erickson et al. (1972) built four more systems. These systems were used in pairs to allow a resting period and restoration of infiltration rate by anaerobic activity. The results obtained using these units on swine and dairy wastes are shown in Table 8. Erickson et al. (1972) indicated there are problems during the winter. An attempt to maintain above-freezing temperatures using electrical heating tape was not successful.

ACKNOWLEDGMENT

This manuscript was prepared immediately following publication of the North Central Regional Research Publication 222, *Livestock Waste Management with Pollution Control* (Miner & Smith, 1975). Appreciation is expressed to the NC-93 committee members, whose efforts contributed greatly to this chapter. Particular appreciation is due Dr. R. J. Smith, Iowa State University, who served as coeditor of the NC-93 publication, along with J. R. Miner.

LITERATURE CITED

American Society of Agricultural Engineers. 1974. Agricultural engineer's yearbook. Am. Soc. Agric. Eng., St. Joseph, Mich.

Babbitt, H. E., and D. H. Caldwell. 1939. Laminar flow of sludges in pipes with special reference to sewage sludge. Univ. Ill. Eng. Exp. Stn. Bull. 319.

Babbitt, H. E., and D. H. Caldwell. 1940. Turbulent flow of sludges in pipes. Univ. Ill. Eng. Exp. Stn. Bull. 323.

Bartlett, H. D., R. E. Bos, and E. C. Wunz. 1973. Dewatering bovine animal manure. ASAE Pap. No. 73-431. Am. Soc. Agric. Eng., St. Joseph, Mich.

Booram, C. V., and R. J. Smith. 1974. Manure management in a 700-head swine finishing unit in the American Midwest: An integrated system incorporating hydraulic manure transport with anaerobic lagoon liquor and final effluent use by corn (*Zea mays*). p. 1089-1097. *In* Water research V. 8. Pergamon Press, Great Britain.

Converse, J. C., C. O. Cramer, R. F. Johannes, and H. J. Larson. 1974. Storage lagoon vs. underfloor tank for dairy cattle manure. ASAE Pap. No. 74-3028. Am. Soc. Agric. Eng., St. Joseph, Mich.

Dale, A. C. 1973. Solids-liquid separation: an important step in the recycling of dairy cow wastes. J. Milk Food Technol. 36:289-295.

Erickson, A. E., J. M. Tiedje, B. G. Ellis, and C. M. Hansen. 1971. A barriered landscape water renovation system for removing phosphate and nitrogen from liquid feedlot waste. p. 232-234. Int. Symp. Proc. Livestock Waste Manage. and Pollut. Abatement. Columbus, Ohio. PROC-271. Am. Soc. Agric. Eng., St. Joseph, Mich.

Erickson, A. E., J. M. Tiedje, B. G. Ellis, and C. M. Hansen. 1972. Initial observations for several medium sized barriered landscape water renovation systems for animal wastes. p. 405-410. *In* Waste management research. Proc. Cornell Agric. Waste Manage. Conf., Rochester, N. Y.

Fairbank, W. C., and E. L. Bramhall. 1968. Dairy manure liquid-solids separation. Univ. Calif. Agric. Ext. Serv. AXT-271.

Fehr, R. L., and R. J. Smith. 1975. Management of a flushing gutter manure removal system to improve atmospheric quality in housing for laying hens. p. 437-440. *In* Managing Livestock Waste, Proc. Third Int. Symp. on Livestock Waste. ASAE Publ. PROC-275. Urbana, Ill.

George, R. M. 1973. Designing gutter flushing systems. Midwest Livestock Waste Manage. Conf. Proc. Agric. Eng. Dep., Iowa State University, Ames, Iowa.

Gilbertson, C. B., T. M. McCalla, J. R. Ellis, O. E. Cross, and W. R. Woods. 1970. The effect of animal density and surface slope on characteristics of runoff, solid wastes and nitrate movement on unpaved beef feedlots. Nebr. Agric. Exp. Stn. Bull. 508.

Gilbertson, C. B., T. M. McCalla, J. R. Ellis, and W. R. Woods. 1971. Methods of removing settleable solids from outdoor beef cattle feedlot runoff. Trans. ASAE 14:899-905.

Glerum, J. C., G. Klomp, and H. R. Poelma. 1971. The separation of solid and liquid parts of pig slurry. p. 345-347. *In* Int. Symp. Proc. Livestock Waste Manage. and Pollut. Abatement. Columbus, Ohio. PROC-271. Am. Soc. Agric. Eng., St. Joseph, Mich.

Graves, R. E., J. T. Clayton, and R. G. Light. 1971. Renovation and reuse of water for dilution and hydraulic transport of dairy cattle manure. p. 341-344. *In* Int. Symp. Proc. Livestock Waste Manage. and Pollut. Abatement. Columbus, Ohio. PROC-271. Am. Soc. Agric. Eng., St. Joseph, Mich.

Gribble, D. J., and H. E. Bennett. 1965. Dairy establishment. U. S. Pat. 3,223,070.

Ham, R. K., W. K. Porter, and J. J. Reinhardt. 1972. Refuse milling for landfill disposal. p. 37-72. *In* Solid Waste Demonstration Projects Proc. U. S. Environ. Prot. Agency, SW-4p. Washington, D. C.

Hart, S. A., J. A. Moore, and W. F. Hale. 1966. Pumping manure slurries. p. 34-38. *In* Nat. Symp. Proc. Manage. of Farm Wastes. SP-0366. Am. Soc. Agric. Eng., St. Joseph, Mich.

Holmes, L. W., D. L. Day, and J. T. Pfeffer. 1971. Concentration of proteinaceous solids from oxidation ditch mixed liquor. p. 351-354. *In* Int. Symp. Proc. Livestock Waste Manage. and Pollut. Abatement. Columbus, Ohio. PROC-271. Am. Soc. Agric. Eng., St. Joseph, Mich.

Howes, J. R. 1966. On-site composting of poultry manure. p. 68-69. *In* Nat. Symp. Proc. Manage. of Farm Animal Wastes. SP-0366. Am. Soc. Agric. Eng., St. Joseph, Mich.

Jones, D. D., A. C. Dale, and D. L. Day. 1971. Aerobic treatment of livestock wastes. Ill. Agric. Exp. Stn. Bull. 737.

Jones, E. E., Jr., G. B. Willson, and W. F. Schwiesow. 1971. Improving water utilization efficiency in automatic hydraulic removal. p. 154-158. *In* Int. Symp. Proc. Livestock Waste Manage. and Pollut. Abatement. Columbus, Ohio. PROC-271. Am. Soc. Agric. Eng., St. Joseph, Mich.

Koelliker, J. K., H. L. Manges, and R. I. Lipper. 1974. Performance of feedlot runoff facilities in Kansas. ASAE Pap. No. 74-4012.

Koelliker, J. K., and J. R. Miner. 1970. Reduction of nitrogen concentrations in swine lagoon effluent by biological denitrification. Purdue Eng. Ext. Ser. 137:472–480.

Koelliker, J. K., J. R. Miner, T. E. Hazen, H. L. Person, and R. J. Smith. 1972. Automated hydraulic waste handling system for 700 head swine facility using recirculated water. p. 249–262. In Waste management research. Proc. Cornell Agric. Waste Manage. Conf. Rochester, N. Y.

Kumar, M., H. D. Bartlett, and N. N. Mohsenin. 1970. Flow properties of animal waste slurries. ASAE Pap. No. 70-911. Am. Soc. Agric. Eng., St. Joseph, Mich.

Loehr, R. C. 1974. Agricultural waste management. Academic Press, New York. p. 500–520.

Loudon, T. L., R. L. Maddox, and C. H. Shubert. 1975. Comparison of design criteria and performance of waste handling systems. ASAE Pap. No. 75-4026. Am. Soc. Agric. Eng., St. Joseph, Mich.

Martin, J. H., Jr., M. Decker, Jr., and K. C. Das. 1972. Windrow composting of swine wastes. p. 159–172. In Waste management research. Proc. Cornell Agric. Waste Manage. Conf. Rochester, N. Y.

Midwest Plan Service. 1971. Reinforced concrete manure tanks, slots, and beams. Midwest Plan Serv. TR-3. Iowa State University, Ames, Iowa.

Midwest Plan Service. 1975. Livestock waste handling facilities. Midwest Plan Serv. Publ. MWPS-18. Iowa State University, Ames, Iowa.

Miner, J. R., and R. J. Smith, ed. 1975. Livestock waste management with pollution control. Midwest Plan Serv. Publ. MWPS-19, North Central Reg. Res. Publ. 222. Iowa State University, Ames, Iowa.

Moore, J. A., R. O. Hegg, D. C. Scholz, and E. Strauman. 1973. Settling solids in animal waste slurries. ASAE Pap. No. 73-438. Am. Soc. Agric. Eng., St. Joseph, Mich.

Ngoddy, P. O., J. P. Harper, R. K. Collins, G. D. Wells, and F. A. Heidar. 1971. Closed system waste management for livestock. U. S. Environ. Prot. Agency Water Pollut. Control Res. Ser. 13040 DKP 06/71.

Pound, C. E., and R. W. Crites. 1973. Characteristics of municipal effluents. p. 49–59. In Proc. of the Joint Conf. on Recycling Municipal Sludges and Effluents on Land. Champaign, Ill. U. S. Environ. Prot. Agency, U. S. Dep. of Agric., and Nat. Assoc. of State Univ. and Land-Grant Coll., Washington, D. C.

Reddell, D. L., P. J. Lyerly, and J. J. Hefner. 1972. Crop yields from land receiving large manure applications. ASAE Pap. No. 72-960. Am. Soc. Agric. Eng., St. Joseph, Mich.

Reed, C. H. 1975. Equipment for incorporating animal manures and sewage sludges into the soil. p. 444–445. In Managing Livestock Waste, Proc. Third Int. Symp. on Livestock Waste. ASAE Publ. PROC-275. Urbana, Ill.

Ross, I. J., J. J. Begin, and T. M. Midden. 1971. Dewatering poultry by centrifugation. p. 348–350. In Int. Symp. Proc. Livestock Waste Manage. and Pollut. Abatement. Columbus, Ohio. PROC-271. Am. Soc. Agric. Eng., St. Joseph, Mich.

Sewell, J. I. 1971. Agitation in liquid manure tanks. p. 135–137. In Int. Symp. Proc. Livestock Waste Manage. and Pollut. Abatement. Columbus, Ohio. PROC-271. Am. Soc. Agric. Eng., St. Joseph, Mich.

Shuyler, L. R., D. M. Farmer, R. D. Kreis, and M. E. Hula. 1973. Environmental protecting concepts of beef cattle feedlot waste management. U. S. Environ. Prot. Agency, Nat. Animal Feedlot Wastes Res. Program, Proj. 21 AOY-05. Ada, Okla.

Smith, R. J. 1971. A prototype to renovate and recycle swine wastes hydraulically. Ph.D. Thesis. Iowa State Univ. (Libr. Contr. Card No. Mic. 72-5258). Univ. Microfilms. Ann Arbor, Mich.

Smith, R. J., and T. E. Hazen. 1967. The amelioration of odor and social behavior in, together with the pollution reduction from, a hog house with recycled wastes. ASAE Pap. No. 67-434. Am. Soc. Agric. Eng., St. Joseph, Mich.

Smith, R. J., T. E. Hazen, and J. R. Miner. 1971. Manure management in a 700-head swine-finishing building: two approaches using renovated waste water. p. 149. In Int. Symp. Proc. Livestock Manage. and Pollut. Abatement. Columbus, Ohio. PROC-271. Am. Soc. Agric. Eng., St. Joseph, Mich.

Smith, R. J., T. E. Hazen, and G. B. Parker. 1973. Waste management. Pork Producers Day Proc. Iowa State Univ. Coop. Ext. Serv. Publ. AS-391.

Sobel, A. T. 1966. Physical properties of animal manures associated with handling. p. 27–32. *In* Nat. Symp. Proc. Manage. of Farm Animal Wastes. SP-0366. Am. Soc. Agric. Eng., St. Joseph, Mich.

Staley, L. M., N. R. Bulley, and T. A. Windt. 1971. Pumping characteristics, biological and chemical properties of dairy manure slurries. p. 142–145. *In* Int. Symp. Proc. Livestock Waste Manage. and Pollut. Abatement. Columbus, Ohio. PROC-271. Am. Soc. Agric. Eng., St. Joseph, Mich.

Staley, L. M., M. A. Tung, and G. F. Kennedy. 1973. Flow properties of dairy waste slurries. Can. Agric. Eng. 16:124–127.

Stewart, B. R., and J. M. Sweeten. 1974. Liquid manure management for swine operations. Tex. Agric. Ext. Serv. MP-1128. 8 p.

Taiganides, E. P., T. E. Hazen, E. R. Baumann, and H. P. Johnson. 1964. Properties and pumping characteristics of hog wastes. Trans. ASAE 7:123, 124, 127, 129.

Taiganides, E. P., and R. K. White. 1971. Automated handling, treatment and recycling of wastewater from animal confinement production unit. p. 146–148. *In* Int. Symp. Proc. Livestock Waste Manage. and Pollut. Abatement. Columbus, Ohio. PROC-271. Am. Soc. Agric. Eng., St. Joseph, Mich.

Taiganides, E. P., and R. K. White. 1972. Automated handling and treatment of swine wastes. p. 331–339. *In* Waste management research. Proc. Cornell Agric. Waste Manage. Conf. Rochester, N. Y.

Toth, S. J., and B. Gold. 1971. Composting. p. 115–120. *In* Proc., Agric. Wastes: Principles and Guidelines for Practical Solutions. Cornell Agric. Waste Manage. Conf. Rochester, N. Y.

U. S. Environmental Protection Agency. 1975. National Pollution Discharge Elimination System (NPDES). Fed. Regist. 20 Nov. 1975. p. 54182–54186.

van Dam, J., and C. A. Perry. 1968. Manure management—costs and product forms. Calif. Agric. 22(December):12–13.

Verley, W. E., and J. R. Miner. 1974. A rotating flighted cylinder to separate manure solids from water. Trans. ASAE 17:518–520.

Willson, G. B. 1971. Composting dairy cow wastes. p. 163–165. *In* Int. Symp. Proc. Livestock Waste Manage. and Pollut. Abatement. Columbus, Ohio. PROC-271. Am. Soc. Agric. Eng., St. Joseph, Mich.

Wise, G. G., and D. L. Reddell. 1973. Water quality of storm runoff from a Texas beef feedlot. ASAE Pap. No. 73-441. Am. Soc. Agric. Eng., St. Joseph, Mich.

Witz, R. L., and G. L. Pratt. 1971. Experimental facilities for studies on beef housing and equipment. Can. Agric. Eng. 13:81–84.

chapter 16

Hand shows depth of rich forest cover. A rich layer of rotting organic matter is just under the top cover of leaves (Photo courtesy of USDA-SCS).

Special Opportunities and Problems in Using Forest Soils for Organic Waste Application[1]

WAYNE H. SMITH, University of Florida, Gainesville, Florida

JAMES O. EVANS, USDA Forest Service, Washington, D. C.

I. INTRODUCTION

The capacity of forested watersheds to renovate and release high quality water to reservoirs has long been recognized. Yet, the potential of forest lands as sites for recycling wastes is often overlooked. The most pondersome problems concerning wastes are the large quantities that must be handled and the toxic substances in many wastes that limit their use in repeated applications to soils, regardless of plant cover. There also may be both human health and cultural restraints against the use of crop plants where these plants from treated areas are to be used for human consumption or by animals that enter the human food supply. These are only some examples of why forests may have a role in the solution of waste disposal problems.

Forests occupy the largest single area of land in the contiguous 48 states of the USA (Table 1). Except for the Great Plains and the southwestern regions of the USA, forests are abundant and well distributed, occurring near major areas where wastes are generated. Even in New Jersey, the most densely populated state, nearly half of the land is in forests. In certain cases (recreation sites, for example) forests may be the only vegetated landscape community or suitable location for waste recycling.

Most products from forests are not consumed by humans, although some game animals that forage in forests may be eaten by humans. Forests possess low human population densities, although they may occur near urban areas. Because of processing and use characteristics, there is no reason to suspect that forest products (lumber, paper, etc.) prepared from trees grown under waste applications would represent a health hazard. Thus direct contact of humans with waste-treated areas is less likely so odors or displeasing esthetic features of waste systems should neither provoke little adverse public reaction nor pose serious health hazards.

Another feature of forests is the relatively long cropping periods associated with forest management. This feature may be more advantageous with solid wastes and dewatered sludge than with liquid effluents since decomposition and nutrient release from the former materials may occur over a long time period. A variety of cropping schemes is possible, ranging from short rotation coppice systems of 5 years or less, to intensively managed forests of the Southeast involving rotations of 20 years or less, to natural stand rotations that may exceed 100 years.

But all features of forests are not favorable to forest land use for waste disposal. A principal factor to consider is public acceptance of disposal in forests since the pristine image of forests may be marred by this use. Some people may regard such use as unacceptable even though they may visit a forest only a few times during their lifetime. Certainly, visible irrigation hardware is not a usual landscape feature, and perhaps changes in the vegeta-

[1]Number 7052 in the Journal Series of the Florida Agricultural Experiment Station, Gainesville, FL 32611. This effort was partially supported by the Center for Environmental Programs, Institute of Food and Agricultural Sciences, University of Florida, Gainesville.

Table 1—Land area of the United States by land type and region†

Land use	Area	Proportion	Region			
			North	South	Rocky Mountains	Pacific Coast
	million ha	%	———————— ha ————————			
Forest land						
Commercial timberland	202.4	22.0	72.0	78.0	24.9	67.4
Unproductive	102.8	10.3	--	--	--	--
Reserved	7.0	0.8	--	--	--	--
Deferred	1.0	0.1	--	--	--	--
Total forest land	305.3	33.2	75.5	85.7	56.0	87.8
Cropland	172.9	18.8	105.4	42.0	15.2	10.4
All other land	441.2	48.0	73.5	79.9	153.6	134.2
Total all land	919.5	100.0	254.3	207.6	224.8	232.4

† From U. S. Forest Service (1973).

tion resulting from waste application may be described by some as "unnatural."

Another problem concerns accessibility, especially on public lands. Here, land use policy and guidelines will have to be developed and articulated to make this use compatible with other resource use requirements (Olson & Johnson, 1973). The article stated further that the National Forest System, USDA, Forest Service, is proceeding with cautious optimism because of the dearth of proven effects, both favorable and potentially adverse. Public land management agencies such as the Forest Service have policies that, where possible, programs should contribute to community development and to improvement of the rural environment in general. When these two criteria are met on the basis of sound research, a clear policy should follow. Private forest lands may be accessible for waste disposal if no adverse effects are apparent on the current land use objectives and if the requirements of the National Environmental Policy Act and local regulatory bodies are met.

Roads or other accesses to forests must be available and cost to the disposal agency should be minimal. Remoteness of forests in some cases makes transport distances so great that their use is impractical. In the East and Northwest, forests are abundant nearby urban areas. For example, in Alachua County, Florida many thousands of hectares of forest land are available within the radius of transport now involved in deliveries to land fills. In St. Johns County, Florida, a governmental official remarked that lands suitable for land fills (i.e., 3 to 4 m of drained profile) were priced at rates for condominium use. But forest land with shallower profiles is abundant in that county where little agriculture occurs except in widely dispersed areas.

Another consideration that may restrict forest land use is the effect of waste on wildlife. Habitats will change under application of wastes (Sopper & Kardos, 1973) and so will wildlife populations. Establishing ownership of mobile resources such as wildlife is often difficult. Hence, adverse public reactions may occur if a prized species for observing or hunting is reduced or one that is threatened or endangered as a resident species is eliminated from the landscape.

Distribution of wastes on forest land poses special problems. Often forest land is steeply sloped and rugged. Furthermore, trees themselves may represent obstructions to distribution systems. Where forest lands are the recycling sites of choice, special systems may need to be devised to obviate certain problems.

II. FOREST SOIL, SITE, AND STAND CHARACTERISTICS AFFECTING USE

A. Soil and Site Properties

Certain properties of forest soils favor their use in waste recycling. First, we shall consider those properties favorable to waste water application. Hydrologists generally regard forest lands as areas with optimum infiltration of water and negligible overland flow (Pierce, 1967). Although several factors influence infiltration, forest soils possess certain characteristics that enhance this soil property for waste water recycling—porous channels due to root and animal activity, incorporated organic matter in the surface horizons, and accumulation of organic debris. As long as organic layers are not disturbed and mineral soils are not exposed, removing the forest cover does not reduce infiltration, but compaction from trampling by cattle, humans, or logging equipment will cause disruption leading to overland flow (Dils, 1953; Hornbeck & Reinhart, 1964). For a northeastern forest, Trimble et al. (1958) estimated infiltration rates for the watershed to range between 35 and 76 cm/hour.

Obviously, forest soils do not have traffic pans generated by frequent passes of machinery. Water that infiltrates forest land percolates through the profile rather freely bringing it quickly into contact with large soil volumes. Percolation rates under various forest types and in soil strata measured in North Carolina ranged from 229 down to 3.8 cm/hour depending on stage of development of the forest and on soil characteristics (Metz & Douglass, 1959). Although infiltration and percolation rates of surface horizons may exceed 50 cm/hour, the total subsoil rarely has percolation rates that high.

Percolation movement is aided in forest soils not only by the decayed-root channels, but also by abundant pores. Bulk density is a useful physical characteristic of soils, especially in determination of pore space through which percolation occurs. The maintenance of porosity in forest soils is clearly demonstrated in Table 2 where a comparison is given for bulk density of several cropped and uncropped soils (Lyon et al., 1952).

Other evidence of the effect of soil disturbance by land use (e.g., agriculture) and its persistence is apparent from the work by Miller (1967). He characterized soils under native woodlands and in forest plantings in abandoned fields in which parent material, slope, texture, and other factors

Table 2—Comparisons of bulk density of cropped and uncropped soils

Soil	Cropped	Uncropped
	g/cm³	
Loam (Pa.)	1.25	1.07
Silt loam (Iowa)	1.13	0.93
Silt loam (Ohio)	1.31	1.05
Avg. 19 soils (Ga.)	1.45	1.14

were essentially constant. The horizon comparisons for a Wilcox silty clay (Table 3) show the effect of prior soil disturbance on bulk density (or soil pore volume). This contrast is typical of several comparisons across soils derived from an array of parent materials in Mississippi.

Parent material can cause wide variations in bulk densities even under similar forest cover through its effect on texture, structure, and other soil properties. An example for two soils formed from different parent materials but under similar forest cover in the Northwest USA (Anderson & Tiedemann, 1970) is given in Table 4. These data also show the high porosity that is possible in some forest soils.

As long as soils are deep, this porous property of forest soils for waste water use is desirable. But we should remember that many forests exist primarily because they occupy adverse sites such as shallow soils, imperfectly drained soils, and steep rocky hillsides unsuitable for other uses. Also forests frequently occur on sands which are not as favorable for mineral retention as permeable soils with substantial colloidal content. The importance of imperfect drainage resulting from shallow soils or clay pans in denitrification has not been established. For these reasons, inherent soil properties per se may be of little meaning when considering the ability of some forest lands to accept water.

Table 3—Bulk density of soil horizons as affected by land use

Soil horizon	Native woodlands	Old field plantation
	g/cm³	
A or Ap	1.09	1.31
B21g	1.29	1.39
B22g	1.32	1.40
B3g	1.37	1.46

Table 4—Bulk density of two forest soils derived from different parent materials

Forest soil	Bulk density, g/cm³
Derived from basalt	0.62
Derived from sandstone	1.03

Many mountainous forest soils are very shallow and steep, thus limiting their use for high rates of effluent recycling. Their abundance, however, makes consideration of low rates over large areas possible. Similarly, many coastal soils are shallow to ground water, which may limit their use and require special regimes.

B. Stand Properties and Ecologic Factors

The capacity of any forest to renovate wastes depends largely on the successional stage of the forest. During early development when the stand is building nutrient-rich tissues, such as a canopy of tree crowns and meristems (cambial and apical), nutrient accumulation will be highest (Smith et al., 1971). In contrast, near-mature or climax forests accumulate lesser quantities because in these forests, "net energy fixation is zero" (materials are only cycling to maintain biomass rather than synthesize storage). This ecological principle is clearly illustrated by data from the report of Sopper and Kardos (1973) on irrigation of a mixed hardwood forest with waste water. They reported only slight increases in diameter growth by older trees (50–70 years), while in younger stands 30 and 50 years old, significant growth increases resulted (69% at 30 years and 40% at 50 years). In another example from this same source, a red pine plantation (see Table 5 for list of common and scientific names) established in 1959 irrigated with effluent at 5.08 cm/week showed a decreased growth rate when compared to a control. But irrigation of an out-planting of red pine (5 years in the field) increased height by 240% over the control.

Native forest vegetation is in equilibrium with natural cyclic processes that affect the pool of nutrients available for growth. Output of nutrients in subsurface flows (overland flow is negligible in undisturbed forests) should seldom exceed input except for special cases (e.g., calcium, where there is a limestone substratum). In fact, forests tend to accumulate, store, and redistribute nutrients. Richards (1962) in Australia and Switzer et al. (1968) in the southeastern United States have reported data showing buildups of N in the soil-plant system during stand development. These capacities to retain nutrients, especially N, may be associated with the organic debris that persists with wide C/N ratios, high cation exchange capacities of the organic materials, and low base saturations (Allison, 1965; Wooldridge, 1970).

Mineral forest soils are often infertile by agronomic standards. Because of generally low pH levels in surface horizons they have the capacity to fix substantial quantities of nutrients. For example, Ballard and Fiskell (1973) have shown that many forest soils under coniferous forests possess high levels of active aluminum that can fix substantial phosphate—an element of concern in many wastes. Unlike agricultural soils, forests have not received lime or annual applications of fertilizer to enrich the profile. Hardwood forest soils and some others of lesser area (especially soils formed from limestone) may have higher base levels than soils under agronomic uses. Wastes usually

Table 5—List of common and scientific names for trees and food plants

	Names		
Common	Scientific	Common	Scientific
Trees			
Maple	*Acer* sp.	Birch	*Betula* sp.
Catalpa	*Catalpa bignonioides* Walt	Red bud	*Cercis canadensis* L.
White cedar	*Chamaecyparis thyoides* (L.) B.S.P.	Dogwood	*Cornus florida*
Hawthorne	*Crataegus*	Arizona cypress	*Cupressus arizonica* Greene
Eucalyptus	*Eucalyptus* sp.	Green ash	*Fraxinus pennsylvanica* Marsh.
Holly	*Ilex* sp.	Walnut	*Juglans* sp.
Red cedar	*Juniperus virginiana* L.	Larch	*Larix* sp.
European larch	*Larix decidua* Mill.	Larch	*Larix leptolepis* (Sieb. Zucc) Gord.
Sweet gum	*Liquidambar styraciflua* L.	Tulip poplar	*Liriodendron tulipifera* L.
Magnolia	*Magnolia* sp.	American olive	*Olea* sp.
White spruce	*Picea glauca* (Moench) Voss	Pine	*Pinus* sp.
Sand pine	*Pinus clausa* (Chapm.) Vasey	Shortleaf pine	*Pinus echinata* Mill.
Slash pine	*Pinus elliottii* var. *elliottii* Engelm	Red pine	*Pinus resinosa* Ait
White pine	*Pinus strobus* L.	Loblolly pine	*Pinus taeda* L.
American sycamore	*Platanus occidentalis* L.	Cottonwood	*Populus* (hybrid)
Aspen	*Populus* (hybrid)	Eastern cottonwood	*Populus deltoides* Bartr
Black cherry	*Prunus serotina* Ehrh	Cherry laurel	*Prunus caroliniana*
Douglas fir	*Pseudtsuga menziesii* (Mirb.) Franco.	Oak	*Quercus* sp.
Saw-tooth oak	*Quercus accutissima* Caruthers	Laurel oak	*Quercus laurifolia* Michx.
Nuttall oak	*Quercus nuttallii* Palmer	Cherry bark oak	*Quercus falcata* var. *pagodaefolia* Ell.
Red oak, northern	*Quercus rubra* L.	Live oak	*Quercus virginiana* Mill.
Black locust	*Robinia pseudoacacia* L.	Willow	*Salix* sp.
Bald cypress	*Taxodium distichum* (L.) Rich.	Chinese elm	*Ulmus parvifolia* Jacq.
Food plants			
Alfalfa	*Medicago sativa* L.	Potato	*Solanum* sp.
Ryegrass	*Lollium multiflorum* Lam.		

contain substantial quantities of elements, especially phosphorus, subject to fixation by both inorganic and organic soil constituents.

A most important forest characteristic, and one frequently overlooked in trials, is the species (or forest type) adaptability to the nature of the waste. Species adapted to moist, fertile conditions should prove more useful in renovation than species adapted to infertile, well-drained soils. A first step in planning waste water disposal in forests is to determine species requirements. By doing this the planner can either adapt the waste water application rate to the forests available nearby or plant forests more suitable to higher rates of application.

Certain forest species cannot tolerate fertile, moist conditions. In Florida, sand pine (*Pinus clausa*) which is adapted to droughty, coarse sands does not persist when planted on moist sites because of susceptibility to root rot under moist conditions (Ross, 1973). Similarly, red pine in New York was found to be adversely affected when grown on wet soils (Stone & Morrow, 1954). Thus, it is not surprising that red pine in the Pennsylvania State University studies responded positively at a low effluent rate but was adversely affected at rates of 5.08 cm/week (Sopper & Sagmuller, 1966).

In Florida, over 30 forest species have been screened under three rates of secondary treated waste water irrigation. First year heights show that under irrigation some species grow more rapidly than others (Table 6). In this study, controls were not included because it is well known that these species will either not grow at all or else grow very poorly on the experimental site (deep, excessively drained sandy soil). In this case, using waste water from the city of Tallahassee, Florida, the forest species responded to water and nutrients.

Settergren et al. (1974) reported first year heights of eight tree species from a study of recreational waste water irrigation in Missouri. They found cottonwood, sycamore, bald cypress, green ash, and maple to be more responsive than oaks and pines under 5 cm/week of irrigation. Based on the modest chemical applications in the recreational area waste water, they concluded that the response was primarily due to water. In trails at Pennsylvania State University, oak and pine responses were small or negative (Sopper &

Table 6—First year growth of selected forest species under waste water irrigation averaged over rates (D. M. Post and W. H. Smith, unpublished report, University of Florida, School of Forest Resources and Conservation, Gainesville)

Hardwoods	Height	Evergreens (and Bald cypress)	Height
	cm		cm
Eastern cottonwood	256	Bald cypress	138
Sycamore	206	Red cedar	83
Green ash	184	Arizona cypress	77
Black locust	180	Loblolly pine	60
Sweetgum	140	Laurel oak	104
Chinese elm	146	Magnolia	61
Tulip poplar	101		

Kardos, 1973). But, in the mixed hardwood stand irrigated in Pennsylvania, it was observed that *Acer* apecies and *Populus* seedlings in the stand grew at least twice as rapidly with irrigation as without. These were naturally occurring hardwoods rather than individual species selected for tolerance to the irrigation environment. These observations from Pennsylvania agree with experimental results reported from Missouri and Florida. In contrast, the Pennsylvania group screened conifers (Sopper & Kardos, 1973) and found larches (*Larix*) and white pine to grow better under irrigation than did red pine or white spruce (*Picea*).

Another important ecologic factor in selecting tree species for recycling wastes other than hydrophytic or xerophytic character is species nutritional requirements. Hardwoods are known to accumulate more nutrients than invader pine species (Rennie, 1955) even though they lose their leaves in winter. Further, certain plants are acidophilic (calciphobic) or calciphilic and these plants react differently to the nutritional regimes. Certain "acid-loving" plants are subject to "lime chlorosis" at elevated pH levels (Grime & Hodgson, 1969). Under near neutral pH conditions these species develop chlorosis, presumably because of a disturbance of iron nutrition. For example, Theobald and Smith (1974) observed that slash pine developed chlorosis correctable with iron sprays at a near neutral pH. This same species developed a similar chlorosis under waste water irrigation (pH near neutral) at Tallahassee, Florida. For the red pine in Pennsylvania where growth declined under irrigation, Sopper and Kardos (1973) stated, ". . .needles of pines being irrigated at the higher rate began to turn yellow"—typical symptoms of iron chlorosis. They suspected boron toxicity but were unable to confirm this. The pH of the effluent used in their studies averaged 8.1.

A related problem with conifers is that conifers appear to prefer NH_4^+ forms of N (McFee & Stone, 1969). Under natural conditions, nitrate reductase, an enzyme essential in NO_3^- metabolism, has a low activity in pine, but it can be induced at an elevated pH (Theobald & Smith, 1974). Analysis of percolate under pine and hardwood under waste water irrigation in Pennsylvania showed that under red pine, percolate remained rich in NO_3^-, especially under the highest irrigation regime, suggesting low removal of NO_3^- by pine trees.

In contrast to pines, legumes thrive at soil pH levels near neutral. For the Florida trials, the two legume tree species in the test, red bud and black locust, were among the most rapidly growing trees—surpassed only by those hardwoods known to prefer moist, fertile sites. Red cedar (*Juniperus*), a calciphile, also grew well under calcium-rich waste water—better than any of the other coniferous species tested except bald cypress. Using these and other criteria provides a way of developing forest tree plantings to fit the characteristics of waste effluents and sludges.

While hardwoods use more nutrients, conifers evaporate more water (Penman, 1967) and thereby leave less for subsurface discharge. This is attributable to their evergreen habit and year around water loss. Baumgartner (1967) examined the energetic bases for differential vaporization and showed

Table 7—Radiation balance for three plant communities

Vegetation type		
Forest	Alfalfa	Potato
langleys cm^{-2} day^{-1}		
505	385	340

that deciduous forests were intermediate between coniferous forests and grasslands in net radiant energy for water loss processes. His calculated radiation balance for a spruce forest, an alfalfa field, and a potato field under clear conditions is given in Table 7.

Based on a value of 100% for alfalfa, forest cover may use 30% more water. This value is in agreement with water yield increases after clearcutting the trees on a forested watershed (Swank & Douglass, 1974). This could mean that an irrigation rate of 3.3 cm/day on forests is effectively equivalent to 2.5 cm/day applied to a crop like alfalfa. Pennypacker et al. (1967) showed that 15 to nearly 30% of the waste water irrigation applied to forest vegetation was intercepted by the tree crown and never reached the forest floor. The structure of forest vegetation also reduces soil freezing such that infiltration is not blocked and irrigation can proceed through more of the year (Sopper, 1968).

III. CHARACTERISTICS OF WASTES AFFECTING USE IN FORESTS

A Current Research Information System (CRIS) search[2] made in August 1974 revealed fewer than a dozen projects nationwide evaluating forests for waste recycling (Table 8). Although some work, especially by private industry, may not be included in the USDA-State Experiment Station-Forestry School file, this finding is an indication of the low intensity of research in this area.

[2]Search conducted for the authors by, Director, Current Research Information System, Cooperative State Research System, U. S. Department of Agriculture, Washington, D. C.

Table 8—Inventory of research projects nationwide in the U. S. Dep. of Agric., state experiment stations, and forestry schools (Current Research Information System [CRIS], USDA, Washington, D. C., 1974 search)

Plant system	Animal waste	Agricultural residues	Sludge	Municipal effluent	Industrial and cannery waste	Solid waste	Total
Agricultural	129	17	25	15	6	4 (2)†	198
Forest	3	0	2	4	1	1	11

† Number in parentheses refers to projects dealing with forest products wastes applied to agricultural soil-plant systems.

A. Municipal Sewage Effluents

Municipal sewage treatment plant waste waters are produced in large volumes, and they often contain elements that may become toxic to plants under prolonged irrigation. In major cities, transportation (via piping) would be a major deterrent to using forests for deposition of effluents. For small towns where urban sprawl is minimal, the treatment plant may be located in or nearby a wooded area. In any case, available forests may prove desirable disposal sites if soil pollution under repeated treatment of agricultural lands proves hazardous. Because of the vastness of forest land acreages, new areas could be treated successively by using portable piping. To do this, a land-owner-lease arrangement would be required, and disposal would be added to collection costs charged to residents. Toward this eventuality, species tolerances and stand loading rates must be researched to assure landowners that no harm will be done to their forest land, that waste water will be adequately renovated, and that benefits may accrue.

Little information has been reported concerning the use of waste water in forested areas. Early work reviewed by Sopper (1968) mainly focused on disposal with rates of several meters per week. Renovation efficiency, ground-water recharge, and environmental protection were seldom examined. The most complete work on this aspect has been in Pennsylvania in the recent decade. In this work, forest species used were not especially suited to the effluent, and the soils were apparently shallow over an impermeable layer that restricted soil volume for root penetration thus causing wind and ice throw of some trees (Sopper & Kardos, 1973). Even with these restraints 5 cm/week of sewage effluent could be applied with satisfactory renovation. In southern California, effluent was required at rates of 7.6 cm or more per week for benefits to accrue, pointing out the role of site factors in choosing application rates (Youngner et al., 1973a, 1973b).

Recycling municipal effluent in woodlands has advantages other than year around irrigation. Irrigation of forests can improve understory vegetation and thereby increase food and cover for birds, rabbits, and deer (Wood et al., 1973).

In Michigan, red pine on a sandy soil (in contrast to the loam soil in Pennsylvania) was irrigated with effluent at rates up to 8.8 cm/week. In early evaluations, needle length and dry weight increased up to 36 and 56%, respectively, when compared to the control. Subsequent evaluations may show that this increase in photosynthetic tissue leads to increased shoot growth. In these tests, renovation efficiency was 83% or better at 60-cm depths with nitrate nitrogen not exceeding 1 ppm. Again, red pine was rather unresponsive to irrigation and fertility; so for this species, effluent irrigation will probably have to be at low rates over large acreages if large effluent volumes are to be recycled.

Fast growing species known to require more nutrients than pine were tested in other studies in Michigan (Sutherland et al., 1974). Cottonwood

Table 9—Growth of cuttings and seedlings during the second growing season

Tree species	Irrigation level		
	None	2.05 cm/week	5.56 cm/week
	Height growth, cm		
Cottonwood hybrid	27	101	112
Aspen hybrid	33	60	69
Green ash	21	38	36
European larch	18	17	28
Japanese larch	27	38	26
Tulip poplar	5	22	21
White cedar	2	10	11
Red oak	1	1	2

grew 35% better and aspen hybrid nearly 25% better than controls. Growth of other species—Japanese larch, tulip poplar, white cedar, and red oak—were affected little in growth but showed improved survival. Slow initiation of accelerated growth was in part attributable to the use of cuttings that require root generation and starting irrigation late in the year. After the second year of effluent irrigation, growth of all species except red oak increased (Table 9). Growth of cottonwood was quadrupled, aspen doubled, and green ash nearly doubled. The best growth was at the highest irrigation rate (about 5.5 cm/week). Green ash, tulip poplar, and white cedar showed a growth improvement but not as much as the other species. A recent report from the experiment in Michigan using oxidation pond effluent (Urie et al., 1975) indicated that growth of the five tree species doubled when irrigated.

At much higher irrigation rates in Florida (Table 6), the same species evaluated in Michigan (cottonwood, green ash, and tulip poplar) grew rapidly under effluent irrigation on a site too dry and infertile for survival without irrigation. Black locust, catalpa, sycamore, sweetgum, and bald cypress (among other species) also showed rapid early growth (more than 2 m/year); oaks and conifers showed substantially smaller growth rates (< 2 m/year). In all tests mentioned above, full evaluation will have to await longer growth periods. Weed control must be practiced by mowing, cultivating, or using herbicides until the tree seedlings rise above the weed level. Two mowings the first year and one the second year proved adequate in a Florida trial (D. M. Post and W. H. Smith, unpublished report, University of Florida, School of Forest Resources and Conservation, Gainesville).

The importance of water and fertility is demonstrated by the Missouri studies (Settergren et al., 1974) and by the observations of Smith and Post in Florida. In Missouri, where the effluent was not highly enriched with nutrients, the investigators concluded that water was the principal cause of the response. In a Florida trial on sandy, infertile soil with effluent that was not highly enriched with nutrients, little response to effluent irrigation occurred—presumably because of inadequate fertility. The same tree species planted the same year under nutrient-rich effluent (Table 6) at a second site grew rapidly under similar irrigation rates (approximately 5.0 cm/week). Most of the 32 species evaluated were not ecologically adapted to either site for survival without irrigation.

The community of St. Charles, Maryland, is an example of a new planned community utilizing spray irrigation as a means of sewage disposal in lieu of treatment facilities. The choice of woodlands ". . .was based on the utility's need for the cheapest disposal system" (Sullivan et al., 1973). In an effort to clear a wooded area, flooding and washout of the soil resulted —forest mulch left undisturbed induced better drainage. Two and one-half to five centimeters of effluent per day were applied but no effort to monitor tree growth was reported. Among other uses, Santee County Water District in California uses waste water to irrigate a tree farm as does Walt Disney World in Florida (Sullivan et al., 1973). Christmas tree production with waste water irrigation in Michigan (Urie, 1975) is under evaluation.

These studies and others now in progress indicate that certain forest species are well suited to sewage effluent and grow rapidly under treatment. Obviously, rates must be adapted to the species and soil restraints. Where lands are too expensive to purchase or too rough to farm, forests may have a role in municipal effluent disposal. Since disposal systems must operate throughout the year, and since forested areas provide better conditions for water infiltration and phosphorus fixation (Sopper & Kardos, 1972), forests are often superior to agricultural lands as disposal sites. In Pennsylvania woodlands, winter application rates of 0.6 cm/hour were accommodated by a sandy loam soil. Since many workers have found agricultural soils unsuited for winter treatment in cold climates, Sopper and Kardos (1972) concluded that a combination of cropland, which has greater absorptive capacities in summer, and forest lands (for winter) will provide the greatest flexibility in operating a disposal system.

B. Rural Sewage Effluents—Recreational Area Wastes, Animal Waste, etc.

Effluent from recreational areas has increasingly become a problem as many are located on sites where the soil is unsuitable for septic tank disposal. Some recreation facilities are on shallow soil; many are near water reservoirs requiring protection. Another unique feature is that recreational waste waters are relatively "clean." The effluent quality of waste water from a recreation area (Settergren et al., 1974) is compared with that from a municipal source (Sopper & Kardos, 1973) in Table 10. Chemical concentrations in the recreational waste water are low even in comparison to a municipality with little industry. Thus, it is apparent that these wastes are quite different and may involve different strategies and rates in their disposal. Some approaches under investigation by the USDA Forest Service include (i) spray irrigation in plantations of fast-growing tree species, (ii) application to peat filterbeds, and (iii) irrigation in natural forests (Urie et al., 1975).

Agricultural wastes are among the most abundant and their disposal one of the major problems facing farmers and scientists. For a variety of reasons these may not be used fully on cropland. When these wastes are applied to

Table 10—Comparison of nutrient concentrations in recreational area sewage effluent with that of a municipality having little industry

Kind of effluent	Element				
	NO$_3$-N	NH$_4$-N	Inorganic P	K	Na
			ppm		
Recreation area at Bennet Springs					
Year 1	0.60	0.81	0.78	3.18	16.2
Year 2	0.13	1.07	†	3.50	7.3
Pennsylvania, municipal effluent	8.9	0.9	2.65	†	28.1

† Not reported.

cropland, runoff can be a serious problem, especially before the area is fully vegetated. Forest buffer strips may be increasingly used to absorb runoff from such areas. A preliminary study in Maryland of a forested strip between a stream and agricultural field receiving manure proved the strip to be effective in preventing runoff from entering the stream (D. C. Wolf, personal communications, University of Maryland, College Park, Md.).

Many animal concentration areas do not have cropland nearby for waste disposal; in some cases the manure causes damage to the crop at high application rates. An example is caged layer hen operations in Connecticut (Stephens & Hill, 1973a, 1973b). Since large trees were not damaged, liquid poultry manure was disposed of by application to white pine plantations 30–35 years old. In contrast to agricultural crops where applications could be made for only a few months each year, the trees could be irrigated over the entire year and the advantage of perennial storage in wood, roots, and bark could be exploited. After 3 years Stephens and Hill (1973b) concluded that 50% of the nitrogen was volatilized in 2 weeks after irrigation. Storage of N in pine needles was fully 58% greater in the heavily manured plots. On a well-drained site, pine basal area (sum of cross sectioned area of all stems) growth was doubled in 3 years while on a poorly drained soil no increased growth occurred. Furthermore, on the drier site where growth was increased, nitrate in ground water was always < 10 ppm NO$_3^-$ when 115 metric tons/ha were applied; on the wet site, levels as high as 80 ppm were recorded. From these studies, Stephens and Hill (1973b) concluded that, on a sustained basis, 1 ha of pine plantations would safely utilize the manure from 1,000 hens. Many farms include woodlands along with their cropland, and when used together, it should be possible to dispose of animal waste near the site of production.

Dairy waste was successfully applied to a stand of mixed hardwoods at Green Valley Farms in Pennsylvania (Reynolds, 1972). The greatest advantage from woodland irrigation was the capacity to operate during winter. Winter irrigation on agricultural fields of rye led to immediate runoff. Trees were not debarked if pressures were sufficiently high (about 70.3 kg/cm^2)

that the spray broke up into small droplets. Pressures near 63 kg/cm² were deleterious to the trees.

A major problem encountered at Green Valley Farms occurred when a path was cleared in the woodlands. In the cleared area water was not absorbed during winter because disturbance caused reduced infiltration. Thus Reynolds (1972) concluded, "for optimum irrigation it is mandatory that no leaf mulch be removed from the woods in any manner." By using the absorptive capacity afforded by forests, Green Valley was able to operate two to three times longer with their acreage and effluent volumes than was possible with cropland. There was no evidence of environmental or health problems.

At Green Valley, numerous tree species were evaluated under the animal waste irrigation. Early appraisals (Reynolds, 1972) showed that certain species did especially well—sycamore, locust, oak, maple, and Japanese cherry. He concluded, ". . .we feel this (tree growing) would be one of the highest yielding crops in dollar revenues." While some trees increased stem diameters up to 70% in 1 year, others could not tolerate heavy watering nor weed competition—the biggest problem in the effort.

C. Industrial Effluents

Industrial effluents vary widely in composition and in disposal difficulty. Cannery wastes possess high levels of organics that demand oxygen for decomposition but are low in the troublesome toxic mineral elements and NO_3^-. An early user of landspreading of industrial effluent found forest soils especially suited for handling cannery waste (Sullivan et al., 1973). Researchers at Seabrook farms in New Jersey found in 1950 that a sandy soil under clover soon became waterlogged when irrigated with 5.1 cm of waste water per week. When the sprinkler was moved 180 m into a woodland no flooding occurred at 20.3 cm/week (nor by irrigating 48 hours at a rate of 2.5 cm/hour). Because of rapid infiltration and percolation, Seabrook handled 2,206 m³/hour on 125 ha of woodland. After 23 years of operation this system has achieved a high degree of water purification without loss of efficiency. However, the original vegetation has been replaced by species adapted to the new environment.

Cannery waste water was applied to forest tree species in Michigan by Rudolph and Dils (1955). Weed growth apparently overtopped the young trees and reduced survival; but the living trees (especially native willow and cottonwood) grew about twice as fast in height as unirrigated trees. Recently, this effort to grow trees was abandoned, although growth rates were outstanding, because the trees interfered with the disposal system (Sullivan et al., 1973). Spraying onto the tree crowns also caused ice to form and damaged the trees. In both the Seabrook, New Jersey, and the Michigan situations, tree cover disappeared mainly because the water distribution systems were not adjusted to the structural characteristics of trees.

Commercial Solvents Corporation in Indiana applied from 7.6 to 25.4 cm daily of fermentation waste to woodlands (Sullivan et al., 1973). Again the woodland was hydrologically suited to handle the water, but the heavy application rates killed the trees. When applied to bare ground the soil sealed quickly.

Where organic filtering is the disposal objective, forestland is especially suited for high hydrologic loadings. Although research on renovation efficiency has been pursued, there is a dearth of work to determine forest species suitable to the environments created or to devise a water handling system that is adapted to the structural characteristics of trees. Obviously, aerial sprinklers, especially with high pressure, are poorly suited to large trees even where winter ice is not a problem.

Pulpmill effluents are like cannery waste waters in that they are rich in organic substances (wood fibers). But, these wastes are unlike cannery wastes since wood fibers possess high C/N ratios that could lead to nitrogen deficiencies (Philip, 1971). Some mills dilute their effluent because the Na content may be as high as 800 mg/liter (Watterston, 1971). In irrigation trials on fine textured soils, Na caused dispersal of the clays and "plugging" of soil pores. In sandy soils where infiltration and percolation were not impeded, the probability of contamination of subsurface water by Na from extended use was high. Kadamki (1971) reported the variability among tree species in their cation concentrations when grown in pots irrigated with pulpmill effluents. Certain eucalyptus species contained high sodium concentrations, while the species of this genus characterized as a halophypte appeared to exclude Na.

Kadamki (1971) did not report growth tolerance under pulpmill effluent nor did Watterston (1971) report the infiltration and percolation characteristics when the pulpmill waste was applied to an undisturbed forest floor with litter mulch in place. As with other effluents, the capacity of a forest system to thrive and renovate the waste water will depend on tolerances of the tree species and on inherent properties of the soil.

D. Sludges

Sludges are more enriched in organics and minerals than effluents, although they typically contain 95+% water (Table 11). Removing water from sludge reduces their volume and makes them easier to handle but at a considerable energy cost (Evans, 1970). Chemical composition of sludges varies widely depending largely upon the industrial activities in the sewage collection area.

The practice of sludge disposal in forests has not been extensive, probably because of limited demand for forest disposal sites and distribution difficulties. Because of the frequency of potentially hazardous substances present in sludge such as certain heavy metals, pesticides, and organic toxicants, forests should be considerably more suitable than food croplands for disposal

Table 11—Range of elemental content in waste water effluents and sludges (Pound &
Crites, 1973; Jorgensen, 1965)

Element	Waste water	Sludge
	ppm	ppm dry weight
Total N	10 to 60	35 to 64 ($\times 10^3$)
Total P	8 to 25	8 to 39 ($\times 10^3$)
K	10 to 40	2 to 7 ($\times 10^3$)
Ca	20 to 120	10 to 50 ($\times 10^3$)
Mg	10 to 50	5 to 30 ($\times 10^3$)
Zn	60 to 750	50 to 500
Cu	40 to 100	250 to 17,000
Ni	--	25 to 8,000
Cr	0 to 10	--
Cd	--	5 to 2,000
B	0 to 0.10	15 to 1,000
Pb	10 to 55	100 to 10,000
Hg	--	1 to 10
As	2 to 8	--

of sludges. Although sludges have been shown to be beneficial to the growth of food crops, a recent evaluation (J. Webber. 1974. Sludge handling and disposal practices in England. Ministry of Agriculture, Fisheries, and Food, Leeds, England) of literature and current practice, led Webber to conclude, ". . .many examples have been encountered over the years of reduced crop growth from this cause (content of metals), especially where sludges from industrial towns have been used." In Webber's review, Zn, Cu, and Ni were most toxic to the plants while Pb, Cd, Hg, and perhaps Cr, were more of a risk to domestic animals and humans eating the crops than to the crops themselves. Because some metal-enriched sludges cannot be applied repeatedly to agricultural soils, and since some foods prepared from them may not be safe, forests seem especially suited for sludge disposal; their products are not eaten by humans, and vast acreages allow for shifting application sites.

Most nonagricultural uses of sludge on land have been for reclamation of disturbed lands such as strip mine spoils in the Shawnee National Forest in southern Illinois (Sutton & Vimmerstedt, 1973). Sludge mixed with acid spoil reduced the acidity and decreased the metals content in percolate (Lejcher & Kunkle, 1973). This treatment improved the spoil medium such that 90% cover developed by the end of the first year, whereas seed germination failed on the untreated spoil site. In other studies, leachate from sludge-treated, acid, strip mine spoil contained lowered levels of all metals at the high sludge rates (Urie, 1976).[3] Only N increased in the leachate as sludge rate increased.

We know of few published reports on sludge applied to undisturbed forest soils, although several studies are underway. However, many forest

---

[3]D. H. Urie. 1976. Unpublished Progress Report, USDA, Forest Service, East Lansing, Mich.

soils are strongly acid and possess the potential to fix sludge borne chemicals in a manner analogous to the spoil materials. Control of pH is fundamental to fixation of elements by soil, so liming may be useful to enhance fixation in some cases (Urie, 1975). A comparison of sludge application on a forest soil is the evaluation of subsoil injection of campground sewage vault waste to a native meadow (Cunningham et al., 1974). This toilet waste from vaults (untreated) possessed titres of viable coliforms. With adequate isolation on level, moderately well-drained soils (water table at 3 m), the soils had adequate filtration capacity to protect ground water from bacterial and virus pollution (Urie, 1976).[3]

Sludge application to mine spoils, to soil contaminated by Cu ore processing, to eroded Piedmont lands, and to borrow pits in the southeastern United States are under study by C. R. Berry and D. H. Marx of the U. S. Forest Service, Athens, Ga. On a severely eroded site, loblolly pine and shortleaf pine growths were best where 17,200 kg/ha of dry sludge were spread on the soil surface and mixed by discing (Berry, 1977). Weed competition probably caused most of the seedling mortality at the high sludge rate (68,800 kg/ha). An important part of this research is the interaction of sludge with mycorrhizal inoculations. In micro-plot studies (Berry & Marx, 1976) using soil from the eroded site, pine with *Pisolithus tinetorius* mycorrhizae grew best at sludge rates of 138 metric tons/ha. Sludge may provide the organic substrate necessary for establishing these symbiotic fungi known to be important in the nutrition and protection of trees against pathogens.

A rather comprehensive study of sludge applications on forest land is underway in Washington state. The first report from this work (H. Riekerk, and D. W. Cole. 1975. Utilization of sewage sludge on forest lands. Agron. Abstr. p. 168.) where large amounts of dewatered sludge were applied to 12 ha of forest land having moderately coarse-textured soils revealed several significant findings. These are summarized as follows: (i) trees planted directly in sludge had poor survival and early growth, (ii) Douglas fir seedlings planted in sludge aged 12 months on the site appeared to do well, (iii) NO_3^- from the sludge moved in significant concentrations into the soil profile and became detectable (up to 30 ppm NO_3^-) in ground water, (iv) fecal coliform, which was initially high in the sludge, was considerably reduced after 30 weeks and none was detected in ground water, and (v) weedy vegetation responded markedly to the sludge application.

Nitrate leaching, N transformations, cation adsorption, loading rates, tree species response, need for other amendments such as lime and/or fertilizers, mycorrhizal inoculations, application techniques, and other problems are a part of the continuing efforts of several studies now in progress.

E. Solid Wastes

Solid wastes other than sludge (such as garbage) represent a substantial disposal problem. The U. S. Bureau of Solid Waste Management has estimated that per capita solid waste production amounts to about 2.3–3.6 kg/

Table 12—First year heights of slash pine and plant biomass production following garbage incorporation in a Florida forest soil

Garbage rate	Broadcast and disced		In furrow	
	Height	Biomass	Height	Biomass
metric tons/ha	cm	kg/ha	cm	kg/ha
112	68	5,070	72	7,000
224	78	7,600	71	14,600
448	65	10,600	56	12,230
Control	40	3,600	--	--

day and that 94% of all land disposal systems (mainly landfills) are unsatisfactory. Most composts have some value to crops (Hortenstine & Rothwell, 1968) but because of bulk, processing and transportation costs, and variable composition, their use has not expanded. In the absence of a market for garbage, incorporation into forest soil during reforestation may be a suitable recycling mechanism.

Garbage compost was applied to an excessively drained sandy soil on Tennessee Valley Authority lands in Florida supporting slash pine (G. W. Bengtson, and J. J. Cornette. 1971. Disposal of composted municipal waste in a young slash pine plantation: effects on soil and trees. Agron. Abstr. p. 117.) to improve water retention. No improvement in growth occurred but neither was growth adversely affected at rates up to 45 metric tons/ha. In sharp contrast, W. H. Smith and D. M. Post (1974. Organic municipal waste disposal in a slash pine plantation. Agron. Abstr. p. 178.) applied garbage compost to a somewhat poorly drained spodosol in Florida; treatment doubled growth of planted slash pine and tripled plant biomass production (Table 12). The garbage (household refuse from which ferrous metals were removed) was coarsely ground by the City of Gainesville Waste Disposal Authority, wetted as necessary with sewage sludge, and composted 5 days. The material was applied to a recently clear-cut forest area and either broadcast and disced into the soil or placed in furrows at rates of 112, 224, and 448 metric tons/ha (wet). Both areas were bedded immediately during summer and planted to slash pine in December.

Although the area was not monitored by soil solution analysis, soil extraction suggested that chemicals released by the garbage did not pass through the spodic horizon. Chemical analysis of plant tissues (ground cover and tree parts) showed that reductions in growth at the high garbage rate (although growth was greater than controls) was because of ground cover competition rather than toxic levels of heavy metals (Table 13).

Because of the perennial nature of forest trees and preparation procedures used in reforestation, ground biodegradable garbage could be incorporated into forest soil and composted in the field—obviating the cost of structures, storage, and handling facilities.

Table 13—Chemical concentrations of selected trace elements in plant tissues from control plots, the high treatment plots, and intermediate treatments where growth was best

	Pine foliage				Ground vegetation			
	Fe	Zn	Mn	Cu	Fe	Zn	Mn	Cu
				ppm				
Control	40	52	241	7.5	306	25	111	4.2
224 metric tons/ha broadcast	31	66	100	5.8	267	88	71	6.7
448 metric tons/ha in furrow	42	66	130	5.4	310	137	76	6.3

IV. TRANSPORT, DISTRIBUTION, LOGISTICS, AND ENGINEERING FACTORS

Land spreading of municipal and industrial wastes involves serious "delivery" problems regardless of the intended plant community. For local forests consisting of species which will not tolerate high effluent loads, extensive piping, trucking, or other types of transport will be required to deliver wastes to the land. Many waste disposing agencies already transport sludge to landfills by surface vehicles. In this case, a series of central but relocatable distribution centers seem advisable for dispersing the sludge among the lands near the center. Each year or two the system will need to be removed to a new area. Because of the danger of soil pollution, continued operation at high loading levels on the same land is not advisable.

Irrigation systems designed for agricultural crops have generally been used in forest irrigation trials. In several cases the tests were abandoned because the water under high pressure physically damaged the trees (Little et al., 1959), or else the trees interfered with the dispersal of water. An area almost totally unresearched is the design of waste handling devices uniquely suited to the characteristics of forests. Perhaps lines of perforated pipe operated at low pressure just above the shrub layer would be much more effective than "rainbird-type" sprinklers designed for low growing agricultural plants. Alternatively, ditch irrigation, underground discharge pipes, or other seepage distribution systems may warrant attention.

V. FOREST MANAGEMENT ALTERNATIVES AND UTILIZATION STRATEGIES

A. Multiple Cropping

Certain deciduous hardwoods may prove to be most desirable for waste application on the basis of growth and nutrient absorption criteria. Unfortunately, hardwoods are unproductive in winter. Certain agronomic crops may be seeded under properly spaced trees at leaf fall and harvested annually by grazing or mechanical methods. Trees could be harvested at longer intervals depending on growth rate. Alternatively, shade-tolerant trees (especially understory species such as dogwood, red bud, holly, American olive, red cedar, Atlantic white cedar, etc.) could be grown in the interrow area beneath deciduous trees for transplanting purposes. Multilayered arrangements may increase water handling capacity for a land unit and also enhance esthetic appeal of the landscape. Erosion from surface flow causing sedimentation problems, which are typical of some agricultural systems, are virtually non-existent in forests where infiltration is rapid.

B. Minirotation Forestry

Certain forest species (especially deciduous hardwoods) grow rapidly and produce satisfactory fiber at an early age. These species could be planted at close spacings, harvested after 2 to 3 years like silage (McAlpine et al., 1966), and the total aboveground biomass could be used for paper and other products (DeBell, 1975). Sycamore, which performed well in the Tallahassee, Florida waste water disposal trials, also has performed well in "silage sycamore" tests. Dry biomass production by sycamore in a minirotation (4 years) without irrigation or fertilizer in Georgia totaled more than 35 metric tons/ha (Saucier et al., 1972). Coppicing, by allowing regeneration of sprouts from the cut stumps, eliminates a number of costly replanting operations; each tree planting can provide about three coppice crops.

Other wetland hardwoods are capable of rapid early growth and production of fibers useful in high quality paper. Cottonwood is an example of fast-growing species which is responsive to fertilizer and abundant water so long as flooding does not occur. Typical growth rates on bottomland sites in Mississippi and southwest Alabama are reported to range from 2 to 3 m/year (Maisenhelder, 1960). In Alabama, the investigators determined that 76% of the height growth of the eight stands could be accounted for by the supply of Ca, K, and P in the tree foliage—important elements contained in wastes (White & Carter, 1968). Waste water irrigation of soils with good internal drainage could result in comparable growth of stands which would be harvestable in 5 years. Poplars, sweetgum, and other hardwoods may grow similarly. Tests now underway in many locations and involving several species will allow for appraisal of productivity and nutrient removal in minirotation forestry (White & Hook, 1975). Belts of such trees can also serve as wind or spray-drift barriers on land supporting low-growing agricultural crops.

C. Explant Production

Until now, discussion has centered on forest tree production for wood fiber and wood products—conventional production forestry. Environmentally aware citizens now recognize the existence of urban forests and their amenity values. Frequently, cities occur in forests and many of the trees now standing are remnants of the original stand. For example, metropolitan Atlanta with a population of 1,500,000 people occupies 0.61 million ha with trees covering 65% of the land (Shirley, 1971). People of Atlanta, like others nationwide, are discovering that trees have values beyond esthetics. Although they are "things of beauty," trees also moderate temperature, wind, and noise; they affect zoning and screen the air of harmful pollutants. Through tree, greenbelt, and flood plain ordinances, many cities are requiring that trees should not only be protected, but should be increased and optimally used. Waste water irrigation in explant production may be a way for municipalities to supply such tree planting stock (entire trees of substantial size lifted with soil root ball intact).

Wastes are potentially useful in nursery production of planting stock for regenerating trees in greenbelts, parks, street liners, and house yards. Tree production at close spacings can lead to large biomass and nutrient yields resulting from whole tree harvest. In fact, such operations may even lift a portion of the nutrient-rich soil along with the tree roots. These operations show great promise not only because they yield marketable trees useful for maintaining urban forests, but also because of total harvest with substantial nutrient export.

Tests conducted mainly at Tallahassee, Florida (Smith & Post, 1973) are evaluating numerous tree species under waste water irrigation to determine those desirable for explanting. Unlike earlier tests where forest habitats developed, these are single-row evaluations suited to lifting and explanting trees to parks, street margins, parking lot perimeters, and home yards.

Several species identified in the tests are also potentially useful for explanting as well as in multiple-cropping schemes. Certain species (e.g., red cedar) should also be evaluated for their potential value in Christmas tree production. Christmas tree production with waste water irrigation is under investigation in Michigan (Urie et al., 1975). Principals in the Green Valley Farms, Pennsylvania waste water project concluded that nursery production at forest trees would generate the maximum economic return among the several cropping systems employed (Reynolds, 1972). The most important benefit, however, is nutrient removal.

D. Management Problems

Minimizing management costs is always desirable, although profit generation may not be the principal objective in land application of wastes. A first step involves development of a low cost, low management water distribution system. Although considerable work has been done on systems for agriculture, little has been done for forest situations. In the final analysis, sprinkler systems commonly used in agriculture may not be as desirable in forests as ditch or trickling pipe systems which could use low cost material and low energy demanding pumps. A recent study in Wisconsin tested several effluent application systems (A. F. Harris. 1975. Evaluations of several methods of applying effluent to forested areas in the winter. Agron. Abstr. p. 166.). Surface and subsurface flooding and sprinklers were employed, but further study is necessary before conclusions can be made concerning the most desirable and useful forest application systems.

In any of the forest tree production schemes discussed, planting and early weed control will constitute the major costs beyond seedling purchase price. Seedling cost is minimal for mass-produced species (currently $7–15/thousand for species such as cottonwood, sweetgum, sycamore, pine, etc.) while the lesser available species will be more costly (currently $7–30/hundred for species such as dogwood, walnut, and birch). When plant composition is controlled early so that the desired trees become dominant, shading often keeps nonmarketable species checked afterwards. These "undesirable" species, however, may constitute the food source for numerous wildlife exporters of the nutrients absorbed during their growth.

Soil compaction leading to poor water infiltration can result if frequent passes with mechanical equipment are required. With tree species, the necessity for use of heavy machinery will not cause serious problems since a ground cover by other plants can be tolerated and because of litter production by the tree community.

After considerable forest floor development, controlled burning may also be a useful management tool. Combustion would volatilize N and S resulting in nutrient reductions and prepare the ground for seeding companion crops if multiple cropping is practiced. The litter, however, has value in aiding and maintaining soil structural integrity and facilitating infiltration. Fortunately, controlled buring does not consume all the organic debris.

Recently developed harvesting procedures and utilization processes now make it possible to use complete trees—roots, stems, branches, and foliage (Koch, 1974). Since most nutrients are in the crown of trees (Smith et al., 1971) such utilization practices will lead to increased nutrient export from sites receiving wastes.

Table 14—Tree species recommended on basis of height growth as low, high, and intermediate responders to chemical and moisture loading from effluent (D. M. Post and W. H. Smith, unpublished results, University of Florida, Gainesville)

Response class to effluent		
Low	Intermediate	High
Slash pine	Tulip poplar	Cottonwood
Cherry laural	Bald cypress	Sycamore
Arizona cypress	Saw-tooth oak	Green ash
Live oak	Red cedar	Black cherry
Holly	Laurel oak	Sweet gum
Hawthorne	Magnolia	Black locust
	Nuttall oak	Red bud
	Cherrybark oak	Catalpa
	Loblolly pine	Chinese elm

Because conditions exist that favor the use of forests for recycling wastes, forest species selection becomes important in the design of the tree production and waste utilization systems. A number of tree species under an array of loading rates are recommended in Table 14. This preliminary grouping was made according to rate of tree height growth following 2 years of effluent irrigation. Final groupings must await longer term observations and nutrient yield determinations.

VI. CONCLUDING SUMMARY

Two important subjects are addressed: (i) potential environmentally acceptable waste disposal procedures; and (ii) the potential for using wastes on forest lands and nurseries safely, economically, and beneficially to aid in the wood fiber supply. The need for recycling wastes and for increasing cellulose fiber production is rapidly increasing. The renewability of forest resources and high biodegradability of forest products are factors favoring the use of wood fiber over that of finite resources such as petrochemicals and fossil fuels.

Research summarized here indicates that waste application to forests and tree plants can contribute to renewing the wood fiber resource and solving waste recycling problems provided proper tree species are selected; wastes are properly adjusted to the ecological requirements of the trees; distribution systems are designed to fit structural characteristics of forests; and the optimum tree production systems are selected, installed, and properly managed.

LITERATURE CITED

Allison, F. E. 1965. Decomposition of wood and bark sawdust in soil, nitrogen requirements, and effects on plants. USDA Tech. Bull. 1332. 58 p.

Anderson, T. D., and A. R. Tiedmann. 1970. Periodic variation in physical and chemical properties of two central Washington soils. U. S. For. Serv. Res. Note PSN-125. 9 p.

Ballard, R., and J. G. A. Fiskell. 1973. Phosphorus retention in coastal plain forest soils. Soil Sci. Soc. Am. Proc. 38:363–366.

Baumgartner, A. 1967. Energetic bases for differential vaporization from forest and agricultural lands. p. 381–390. In W. E. Sopper and H. W. Lull (ed.) Int. Symp. on Forest Hydrology. Pergamon Press Ltd., Oxford.

Berry, C. R. 1977. Growth of pine and associate vegetation on an eroded forest site amended with dried sewage sludge. U. S. For. Serv., SE For. Exp. Stn. Paper. (In press.)

Berry, C. R., and D. H. Marx. 1976. Sewage sludge and Pisolithus tinctorius ectomycorrhizae: their effect on growth of pine seedlings. For. Sci. 22(3):351–358.

Cunningham, R., L. Tluczek, and D. H. Urie. 1974. Soil incorporation shows promise for low cost treatment of sanitary vault wastes. U. S. For. Serv. Res. Note NO-181. 3 p.

DeBell, D. S. 1975. Short-rotation culture of hardwoods in the Pacific Northwest. Iowa J. Res. 49:345–352.

Dils, R. E. 1953. Influence of forest cutting and mountain farming on some vegetation, surface soil, and surface runoff characteristics. U. S. For. Serv., SE For. Exp. Stn. Pap. 24. 55 p.

Evans, J. O. 1970. The soil as a resource renovator. Environ. Sci. Technol. 4:732–735.

Grime, J. P., and J. G. Hodgson. 1969. An investigation of the ecological significance of lime chlorosis by means of large scale comparative experiments. p. 67-100. *In* I. H. Rorison (ed.) Ecological aspects of mineral nutrition of plants. Blackwell Scientific Publ., Oxford.

Hornbeck, J. W., and K. G. Reinhart. 1964. Water quality and soil erosion as affected by logging in steep terrain. J. Soil Water Conserv. 19:23-27.

Hortenstine, C. C., and D. F. Rothwell. 1968. Garbage compost as a source of plant nutrients for oats and radishes. Compost Sci. 9:25.

Jorgensen, J. R. 1965. Irrigation of slash pine with papermill effluents. Div. of Eng., Louisiana State Univ. Res. Bull. No. 8.

Kadamki, K. 1971. Accumulation of sodium in potted soil irrigated with pulpmill effluents. Consultant 16:93-94.

Koch, P. 1974. Harvesting southern pine with taproots can extend pulpwood resource significantly. J. For. 72:266-268.

Lejcher, T. R., and S. K. Kunkle. 1973. Restoration of acid spoil banks with treated sewage sludge. p. 184-199. *In* W. E. Sopper and L. T. Kardos (ed.) Recycling treated municipal wastewater and sludge through forest and cropland. Penn. State Univ. Press, University Park, Pa.

Little, S., H. W. Lull, and I. Remson. 1959. Changes in woodland vegetation and soils after spraying large amounts of wastewater. For. Sci. 5:18-27.

Lyon, T. L., H. O. Buckman, and N. C. Brady. 1952. The nature and property of soils. The MacMillan Co., New York. 591 p.

McAlpine, R. G., C. L. Brown, A. M. Herrick, and H. E. Ruark. 1966. "Silage" sycamore. For. Farmer 26:6-7, 16.

McFee, W. W., and E. L. Stone. 1969. Ammonium and nitrate as nitrogen sources for *Pinus radiata* and *Picea glauca*. Soil Sci. Soc. Am. Proc. 33:812-817.

Maisenhelder, L. C. 1960. Cottonwood plantations for southern bottom lands. U. S. For. Serv., South. For. Exp. Stn. Occas. Pap. 179. 24 p.

Metz, L. J., and J. E. Douglass. 1959. Soil moisture depletion under several Piedmont cover types. U. S. For. Serv. Tech. Bull. No. 1207. 23 p.

Miller, W. F. 1967. Physical and chemical properties of forested soils. Miss. Agric. Exp. Stn. Bull. 734. 112 p.

Olson, O. C., and E. A. Johnson. 1973. Forest Service policy related to the use of national forestlands for disposal of wastewater and sludge. p. 435-437. *In* W. E. Sopper and L. T. Kardos (ed.) Recycling treated municipal wastewater and sludge through forest and cropland. Penn. State Univ. Press, University Park, Pa.

Penman, H. L. 1967. Evaporation from forests: a comparison of theory and observation. p. 373-389. *In* W. E. Sopper and H. W. Lull (ed.) Int. Symp. of Forest Hydrology. Pergamon Press Ltd., Oxford.

Penneypacker, S. P., W. E. Sopper, and L. T. Kardos. 1967. Renovation of wastewater through irrigation of forest land. J. Water Pollut. Control Fed. 39:285-296.

Philip, A. H. 1971. Disposal of insulation board effluent by land irrigation. 2 vol. U. S. Environ. Prot. Agency, EPA 660/2-73-006a. Washington, D. C.

Pierce, R. S. 1967. Evidence of overland flow on forest watersheds. p. 247-252. *In* W. E. Sopper and H. W. Lull (ed.) Int. Symp. on Forest Hydrology. Pergamon Press Ltd., Oxford.

Pound, C. E., and R. W. Crites. 1973. Wastewater treatment and reuse by land application. 2 vol. U. S. Environ. Prot. Agency, EPA 660/2-73-006. Washington, D. C.

Rennie, P. J. 1955. The uptake of nutrients by mature forest growth. Plant Soil 7:49-95.

Reynolds, B. J. 1972. Land treatment by spray irrigation: An ecological blueprint for today. Benjamin J. Reynolds, Avondale, Pa.

Richards, B. N. 1962. Increased supply of soil nitrogen brought about by pines. Ecology 43:538-541.

Ross, E. W. 1973. Important diseases of sand pine. p. 199-206. *In* Proc. Sand Pine Symp. U. S. For. Serv. Gen. Tech. Rep. SE-2.

Rudolph, V. J., and R. E. Dils. 1955. Irrigating trees with cannery wastewater. Mich. Agric. Exp. Stn. Quart. Bull. 37(3):407-411.

Saucier, J. R., A. Clark, III, and R. G. McAlpine. 1972. Aboveground biomass yields of short-rotation sycamore. Wood Sci. 5:1-6.

Settergren, C. D., J. A. Turner, and W. F. Hansen. 1974. The use of sewage effluent irrigation techniques at large recreational developments. p. 272–282. In Proc. Soc. Am. For., Forestry Issues in Urban America. Soc. of Am. For., Washington, D. C.

Shirley, A. R. 1971. Social and legal factors affecting environmental forestry programs. p. 95–99. In Symp. on the Role of Trees in the South's Urban Environment. Univ. of Georgia, Athens.

Smith, W. H., L. E. Nelson, and G. L. Switzer. 1971. Development of the short system of loblolly pine. II. Dry matter and nitrogen accumulation. For. Sci. 17:55–62.

Smith, W. H., and D. M. Post. 1973. Wastewater disposal in forests and the production of forest plants. p. 102–103. In Proc., 1973 workshop: Landspreading municipal effluent and sludge in Florida. Inst. of Food and Agric. Sci., Coop. Ext. Serv., Gainesville.

Sopper, W. E. 1968. Wastewater renovation for reuse: key to optimum use of water resources. Water Res. 2:471–480.

Sopper, W. E., and L. T. Kardos. 1972. Effects of municipal wastewater disposal on the forest ecosystem. J. For. 70:540–545.

Sopper, W. E., and L. T. Kardos. 1973. Vegetation responses to irrigation with treated municipal wastewater. p. 271–294. In W. E. Sopper and L. T. Kardos (ed.) Recycling treated municipal wastewater and sludge through forest and cropland. Penn. State Univ. Press, University Park, Pa.

Sopper, W. E., and L. J. Sagmuller. 1966. Forest vegetation growth responses to irrigation with municipal sewage effluent. p. 639–647. Proc. First Am. Soil Conserv. Congr. San Paulo, Brazil.

Stephens, G. R., and D. E. Hill. 1973a. Utilizing liquid poultry waste in forests. Conn. Agric. Exp. Stn. Spec. Soils Bull. 31. 7 p.

Stephens, G. R., and D. E. Hill. 1973b. Utilizing liquid poultry wastes in woodlands. p. 234–242. In Proc. Int. Conf., Land for Waste Management, Ottawa, Canada.

Stone, E. L., and R. R. Morrow. 1954. A malady of red pine on poorly drained sites. J. Forest. 52:104–114.

Sullivan, R. H., M. M. Cohn, and S. S. Baxter. 1973. Survey of facilities using land application of waste water. U. S. Environ. Prot. Agency, EPA 430/9-73-006. Washington, D. C.

Sutherland, J. C., J. H. Cooley, D. G. Neavy, and D. H. Urie. 1974. Irrigation of trees and crops with sewage stabilization pond effluent in southern Michigan. p. 295–313. In Proc. Wastewater Use in the Production of Food and Fiber. U. S. Environ. Prot. Agency, EPA 660/2-74-041. Washington, D. C.

Sutton, P., and J. P. Vimmerstedt. 1973. Treat stripmine spoils with sewage sludge. Ohio Rep. 58:121–123.

Swank, W. K., and J. E. Douglass. 1974. Streamflow greatly reduced by converting deciduous hardwood stands to pine. Science 185:857–859.

Switzer, G. L., L. E. Nelson, and W. H. Smith. 1968. The mineral cycle in forest stands. p. 1–19. In Forest fertilization—theory and practice. Tennessee Valley Authority, Knoxville, Tenn.

Theobald, W. N., and W. H. Smith. 1974. Nitrate production in two forest soils and nitrate reduction in pine. Soil Sci. Soc Am. Proc. 38:668–672.

Trimble, G. R., Jr., R. S. Sartz, and R. S. Pierce. 1958. How type of soil forest affects infiltration. J. Soil Water Conserv. 13:81–82.

Urie, D. H. 1975. Nutrient and water control in intensive silviculture on sewage renovation areas. Iowa J. Res. 49:313–317.

U. S. Forest Service. 1973. For Resour. Rep. No. 20. Washington, D. C. p. 9.

Watterston, K. G. 1971. Water quality in relation to fertilization and pulpmill effluent disposal. Consultant 16:91–92.

White, E. H., and M. C. Carter. 1968. Relationships between foliage nutrient levels and growth of young natural stands of Populus deltoides Bartr. p. 283–294. In Tree growth and forest soils. Oregon State Univ. Press, Corvallis, Oreg.

White, E. H., and D. D. Hook. 1975. Establishment and regeneration of silage plantings. Iowa J. Res. 49:287–296.

Wood, G. W., D. W. Simpson, and R. L. Dressler. 1973. Effects of spray irrigation of forests with chlorinated sewage effluent on deer and rabbits. p. 311–323. In W. E. Sopper and L. T. Kardos (ed.) Recycling treated municipal wastewater and sludge through forest and cropland. Penn. State Univ. Press, University Park, Pa.

Wooldridge, D. C. 1970. Chemical and physical properties of forest litter layers in central Washington. p. 327–338. *In* Tree growth and forest soils. Oregon State Univ. Press, Corvallis, Oreg.

Youngner, V. B., W. D. Kesner, A. R. Berg, and L. R. Green. 1973a. Ecological and physiological implications of greenbelt irrigation with reclaimed wastewater. p. 396–407. *In* W. E. Sopper and L. T. Kardos (ed.) Recycling treated wastewater and sludge through forest and cropland. Penn. State Univ. Press, University Park, Pa.

Youngner, V. B., T. E. Williams, L. R. Green, and R. W. Shultz. 1973b. Ecological and physiological implicaitons of greenbelt irrigation. Program report of the Maloney Canyon Project. Univ. of California, Riverside (U. S. For. Serv. cooperating). 105 p.

chapter 17

Tractor with front-end manure spreader works on a field in Indiana (Photo courtesy of Allis-Chalmers Mfg. Co.).

Land Utilization and Disposal of Organic Wastes in Cool Subhumid and Humid Regions

L. R. WEBBER and E. G. BEAUCHAMP
University of Guelph, Guelph, Ontario, Canada

I. INTRODUCTION

A measure of man's ability to feed himself is manifested in his manipulation and management of the environment. The primary function of land must be the production of food, feed, and fiber but as the earth's population continues to increase, more and more pressure is being applied to use land for disposal of many kinds of waste. If we accept the premise that expansion of agriculture into new lands has about ended, then man's best management of land currently under use is crucial.

This symposium has been structured for the express purpose of evaluating the state of our knowledge in using land, primarily agricultural land, as the ultimate sink for organic wastes and waste waters. Three broad climatic regions are to be considered in terms of waste disposal. This approach appears timely because climate dominates the environments of soil, air, and water. The ultimate fate of chemical and biological materials when applied to soil depends upon the physical, chemical, and biological properties of the soil.

II. FEATURES OF COOL SUBHUMID AND HUMID REGIONS

For this discussion, the cool subhumid and humid regions for North America are arbitrarily taken to include those agricultural areas lying north of the 10C (50F) average annual isotherm and south of the permafrost in Canada. This region includes all or part of the northeastern states, New York, Pennsylvania, Ohio, Michigan, Wisconsin, Minnesota, the Dakotas, Montana, Washington, Oregon, and Canada. The semiarid and cool portions of the southwest and north-central USA and southern Alberta and Saskatchewan are excluded (Fig. 1).

Climatic parameters cover a very wide range of conditions. In the Peace River country in northern Alberta the mean annual temperature is 4C (25F) and annual precipitation is 32 cm (13 inches), with 19 cm (7.5 inches) occurring in the growing season from May to September. Throughout the western plains of the USA and Canada, precipitation is generally < 50 cm (20 inches) with pronounced moisture deficits. In parts of Washington state, precipitation increases to values in excess of 250 cm (100 inches). Length of growing season or cumulative number of degree-days (temperature > 5C) decreases as one goes from south to north. A summary of selected climatic parameters is given in Table 1.

Obviously the region includes many kinds of soils. The northern limit of potential agricultural land is the Peace River country, with an estimated area of 4 million ha in northern Alberta. The soils of this region are dominantly Cryoboralls (Gray Luvisols or Gray Wooded). As one proceeds southerly the soils include Boralls (Black, Dark Brown, and Brown Chernozems). In the eastern states and Canada, the soils are dominantly Udalfs (Gray Brown Luvisols), Aquolls (Gleysols), Eutrochrepts (Brunisolic), and

Fig. 1—The cool subhumid and humid regions of North America are considered to occur between the permafrost on the north and the 10C average annual isotherm on the south. The area includes a semiarid cool region as shown by the dotted line.

Humods and Orthods (Podzols). All drainage classes are found in the region from well-drained mineral soils to the muskegs and bogs of Minnesota, Wisconsin, and northern Canada.

Naturally, the many kinds of soils and the range in climatic variables provide an environment for a wide range of agricultural crops. Throughout the region, the principal crops are spring and fall-seeded grains, grasses, and legumes. Corn (*Zea mays* L.) for silage and grain is concentrated primarily in the North-central States, New York, Pennsylvania, and southern Ontario. Limited acreages of corn are found in the Fraser Valley, B. C., southern Manitoba and southeast Quebec, and in some states in the northeastern USA.

Table 1—Selected climatic parameters for the cool subhumid and humid regions of North America

	Range of values, southern boundary states of the USA to Peace River, Canada
Latitude, north	40 to 55°
January mean temperature	−7 to −14C (20 to 6F)
July mean temperature	21 to 16C (70 to 60F)
Frost-free days	160 to 80
Start of growing season daily mean >5C	1 May to 1 June
End of growing season	1 October to 1 September
Average annual precipitation	30 to 250 cm
Frost penetration	few cm to few meters

III. SOIL-CLIMATE PROBLEMS

A. Natural Soil Drainage

The kind of microbial degradation which occurs when organic wastes are incorporated into the soil depends significantly on soil drainage. Soils that are naturally well drained normally provide an aerobic environment for waste decomposition. In the presence of free oxygen, more or less complete oxidation of carbonaceous compounds is achieved.

In the northern climatic region, a significant portion of the land mass is occupied by poorly drained organic and mineral soils. The water table may be permanently high as in undrained organic soils, bogs, peats, and muskeg. High water tables contribute to an oxygen-deficient medium, with the result that decomposition of applied organic wastes is retarded.

Under anaerobic conditions the breakdown of organic materials which leads to the release of NH_4^+ proceeds at a slower rate than in a well-drained soil. Mineralization of N does not proceed beyond the NH_4^+ stage because oxygen is necessary for the microbial conversion of NH_4^+ to NO_3^- (Patrick & Mahapatra, 1968).

The disposal and utilization of organic wastes containing N on soils without contributing to environmental pollution involves the fate and ulti-mate sink of NH_4^+-N and NO_3^--N. Olsen et al. (1970) determined exchange-able NH_4^+-N and NO_3^--N in the soil solution under four soil conditions: un-disturbed, cultivated, poorly drained, and well-drained barnyards. The high content of NH_4^+-N (21 to 207 kg N ha^{-1} 15 cm^{-1}) at all depths in the poorly drained soils indicated that nitrification was inhibited. The lowest accumula-tion of NH_4^+-N was found in the well-drained cultivated soils and apparently nitrification was operative, as the greatest amount of NO_3^--N was also found in these profiles.

Under reducing conditions encountered in waterlogged organic soils and in mineral soils with high groundwater levels certain phosphorus compounds tend to become soluble and could appear in the drainage effluents. It has been demonstrated that under aerobic soil conditions phosphorus is rendered immobile by fixation or precipitation but if the system is placed under anaerobic conditions, phosphorus may be released (Burns & Ross, 1972; G. C. Roberts. 1972. Sewage phosphorus retention and movement under aerobic and anaerobic soil conditions. M.S. Thesis. University of Guelph, Guelph, Ontario, Canada).

Tilstra (1972) investigated the feasibility of disposing waste water ef-fluents on peat in the Detroit Lakes area of Minnesota. When the effluent was applied to the waterlogged peat and presumably under anaerobic condi-tions, P fixation and N removal were not significant. Lysimeters were con-structed using the peat and, when operated under aerobic conditions and artificially drained, P was retained and thus removed from the applied ef-fluents. Nitrification occurred under aerobic conditions in the lysimeters.

The authors concluded that the Detroit Lakes peat field, with its high water-table and anaerobic soil environment, was not a favorable waste water effluent disposal site from either a hydrological or nutrient removal standpoint.

In southern Ontario, two studies were conducted on the nutrient quality of the waters draining from cultivated and uncultivated organic deposits. The loss of nutrients in runoff water from an undeveloped portion of the Holland Marsh was difficult to estimate (Nicholls & MacCrimmon, 1974). They speculated that nutrients in the major portion of the undeveloped marsh were locked into a cycle of wintertime leaching from frozen vegetation followed by spring and summer assimilation. The undeveloped marsh would thus contribute a comparatively small amount of P and NO_3^--N during the spring runoff relative to that lost from the cultivated marsh. The nutrient data from cultivated portions of the Holland Marsh indicated that four to five times more P and 40 to 50 times more NO_3^--N were contained in the runoff waters than from the undeveloped areas. The importance of overwinter leaching of phosphorus from frozen marsh vegetation was indicated. Nicholls and MacCrimmon (1974) concluded that 1.56 kg total P/ha and 4.15 kg NO_3^--N/ha were lost in the drainage waters from the cultivated marsh, and were related to chemical fertilization practices and the oxidizing and nitrifying conditions in organic soils.

Miller (1973) measured the nutrient content of tile drainage waters from four sites in the Erieau Marsh cropped with onions (*Allium cepa* L.) and carrots (*Daucus carota* L.). The highest concentrations of nutrients occurred during periods of greatest drainage in the spring and fall. The average weekly NO_3^--N concentration varied from 1.0 to over 50 mg/liter. The soluble orthophosphate and total P concentrations ranged from 0.1 to 8.9 and 0.1 to 11.7 mg P/liter, respectively. It was estimated that 4,272 kg P/year (21 kg P ha^{-1} year^{-1}) entered Rondeau Harbour and contributed significantly to the deterioration of the surface water. Miller (1973) cited instances of fertilizer applications that were about 10 times greater than that recommended by soil tests, particularly for P. It was suggested that the overfertilization with P could have contributed to the deterioration of water quality in Rondeau Harbour.

Over a 3-year period, natural soil core lysimeters were used at Guelph, Ontario to study the NH_4^+-N and NO_3^--N concentrations in percolates following surface additions of liquid poultry manure (Bielby et al., 1973). Based on total NO_3^--N discharged, nonmanured lysimeters discharged about 30% of the nitrate-nitrogen in the spring prior to corn planting, 20% during the growing season, and 50% in the fall after corn harvest. Those lysimeters that received N from poultry manure discharged about 10% of the NO_3^--N prior to corn planting and 60 to 84% after corn harvest. Peak concentrations of NO_3^--N in the percolates tended to occur during the period after corn harvest and before winter freezeup. Bielby et al. (1973) concluded that for mineral soil conditions in southern Ontario, the maximum spring application of poultry manure should be somewhat < 550 kg N/ha on corn land. The fall application of manure on stubble land following corn, soybeans (*Glycine max* [L.] Merr.), or spring grain could not be recommended.

The importance of natural soil drainage conditions in relation to organic waste disposal on land appears to relate to the oxygen status of a soil and associated microbial activity. Under well-drained conditions, many carbonaceous compounds would be oxidized to simple stable components, NO_3^--N could occur in excess of plant requirements, and P would be immobilized. There is the ever-present hazard of applying wastes in excess which would result in the discharge of contaminants. While research has been undertaken on waste disposal on poorly drained soils, many questions remain unanswered: Can wet soils be managed to enhance denitrification, and what is the magnitude and significance of the release of mobile P compounds? The release of plant nutrients to the environment appears to be a major problem on organic deposits in the cool subhumid and humid regions, and may require special soil and crop management practices.

B. Watershed Discharges

During the winter season, most watersheds in Canada and northern USA are characterized by frozen, snow-covered surfaces. Under these conditions, runoff containing sediment or nutrients is limited or practically nonexistent. However, peak values are usually reached in the spring during thawing.

In Ontario, maximum flow discharges from the Grand River and Bay of Quinte watersheds occur predominantly from December through April (MacCrimmon & Kelso, 1970; Johnson & Owen, 1971). In western Canada, peak discharge from the Saskatchewan River occurs somewhat later in the year, April to September (Stichling, 1974). In these studies, maximum quantities of N, P, and sediments in the runoff occur when hydrographs show peak volume discharges.

Peak concentrations of N and P from vegetable-producing organic deposits in southern Ontario have been reported to occur in late fall and in the spring (Nicholls & MacCrimmon, 1974; Miller, 1973).

C. Agricultural Runoff and Erosion

Since peak quantities of water, nutrients, and sediment from watersheds in the northern regions are, in general, coincident with spring thaws and melting periods, it is important to relate land application of wastes to time of year and cultural practices.

Data involving land disposal of animal wastes illustrate the magnitude of potential losses in runoff and erosion. Since 1970, Ketcheson (1973) has investigated the effect of weather conditions, soil cover, and depth of frost in soil on runoff and erosion associated with the winter spreading of livestock manures. When manure was applied in January, winter losses of nutrients (N and P) tended to be as high as those from summer runoff. Corn stover left on the soil surface with an application of manure effectively reduced water, soil, and nutrient losses for both summer and winter periods. Runoff during May to October was less than in the winter period (November to April) but soil losses were greater during the summer period with high intensity rainfalls.

Many farmers in northern states dispose of animal waste by winter spreading on frozen or snow-covered ground which constitutes a hazardous practice on sloping land (Holt et al., 1973). Studies at Morris, Minn., show that 18% of the N and 8% of the P were lost from a 45 metric ton/ha (20-ton) application of dairy manure spread on snow-covered alfalfa (*Medicago sativa* L.) land with 8% slope. Holt et al. (1973) noted that surface water quality apparently is more adversely affected by winter application of manure on hay land than on fall-plowed land. In the same study, manure plowed under in the fall contributed no more nutrients to spring runoff than the non-manured plot.

An early report on Lake Erie singled out runoff from agricultural land as a major source of nutrient and silt pollution of the lake. It was estimated that runoff removed about 0.4 kg P/ha of watershed and deposited the nutrient in the lake (U. S. Dep. of Interior, 1968). This estimate is considerably greater than other reported values (Biggar & Corey, 1969; Johnson & Owen, 1971; Timmons et al., 1970). There are still many unanswered questions relating the significance of agricultural runoff to water quality.

D. Freezing and Thawing of Vegetation

When soils are frozen, snow-covered, or when midwinter rains occur surface runoff water will contain small amounts of dissolved nitrogen and phosphorus. The release of these nutrients is attributed to the desiccation of plant material by air drying or freezing.

Studies with desiccated vegetation have demonstrated that virtually all plant P was removed by leaching with water and over 70% was in the orthophosphate form. Average loss of dissolved orthophosphate from land covered with 50% corn, 25% oats (*Avena sativa* L.), and 25% hay was about 0.09 kg P ha^{-1} year^{-1} (Timmons et al., 1968, 1970). In western Minnesota, the average 2-year snowmelt runoff accounted for 83 to 100% of the total runoff from different crops (Holt et al., 1973). Another report by Holt (1969) who investigated the sources of P to Big Stone Lake, concluded that runoff at a rate of 0.09 kg P/ha would result in a P concentration in the lake above the usually accepted threshold value (0.01 mg/liter) for algal growth.

IV. MICROBIAL ACTIVITY AND TEMPERATURE

In cool subhumid and humid regions, temperature is an environmental factor affecting the microbial degradation of soil-applied wastes. The degradation rate of organic wastes is considered to increase with ambient temperature but under field conditions the soil is known to undergo diurnal and seasonal fluctuations. Some research involving microbial activity and temperature changes has been reported.

Mahendrappa et al. (1966) tested the hypothesis that dissimilar nitrifying capacities of soils were due to natural selection and a gradual adjustment of nitrifying bacteria to the climatic environment in which the soil was located. Their data indicate that the climatic zone influences nitrification rate. Nitrification was more rapid in soils from the northern section of western USA at cooler temperatures (20 and 25C) than at 35 and 40C, and the adverse effects of high temperatures were never completely overcome. It was stated that, considering temperature only, natural selection and adjustment of the organisms were governed in part by the temperature of the region where the soil was formed.

In parts of the region under consideration the soils usually freeze and thaw several times between fall and early spring. In laboratory studies, Biederbeck and Campbell (1973) found that an upward shift in temperature enhanced microbial numbers and the production of inorganic N, but when the temperature was lowered there was a marked decrease in the microbial population and a transient flush in inorganic N production. The killing-off of microbial cells was more extensive under fluctuating than under constant temperature conditions. These observations were supported by field studies where it was observed that, with the onset of the first cold spell in early fall, or with the occurrence of a cold spell in late spring, there was a concomitant sudden flush in NO_3^--N production. It was suggested by Biederbeck and Campbell (1973) that during transient cold spells microbial cells are killed and their protoplasm then serves as a source of readily available N for the surviving and adapting microflora.

Thus, it appears that on a regional basis there is some selection and adaptation of soil microorgansims to soil temperature and that fluctuating low temperatures are highly lethal to segments of the microbial population. The results indicate possible causes for transient flushes of inorganic nitrogen at a time when probably not expected. While it is not feasible to control fluctuations in soil temperature, this research may help to explain anomalies in field plot data.

V. UTILIZATION OF WASTES

A. Carbon-nitrogen Relationships

When organic materials from crop residues, wastes, or waste waters are incorporated with the soil, a multitude of reactions occur as the materials undergo decomposition. Many research projects and technical papers have dealt with the immense topic. In this discussion we are concerned primarily with the carbon-nitrogen relationship that may be uniquely modified by the climatic environment in the cool subhumid or humid region.

If an organic waste low in N is added to a soil, the microbial population assimilates the available NH_4^+ and NO_3^- in the soil. The organisms degrade the organic material to obtain energy; the loss of organic material is reflected in the production of carbon dioxide and with little loss of N as the C/N ratio narrows. Such processes indicate a relationship between the C/N ratio of added material and the quantity of inorganic N in the soil. Generally research has supported Bartholomew's conclusion (1965) that inorganic N levels in a soil increase when the C/N ratio of the added material is lower than 25 and that levels are lowered when the C/N ratio is above 35.

With a wide C/N ratio (> 35) the supply of energy for microbial activity is optimal and inorganic N tends to be immobilized. Net immobilization involves the assimilation of inorganic N in proportion to an increase in the microbial population enhanced by an excess of available energy. Available N is rapidly used by the organisms for metabolic processes, primarily the production of cell protein. Stojanovic and Broadbent (1956) observed that immobilization is of significant magnitude even in the decomposition of a high N material such as corn leaves.

Nitrogen that is immobilized during the decomposition of grain straw is not immediately available for plant use but may become available slowly with time. However, the immobilized N may never be available with the same recovery efficiency as fertilizer N applied for a crop (Stewart et al., 1963; Smith & Douglas, 1971; Tyler & Broadbent, 1958). It should be noted that the rate and extent of immobilization of inorganic soil N decreases with increasing plant residue size or decrease in contact between residues and soil during decomposition (Smith & Douglas, 1971; Sims & Frederick, 1970).

The addition of easily decomposable materials, such as green manuring crops, stimulates the organisms to decompose the added organic matter and also some fractions of the more resistant organic matter. The addition of sudangrass (*Sorghum vulgare* var. *sudanense* Hitchc.) increased the total of CO_2 production by a factor of three and inorganic N twofold. Data of this nature indicate that green manure crops are not likely to increase the soil organic matter content (Broadbent, 1947).

B. Application of High-carbon Wastes on Land

Considering the plant nutrients and the carbon contained in domestic organic wastes and sewage sludges it seems feasible that these wastes should be incorporated directly into soils. The literature contains at least two reports in which noncomposted domestic organic solid waste was mixed with soils under field conditions (Volk & Ullery, 1972[1]; Webber & King, 1974).

In the Boardman, Oregon project up to 900 metric tons/ha of shredded solid waste plus liquid sewage sludge were added to sandy soils (Volk &

---

[1]J. V. Volk, and C. A. Ullery. 1972. Disposal of municipal wastes on sandy soils. Report to Boeing Company. Dep. Soil Sci., Oregon State Univ., Corvallis.

Ullery, 1972).[1] The annual precipitation is normally < 37 cm (15 inches) but with irrigation the area has a potential for agricultural crop production. Fescue (*Festuca eliator* L.) and alfalfa forage were harvested three times during the summer. The first-year yield was slightly reduced with a waste application of 224 metric tons/ha but the yield was equivalent or higher than that from the check plots at the second and third cuttings. It was suggested that waste application to sandy soils should not exceed 450 metric tons/ha unless significant yield reductions in the first-year fescue and alfalfa are acceptable. Extensive laboratory studies indicated that the soil physical properties had changed markedly by the addition of the shredded waste: soil bulk density was decreased significantly; moisture retention by the soil was increased at any given tension, particularly during the early stages of decomposition; and wind erosion losses were reduced 88% by the application of 450 metric tons/ha.

In a Guelph study, shredded solid waste was spread on the soil surface and plowed under in August (Webber & King, 1974). Liquid, anaerobically digested sludge or liquid poultry manure was added with some solid waste treatments to the plowed soil, allowed to dry, and incorporated by discing. The plots were left undisturbed for 6 weeks, then disced and seeded to winter rye (*Secale cereale* L.) in September. Samples of the rye were obtained in November. In May, the rye plants were incorporated by discing and the plots seeded to corn.

The C/N ratios of the solid waste, digested sludge, and liquid poultry manure were 65:1, 4.8:1, and 2.3:1, respectively. The oxidizable carbon was determined by treating samples with an excess of potassium dichromate in sulphuric acid (Jackson, 1958). The total N, except for NO_3^--N was estimated by an automated procedure (Thomas et al., 1967). The field treatments are summarized in Table 2.

It was speculated that the NO_3^--N concentration in the rye when harvested in November several weeks after the first fall frost, would reflect the level of soil NO_3^--N available for plant growth at that time of year.

Table 2—The amounts of carbon and nitrogen added to field plots using solid waste, sludge, and poultry manure

Treatment	Amount added	
	Carbon	Nitrogen
	metric tons/ha	kg/ha
C—control, no wastes added	--	--
W—solid waste, 280 metric tons/ha	70	1,070
S—sludge, 2.3 cm/ha	2.9	610
WS—combination, W plus S	23	1,680
2(WS)—solid waste, 560 metric tons/ha plus sludge, 4.6 cm/ha	146	3,360
WM—solid waste, 280 metric tons/ha plus poultry manure, 1.4 cm/ha	73	1,700

If the upper concentration of NO_3^--N in forage that may be fed to or grazed by livestock is 0.3% (Hanway et al., 1963), then only the rye forage from the solid waste (W) and solid waste plus sludge 2(WS) treatments could be considered safe for livestock consumption. The highest NO_3^--N concentration (0.98%) was found in forage grown on plots receiving only digested sludge. The C/N ratio of the solid waste and the NO_3^--N uptake by the rye would indicate that N released from the waste by microbial decomposition was immobilized and not available to the rye crop.

The distribution of NO_3^--N in the soil to a depth of 75 cm after corn harvest is shown in Fig. 2. During the period from corn planting to harvesting (25 May to 3 October), rainfall amounted to 25.3 cm or a monthly average of 7.9 cm with a low of 5.9 cm in August and a high of 10 cm in July. The data in Fig. 2 indicate that significant amounts of NO_3^- remained in the soil after corn harvest. Under most treatments < 50% of the NO_3^--N in the

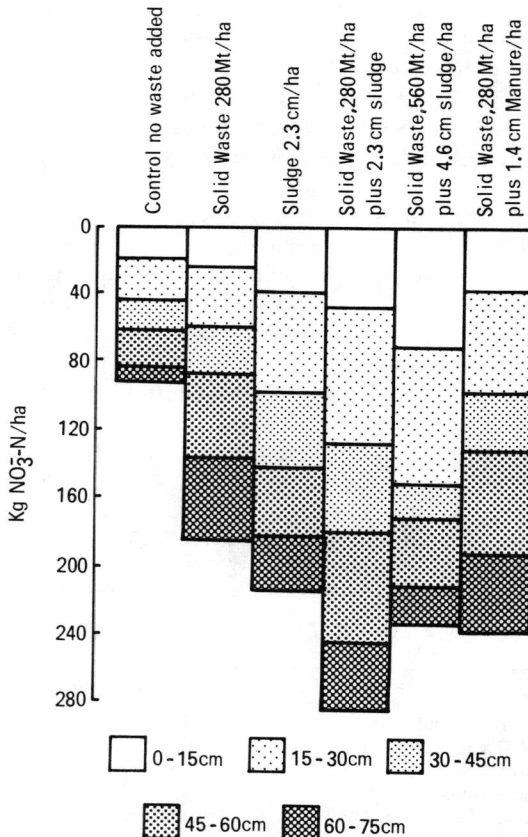

Fig. 2—The distribution of nitrate-nitrogen in the soil with depth following the application of various wastes and after corn harvest.

profile was found in the 0- to 30-cm depth which would indicate an appreciable movement (Webber & King, 1974). It was apparent that 14 months after waste incorporation, NO_3^--N was continuing to be released from the wastes and in quantities that could lead to ground-water contamination.

In principle soil incorporation of high-carbon wastes with high-nitrogen wastes (sludges, manures) appears feasible, but the practice requires further research before it can be promoted on an extensive scale. Rates of waste applications and combinations have to be determined in relation to kind of soil, crop requirement for N, and crop rotation. Thought must be given to cultural practices that utilize the N remaining in a soil after corn harvest. The interseeding of the corn with winter rye has been proposed, but utilizing the forage poses problems because of the possible high content of nitrate-nitrogen.

C. Utilization of Sewage Sludges

Throughout Canada and the USA, the disposal of digested sludges on agricultural land is receiving many man-years of research. While there has been a long history of using sludge, the research has been primarily concerned with the fertilizing value of the waste in crop production. In addition to the nutrient value of sludge, research is being directed to the potential hazard of soil contamination by heavy metals, nonmetals, and organic substances.

In Ontario, research on the agricultural utilization of sludge is concerned with the ultimate sink for N at various rates of application and, of course, the heavy metal problem.

An earlier publication suggested that for corn and hay crops in Ontario the maximum N to be applied as in livestock manures was in the order of 300 kg/ha (Webber & Lane, 1969). Subsequent research using anaerobically digested sewage sludge has indicated that on a well-drained loam soil, yields of corn grain and stover were not significantly increased by sludge applications rates in excess of 1.25 cm/ha (400 kg N/ha). At the conclusion of the experiment, 6 to 10% of the N supplied by the 1.25 cm/ha sludge treatment remained in the soil and on the soil surface in residual solids. After corn harvest, it was estimated that about 97 kg NO_3^--N/ha remained in the soil to a depth of 90 cm. By the following spring, this value had not changed significantly, that is, 90 kg NO_3^--N/ha were found in the soil (Stewart & Webber, 1973). These data suggest that application of excessive N as sludge may result in an accumulation of inorganic N at a time of the year when crops cannot utilize it.

As an example of heavy metal problems, it is generally accepted that Cd in our environment poses a health problem to humans and animals. For Ontario, health authorities suggest that the maximum allowable concentration of Cd in drinking water be set at 0.01 mg/liter; zero content is preferred.

As yet, regulatory agencies in the USA and Canada have not set levels for maximum dietary intake of cadmium by humans; a set dietary level would have to be conditioned by many variables.

Published data indicate that certain crops accumulate Cd from Cd-fortified substrates (Page, 1974). The leaf portion of some plants accumulated substantial amounts of Cd from sewage-treated soils. Chicago sludge was applied to a Blount silt loam soil for 3 consecutive years (Hinesly et al., 1972). The authors noted that corn plants did not accumulate toxic levels of trace elements, even under acid soil conditions which should favor element mobility and plant availability.

It has been postulated that Cd in a neutral or alkaline soil would form highly insoluble compounds or would be strongly complexed with organic matter. A concomitant increase of Zn with Cd has been reported as advantageous in suppressing a health hazard that might exist by the ingestion of plant materials (e.g., soybeans) containing above-normal levels of Cd (Jones et al., 1973). Similarly, the soil concentration of Cd per se may not determine the amount accumulated by plants but rather the Zn/Cd ratio (Chaney, 1973). Chumbley (1971) reported that the toxic effects of certain metals to plants were additive. It was suggested that it may be permissible to add to an uncontaminated soil, a Zn equivalent to 250 ppm (in topsoil) over a 30-year period. The Zn equivalent was calculated by totalling the Zn concentration plus twice Cu plus eight times the Ni concentration.

VI. CONCLUSIONS

The management of soils for the utilization of organic wastes and waste waters poses unique problems in cool subhumid and humid regions such as the northern tier of the United States and Canada. The assimilative capacity of a soil is directly related to the microbial, chemical, and physical properties of a soil which in turn are conditioned by the temperature, moisture regime, and length of growing season in a region.

One must assume that the assimilative capacity of a soil is finite. Where land is used for waste disposal, the capacity of a soil to accept, degrade, and render innocuous applied wastes must be determined. If we propose to use land as a receptor of wastes, then our ultimate objective must be to utilize these wastes as a nutrient source or dispose of them on soils in such a manner that the practice does not impair the many aspects of our environment or the quality and quantity of food and feed produced. At the same time, we must always strive to achieve the full economic potential of these wastes.

ACKNOWLEDGMENT

The authors gratefully acknowledge the financial aid in support of this research by the Ministry of the Environment, Province of Ontario, Canada.

LITERATURE CITED

Bartholomew, W. V. 1965. Mineralization and immobilization of nitrogen in the decomposition of plant and animal residues. *In* W. V. Bartholomew and F. E. Clark (ed.) Soil nitrogen. Agronomy 10:285-306. Am. Soc. of Agron., Madison, Wis.

Biederbeck, V. O., and C. A. Campbell. 1973. Soil microbial activity as influenced by temperature trends and fluctuations. Can. J. Soil Sci. 53:363-376.

Bielby, D. G., M. H. Miller, and L. R. Webber. 1973. Nitrate content of percolates from manured lysimeters. J. Soil Water Conserv. 28:124-126.

Biggar, J. W., and R. B. Corey. 1969. Agricultural drainage and eutrophication. p. 405-445. *In* Eutrophication: causes, consequences, correctives. Nat. Acad. of Sci., Washington, D. C.

Broadbent, F. E. 1947. Nitrogen release and carbon loss from soil organic matter during decomposition of added plant residues. Soil Sci. Soc. Am. Proc. 12:246-249.

Burns, N. M., and C. Ross. 1972. Oxygen nutrient relationships within the Central Basin of Lake Erie. p. 85-119. *In* N. M. Burns and C. Ross (ed.) Project Hypo: an intensive study of Lake Erie Central Basin hypolimnion and related surface water phenomena. Canada Centre for Inland Waters, Burlington, Ontario, Canada.

Chaney, R. L. 1973. Crop and food chain effects of toxic elements in sludges and effluents. p. 129-141. *In* Recycling municipal sludges and effluents on land. Proc. Joint Conf., 9-13 July, Champaign, Ill. Nat. Assoc. State Univ. and Land Grant Coll., Washington, D. C.

Chumbley, C. G. 1971. Permissible levels of toxic metals in sewage used on agricultural land. Min. of Agric., Fish., and Food, ADAS, Wolverhampton, England.

Hanway, J. J., J. B. Herrick, T. L. Willrich, P. C. Bennett, and J. T. McCall. 1963. The nitrate problem. Spec. Rep. No. 34. Iowa State Univ., Ames, Iowa. 20 p.

Hinesly, T. D., R. L. Jones, and E. L. Ziegler. 1972. Effects on corn by applications of heated anaerobically digested sludge. Compost Sci. 13:26-30.

Holt, R. F. 1969. Runoff and sediment as nutrient sources. Water Resour. Res. Center, Univ. of Minnesota, St. Paul. Bull. 13.

Holt, R. F., H. P. Johnson, and L. L. McDowell. 1973. Surface water quality. p. 141-156. *In* Proc. Conf. on Conservation Tillage. Soil Conserv. Soc. of Am., Ankeny, Iowa.

Jackson, M. L. 1958. Soil chemical analysis. Prentice-Hall Inc., Englewood Cliffs, N. J.

Johnson, M. G., and G. E. Owen. 1971. Nutrients and nutrient budgets in the Bay of Quinte, Lake Ontario. J. Water Pollut. Control Fed. 43:836-853.

Jones, R. L., T. D. Hinesly, and E. L. Ziegler. 1973. Cadmium content of soybeans grown on sewage-sludge amended soil. J. Environ. Qual. 2:351-353.

Ketcheson, J. W., and E. L. Dickson. 1973. Runoff and erosion during winter and spring. p. 52-53. *In* L. R. Webber (ed.) Progress report. Dep. of Land Resour. Sci., Univ. of Guelph, Guelph, Ontario, Canada.

MacCrimmon, H. R., and J. R. M. Kelso. 1970. Seasonal variation in selected nutrients of a river system. J. Fish. Res. Board Can. 27:837-846.

Mahendrappa, M. K., R. L. Smith, and A. T. Christiansen. 1966. Nitrifying organisms affected by climatic regions in western United States. Soil Sci. Soc. Am. Proc. 30: 60-62.

Miller, M. H. 1973. The contribution of plant nutrients from agricultural lands to water supplies. p. 47-52. *In* L. R. Webber (ed.) Progress report. Dep. of Land Resour. Sci., Univ. of Guelph, Guelph, Ontario, Canada.

Nicholls, K. H., and H. R. MacCrimmon. 1974. Nutrients in subsurface and runoff waters of the Holland Marsh, Ontario. J. Environ. Qual. 3:31-35.

Olsen, R. J., R. F. Hensler, O. J. Attoe, and S. A. Witzel. 1970. Comparison of inorganic nitrogen contents of undisturbed, cultivated, and barnyard soil profiles in Wisconsin. Soil Sci. Soc. Am. Proc. 34:699-701.

Page, A. L. 1974. Fate and effects of trace elements in sewage sludge when applied to agricultural lands. A literature review study. Prepared for Office of Res. and Develop., U. S. Environ. Prot. Agency, Cincinnati, Ohio. EPA Technol. Ser., 670/2-74-005. 98 p.

Patrick, W. H., Jr., and I. C. Mahapatra. 1968. Transformation and availability to rice of nitrogen and phosphorus in waterlogged soils. Adv. Agron. 20:323-356.

Sims, J. L., and L. R. Frederick. 1970. Nitrogen immobilization and decomposition of corn residue in soil and sand as affected by residue particle size. Soil Sci. 109:355–361.

Smith, J. H., and C. L. Douglas. 1971. Wheat straw decomposition in the field. Soil Sci. Soc. Am. Proc. 35:269–272.

Stewart, B. A., D. D. Johnson, and L. K. Porter. 1963. The availability of fertilizer nitrogen immobilized during decomposition of straw. Soil Sci. Soc. Am. Proc. 27:656–659.

Stewart, N. E., and L. R. Webber. 1973. Land disposal of anaerobically treated sewage sludge. p. 108–117. In W. J. Eden (ed.) Proc. Nat. Conf. on Urban Engineering Terrain Problems. Tech. Memo No. 109. Nat. Res. Counc., Ottawa, Canada.

Stichling, W. 1974. Sediment loads in Canadian rivers. Tech. Bull. 74. Environ. Canada, Water Resour. Branch, Ottawa, Canada.

Stojanovic, B. J., and F. E. Broadbent. 1956. Immobilization and mineralization rates of nitrogen during decomposition of plant residues in soil. Soil Sci. Soc. Am. Proc. 20:213–218.

Thomas, R. L., R. W. Sheard, and J. R. Moyer. 1967. Comparison of conventional and automated procedures for nitrogen, phosphorus and potassium analysis of plant material using a single digestion. Agron. J. 59:240–243.

Tilstra, J. R., K. W. Malueg, and W. C. Larson. 1972. Removal of phosphorus and nitrogen from wastewater effluents by induced soil percolation. J. Water Pollut. Control Fed. 44:796–805.

Timmons, D. R., R. E. Burwell, and R. F. Holt. 1968. Loss of crop nutrients through runoff. Minn. Sci. 24. 3 p.

Timmons, D. R., R. F. Holt, and J. J. Latterell. 1970. Leaching of crop residues as a source of nutrients in surface runoff water. Water Resour. Res. 6:1367–1375.

Tyler, K. B., and F. E. Broadbent. 1958. Nitrogen uptake by ryegrass from three tagged ammonium fertilizers. Soil Sci. Soc. Am. Proc. 22:231–234.

U. S. Department of Interior. 1968. Lake Erie report. U. S. Dep. of Interior, Fed. Water Pollut. Control Admin., Great Lakes Region, Washington, D. C. 107 p.

Webber, L. R., and L. D. King. 1974. Recycling urban wastes through farm soils. Trans. ASAE. 17:530–532.

Webber, L. R., and T. H. Lane. 1969. The nitrogen problem in the land disposal of liquid manure. p. 124–130. In Animal waste management. Cornell Univ. Conf. on Agricultural Waste Management. Ithaca, N. Y.

chapter 18

Runoff water from this Lubbock, Tex. feedlot is diverted into holding ponds near top left of picture (Photo courtesy of *Soil Conservation Magazine*/SCS).

Land Utilization and Disposal of Organic Wastes in Arid Regions[1]

W. H. FULLER and T. C. TUCKER
University of Arizona, Tucson, Arizona

I. INTRODUCTION

One of the most characteristic features of an arid region is its lack of a fixed boundary. Margins cannot be clearly defined because they expand during dry years and contract in wet periods. Arid lands are generally considered to occupy climatic locations having up to 25 cm of rain per year; semiarid lands are those which receive between 25 and 50 cm of rain. Rainfall limits are arbitrary, however, since there are arid areas that may receive up to 50 cm, such as those in some equatorial regions.

The various climatic zones, based more on water supply and need than on rigid rainfall parameters, are shown in Fig. 1 (Merrell et al., 1967). The heart of the arid region in the United States is represented by Arizona, Nevada, New Mexico, Utah and parts of California, Oklahoma, and Texas, although arid lands occur in parts of nearly every state west of the Mississippi River.

Aridity is more of a comparison between water *supply* and water *need*. Water supply is relatively easy to measure. Water need refers primarily to potential evapotranspiration and is more difficult to evaluate. Aridity thus has been expressed as some function of temperature and rainfall and is related to observed responses of the environment to climate. Generally, under arid conditions vegetation is sparse, short, and develops as individual plants. Organic matter accumulates slowly if at all. Erosion, runoff, and evaporation seriously influence the effectiveness of rainfall in supporting plants and in leaching salts from the surface horizons.

Chemically, arid land soils are neutral to alkaline in pH and high in salts, the most notable of which are calcium carbonate (lime) and calcium sulfate (gypsum). Salts of Na, K, and Mg are less prominent, but they can be a problem in local areas. Elements toxic to plants and animals such as B, Li, and Se may be found in isolated spots and accumulate in soils where irrigation waters are naturally contaminated. Excessive chlorides and bicarbonates adversely influence the quality and productivity of some crops. Thus, some soils ($\sim$ 25%) of arid climates are naturally polluted with salts not normally present in excess in soils of humid climates.

In comparison with humid climates, physical weathering in arid climates often dominates over chemical weathering. An important feature of arid regions is the wide diurnal temperature change which causes rock fracturing and mineral fragmentation. The relatively constant wind action results in abrasion and wearing away by shifting debris, sand, and dust.

Biologically, soils of arid lands do not lack for an active microbial flora (Cameron & Blank, 1965; Fuller, 1975) whose activity is not inhibited by high temperatures on the surface; but it is retarded by a lack of moisture. Arid soils teem with an abundant and varied microbial population which is

---

[1]Contribution from the University of Arizona Agricultural Experiment Station; Soils, Water, and Engineering. Western Regional Research Project W-124. Journal Series Paper No. 1342.

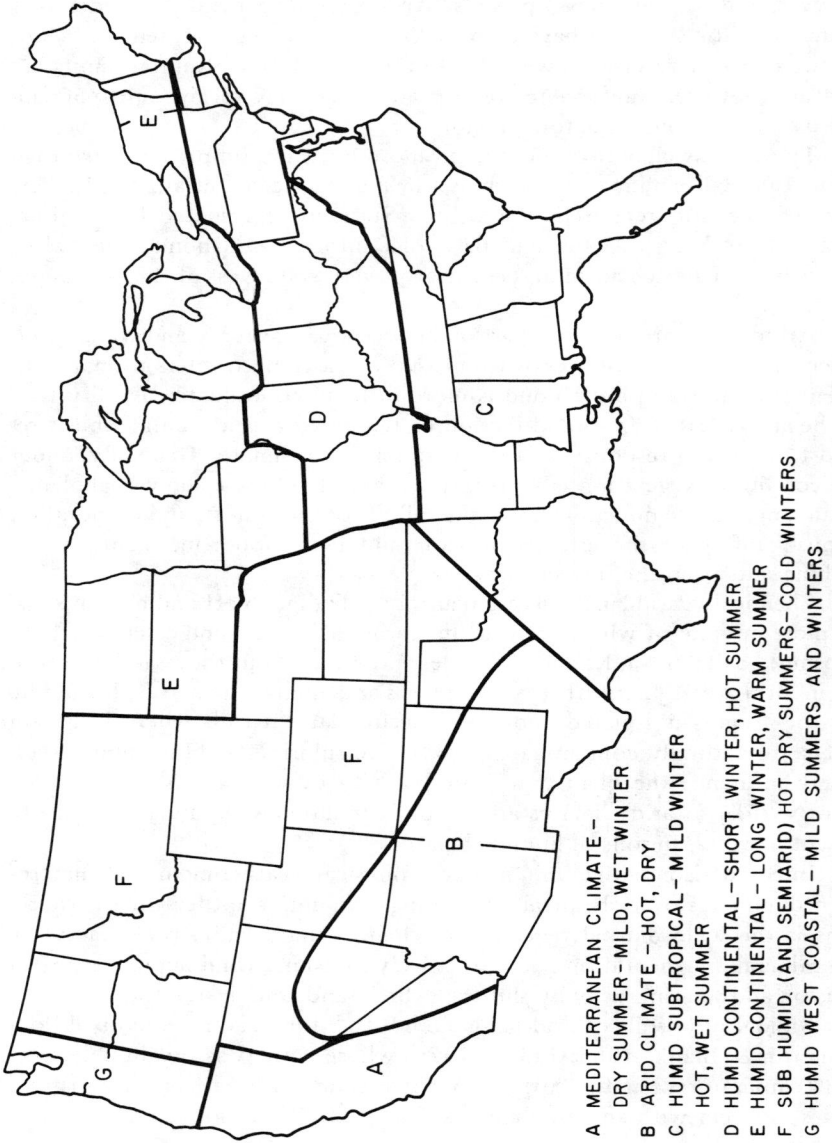

A MEDITERRANEAN CLIMATE –
   DRY SUMMER – MILD, WET WINTER
B ARID CLIMATE – HOT, DRY
C HUMID SUBTROPICAL – MILD WINTER
   HOT, WET SUMMER
D HUMID CONTINENTAL – SHORT WINTER, HOT SUMMER
E HUMID CONTINENTAL – LONG WINTER, WARM SUMMER
F SUB HUMID (AND SEMIARID) HOT DRY SUMMERS – COLD WINTERS
G HUMID WEST COASTAL – MILD SUMMERS AND WINTERS

Fig. 1—Generalized climatic zones for land application.

as adequate as that in other soils for biodegradation and transformation of matter when energy sources and moisture become available. Even algae resist heat and drought and are a prominent feature of North America deserts (Cameron & Blank, 1965; Fuller, Cameron & Raica, 1960).

Irrigating and disposing of aqueous wastes and solid waste leachates in arid soils can pose a serious problem by transporting pollutants if water reaches the capillary fringes of underground sources before purification occurs. Residents in some arid lands depend on underground water resources for survival. For example, the city of Tucson, Arizona, obtains *all* of its water from subsurface storage. Prevention of underground water contamination is critical and essential to the persistence of life in most arid lands. Of course, purification could be achieved at some expense, but then the water costs could not compete with those of humid regions, particularly for food and fiber production (Fuller, 1975; Tucker & Fuller, 1971).

II. NATURE OF WASTES

A. Animal Wastes

Animal manures in arid regions probably differ from those in humid regions more than other waste materials. Moisture content of waste and also the leaching associated with rainfall are less prominent in arid than in humid lands. The custom of housing animals in exposed and unpaved corrals and feedlots provides ready opportunity for sand and soil material to collect along with manure. It is not unusual for steer and horse corral manure to contain 50 to 65% sand and ash, much of which originates from the dirt floors. Some lots are deliberately located on caliche (lime) ridges of higher elevations where drainage is favorable. Interaction of the animal excreta with calcareous soils adds to the salt burden of already salty refuse, thereby reducing its value for fertilizing cropland. Exposed salt (NaCl) licks and high salt content of "free-choice" feeds contribute still more salt. Because arid lands already are abundantly supplied with soluble salts, the value of animal manures high in salt is downgraded in a climate where leaching by natural precipitation is minimal. Some typical values of salts and nutrients in feedlot manures near Phoenix, Arizona (Abbott, 1968; Steffger, 1974), are provided in Table 1. Comparisons in composition of various feedlot wastes in arid and humid regions may be found in a Western Regional Research Committee (W-124) guidelines publication (Meek et al., 1975).

The year-round favorable climate in arid regions for horseback riding, horse racing, and breeding programs has greatly contributed to a large accumulation of manure in and on the outskirts of most cities. Amounts range from a few hundred metric tons to several million metric tons near large population centers such as Los Angeles, California.

Table 1—Some chemical characteristics of feedlot manure from the
Phoenix, Arizona, area†

Sample	pH	Moisture	Ash	Total soluble salts	Na
			%		
Feedlot 1	6.7	21	44	6.6	0.6
Feedlot 2	7.4	24	29	10.8	0.87
Feedlot A	7.3	42	38	10.6	0.6
Feedlot B	7.1	31	47	8.0	0.8
Feedlot C	8.0	25	60	9.5	1.2
Feedlot D	6.8	20	56	4.2	0.5
Feedlot E	7.8	24	62	7.2	0.8
Four feedlots					
Range	6.5-8.1	28-58	9-38	4.4-8.7	0.3-0.7
Average	7.3	40	20	6.1	0.5

† All figures reported on an oven-dry basis.

B. Plant Wastes

The relatively high concentration of agriculturally oriented industries in arid regions emphasizes problems associated with plant wastes (e.g., canning, crop, and range waste). Fortunately, much of the range and forestry waste problems are being solved by the accelerated demand of homeowners and nurserymen for bark, chips, shreds, sawdust, etc., to fortify arid soils impoverished of organic matter. Disposal of crop wastes such as fruit and nut tree trimmings, hulls, and pits continues to pose numerous problems. Perishable canning wastes are seasonal and present problems of waste water disposal within the permissible limits of pollution of underground water resources. Canning wastes also may contain pesticide residues which prohibit their disposal as animal feed. Algal growth in irrigation canals and ponds is a serious organic waste problem in warm desert regions.

C. Municipal Solid Wastes

Residential solid wastes in arid regions are drier and higher in cellulosic yard trimmings throughout the year, because of the year-round growing weather, than are those of humid regions. An example of mixed refuse of average composition from Oakland, California (B. C. Petrucci, personal communications), and Phoenix, Arizona (W. H. Fuller, unpublished data), is given in Table 2. Compared with municipal refuse of western Europe, that from western USA arid climates is lower in proteinaceous and readily available carbohydrate substances (Fuller, 1961). The characteristics of accumulated dirt (soil material) in solid wastes, in general, will be related to the nature of the native soil. Thus, dirt particles in refuse from arid regions are expected to

Table 2—Distribution of selected municipal solid waste constituents comparing
city and county disposals

Solid waste	City		County	
	Oakland, Calif.†	Phoenix, Ariz.‡	Sacramento Co., Calif.†	Pima Co., Ariz.‡
			%	
Cans and metals	7.4	7.1	6.4	8.0
Glass	10.0	8.9	9.8	9.7
Paper	38.0	42.5	26.0	39.2
Organics and yard trim	31.0	27.2	29.0	28.5
Plastics	5.0	6.2	1.3	4.8
Textiles (rags)	2.5	3.1	1.1	1.6
Wood	2.5	1.2	0.8	1.0
Tires	2.2	1.0	1.0	1.0
Other (ashes, dirt, etc.)	1.4	2.8	24.6	6.2

† Private communication from Benjamin Petrucci, Sacramento, Calif.
‡ Unpublished data, W. H. Fuller, Univ. Ariz., Tucson.

be coarse, i.e., contain more sands than in more humid regions where silts and clays predominate. Constant wind movement, characteristic of desert areas, is an important factor contributing to dirt content of solid wastes. Other than the varying amounts of water, salts, plant residues, and sand, municipal solid wastes probably are not greatly differentiated, regionally, in the United States.

D. Sludges

Like municipal solid waste, sludges vary considerably *within* a selected climatic region, and *among* different regions. However, sludges are drier and higher in dirt and sand in arid than in humid regions. Where sludge is dried in open basins, microbial decomposition will not have proceeded as far in arid as in humid regions where sludge is wet or stored in deep, lagoon-like ponds throughout the year. Sludges from cities and urban areas, except where certain manufacturing industries develop around the cities, probably are generated from wastes which are similar in composition. For example, the sludge from Tucson and Phoenix, Arizona, Table 3 (A. D. Day, personal communication), reflects a less intensive, nonagricultural industry, than that of Michigan (Page, 1974). Sludges thus relate to the type of industries attracted within a specific climatic area. Under this same reasoning, the large manufacturing industries in humid temperate and coastal regions would be expected to contain a higher concentration of potentially hazardous trace elements[2] (As, Be, Cd, Cr, Cu, Hg, Ni, Pb, Se, V, and Zn) than the less industry-oriented arid regions. The method of handling, efficiency of the

---

[2] *Trace elements* is the term used here to identify those selected elements which may be found in biological tissues or cells in amounts usually considered to be trace as opposed to more macrolevels of N, P, K, H, O, C, Ca, Na, etc.

Table 3—Comparisons of the average composition of dried sewage sludge from
Tucson and Phoenix, Arizona

	1970	
Constituents	Tucson samples†	Phoenix samples†
H_2O-soluble salt, ppm	9,493	7,167
Total N, %	1.4	1.8
NO_3-N, ppm	19	102
Total P, %	1.5	1.5
CO_2-soluble P, ppm	100	157
K, %	0.4	0.4
Ca, %	4.4	3.4
Mg, %	3.6	5.2
Na, ppm	1,480	1,575
B, ppm	2.6	2.5
Cd, ppm	21	50
Cu, ppm	524	712
Fe, %	18,925	24,475
Pb, ppm	213	254
Mn, ppm	281	295
Zn, ppm	2,808	4,150
pH	6.8	6.5

† Data supplied through the courtesy of Dr. A. D. Day, Univ. of Ariz., Tucson.

treatment system, and modernization of the system also determine the nature of the sludge product (Page, 1974).

E. City Effluent

Treated municipal waste waters in arid regions should not differ significantly from those in other United States regions, except as specific industries use city facilities. Sewage effluent from municipal sources is valued in
arid lands for the water content as well as for N–P–K fertilizing qualities
(Day et al., 1962; Day et al., 1963; Day et al., 1972). Fortunately, only
rarely are noxious industrial wastes disposed of in municipal sewage facilities
in such quantities that the water cannot effectively be used for irrigation of
nonedible crops, golf greens, or recreation landscapes. The quantity of N, P,
and K varies from city to city. Generally, however, the N content (20 to 30
ppm) is higher than is usually necessary for favorable crop production if all
the water needs are met by the effluent. Some characteristic municipal waste
water analyses are provided in Table 4 from Tucson (Day et al., 1963), where
industrial waste contributions are minimal.

F. Industry Waste Streams

Hazardous waste generating industries in the United States center more
in humid than in arid regions. On the other hand, the fossil fuel industry
may be highly concentrated in arid regions such as Texas. Certain arid re-

Table 4—Time comparisons of sewage effluent composition from Tucson, Arizona, sewage plant

Constituent	December 1956	Avg. 1964-65	October 1974	February 1975
		mg/liter		
TDS†	819	691	614	674
Total N	24	26	20	21
NO$_3$-N	2	1	0.5	0.1
NH$_3$-N	--	21	18	17
P	9	10	7	7
K	11	--	12	13
Ca	36	68	60	78
Mg	16	11	11	17
Na	152	72 (Na+K)	107	92
CO$_3$	0	0	0	0
HCO$_3$	378	149	224	315
Cl	92	72	83	85
SO$_4$	105	98	141	152
pH	--	--	6.5	7.5
Hardness, grains per gallon	9	16	18	16

† Total dissolved solids.

gions (e.g., the desert Southwest United States) support abundant agricultural industries. Wastes from agricultural products characteristically contain highly carbonaceous, low proteinaceous materials. Since N is the single nutrient element most limiting to crop production in arid lands, disposal of industrial wastes on land creates special management problems.

III. CURRENT STATUS OF WASTE MANAGEMENT IN ARID SOILS

At present, organic waste and waste water management in arid region soils of the United States is based on limited research. For example, despite the use of Tucson municipal waste water for crop irrigation for over 15 years, Day (1973) stated, "The maximum amount of treated municipal wastewater that may be applied to a given soil on which a specific crop is grown is not known." He further lists a number of fundamental management practices which require intensive research before guidelines can be established with a high degree of certainty. On the other hand, because of the great need for use of all water resources in arid regions and the favorable soil moisture conditions, the waste water management program is advancing rapidly, having had encouragement from agricultural programs aimed at salt control for permanent agricultural production. In view of this development, the quotation from Day (1973) appears to be an overstatement of the problem.

One of the most successful municipal solid waste recycling and composting operations, technically speaking, was initiated in Phoenix (Fuller, Johnson & Sposito, 1960; Fuller, 1966) by the Arizona Biochemical Corporation in the early sixties. The operation, which closely resembled the Dano Process of western Europe, marketed compost of high agricultural value (Fuller, Johnson & Sposito, 1960; Fuller, 1966; Fuller et al., 1967). At that time, society in the United States had not yet awakened to the need for more advanced waste disposal methods, and the plant was closed down because technology had outstripped society's capacity for economic acceptance.

Equipment necessary for deep incorporation of large quantities of organic and shredded solid wastes is highly essential for successful land loading of bulky nonliquid materials (Stout, 1974). Land loading of animal manure, however, does not suffer from equipment need as much as that of municipal and some industrial wastes.

The current program of animal waste utilization in soils is much better developed than utilization of other wastes. Guidelines describing effective and efficient manure utilization in the western region, United States, for example, have been well established (Meeks et al., 1975).

Management of sanitary landfill leachate is even less developed than that of municipal waste waters and sludges. This is also true for hazardous industrial waste "streams." Research now in progress at the University of Arizona is just beginning to relate movement of selected hazardous trace elements (As, Be, Cd, Cr, Cu, Hg, Pb, Se, V, and Zn) contained in landfill leachates with soil parameters in laboratory columns. Industrial waste stream conditions are being simulated by "spiking" the leachates with each of the above selected trace elements at values well above those found in municipal landfill leachates.

A. Airborne Dust

The sparse vegetation, and further devegetation by cattle grazing and cultivation, combined with the general tendency toward accelerated wind activity add to the pollution hazard in arid climatic regions. The dust storm is a common part of life in deserts and plains; dust from land surfaces may contain pesticides (Stubblefield & Smith, 1964). Dust may originate also from cotton gins, alfalfa dehydrators, and beef-cattle feedlots. Smoke from burning grain stubble and crop trash also reduces visibility and irritates respiratory organs of human beings and other animals. Burning to rid the land of organic agricultural wastes (orchard pruning, straw, range weeds, grass, and shrubs) is still a practice too often employed in arid lands.

IV. ENERGY IMPLICATION

The value of municipal waste water from the standpoint of energy conservation is significant. For the N and P contained in 1,234 m^3 (1 acre-foot) of waste water, the energy required to manufacture the equivalent amounts in fertilizer is equal to the energy in 60.5 liters of gasoline (Table 5). There-

Table 5—Energy evaluation for pumping 1,234 m^3 (1 acre-foot) of water for irrigation use

Lift	kW·h†	Joules (J)	British thermal unit (Btu)	Gasoline equivalent†
m				liters
30.5	204	733,657,701	695,542	21.2
60.9	408	1,466,366,083	1,390,184	42.4
91.4	612	2,200,973,104	2,086,626	63.6
121.9	816	2,934,630,806	2,782,168	84.8
152.3	1,020	3,668,288,508	3,477,710	106.0

† Values assume 50% pumping efficiency at all depths.

fore, each 1,234 m^3 of waste water used for irrigation instead of ground water pumped from a 91-m depth could result in an energy savings equal to 124 liters of gasoline (Table 5). (Also see Steffger, 1974.) The cities of Phoenix and Tucson could, in 1985, save energy equal to 10.24 × 10^6 and 3.15 × 10^6 liters of gasoline for nutrients and 10.75 × 10^6 and 3.30 × 10^6 liters, respectively, for pumping if the total sewage effluent were used for irrigation. These figures assume an average composition of N and P from past analyses and a 91-m lift of irrigation water.

The fertilizer value of dried sewage sludge adds to the total energy conservation possible (Table 6). Of that efficiently utilized by plants, the N value should be equal to approximately 5% of the N in waste water, i.e., $\sim$ 512,000 liters of gasoline for Phoenix, and the P value between 10 and 15% of the total P in waste water (Table 6). Similarly, animal wastes can provide source constituents useful in energy conservation (Steffger, 1974; Stout, 1974).

Other, more subtle implications of energy relations in an arid region concern the high net radiation received (Selitch, 1974). The availability of energy from the sun throughout the year for plant growth and biological decomposition and transformation of mineral constituents must be considered as a part of any program of soils for management and utilization of organic

Table 6—Municipal waste water, sludge, nitrogen, and phosphorus levels for Phoenix and Tucson, Arizona, and predicted levels for 1975 and 1985

Year	Sample source	Phoenix			Tucson		
		Amount	N	P	Amount	N	P
				metric tons			
1963	Waste water	45.9 × 10^6	1,100	372	25.5 × 10^6	611	207
	Sludge	2,757	49	40	2,776	40	42
1973	Waste water	121.8 × 10^6	2,916	987	45.2 × 10^6	1,083	367
	Sludge	7,309	130	104	3,629	52	54
1975	Waste water	131.0 × 10^6	3,137	1,061	49.3 × 10^6	1,179	399
	Sludge	7,320	140	112	3,951	56	59
1985	Waste water	208.3 × 10^6	4,987	1,688	64.0 × 10^6	1,533	519
	Sludge	12,500	222	179	5,136	73	77

wastes and waste water. For example, bacterial reaction rates normally double with every 10C rise in temperature up to optimum limits. Evaporation and evapotranspiration in hot, dry climates obviously affects the energy efficiency of water application systems.

V. MAJOR PROBLEMS

A. Volume, Place, and Time Economy

Many small communities in arid lands lack sufficient economic resources to support adequate sewage and solid-waste disposal plants. Four factors responsible for these problems are

1) The volume of industrial waste and waste water is not large enough to make recycling, recovery, and reuse of their constituents economically feasible. Coordination of the inhabitants toward costly disposal programs often is more difficult to attain than in large cities that have paid managers.

2) Communities often are widely scattered, being located great distances from markets for recycled wastes (glass, rags, paper, aluminum, manure, and even water from municipal sewage and waste water processing plants). Transportation costs do not always provide economical irrigation utilization of waste waters on the land. Transportation of segregated solid waste-recyclable material in the 1960's from Phoenix to Los Angeles, for example, where it could be used was more costly than it was worth. Cost of recycling of low value materials, however, can change with changing economics which would make recycling a possible disposal method for parts of the arid regions.

3) Seasonal volumes of wastes vary in agricultural-dominated regions. Time economy can make it difficult for most effective disposal; as with some canning wastes, crop residues, and waste waters.

4) Disposing of potentially hazardous pollutants on the "wide-open" ranges of the West rather than on high-valued municipal property or irrigated farmland of more densely populated areas is an attractive thought. A problem, however, centers around volume, place, and time economy, which seldom favors wide-open space disposal.

B. Hazardous Constituents

Special problems associated with the widespread use of pesticides in intensive irrigated agriculture may be a distinguishing feature of arid regions. Accumulation in soils of hazardous trace elements contained in pesticides has been reported (Stubblefield & Smith, 1964). The elements include As, Cd, Cr, Cu, Pb, Hg, Se, and Zn. Thus, like most sludges, those from arid regions also contain some potentially hazardous elements (except for B and V) (Page, 1974). The high concentration of Cu, Pb, Zn, and Fe mining and smelting activity in the arid region also poses problems of trace element, as well as sulfur-oxides contamination from emissions from mining stack gas sources. Most of these problems require further verification and definition to evaluate their long term hazard to the surrounding environment and underground water resources.

Because of the tendency for high evaporation in arid regions, disposal by spraying liquid waste on vegetation can cause serious trace element foliage burn and toxicity problems. Potentially hazardous trace elements such as As, Be, Cd, Cr, Cu, Hg, Ni, Pb, Se, and Zn in waste waters may accumulate to harmful levels by repeated wetting and drying of the spray. Some crop plants are known to develop common salt burn even from the spray of high quality irrigation water. For example, chloride burn is common in citrus (*Citrus* sp.) where spray application is used.

C. Soil Contamination

Soil is a basic resource which, in addition to its many functions in the support of life, converts or inactivates the multiplicity of wastes which accumulate around human society. Rarely has the soil been researched to evaluate its capacity to withstand contamination from its ever-increasing intensified use by a rapidly expanding society. How long can the soil respond favorably to continued dumping of pesticides, human and animal manures, city effluents, sludges, industrial waste streams, garbage, trash, and landfill leachate without becoming contaminated to the extent that it poses a threat to food production and underground water sources? The problem needs constant evaluation and monitoring.

Potential pollution hazards in a region are related to the nature, kind, and amounts of urban and industrial discharges. The obvious soil contamination problems, thus, are highly localized. In an arid region some of the most prominent sources of soil contamination are: (i) pesticide application to cropland and wind drift, (ii) pesticide overapplication to target crops and persistence in the soil, (iii) mine smelter deposits of hazardous trace elements from stack emissions, (iv) cannery waste disposal, and (v) disposal and use of low quality (high salt) animal wastes.

Organic waste and sludge applied to the dry soil of arid regions require additional irrigation water to carry decomposition to completion, since rainfall is not adequate to do the job as it is in the humid regions. Soils of arid regions can be overloaded easily with organic wastes if sufficient moisture is not available for decomposition. Even spraying of animal waste lagoon effluents onto land can build up high levels of organic material. When a summer rainy period occurs, an intense fly population can develop. In one such instance, hoards of flies resulted in the closing of schools and businesses (private communication, Dr. Bartley Cardon, Arizona Feed, Tucson, Ariz.).

D. Underground Water Quality

Inhabitants of arid regions maintain a more watchful eye on the underground water quality than inhabitants of other regions. Ground water constitutes the only source of domestic and irrigation water for much of the desert regions. To prevent contamination, even under conditions of normal recharge, requires constant vigilance. Application of organic wastes and waste waters to the soil, therefore, certainly must be accompanied by an assurance that potentially hazardous constituents will not migrate to the capillary fringes of the ground water. With so little known about migration rates through soils, even disposal in some remote arid area may constitute a threat to water quality at some future date. The research of the Flushing Meadows Project near Phoenix (Bouwer, 1973) indicates that renovation of municipal sewage effluent by passing it through coarse-textured soil in the field is possible on a practical basis when managed wisely.

E. Crop Quality

Perhaps the most perplexing problem confronting management and utilization of waste waters, where they are needed for irrigation in arid regions, is in the evaluation of hygienic effects on food crops. For example, Tucson sewage water has been used successful to irrigate cotton (*Gossypium hirsutum* L.) and cereal grains for over 15 years. The water is not permitted for use on crops where contact is made with the edible portion [e.g., lettuce (*Lactuca sativa* L.) and carrots (*Daucus carota* L.)].

F. Underdeveloped Management Practices

As in most other climatic regions in the United States, management of soils in arid regions for the utilization of organic waste and waste waters is poorly developed. Each region must have its own management practices, and should not expect to depend entirely on those developed for another climatic zone where climate and soil conditions are vastly different (Fuller, 1975; Tucker & Fuller, 1971).

Recent population growth rates in arid regions of the United States greatly exceed those in the rest of the country. As growth continues, management of waste will become increasingly more complicated (Bouwer, 1970). Some encouragement and comfort can be derived from the fact that there is a vast amount of knowledge developed in soil and water management to maintain a permanent agricultural industry. The same knowledge can be helpful in planning for reuse or recycling of waste waters. The high value of water in an arid region, along with the need for fertilizer nutrients, provide economic incentives for development of systems for utilization of organic wastes and waste waters on soils.

VI. USE AND MANAGEMENT OF SOILS

A. Animal Wastes

Suggestions and recommendations for use and management of soils with respect to animal waste in arid regions are outlined in several publications (Fuller, 1975; Tucker & Fuller, 1971) and in a report of the Western Regional Research Committee W-124 entitled "Guidelines for Manure Utilization in the Western Regions, USA" (Meek et al., 1975). The recommendations are developed primarily for the beneficial use of manures on soils in crop production and must involve special management practices to avoid the following problems: (i) excessive salt hazard, (ii) high loading rates, (iii) accumulation of unfavorable levels of trace and heavy metals, (iv) potential disease problems, and (v) high levels of Na and K in proportion to Ca plus Mg.

B. Municipal Sewage Effluent

Municipal sewage effluent is highly valued in arid regions for its nutrient and water content. Experimentation in the use of municipal sewage effluent can be grouped in three categories:

1) Direct use of effluent for flood irrigation of food and nonfood crops (Day et al., 1962; Day, 1973). The general low trace element content of effluents in most cities of the arid regions, the tendency of trace elements to concentrate in the sludge, and the neutral-to-alkaline characteristic of the soils minimize the migration hazard of these elements to underground waters.

2) Irrigation can renovate municipal waste water (Bouwer, 1970; Bouwer et al., 1972; Bouwer, 1973). The biological health hazard associated with recharge of underground water to be cycled for human consumption and pretreatment of the waste water remains to be defined more clearly in an expanded use of waste water by this type of management. Sprinkler irrigation with chlorine-treated effluent on golf-course turf is becoming a widespread practice in arid regions. Some examples may be found in Arizona at Mesa (G. V. Johnson, personal communication) and Tucson (Johnson, 1973), in Nevada at Las Vegas (R. A. Young, personal communications), and in California at Santee (Merrell et al., 1967).

3) Soil-treated municipal effluents can augment ground water for recreation areas. At Santee, California, renovated water moves laterally through a series of four ponds, which initially were used for holding, for boating and fishing, and finally for swimming purposes (Merrell et al., 1967).

4) The potential for municipal effluent as a water source, fertilizer, and

possible soil conditioner only waits to be utilized. Such problems as B excesses must be overcome, however, since B is substituted more and more for such elements as P in detergent manufacturing.

When municipal waste water has been used on a continued basis as the sole source of irrigation water for field crops in arid regions, excessive amounts of nutrients often have been applied. Nitrogen, in particular, tends to accumulate with time, unfavorably affecting plant behavior, yield, and quality of certain field crops, and poses a potential ground-water pollution hazard. Except for small grain, P accumulation has not been shown to cause detrimental effects. Therefore, a practical management system of waste water utilization should consider first its nutrient content in relation to the specific crop requirements. In supplying the nutrients required by the crop, the amount of waste water used will represent an equivalent reduction in the requirement for normal sources of irrigation water. In an arid region, this will usually mean a saving from one-third to two-thirds of the normal source of pump or surface water. This situation is in sharp contrast to the limitation of effective water use in humid regions.

C. Municipal Sludges

One of the most outstanding soils application practices with municipal sludge may be found in the Phoenix, Arizona area, where sun-dried sludge has been incorporated as a filler in mixed fertilizer production for over 25 years. The economical use of radiant energy for drying sludge in a series of shallow ponds makes the management of sludge in a hot, dry region attractive. Application to municipal park and golf turf is a well-established practice in Tucson and Phoenix. Arid land soils, so notoriously low in organic matter, respond well to such use. The implication of possible migration of trace elements through the soil and possible accumulation in underground water sources, however, has not received attention until recently.

D. Municipal Solid Waste

Composted organic waste from municipalities can be used effectively on arid region soils for soil conditioning (Fuller, Johnson & Sposito, 1960; Fuller & Bosma, 1965; Fuller et al., 1967). The small amounts of N and high C content require an input of N to prevent N deficiencies. Loading rates of municipal compost, however, remain to be worked out. Supplemental N application to the soil at the different loading rates is poorly understood with respect to *method* and *amount* of addition. The effect of long term accumulation of metals and very slowly biodegradable solids on soil properties needs to be investigated.

E. Industrial Organic Wastes and Waste Waters

Safe disposal of industrial organic wastes in soils of arid regions offers numerous challenges. The problems are so diverse that space does not permit detailed discussion here. The whole field of potentially hazardous trace element and organic compound contamination of soil presents unique disposal problems. One paper pulp mill in Arizona takes advantage of the climatic factors and low population density by disposing of waste waters and sulfite liquors in a dry lake bed of great clay depth. Rarely is there a combination of factors so favorable for disposal.

F. Air Pollution Byproducts

Air pollution byproducts as may be represented by pesticides, again are so diverse as to prohibit in-depth discussion here. Organic air pollutants become soil pollution problems as they deposit from air and water transport onto the soil. Air pollutants require special attention as hazardous constituents with a wide variety of individual characteristics, from alkali dust to allergens. Some can actually be toxic, as in the episode of tetrachlorodibenzodioxin (TCDD) in the horse arenas (Carter et al., 1975). Dust is a common arid region air pollutant. Waste oil sludge frequently is used for its control, particularly on roads, horse riding areas, and recreation tracks. Contaminated waste oil sludge, from a salvage oil company, sprayed on three horse arenas (Carter et al., 1975) and a road proved to be toxic to birds, small animals, and horses to the extent of being lethal to some of the life frequenting the sprayed areas. A 6-year-old girl who played in one arena became seriously ill. The toxic compound which contaminated the oil was identified as TCDD (Carter et al., 1975).

On the other hand, some air pollutants have been observed to be assets, such as the classical burning of tar pits and the resultant correction of sulfur deficiency on bordering lands.

ACKNOWLEDGMENT

Some of the research and manuscript preparation was supported by the Western Regional Research Project W-124 and the U. S. Environmental Protection Agency, Solid and Hazardous Waste Research Division, Municipal Environmental Research Lab, Contract 68-03-0208, Cincinnati, Ohio.

LITERATURE CITED

Abbott, J. L. 1968. Use animal manure effectively. Univ. Ariz. Agric. Exp. Stn. Bull. A-55.

Bouwer, H. 1970. Groundwater recharge design for renovating wastewater. J. Sanit. Eng. Div. Am. Soc. Civ. Eng. 96:59–74.

Bouwer, H. 1973a. Design and operation of land treatment systems for minimum contamination of groundwater. Proc. Int. Symp. on Underground Waste Management and Artificial Recharge. New Orleans, La. September 1973. Sponsored by Am. Assoc. Petrol. Geol., U. S. Geol. Surv., and Int. Assoc. Hydrol. Sci.

Bouwer, H. 1973b. Land treatment of liquid waste: The hydrologic system. p. 103–112. In Recycling municipal sludges and effluents on land. Proc. Joint Conf., 9–13 July, Champaign, Ill. Nat. Assoc. State Univ. and Land-Grant Coll., Washington, D. C.

Bouwer, H., R. C. Rice, E. D. Escorcega, and M. S. Riggs. 1972. Renovating secondary sewage by groundwater recharge with infiltration basins. U. S. Environ. Prot. Agency. Water Pollut. Control Res. Ser. Proj. No. 16060 DRV. U. S. Government Printing Office, Washington, D. C. 101 p.

Cameron, R. E., and G. B. Blank. 1965. Soil studies—Microflora of desert regions. VIII. Distribution and abundance of desert microflora. p. 193–201. In Space Programs Sum. No. 37–34, Vol. IV. Jet Propulsion Lab., Pasadena, Calif.

Carter, C. D., R. D. Kimbrough, J. A. Little, R. E. Cline, M. J. Zack, Jr., W. F. Bartel, R. E. Koehler, and P. E. Phillips. 1975. Tetrachlorodibenzodioxin: an accidental poisoning episode in horse arenas. Science 188:738–740.

Day, A. D. 1973. Recycling urban effluent on land using annual crops. p. 155–160. In Recycling municipal sludges and effluents on land. Proc. Joint Conf., 9–13 July, Champaign, Ill. Nat. Assoc. State Univ. and Land-Grant Coll., Washington, D. C.

Day, A. D., A. D. Dickerson, and T. C. Tucker. 1963. Effects of city sewage effluent on grain yield and grain malt quality of fall-sown, irrigated barley. Agron. J. 55:317–318.

Day, A. D., J. L. Stroehlein, and T. C. Tucker. 1972. Effects of treatment plant effluent on soil properties. Water Pollut. Control Fed. 44:372–375.

Day, A. D., T. C. Tucker, and M. G. Vavich. 1962. Effects of city sewage effluent on the yield and quality of grain from barley, oats, and wheat. Agron. J. 54:133–135.

Fuller, W. H. 1961. Composting city refuse. Int. J. Milk Food Technol. 24:385–389.

Fuller, W. H. 1966. New organic pelleted compost. Compost Sci. 6(4):30.

Fuller, W. H. 1974. Desert soils. p. 31–101. In G. W. Brown (ed.) Desert biology. Vol. 2. Academic Press, New York.

Fuller, W. H. 1975. Management of southwestern desert soils. The Univ. of Ariz. Press. Tucson.

Fuller, W. H., and S. Bosma. 1965. The nitrogen requirement of some municipal compost. Compost Sci. 6(2):26–32.

Fuller, W. H., R. E. Cameron, and N. Raica, Jr. 1960. Fixation of nitrogen in desert soils by algae. Int. Soil Sci. Soc. Proc. 2:617–624.

Fuller, W. H., E. W. Carpenter, and M. F. L'Annunziata. 1967. Evaluation of municipal waste compost for greenhouse potting purposes. Compost Sci. 8(2):16–21.

Fuller, W. H., G. V. Johnson, and G. Sposito. 1960. Influence of municipal compost on plant growth. Compost Sci. 1(3):16–19.

Johnson, G. V. 1973. Irrigation recreational turfgrass with sewage effluent. Prog. Agric. 15:8–9. Univ. of Ariz. Agric. Exp. Stn., Tucson.

Meek, B., L. Chesnin, W. Fuller, R. Miller, and D. Turner. 1975. Guidelines for manure utilization in the western region, USA. Rep. Western Reg. Comm. W-124. Wash. State Univ., Coll. Agric. Res. Center Bull. 814.

Merrell, J. C. et al. 1967. The Santee recreational project—Santee, California, final report. FWPCA, U. S. Dep. Int., Cincinnati, Ohio.

Page, A. L. 1974. Fate and effects of trace elements in sewage sludge when applied to agricultural lands. A literature review study. U. S. Environ. Prot. Agency, Nat. Environ. Res. Center, Cincinnati, Ohio.

Seabrook, B. L. 1973. Land application of wastewater with a demographic evaluation. p. 9-24. *In* Recycling municipal sludges and effluents on land. Proc. Joint Conf., 9-13 July, Champaign, Ill. Nat. Assoc. State Univ. and Land-Grant Coll., Washington, D. C.

Selitch, I. 1974. Improving the energy conversion in agriculture. p. 37-50. *In* D. E. McCloud (ed.) A new look at energy sources. Am. Soc. of Agron., Madison, Wis.

Steffger, F. W. 1974. Energy from agricultural products. p. 23-25. *In* D. E. McCloud (ed.) A new look at energy sources. Am. Soc. of Agron., Madison, Wis.

Stout, P. R. 1974. Agriculture's energy requirements. p. 13-22. *In* D. E. McCloud (ed.) A new look at energy sources. Am. Soc. of Agron., Madison, Wis.

Stubblefield, T. M., and A. H. Smith. 1964. A survey of the production and marketing of cattle manure in Arizona. Univ. Ariz. Agric. Exp. Stn. Bull. A-36.

Tucker, T. C., and W. H. Fuller. 1971. Soil management: humid vs. arid areas. *In* W. G. McGinnies, B. J. Goldman, and P. Paylore (ed.) Food fiber and the arid land. The Univ. Ariz. Press, Tucson.

Volk, V. V., and C. H. Ullery. 1972. Disposal of municipal waste on sandy soils. Rep. to the Boeing Co. Oregon State Univ., Corvallis.

Wadleigh, C. H. 1968. Wastes in relation to agriculture and forestry. USDA Misc. Publ. No. 1065. U. S. Government Printing Office, Washington, D. C.

chapter 19

Land Utilization and Disposal of Organic Wastes in Hot, Humid Regions

GRANT W. THOMAS, University of Kentucky, Lexington, Kentucky

I. INTRODUCTION

Two impossible tasks are encountered in discussing special problems of the hot, humid region with respect to waste disposal in soils. First, identification of problems that are unique to the region; and second, determining the importance of these problems. Instead, I have tried to mention problems which are important to the hot, humid region, but which undoubtedly overlap into other climates, and I have tried to make the best guess about practical problems based on soil characteristics, temperature, rainfall, and evapotranspiration in the region. These guesses are based on general principles applicable to all soils; in most cases, they have not been tested with wastes of the types or amounts now being used.

The subjects to be considered are: (i) the influence of rainfall and evapotranspiration on water status of soils and streams, (ii) the effect of annual and seasonal temperatures in the region on nitrification and organic matter oxidation, (iii) soil characteristics, with emphasis on soil groups which appear to show the greatest problems with regard to waste disposal, and (iv) specific problems of wastes produced in the region.

II. RAINFALL VS. EVAPOTRANSPIRATION

Rainfall in the southeastern USA is much higher than in the Northeast, but streamflow, which is a good measure of excess water, does not average much higher. Crowther (1930) estimated that for each 1C increase in annual temperature there must be an increase of 3.4 cm of precipitation in order to maintain the water balance. This estimate may be slightly low according to Table 1, which shows the rainfall, mean annual temperature, streamflow, and streamflow estimated from reservoir or open pan evaporation multiplied by a factor of 0.7.

Streamflow in the Genessee river in western New York is 39 cm (15.32 inches/year) with rainfall of 89 cm (35 inches) compared to a streamflow of 48 cm (18.73 inches/year) in the Savannah river in Georgia with 127 cm (50 inches) of rainfall per year. Streamflow in the Suwanee river in Florida is

Table 1—Annual rainfall, mean annual temperature, and yearly streamflow for several rivers (Todd, 1970)

River	Location	Yearly rainfall	Mean annual temperature	Yearly streamflow	Calculated streamflow[†]
		cm	°C	cm	
Suwannee	Florida	139	21.0	24.8	27.4
Savannah	Georgia	127	16.0	47.6	39.9
Tennessee	Tennessee	122	15.5	54.6	52.6
Kentucky	Kentucky	114	13.5	44.1	41.4
Genessee	New York	89	8.5	39.0	35.5

† Calculated from reservoir or open pan evaporation × 0.7 subtracted from rainfall.

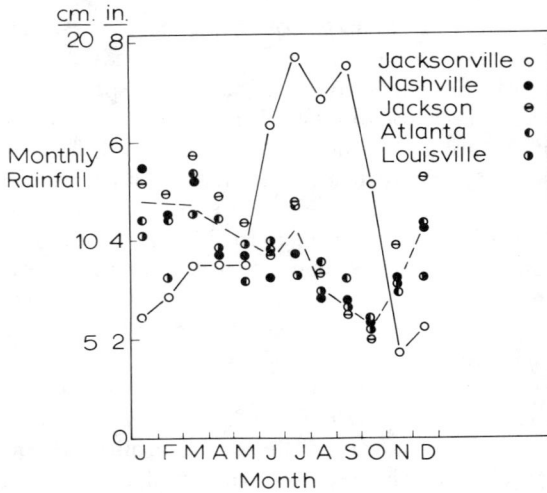

Fig. 1—Monthly rainfall for several southeastern locations (Todd, 1970).

abnormally low, not only because of the high mean temperature but also because a higher proportion of the rainfall falls during the hot summer months.

The water status of an area depends not only on rainfall and its distribution, but on evapotranspiration rate (ET). Since potential ET is directly related to radiation, and closely correlated to temperature, ET would be expected to be great in a hot, humid region. *Potential* evapotranspiration is even higher in hot, arid regions but *actual* evapotranspiration is lower because of absence of water.

Even in a humid region, there are periods during which ET is low compared to potential ET. This situation exists over most of the southeastern USA during late summer and early fall, for the reasons that plants have extracted most of the available soil water and rainfall is low during these periods. The results are low humidities, delightful fall weather, and disappointing crop yields. Figure 1 shows monthly rainfall for several regions in the Southeast. Except for Florida, these data show low precipitation during late summer and early fall and high rainfall during December and January. The effect of this on streamflow is shown in Fig. 2 and 3. The stream shown in Fig. 2 (Cave Creek, Ky.) flows from a small agricultural watershed where recharge is rapid and flow ceases during late summer and early fall. Figure 3 shows Tennessee river flow at Paducah, Ky., which, despite construction of many dams, shows the same tendency. Low water flow during the months of July through November for this huge watershed coincides with a deficiency in soil water.

During the period May through October, it is possible to apply excess water to the soil without causing much increase in stream flow. In other words, actual ET can approach potential ET if the soil is kept continuously wet. An estimate of the amount of water that could be applied to soils at

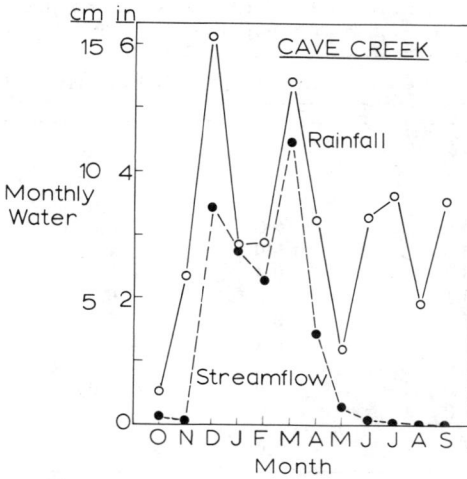

Fig. 2—Rainfall and streamflow for Cave Creek, Fayette Co., Ky. (U. S. Geological Survey, 1970–1973).

several locations in the Southeast without increasing leaching is given in Table 2. Due to a different rainfall pattern, the times of year that water can be added in Florida are not the same as for the other locations.

The values given in Table 2 must be considered minimum under average soil conditions. In well-drained soils, the amount of water added can be doubled or tripled in some months; however, the great excess of water during the months of December through March (except in Florida) precludes the use of much additional water. The numbers in Table 2 apply best to a permanent sod; if the land is to be used for cultivated crops, the amount which could be applied at planting and harvesting would be much lower.

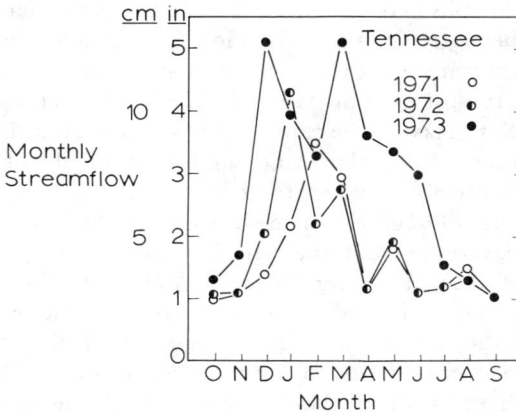

Fig. 3—Streamflow for the Tennessee River at Paducah for 3 years (1971, 1972, 1973) (U. S. Geological Survey, 1970–1973).

Table 2—Estimated water that can be added to soils on a monthly basis without increasing deep drainage

Month	Louisville, Ky.	Nashville, Tenn.	Atlanta, Ga.	Jacksonville, Fla.
			cm	
January	0	0	0	4.6
February	0	0	0	3.8
March	0	0	0	3.8
April	0.5	2.0	0.5	5.8
May	3.6	5.6	7.1	8.4
June	6.4	9.4	7.9	2.5
July	9.1	8.9	6.1	0
August	7.1	8.9	6.6	0.5
September	5.6	6.1	6.1	0
October	2.0	3.1	3.3	0.5
November	0	0	0	5.1
December	0	0	0	1.3
Yearly total	34.3	44.0	37.6	36.3

The effects of this seasonal variation in soil water on disposal of wastes containing large amounts of water are that either large areas of land must be covered very lightly or that considerable waste storage capacity must be provided. During the wetter months, both approaches could be used. During wet periods, moderate water applications can be made on well-drained soils, but extra storage capacity would be required for very wet periods. The storage capacity must be much larger when soil drainage problems are encountered.

An important reaction affected by soil water content is that of nitrification. It has been observed (Alexander, 1965) that nitrification proceeds at a rapid rate after dry soils are rewetted. Apparently this is a result of a new surge of bacterial growth. The situation with respect to water in southeastern soils is that they are nearly air-dry in the surface layer in late summer. When fall rains rewet the soils in September and October, particularly where cultivated crops have been grown, there is almost no transpiration by the crop. Soil temperatures remain rather high through November and rapid nitrification occurs. A typical example where N had been added to the surface of a bare soil in Kentucky at the rate of 336 kg/ha is given in Fig. 4. During this particular year (1971), almost no rain fell on the plots from late August through early September. When the rains began in late September nitrification of more than 300 kg/ha remained. Chloride (as KCl), added as a check, showed some variation but the rate of winter leaching loss was identical to that for NO_3^--N. However, there was only a small increase in the amount of Cl recovered in the profile as compared with the increase in NO_3^--N at the 2 December sampling. Figure 4 shows both the enormous capacity for nitrification after initiation of fall rains and the heavy losses that are apt to occur during the winter. Also illustrated is the importance of applying organic wastes to soils on which an active crop (e.g., grass or a small grain) will be growing during the winter.

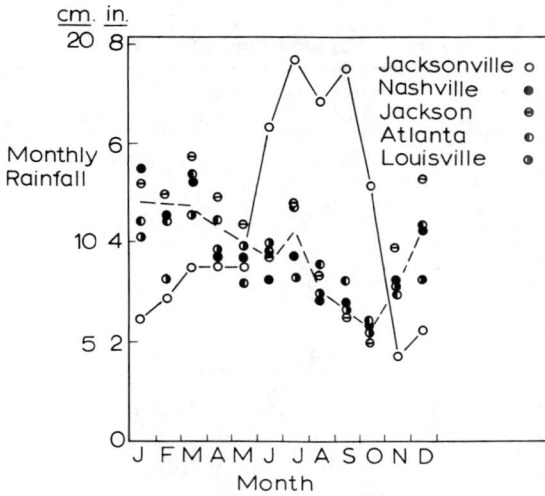

Fig. 4—The quantities of NO$_3$-N and Cl in a 90-cm profile of Maury silt loam showing the rapid increase in NO$_3$-N during the fall, fallow period followed by winter leaching. Dashed lines represent the quantities of NO$_3$-N and Cl added to the soil.

III. TEMPERATURE AND ITS EFFECTS

Mean annual temperatures for the southeastern USA are closely related to latitude, except in the Appalachian region, and they range from 10 to 21C.

The casual visitor to Atlanta, Ga., in July or to Albany, N. Y., in January might conclude that typical weather at the two locations is more different than is actually the case. This is shown in Fig. 5, where mean monthly temperatures for several locations along a north to south transect of the eastern USA are given. In addition to the general seasonal trend at all locations, and the increasing temperatures with decreasing latitudes, it should be

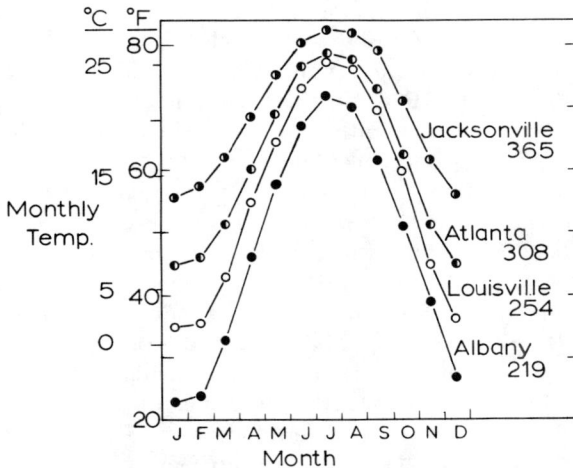

Fig. 5—Monthly mean temperatures for several locations (Todd, 1970).

Table 3—Ramann's (1911) weathering index for several cities based on a value of 1.0
for Albany, N. Y. (quoted by Jenny, 1941)

Place	Average temperature		Relative dissociation of water	No. of days	Relative weathering
	°C	°F			
Albany	8.7	47.6	1.6	219	1
Cleveland	9.6	49.2	1.65	232	1.09
Philadelphia	11.9	53.5	1.75.	256	1.28
Louisville	13.2	55.7	1.95	254	1.42
Nashville	15.5	60.0	2.1	290	1.74
Atlanta	16.3	61.4	2.2	308	1.93
Jackson	18.6	65.5	2.4	321	2.20
Jacksonville	20.8	69.5	2.6	354	2.63

noted that wintertime temperature differences between north and south are
greater than are summertime temperatures. For example, the difference between Albany, N. Y., and Jacksonville, Fla., in July is 4C (10F); the January
difference is 18.3C (33F). Next to the name of each city in Fig. 5 is the
number of days that temperature is above freezing. Notice that Albany is
above freezing little more than half the year compared to Jacksonville, which
is essentially always above freezing.

An interesting approach to the cumulative effect of temperature and
number of days above freezing was given by Ramann (1911) (quoted by
Jenny, 1941). Results for several locations are given in Table 3. The calculation is based on the change in dissociation of water with increasing temperature multiplied by the number of days that the temperature is above freezing.
Giving Albany a value of 1.0, the relative "weathering" is given for all locations.

As a check on this calculation, the Van't Hoff temperature rule (Jenny,
1930) states that for every 10C rise in temperature the velocity of a chemical
reaction increases by a factor of two or three. The mean annual temperature
at Jackson, Miss., is 10C higher than for Albany, N. Y., and its "weathering"
factor of 2.2 follows this relationship rather well.

Biological reactions also have Van't Hoff Q_{10} values of two or three.
For example, Sabey et al. (1956), in their much misquoted paper, showed a
Q_{10} slightly over 2 for nitrification up to 25C in field soils. Admitting that
soil to soil variation is astounding (Anderson, 1960; Anderson & Boswell,
1964), this means that there is an opportunity for twice the nitrification to
occur at Jackson as at Albany if there are nitrogenous compounds which can
be oxidized. When waste applications to soils are made so that substrate is
not the limiting factor, increases of this magnitude will be expected.

Supposing that values for temperature dependence on nitrification are
reasonable, relative nitrification can be calculated based on monthly surface
soil temperature at various locations. These calculations were done for
Blacksburg, Va., Lexington, Ky., and Gainesville, Fla. Mean monthly soil
temperature at a 10-cm depth was determined from data published by Fribourg et al. (1967) and the relative rate of nitrification for each month was
determined using the relationship of Sabey et al. (1959). Total nitrification

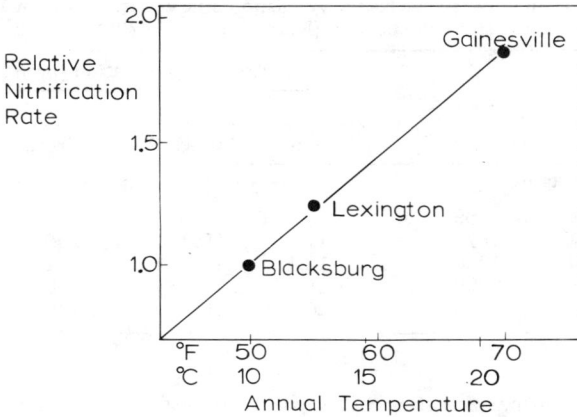

Fig. 6—Relative nitrification rate computed from data of Sabey et al. (1959) and based on temperature data of Fribourg et al. (1967).

for the year was determined and plotted against mean yearly temperature which gave a straight line with a temperature coefficient (Q_{10}) of 1.8 (Fig. 6).

Many years ago, Jenny (1930) published data showing the effect of temperature (T) on organic matter (OM) in soils along the east coast of the U. S. He found the following equation fit data from New York to Florida:

$$OM = 6.50 \exp{(-0.104\ T_{8C}^{20C})}.$$

This equation was designed only to show that existing organic matter in soils is related to temperature by the decay equation. However, relative indices of decomposition rate can be calculated from this equation for a basis of comparison. This was done in Fig. 7. Several locations in the southeastern U. S. are compared with Albany, N. Y. Relative decomposition rates range from 1.0 at 8C to 3.5 at 20C. The value for Q_{10} is approximately 3.

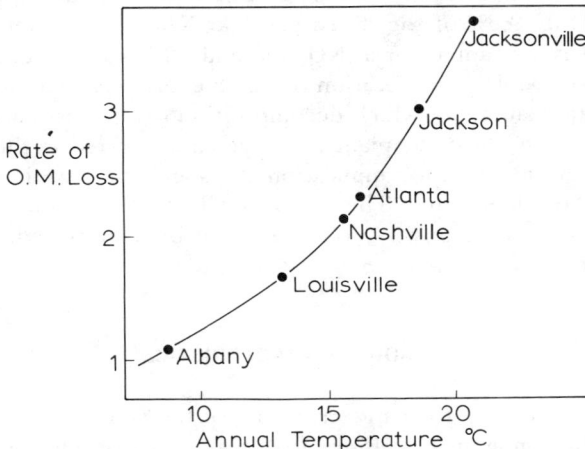

Fig. 7—Computed relative organic matter loss based on Jenny's equation (Jenny, 1930).

Table 4—The relationship between rye forage yield and tissue concentration of Zn and Cu in limed and unlimed soils (King & Morris, 1972b)

Total sludge added	Rye forage yield		Tissue concentration (ppm)			
			Cu		Zn	
	Unlimed	Limed	Unlimed	Limed	Unlimed	Limed
cm	kg/ha			ppm		
0	300	269	--	--	--	--
6.93	1180	1570	11.0	10.2	150	106
13.75	1540	1960	12.1	11.5	232	186
20.0	1650	2090	14.5	12.0	340	251
40.0	390	900	20.0	16.0	775	579

It should be emphasized that this apparent temperature coefficient does not necessarily apply to *added* organic wastes. Perhaps a better estimate of waste breakdown would be the potential nitrification shown in Fig. 6 which has a temperature coefficient of 1.8. In any case the mean annual temperature appears to be a useful measurement of relative activity and does *not* have to be corrected for the number of days below freezing.

The importance of increased carbon loss in hot, humid regions has not been studied intensively. However, a fairly comprehensive study of municipal sludge application to Cecil sandy clay loam at Athens, Ga., was reported by King and Morris (1972a, b, c). This study is of special interest with regard to oxidation of organic matter at high temperatures because the effect of the sludge treatments on rye (*Secale cereale* L.) growth was measured *after* sludge application stopped (Table 4). At the high residual rate of sludge, rye yield dropped drastically. Whether this was caused by 775 ppm Zn or 20 ppm Cu in the plant tissue is not clear. What is clear is that it is possible to overload soil systems with metals that reduce plant growth. It is also of interest that the high treatment had a NO_3^--N content of around 1,000 kg/ha during the fall after the last application. Obviously this type of residual NO_3^--N is not acceptable. Only 9.3% of the added 9,100 kg N/ha was recovered in plants, about 10% was accounted for as NO_3^--N, and 80% was still unmineralized. Even with appreciable denitrification this is an enormous load on a soil.

This is an example in which decomposition of organic matter can have adverse effects on soils and plants. As yet unmeasured is the effect the metals may have after sludge application stops and the organic matter level has returned to a low value. The heavy metal situation is uncertain at this time because this decomposition effect has not been measured. This area is discussed further in section V.

IV. SOIL CHARACTERISTICS

It is a difficult task to characterize soils of the Southeast by any quantitative method. While most soils are acid, many are not; most are low in organic matter, many are very high; cation exchange capacity ranges from the lowest to the highest in the United States.

There is one thing that can be said, however, and that is that parent material and topographic position show more marked effects on soils in the hot, humid region than do the other soil-forming factors. In addition, a higher proportion of soils are formed from consolidated material than is typical in other parts of the country, and the relation between parent rock and soil is quite distinctive in terms of both soil morphology and landscape appearance.

The main emphasis of this section will be on groups of soils which are liable to present problems in waste application, particularly when water is an important part of the waste. The reasons for the waste and water disposal problems are outlined for each soil grouping.

Figure 8 shows the steep, shallow soils of Virginia, the Carolinas, Kentucky, and Tennessee. These soils, typically, are not easily accessible for waste disposal; the steep slopes lose some water from surface runoff, but the most serious problem is the inadequate depth of soil for filtration of water to occur. Nearly all the stream flow in the area is a result of soil water interception by impermeable rock layers. Capacity of the soils to handle additional water is limited to dry periods in the summer.

The limestone-derived soils have a specific problem with regard to disposal of contaminated, watery waste. Whereas the soils themselves tend to be well drained and to have desirable structural properties, the channels between soil and underlying rock are large and frequent. The result is that unfiltered water, solid waste, soil, and occasionally calves and tractors can move

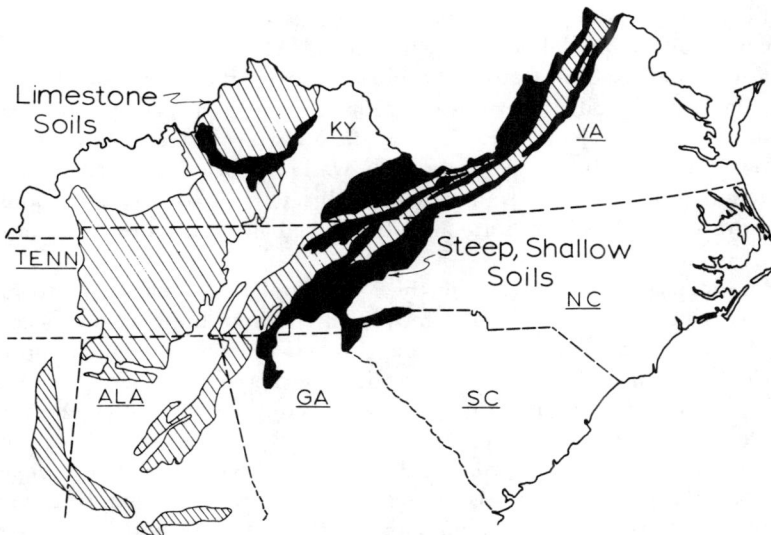

Fig. 8—Soil association map showing areas of steep, shallow soils and limestone soils in the southeastern USA (Buol, 1973).

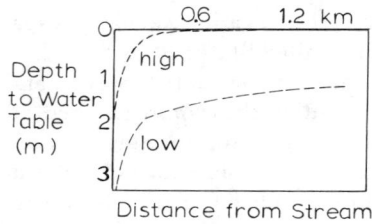

Fig. 9—The relation between distance from streams and high and low water table depths for the Middle Coastal Plain of North Carolina (Daniels et al., 1971).

directly into the cavernous rock. Once in the solution channels of the limestone, the contaminants move in underground streams and can appear in well and spring water or break out into surface water. For this reason, the underground rock status of limestone areas where large scale waste deposition is to be practiced must be known. Much of the area is useable for waste disposal but extreme caution must be exercised in this region.

The coastal plain is not delineated on this map (Fig. 8) since the problem there is quite complex. In the middle and lower coastal plain of the Atlantic coast a large proportion of the soils have water tables of varying depths. Near the streams, water tables are very deep. Figure 9 shows the relationship between high and low water tables and distance from streams in the Middle Coastal Plain of North Carolina (Daniels et al., 1971). In the broad flats between the streams, land surfaces are suitable for waste disposal but water tables rise near the surface during part of the year. Thus, the effect of the water table is to increase the chance of ground-water contamination and to reduce the time during the year when waste can be applied.

Figure 10 is a map of the Mississippi delta region with alluvial, loessial, and marsh soils being delineated. The alluvial soils range from excellent to poor for waste disposal, but nearly all suffer from the limitation that they are liable to be flooded every year. Judging from past history, chances of flooding are greater than 1 year in 5 in the Mississippi river valley (Hoyt & Langbein, 1955). Obviously, large scale waste application is hazardous on alluvial soils since by definition they are subject to flooding. Marsh soils at the mouth of the Mississippi are already flooded and cannot be used for disposal of organic wastes.

The loessial soils that follow the Mississippi river basin tend to have fragipans which restrict the water movement. Where the pans are deep ($\geq$ 75 cm), excess water is present during the period from November through June, leaving 4 months when wastes could be applied. Where the pan is close to the surface (about 30 cm), there is a continuous problem of water management after every rain and a tendency for the soils to remain saturated from December through April. The solution to the problem for organic waste disposal in loessial soils is to use soil survey information in such a way that disposal is only practiced on soils without a pan or with a very weak fragipan.

Florida (Fig. 11) offers the most exacting challenge from the standpoint

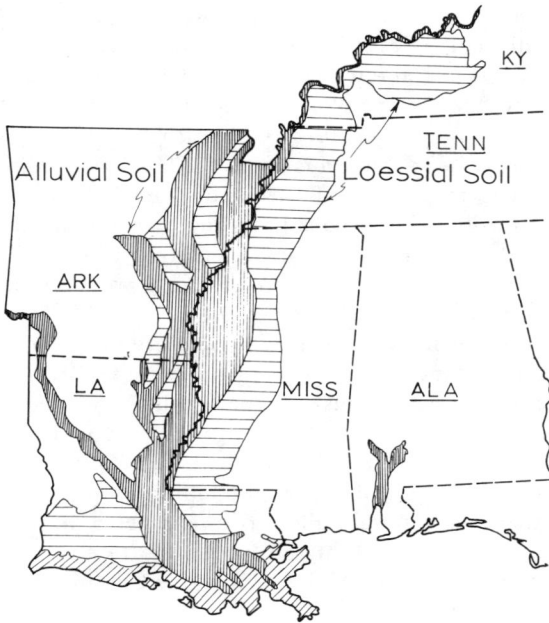

Fig. 10—Alluvial, loessial, and marsh soils associated with the Mississippi River (Buol, 1973).

of soil limitations of any state in the southeastern USA. At first glance, Florida is seen as a region of deep, sandy soils that offers unlimited possibilities for waste and water disposal. The sand is only a camouflage which hides the soil behavior.

Problems to contend with include swamps such as the Everglades where the water table tends to lie above the land surface. Obviously, these areas are

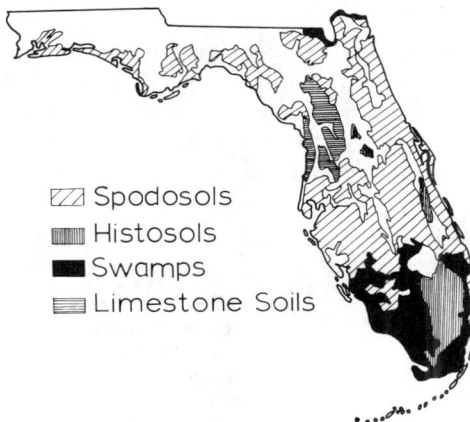

Fig. 11—Problem soils of Florida (Buol, 1973).

not useable. Another problem is the histosols (organic soils) where water
management is difficult at best. These areas have several drawbacks as sites
for waste disposal. First, the water balance is delicate; second, much of the
area is devoted to food crops which cannot be irrigated safely with waste;
and third, there is a marl layer directly beneath the organic soil which presents the same problems in water quality as the limestone region.

By far the largest portion of the sandy soils in Florida are spodosols
(ground-water podzols). These soils have one or more metallic-organic pans
near the surface which slow water drainage greatly. The use of these soils for
waste disposal is restricted, although many such areas are being used with
fair success over most of the year.

Finally, in north-central Florida, an apparently desirable soil area occurs
over cavernous limestone, with the same problems of contamination that
other limestone regions have. This area surrounds Gainesville.

Much of the Piedmont, the Upper Coastal Plain, deeper soils of the
mountains, and the mid-south offer excellent disposal sites from the standpoint of water handling, ease of access, and limited slope (Peele et al., 1973).
Fortunately, these areas contain many of the larger cities in the South, including Atlanta, Richmond, Birmingham, Charlotte, and Jackson. Other
cities, such as Norfolk, Miami, Mobile, and New Orleans, have severe problems because of high water tables.

V. SPECIFIC PROBLEMS OF ORGANIC WASTES

All organic wastes contain some of the essential plant nutrients and all
offer some advantage to crop growth. All of them could potentially cause
NO_3^- contamination of ground and surface waters, and degrade runoff water
with organic compounds and phosphorus. All have an odor problem and all
are too bulky to be worth handling costs based on nutrient content alone.
Some specific problems are given in Table 5. Only wastes of nationwide
interest are shown.

Sewage sludge and sewage effluent present several severe problems.
Pathogens are potentially the most serious from a health standpoint. Soil
disposal offers both the opportunity of improved water quality, if effective
filtration takes place, and also the possibility of widespread contamination of
air, soil, surface, and ground water if poor sites are used. Sewage sludge also

Table 5—Some specific problems with several types of wastes

Type of waste	Notable problems
Sewage sludge	Pathogens, heavy metals, transportation
Sewage effluent	Pathogens, transportation, water
Dairy manure	Storage, water
Beef manure	Transportation
Hog manure	Odors, storage
Poultry manure	Transportation

Table 6—Heavy metals in three sewage sludges

Element	Miami, Fla.[†]	Atlanta, Ga.[‡]	Athens, Ga.[§]
		mg/kg in dry sludge	
Cd	>59	165	--
Cu	500	636	520
Zn	1,780	11,812	2,825
Ni	241	183	26

[†] R. L. Chaney, 1974, unpublished data.
[‡] F. C. Boswell, 1973, unpublished data.
[§] King and Morris (1972a).

has the unique problem of heavy metals complexed by organic compounds. These metals (Ni, Cd, Cu, Zn) are often very high in wastes from industrial cities (Chaney, 1973; Page, 1974); Zn and Cu are high in sludges from small cities (King & Morris, 1972a). Apparently, Zn and Cu will always be high from the erosion of pipes, even in a residential community. Values for three locations are given in Table 6.

The problem of heavy metals could be minimized if the wastes were applied at low rates. The actual philosphy, which is a result of short term economics, is to spread a thick layer of waste over as little soil as possible. This causes problems for both NO_3^- and heavy metals as shown by King and Morris (1972b, c). Chaney (1973) and Page (1974) reviewed the heavy metal problems of sewage sludges and both concluded that problems with Cd are particularly serious from a human health standpoint and that Cu and Zn are potential problems from a soil sterilization point of view.

In addition to the biological problems, transportation of both sludge and effluent is expensive and the water disposal problem of effluent is great during the wetter parts of the year. Again, the usual temptation is for cities to dump large amounts too close to residential areas on marginal soils, and at rates much too high for safety.

Dairy manure from milking parlors, loafing sheds, and alleyways is generally washed down with large quantities of water, resulting in a solid content of about 2%. In an experimental system at the University of Tennessee (Barker & Sewell, 1972),[1] a 1.6-ha (4-acre) land surface of limestone-derived soil was used to accommodate the washing from a 0.4-ha (1-acre) concrete area. The storage tank had a capacity for 6.5 cm (2.5 inches) of runoff so that under conditions of very heavy rainfall, irrigation had to be carried out during the rain. The results with this system show a considerable worsening of both ground-water and runoff water quality as regards coliform bacteria and nitrate. However, this must be balanced with the alternatives of either very extensive spreading at high cost or using a solid removal system.

Beef manure in hot, humid regions is a problem only in relation with

---

[1]J. C. Barker, and J. I. Sewell. 1972. Effects of spreading manure on ground water and surface runoff. Preprint of paper presented at the meetings of Am. Soc. of Agric. Eng. June 1972.

confined cattle feeding installations, which are not common. In general, transportation during the wet portion of the year (December through March) is the only major problem.

Hog manure in confined systems is a problem not unlike dairy manure except that less waste is added and the systems are smaller. The general problems are odor and sufficient storage to handle the slurry during the wet winter periods. Periods of soil wetness are shorter than in the more northerly areas but the trafficability of the soil is worse due to heavier rains in December and January and higher temperatures during the winter.

Poultry manure, largely a southern product, causes two separate problems: (i) distribution over a sufficiently large area to prevent problems of soil impermeability (L. E. Hileman, 1973, personal communication) and (ii) fat necrosis of cattle grazed on fescue pastures where poultry manure is applied (Wilkinson et al., 1971) at very high levels. The obvious solution is lighter applications of poultry manure to more land. This route has become more attractive economically because of the recent rise in the value of poultry manure as fertilizer.

In the author's opinion, the problems of organic waste disposal on soils can be handled using principles which have been developed over the years for the beneficial use of beef and dairy cattle manure, namely applications at rates which supply adequate nutrients for crop growth. This increases the transportation costs, but little environmental damage results, either on a short or long term basis.

One type of waste which appears to have serious problems over the long run, even at low application rates, is sewage sludge. The chief problem for continued usage of the same land is the accumulation of toxic heavy metals. As long as high organic matter applications are continued, the problem may not become acute, but once organic matter begins to decompose, the metals are apt to become more available.

This problem has been shown for Cu in data cited by Page (1974) and no elaboration will be made here except to point out that in hot, humid regions decomposition of organic matter is likely to occur at rates two to three times that in cold regions, with a result that heavy metal toxicities may be more severe. This, of course, has not been proven, but the principles discussed in section III lead to that conclusion.

Perhaps the most serious general restriction when applying wastes to soils in hot, humid regions is that of water management (for wastes high in water). The extremes in water content of the soil (section II) are such that during the winter and spring much less water can be applied than during the summer and early fall (except in Florida).

The magnitude of the problem varies with soil type and a considerable range exists. In any event, systems designed for optimum summer use will not perform during the winter regardless of soil type. There seem to be only two answers for this problem: (i) the use of very large land areas for winter application, and (ii) the use of extensive storage lagoons so that application can be made primarily during the summer and fall.

LITERATURE CITED

Alexander, M. 1965. Nitrification. *In* W. V. Bartholomew and F. E. Clark (ed.) Soil nitrogen. Agronomy 10:307-343. Am. Soc. of Agron., Madison, Wis.

Anderson, O. E. 1960. The effect of low temperatures on nitrification of ammonia in Cecil sandy loam. Soil Sci. Soc. Am. Proc. 24:286-289.

Anderson, O. E., and F. C. Boswell. 1964. The influence of low temperature and various concentrations of ammonium nitrate on nitrification in soils. Soil Sci. Soc. Am. Proc. 28:525-529.

Buol, S. W., ed. 1973. Soils of the southern states and Puerto Rico. South. Coop. Ser. Bull. 174.

Chaney, R. L. 1973. Crop and food chain effects of toxic elements in sludges and effluents. p. 129-141. *In* Recycling municipal sludges and effluents on land. Proc. Joint Conf., 9-13 July, Champaign, Ill. Nat. Assoc. State Univ. and Land-Grant Coll., Washington, D. C.

Crowther, E. M. 1930. The relationship of climatic and geological factors to the composition of soil clay and the distribution of soil types. Proc. R. Soc. London B107: 1-30.

Daniels, R. B., E. E. Gamble, and L. A. Nelson. 1971. Relations between soil morphology and water table levels on a dissected North Carolina Coastal Plain surface. Soil Sci. Soc. Am. Proc. 35:781-784.

Fribourg, H. A., R. H. Brown, G. M. Prine, and T. H. Taylor. 1967. Aspects of the microclimate at five locations in the southeastern United States. South. Coop. Ser. Bull. 124.

Hoyt, W. G., and W. B. Langbein. 1955. Floods. Princeton Univ. Press, Princeton, N. J.

Jenny, H. 1930. A study on the influence of climate upon the nitrogen and organic matter content of the soil. Missouri Agric. Exp. Stn. Res. Bull. 152. 66 p.

Jenny, H. 1941. Factors of soil formation. McGraw-Hill Book Co., Inc., New York, N.Y.

King, L. D., and H. D. Morris. 1972a. Land disposal of liquid sewage sludge: I. The effect on yield, in vivo digestibility and chemical composition of Coastal bermudagrass (*Cynode dactylon* L. Pers.). J. Environ. Qual. 1:325-329.

King, L. D., and H. D. Morris. 1972b. Land disposal of liquid sewage sludge: II. The effect on soil pH, manganese, zinc, and growth and chemical composition of rye (*Secale cereale* L.). J. Environ. Qual. 1:425-429.

King, L. D., and H. D. Morris. 1972c. Land disposal of liquid sewage sludge: III. The effect on soil nitrate. J. Environ. Qual. 1:442-446.

Page, A. L. 1974. Fate and effects of trace elements in sewage sludge when applied to agricultural lands. Environ. Prot. Agency Technol. Ser. EPA-670/2-74-005.

Peele, T. C., H. P. Lynn, C. L. Borth, and J. N. Williams. 1973. South Carolina guidelines for land application of animal waste. South Carolina Agric. Exp. Stn. Bull. 570. 18 p.

Ramann, E. 1911. Bodenkunde. Verlag Julius Springer, Berlin, Germany.

Sabey, B. R., W. V. Bartholomew, R. Shaw, and J. Pesek. 1956. Influence of temperature on nitrification in soils. Soil Sci. Soc. Am. Proc. 20:357-360.

Sabey, B. R., L. R. Frederick, and W. V. Bartholomew. 1959. The formation of nitrate from ammonium in soils: III. Influence of temperature and initial population of nitrifying organisms on the maximum rate and delay period. Soil Sci. Soc. Am. Proc. 23:462-465.

Todd, D. K. 1970. The water encyclopedia. Water Information Center, Port Washington, N. Y. 589 p.

U. S. Geology Survey. 1970-1973. Water resources data for Kentucky, Part 1: Surface water records. U. S. Geol. Surv., Louisville, Ky.

Wilkinson, S. R., J. A. Stuedemann, D. J. Williams, J. B. Jones, R. N. Dawson, and W. A. Jackson. 1971. Recycling broiler house litter on tall fescue pastures at disposal rates and evidence of beef cow health problems. p. 321-324. *In* Livestock Waste Management and Pollution Abatement Proceedings of Symposium held at Ohio State University. ASAE Proc. 271, St. Joseph, Mo.

chapter 20

Water heated to calculated power plant condenser discharge temperatures is used in open field and greenhouse soil warming studies (Photo courtesy of the Tennessee Valley Authority, Muscle Shoals, Ala.).

Special Problems and Opportunities in Use of Waste Heat for Soil Warming

DAVID A. MAYS, Tennessee Valley Authority, Muscle Shoals, Alabama

I. INTRODUCTION

When the word waste is used most people have traditionally thought of the useless solid and liquid byproducts resulting from manufacturing and the biological processes of living organisms. These assorted wastes have caused concern only when their volumes were great relative to the size of the environmental sinks into which they were discharged. Interest in this and similar symposia suggests that a level of urbanization and industrialization has been reached where the volumes of waste produced are so great that they can no longer be ignored.

Waste heat is that portion of the heat created by the decomposition of conventional or nuclear fuels which is not beneficially used in the intended manufacturing or power production process and is discharged to the environment. Some of this heat occurs as hot air or gases emitted from stacks, some is radiated directly to surrounding air by warm machinery and boiler jackets, and some occurs in cooling water circulated through machinery or in heated waste water. However, the largest amount by far is found in the condenser cooling water used by fossil and nuclear fueled power plants. Concern for waste heat has arisen primarily because of the proliferation of these power plants in the last 25 years.

A few industries are recirculating part of their stack gas heat to warm work areas, and some radiated heat may be used for this purpose. However, these heat sources appear to have little or no application in crop production and will not be considered again.

Warren (1969) estimated that power generation accounts for 80% of the total cooling water used in the United States and for almost one-third of all water used. Heated water discharge from power plants amounts to 50 trillion gallons annually or about 15% of the total flow of U. S. rivers and streams (L. Barry Goss, 1975. The effects of varying rates of temperature changes on survival of *Daphnia tulex* Ledig and *Daphnia magna* Straus. M.S. Thesis. University of Tennessee, Knoxville).

Because thermal power plants account for the bulk of waste heat production, and since they are concentrated in a few locations relative to other industries, the remainder of this discussion will be oriented toward waste heat generated from this source. However, comments on the agricultural usefulness of waste heat should be applicable to any hot water source subject to modifications necessary because of temperature and volume.

II. CHARACTERIZATION OF WASTE HEAT FROM POWER PLANTS

A. Sources and Amounts

Since this discussion is limited to waste heat from power plants the source can be described quite simply. The efficiency of present day thermal power plants in converting heat to useful electricity lies in the 30 to 40%

range (Boersma, 1970). Nuclear plants reject 6,500 Btu of heat for each kW of power produced, while the more efficient fossil fuel plants reject about 4,600 Btu (Warren, 1969). It is expected that heat rejection per kW produced will be reduced by about 10% in the 1980's and by about 50% by the year 2000. [British thermal units (Btu) may be converted to joules (J) by multiplying the Btu measurement by 1,054.8.]

Based on the assumption that power generation requirements will double every 10 years for the next few decades, total thermal generating capacity will be 530,000 MW by 1980 and 1,033,000 MW by 1990 (Federal Power Commission, 1971). These production rates will result in waste heat amounts of 5,914 and 12,773 trillion Btu's, respectively, in 1980 and 1990. However, the present fuel supply and price situation in 1975, if it continues, may result in a somewhat slower growth rate.

While nuclear power capacity is now $< 10\%$ of the total production, it is expected to increase to 22% by 1980 and 40% or more by 1990 (Warren, 1969). During this period, fossil fuel power will decline to 59% of the total in 1980 and 44% in 1990. Other writers using somewhat different assumptions or base data have arrived at slightly higher or lower projections, but the ones mentioned here seem adequate to indicate the magnitude of the waste heat resource.

B. Thermal Characteristics of Cooling Water

With present technology, all of the potentially useful waste heat for agriculture from power plants is in condenser cooling water; however, several workers are beginning to study the feasibility of direct removal of steam from turbines for heating purposes.

The water actually used to make steam in both fossil fuel and nuclear power plants is in a closed system. After maximum energy is extracted by the turbine, the steam is condensed to water before being recirculated and reheated. The condenser is, in turn, cooled by water at ambient temperature which is pumped through it from a river, lake, or ocean, or by water in a closed-loop system with cooling towers, ponds, or other methods for heat dissipation to the atmosphere.

The thermal characteristics of the condenser water depend on geography, season of the year, and design of the power plant. Where once-through cooling is used, incoming water temperatures may range from a few degrees above freezing at northern locations in the winter to temperatures of 30C or above at southerly locations in midsummer. As cooling water passes through the condenser, temperature increases may range from 5 to 10C for coal-fired plants to somewhat higher for nuclear plants, depending on design. Thus, discharge water temperatures with once-through cooling could range from 10 to 50C, so general statements of the usefulness of such heated water may have little meaning. Condenser water discharge temperatures in closed-loop cooling tower or pond systems are usually higher than with once-through cooling, but the extreme range of variability still exists. Unfortunately, condenser water temperatures are low in the winter and high in the summer, while we would wish the reverse were true.

C. Present and Potential Fate of Waste Heat

Until fairly recently most thermal generating plants were sited on rivers, lakes, or salt water bodies sufficiently large so that once-through cooling was used and the waste heat was discharged to the natural body of water.

With the development of larger plants, the need for additional generation sites on smaller waterways, and the increased concern for the possibility of environmental changes caused by elevated river temperatures, closed-loop condenser systems using cooling towers have been developed. Cooling towers now being used are of the wet, natural- or forced-draft type which may result in icing or fogging problems in the area and in fairly large evaporative losses of water. Dry cooling towers which could eliminate these problems have been proposed but are considered costly and of unknown efficiency. Man-made cooling ponds can be used, but they have the disadvantage of requiring 400 to 1,200 ha of what is often prime farm land for a 1,000-MW plant, and they may produce the same icing and fogging problems as towers.

Because of the low temperatures of these waste heat sources, profitable industrial uses have not been identified (Claude J. Powell, personal communication). Use for residential or commercial heating is not considered feasible unless heat pumps are used to upgrade the quality of the heat. Also, the large residential developments needed to utilize the available heat are not expected to be located near most power plants.

Several biological uses for waste heat have been proposed because biological processes are often hastened by fairly small increases in temperature which could be achieved with this heat source. The possible agricultural uses which have been identified include soil heating for crop production, heating and cooling of greenhouses and livestock shelters, warming of waste treatment facilities to enhance the growth rate of algae used for recycling, and irrigation and frost protection for fruit and vegetable crops. Some of these are consumptive uses of heat and might substitute for part of the cooling tower investment, but others are not. Since soil warming is the only use which is within the scope of this paper further discussion will be restricted to that subject.

III. PHYSICAL AND ENGINEERING ASPECTS OF SOIL HEATING

Utilization of waste heat for soil warming has the advantage of dissipating heat in a manner which does not appear to cause environmental harm, while at the same time serving as a production resource for enhanced growth of crops.

A. Thermal Characteristics of Soils

Thermal characteristics of soils, including specific heat, conductivity, and effects of soil moisture on heat transfer, have been studied by Smith and Byers (1938), DeVries (1963), Nakshabandi and Kohnke (1965), and others, although thermal water usage was not the ultimate objective at that time. These authors found that the thermal characteristics of most soil minerals are similar, with most differences among soils being due to variations in organic matter content, moisture levels, bulk density, and porosity.

DeVries (1963) found that specific heat is about 0.20 for mineral soils but that it is considerably higher for organic soils. It appears to this author that specific heat of a given soil would be important only in determining the time required for initial warmup and would have little effect in the long term maintenance of a uniform temperature.

Heat conductivity seems to be the most important soil characteristic, since it determines the rate of lateral heat movement away from pipes, to deep layers, and to the surface where it can be radiated to the air.

Moisture content seems to be the most important factor causing variable conductivity among soils, since heat is not only conducted by water but moves through the soil with the liquid and vapor phases. Nakshabandi and Kohnke (1965) reported that the thermal conductivity of soil water is 30 times that of the air spaces but considerably less than that of the soil particles. According to DeVries (1963), conductivity changes only slightly between saturation and field capacity but decreases rapidly at lower moisture levels. Nakshabandi and Kohnke (1965) agreed with Smith and Byers (1938) that soil type has little effect on heat conductivity of dry soils and that there is little increase in conductivity until moisture content reaches the hygroscopic point.

Smith and Byers (1938) reported that heat conductivity increases with bulk density and decreases with increased porosity. There are conductivity differences among soil horizons, with the A and B horizons usually having lower conductivities than the C horizons. This is likely related directly to bulk density and porosity differences among horizons.

B. Methods of Warm Water Distribution

Effective heat dissipation or soil warming requires that the waste heat be brought into contact with the soil. Wierenga et al. (1970) investigated the possibilities of direct application of warmed water to soil. While this was an effective way of dissipating heat from the water, it did not result in warming the soil. They found that flood irrigating with water at 4.1 to 21.6C resulted in lowering the soil temperature because of increased evaporation. The water temperature differences resulted in only small differences in soil temperature for a short time after irrigation. In a similar experiment, Cline et al. (1969)

sprinkler irrigated with 50C water and found that droplets cooled to a temperature safe for plants by the time they reached the soil. While this was not a satisfactory way of warming soil, it showed that thermal water from power plants can be safely used for sprinkler irrigation. Miller (1970) reported on a similar use of thermal water for frost protection of fruit crops and irrigation of vegetables, fruits, and ornamental plantings.

Aside from the evaporative cooling problem pointed out by Wierenga et al. (1970), it appears that an attempt to warm soil by direct application of hot water would result in a saturated soil before any meaningful amount of heat had been added. Because of these problems the only option appears to be the use of a pipe grid to conduct heated water through the soil. This aspect of soil warming has been the subject of several research efforts, including field experimentation and computer modeling.

deWinter (1970)[1] performed an extensive economic analysis to compare copper vs. polyethylene tubing for soil warming, apparently with the objective of obtaining maximum heat dissipation in a system designed both for heat dissipation and soil warming for crop production. Copper appeared to be competitive with polyethylene only if extremely small diameter copper pipes in the 0.625-cm range were used. This size would put severe constraints on length of run and power requirements for pumping water.

Copper pipe 1.25 cm in size was used in a Minnesota field experiment (Allred et al., 1973). Most if not all other investigators including Skaggs et al. (1973) in North Carolina, the Interdisciplinary Research Team at Penn State (1974), and Berry and Miller (1974) in Oregon used polyethylene or polyvinyl chloride (PVC) pipe.

It seems that polyethylene pipe will be the most desirable for the lateral lines in a soil heating grid, since it is cheap and flexible enough to be installed with the cable-laying machine; ditching is required to install PVC pipe. Although it is technically a cold water pipe, polyethylene will tolerate temperatures in excess of 40C, which is above the optimum soil temperature for plant growth. It is likely that PVC pipe would be used for headers and larger supply lines, since excavation would be required to allow connection of the laterals.

C. Pipe Grid Designs

The design characteristics of a satisfactory water distribution grid for a large acreage have yet to be proven in actual field experimentation. A number of computer modeling exercises have been carried out to estimate heat dissipation rates, either to determine the maximum acreage which could be heated from a given sized power plant, or the minimum acreage needed for successful heat dissipation.

---

[1]F. deWinter. 1970. Underground heating tubes in agriculture; copper versus polyethylene. Report on study prepared for Copper Develop. Assoc., Inc., New York, N. Y. 60 p.

Perhaps the most widely quoted model is one developed by Kendrick and Havens (1973) to calculate the effects of several variables in a soil heating system. Their model assumes constant soil temperature, thermal conductivity, uniform radial water temperature in pipes, and radial heat transfer away from pipes by conduction, even though these factors are not necessarily constant. They determined that closely spaced, shallowly buried pipes would require the least land area for dissipation of a given amount of heat but, of course, this would likely not lead to the most efficient use of heat for crop production.

A Penn State Interdisciplinary Research Team (1974) combined a small field experiment with computer modeling and determined that 1,821 ha would be required to dissipate the heat from a 1,000-MW power plant with 5-cm pipe buried 30 cm deep and 60 cm apart. The research team reported that this system is less expensive than dry cooling towers but more expensive than wet towers.

Skaggs et al. (1973) conducted a similar experiment in North Carolina using 2.5-cm polyethylene pipe buried 50 cm deep and 50 cm apart, with the water temperature maintained at 37.5C. They found that both surface and subirrigation increased heat transfer from pipes and that the surface temperature of a moist heated soil was 3.8 to 5.0C lower at midday than a similarly heated dry soil. They calculated that heat dissipation to the soil from a 1,000-MW plant in North Carolina would require 1,255 ha in the winter and 4,251 ha in the summer with irrigation or 5,263 ha without irrigation.

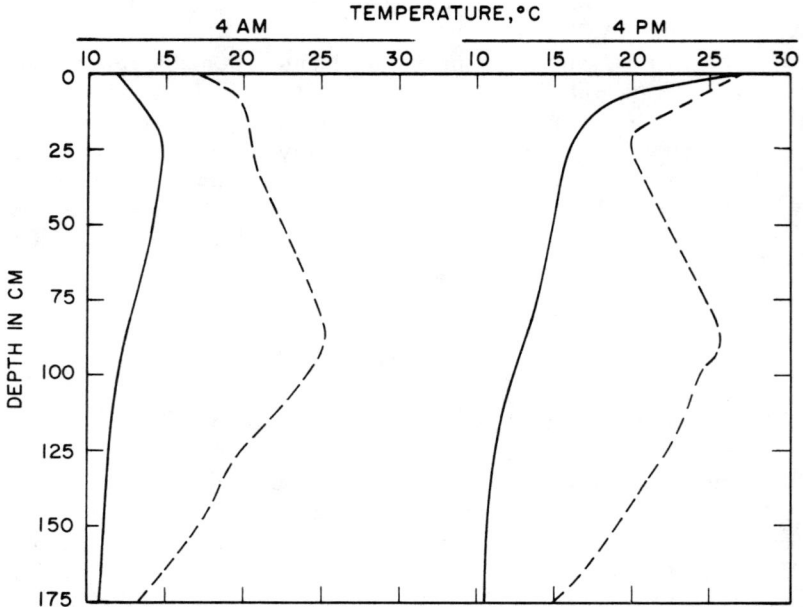

Fig. 1—Soil temperature as a function of depth on 31 May at Corvallis, Oregon. Measurements in verticle profile over 25C heating cable at 90-cm depth (broken line) and in control profile (solid line) (Boersma, 1970).

Extensive equations useful in characterizing various aspects of heat transfer in soils can be found in the several references cited in this and the previous section.

D. Temperature Profiles in Soils

Research with electrical cables has shown that heat was radiated rather uniformly around a heat source in soil (Fig. 1, Rykbost & Boersma, 1973) and that temperature dropped as the distance from the heat source increased. Soil temperature decreased rather rapidly (Fig. 2) above the 90-cm deep heat source, resulting in surface temperatures below the optimum for good response, particularly from young, shallow-rooted plants. Kendrick and Havens (1973) published similar profile descriptions which were developed mathematically. Theoretically the most uniform temperature throughout the heated profile occurs when lateral spacing and buried depth are equal. How-

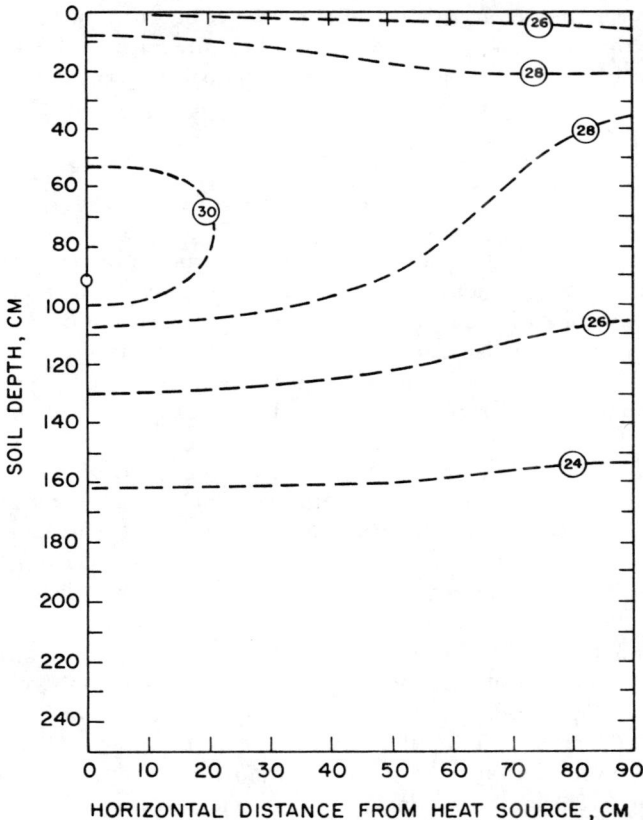

Fig. 2—Soil temperature profile as affected by verticle and horizontal distance from a cable heat source (Boersma, 1970).

ever, from a practical standpoint one might not strive for maximum uniformity, because of mechanical and economic constraints on deep placement or increased materials cost involved in close spacing if pipes are shallow.

Pipes should probably be buried at least 45 cm and preferably 60 cm deep to assure protection from physical damage, but it seems that placement much deeper than 60 cm might result in excessive power requirements in many soils and also in undesirably cool surface layers. Problems associated with lack of lateral uniformity in temperature profiles could be alleviated by adjustments in row spacings and plant geometry.

IV. CROP RESPONSES ON WARMED SOIL

Fairly extensive research to measure crop responses to heated soil has been conducted in Oregon, Minnesota, Maryland, and Alabama. With exception of the Maryland work, these experiments were oriented specifically toward using waste heat from power plants as a crop production resource. Over the years numerous other studies on the effects of root temperature on plant growth have been conducted primarily for the purpose of studying temperature effects on physiological and morphological development of plants and to help explain differences in seasonal growth responses.

A. Results in Oregon

Experiments to evaluate waste heat as a production resource were begun in 1969 at Corvallis, Oregon, and were reported by Rykbost et al. (1974). Heating was done with electrical cables with a capacity of 65.6 W per linear meter buried 92 cm deep and 183 cm apart. Temperature profiles in this installation were reported in the previous section.

A number of forage crops, including sudangrass (*Sorghum vulgare sudanense* L.), a sorghum-sudangrass hybrid (*Sorghum bicolor* [L.] moench × *Sorghum vulgare sudanense* L.), annual ryegrass (*Lolium multiflorum* Lam.), and tall fescue (*Festuca arundinacea* Schreb.), were compared on heated and unheated plots by Rykbost et al. (1974). The sudangrass and sorghum-sudan hybrid were both grown as summer annuals and cut four times. Soil heating increased total yield of sudangrass by 79% and of sorghum-sudangrass hybrid by 57%. Yield increases due to heating for the winter annuals, ryegrass and crimson clover (*Trifolium incarnatum* L.) were 69 and 64%, respectively; tall fescue yields were increased only by 19%.

Field corn (*Zea mays* L.) showed total dry matter and grain yield increases of 20% or more in 3 out of 4 years (Table 1), while soybeans (*Glycine max* L.) showed yield increases of 60% or more one year and 10% or less another year. The authors concluded that soil heating probably would be justified for corn grain production but not for corn silage or for soybeans grown for grain or silage.

Table 1—Effect of soil heating on corn grain and silage yields in Oregon
(Rykbost et al., 1974)†

Year	Plant component	Yield		Increase
		Unheated	Heated	
		— metric tons/ha —		%
1969	Total	12.3	17.9	45
	Grain	7.2	9.6	34
1970	Total	22.7	23.8	5
	Grain	11.7	13.7	17
1971	Total	14.3	19.3	35
	Grain	4.8	7.6	57
1972	Total	13.2	15.8	20
	Grain	7.6	9.2	22

† Total yields were on dry basis while grain yields were calculated at 15.5% moisture.

Several vegetable crops including bush beans (*Phaseolus vulgaris* L.), lima beans (*Phaseolus linatus* L.), tomatoes (*Lycopersicon esculentum* Mill.), broccoli (*Brassica oleracea,* var. Italica), and peppers (*Capsicum annum*) were grown on heated and unheated plots for periods of 1 to 3 years. The greatest number of comparisons were made with bush beans, comparing several planting dates and fertility levels each year. Table 2 shows the average response to heat for these vegetables. With bush beans, results varied from yield reductions at low fertility to increases as great as 85%, with the average being a modest 19%. Broccoli showed the greatest response—113%. With multiple harvest crops like broccoli and peppers, greatest responses were evident at early harvests.

Strawberries (*Fragaria virginiana* Duchesne) showed marked yield increases to soil heating, but maturity was not hastened.

Economic analyses performed for some of the crops indicated that soil heating would not be feasible in the Willamette Valley for forage crops, but would be for fresh market bush beans and perhaps strawberries. All crops in the trials were not evaluated for economic potential.

Berry and Miller (1974) conducted soil heating research at Springfield,

Table 2—Effect of soil heating on vegetable yields at Corvallis, Oregon
(Rykbost et al., 1974)

Crop	Years of data averaged	Yield		Increase
		Unheated	Heated	
		— metric tons/ha —		%
Bush beans	3	15.6	18.6	19
Lima beans	2	4.9	6.3	28
Tomatoes	3	7.0	9.5	36
Broccoli	1	2.3	4.9	113
Peppers	1	6.3	8.9	41

Table 3—Effect of soil heating on vegetable yields at Springfield, Oregon
(Berry & Miller, 1974)

Crop	Yield		Increase
	Unheated	Heated	
	— metric tons/ha —		%
Tomato (Avg. of two varieties)	136.2	136.2	0
Sweet corn (ears)	13.0	16.6	28
Cantaloupe	10.5	5.3	−49
Acorn squash	55.7	62.9	13

Oregon, using a 6.25-cm diameter PVC pipe buried 60 cm deep with a 150-cm spacing. They grew tomatoes, sweet corn (*Zea mays saccharata*), asparagus (*Asparagus officinalis* L.), rhododendron (*Rhododendron* sp.), cantaloupes (*Cucumis melo* L.), and acorn squash (*Cucurbita maxima* Duchesne) with and without soil heating.

Table 3 shows that there were no yield differences with tomatoes, while cantaloupe yields were decreased by 49%; sweet corn yields were increased by 28% and squash by 13%. Asparagus growing on heated soil had larger crowns, made more top growth, and produced a greater harvest of spears than that grown without heat. Two of six rhododendron varieties showed significant height incresaes on heated soil, while five of the six showed marked increases in lateral spread. The rhododendron results indicated that potential to reduce nursery growing time of some rhododendron varieties by a year.

Berry and Miller (1974) also constructed a plastic greenhouse over heated soil at Springfield and grew vegetables with no heat except that from the soil and the sun. Inside the greenhouse, heated soil temperatures averaged 5.3C higher at the 15-cm depth than unheated control soil from May through January. At the 30-cm depth, soils in the greenhouse were 1.7 to 3.2C warmer than heated soils at the same depth outside the greenhouse.

Leaf lettuce (*Lactuca sativa* L.), tomatoes, and cucumbers (*Cucumis sativus* L.) were successfully grown this way during several periods of the year when outside temperatures were not extreme. In December 1972, when the ambient air temperature fell to −20C, the inside air temperature fell to about −7.5C at the 150-cm height and froze tomato and cucumber plants; however, lettuce was not damaged.

B. Results in Minnesota

Allred et al. (1973) grew white potatoes (*Solanum tuberosum* L.) on heated soil at Elk River, Minnesota. Heating was started on 1 March using 40 to 45C water in 1.25-cm copper pipes buried 90 cm apart and 30 cm deep. At that time the soil was frozen to a depth of 90 cm and had 20 cm of snow cover. The upper 60 cm of soil was warm enough to plant after 10 days. Potatoes were planted on 19 and 26 March and emerged 10 and 13

April on heated soil; planting was delayed until 11 April on unheated soil and potatoes did not emerge until 7 May. During April and May, temperatures on heated soils were 9 to 12C above ambient soil temperature; however, differences were minimal in midsummer.

Even though top growth of emerged plants was severely damaged by frost three times in April and early May, potatoes on heated plots reached maximum yield 3 weeks earlier than those on unheated plots; however, total yields were similar with both treatments. A crop grown on the heated soil in the fall produced only a modest yield; no second crop was planted on unheated soil. The authors attributed the low fall yield to cool air temperatures, but it seems possible that the effect of shorter days and lower light intensities may have limited yields too.

C. Results in Maryland

Several experiments were carried out at College Park, Maryland, to study the effects of soil temperatures ranging from 10 to 32C on growth responses of a variety of plant species. Hawes and Decker (1971) found that cool-season grasses maintained the highest carbohydrate reserves with a 10C soil temperature but that yields were highest at 21C and at ambient temperature; these species were under a definite stress with a 32C soil temperature. On the other hand, bermudagrass (*Cynodon dactylon* L.), a warm-season species, flourished with a 32C soil temperature. High soil temperatures also enabled established bermudagrass to stay green until mid-December and eliminated winterkilling in newly planted bermudagrass.

Decker (1973) reviewed earlier work showing that root morphology is modified by soil temperature and that root and tiller formation by cool-season grasses is decreased at high temperatures. His own work showed that increasing the soil temperature hastened heading of orchardgrass (*Dactylis glomerata* L.) and timothy (*Phleum pratense* L.) but decreased total growth and tiller production of orchardgrass. Temperature responses appeared to be related to place of origin of a particular strain.

Decker and Walker (1972, 1973) found large soil temperature effects on corn growth. When early season soil temperature was below 15C, very little growth occurred; total season production at 10C was minimal (Table 4).

Table 4—Effect of soil temperature on corn yield in Maryland (Decker & Walker, 1974)

Soil temperature	Yield		Relative yield	
	Grain	Total plant	Grain	Total plant
	— g dry matter/plant† —		— % —	
10C	14.1	69.1	14	25
21C	84.7	239.0	85	85
Ambient	99.2	280.3	100	100
32C	79.8	260.2	80	93

† Yields are averaged for three fertility levels.

Early plant growth was greatest at 32C but these plants quit growing 2 weeks earlier than those at 21C and were 50 to 60 cm shorter. With low fertility, grain yields increased with each incremental increase in soil temperature,,but where nutrients were adequate best grain yields were obtained at 21C and at ambient soil temperatures. Late varieties performed relatively better at high soil temperature and early varieties were best at low soil temperatures.

With the exception of bermudagrass, all of the crops evaluated in Maryland grew best at soil temperatures likely to be achieved using waste heat from condenser water.

D. Results at Muscle Shoals, Alabama[2]

Since 1970, soil heating investigations have been carried out at Muscle Shoals, Alabama, using electric cables as the heat source. There are minor engineering differences in the two soil heating systems used, but in general terms the cables are buried about 30 cm apart and 25 to 30 cm deep and provide 110 W of heating capacity per m^2. This provides sufficient heat to achieve a temperature rise of 8 to 10C above ambient soil temperature. In some experiments, an irrigation variable was included; in others the entire area was irrigated uniformly. Soil fertility was maintained at a level considered to be optimum for the crops being grown. Attempts have been made to grow something in every season of the year both outdoors and in the greenhouse.

In 1971 and 1972 sweet corn, green beans, and summer squash (*Cucurbita moschata,* Duchesne) were planted in early April when ambient soil temperatures were below 15C. Thermostats controlling heated plots were set at 30C; however, that temperature could not actually be reached until late April when the ambient soil temperature reached about 20C. Ambient temperature did not approach the heated soil temperature closely until mid-June.

The added heat hastened emergence of all crops by about 4 days, and early plant growth was greatly enhanced. The first harvests were only 4 to 7 days earlier, but a greater proportion of the total yield from heated plots removed at the first picking indicated that a slight early market advantage should occur.

Table 5 shows that the effects of heat or irrigation alone were similar in magnitude and that the two inputs added together resulted in total yield increases of 131% with squash, 178% with beans, and 268% with sweet corn. Responses to heat were greater with corn and beans than with squash because the squash continued to produce fruit throughout the midsummer period when ambient soil temperatures were sufficient for maximum growth.

The addition of heat to the soil increased the evapotranspiration rate so that the irrigation water requirement was 6 to 8 cm greater per growing sea-

[2]Unpublished data from Forage and Field Res. Rep. No. 7–11, Soils and Fertilizer Res. Br., Tennessee Valley Authority, Muscle Shoals, Ala.

Table 5—Effect of soil heating on vegetable yields (2-year avg.) at Muscle Shoals, Alabama, 1971–72

Treatment	Sweet corn		Green beans		Summer squash	
	Yield	Increase	Yield	Increase	Yield	Increase
	metric tons/ha	%	metric tons/ha	%	metric tons/ha	%
Heat plus irrigation	19.5	268	18.6	178	68.6	131
Irrigation only	11.3	113	12.5	86	49.5	66
Heat only	11.6	118	12.7	90	41.2	39
No heat or irrigation	5.3	0	6.7	0	29.7	0

son. This greater evapotranspiration meant also that during certain periods when rainfall was limited the addition of heat without irrigation water resulted in a net decrease in plant growth rate and yields.

The slightly earlier maturity possible with soil heating can aid crops in avoiding a drought stress which would ordinarily occur during late growth. On the other hand, heating can cause a crop to mature during a dry period; a rain occurring a few days later could save a similar crop on unheated soil. However, until long range rainfall occurrence can be predicted with greater accuracy than is now possible, knowledge of these effects probably is not useful in planning crop management systems.

In a subsequent experiment, several vegetables were grown with no added heat and at thermostat settings of 21, 27, and 32C. These specified temperatures do not represent actual season-long temperature regimes because sufficient heating capacity was not available to reach the highest temperatures at the beginning of the growing season, and the no heat, 21, and 27C beds had similar temperatures in midsummer from solar heat inputs.

Results in Table 6 clearly show that all vegetables except tomatoes gave a positive yield response to at least some heat input; all except cantaloupe and okra (*Hibiscus esculentus* sp.) showed a negative response to the 32C soil temperature. This negative response was particularly striking with tomatoes and peppers, both of which barely remained alive throughout the season with 32C soil temperatures.

Table 6—Effect of different soil temperature regimes on vegetable yields at Muscle Shoals, Alabama, 1973

Crop	Yield			
		Thermostat settings		
	No heat	21C	27C	32C
		kg/plot		
Summer squash	39.6	96.4	97.7	83.8
Cantaloupe	20.9	99.2	124.9	144.8
Okra	145.2	181.3	--	182.8
Tomato	177.7	179.9	162.6	75.0
Sweet pepper	123.6	167.4	170.1	66.9
Green beans	49.3	61.8	38.9	37.6

Table 7—Effect of soil heating on field crop yields at Muscle Shoals, Alabama, 1973

	Yield		Increase
	Unheated	Heated	
	kg/plot		%
Corn stalks	58.1	67.5	16
Corn grain	71.7	74.6	4
Soybean stalks	42.7	46.2	8
Soybean grain	29.7	29.9	1
Seed cotton (Seed plus lint)	8.6	12.3	43

These data indicate that power plant discharge water with temperatures as low as 25 to 30C could be used successfully to enhance crop growth, providing that adequate heat transfer from the water to soil can be achieved. Also, the data show rather clearly that if crop production is an objective, field soils cannot be used for dissipation of waste heat during the midsummer period in the southern United States when discharge water temperatures are in excess of 35 to 40C. However, this conclusion may not be valid for northern climates where water temperatures are lower.

The three most common Tennessee Valley field crops, cotton (*Gossypium hirsutum* L.), soybeans, and corn, were grown for 2 years on soil warmed to 29C. Of these only cotton appeared to show a level of response that might indicate real economic potential (Table 7).

Cotton did not respond to extremely early planting, even on heated soil, because air temperatures below about 10C were damaging to newly emerged plants. Corn, however, might have shown a greater response if planted before the early May date used in these tests.

Early growth of soybeans was enhanced on heated plots but seed yields were not increased. Much research has shown that climatic conditions during pod-set and filling are more important to yield than are early season conditions. There is no need for early planting of southern type soybeans, as they flower and fruit primarily in response to day length rather than plant age. Naturally high soil temperatures coincide with the normal fruiting period of soybeans in the South; however, Oregon results indicate that soybeans do respond to soil heating in cool climates.

Thus far, little increase in crop production during the cool fall period has been found. Turnip greens (*Brassica rapa* L.), rye (*Secale cereale* L.), and ryegrass have been grown in the fall months with and without heat but thus far yield increases have been minimal or nonexistent. This may be because weather conditions at Muscle Shoals are such that soils normally stay warm until the day length and light intensity begin to limit yields.

As reported by Berry and Miller (1974), greenhouse vegetable production in warmed soil appears to be feasible. A plastic greenhouse was erected over heated soil at Muscle Shoals and used with no supplementary air heating. The production cycle currently being studied includes tomatoes, hybrid Dutch cucumbers, and sweet peppers in the spring and fall seasons, with a midwinter crop of broccoli, cauliflower (*Brassica aleracea* L.), or bibb lettuce, and no midsummer crop.

Spring crop yields were about 4.5 kg/plant for tomatoes, 10 kg for cucumbers, and 1.5 kg for peppers. Fall yields were considerably less, particularly for tomatoes and cucumbers, but some of this decrease appeared to be related to disease and insect problems which better management should control. Tomatoes grew and yielded best at a soil temperature of about 21C while peppers and cucumbers grew better at 27C.

The winter crops grew well and had an apparent potential to produce about 45,000 heads of broccoli or cauliflower and about 150,000 heads of bibb lettuce per ha. Soil temperatures as low as 18C were satisfactory for these crops. The data showed that this production system might have some economic potential if one already had a heated field available since additional capital and heating costs would be low.

Cauliflower, broccoli, and peppers are somewhat lower in value than crops normally grown in the greenhouse, but the labor requirements are very low relative to tomatoes or cucumbers. Transplanting and harvesting are the only operations required, at least part of which could be mechanized in large houses.

V. DISCUSSION AND CONCLUSIONS

Only a few parts of the waste heat-soil warming relationship are clear at this time. It is well documented that tremendous amounts of waste heat are already available in power plant condenser water, and to a much smaller degree, in other industrial cooling waters. Although rate of growth may differ somewhat from current projections, it seems certain that the end of this century will see a great increase in the amount of power generated and thus the amount of heat rejected to cooling water. There also seems to be ample evidence that many crop plants have the capacity for large growth and yield responses to increased soil temperatures under rather diverse climatic conditions.

Questions which have not been answered adequately include: (i) those concerning capital and operating costs and length of life of soil heating systems; (ii) which plant species have the greatest profit potential and whether major response differences exist among varieties of the same crop species; (iii) which geographical areas show the greatest promise for profitable use of waste heat; and (iv) whether heat dissipation in normal soil heating operations is of financial consequence in power operations. Because of the great number of unknowns it is very difficult to gain sufficient perspective so that one can predict any details of an efficient soil heating facility.

Related external factors such as food supply and price, availability, and price of various agricultural inputs including labor, and the possible cost to a power plant of waste heat use also are likely to influence the adoption of soil heating as a production resource.

There is a need for a great amount of both basic and applied research if soil heating is to become feasible. The basic questions are primarily the whys and hows of individual plant response to different soil temperatures. From an applied research standpoint, much information is needed to identify species and varieties most likely to give profitable response. Information specific to heated soil conditions is needed on almost all phases of management including fertilization, spatial arrangement of plants, irrigation requirements, changes in traditional seasonal growth patterns, and geographical location.

Although there seem to be a great many unanswered questions at this point there is no cause for despair. It is very difficult to visualize any other new management input which, when added to existing techniques, could increase crop yield as much as soil heating has done in many of the experiments discussed in this paper. This single factor, if no other, makes continued research and development on soil heating imperative.

LITERATURE CITED

Allred, E. R., P. E. Read, and J. R. Gilley. 1973. Use of waste heat for soil warming and irrigation in northern climates. *In* Proc. Am. Soc. of Agric. Eng. Winter Meeting. Chicago, Ill., 11–14 Dec. 1973. ASAE Pap. No. MI 93–254.

Berry, J. W., and H. H. Miller. 1974. A demonstration of thermal water utilization in agriculture. Environ. Prot. Agency Technol. Ser., EPA 660/24-74-011.

Boersma, L. 1970. Warm water utilization. p. 74–107. *In* S. P. Mathur and R. Steward (ed.) Proc. Conf. on Beneficial Uses of Thermal Discharges, 17–18 Sept. 1970. N. Y. State Dep. of Environ. Conserv., Albany, N. Y.

Cline, J. F., M. A. Wolf, and F. P. Hungate. 1969. Evaporative cooling of heated irrigation water by sprinkler application. Water Resour. Res. 5:401–406.

Decker, A. M. 1973. Plant environmental studies with controlled soil temperature field plots. *In* Plant morphogenesis as the basis for scientific management of range resources. Proc. Workshop of the United States-Australia Rangeland Panel, Berkeley, Calif., January 1974. USDA Misc. Publ. No. 1271. 232 p.

Decker, A. M., and J. M. Walker. 1972. Responses of corn to controlled field soil temperatures and fertilizer rates. Maryland Agric. Exp. Stn., Dep. of Agron., Progr. Res. Rep. Crops Soils Res., 1971. Vol. 5:109–114.

Decker, A. M., and J. M. Walker. 1973. Early response of medium and late corn varieties to controlled field soil temperature. Maryland Agric. Exp. Stn., Dep. of Agron., Progr. Res. Rep. Crops Soils Res., 1972. Vol. 6:85–88.

DeVries, D. A. 1963. Thermal properties of soils. p. 210–235. *In* W. R. van Wijk (ed.) Physics of the plant environment. North Holland Publishing Co., Amsterdam.

Federal Power Commission. 1971. The 1970 national power survey. Part I. U. S. Government Printing Office, Washington, D. C.

Hawes, D. T., and A. M. Decker. 1971. Plant environmental studies with turf. Maryland Agric. Exp. Stn., Dep. of Agron., Progr. Res. Rep. Crops Soils Res., 1970. Vol. 4: 64–69.

Interdisciplinary Research Team. 1974. An agro-power-waste water complex for land disposal of waste heat and waste water. Inst. for Res. on Land and Water Resour., The Pennsylvania State Univ., University Park, Pa.

Kendrick, J. H., and J. A. Havens. 1973. Heat transfer models for a subsurface, water pipe, soil warming system. J. Environ. Qual. 2:188–196.

Miller, H. H. 1970. The thermal-water horticultural demonstration project at Springfield, Oregon. p. 62–69. *In* S. P. Mathur and R. Stewart (ed.) Proc. Conf. on Beneficial Uses of Thermal Discharges, 17–18 Sept. 1970. N. Y. State Dep. of Environ. Conserv., Albany, N. Y.

Nakshabandi, G. A., and H. Kohnke. 1965. Thermal conductivity and diffusivity of soils as related to moisture tension and other physical properties. Agric. Meteorol. 2: 271-279.

Rykbost, K. A., and L. Boersma. 1973. Soil and air temperature changes induced by sub-surface line heat sources. Oregon Agric. Exp. Stn. Spec. Rep. 402. 105 p.

Rykbost, K. A., L. Boersma, H. J. Mack, and W. E. Schmisseur. 1974. Crop response to warming soil above their natural temperatures. Oregon Agric. Exp. Stn. Spec. Rep. 385. 98 p.

Skaggs, R. W., C. R. Willey, and D. C. Sanders. 1973. Use of waste heat for soil warming. In Proc. Am. Soc. of Agric. Eng. Winter Meeting, Chicago, Ill., 11-14 Dec. 1973. ASAE Pap. No. 73-3530.

Smith, W. O., and H. G. Byers. 1938. The thermal conductivity of dry soils of certain of the great soil groups. Soil Sci. Soc. Am. Proc. 3:13-19.

Warren, F. H. 1969. Electric power and thermal output in the next two decades. p. 21-46. In M. Eisenbud and G. Cleason (ed.) Electric power and thermal discharge. Gordon and Beach, New York.

Wierenga, P. J., R. M. Hagan, and D. R. Nielsen. 1970. Soil temperature profiles during infiltration and redistribution of cool and warm irrigation waters. Water Resour. Res. 6:230-238.

chapter 21

Organic materials are broken down in an aeration basin at the Durham Facilities of the Unified Sewerage Agency of Washington County, Oreg. (Photo courtesy of Stevens, Thompson & Runyan, Inc., Engineers/Planners).

Odors and Emissions from Organic Wastes[1]

A. R. MOSIER, USDA-ARS, Fort Collins, Colorado

S. M. MORRISON and G. K. ELMUND
Colorado State University, Fort Collins, Colorado

I. INTRODUCTION

A. The Problem of Odors and Emissions

Malodorous emissions from organic wastes can have detrimental esthetic and economic effects on a community, as well as on its residents' mental and physical health. Public opinion surveys frequently identify malodors as the air pollutant of greatest concern (Osag & Crane, 1974). There is no apparent relationship between odors and a specific organic disease, or toxicity of a gas. However, odors have been documented to incite allergic responses, poor appetite, lower water consumption, impaired respiration, nausea and vomiting, insomia, and mental stress. The subjective nature of the physiological effect of odors is demonstrated by its dependence on an individual's attitude, disposition, and even the time of day (Petri, 1961). Economically, odors can stifle growth and development of a community: tourism decreases, property values decline, and tax revenues and payrolls diminish (Sullivan, 1969).

Unfortunately, there is no direct approach to many odor problems. Assessment of individual situations is subjective and clearly objective methods are not available. The olfactory sense varies among people, though most can distinguish odor quality and intensity. Individuals disagree as to the objectionability of an odor. Persons responsible for an odor source may deem it as merely the smell of success, while their downwind neighbors may consider it highly objectionable and deleterious.

Many of the chemicals considered responsible for malodors from organic wastes can be sensed at very low concentrations. Thus malodors are commonly one of the first manifestations of air pollution and evoke emotional reactions. As urban areas encroach further into previously isolated agricultural lands, controlling malodorous emissions from organic wastes becomes more important, not only physiologically and economically, but also psychologically and sociologically.

Currently, with no federal regulations for the control of odors, state and local pollution control agencies have been charged with developing odor control regulations. In 1971, 203 state and local air pollution control agencies existed (Copley International Corporation, 1973). As a result, odor control regulations vary geographically. Leonardos (1974) recently reviewed odor-control regulations throughout the United States and he grouped the regulations of the 52 major agencies into nine classifications; the most widely used are air pollution nuisance regulations, objectionability criteria, scentometer based regulations, and highest and best practicable control. None of the control regulations have been completely satisfactory. Leonardos (1974) attributed the difficulty to the inability to measure odor objectively and reliably.

[1]Contribution of Agricultural Research Service, USDA, in cooperation with Colorado State University Experiment Station, Fort Collins, CO 80523. Scientific Journal Series No. 2045.

Local odor control regulations are important if one is concerned with the operation of an industry with a potential for accumulating organic materials. Willrich and Miner (1971) cited instances of litigation based on the law of private nuisance or the common-law concept of trespass brought against livestock and poultry producers because of offensive odor emissions.

B. Scope of this Chapter

In this chapter we wish to acquaint the reader with the subjective nature of the problems of odors and emissions from organic wastes and describe attempts to objectively study these problems. We will define the nature of the principle organic wastes, describe microbial and biochemical processes involved in the production of volatile compounds, relate their chemical nature, and discuss analytical methods employed to determine their nature. We will also suggest how volatile chemicals are dispersed from a source and how human olfaction responds. In addition, we discuss methods employed to limit nuisance odors and emissions. We have emphasized organic wastes derived from livestock production, but the biochemical transformations involved in odor production apply to other biodegradable organic waste. Our discussion of the dispersal and chemical analysis of organic chemicals is also applicable to other waste sources. We will emphasize N emissions from organic wastes, primarily because we have had the most experience with this class of chemicals and because N is also a major plant nutrient which may also pollute surface and ground waters.

II. THE NATURE OF ODORS AND EMISSIONS

A. Chemical Characteristics of Volatile Emissions

A large number and variety of organic chemicals are evolved from animal wastes and sewage (Table 1); unfortunately, many are malodorous. Esthetically, they can create problems and in high concentrations, many are also toxic (Lillie, 1969). The compounds considered malodorous (those containing sulfur, reduced nitrogen, and carboxyl moieties) evolve from most animal manures and sewage (Miner, 1973; Mosier et al., 1973; Rains et al., 1973; Warden, 1972[2]). Odor from anaerobically digested sewage sludge has been described as "earthy," similar to crude oil. Persons living near areas where sludge is applied to land or held in lagoons frequently complain of objectionable odors similar to ammonia (Thorne et al., 1975).

Rendering plants create odors similar to animal feeding and sewage treatment facilities. Poultry rendering emissions are predominantly aldehydes and small amounts of dialkyl disulfides. The odors from dry rendering of dead

---

[2] T. H. Warden. 1972. Volatile pollutants from feedlots. M.S. Thesis. Colorado State University, Fort Collins, Colo.

stock, beef offal, and slaughterhouse trimmings contain various carbonyl components with traces of sulfurous and nitrogenous compounds. Fish meal processing emits mainly amines, most notably trimethylamine and putrescine. Feather cooking and drying generate large quantities of acrolein, acetaldehyde, methyl mercaptan, diethylamine, n-propylamine, NH_3, and H_2S (Bethea et al., 1973). These odorous materials are produced in a confined area, and unlike open animal feeding operations, some form of control can and has been applied.

The general characteristics of odor and emission problems associated with meat, poultry, and seafood processing were reviewed by Jones (1974). Odors and emissions from the fruit and vegetable processing industry are apparently not a major source of air pollution (Jones, 1973).

Kraft wood pulping operations are also a source of malodorous volatiles. The process involves cooking wood chips with a solution of sodium sulfite and sodium hydroxide, followed by several recovery steps (Sullivan, 1969). Volatile chemicals identified from aqueous kraft mill effluents include H_2S, methyl mercaptan, dimethyl sulfide, dimethyldisulfide, methanol, ethanol, acetone, methyl isobutyl ketone, α-pinene, Δ^3-carene, camphene, limonene, cineole, and α-terpenol plus additional unidentified compounds (Hrutfiord & McCarthy, 1967).

B. Dust Emissions

Dust is another form of emission associated with animal feeding operations and waste application to land. Dust is a particular problem in arid and semiarid climates where cattle feeding is extensively practiced and dust-caused respiratory distress has been noted (Azevedo et al., 1974). Little information on the nature of the dust from animal feeding operations has been reported. Dusts appear of greatest concern to the health of poultry and zoo animals (Lillie, 1969). Koon et al. (1963) reported that dust in housed poultry operations clogged ventilation equipment. The dust from caged laying hens consisted of two distinct types of particulates. The majority of the material was flat, flaky, and cellular. These particles, 1 to 450 μm in diameter, were identified as a mixture of feed and skin debris from birds epidermis. The other material consisted of long, cylindrical particles with nodes and internodes; these were identified as broken feather barbules with an average diameter of 4 μm. The dust, which contained approximately 92% dry matter, was composed of 60% crude protein, 9% fat, 4% cellulose, and 19% ash (Koon et al., 1963). Dust production has been reported to be inversely proportional to the relative humidity of the poultry house (Lillie, 1969). Burnett (1969a) reported that particulate matter transported odor in high density poultry houses. Gas chromatographic and olfactory analysis of volatiles from a water slurry of poultry dust indicated the presence of odoriferous compounds.

Azevedo et al. (1974) showed that manure dusts could be identified in

Table 1—Citation of chemicals identified as volatiles from cattle, poultry, and swine wastes and sewage

Chemical	Cattle	Poultry	Swine	Sewage
		Waste source		
	——— Bibliographic reference† ———			
Alcohols				
Methanol	3	--‡	8	--
Ethanol	3	--	8	--
n-Propanol	12	--	8	11
Isopropanol	3	--	8	11
n-Butanol	--	--	8	11
Isopentanol	--	--	8	--
Carbonyl-containing				
Acetic acid	--	2,4	--	11
Propionic acid	--	2,4	--	11
n-Butyric acid	11	2,4	--	--
Isobutyric acid	--	4	--	11
n-Valeric acid	--	4	--	--
Isovaleric acid	--	4	--	--
Formaldehyde	--	--	--	--
Acetaldehyde	3,12	--	7,8	11
Propionaldehyde	3	--	7,8	11
n-Butryaldehyde	--	--	7	--
Isobutryaldehyde	--	--	8	--
n-Valeraldehyde	--	--	8	--
n-Hexaldehyde	--	--	7	--
n-Octaldehyde	--	--	8	--
n-Decaldehyde	--	--	8	--
Ethylformate	3	--	--	--
Methylacetate	3	--	--	--
Isopropylacetate	3	--	--	--
Isopropylpropionate	3	--	--	--
Isobutylacetate	3	--	--	--
Acetone	--‡	--	7	--
2-Butanone	11	--	7	--
3-Pentanone	--	--	7	--
2,3-Butanedione	--	4	--	--
3-Hydroxy-2-Butanone	--	4	--	--
Nitrogen-containing				
Methylamine	10	--	--	--
Dimethylamine	10	--	--	--
Trimethylamine	12	--	--	--
Ethylamine	10,12	--	9	11
Triethylamine	--	--	9	11
n-Propylamine	10	--	--	--
Isopropylamine	10	--	--	--
n-Butylamine	10	--	--	--
n-Amylamine	10	--	--	--
n-Hexylamine	--	--	--	11
Diisopropylamine	--	--	--	11
Dibutylamine	--	--	--	11
Diisobutylamine	--	--	--	11

(continued on next page)

Table 1—continued

Chemical	Cattle	Poultry	Swine	Sewage
		Waste source		
		Bibliographic reference†		
Ammonia	10,6	4	7,8	11
Indole	3	4	--	--
Skatole	3	4	--	--
Sulfur-containing				
Hydrogen sulfide	1,3,5	1,4	8	11
Carbonyl sulfide	1,5	--	--	--
Dimethyl sulfide	1,12	1,4	--	--
Diallyl sulfide	--	--	--	11
Carbon disulfide	1	--	--	--
Dimethyl disulfide	1	1	--	--
Methanethiol	1,13	1,4	--	11
Ethanethiol	--	4	--	11
Propanethiol	--	4	--	11
t-Butylthiol	--	--	--	11
t-Amythiol	--	--	--	11
Simple Organics				
CO_2	5	--	8	11
Methane	5	--	8	11

† Bibliographic references:
1) W. L. Banwart, and J. M. Bremner. 1974. Identification of sulfur gases evolved from animal manures. Am. Soc. Agron. Abstr. p. 27.
2) Bell (1970).
3) Bethea and Narayan (1972).
4) Burnett (1969b).
5) Elliott and Travis (1973).
6) Elliott et al. (1971).
7) Hartung et al. (1971).
8) Merkel et al. (1969).
9) Miner and Hazen (1969).
10) Mosier et al. (1973).
11) Rains et al. (1973).
12) Warden (1972)[2].
13) White et al. (1971).
‡ Chemical not identified from this source.

the presence of background dusts using alpha-excited X-ray spectroscopy. High ratios of P, S, Cl, K, and Ca to Si were characteristic of the > 2 μm particle size, and the < 2 μm particles showed relatively high Cl, K, and Ca to Si ratios. They reported that feedlot dusts consisted mainly of large particles (> 17 μm) that dropped out rapidly and locally. Feedlot dusts were not distinguished from background dust 750 m downwind of the lot. Luebs et al. (1974) reported that using filters on their ammonia collection impingers reduced distilled N measurements by an average of 16%. Thus, NH_3 and probably other volatile N compounds were either adsorbed onto the filter or attached to dust particles collected on the filter.

C. Principal Gaseous Emissions from Animal Wastes

Ammonia is the predominant basic N compound volatilized from cattle feedlots (Mosier et al., 1973). Hutchinson and Viets (1969) showed that sufficient NH_3 was volatilized from cattle feedyards to contribute significantly to the N enrichment of surface waters in northeastern Colorado. An exposed dilute acid surface absorbed 20 times more NH_3 near feedlots than those at control sites. There was no apparent seasonal influence on the amount of NH_3 collected. Weekly fluctuations were a function of the moisture status of the feedlot surfaces. Elliott et al. (1971) placed acid traps next to a feedlot and 0.8 km away, and collected 148 and 16 kg basic N ha^{-1} $year^{-1}$, respectively. Luebs et al. (1974) measured distillable-N concentrations of 540 $\mu g/m^3$ at the downwind corral fence of a 600-cow dairy, 190 $\mu g/m^3$ 0.8 km upwind from the animals, and 6 $\mu g/m^3$ 11.2 km upwind from the dairy area during a 24-hour collection period. Denmead et al. (1974) reported that a substantial part of the N lost from grazed pastures can be attributed to NH_3 volatilization. The average daily flux density of N from a pasture grazed by sheep was 0.126 kg/ha.

Hydrogen sulfide is another major volatile from organic wastes (Table 1). Hydrogen sulfide was found at nine times the ambient concentration over cattle waste ponds (R. E. Luebs, and A. E. Laag. 1974. Hydrogen sulfide emission from dairy animal waste. Am. Soc. Agron. Abstr. p. 33). On the downwind side of a pond receiving wastes from a 600-cow dairy, H_2S concentrations of 181 and 66 $\mu g/m^3$ were measured at 10 and 100 cm above the ground, respectively. Hydrogen sulfide odor could be detected 50 m downwind from the waste pond. Sprinkling dairy waste effluent on land increased the downwind detection distance to 320 m.

Methane, CO_2, NH_3, and H_2S are the gases evolved from organic waste in greatest concentration (Taiganides & White, 1969; Muehling, 1970). Although methane is not considered toxic, concentrations can be sufficiently high in an improperly ventilated confined animal feeding enclosure to produce an asphyxiating atmosphere and explosive conditions (McAllister & McQuitty, 1965). Carbon dioxide, like methane, is not highly toxic but may contribute to O_2 deficiency in unventilated confinement (Muehling, 1970). Air containing 50-100 ppm NH_3 did no apparent harm to humans when inhaled for several hours (Lillie, 1969). Stombaugh et al. (1969) reported prolonged exposure of pigs to air containing 100 to 150 ppm NH_3 caused excessive nasal, lacrimal, and mouth secretions, and adversely affected pig feed consumption and average daily weight gain. Hydrogen sulfide is one of the more toxic gases associated with liquid manure storage. Human exposure to 20-150 ppm concentrations causes severe irritation to eyes and respiratory tract, while exposures to 500 ppm for 30 min affects the nervous system (Lillie, 1969). Concentrations of 800 to 1,000 ppm cause immedaite unconsciousness and death through respiratory paralysis unless artificial respiration is immediately applied (Muehling, 1970). Lillie (1969) documented instances where swine were killed when manure storage tanks located in under-floor pits were agitated; H_2S was implicated as the causal agent. Since H_2S does not collect in open feeding areas because of air dilution and oxidation to SO_2, potential acute toxicity of H_2S is important only in confined feeding operations during agitation of anaerobic underhouse waste collection pits where house ventilation is poor. In the atmosphere, H_2S has a reported residence time of less than a day (Kellogg et al., 1972).

D. Microbial Aerosols

During many agricultural activities, both pathogenic and nonpathogenic microorganisms can be dispersed to the environment as aerosols. Pathogenic microorganisms transmitted via aerosols include anthrax (Albrink, 1961) histoplasmosis (Campbell, 1957), and Q-fever (Dyer, 1949).

Aerosols containing microorganisms can also be produced during sewage treatment processes like those using trickling filters (Adams & Spendlove, 1970) and activated sludge (Napolitano & Rowe, 1966). Using an in vitro system, Poon (1968) evaluated the survival of *Escherichia coli* in aerosols under various environmental conditions. Maximal death rates were correlated with high temperatures and low relative humidity and consequent high evaporation rates. Addition of sodium chloride or glucose to the aerosol apparently affected the evaporation rate and reduced the death rate of the microbes. *Serratia marcescens* has been used as a tracer organism to demonstrate production of aerosols from an activated sludge plant (King et al., 1973). In these studies, increased microbe death rates were correlated with high temperatures and low relative humidities. Under their experimental conditions, species of *Bacillus* were the predominant microbes both upwind and downwind of the activated sludge plant. Downwind of the plant, species of *Alcaligenes* and *Achromobacter* were the second and third most predominant microorganisms recovered, respectively. In addition to temperature and relative humidity, other factors may affect the survival of microbial aerosols including the type of microorganisms, wind velocity, air temperature stability, particle size, and solar radiation (Spendlove, 1973).

Will et al. (1974) studied transmission of pathogenic microorganisms in aerosols from a model Pasveer oxidation ditch. Results suggested that survival and virulence of a pathogen inoculated into manure slurries depended, in part, on the type of microorganism. Viable cells of *Salmonella typhimurium* were recovered (more readily than cells of *Leptospira* sero-type *pomona*) from aerosols of inoculated manure slurries. In addition, turkey poults exposed to *S. typhimurium* in aerosols from manure slurries developed infection, whereas hamsters (*Mesocricetus auratus*) exposed to aerosols of *L. ponoma* did not develop infection. Goodrich et al. (1974) studied aerosol production from an oxidation ditch below a slotted floor in an enclosed beef cattle housing unit. Under field conditions, the oxidation ditch did not contribute significant quantities of microbial aerosols as compared with the contribution of the animals themselves. The wet floor in the buildings apparently suppressed aerosol formation. However, usually high and potentially hazardous levels of microbial aerosols existed within the barn during cleaning periods with a dry floor.

III. HUMAN RESPONSE TO ODOROUS CHEMICALS

A. Human Olfaction

In spite of recent analytical advances, the human nose remains the only reliable odor detector. Human olfactory response to odors may vary from person to person, but all healthy people are usually aware of odors and generally agree that some odorous compounds are obnoxious (Sullivan, 1969). A detailed description of human olfaction can be obtained from Moncrieff (1967) and Amoore (1970).

B. Characteristics of Odor

Characterization of an odor normally depends on intensity (strength) and quality (a verbal description of the odor). Experiments have shown that an average observer can distinguish between three intensities—weak, medium, and strong, while a trained observer can distinguish five degrees of intensity (Sullivan, 1969). The sensing of odor intensity varies according to the Weber-Fechner psychophysical law which states that the intensity of the sensation is proportional to the logarithm of the strength of the stimulus. The odor threshold concentration (the minimum concentration which will result in the stimulation of the olfactory nerves) varies per person. Therefore, odor panels are used to determine odor threshold concentrations. In addition, odor threshold measurements depend largely on the purity of the odorant. As a result, reported threshold concentrations of odorants vary. Table 2 lists odor threshold values for chemicals considered to be associated with animal feeding and municipal waste.

1. ODOR INTENSITY

The intensities of components of a mixture may be independent, counteractive, additive, or synergistic (Sullivan, 1969). Guadnagni et al. (1963) concluded that odorant chemical interactions are mostly additive. Though many chemical combinations were tried, none approximated the mixture of odorants emitted from animal wastes.

Odor intensity can be measured quantitatively. The most commonly used method is that of vapor dilution. Odor strength can be defined as the number of dilutions with odor-free air necessary to reduce the odor to the threshold concentration (the threshold odor number). The odor intensity

Table 2—Odor threshold and quality description of chemicals considered to be important to organic waste odors

Chemical	Odor threshold	Odor description
	ppm	
	Carbonyl-containing	
Acetaldehyde	0.21 §	Green sweet §
Propionaldehyde	0.0095 ¶	--†
3-Hydroxy-2-butanone	--†	Butterlike ‡
Acetic acid	1.0 §	Vinegarlike,‡ sour §
Propionic acid	20.0 ¶	Pickle-like ‡
2-Methylpropionic acid	8.1 ¶	Sweat-like ‡
Butyric acid	0.001 §	Sour, § rancid ‡
	Nitrogen-containing	
Methylamine	0.021 §	Fishy, § ammoniacal ‡
Dimethylamine	0.047 §	Fishy §
Trimethylamine	0.00021 §	Fishy §
Ethylamine	--†	Fishy §
Skatole	0.019 ¶	--†
Ammonia	46.8 §	Ammoniacal ‡ ¶
	Sulfur-containing	
Methanethiol	0.0021 §	Skunk,# foul††
Ethanethiol	0.001 §	Onion-like,‡ skunk#
Propanethiol	0.00074 #	Onion-like,‡ skunk#
t-Butylthiol	0.00009 #	--†
Dimethyl sulfide	0.001 §	Rotten cabbage#
Diethyl sulfide	0.003 #	Rotten cabbage#
	0.00047 §	
Hydrogen sulfide	0.0072 #	Eggy sulfide, § foul††
	0.072 #	

† No data available.
‡ Burnett (1969b).
§ Leonardos et al. (1969).
¶ Stahl (1973).
# Sullivan (1969).
†† White et al. (1971).

index (OII) is another term for the quantitative description of odor intensity. The OII is the number of times an odorant must be diluted by half with odor-free medium to reach the odor threshold. The methods used for odor panel determination of the odor intensity and quality of an odorant are described by Leonardos et al. (1969), Miner (1973), and Sullivan (1969). Detailed technology for the standardizing of odor measurement can be obtained from their references.

2. ODOR QUALITY

Odor quality is a verbal description of an odor. The quality is commonly described by associating an unfamiliar with a familiar odor. Many different systems for categorization have been developed, most applicable to trained odor panelists. Miner (1973) and Sullivan (1969) listed several of these classifications. A classification system using familiar odors for comparison has been used by untrained odor panels to attempt to describe the odorous components of animal wastes. To describe the odors from dairy cattle waste, White et al. (1971) used the terms: foul, sweetish, acetate, nutlike, pungent, and musty. Burnett (1969b) used the following terms to describe poultry manure odor: rotten egg, rotten cabbage, onion-like, putrid, butter-like, and garlic. Table 2 lists the characteristic odors for different volatile chemicals identified from animal and sewage wastes. There is no straight-forward method to quantify odor quality.

3. ODOR ACCEPTABILITY

The intensity and quality of an odor govern whether it is acceptable or unacceptable. Obnoxious odors may become acceptable at very low concentrations. At high intensities, perfumes can be unacceptable (Sullivan, 1969). According to Moncrieff (1966), the chemical classes with the least acceptable odors are: mercaptans, sulfides, disulfides, amines, and aldehydes.

C. Characteristics of Odors from Organic Wastes

The quality of animal waste and sewage sludge odors has been measured and related to chemicals separated from the mixture of volatiles from these sources (Burnett, 1969b; Merkel et al., 1969; Rains et al., 1973; White et al., 1971). Mercaptans, amines, and acids have been most strongly implicated as the offensive odorants. Due to the complex mixture of gases evolved from animal feeding or sewage treatment operations, the true nature of the odor evolved has not been resolved. Many of the individual odorants are unacceptable, but at times the concentration of these odorants is considered to be individually below the threshold. Guadnagni et al. (1963) indicated that many chemicals, when mixed, interacted additively on the olfactory threshold response. The combined effect on the human response of mercaptans, sulfides, disulfides, amines, acids, alcohols, aldehydes, ketones, and hydrocarbons evolved from a waste source must be great. The characteristic feedlot odor can usually be detected several miles from the source.

IV. CHARACTERISTICS AND BIOLOGICAL DECOMPOSITION OF ANIMAL WASTES

A. Characteristics and Factors Affecting the Decomposition of Animal Wastes

Both the chemical composition and management of livestock manures influence microbial decomposition of the waste and, thus, odor production. The chemical characteristics of livestock manures are highly variable, due in part to the variations in feeding practices in different regions and to waste management practices. McCalla et al. (1970) noted that approximately 90% of manure dry matter is undigested organic material of which 60 to 75% is potentially biodegradable. Stewart (1970) suggested that as much as 90% of the urea N from feedlot waste may be volatilized to the atmosphere. This loss influences the carbon-nitrogen balance in the waste as well as the rate of microbial activity. Chemical analyses of manures in various stages of decomposition will aid in identifying decomposition processes which contribute to odor production. However, even detailed chemical analyses may not indicate the actual accessibility of these substrates for microbial growth.

As described by Alexander (1965), factors which may limit the biodegradability of organic materials include: (i) the inaccessibility of the substrate to microorganisms and their enzymes; (ii) the absence of some factors essential for microbial growth such as water or a suitable electron acceptor; (iii) the toxicity of the environment such as extremes in temperature, pH, or salt concentrations; (iv) the inactivation of the requisite enzymes by adsorption or by noncompetitive inhibitors; (v) the structural characteristics of the molecule which prevent the enzymes from acting; and (vi) the inability of the microbial community to metabolize the substrates because of some physiological inadequacy. To a limited degree, microorganisms are capable of genetic changes and of adaptive responses to selective environmental pressures. The complex carbohydrates and proteins present in agricultural wastes are attacked in part by various extracellular enzymes such as hydrolases which will release substrates such as sugars and amino acids in a more metabolizable form. The environmental conditions which favor or inhibit the formation of these enzymes will also affect the rate of manure or waste decomposition. For example, in the presence of glucose or other readily metabolized carbon sources, the rate of synthesis of certain enzymes, especially those of catabolic metabolism, will be reduced. This phenomenon is called catabolite repression (Paigen & Williams, 1970). It has also been observed that, when microorganisms are growing in the presence of two readily metabolized carbon sources, the microbes will defer the formation of enzymes for using one of the carbon sources until the other has been exhausted from the subtrate; this is referred to as diauxic growth. There are also population interactions which occur between microbial types and their environment under both aerobic (Waksman, 1941; McCalla & Haskins, 1964) and anaerobic conditions (Klugelman & Chin, 1971). Interactions between microbial populations are discussed in detail by Brock (1966) and Alexander (1971).

In the food processing industries, odors can be generated not only from microbial growth on waste materials but from microbial spoilage of the product as well. Microbial contamination and spoilage of food products can range from the readily detectable sour milk or off-odor meats and fish to the organoleptically undetectable but hazardous presence of botulinum toxin. Odors can also be released during the processing of the food itself, i.e., cooking, aging. Cost, as well as the chemical nature of some food processing wastes, can make byproduct recovery more attractive than treatment for disposal. The variability in concentration and rate of discharge of meat processing wastes can necessitate physical and/or chemical pretreatment as well as load balancing prior to aerobic and anaerobic treatment (Dart, 1974). Reviews concerning the characteristics and treatment of meat, seafood, and poultry processing wastes (Jones, 1974) as well as fruit and vegetable processing wastes (Jones, 1973) are available.

B. Biological Decomposition of Polysaccharides

Waksman (1952) described the complex microbial decomposition of organic matter:

> The plant and animal residues do not decompose as a whole. The various chemical constituents are attacked at different rates. The sugars, starches, some of the hemicelluloses, and some of the proteins undergo a most rapid decomposition by a great variety of microorganisms. The cellulose, certain hemicellulose, and some of the fats, oils and other plant constituents are decomposed more slowly and, commonly, by specific organisms. The lignins and some of the waxes and tannins are most resistant to decomposition; some of the lignins may even affect the decomposition of the proteins by rendering the latter more resistant to attack.

The rate of polysaccharide decomposition and the products formed during that process depend on various factors. These include the participating microorganisms, the nature of the polysaccharides, the availability of nitrogen, and especially important in relation to odor production, the oxygen potential. The microorganisms demonstrated to participate directly in cellulose decomposition include aerobic mesophilic bacteria (Han & Callihan, 1974), aerobic thermophilic bacteria (Allen, 1953), actinomycetes and myxobacteria (Waksman, 1927), anaerobic bacteria (Hungate, 1950; McBee, 1950), and fungi (Jurasek et al., 1967). The microorganisms present in fresh manures may participate in the initial stages of decomposition. However, the manure is rapidly inoculated with other microorganisms from the soil and air which also participate in the decomposition process (Rhodes & Hrubant, 1972; Thayer et al., 1974). The moisture concentration, pH, and temperature will in part determine the selection of microbial types which actively utilize polysaccharides. In general, the growth of aerobic bacteria on cellulosic substrates is enhanced by moderate temperature (20-28C), a neutral or slightly alkaline pH (6-9), and a water content between 50 and 75%. Moderately acidic conditions, a moisture content between 50 and 75%, temperatures between 45 and 55C, and the availability of sufficient oxygen favor the development of fungi and actinomycetes. A high moisture content (80 to 95%), limited oxygen, and temperatures between 25 and 37C select for anaerobic cellulolytic microorganisms (Waksman, 1952). The rate of decay will also be affected by the relatively complex interactions that occur between microbial populations participating in the decomposition process.

The recalcitrance of cellulose to enzymatic hydrolysis is also due in part to apparent crystalline regions in its structure as well as its association with lignin (Norkrans, 1967). The microbial decay of lignin appears to be primarily an aerobic process (Gottlieb & Pelczar, 1951). Since ruminants anaerobically ferment the readily available polysaccharides present in their feed and lignin is only partially degraded (Hungate, 1966), the lignin tends to become concentrated in the manure (Waksman & Starkey, 1931). A concise review concerning the microbial degradation of lignin is provided by Oglesby et al. (1967). Other research (Hattingh et al., 1967; Kotzé et al., 1968; Elmund, Morrison, and Grant, 1971; Bomar & Schmid, 1973; Mandels et al., 1974; Han & Callihan, 1974) suggests that the limited accessibility of cellulose to microorganisms may reduce the rate of decomposition.

Cellulose, starch, hemicelluloses, and pectins are hydrolyzed to various saccharides and are metabolized both aerobically and anaerobically to a common intermediate, pyruvate. The fate of pyruvate in relation to odor production depends primarily on the availability of oxygen. Under aerobic conditions, oxygen acts as the terminal electron acceptor and pyruvate is oxidized via the citric acid cycle to form carbon dioxide, NADH, and FADH. Under anaerobic conditions, organic molecules act as the terminal electron acceptors to form lactate, propionate, butyrate, acetate, and formate as well as ethanol, acetone, isopropanol, butanol, acetoin, 2,3-butanediol, and CO_2. The organic acids, alcohols, etc., produced during anaerobic glycolysis are subsequently released to the environment either to be metabolized anaerobically (Toerien & Hattingh, 1969), aerobically (Wegener et al., 1968), or volatilized to the atmosphere. Table 3 gives the relationships among some genera of anaerobic microorganisms and selected volatile products of their metabolic activities.

Table 3—The relation of various anaerobic microorganisms to the products of the metabolic activities[†]

Bacterial genus	Formate	Acetate	Ethanol	Acetone	Lactate	Propionate	Propanol	Butyrate	Butanol	Isobutyrate	Succinate	Valerate	Isovalerate	Caproate	Isocaproate
Clostridium	X	X	X	X	X	X	X	X	X	X	X	X	X	X	X
Peptococcus	X	X	X		X	X		X		X	X		X		X
Fusobacterium	X	X			X	X		X	X	X					
Bacteriodes	X	X			X	X		X		X	X		X		
Eubacterium	X	X	X			X	X	X	X	X	X		X		
Propionibacterium	X	X			X	X					X		X		
Butyrivibrio	X	X			X			X			X				
Selenomonas		X									X				
Peptostreptococcus		X	X		X	X		X		X	X		X		X
Escherichia	X	X	X	X							X				

† Derived from Holdemand and Moore (1973).

Reim et al. (1973)[3] demonstrated that the relative quantities of volatile fatty acids (VFA's) produced from anaerobically incubated batch samples of feedlot waste were butyrate > propionate > valerate > isovalerate. Addition of glucose to the manure samples increased growth of anaerobic microorganisms and altered the relative amounts of individual VFA's. Supplementing the manure with peptone also increased microbial growth but did not alter the VFA pattern. Furthermore, quantities of straight chain VFA's, from C_2 to C_6, were 20 to 40 times higher in fresh feedlot waste than in accumulated pen wastes (Reim & Morrison, 1974).[4] The authors suggested that the feed ration, moisture, and incubation temperature are the primary determinants for the amounts of VFA's that will accumulate in feedlot wastes.

The production of volatile fatty acids, alcohols, etc. is referred to as the acid-forming phase of methane formation. The biochemistry (McBride & Wolfe, 1971; Wolfe, 1971), microbiology (Toerien & Hattingh, 1969), and the nutritional requirements of methanogenic bacteria (Bryant et al., 1971) have been reviewed. The relative concentrations of the various volatile fatty acids produced during the acid-forming phase may reflect the balance of methane formation. McCarty et al. (1963) observed that relatively high concentrations of acetic and propionic acids indicated an unbalanced fermentation, whereas formic and butyric acids were utilized rapidly in methane formation. However, the preferred substrates for methanogenic bacteria appear to be hydrogen plus CO_2, formic, and acetic acids (Zeikus, 1974).

C. Biological Decomposition of Nitrogenous Materials

Decomposition of nitrogenous organic materials such as protein, urea, and uric acid is accomplished by aerobic, anaerobic, and facultative microorganisms. Both the aerobic and anaerobic decay of organic N results in the release of NH_3 to the environment. However, in the presence of available carbohydrates, more NH_3 will be incorporated into microbial biomass and less ammonia will be lost to the atmosphere (Waksman, 1952; Weiner & Rhodes, 1974). Anaerobic decomposition of nitrogenous materials results in the formation of volatile amines, mercaptans, and hydrogen sulfide in addition to NH_3.

Uric acid is excreted by birds and reptiles and is degraded via allantoic acid to form urea and glyoxylate. Barker and Beck (1942) studied the anaerobic decay of uric acid by clostridia isolated from various sources, including a heavily manured garden. Some of the isolates metabolized uric acid as a sole source of N, C, and energy. In addition to NH_3 and CO_2, acetic

---

[3]R. L. Reim, M. G. Petit, and S. M. Morrison. 1973. Volatile fatty acid production in cattle feedlot wastes. p. 55. Am. Soc. Microbiol. Abstr. No. G-177.

[4]R. L. Reim, and S. M. Morrison. 1974. Fresh cattle wastes as a substrate for volatile fatty acid production. p. 55. Am. Soc. Microbiol. Abstr. No. G-72.

acid was produced from uric acid substrates. Facultative anaerobic species of *Pseudomonas* were isolated from poultry litter, manure, and nearby soil. These organisms metabolized uric acid, both aerobically and anaerobically, to form NH_3 and CO_2 (Bachrach, 1957). Enzymes which degrade uric acid are induced in the presence of substrate in aerobes such as *Bacillus subtilis* and facultative bacteria such as *Enterobacter aerogenes* and *Pseudomonas aeruginosa* (Rouf & Lomprey, 1968). Burnett and Dondero (1969) observed the aerobic and anaerobic decomposition of uric acid in dried poultry waste. Ammonia production occurred maximally between 4 and 8 days of incubation, whereas amine production followed at days 8 through 16. Both NH_3 and amine production were correlated with a decrease in the concentration of uric acid. Urease, which catalyzes the hydrolysis of urea to ammonia and CO_2, is an inducible enzyme system in several genera of microorganisms. These include *Proteus, Pseudomonas, Bacillus,* and the strict anaerobe, *Clostridium* (Stephenson, 1966).

Amino acids from the hydrolysis of protein can be assimilated by the microorganisms and used either biosynthetically to form new amino acids and proteins or used as a source of energy during carbohydrate-limited growth. The microbial decomposition of amino acids in the absence of fermentable carbohydrate can result in the formation of organic acids, amines, NH_3, H_2S, and mercaptans.

Amino acids can be either deaminated or decarboxylated. Bacterial deaminases are produced in response to growth in neutral or alkaline environments and result in the release of NH_3 and an organic acid (Stephenson, 1966). Oxidative deamination results in the release of NH_3 and α-keto acid. Strict anaerobes and facultative anaerobes can reductively deaminate amino acids to NH_3 and the corresponding saturated fatty acid (Cohen, 1949). *Clostridium sporogenes* obtains energy from the unique process of mutual oxidation/reduction of pairs of amino acids. This is called the Stickland reaction and has been reviewed by Nisman (1954). *Clostrodium propionicum* is able to metabolize serine, alanine, and threonine by a mutase reaction; for example, alanine is deaminated as follows:

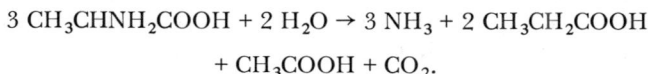

$$3 \; CH_3CHNH_2COOH + 2 \; H_2O \rightarrow 3 \; NH_3 + 2 \; CH_3CH_2COOH$$
$$+ \; CH_3COOH + CO_2.$$

The S-containing amino acid, cystine, is decomposed after preliminary reduction to cysteine through a hydrogenase reaction. Subsequent to initial reduction, cysteine can be reduced and deaminated to form acetic acid, NH_3, H_2S, and formic acid. The amino acid tryptophan is the precursor of both indole and skatole. The deamination of tryptophan forms α-indolpropionic acid, which can react subsequently through a tryptophanase enzyme to form indole and propionic acid. Tryptophan also can be deaminated and deacetylated to form skatole. The presence of a fermentable carbohydrate has been reported to repress the formation of deaminases independent of the environmental pH (Stephenson, 1966). These fermentable carbohydrates may not be continually in excess or even available in an anaerobic manure system.

The decarboxylation of amino acids results in the production of amines and carbon dioxide. Microorganisms capable of producing amino acid decarboxylases include *Escherichia coli, Streptococcus faecalis,* and *Proteus vulgaris* as well as members of the genera *Bacillus* and *Clostridium.* Bacterial decarboxylases have been demonstrated for amino acids with strong polar groups at the end of the molecule distal from the carboxyl group. These include arginine, lysine, ornithine, glutamic acid, histidine, and tyrosine. Decarboxylation of these amino acids results in the formation of malodorous putrescine and cadaverine, α-aminobutyrate, histamine, and tyramine, respectively. In contrast to the deaminases, bacterial decarboxylases have an optimum pH range of 3.5 to 4.5. The metabolic significance of amino acid decarboxylation as well as the formation pathways for some of the short chain aliphatic amines found in decomposing manures remains to be elucidated. Tornabene (T. G. Tornabene, personal communication) suggested that the phosphatidyl-ethanolamine residues of the bacterial membrane glycolipids may interact with the methyl group of methionine during cell aging to form some of the more complex volatile amines.

D. Biological Formation of S-containing Volatiles

Hydrogen sulfide can be derived from the reduction of the sulfhydryl (-SH) groups of the amino acids cysteine and methionine as well as through sulfate reduction which is accomplished by strict anaerobes usually of the genus *Desulfovibrio* (Postgate, 1965). *Desulfovibrio* have been isolated from the sheep rumen (Huisingh et al., 1974) and from poultry manure (Burnett & Dondero, 1969). The microbial metabolism of inorganic sulfur compounds has been reviewed by Peck (1962) and Trudinger (1969). *Clostridium nigrificans* also can use sulfate as a terminal electron acceptor to form hydrogen sulfide. The biological oxidation of H_2S to thiosulfate and sulfur can occur both aerobically and anaerobically. Aerobic oxidation is accomplished by colorless sulfur bacteria such as *Thiobacillus.* Anaerobic oxidation involves the activities of the photosynthetic purple and green sulfur bacteria. Methyl and ethyl mercaptans may accompany H_2S as products of anaerobic protein decomposition (Waksman, 1927). Mercaptans were formed from l-cystine by *Proteus vulgaris* and *Escherichia coli,* independent of the presence of an available source of carbohydrate (Kondo, 1922). Francis et al. (1973) reported the production of methyl mercaptan, dimethyl mercaptan, and N-butyl mercaptan when soil was supplemented with methionine. However, when cysteine was added to the soils tested, there was no detectable conversion of the added substrate to volatile organic sulfur compounds. The microbial formation of volatile sulfur compounds has been reviewed by Kodata and Ishida (1972).

E. Role of Dietary Supplements on Manure Decomposition

The incorporation of antibiotics, hormones, metal salts, etc. into the diets of livestock animals has been reported to affect the decomposition process in the excreted wastes. Clark (1965) observed that 5-day biochemical oxygen demand (BOD) tests of swine wastes were erratic and did not correlate with the strength of the waste as determined by other methods. The feed the animals received contained large quantities of the antibiotics, penicillin, chlortetracycline, bacitracin, and terramycin. The supplementation of feeds with antibiotics and sulfa drugs apparently caused a 75% decrease in standard plate counts of lagoon wastes samples as well as a 70% decrease in the volatile acid concentration of the waste. The absence of phytoplankton in lagooned waste was correlated with the excretion of copper sulfate from swine receiving a dietary supplement. Morrison et al. (1969) observed that a variety of microorganisms isolated from in situ and stockpiled manures from feedlot cattle receiving dietary chlortetracycline were less efficient in metabolizing the water soluble manure constituents than similar microorganisms isolated from pasture manures. Significant quantities of biologically active chlortetracycline were recovered from the cattle feedlot manure. When samples of these manures were incubated at 37, 28, and 4C, the antibiotic had a half-life of 7 days of 37C and > 30 days at both 28 and 4C. Dietary antibiotic supplementation may alter the digestive processes in feedlot cattle resulting in manures less biodegradable than those from animals not receiving antibiotic supplementation (Elmund et al., 1971). Nearly 93% of the ^{14}C-diethylstilbesterol fed to sheep was recovered in the feces (Aschbacker, 1971). Thayer et al. (1974) reported that diethylstilbesterol in cattle feedlot waste may inhibit the growth of microorganisms participating in the decomposition process. Additional research is needed to determine the specific effects of dietary supplements of antibiotics, hormones, insecticides, antifungal agents, etc. on the decomposition of livestock wastes in relation to the microbial populations involved and the generation of odors. Finding suitable control animals may prove to be a difficult task.

V. DISPERSAL OF CHEMICALS FROM THE SOURCE

A. Movement of Chemicals

The movement of volatile chemicals from a source may be pictured as in Fig. 1. Volatile chemicals dispersed into the atmosphere follow several pathways in contributing, not only to a deterioration of air quality, but also to soil and water pollution. Airborne pollutants may react with other chemicals, like NH_3 with SO_4^{2-} or NO_3^-, and water to form $(NH_4)_2SO_4$ or NH_4NO_3 aerosols (Alexander, 1971; Gordon & Bryan, 1973; Viets, 1974). The chem-

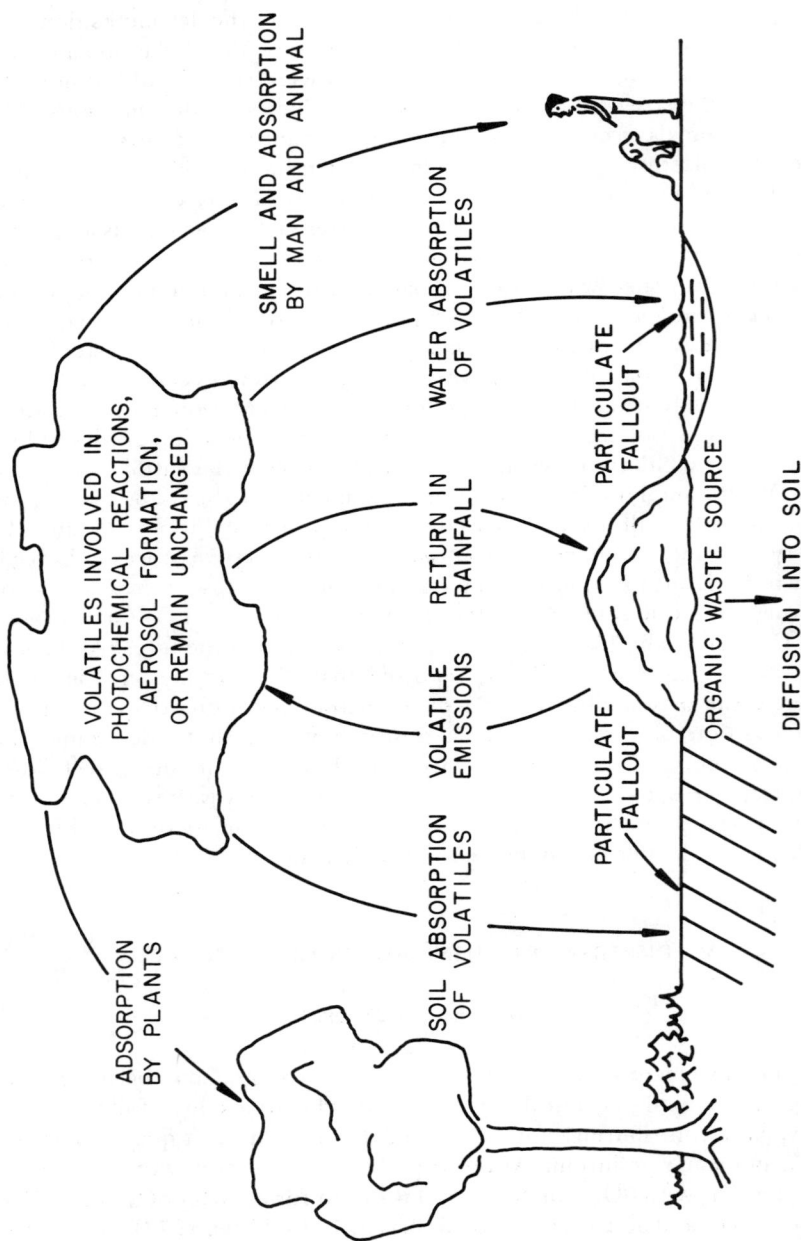

Fig. 1—Dispersion of volatile emissions from an organic waste source.

icals may be absorbed from the atmosphere by water (Hutchinson & Viets, 1969), or returned to the earth's surface with precipitation (Hoeft et al., 1972) or particulate fallout (Kellogg et al., 1972). Airborne chemicals may be absorbed by plants (Heichel, 1973; Porter et al., 1972), onto soil (Bohn, 1972; Heichel, 1973; Abeles et al., 1971), by animal life (Bohn, 1972), and when present in sufficient amounts, may be detected by man. Volatiles from an organic source can also diffuse into the underlying soil (Elliott & McCalla, 1971).

Using an open, unsurfaced, high density cattle feedlot such as those in northern Colorado as an example, let us follow the possible fate of an NH_3 or an amine molecule produced in the manure pack from the anaerobic bacterial digestion of manure protein. The feedlot profile may be described as a mixture of animal feces, urine, and soil overlying the original surface upon which the lot was constructed. The manure layer is commonly comprised of three layers: (i) the aerobic surface; (ii) the manure pack (varying from a few to several centimeters, depending upon the "housekeeping" practices of the feedlot operation) which is anaerobic below the surface; and (iii) a hard-packed, "humified" layer contacting the soil, which is essentially impervious to water (Elliott & McCalla, 1971; Mosier et al., 1972; Mielke et al., 1974). In a high density feedlot, little water moves down through the soil profile, so movement of organic chemicals in solution downward is of little importance in an established feedlot (Mosier et al., 1972). The pH of the manure pack surface is slightly basic due to the large amount of NH_3 produced from the hydrolysis of urea in cattle urine; hence, NH_3 is rapidly volatilized. Stewart (1970) found in a soil column study that the N in cattle urine was 80 to 90% volatilized as NH_3 from dry soil. However, in moist soil, 25% of the N was volatilized as NH_3 while the remainder was nitrified.

The volatile chemicals produced from manure digestion may diffuse from the manure upward into the atmosphere or down into the soil profile. In the soil, ammonium ions formed by the reaction of NH_3 with water can be tightly bound in soil clays (McCalla et al., 1970) or react with organic anions in the soil (Stevenson & Wagner, 1970). Ammonium can be bacterially nitrified to NO_3^- where it can potentially serve as a source of mobile N, though most research indicates that NO_3^- mobility is not a large problem under high density cattle feedyards (Ellis et al., 1975; Viets, 1974). The NH_3 that diffuses into the air may be carried away from the source by the prevailing air currents (Luebs et al., 1974) and interact as described above. The movement of other chemicals from feedlots has not been studied.

Plants can also serve as sinks for chemicals volatilized from organic wastes. Hutchinson et al. (1972) and Porter et al. (1972) showed that plants can absorb significant quantities of NH_3 from the air even at natural, low atmospheric concentrations. Sulfur dioxide produced from H_2S in the atmosphere (Kellogg et al., 1972) is also absorbed by plants (Faller, 1970). The extent of absorption by plants and adsorption onto soil of other airborne compounds could be significant but has not been substantiated.

The soil is also a sink for atmospheric contaminants. Inman et al.

(1971) showed that soil can remove CO_2 from the atmosphere and that removal is dependent upon soil microbial activity. Ethylene, SO_2, and NO_2 may be removed from air by microbial or chemical reactions in soil (Abeles et al., 1971). Turner et al. (1973) found that ozone in the atmosphere can be taken up by soil. Moist, nonacid soils (Bohn, 1972) absorbed SO_2 rapidly and converted it to H_2SO_4. The absorption capacity of sand, clay, and an acid soil for H_2S increased with higher soil pH and the presence of transition metal cations. Soil absorbs mercaptans at rates increasing with the size of the microbial population. Ethanethiol, ethyldisulfide, thioacetic acid, cystine, and the various compounds in sewer gases are absorbed by soils (Carlson & Leiser, 1966). Acetic, butyric, and lactic acids are taken up and oxidized rapidly. Malo and Purvis (1964) observed that soils in New Jersey absorb NH_3 from the air.

The dispersal of organics, NH_3, and H_2S into the environment is increased when sewage sludge or animal manures are removed from their collection site to be applied to land. The release of H_2S and other strong odors has been noted when animal manures collected in pits under confined feeding houses have been disturbed (Muehling, 1970). The feedlot manure pack or the collection pit under swine can be an essentially closed system which tends to be self-confining. Breaking the "seal" and exposure of the ensuing volatiles to air produces an initial high concentration of the anaerobically produced volatiles. The emission of malodors from spreading of animal manures is well known (Loehr, 1974; Ludington, 1971). Analytical measurements of the chemicals emitted during spreading have not been made, but qualitatively we assume that the chemicals are similar to those detected near animal feeding and sewage treatment operations. Dispersal may change after manures are applied to the land due to runoff or movement into the soil profile with water. Losses of 61 to 99% of the total ammoniacal N were detected 25 days after field application of dairy cattle manure (D. A. Lauer, D. R. Bouldin, and S. D. Klausner. 1974. Ammonia volatilization from dairy manure spread on the soil surface. Am. Soc. Agron. Abstr. p. 31.).

Adding large amounts of manure to land stimulates the growth of saprophytic bacteria, fungi, and actinomycetes. Numbers of aerobic, cellulolytic bacteria, protozoa, and actinomycetes, and CO_2 production are increased by manure addition. Under aerobic conditions, the manure is decomposed to CO_2, NO_3^-, $H_2PO_4^-$, SO_4^{2-}, and microbial cells, but under anaerobiosis, the production of odorous organics like those emitted from the feedlots may result (Table 1).

B. Important Factors Controlling Emission of Volatile Chemicals

Important factors controlling the emission of gases from field-applied manures, as well as from the animal feeding site or sewage treatment plant, are pH, oxidation-reduction potential, moisture, and temperature.

1. EFFECT OF pH

A molecule in its ionized state has no vapor pressure, so only molecules in the free state are odorous. The pH of an aqueous solution controls the relative amount of ionized or free molecular species according to the familiar expression, pH = pK_a + log (base)/(acid). A dissolved gas is lost from solution when its partial pressure in the atmosphere over the solution falls below the level dictated by its Henry's solubility constant. Thus, for a basic volatile compound like an amine, the odor produced from a solution of the compound will increase as the pH of the solution increases until a solution/ atmosphere equilibrium is reached. The effect of pH on NH_3 volatility is demonstrated by equations developed to predict the rate of NH_3 desorption from chicken manure slurries (Hashimoto & Ludington, 1971).

2. EFFECT OF OXIDATION-REDUCTION POTENTIAL

The redox potential (Eh) of a system affects the type of chemical (Engler & Patrick, 1973; Whisler et al., 1974) and biological (Alexander, 1961) reactions that occur. Takai and Kamura (1966) outlined the succession of events which may occur in waterlogged soils or in those receiving excessive loads of organic waste. The early period of incubation is characterized by the disappearance of molecular O_2 in the redox potential range of +600 to +400 mV. Aerobic microbial metabolism is functional in this range. The disappearance of NO_3^- and the formation of Mn^{2+} and Fe^{2+} occurs in the +400 and +100 mV range where facultative anaerobic organisms predominate and organic acid accumulation begins. Later, sulfides form and organic acids rapidly accumulate at redox potentials of 0 to –150 mV. At redox potentials of –150 to –220 mV, obligate anaerobic organisms are the active metabolic species, organic acid concentrations decrease, and H_2 and CH_4 are formed.

3. EFFECT OF MOISTURE

Water content of a system plays an important role in determining available oxygen, redox potential, and microbial activity. The elimination of water in confined feeding waste collection is not practical; thus, aerators have been developed (Jones et al., 1970). The moisture content of an open feedlot cannot be well controlled, but there are management practices that will help solve the problem (Shuyler et al., 1973). The rate of soil-applied organic residue decomposition can be affected by the interaction of soil moisture and microorganisms. Maximum growth and activity of soil bacteria occur at high water contents but noticeably decreases at about 3 bars tension. Soil fungi tend to thrive in soils of much lower water potentials where bacteria are less active (Alexander, 1961).

4. EFFECT OF TEMPERATURE

Most reaction rates are, to a point, increased as the temperature of the particular environment increases. Temperature variations can alter the species composition of the soil microflora. Maximum residue decomposition rates in soil generally occur in the range of 30 to 35C, although various organisms have different optima for maximum growth and activity (Alexander, 1961). Anaerobic animal waste storage lagoons are relatively odorless during the cold seasons but become malodorous as the temperature increases. Adriano et al. (1974) reported that higher moisture and temperature caused higher N losses from land applied manure. Most of these losses were during the early stages of incubation, apparently through volatilization of NH_3.

VI. COLLECTION AND ANALYSIS OF ODORS AND VOLATILE EMISSIONS FROM ANIMAL WASTES

A. Problems of Collection and Analysis

Identifying organic chemicals volatilized from animal feeding operations and sewage treatment facilities has been attempted by numerous researchers (Bethea & Narayan, 1972; Burnett, 1969b; Elliott & Travis, 1973; Miner & Hazen, 1969; Mosier et al., 1973; Rains et al., 1973; Warden, 1972; White et al., 1971). Efforts to identify all of the compounds evolved have had only marginal success, although specific classes of compounds have been identified (Banwart & Bremner, 1974b; Elliott & Travis, 1973; Hartung et al., 1971; Mosier et al., 1973). The major stumbling block appears to be collecting sufficient amounts of each of the numerous chemicals emitted from a source. Numerous analytical instruments are presently available with adequate sensitivity, but the appropriate collection methods have not been developed that accurately represent the organics evolved from the source.

B. Collection

1. COLLECTION PROBLEMS

Difficulty in collection and analysis lies in the very low concentration of substances volatilized compared to other atmospheric constituents such as water. Obtaining sufficient amounts of the individual compounds present in the atmosphere without initiating reactions of the collected chemicals with each other or with the collection or analytical apparatus is a major problem. Research of Mosier et al. (1973), who attempted to identify the basic N compounds emitted from beef cattle feedlots, illustrates this point. They found that aliphatic amines could be easily collected, concentrated, and stored in dilute sulfuric acid. However, the decomposition of heterocyclic N compounds is acid catalyzed and NH_3, which is present in much larger concentrations (100 to 1,000 times) than any of the individual amines, interferes with most qualitative and quantitative methods specific for aliphatic amines. Gas chromatography (GLC) using a flame ionization detector (FID) is satisfactory for ng/μliter concentrations of amines, as the detector is insensitive to NH_3. Pentafluorobenzamide derivatives of the basic N compounds were made and electron capture (EC) detection was used with some success. However, no gas chromatographic column system was found to completely resolve the NH_3 derivative from all amine derivatives.

2. LABORATORY COLLECTIONS

Because of the low atmospheric concentrations of odorous chemicals near animal feeding operations, wastes have been brought into the laboratory, and the gases emitted from the materials analyzed after a concentration step. Miner and Hazen (1969) used acetic acid traps in an attempt to collect the basic volatiles from swine manure. Merkel et al. (1969) salted out the volatiles over a sample of swine manure, and used fractional condensation (ice water, dry ice-ethanol, and liquid N_2) to collect the volatiles from a slurry of swine manure. Bethea and Narayan (1972) used a series of solvent traps to selectively absorb the volatiles from the headspace above samples of aerobically maintained cattle manure. White et al. (1971) employed an equilibrium sampling method to collect the volatiles from the headspace above dairy cattle manure in the laboratory. Warden (1972)[2] evaluated several methods of volatile collection which included "equilibrium" headspace sampling, three methods of cryogenic concentration [cold traps similar to those used by Merkel et al. (1969), precolumn collection, and GC on-column collection], activated charcoal collection, and another cryogenic trap comprised of a 1-liter FEP Teflon bottle modified with side ports, which contained silanized Pyrex beads for sampling headspace vapors over cattle manure. Burnett (1969b) used a cryogenic (acetone-dry ice) collection in a GC precolumn of the volatiles from aqueous slurries of poultry manure. Mosier et al. (1973) used $0.01N$ H_2SO_4 traps to collect the basic compounds volatilized from beef cattle manure in the laboratory. They noted that the chemical content of the manure headspace gases collected were very different from the compounds collected in similar traps located near a cattle feedlot.

Obtaining a volatile sample truly representative of the organic source remains a problem in laboratory work. The headspace analysis of "equilibrium" vapors in a closed container of animal excreta is simple and rapid, but suffers from interference due to vapor pressure predominance by major components and low sample concentrations of low vapor pressure volatiles (Warden, 1972[2]; Johnson & Nursten, 1971). Cryogenic collection has the advantages of increased concentration and minimization of side reactions, but high water concentrations require desiccants which can absorb the compounds of interest (Williams, 1965). Solvent trains designed to collect fractions based upon solubility were functional for qualitative analysis, but incomplete elution of functional group compounds not specific for this collection trap frequently can occur (Mosier et al., 1973). For example, small amounts of alcohols remain in the dilute acid traps and their presence interfered with analytical methods for amines.

3. FIELD COLLECTION

Warden (1972) used the cryogenic collection described above to concentrate cattle feedlot atmosphere samples, while Mosier et al. (1973) used dilute sulfuric acid to collect basic compounds volatilized from cattle feedyards. Hartung et al. (1971) sampled swine-building atmospheres for carbonyl compounds by pumping the confinement atmosphere through tubes containing Celite or silica gel impregnated with dinitrophenyl hydrazine. Fractional condensation and selective absorption were used by Merkel et al. (1969) to concentrate the organics from swine building atmospheres, and

Table 4—Methods used to collect chemicals volatilized from animal wastes

Chemical group	Collection method	Reference
Acids	Cryogenic collection in GLC column	Burnett (1969b)
	Manure extract	Merkel et al. (1969)
Alcohols	Propylene glycol trap	Bethea and Narayan (1972)
		Merkel et al. (1969)
	Cryogenic collection	Warden (1972)[2]
Aldehydes, esters, and ketones	Propylene glycol trap extracted with CCl_4	Bethea and Narayan (1972)
		Merkel et al. (1969)
	Silica gel impregnated with dinitrophenyl-hydrazine	Hartung et al. (1971)
	Cryogenic collection	Warden (1972)[2]; Burnett (1969b)
Amines	1.2N HCl trap	Bethea and Narayan (1972)
		Merkel et al. (1969)
	0.01N H_2SO_4 trap	Mosier et al. (1973)
	5% acetic acid trap	Miner and Hazen (1969)
Mercaptans and sulfides	$HgCl_2$ + $Hg(CN)_2$ traps	Bethea and Narayan (1972)
		Merkel et al. (1969)
	Equilibrium vapor method	White et al. (1971)
Nitrogen heterocyclics	Extraction and steam distillation of manure	Burnett (1969b)
All groups	Cryogenic collection	Warden (1972)[2]; Burnett (1969b)
	Equilibrium vapor method	White et al. (1971)

Miner and Hazen (1969) used acetic acid traps to collect the basic compounds.

Rains et al. (1973) reported the collection of volatile organics over a municipal sewage treatment facility during thickening of raw and secondary sewage sludge. A styrofoam dome covering a sludge thickener controlled atmospheric conditions and concentrated odors. Volatiles were collected by cryogenic collection (dry ice-acetone).

Table 4 lists the methods used to collect the different classes of compounds characterized as components of the organic chemicals volatilized from animal wastes.

C. Analytical Methods

The principal analytical tool used to identify specific organic volatiles has been gas chromatography, more specifically, gas-liquid chromatography (GLC), because this instrument facilitates separation of individual compounds and detection of very small quantities of organic materials. Choice of detectors is important: the hydrogen flame ionization detector (FID) is sensitive to oxidizable organic C in the ng range; the electron capture detector (EC) is useful for chemicals containing electron-capturing elements and has a sensitivity in the pg range. Other detectors are specific for individual elements like the flame photometric detector (FPD), specific for S and P, and the microcoulometric detector that is specific for N. The FPD and the microcoulometric detectors are useful for sample concentrations and in the ng to pg range. Table 5 lists the column and detectors used to analyze the organics identified as volatiles from animal wastes, municipal wastes, and the other sources.

To date, limited use of the combined GLC-mass spectrometer has been made to identify organic compounds volatilized from organic wastes. The high cost of the instrumentation and necessary specialized expertise largely contribute to its lack of use.

Much has been accomplished toward identifying the organic flavor components of fruits and beverages (Schultz et al., 1971; Timmer et al., 1971) and the analysis of human urine for uncommon metabolites (Zlatkis et al., 1973) by GLC-mass spectrometric analysis. These methods should have direct applications to the area of volatile organic chemical analysis of agricultural and municipal wastes. Abramson (1972) and McFadden (1973) described the versatility of this instrumental method.

A major criticism of the analytical work in the field of agricultural organic waste volatiles is that most of the work does not provide corroborating evidence to support the claimed identifications. Of the reports of compound identification from animal wastes, only three (Elliott & Travis, 1972; Mosier et al., 1973; Warden, 1972[2]) provided supplementary analytical data which supported their identifications. The majority of the identifications were probably correct, but the methods used leave room for doubt.

Table 5—Gas chromatographic and other methods of analysis of odors and volatiles

Chemical group for which analysis designed	Column	Detector	Reference
Acids	AW Chromosorb W + 10% SP-1200 + 1% H_3PO_4	FID	Ottenstein and Bartley (1971)
	AW Chromosorb W + 10% Carbowax 20M	FID	Burnett (1969b)
	Graphitized carbon + 0.5% H_3PO_4 + 3% PEG 20M	FID	DiCorcia and Samperi (1974a)
Alcohols	Porapak Q	FID	Merkel et al. (1969)
Amines	Chromosorb 103 (injection port packed with Ascarite)	FID	Mosier et al. (1973); Andre and Mosier (1973)
	Gaschrom R + 28% Penwalt 223 + 4% KOH	FID	Mosier et al. (1973); Andre and Mosier (1973)
	Anakrom SD + 10% Igepal CO 880	EC	Mosier et al. (1973)
	Graphitized carbon + 0.8% KOH + 5% PEG-20M	FID	DiCorcia and Samperi (1974b)
	Chromosorb W + 10% amine 220 + 10% KOH	FID	Umbreit et al. (1969)
Mercaptans and sulfides	Graphon + 0.5% H_3PO_4 + 0.3% Dexsil	FPD	Bruner et al. (1972)
	Chromosorb T + 12% Polyphenyl ether + 0.5% H_3PO_4	FPD	Banwart and Bremner (1974)
	Chromosorb G + 10% Triton X 305 + 0.5% H_3PO_4	FPD	Ronkainen et al. (1973)
Nitrogen heterocyclics	AW DMCS Chromosorb W + 5% SW-30	FID	Burnett (1969b)
Multigroup analysis	Chromosorb P + 10% Carbowax 20M AW-DMCS	FID	White et al. (1971)
	Porasil S, C + Durapak-Carbowax 400	FID	Warden (1972)[2]

Other Analytical Methods

Hydrogen sulfide	$AgNO_3$ impregnated filter paper, analyzed fluorometrically		Natusch et al. (1974)
	$Pb(OAc)_2$ + 5% acetic acid impregnated filter paper analyzed spectrophotometrically		Okita et al. (1971)

D. Field Analytical Methods

Few sensitive qualitative or quantitative analytical methods, short of instrumentation in a complete mobile laboratory, have been developed for the in-field collection and analysis of odors and emissions. There are a few methods that could have utility for quick field analysis of specific major volatiles when accuracy is not important. Moum et al. (1969) described a method for measuring atmospheric NH_3 which utilized the equilibrium relationship of NH_3 between the atmosphere and contacting water surfaces. In their procedure, pH test paper was moistened with distilled water and exposed to air for 15 sec. The pH is noted and the NH_3 content of the air can be calculated. Hydrogen sulfide can be collected in the field by passing a measured volume of air through solutions of zinc acetate (American Public Health Association, 1971) or cadmium hydroxide. The H_2S reacts to form a zinc or cadmium sulfide suspension. The sulfide content can be measured by titration or colorimetric methods applicable for field use. Other methods of H_2S collection were reported by Natusch et al. (1974) who used silver nitrate-impregnated filter paper, and Okita et al. (1971) who described a lead acetate-acetic acid-impregnated filter paper. These two methods describe the use of sensitive fluorometric or spectrophotometric analytical methods which should be adaptable to field analysis with the probable sacrifice of sensitivity.

E. Olfactory Analysis

Several researchers have correlated GLC analyses of organic volatiles from wastes with olfactory sensation. The use of olfactory panels has been helpful to substantiate that some of the chemicals separated and identified by GLC constituted a portion of the odor from organic wastes. White et al. (1971) developed a method of "equilibrium" sampling of anaerobic dairy cattle manure. Using a column splitter to direct the injected gas sample to both the detector and a sniffing port, GLC, and olfactory analysis of volatiles was accomplished. Kovat indices were used to characterize the chromatogrammed volatiles, and the odor of each chromatogram peak, where there was an amount injected above the odor threshold, was determined by a person. Characteristic odor signatures for volatiles from dairy animal wastes under anaerobic conditions were obtained. The contribution of each particular compound to the odor perceived was not determined. Dimethyl sulfide was considered the principal component of anaerobic dairy waste odor. Aeration reduced or eliminated the sulfur compounds tentatively identified.

Burnett (1969b) trapped volatile odorous substances from liquid poultry manure in short sections of GLC columns held at −78C. The individual volatile components were separated by GLC and identified by correspondence between relative retention time and the odor of the peak for the unknown and authentic compounds. Sulfur compounds, organic acids, and skatole were implicated as important malodorous components. Bell (1970) using GLC and a simple column chromatographic method related the total amount of fatty acids (acetic + propionic + butyric acid) to the odor of liquid poultry manure. He concluded that if the fatty acid content of the liquid manure could be kept below 0.1%, then the odor would be maintained within an "acceptable" range. He contended that the volatile fatty acids were not necessarily the sole source of offensive odor but that their presence indicated the extent of anaerobic metabolism. Measures designed to prevent the accumulation of fatty acids, i.e., anaerobiosis, would prevent the accumulation of other compounds which contribute to the obnoxious odor of putrid poultry manure.

Merkel et al. (1969) used the selective solvent absorption train to observe the effect of the various absorbents on swine manure odor. Correlation of GLC analysis of individual absorption traps with olfactory judgment indicated that the major odor constituents in the atmosphere of swine buildings belonged to the amine and sulfide groups while the alcohols and carbonyls were judged unimportant. The scentometer (Barnebey-Cheney Co., 1968) has been used extensively for odor evaluation (Miner, 1973) and is used as a standard odor measuring device on which odor limits are based in some areas (Leonardos, 1974).

Odor evaluations of individual chemicals isolated by GLC from sewage digesters indicated that propyl amine, tert-butyl mercaptan, ethyl amine, and acetaldehyde were major contributors to malodors. Other compounds that were minor contributors were alcohols and organic acids (Rains et al., 1973).

VII. METHODS OF LIMITING NUISANCE ODORS AND EMISSIONS

A. General Considerations

The obvious solution to odors produced from the emission of nuisance substances is to prevent their formation by eliminating anaerobic conditions in the feeding or treatment site. Since all biological processes depend upon temperature, pH, oxidation-reduction potential, water, and available nutrients, manipulation of one or more of these factors seems an easy solution. Unfortunately, the control of odors and emissions from large concentrations of organic waste is not simple and direct. No totally satisfactory solution has yet been found that is both economically and technologically feasible. Methods of controlling nuisance odors and emissions include source management, animal diet, and chemical treatment.

B. Source Management

The initial step in the management of a source of odors and emissions is the location of the site. Proximity to residential areas, businesses, and public gathering places, soil type, topography, and prevailing winds are important factors to consider when locating an animal feeding operation (Shuyler et al., 1973). Miner (1970) recommended locating a livestock feeding operation at least 4.8 km from an urban area. Land which is too flat may be poorly drained, resulting in sloppy pen conditions which can lead to odor production (Shuyler et al., 1973).

1. OPEN CATTLE FEEDLOTS

For both established and new feedlots, good housekeeping is the basic method available for odor limitation. According to Moorman (1965), odors can be significantly reduced by minimizing conditions which aid in their production: (i) poor drainage, allowing water to stand for long periods of time; (ii) spilled feed wasted from feedbunks or around feed mills; (iii) improper disposal of dead animals; (iv) accumulation of manure in feeding pens; and (v) manure disposal operations improperly managed. The major problems usually result from accumulation of manure in the pens and manure disposal (Moorman, 1965). Miner (1970) suggests that animal feeding areas and pens should be kept as dry as possible to minimize anaerobic manure decomposition. As warm-bodied cattle provide a site for accelerated bacterial growth and odor production, manure management should be designed to prevent dirty, manure-covered animals. Slopes of 4 to 6% are generally preferred for unsurfaced lots for adequate drainage. Mounding of manure in the feeding pen is frequently practiced to provide animals a dry place to lie down if drainage is marginal. Frequent removal of straw, wood chips, or other fibrous material is suggested to minimize manure moisture content and anaerobic conditions. Proper maintenance of cattle watering equipment is also important (Miner, 1970).

Once manure has been removed from feeding pens, a controlled system of handling should be followed. Satisfactory methods being used include properly constructed compost piles (Wells et al., 1969), dehydration (Moorman, 1965), lagooning, spreading on fields, and land incorporation (Shuyler et al., 1973). Immediate utilization of manure when cleaning pens can be a problem. Under circumstances where land application or other use is not immediately possible, stockpiling is practiced. Moorman (1965) recommended placing the stockpiles in a sparsely populated area so that little odor problems result. The pile should be located so runoff from the site is minimal (Shuyler et al., 1973).

2. CONFINED SWINE AND CATTLE FEEDING

In areas of high rainfall, confined cattle and swine feeding on slotted floor buildings is practiced. This practice requires a different approach, i.e., liquid/slurry waste technology. Shuyler et al. (1973) define liquid/slurry waste management systems as those which handle material having a moisture content of 85% or greater. Such systems typically perform four functions: (i) waste collection and transport to storage; (ii) storage and treatment; (iii) transport from storage; and (iv) reuse, further processing, or disposal. The liquid/slurry may be collected and held in a pit under a slotted floor or fluid transported to a holding pond. The stored material can be treated either anaerobically or aerobically. The anaerobic treatment may have cost and handling advantages, and the nutrient quality of anaerobically digested slurry is greater than aerobically treated slurry. Unfortunately, odor problems associated with handling anaerobically treated wastes are great. Jones et al. (1970) are proponents of aerobic treatment of livestock waste. Though the cost and difficulty of aeration may be significant, the "odorless" method of waste treatment may be advantageous. Day et al. (1971), Jones et al. (1970), and Shuyler et al. (1973) outlined several methods that have been developed to handle and treat liquid/slurry animal wastes. Anaerobic lagoon storage and treatment of liquid/slurry animal wastes have been used for years. The anaerobic lagoon serves as both a storage and treatment facility where a considerable amount of solids break down to liquid during storage. The odors from lagoons have been described as nonoffensive to highly offensive. The offensive odor production is related to an overload of organics in the system which causes an imbalance between the production of nonodorous methane to the production of odorous gases by bacteria (Miner, 1973).

3. POULTRY PRODUCTION

Malodors from housing poultry operations are produced during the anaerobic storage of manure and are released during storage and spreading. The major sources of odors from poultry production are house ventilation air, solid and liquid loading areas, storage tanks, and the land after field application (Ludington, 1971). Ludington et al. (1971) reported that malodors from poultry can be limited by: (i) frequent cleaning and spreading of manure; (ii) in-house moisture removal; (iii) aeration of liquid manure handling systems; (iv) plow-furrowing cover of field application; and (v) injection of liquid manure. Herr (1970) described a mechanical method for the under-cage drying of poultry manure which aids in the limitation of odor production. Bell (1970) described a method for the aeration of liquid poultry manure which serves as a stabilization process as well as odor control process. Ludington et al. (1971) concluded that management systems which do not allow manure buildup, or remove moisture from the manure, effectively limited in-house odor production. He ranked adequately aerated, liquified manure above moisture removal, frequent cleaning, and chemical treatment as the best methods of controlling odors for ventilation air, loading area and storage, and land spreading. He concluded that "no matter how carefully the method of handling is chosen, if good management and good housekeeping are not followed, the system will fail."

C. Chemical Treatment of Wastes and Malodorous Air

Several chemicals have been tested to limit odor from animal wastes or to treat malodorous air emanating from the wastes. None of the products tested have been totally successful in limiting odors. Wilmore (1972) noted that "no product eliminates all odors; some are more effective than others; some don't work at all." A number of the commercially available chemicals for odor treatment are listed by Miner (1973).

1. TREATMENT OF WASTES

Hydrogen peroxide has been tested for a number of treatment applications to oxidize the odorous compounds in a manure slurry. A 100-ppm dosage of H_2O_2 was effective in reducing odors from dairy cattle and swine manure slurries (Miner, 1973). Aerial H_2S over the swine manure slurry was reduced from 10 to 0 ppm immediately after treatment (Miner, 1973). Kibbel et al. (1972) reported that 100, 150, and 175 ppm H_2O_2 significantly reduced the odors evolved from the spreading of chicken manure slurry on land.

Potassium permanganate is reported to be effective in limiting odors from southern California cattle feedlots. The cattle feedlot surface was sprayed with a 1% solution of $KMnO_4$ (Faith, 1964). The residual effects of Mn on the characteristics of the waste are not known.

Hydrated lime applied to hog manure slurries reduced H_2S and CO_2 production. Applied in liquid manure collection pits at the rate of 0.16 pounds/day per 100-pound hog, hydrated lime limited odor production by inactivating anaerobic bacteria. In the same study, Hammond et al. (1968) found that 0.1 pounds/day per 100-pound hog of active Cl was an effective odor-reducing agent. Cost was a limiting factor for this treatment.

Seltzer et al. (1969) report that application of 1 g paraformaldehyde flakes/100 g poultry manure limited NH_3 odor for 9 days and also limited H_2S evolution. This agent may be of limited applicability because of its toxicity to animals if ingested, and because paraformaldehyde decomposes to formaldehyde.

2. TREATMENT OF ATMOSPHERIC ODORANTS

Chemicals used to treat atmospheric odorants commonly are grouped according to their method of malodor limitation. These groupings include odor masking, counteractants, deodorants, and digestive deodorants. The odor-masking agents are aromatic oils used to cover malodors. According to Young (1972), these agents have had only limited success because the odor of the agents may be considered malodorous to some people. Counteractants are generally mixtures of aromatic oils selected to counteract the odor components in the waste (Miner, 1973). This is the method that Young (1972) found most successful. This method is based on the phenomenon that certain odors neutralize the malodor so that neither odor is perceptible. Another group, deodorants, are a formulation designed to eliminate malodors without imparting a covering odor. The digestive deodorants have not proven very successful (Young, 1972). These agents combine digestive enzymes, aerobic, and anaerobic bacteria to create a digestive process to eliminate the odor. This method is similar to composting.

Burnett and Dondero (1970) evaluated several masking agents, counteractants, deodorants, and digestive deodorants on the malodors from chicken manure. The masking agents and counteractants were found most effective; deodorants were moderately effective, and digestive deodorants least effective. The chemical treatments were effective for limited time periods which would indicate that chemical treatment of wastes might be applicable to chemical addition to stored manure immediately before land spreading.

3. ANIMAL FEEDING

The literature documenting the effect of food additives on animal waste odor is very limited. Ingram et al. (1973) reported feeding young pigs a commercial bacterial culture, *Lactobacillus acidophilus,* a lyophilized yeast culture, or activated charcoal in a basal grain sorghum-soybean [*Sorghum bicolor* Moench.-*Glycine max* (L.) Merr.] meal diet. Their study showed that adding the yeast culture or the bacterial culture to the pig diet reduced the indole and skatole content of the young pig feces; however, the olfactory panelists could not differentiate between odors produced by the treated compared with untreated excrement. The use of sagebrush (*Artemisia* sp.) as a cattle feed additive was reported to reduce feedlot odors during limited feeding trials at Colorado State University. The effectiveness of the additive was attributed to the volatile oils being carried through the manure (Eller & Matsushima, 1972).

D. Odor Limitation from Sewage Treatment and Land Application

Rains et al. (1973) studied air dilution, activated carbon adsorption, and chlorine oxidation as methods to limit malodorous emissions from a dome-covered sewage sludge thickening unit. Air dilution using cyclic operation of an exhaust fan was effective only when atmospheric conditions were conducive to odor dissipation. Passing the vapors through activated carbon filters did not completely eliminate the odor. A 1.5 mg/liter solution of Cl was effective in removing all odors from vapor samples bubbled through the solution. Odors from liquid sludge stored in lagoons may be reduced by minimizing surface area and reducing storage time. Injecting liquid sludge into soil is preferred to surface sprinkling for odor control during land application (Thorne et al., 1975). Municipal sewage treatment works in England irrigate land with liquid digested sludge without associated odor and fly problems

(Seabrook, 1973).   Sewage waste water has been made acceptable for golf course sprinkle irrigation by chlorination (Merz, 1959).

E. Odor Control from Rendering Plants

The chemicals emitted from the rendering of inedible animal products are similar to those from animal feeding and sewage sources.   Control of rendering plant emissions is potentially simpler than that of the other sources discussed, since the rendering is performed in a housed, closed system.   Currently used odor control systems are condensation followed by incineration, scrubbing, or a combination of these.   Other control methods used with varying degrees of success are catalytic combustion, adsorption, and ozonation.   No odor control system, according to Bethea et al. (1973), can be effective, unless the renderer practices good housekeeping and proper materials management.

VIII. SUMMARY

The biochemical transformations accompanying the growth and metabolism of microorganisms on animal wastes under anaerobic conditions can lead to the formation of malodorous products.   Table 6 gives the relationship between some odor constituents, like volatile fatty acids, amines, and mercaptans, and microbial metabolic activities which may contribute to their formation.   In developing and evaluating techniques for modifying or controlling odor production, various carbonaceous and nitrogenous waste constituents can be interconverted during metabolism via intermediates, like pyruvate, aspartate, and glutamate, to form cell biomass as well as metabolic end products.   In addition to the limiting factors, like water, pH, or redox potential, nonbiodegradability or inaccessibility of the growth substrate may potentiate metabolic shifts or selective pressures contributing to odor formation.

Table 6—Relation of odor constituent to microbial metabolic activity

Odor constituent	Microbial metabolic activity
Volatile fatty acids; ketones	Anaerobic carbohydrate fermentation and amino acid deamination
Alcohols	Anaerobic carbohydrate fermentation
Amines	Amino acid decarboxylation; phosphatidy-lethanolamine-methionine interaction
Mercaptans	Decarboxylation of cysteine, methionine
Hydrogen sulfide	Aerobic and anaerobic catabolism of cysteine and methionine
Ammonia	Aerobic and anaerobic amino acid deamination; denitrification; urea and uric acid decomposition

Fifty-seven different organic chemicals, as well as NH_3, N_2S, CS_2, and COS, have been identified as volatiles from cattle, poultry or swine feeding wastes, and sewage treatment plants. The chemical classes represented by these organics include acids, alcohols, aldehydes, amines, ketones, N hetero-cyclics, and reduced sulfur compounds. Ammonia and H_2S contribute the greatest amount of reduced N and S to the atmosphere during organic waste decomposition. Many of the volatilized chemicals are toxic to animals in high concentrations, but usually the concentrations are low and are not con-sidered harmful. Several identified chemicals, particularly the sulfur-con-taining and amine compounds, are considered malodorous.

The dispersal patterns of volatile chemicals from organic waste sources have not been fully investigated. Theory and limited research suggest that volatiles may be widely dispersed from a source, depending upon atmospher-ic conditions. Volatile chemicals can exist in the atmosphere in their free state or combine with water, other chemicals, and/or particulate fallout. Air-borne gaseous chemicals can be absorbed directly by plants, be adsorbed on-to soil, and if present in sufficient amounts, be detected by man. Volatiles from an organic source may diffuse into the soil beneath the source where they can be fixed, incorporated into biomass, or leached into the ground water.

Collection and analysis of volatile chemicals dispersed from an organic waste source can be an arduous task. Since most organic chemicals volatilized during decomposition are found in the atmosphere at relatively low concen-trations compared to the major atmospheric components, collection of suf-ficient amounts of these chemicals for analysis by commonly practiced methods requires sampling large volumes of air. At times major atmospheric constituents interfere with collection and analytical methods. Laboratory sampling of waste head-gas, commonly practiced to identify chemicals volatilized from the decomposing wastes, provided useful information, but may not be the most accurate representation of the volatile decomposition products or organic wastes in situ. Gas-liquid chromatography has been the method most frequently used to identify individual volatiles. Unfortunately, the coupling of the GLC and the mass spectrometer has not been exploited to identify volatiles from the different organic waste sources. Organoleptic analysis accompanying GLC separations has been helpful in distinguishing odorous volatiles in gases evolved from organic wastes.

The problems associated with volatile emissions from decomposing or-ganic wastes represent a complex mixture of human emotion, management, economics, and analytical complications. The problems are accentuated by the diverse effects of volatile emissions on people and our inability to meas-ure these substances objectively and reliably. Many questions such as the quantitative and qualitative loss to the atmosphere of organic chemicals from wastes during land application, movement of volatiles and particulates from the source, role of particulates in odor transport and detection, and the quantitative analysis of the various volatiles identified as emissions have not been satisfactorily answered. One specific problem is the loss of N from organic wastes through NH_3 and organic N. As yet, we do not know how much N is lost initially from a source, the subsequent losses during land application, or how the N is dispersed, although limited research suggests that the losses are large.

No instrumental method has been developed to replace the human nose for the qualitative analysis of odorants. The extremely low threshold limits of human olfactory detection of individual and groups of chemicals presents a formidable problem when attempting to control odors from organic wastes. The quality and intensity of individual odorants and the relationship between odor detection of some mixtures of odorant chemicals has been determined; however, the relationship between the complex mixture of gases identified as volatiles from organic wastes and the odors perceived by human olfaction is as yet unknown.

No technologically and economically feasible methods of odor prevention from large concentrated sources of organic wastes have been developed. Odor control problems are accentuated by the large variety of odorous chemicals emitted. Since the majority of the obnoxious malodorous emissions evolved from organic waste digestion are produced during incomplete anaerobic digestion, management efforts can be directed to limit the development of conditions for their production. Management of the source is the principle tool used to combat odor production. Through site location, drainage, and efficient handling of organic wastes, offensive odor formation can be reduced. Chemical and atmospheric treatment of wastes is expensive and has not been generally successful. Limited odor amelioration has been attained by chemical treatment of wastes during handling operations.

LITERATURE CITED

Abeles, F. B., L. E. Craker, L. E. Forrence, and G. R. Leather. 1971. Fate of air pollutants: Removal of ethylene, sulfur dioxide, and nitrogen dioxide by soil. Science 173:914-916.

Abramson, F. P. 1972. Application of mass spectrometry to trace determination of environmental toxic materails. Anal. Chem. 44:28A-34A.

Adams, A. P., and J. C. Spendlove. 1970. Coliform aerosols emitted by a sewage treatment plant. Science 169:1218-1220.

Adriano, D. C., A. C. Chang, and R. Sharpless. 1974. Nitrogen loss from manure as influenced by moisture and temperature. J. Environ. Qual. 3:258-261.

Albrink, W. S. 1961. Pathogenesis of inhalation anthrax. Bacteriol. Rev. 25:268-273.

Alexander, M. 1961. Introduction to soil microbiology. John Wiley and Sons, Inc., New York. 472 p.

Alexander, M. 1965. Biodegradation: Problems of molecular recalcitrance and microbial fallibility. Adv. Appl. Microbiol. 7:35-80.

Alexander, M. 1971. Microbial ecology. John Wiley and Sons, Inc., New York. 511 p.

Allen, M. B. 1953. The thermophilic aerobic sporeforming bacteria. Bacteriol. Rev. 17: 125-173.

American Public Health Association. 1971. Standard methods for the examination of water and wastewater. 13th ed. APHA, Washington, D. C.

Amoore, J. E. 1970. Molecular basis of odor. Charles C. Thomas, publ., Springfield, Ill. 200 p.

Andre, C. E., and A. R. Mosier. 1973. Precolumn inlet system for the gas chromatographic analysis of trace quantities of short-chain aliphatic amines. Anal. Chem. 45: 1971–1973.

Aschbacker, P. W. 1971. Excretion of ^{14}C-diethylstilbesterol by sheep. J. Anim. Sci. 33:248.

Azevedo, J., R. G. Flocchini, T. A. Cahill, and P. R. Stout. 1974. Elemental composition of particulates near a beef cattle feedlot. J. Environ. Qual. 3:171–174.

Bachrach, U. 1957. The aerobic breakdown of uric acid by certain *Pseudomonads*. J. Gen. Microbiol. 17:1–11.

Banwart, W. L., and J. M. Bremner. 1974. Gas chromatographic identification of sulfur gases in soil atmospheres. Soil Biol. Biochem. 6:113–115.

Barker, H. A., and J. V. Beck. 1942. *Clostridium acidic-urici* and *Clostridium cyclindrosporum*, organisms fermenting uric acid and some other purines. J. Bacteriol. 43: 291–304.

Barnebey-Cheney Co. 1968. Scentometer: An instrument for field odor measurement. Instruction Sheet 9-68. 835 N. Cassady Ave., Columbus, Ohio.

Bell, R. G. 1970. Fatty acid content as a measure of the odor potential of stored liquid poultry manure. Poult. Sci. 49:1126–1129.

Bethea, R. M., B. N. Murthy, and D. F. Carey. 1973. Odor controls for rendering plants. Environ. Sci. Technol. 7:504–510.

Bethea, R. M., and R. S. Narayan. 1972. Identification of beef cattle feedlot odors. Trans. ASAE 15:1135–1137.

Bohn, H. L. 1972. Soil absorption of air pollutants. J. Environ. Qual. 1:372–377.

Bomar, M. T., and S. Schmid. 1973. Control of the bacterial breakdown of cellulose. Process Biochem. 8:22–23.

Brock, T. D. 1966. Principles of microbial ecology. Prentice-Hall, Inc., Englewood Cliffs, N. J. 306 p.

Bruner, F., A. Liberti, M. Possanzini, and I. Allegrini. 1972. Improved gas chromatographic method for the determination of sulfur compounds at the ppb level in air. Anal. Chem. 44:2070–2074.

Bryant, M. P., S. F. Tzeng, I. M. Robinson, and A. E. Joyner, Jr. 1971. Nutrient requirements of methanogenic bacteria. Adv. Chem. Ser. 105:23–40.

Burnett, W. E. 1969a. Odor transport by particulate matter in high density poultry houses. Poult. Sci. 48:182–184.

Burnett, W. E. 1969b. Air pollution from animal wastes, determination of malodors by gas chromatographic and organoleptic techniques. Environ. Sci. Technol. 3:744–749.

Burnett, W. E., and N. C. Dondero. 1969. Microbiological and chemical changes in poultry manure associated with decomposition and odor generation. p. 271–291. *In* Animal waste management. Cornell Univ. Conf. on Agricultural Waste Manage., Syracuse, N. Y.

Burnett, W. E., and N. C. Dondero. 1970. Control of odors from animal wastes. Trans. ASAE 13:221–224.

Campbell, C. C. 1957. A family outbreak of histoplasmosis. II. Epidemiological studies. J. Lab. Clin. Med. 50:841–848.

Carlson, D. A., and C. P. Leiser. 1966. Soil beds for the control of sewage odors. J. Water Pollut. Control Fed. 38:829–840.

Clark, C. D. 1965. Hog waste disposal by lagooning. J. Sanit. Eng. Div. Am. Soc. Civ. Eng. 91:27–41.

Cohen, G. N. 1949. Nature et mode de formation des acides volatils trouves dans les cultures de bacteries anaerobes strictes. p. 471–551. *In* Le role des anaerobies dans la nature. 2nd Congres Int. Des Microbiologistes De Langue Francaise, Bruxelles 23-27 Mai, 1949. UNESCO.

Copley International Corporation, La Jolla, Calif. 1973. A study of the social and economic impact of odors—Phase III. Environ. Prot. Agency Rep. No. EPA-650/5-73-001. U. S. Government Printing Office, Washington, D. C.

Dart, M. C. 1974. Treatment of waste waters from the meat industry. Process Biochem. 9:11–14.

Day, D. L., D. D. Jones, J. C. Converse, A. H. Jensen, and E. L. Hansen. 1971. Oxidation ditch treatment of swine wastes. Agric. Eng. 52:71–73.

Denmead, O. T., J. R. Simpson, and J. R. Freney. 1974. Ammonia flux into the atmosphere from a grazed pasture. Science 185:609-610.

Di Corcia, A., and R. Samperi. 1974a. Determination of trace amounts of C_2-C_5 acids in aqueous solutions by gas chromatography. Anal. Chem. 46:140-143.

Di Corcia, A., and R. Samperi. 1974b. Gas chromatographic determination at the parts-per-million level of aliphatic amines in aqueous solution. Anal. Chem. 46:977-981.

Dyer, R. E. 1949. Q fever: History and present status. Am. J. Public Health 39:471-477.

Eller, B. R., and J. K. Matsushima. 1972. Sagebrush in feedlot rations. Colorado State Univ. Exp. Stn. Publ. No. 925. p. 11.

Elliot, L. F., and T. M. McCalla. 1971. Air pollution from agriculture. p. C1-C6. In Proc.: Exploring Nebraska's pollution problems. 22 Apr. 1971. Univ. of Nebraska Ext. Serv., Lincoln, Nebr.

Elliot, L. F., G. E. Schuman, and F. G. Viets, Jr. 1971. Volatilization of nitrogen-containing compounds from beef cattle areas. Soil Sci. Soc. Am. Proc. 35:752-755.

Elliot, L. F., and T. A. Travis. 1973. Detection of carbonyl sulfide and other gases emanating from beef cattle manure. Soil Sci. Soc. Am. Proc. 37:700-702.

Ellis, J. R., L. N. Mielke, and G. E. Schuman. 1975. The nitrogen status beneath beef cattle feedlots in eastern Nebraska. Soil Sci. Soc. Am. Proc. 39:107-111.

Elmund, G. K., S. M. Morrison, and D. W. Grant. 1971. Enzyme facilitated microbial decomposition of cattle feedlot manure. p. 174-175. In Proc. Int. Symp. Livestock Wastes, Ohio State Univ., Columbus, Ohio. Am. Soc. of Agric. Eng., St. Joseph, Mich.

Elmund, G. K., S. M. Morrison, D. W. Grant, and Sr. M. P. Nevins. 1971. Role of excreted chlortetracycline in modifying the decomposition process in feedlot waste. Bull. Environ. Contam. Toxicol. 6:129-132.

Engler, R. M., and W. H. Patrick, Jr. 1973. Sulfate reduction and sulfide oxidation in flooded soil as affected by chemical oxidants. Soil Sci. Soc. Am. Proc. 37:685-688.

Faith, W. L. 1964. Odor control in cattle feedyards. J. Air Pollut. Control Assoc. 94:459-460.

Faller, N. 1970. Effects of atmospheric SO_2 on plants. Sulphur Inst. J. 6:5-7.

Francis, A. J., J. Adamson, J. M. Duxbury, and M. Alexander. 1973. Life detection by gas chromatography—mass spectrometry of microbial metabolites. p. 485-488. In T. Rosswall (ed.) Modern methods in the study of microbial ecology. No. 17. Bull. from the Ecol. Res. Comm. Swed. Nat. Res. Counc. N.F.R. Publ., Sweden.

Goodrich, P. R., S. L. Spier, S. L. Diesch, and L. A. Will. 1974. Microbial aerosol monitoring of a beef housing oxidation ditch. Am. Soc. Agric. Eng. SP-0174. p. 182-188.

Gordon, R. J., and R. J. Bryan. 1973. Ammonium nitrate in airborne particles in Los Angeles. Environ. Sci. Technol. 7:645-647.

Gottlieb, S., and M. J. Pelczar, Jr. 1951. Microbial aspects of lignin degradation. Bacteriol. Rev. 15:55-102.

Guadnagni, D. G., R. G. Battery, S. Okano, and H. K. Burrn. 1963. Additive effect of sub-threshold concentrations of some organic compounds associated with food aromas. Nature 200:1288.

Hammond, W. D., D. L. Day, and E. L. Hansen. 1968. Can lime and chlorine suppress odors in liquid by manure? Agric. Eng. 49:340-343.

Han, Y. W., and C. D. Callihan. 1974. Cellulase fermentation: Effect of substrate pretreatment on microbial growth. Appl. Microbiol. 27:159-165.

Hartung, L. D., E. G. Hammond, and J. R. Miner. 1971. Identification of carbonyl compounds in a swine-building atmosphere. p. 105-106. In Proc. Int. Symp. Livestock Wastes. Ohio State Univ., Columbus, Ohio. Am. Soc. of Agric. Eng., St. Joseph, Mich.

Hashimoto, A. G., and D. C. Ludington. 1971. Ammonia desorption from concentrated chicken manure slurries. p. 117-121. In Proc. Int. Symp. Livestock Wastes. Ohio State Univ., Columbus, Ohio. Am. Soc. of Agric. Eng., St. Joseph, Mich.

Hattingh, W. H. J., J. P. Kotzé, P. G. Theil, D. F. Toerien, and M. L. Siebert. 1967. Biological changes during the adaptation of an aerobic digester to a synthetic substrate. Water Res. 1:255-277.

Heichel, G. H. 1973. Removal of carbon monoxide by field and forest soils. J. Environ. Qual. 2:419-423.

Herr, G. H. 1970. Under-cage manure drying system solves odor problem. Poult. Dig. 29:476–469.

Hoeft, R. G., D. R. Keeney, and L. M. Walsh. 1972. Nitrogen and sulfur in precipitation and sulfur dioxide in the atmosphere in Wisconsin. J. Environ. Qual. 1:203–208.

Holdemand, L. V., and W. E. C. Moore, ed. 1973. Anaerobe Laboratory manual. 2nd ed. VPI Anaerobe Lab., Virginia Polytechnic Inst. and State Univ., Blacksburg, Va.

Hrutfiord, B. F., and J. L. McCarthy. 1967. SEKORI: Volatile organic compounds in kraft pulp mill effluent streams. Tappi 50:82–85.

Huisingh, J., J. J. McNeill, and G. Matrone. 1974. Sulfate reduction by a *Desulfovibrio species* isolated from sheep rumen. Appl. Microbiol. 28:489–497.

Hungate, R. E. 1950. The anaerobic mesophilic cellulolytic bacteria. Bacteriol. Rev. 14: 1–50.

Hungate, R. E. 1966. The rumen and its microbes. Academic Press, New York. 490 p.

Hutchinson, G. L., R. J. Millington, and D. B. Peters. 1972. Atmospheric ammonia: Absorption by plant leaves. Science 175:771–772.

Hutchinson, G. L., and F. G. Viets, Jr. 1969. Nitrogen enrichment of surface water by absorption of ammonia volatilized from cattle feedlots. Science 166:514–515.

Ingram, S. H., R. C. Albin, C. D. Jones, A. M. Lennon, L. F. Tribble, L. B. Porter, and C. T. Gaskins. 1973. Swine fecal odor as affected by feed additives. J. Anim. Sci. 36:207.

Inman, R. E., R. B. Ingersoll, and E. A. Levy. 1971. Soil: A natural sink for carbon monoxide. Science 172:1229–1231.

Johnson, A. E., and H. E. Nursten. 1971. Methods of assessing the odors of glues (and gelatine). J. Sci. Food Agric. 22:149–155.

Jones, D. D., D. L. Day, and A. C. Dale. 1970. Aerobic treatment of livestock wastes. Bull. 737. Univ. Ill. Agric. Exp. Stn. 155 p.

Jones, H. R. 1973. Waste disposal control in the fruit and vegetable industry. Pollut. Technol. Rev. No. 1. Noyes Data Corporation, Park Ridge, N. J. 261 p.

Jones, H. R. 1974. Pollution control in meat, poultry and seafood processing. Pollut. Technol. Rev. No. 6. Noyes Data Corporation, Park Ridge, N. J. 263 p.

Jurasek, L., J. R. Colvin, and D. R. Witaker. 1967. Microbiological aspects of the formation and degradation of cellulosic fibers. Adv. Appl. Microbiol. 9:131–170.

Kellogg, W., R. D. Cadle, E. R. Allen, A. L. Lazrus, and E. A. Martell. 1972. The sulfur cycle. Science 175:587–596.

Kibbel, W. H., Jr., C. W. Raleigh, and J. A. Shephard. 1972. Hydrogen peroxide for pollution control. Proc. 27th Purdue Ind. Waste Conf. Purdue Univ., West Lafayette, Ind. 24 May 1972.

King, E. D., R. A. Mill, and C. H. Lawrence. 1973. Airborne bacteria from an activated sludge plant. J. Environ. Health 36:50–54.

Klugelman, I. J., and K. K. Chin. 1971. Toxicity, synergism, and antagonism in anaerobic waste treatment processes. Adv. Chem. Ser. 105:55–90.

Kodata, H., and Y. Ishida. 1972. Production of volatile sulfur compounds by microorganisms. Annu. Rev. Microbiol. 26:127–138.

Kondo, M. 1922. Uber die bitdung des mercaptans aus l-cystine dirch bacterien. Biochem. Z. 136:198–202.

Koon, J., J. R. Howes, W. Grub, and C. A. Rollo. 1963. Poultry dust: Origin and composition. Agric. Eng. 44:608–609.

Kotzé, J. P., P. G. Thiel, D. F. Toerien, W. H. J. Hattingh, and M. L. Siebert. 1968. A biological and chemical study of several anaerobic digesters. Water Res. 2:195–213.

Leonardos, G. 1974. A critical review of regulations for the control of odors. J. Air Pollut. Control. Assoc. 24:456–468.

Leonardos, G., D. Kendall, and N. Barnard. 1969. Odor threshold determinations of 53 odorant chemicals. J. Air Pollut. Control Assoc. 19:91–100.

Lillie, R. J. 1969. Air pollutants affecting the performance of domestic animals. Agric. Handb. 380, U. S. Dep. of Agric. p. 1–7.

Loehr, R. C. 1974. Agricultural waste management. Academic Press, New York and London. 576 p.

Ludington, D. C. 1971. Odors and their control. p. 103–136. *In* D. C. Ludington (ed.) Agricultural wastes: Principles and guidelines for practical solution. Cornell Univ. Conf. on Agric. Waste Management, Ithaca, N. Y.

Ludington, D. C., A. T. Sobel, and B. Gormel. 1971. Control of odors through manure management. Trans. ASAE 14:771-780.

Luebs, R. E., K. R. Davis, and A. E. Laag. 1974. Diurnal fluctuation and movement of atmospheric ammonia related gases from dairies. J. Environ. Qual. 3:265-269.

McAllister, J. S. V., and J. B. McQuitty. 1965. Release of gases from slurry. Record of Agric. Res. (Min. of Agric., N. Ireland) Vol. XIV, Part 2, p. 73.

McBee, R. H. 1950. The anaerobic thermophilic cellulolytic bacteria. Bacteriol. Rev. 14:51-64.

McBride, B. C., and R. S. Wolfe. 1971. Biochemistry of methane formation. Adv. Chem. Ser. 105:11-22.

McCalla, T. M., L. R. Frederick, and G. L. Palmer. 1970. Manure decomposition and fate of breakdown products in soil. p. 241-255. In T. L. Willrich and G. E. Smith (ed.) Agricultural practices and water quality. Iowa State Univ. Press, Ames, Iowa.

McCalla, T. M., and F. A. Haskins. 1964. Phytotoxic substances in soil microorganisms and crop residues. Bacteriol. Rev. 28:181-207.

McCarty, P. L., J. S. Jeris, and W. Murdoch. 1963. Individual volatile acids in anaerobic treatment. J. Water Pollut. Control Fed. 35:1501-1516.

McFadden, W. H. 1973. Techniques of combined gas chromatography/mass spectrometry: Applications in organic analysis. John Wiley and Sons. New York. 459 p.

Malo, B. A., and E. R. Purvis. 1964. Soil absorption of atmospheric ammonia. Soil Sci. 97:242-247.

Mandels, M., L. Hontz, and J. Nystrom. 1974. Enzymatic hydrolysis of waste cellulose. Biotechnol. Bioeng. 16:1471-1493.

Merkel, J. A., T. E. Hazen, and J. R. Miner. 1969. Identification of gases in a confinement swine building atmosphere. Trans. ASAE 12:310-315.

Merz, R. C. 1959. Waste water reclamation for golf course irrigation. J. Sanit. Eng. Div. Am. Soc. Civ. Eng. 85:79-85.

Mielke, L. N., N. P. Swanson, and T. M. McCalla. 1974. Soil profile conditions of cattle feedlots. J. Environ. Qual. 3:14-17.

Miner. J. R. 1970. Raising livestock in the urban fringe. Agric. Eng. 13:702-703.

Miner, J. R. 1973. Odors from livestock production. Agric. Eng. Dep., Oregon State Univ. Corvallis, Oreg. 127 p.

Miner, J. R., and T. E. Hazen. 1969. Ammonia and amines: Components of the swine building odor. Trans. ASAE 12:772-774.

Moncrieff, R. W. 1966. Odor preferences. John Wiley, New York. p. 89.

Moncrieff, R. W. 1967. The chemical senses. 3rd ed. Leonard Hill, London. p. 44.

Moorman, R., Jr. 1965. Controlling odors from cattle feed lots and manure dehydration operations. J. Air Pollut. Control Assoc. 15:34-35.

Morrison, S. M., D. W. Grant, Sr. M. P. Nevins, and K. Elmund. 1969. Role of excreted antibiotic in modifying microbial decomposition of feedlot waste. p. 336-339. In R. C. Loehr (ed.) Animal waste management. Cornell Univ. Conf. on Agric. Waste Manage., Ithaca, N. Y.

Mosier, A. R., C. E. Andre, and F. G. Viets, Jr. 1973. Identification of aliphatic amines volatilized from cattle feedyard. Environ. Sci. Technol. 7:642-644.

Mosier, A. R., K. Haider, and F. E. Clark. 1972. Water soluble organics substances leachable from feedlot manure. J. Environ. Qual. 1:320-323.

Moum, S. G., W. Seltzer, and T. M. Godlhaft. 1969. A simple method of determining concentrations of ammonia in animal quarters. Poult. Sci. 46:347-348.

Muehling, A. J. 1970. Gases and odors from stored swine wastes. J. Anim. Sci. 30:526-531.

Napolitano, P. J., and D. R. Rowe. 1966. Microbial content of air near sewage treatment plants. Water & Sewage Works 113:480-483.

Natusch, D. F. S., J. R. Sewell, and R. L. Tanner. 1974. Determination of hydrogen sulfide in air—an assessment of impregnated paper tape methods. Anal. Chem. 46:410-415.

Nisman, B. 1954. The Stickland reaction. Bacteriol. Rev. 18:16-42.

Norkrans, B. 1967. Cellulose and cellulolysis. Adv. Appl. Microbiol. 9:91-130.

Ogelsby, R. T., R. F. Christman, and C. H. Driver. 1967. The biotransformations of lignin to humus—facts and postulates. Adv. Appl. Microbiol. 9:171-184.

Okita, T., J. P. Lodge, Jr., and H. D. Axelrod. 1971. Filter method for the measurement of atmospheric hydrogen sulfide. Environ. Sci. Technol. 5:532–534.

Osag, T. R., and G. B. Crane. 1974. Control of odors from inedibles—rendering plants. U. S. Environ. Prot. Agency. EPA-450-/1-74-006.

Ottenstein, D. M., and D. A. Bartley. 1971. Separation of free acids C_2-C_5 in dilute aqueous solution column technology. J. Chromatogr. Sci. 9:673–681.

Paigen, K., and B. Williams. 1970. Catabolite repression and other control mechanisms in carbohydrate utilization. p. 252–324. In A. H. Rose and J. F. Wilkinson (ed.) Advances in microbial physiology. Vol. 4. Academic Press, London.

Peck, H. D., Jr. 1962. Comparative metabolism of inorganic sulfur compounds in microorganisms. Bacteriol. Rev. 26:67–94.

Petri, H. 1961. The effect of hydrogen sulfide and carbon disulfide. Staub 21:64.

Poon, C. P. E. 1968. Viability of long-stored airborne bacterial aerosols. J. Sanit. Eng. Div. Am. Soc. Civ. Eng. 94(SA 6):1137–1146.

Porter, L. K., F. G. Viets, Jr., and G. L. Hutchinson. 1972. Air containing nitrogen-15 ammonia: Foliar absorption by corn seedings. Science 175:759–761.

Postgate, J. R. 1965. Recent advances in the study of the sulfate reducing bacteria. Bacteriol. Rev. 29:425–441.

Rains, B. A., M. J. DePrimo, and I. L. Groseclose. 1973. Odors emitted from raw and digested sewage sludge. EPA-670/2-73-098. Grant No. WPRD 23-01-68.

Rhodes, R. A., and Hrubant. 1972. Microbial population of feedlot waste and associated sites. Appl. Microbiol. 24:369–377.

Ronkainen, P., J. Denslow, and O. Leppänen. 1973. The chromatographic analysis of some volatile sulfur compounds. J. Chromatogr. 11:384–390.

Rouf, M. A., and R. F. Lomprey, Jr. 1968. Degradation of uric acid by certain aerobic bacteria. J. Bacteriol. 96:617–622.

Schultz, T. H., R. A. Flath, and T. R. Mon. 1971. Analysis of orange volatile with vapor sampling. J. Agric. Food Chem. 19:1060–1065.

Seabrook, B. L. 1973. Irrigating with liquid digested sludge. Compost Sci. 14:26–27.

Seltzer, W., S. G. Moum, and T. M. Goldhaft. 1969. A method for the treatment of animal wastes to control ammonia and other odors. Poult. Sci. 48:1912–1918.

Shuyler, L. R., D. M. Farner, R. D. Kresi, and M. E. Hula. 1973. Environment protecting concepts of beef cattle feedlot wastes management. Project No. 21AOY-05, Program Element 1B2039, U. S. Environ. Prot. Agency, Corvallis, Oreg.

Spendlove, J. C. 1973. Industrial agricultural and municipal microbial aerosol problems. Dev. Ind. Microbiol. 15:20–27.

Stahl, W. H., ed. 1973. Compilation of odor and taste threshold values data. ASTM Data Series DS48. Am. Soc. Test. Mater., Philadelphia, Pa. 249 p.

Stevenson, F. J., and G. H. Wagner. 1970. Chemistry of nitrogen in soils. p. 125–151. In T. L. Willard and G. E. Smith (ed.) Agricultural practices and water quality. Iowa State Univ. Press, Ames, Iowa.

Stephenson, M. 1966. Bacterial metabolism. 3rd ed. MIT Press, Cambridge, Mass. 398 p.

Stewart, B. A. 1970. Volatilization and nitrification of nitrogen from urine under simulated cattle feedlot conditions. Environ. Sci. Technol. 4:579–582.

Stombaugh, D. P., H. S. Teague, and W. L. Roller. 1969. Effects of atmospheric ammonia on the pig. J. Anim. Sci. 28:844–847.

Sullivan, R. J. 1969. Air pollution aspects of odorous compounds. Prepared for the Nat. Air Pollut. Control Admin. Consum. Prot. and Environ. Health Serv. Dep. of Health, Educ., and Welfare (Contract No. PH-22-68-25).

Taiganides, E. P., and R. K. White. 1969. The menace of noxious gases in animal units. Trans. ASAE 12:359–367.

Takai, Y., and T. Kamura. 1966. The mechanism of reduction in waterlogged paddy soil. Folia Microbiol. 11:304–313.

Thayer, D. W., Sr., P. Lewter, J. Barker, and J. J. J. Chen. 1974. Microbiological and chemical survey of beef cattle waste from a nonsurfaced feedlot. Bull. Environ. Contam. Toxicol. 11:26–32.

Thorne, M. D., T. D. Hinesly, and R. L. Jones. 1975. Utilization of sewage sludge on Agricultural land. Univ. of Ill. Dep. of Agron. Fact Sheet SM-29. p. 6.

Timmer, R., R. ter Heide, P. J. deValois, and H. J. Wobben. 1971. Qualitative analysis of the most volatile neutral components of reunion granium oil (Pelagonium roseum Bourbon). J. Agric. Food Chem. 19:1066–1068.

Toerien, D. F., and W. H. J. Hattingh. 1969. Anaerobic digestion: I. The microbiology of anaerobic digestion. Water Res. 3:385–416.

Trudinger, P. A. 1969. Assimilatory and dissimilatory metabolism of inorganic sulfur compounds by micro-organisms. p. 111–158. In A. H. Rose and J. F. Wilkinson (ed.) Advances in microbial physiology. Vol. 3. Academic Press, Inc., London.

Turner, N. C., S. Rich, and P. E. Waggoner. 1973. Removal of ozone by soil. J. Environ. Qual. 3:259–264.

Umbreit, G. R., R. E. Nygren, and A. J. Testa. 1969. Determination of traces of amine salts in water by gas chromatography. J. Chromatogr. 43:25–32.

Viets, F. G. 1974. Fate of nitrogen under intensive animal feeding. Fed. Proc. 33:1178–1182.

Waksman, S. A. 1927. Principles of soil microbiology. The Williams and Wilkins, Co., Baltimore, Md. 897 p.

Waksman, S. A. 1941. Anatagonistic relations of microorganisms. Bacteriol. Rev. 5:231–291.

Waksman, S. A. 1952. Soil microbiology. John Wiley and Sons, Inc., New York. 356 p.

Waksman, S. A., and R. L. Starkey. 1931. The soil and the microbe. John Wiley and Sons, Inc., New York.

Wegener, W. S., H. C. Reeves, R. Rabin, and J. Ajl. 1968. Alternate pathways of metabolism of short chain fatty acids. Bacteriol. Rev. 32:1–26.

Weiner, B. A., and R. A. Rhodes. 1974. Fermentation of feedlot waste filtrate by fungi and streptomycetes. Appl. Microbiol. 28:845–850.

Wells, D. M., R. C. Albin, W. Grub, and R. F. Wheaton. 1969. Aerobic decomposition of solid wastes from cattle feedlots. p. 58–62. In R. C. Loehr (ed.) Animal waste management. Cornell Univ. Conf. on Agric. Waste Manage., Ithaca, N. Y.

Whisler, F. D., J. C. Lance, and R. S. Linebarger. 1974. Redox potential in soil columns intermittently flooded with sewage water. J. Environ. Qual. 3:68–74.

White, R. K., E. P. Taiganides, and G. D. Cole. 1971. Chromatographic identification of malodors from dairy animal waste. p. 110–113. Proc. Int. Symp. Livestock Wastes, Ohio State Univ., Columbus, Ohio.

Will, L. A., S. L. Diesch, B. S. Pomeroy, S. L. Spier, and P. R. Goodrich. 1974. Aerosol dispersion of pathogens from a model oxidation ditch. Am. Soc. Agric. Eng. SP-0174. p. 176–181.

Williams, I. H. 1965. Gas chromatographic techniques for the identification of low concentrations of atmospheric pollutants. Anal. Chem. 37:1723–1732.

Willrich, T. L., and J. R. Miner. 1971. Litigation experiences of five livestock and poultry producers. p. 99–101. In Proc. Int. Symp. Livestock Wastes, Ohio State Univ., Columbus, Ohio.

Wilmore, R. 1972. Manure deodorants . . . how well do they work? Farm J. 92:22, 38.

Wolfe, R. S. 1971. Microbial formation of methane. p. 107–146. In A. H. Rose and J. F. Wilkinson (ed.) Advances in microbial physiology. Vol. 6. Academic Press, Inc., London.

Young, J. 1972. Dust and odor problems of the feedlot. p. 81–87. In Control of agriculture-related pollution in the Great Plains. Seminar, 24–25 July 1972, Lincoln, Nebr. Great Plains Agric. Counc. Publ. No. 69.

Zeikus, J. G. 1974. Biology of methanogenic bacteria. Am. Soc. Microbiol. News 40:847–849.

Zlatkis, A., H. A. Lichtenstein, A. Tishbee, W. Bertsch, F. Shunbo, and H. M. Liebich. 1973. Concentration and analysis of volatile urinary metabolites. J. Chromatogr. Sci. 11:299–302.

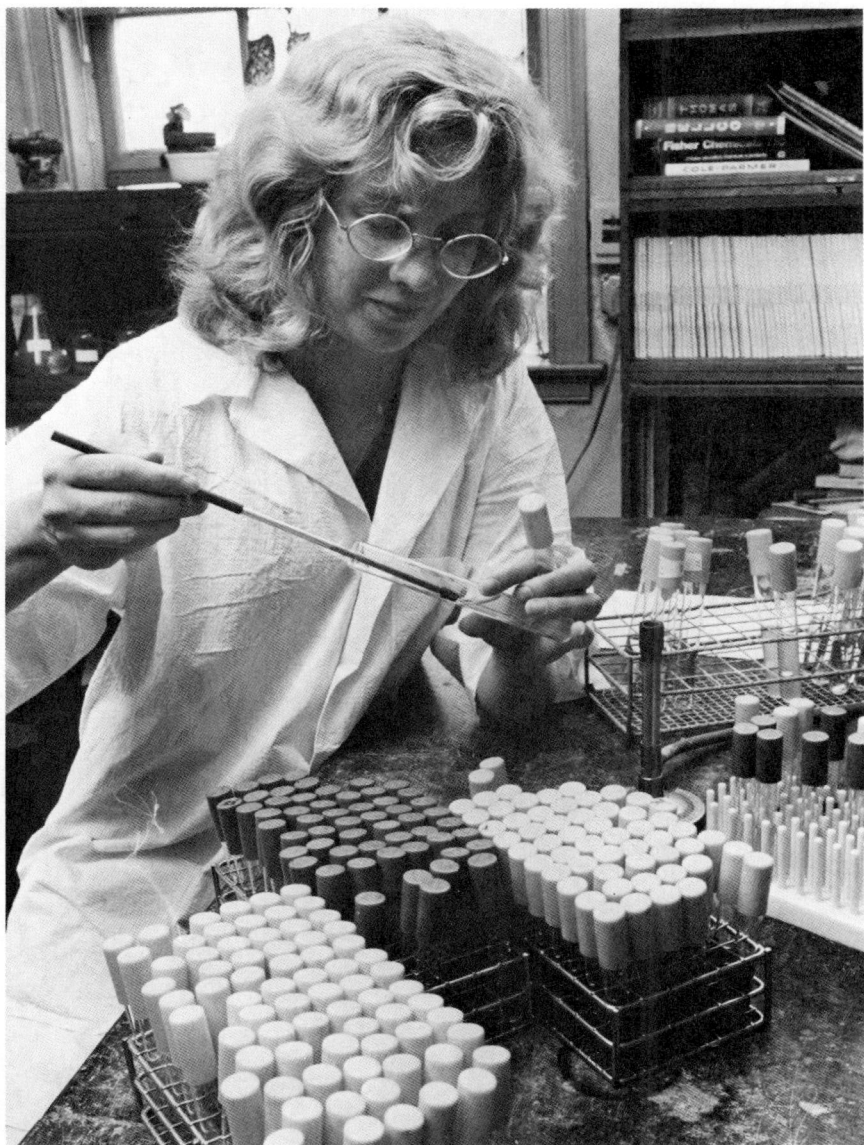

chapter 22

Microbiologist analyzes compost samples for fecal coliform and salmonella to insure that composting had destroyed such pathogens (Photo courtesy of USDA).

Pathogen Considerations for Land Application of Human and Domestic Animal Wastes

J. D. MENZIES, USDA-ARS, Beltsville, Maryland

I. INTRODUCTION

Municipal sewage wastes and farm animal manures, do contain disease agents. The numbers of these pathogens vary widely depending on the health of the populations producing the wastes and how these wastes are subsequently handled or treated. Whether this pathogen contamination is a significant hazard to human and domestic animal health is important in evaluating systems for use of these wastes on agricultural land. This is not easy to determine, because even if assay methods were simple and accurate (which they are not) the detection of pathogens alone does not evaluate risk. One has to consider the probabilities of the pathogens moving from these wastes to susceptible hosts and this must be judged against the inevitable background level of pathogens in the environment. This paper attempts to assess the relative risks based on the knowledge of life cycles and survival capabilities of different types of pathogens, based on historic epidemiological experience, and compared with our options for disposing of these wastes in other ways.

The return of organic wastes to the land is part of the natural cycle. In "uncivilized" nature, the soil decomposes these residues for recycling and decontaminates them of any disease organisms that they may contain. One does not need to be a dedicated organic farmer to recognize the fundamental importance of both these functions. No system of waste management that either avoids or abuses this cycle can be considered permanent.

Sustained productivity of our land requires the highest level of organic waste recycling that we can devise. In our modern agriculture, we have broken this cycle by transporting the organic yields of our farms to remote centers of processing or consumption without establishing a network that returns the residues to the land. This system works for awhile because we use fossil fuel and unrenewable chemical deposits to substitute chemical fertilizers for the lost nutrients. No one supposes that this can go on forever, especially with the escalating costs in energy and the degradation of environmental quality resulting from some of the popular methods for destructive disposal of the wastes.

This point is stressed to emphasize that we cannot consider a solution to the pathogen hazards of waste recycling without recognizing the essential need for recycling. To eliminate the pathogen hazard by destroying the wastes is not an acceptable solution. The problem is to develop land application systems for the waste that strike an acceptable balance between the benefits derived from using the wastes and the potential risks to human and animal health.

II. THE PATHOGENS

Organic wastes from humans or domestic animals inevitably contain pathogens. Each disease agent has its own life cycle that usually includes a saprophytic or nonhost period. It is in this nonhost stage that transmission from one host to another occurs. It is also in this stage that destruction of pathogens is greatest. Survival away from the host is so rare that millions of pathogen propagules usually must be produced for each successful infection. The ability of pathogens to survive in the outside environment, especially in soil, air, and water, varies greatly among kinds and species. Fortunately for the health of man and animals, only a few of the hundreds of disease agents have high enough survival rates in soil and water that they are part of the waste recycling problem.

Many studies and reviews have been published listing the human pathogens that can occur in sewage wastes. Burge reviewed the published literature in 1973 (1974a, 1974b). Other reviews have been made by Foster and Engelbrecht (1973), Grabow (1968), Dunlop (1968), and Sepp (1971). These reviewers list many other reports of detection of various pathogens in sewage and sludges. It appears that only a few pathogens are important enough to require special attention.

Among the intestinal parasites, *Ascaris lumbricoides,* a round worm or nematode, is most frequently mentioned as a potential problem in human health (Dunlop, 1968; Sepp, 1971). The eggs or ova of this pest are excreted in the feces of infested individuals and have been shown to survive to some degree after sewage treatment, including anaerobic digestion. These ova are quite resistant to destruction in the soil and may persist there for several years. The ova must be ingested in order to parasitize.

Another intestinal parasite, the protozoan *Entamoeba histolytica*, which causes amoebic dysentery, produces cysts that are also voided in the feces of infected individuals. These cysts, like the ova of *Ascaris,* are resistant propagules that presumably can survive sewage treatment and could contaminate soil where sewage is spread. The published data, however, indicate that they usually survive for only a few days in soil (Beaver & Deschamps, 1949).

The bacterial diseases of man that are likely to be considered possible hazards in use of sewage wastes are those that are ordinarily spread by water. In other words, infectious bacterial diseases such as those that require special insect or animal vectors, or direct person-to-person contact and do not infect through water, are not likely to survive and infect via waste on land. The water-borne bacterial diseases concerning public health include the enteric fevers typhoid and paratyphoid, cholera, dysentery, gastroenteritis, and leptospiral jaundice (Holden, 1970).

Typhoid and paratyphoid fevers are caused by *Salmonella typhi* and *S. paratyphi*, respectively. These fevers usually result from drinking contaminated water. The causal bacteria, however, may contaminate food directly from an infected or carrier individual, or indirectly through contamination from irrigation water on fresh vegetables.

Gastroenteritis, or food poisoning, is caused by a number of species of *Salmonella,* notably *S. typhimurium.* Ingestion of contaminated food on which this organism has multiplied is the common source of this ailment. In water, relatively massive doses of bacilli will be required to produce symptoms, far larger than those required to produce enteric fever or cholera

(Holden, 1970). Bacillary dysentery, caused by *Shigella* sp., is a water borne intestinal disease with characteristics similar to salmonellosis.

Another class of bacterial diseases generally found only in injured or weak individuals, include enteric infections with strains of *Escherichia coli, Pseudomonas aeruginosa,* and *Klebsiella.* These pathogens can occur in human wastes and are at least potential hazards in sewage, especially waste waters.

Strains of these organisms are widespread and their presence in soil or water in low numbers need not be related to sewage pollution. For example, Green et al. (1974), recently showed that *P. aeruginosa* could be detected in 24% of soil samples taken from vegetable growing areas in California where sewage was not involved. Under very favorable conditions of humidity and temperature this organism can colonize on the surface of leafy vegetables. They concluded that such contamination is usually not sufficient to induce illness in healthy adults but may do so if the raw vegetables are fed to debilitated persons as, for example, in hospitals. Numerous other food-borne disease agents probably fit this pattern.

Some of the famous bacterial diseases of past generations are relatively minor problems in waste management in the United States today because they have been almost eliminated, or because the population is being routinely immunized. These major diseases of the past include cholera, plague, tetanus, and anthrax.

Among the human virus diseases, the enteric group is the most important in waste management. Enteric viruses include the causal agents of polio and viral hepatitis. The polio virus is readily detected in sewage and has been shown to be present in even digested sludge and chlorinated effluent water (Clarke & Kabler, 1964). Although the polio virus is liberated via the intestine, there is no evidence that water or soil pollution is a vector. It is generally believed that polio is spread among children by direct fecal contamination (Holden, 1970). In any case, the almost complete elimination of polio in the United States by vaccination reduces to nonsignificance any hazard of this disease being spread by land application of treated sewage wastes.

The infectious hepatitis agent is presumed to be virus-like, but has not yet been isolated in the laboratory and no assay methods have been developed to determine its populations in waste. According to Grabow (1968), it is very difficult to demonstrate directly whether the hepatitis virus is present even in water. We have to depend upon the indirect evidence of fecal coliforms to indicate water pollution, or upon actual disease outbreak.

Numerous other viruses such as echo virus, coxsackie virus, and reo virus cause short-lived summer disease with symptoms of fever, headache, vomiting, and diarrhea. Grabow (1968) in his review cites numerous instances where these viruses have also been detected in sewage after conventional treatment.

The viruses, unlike some bacteria (*Salmonella* and *Shigella*) can cause infection at very low titer. Berg (1971) believes that the smallest amount of virus that can now be detected in water is sufficient to cause infection in man. Their demonstrated ability to survive sewage treatment and their low inoculum requirements make the viruses at least theoretically important in the health aspects of waste disposal.

Animal manures or barn wastes are recognized as sources of animal disease agents, and livestock management practices have been developed to handle that problem. Recycling of animal manures directly in pasturing or by spreading manure from barns and feedlots on the land is an historically accepted practice. As with human sewage and human pathogens, the number of disease agents affecting animals that specifically become important problems in waste disposal on land is small. In this discussion, mention will be made only of animal pathogens that are considered to be a hazard to animals on pasture treated with sludges or manures.

Several species of *Salmonella* either from manure or sewage can constitute health hazards to animals. There are reported cases of salmonellosis, especially in calves, from grazing on freshly sludged pastures. *S. dublin* and *S. typhimurium* are common contaminants of animal manure and can cause disease in both man and animals (Jensen & Mackey, 1971). Accidental massive pollution of pasture from raw sewage was blamed for an outbreak of *S. aberdeen* in dairy cattle (Bicknell, 1972).

Manure on pasture, and possibly sludge on pasture, can contaminate the forage with cysts and eggs of roundworms or tapeworms. The transmission of such intestinal parasites, especially from adults to young animals, is a widely recognized hazard when the young are allowed to pasture with the parents.

In the case of serious quarantinable illnesses of livestock, such as foot-and-mouth disease, exotic Newcastle disease, and anthrax, the barn wastes as well as infected animals are usually restricted from land spreading, treated chemically, deep buried, or incinerated (Jensen & Mackey, 1971).

III. ROUTE TO TARGET HOSTS

It was mentioned earlier that one of the natural functions performed by soil was to decontaminate waste material. It would follow that recycling wastes to the land would be a sanitizing practice, rather than a hazard. The concern, however, is that pathogens may escape this destructive action of the soil or survive it long enough to complete the cycle back to man.

One of the most frequently cited possibilities of human inoculation from wastes is by way of aerosols or sprays produced when liquid wastes are spread on land by sprinkling. These aerosols may carry bacteria or viruses for a few hundred meters from a disposal site, but there is no claim that airborne pathogens become widely dispersed like smog or radioactive fallout.

Pathogens also can be recycled to hosts in the process of recycling waste to land by direct contamination of food crops. The best known example is the use of raw sewage in truck crop production as is practiced in some parts of the world. Even if the pathogens enter the soil, there is some danger of contamination—especially on root crops or raw vegetables exposed to water splashed from the soil surface. Sprinkler irrigation of sewage effluent or sludges, liquid manure, or feedlot lagoon effluent is a process for directly contaminating food and feed crops. This route back to the host is the most likely to succeed; but, on the other hand, it is the easiest to block as will be described below.

A third avenue of pathogen movement to hosts is the water supply. When pathogen-containing wastes are allowed to pollute streams, lakes, or ponds used as domestic water sources or for contact recreation, the pathway to man is direct. Even when wastes are spread on land away from water courses, however, surface runoff may carry pathogens into waterways before the destructive forces in the soil can kill them. The likelihood of pathogens, even viruses, moving down through the soil to underground water supplies is much less; but, it still must be considered a possibility.

Finally, the spreading of wastes on land requires workers to load, haul, spread, and incorporate the waste into the soil. There are many opportunities for direct contamination of these people, especially if they are dealing with spray application or dry wastes that may be dusty. Both dusts and aerosols may be inhaled, and these carriers may reach nearby neighborhoods.

IV. FACTORS AFFECTING SURVIVAL

After waste has been spread on the soil surface or incorporated, numerous soil and environmental processes work to destroy any pathogenic agents that may be present. Viruses do not multiply outside of living hosts; neither do most pathogenic bacteria and parasites. Their survival depends entirely on resisting killing or inactivation. On the soil surface, high temperature, ultraviolet radiation, and desiccation all are lethal to pathogens if the exposure and intensity are sufficient. In the soil, the natural microflora, adapted to metabolize almost any organic substance, will gradually attack the dormant pathogens. There are believed to be many antibiotic, antagonistic, or predatory activities going on in soil against "alien" organisms such as human pathogens. Over a long period of time, these forces, plus the nonbiotic factors of freezing and thawing, flooding and drying, and aeration and anaerobiosis, will further reduce the pathogen population.

Even if the pathogens should be able to persist in the generally hostile environment of the soil, they may be effectively trapped from ever having a chance to reinfect man. Once a disease agent is buried a few inches in the soil, it is protected from aerosol dispersal or surface runoff and can only move very slightly through the fine pores of the soil. If such pathogens survive to be turned up on the soil surface by cultivation, drying and solar radiation will soon destroy them.

V. PATHOGEN POPULATIONS IN WASTE

Many studies and reports show that any of the above-mentioned pathogens can be detected in sewage or waste water (Burge, 1974a, 1974b; Dunlop, 1968; Foster & Engelbrecht, 1973; Sepp, 1971; Wiley, 1972). These data were obtained by painstaking techniques that, in many cases, were quite inefficient. One assumes that there must be a much higher population in the sample than was revealed by the tests. How much of the pathogen population remains undetected, however, is often only a guess. Most microbiologists working with virus assays from waste samples would probably say, that for every virus propagule detected by their tests there may be 10 to 100 not detected.

This difficulty in directly assaying for pathogenic agents has led to the more convenient procedure of testing for pollution indicator organisms. In public health circles, the coliform bacteria, and more recently the fecal coliform bacteria, have been selected as indicators of fecal pollution. Fecal pollution, in turn, presupposes pathogens. The assumption is made that there is a more or less direct correlation between numbers of fecal coliforms and numbers of pathogens. This is probably true for water, but, we really do not know what this relation is with sludges. The fecal coliforms are so numerous in these wastes that basing estimates of pathogen survival on the survival of coliforms may be quite erroneous.

No attempt will be made here to catalogue the available data on pathogen populations in waste. These figures can be found in the reviews listed in the Introduction. Instead, it will be taken as proved that the pathogens may be there and the discussion will turn to the question of hazard. This brings up the epidemiological evidence of disease spread from use of sewage and manure on land.

VI. EPIDEMIOLOGICAL HISTORY

The present-day, almost emotional concern with environmental quality and the protection of human and animal health has resulted in two strongly held points of view. One is that, if an organism or agent known to cause disease or injury is present in some situation or substance in the environment, it must be removed. The other point of view is that mere detection is not sufficient; some statistically reliable showing of injury should be required before banning the use of a waste, a pesticide, or a manufacturing process. This is the so-called requirement to "produce a dead body." These two extreme views are becoming more polarized as arguments involve law suits, injunctions, and regulatory actions.

When one reviews the literature on disease aspects of using municipal wastes on land in the United States, one finds successful demonstration of the presence of pathogens but almost no evidence linking human disease to this presence. Most authors refer to the "hazard" or the "risk," but do not address this question in any epidemiological way.

There is probably better evidence that domestic animal diseases can be spread by recycling contaminated barn manures onto pasture. The grazing animal is consuming food without the benefits of prior harvesting, processing, cleaning, or cooking. However, before condemning the spreading of manure on pasture, one needs to reflect that the act of pasturing itself is inevitably accompanied by defecation by the grazing animal.

A realistic approach to the question of relative risk in this country has to take into account the general health and sanitation practices prevailing. One does not need further documentation to accept the fact that human disease is spread by uncontrolled use of raw sewage on food crops, especially from populations with a high incidence of intestinal parasites and enteric diseases. In such populations, poor sanitation, ignorance, and lack of effective waste handling systems perpetuate the disease problem. This is not generally true in the United States and in the few situations where it may be true there is no excuse for not correcting the problem.

What we are talking about today, however, is the land application of wastes from a generally healthy population, collected in sanitary sewage systems, and variously treated or stabilized before use. According to the scant epidemiological evidence in this country, the segments of our population living near sewage plants or on farms where treated sludges have been applied are just as healthy as the rest of the population. This is also true of sewage plant workers or those engaged in hauling and spreading sewage on land (Glass & Jenkins, 1962; Hanks, 1967).

One frequently hears calls for more epidemiological evidence on this point. Such studies are extremely expensive, difficult to conduct properly, and may take years to complete. Common experience strongly indicates a lack of any significantly greater incidence of disease associated with sewage treatment handling. Therefore, one would be surprised if more detailed studies were to reveal a serious problem. If the general public does not now recognize a problem, it is probably not very great. Under these circumstances, and while further epidemiologic studies are done, it would seem justifiable to continue to use what appears to be safe procedures until evidence is presented that these are inadequate to protect the public health.

VII. PROPOSED REASONABLE SAFEGUARDS

If agricultural interests expect to get general agreement on use of wastes on cropland, they should be willing to adopt some basic guidelines for health protection. There will also be needs for other guidelines based on toxic metals, nitrogen, etc., but to deal with the pathogen question, the following safeguards are proposed.

A. Municipal Wastes

1) Present regulations protecting domestic water supplies should be maintained. Protecting household wells from surface drainage and from polluted ground water is a basic requirement. The best protection should be based on setting safe distances from waste accumulation or disposal areas rather than on any monitoring of pathogen movement into ground water. Safe distances recommended by sanitary engineers and health authorities probably offer more protection than ground-water monitoring. Testing of well waters for evidence of pollution (the coliform tests) of course should be continued. If coliforms reach a water supply, more dangerous viruses probably will also.

2) In using sewage sludges on land, properly operated, anaerobic digestion, secondary activation, lagooning, or liming to pH 11.5 or above should be sufficient treatment for reducing pathogens to a safe level. This proposal is based on lack of evidence that highway transport or field spreading of sludges treated in these ways have been hazardous to health.

3) Untreated raw sludges and raw effluent water should not be used in food crop production until we know more about this particular hazard. There are safe ways for disposing of raw sludge by burial in trenches or landfills, and of raw effluent by surface flooding or soil injection, but these sites should be closed from public access, sufficiently isolated, and not used for crop production. The hazards in using these raw sludges may still be unproven, but agriculturalists would be wise to avoid any possible health question that might adversely affect the freedom to use treated or stabilized sludges.

4) Composted sludges should be considered to be sufficiently free of pathogens to be used for crop and food production. Normal compost temperatures of 50 to 70C over several days exposure are lethal to most bacteria and viruses. There is a statistical chance of a pathogen surviving composting, but populations are reduced so low that it would be difficult to show them to be higher than in the normal environment.

5) Raw sewage effluent should be safe for sprinkling on land where public access is controlled and suitable buffer zones are provided to take care of aerosol drift. Such treated land should not be used for cultivated crops but can be used for grazing.

6) Effluent from properly operated secondary treatment plants or lagoon systems should be considered safe to use on grains, forage, and other animal feed crops. Such effluent would be acceptable also on parks and golf courses when applications are made at night or other times when public access is prevented. Some states may require chlorination or other disinfection in this situation. More research is required here because it may be that chlorination is neither effective nor needed, and in fact may create hazards we do not know about.

7) Sewage effluent, even after treatment or disinfection, should not be

sprinkled on human food crops. There is no need to take the risk of some failure of the treatment process or to offend the aesthetic sensibilities of the consumer.

8) Any surface application of sewage sludges or effluent should be done with reasonable precautions to prevent surface runoff into open water. Such precautions should include a ban on sludge spreading on frozen ground unless runoff is controlled and perhaps a buffer strip of vegetation along streams should be required.

B. Farm Animal Wastes

As noted above, the recycling of animal manures to the land has been an integral part of livestock management since animals were first domesticated. The only new factor in manure management in the United States today is the generation of great quantities of manure at centralized feedlots. The modern feedlot operator probably does a better job of maintaining the health of the animals in his lots than does the small farmer. The pathogen content of the manures is likely less. Stockpiling and storage undoubtedly help to reduce pathogen populations in the manure. The modern farm animal operation in general is likewise geared to disease control. Consequently, the standard practice of spreading manure on land appears to present no serious hazard.

Spray irrigation of liquid manure, sewage sludges, and sewage or lagoon effluents on pastures appears to increase the animal health hazard. But here again, experience indicates that this risk is small. A number of studies have shown that it is a good precautionary practice to restrict animals from sprayed pastures for a week or 10 days of sunny weather after application. This exposure to heat, ultraviolet radiation, and drying is generally lethal to *Salmonella,* fecal coliforms, and probably viruses.

VIII. CONCLUSION

The health questions surrounding use of municipal wastes on land will probably never be answered to everyone's satisfaction. It seems reasonable to admit that pathogens can occur in these wastes, so they should be treated with reasonable care. But we must also accept the lack of any epidemiologic evidence of health damage to our population. Since these wastes are needed on land, we should proceed to use them with restrictions such as just outlined. If these uses of this valuable resource should later be shown to be causing disease problems, health authorities must impose restrictions. If such added restrictions are imposed, health agencies should have to show that such restrictions do not force alternative waste disposal methods that are equally hazardous.

A recent report prepared for the Council for Agricultural Science and Technology (CAST Report No. 41, Feb. 1975, Utilization of animal manures and sewage sludges in food and fiber production, Dept. of Agronomy, Iowa State Univ., Ames, IO 50010) states "utilization of urban and animal wastes is probably impeded to a greater extent by the fear of disease than by the actual disease hazard involved. Information from field tests suggests that the hazards from pathogens are more imaginary than real. Irrigation of soil with liquid digested sludge is accepted in Great Britain, Germany, and France, where more than 100 years of practice in sewage-farm irrigation has produced no epidemics of cattle or animal disease."

Thomas (1974), in his book *The Lives of a Cell*, discusses our modern propensity to overreact to "germs." He says, "Watching television you'd think. . .that we live in a world where the microbes are always trying to get at us, to tear us cell from cell, and we only stay alive through diligence and fear. . . These are paranoid delusions on a societal scale." There is no need to be paranoid about using municipal or animal wastes on land. It is a natural process that, with prudent management, can be compatible with high public health standards.

LITERATURE CITED

Beaver, P. C., and G. Deschamps. 1949. The viability of *Entamoeba histolytica* cysts in soil. Am. J. Trop. Med. 29:189.

Berg, G. 1971. Integrated approach to the problem of viruses in water. J. Sanit. Eng. Div.; Proc. Amer. Soc. Civ. Eng. SA6, 867.

Bicknell, S. R. 1972. *Salmonella aberdeen* infection of cattle associated with human sewage. J. Hyg. 70:121–126.

Burge, W. D. 1974a. Pathogen considerations. p. 37–50. *In* Factors involved in land application of agricultural and municipal wastes. Spec. Pub., USDA-ARS.

Burge, W. D. 1974b. Health aspects of applying sewage wastes to land. Section II. Proc. Conference-Workshop, "Educational Needs Associated with Utilization of Wastewater Treatment Products on Land." Cooperative Extension Service, Michigan State University, East Lansing. Sept. 23–26, 1974.

Clarke, N. A., and P. W. Kabler. 1964. Human enteric viruses in sewage. Health Lab. Sci. 1:44.

Dunlop, S. G. 1968. Survival of pathogens and related disease hazards. p. 107–122. *In* C. W. Wilson and F. E. Beckett (ed.) Municipal sewage effluent for irrigation. Louisiana Tech Alumni Foundations, Ruston, La.

Foster, D. H., and R. S. Engelbrecht. 1973. Microbial hazards in disposing of waste water on soil. p. 247–270. *In* W. E. Sopper and L. T. Kardos (ed.) Recycling treated municipal wastewater and sludge through forest and crop land. Penn. State University Press, University Park, Pa.

Glass, A. C., and K. H. Jenkins. 1962. Statistical summary of 1962 inventory of municipal waste facilities in the United States. Publ. No. 1165. U. S. Public Health Serv.

Grabow, W. O. K. 1968. The virology of wastewater treatment. p. 675–701. *In* Water Research, Vol. 2. Pergamon Press.

Green, Sylvia K., M. N. Schroth, J. J. Cho, S. D. Kominos, and W. D. Vitanza-Jack. 1974. Agricultural plants and soil as a reservoir for *Pseudomonas aeruginosa*. Appl. Microbiol. 28:987–991.

Hanks, T. G. 1967. Solid waste/disease relationships. Publ. 999-UIH-6. U. S. Public Health Serv.

Holden, W. S. 1970. Water treatment and examination. Williams & Wilkins, Baltimore, Md.

Jensen, R., and D. R. Mackey. 1971. Diseases of feedlot cattle. Ed. 2. Lea & Febiger, Philadelphia, Pa.

Sepp, E. 1971. The use of sewage for irrigation. Bur. Sanit. Eng., Calif. Dept. Public Health, Los Angeles. (Publication may be obtained from Water Sanitation Dep., Dep. of Public Health, Berkeley, Ca.)

Thomas, Lewis. 1974. The lives of a cell. Viking Press, N. Y.

Wiley, J. S. 1972. Pathogen survival in composting municipal wastes. J. Water Pollu. Control Fed. 34:80–90.

chapter 23

Problems and Need for High Utilization Rates of Organic Wastes[1]

W. E. LARSON, USDA-ARS, St. Paul, Minnesota

G. E. SCHUMAN, USDA-ARS, Cheyenne, Wyoming

I. INTRODUCTION

Organic wastes and waste waters can sometimes be applied on land at rates exceeding normal crop nutrient requirements if the site is selected and managed properly. Large amounts of farm manures and sewage wastes have been used on mine spoil banks, dredge materials, sand dunes, covered sanitary landfills, former factory waste lagoons, lands marginal for crop production, and lands overlying confined aquifers. Crops have been successfully established on these areas, although adequate monitoring for environmental changes was not always done. Large amounts of wastes also can be used on productive lands.

Safe use of large amounts of wastes per unit area requires special soil and geologic conditions and/or management of the waste. This may not be possible at all locations. It will often require some land modification and a high level of management.

The purpose of this chapter is to describe situations where large amounts of organic wastes and waste waters can be used successfully without environmental deterioration and to discuss some of the requirements for site selection, needed properties of the waste, and management of the area.

II. NEEDS FOR HIGH UTILIZATION OF WASTES

If transportation distance is great, it may be desirable to develop high-rate utilization facilities close to the waste source. Liquid sludges and waste waters usually are transported by truck, ship, rail, or pipeline. Solid wastes are moved by rail, truck, or ship.

The Metropolitan Sanitary District of Greater Chicago barges daily 5,680 m³ of sewage sludge from Chicago to Liverpool, Ill., on the Illinois River. From the dock the sludge is piped 18 km in 50-cm diameter underground pipe to large storage reservoirs at the land site. The reservoirs cover a total of 73 ha and are 9 to 15 m deep. Liquid sludge also is hauled from Chicago by unit trains to a site near Arcola, Ill., a distance of about 225 km (Bauer & Sheaffer, 1974). At Melbourne, Australia, 5,450 m³ of raw sewage are transported daily by gravity in open ditches 37 km to a land utilization site. Sewage sludges are sometimes vacuum filtered to reduce transportation costs. Composts, dried sludges, and manures also are used in the United States.

Riddell and Cormack (1966) concluded that for cities of 10,000 population land utilization of sewage sludge was more economical, if the transport distance was less than 46 km, than was incineration (ash to landfill) or dewatering and transport to a landfill. The economical distance for land utiliza-

[1]Contribution from the Soil and Water Management Research Unit, North Central Region, ARS, USDA, St. Paul, Minn., in cooperation with the Minnesota Agricultural Experiment Station, Paper no. 9326, Sci. J. Ser.; and the Nebraska Agricultural Experiment Station, Paper no. 5022, J. Ser.

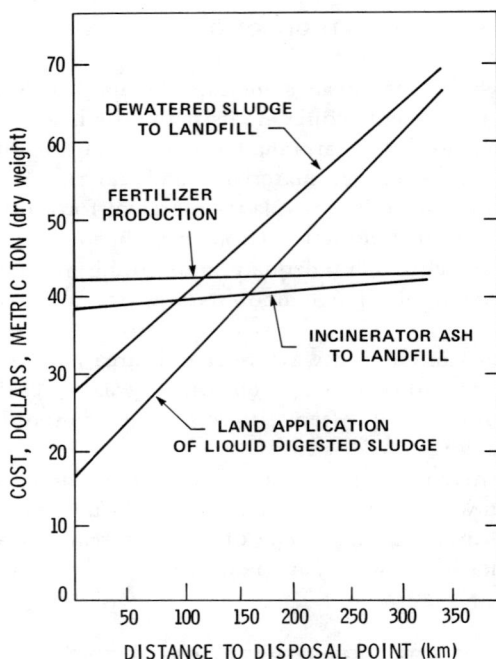

Fig. 1—Cost of disposal of sewage sludge by various methods for city of 1,000,000 (from Riddell and Cormack, 1966).

tion was 170 km for cities of 1 million population (Fig. 1). As shown in Figure 1, land application costs increase from $16 to $46 per dry ton as the transport distance increases from 0 to 200 km. While Riddell and Cormack's study was presented in 1966, the relative costs probably are not much different today.

Tank trucks are more convenient and probably cheaper for transportation of sludge from smaller cities but pipelines are more economical for larger cities with larger volumes. Unit trains also have been used for large volumes.

The cost of land application of solid materials, like garbage, paper products, food processing, and animal wastes is usually directly related to the bulkiness of the material and transportation distance. Transportation costs may be reduced by compacting bulky materials into bales, although the total costs may not be reduced.

Method of field application also becomes a sizeable economic factor for most organic wastes. For most liquid sludges, some type of land surface application with irrigation equipment is cheaper than soil injection. However, for many conditions, soil injection or soil covering is usually more socially acceptable.

With more strict environmental laws, the cost of land preparation will probably increase for many sites and be a significant fraction of the total utilization cost. Lack and cost of suitable land within reasonable transportation distances, high cost of site preparation per unit area, and difficulty in developing socially acceptable field application procedures are factors which make high utilization rates economically desirable.

III. LANDSCAPE AND SOIL SITUATIONS

In using high loading rates of wastes, one must consider the landscape (that portion of the earth's surface that is visible), together with the soil and underlying layers. The landscape can be considered the surface transport system, and the soil can be considered the internal transport system. Their interaction determines the effectiveness of a waste utilization or disposal system.

The soil profile horizons may range in texture from sands to clays. Coarse-textured materials on the soil surface are advantageous in waste utilization systems because they are easier to till and manage than are fine-textured materials.   Coarse-textured soils usually have high saturated hydraulic conductivities and high water intake rates unless a seal develops at the soil surface. Fine-textured soils have low saturated conductivities, small pores that readily clog with organic wastes, and a low volume of air-filled pores. All of these characteristics make them unsuitable for many high-rate waste utilization systems.

Many soils have subsurface horizons that are less permeable then their surface horizons. Examples of horizons with less permeable subsoils than surface horizons are argillic, matric, placic, duripan, and fragipan. Examples of restricting dense materials in the C horizon or below are dense glacial till, shale, siltstone, and residium overlying limestone. Subsurface horizons that restrict water flow can be caused also by vehicular traffic. Where these layers are present, much of the water moves downward to the slowly permeable layer and then may move laterally downslope.

Figure 2 illustrates six types of landscape and soil conditions for consideration in waste utilization. If wastes are applied to Condition I, Open landscape, deep permeable soil, precautions may be necessary to limit surface water runoff as well as movement of contaminated water into deep layers and ground water. On the other hand, deep permeable soils favor rapid infiltration and provide a large amount of soil surface area to interact with contaminants.   A closed landscape with a deep permeable soil eliminates concern about spread of contaminants outside the watershed by runoff and will favor deep percolation.

An open landscape with permeable surface and slowly permeable subsurface has the added hazard to those for Condition I that runoff is likely to be increased and movement through the profile will be decreased. Lateral flow above the constricting layer may be increased. Because of the constricting layer, downward movement may be intercepted with tile drains.

In a closed landscape with a slowly permeable soil, runoff is collected at the low point in the watershed and lateral flow above a constricting layer

I   OPEN – DEEP PERMEABLE SOIL   II   CLOSED – DEEP PERMEABLE SOIL

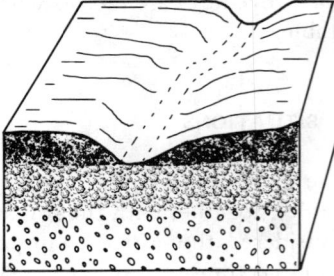

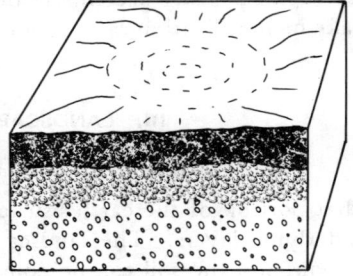

III   OPEN – PERMEABLE SURFACE –
      SLOWLY PERMEABLE SUBSURFACE   IV   CLOSED – SLOWLY PERMEABLE SOIL

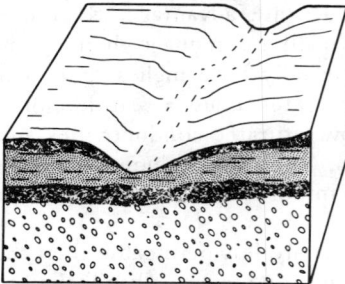

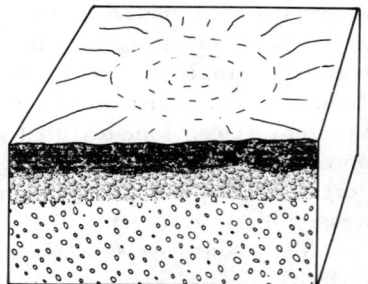

V   OPEN – SLOWLY PERMEABLE   VI   CLOSED – SLOWLY PERMEABLE
    SURFACE AND SUBSURFACE   SURFACE AND SUBSURFACE

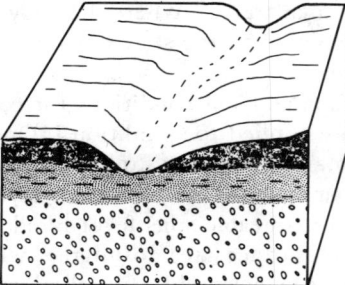

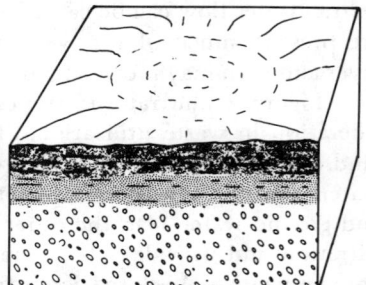

Fig. 2—Hydrologic conditions for consideration in waste utilization.

may also go toward the low point. This may be an advantage in that water can be intercepted and collected for further renovation.

Usually Conditions V and VI are less favorable for high rates of waste utilization than Conditions I through IV; but, this may not always be true, depending on the particular waste and management scheme. Overland flow techniques are well adapted to Conditions V and VI.

The six landscape and soil conditions can be engineered to create many soil and landscape configurations favorable to high rates of waste utilization. The open landscapes (Conditions I, III, and V) can be converted into closed landscapes by use of water detention structures. Structures such as agricultural terraces may be used in any of the landscape conditions to slow water runoff and increase water movement into the soil. Tile-outlet terraces offer the possibilities both for increasing water movement through the surface soil and intercepting the flow with tile drains (Phillips, 1969). If a water table is present or if tile drainage systems can be used, water may be intercepted and used for other purposes.

A number of examples can be given to illustrate innovative engineering designs that may be applicable to high-rate waste systems. At Treynor, Iowa, water behavior was compared in terraced and unterraced watersheds. The soils were developed from deep permeable loess overlying glacial till (Typic Hapludolls, Typic Udorthents, and Cumulic Hapludolls). Slopes on the watershed ranged from 2 to 4% on the ridges and in the valleys and from 12 to 18% on the hillsides. Both the terraced and unterraced watersheds were cropped to corn. Water flow was measured as surface runoff and as subsurface flow that seeped into streams at the base of the watershed. The data in Table 1 illustrate that terracing did not greatly affect the total flow (subsurface and surface) out of the watershed. However, 78% of the total flow was subsurface on the terraced watershed compared with 65% subsurface flow on the unterraced watershed (R. E. Burwell, soil scientist, North Central Watershed Research Center, 207 Business Loop 70 East, Columbia, MO 65201. Personal communication).

Hanway and Laflen (1974) found that tile-outlet terraces at four locations in Iowa on well-fertilized soil for corn production reduced soil erosion and loss of nutrients that are sorbed on the soil particles, but did not reduce the concentrations of soluble nutrients in the runoff. The total soluble nutrients may have been reduced, however, because of greater infiltration into the soil. Average annual inorganic N concentrations in the runoff were 4 ppm or less from three of the four sites and 11 ppm from one site. Annual average inorganic P concentrations in surface runoff ranged from 0.013 to 2.204 ppm and in tile drainage from 0.004 to 0.018 ppm.

In many high waste utilization systems, it may be desirable to have a closed landscape, with areas within the watershed for initial and final treatment. Such a study is now underway at Rosemount, Minn. (Larson et al., 1973). In this study, a Condition III landscape was converted into a closed

Table 1—Annual stream flow discharges of water from agricultural watersheds cropped to corn (*Zea mays* L.) Treynor, Iowa, 1964–1973 (from Burwell et al., 1975).

Treatment	Precipitation, cm/yr	Total stream flow, cm/yr	% of total flow	
			Subsurface flow	Surface flow
Unterraced	84.4	16.7	65	35
Terraced	84.6	19.4	78	22

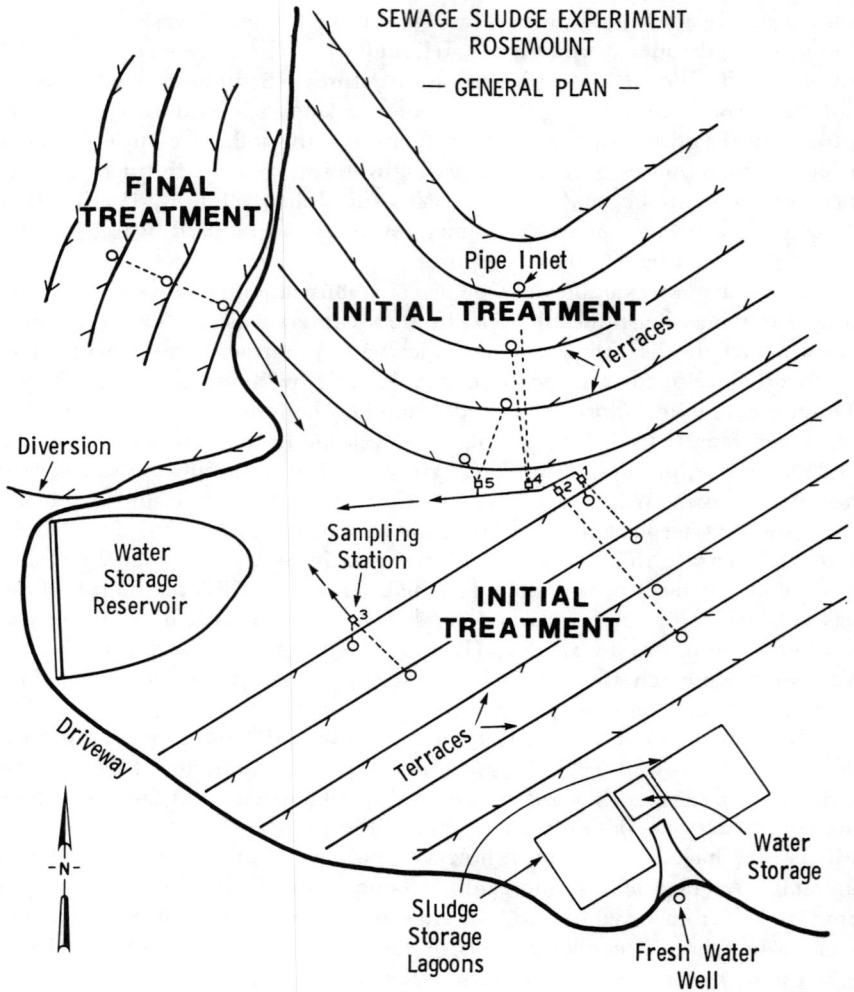

Fig. 3—Runoff control design for a watershed receiving sewage sludge at Rosemount, Minn. (from Larson et al., 1973).

landscape by construction of an earthen dam at the watershed base (Fig. 3). The level detention terraces constructed have surface pipe inlets to slow run-off during intense rains and to enhance infiltration. Sewage sludge is applied to much of the terraced area. Water runoff is directed to the storage reservoir through the pipe system with inlets in the terrace channels. Water moving laterally along the slowly permeable loess-glacial till interface can seep into the waterway channel and is directed into the reservoir. Final treatment is achieved by using water in the reservoir for irrigating a desig-nated area of the watershed.

It is often desirable to reclaim large amounts of waste waters on land

and to intercept and use the water for irrigation or industrial uses. In such cases reclamation regulations may be less stringent than if the water is returned to natural streams.

The Flushing Meadows project at Tempe, Ariz., consists of six parallel recharge basins, 7.1 by 213 m each, and 7.1 m apart, that receive sewage effluent. Constant water depth (usually 20 cm) is maintained during inundation. The soil profile consists of about 1 m of fine, loamy sand (hydraulic conductivity of 5.4 m/day) underlain by an irregular succession of sand and gravel layers to a depth of 75 m where a clay layer begins. The ground water table is at a depth of about 5 m. Observation wells were drilled at various sites within and around the project area. This system has been found effective in renovating sewage effluent (Bouwer, 1973).

As a followup to the Flushing Meadows project, the city of Phoenix, Ariz., is developing a project whereby four 4-ha rapid infiltration basins will be flooded with sewage effluent. The water will be reclaimed by a network of pumps within the basins. Water pumped from the aquifer will then be placed in irrigation ditches and purchased by an irrigation company.

Wherever aquifers are confined or where facilities can be constructed for minimizing movement of the reclaimed water outside the recharge system, great potential exists for use of organic waste waters or wastes. Bouwer (1970) discussed possible designs with a pumping network for limiting spread of the reclaimed water outside the recharge system.

Sometimes it may be practical to construct confined aquifers. An impermeable layer of clay, asphalt, or other material may be laid down and covered with permeable sands or gravels. The partially reclaimed waters can then be recovered by pumping and used for irrigation, industrial cooling, or other uses. This technique may be particularly economical if the aquifer is constructed at the same time as the industrial plant.

Overland flow techniques with short flow distances also may be feasible if waste waters need only partial reclamation. For example, water may be distributed over the surface of slowly permeable soils and the runoff caught in a basin at the base of the watershed for agricultural or industrial use. Overland flow techniques are more efficient in removing N than in removing P from sewage waste waters (Carlson et al., 1974).

In some situations, overland flow (initial treatment) can be combined with crop infiltration (final treatment) for reclaiming waste waters or combinations of waste waters and solid organic wastes.

IV. HYDRAULIC LOADING

Hydraulic properties of soil are important considerations for land disposal of wastes because they influence the amount of water runoff and the amount leached through the soil profile. In addition, water retention characteristics of the soil will influence aeration status of the soil and thus may affect microbial and plant growth.

For land application of wastes, a moderately high but not excessive soil hydraulic conductivity usually is desired. Excessively rapid transport through the profile may not allow sufficient time for the organic materials to biodegrade, for nutrients to be absorbed by plants, or for ion exchange or other chemical reactions to take place between waste components and the soil.

When liquid wastes are applied on the soil, the surface of the soil often "seals" and has much lower hydraulic conductivity than underlying layers. The seals may develop because of soil pore clogging by organic materials or because of soil particle sorting into a thin, dense layer by action of falling water drops. In either case, infiltration rate may be reduced materially and the soil may have to be tilled or thoroughly dried to increase the infiltration rate (Bouwer, 1973).

The effect of surface seal on infiltration rate into an Ida silt loam (Typic Udorthents) is shown in Fig. 4. Curve A is the computed infiltration rate assuming that soil conditions were uniform throughout the wetted layer. Curve B shows the rate assuming that a seal forms during rainfall and that the decreasing saturated hydraulic conductivity reaches a value of 0.20 cm/ hour at the end of 2 hours. Curve C shows the rate assuming that at the end of 2 hours saturated conductivity of the seal is 0.02 cm/hour.

Rate of infiltration into a soil sealed at the surface depends not only on hydraulic conductivity of the seal, but also on the hydraulic conductivity-matric potential-water content relations in the underlying soil. If hydraulic conductivity of the underlying soil drops rapidly as suction increases from zero, the potential gradient through the seal will be relatively large and in-

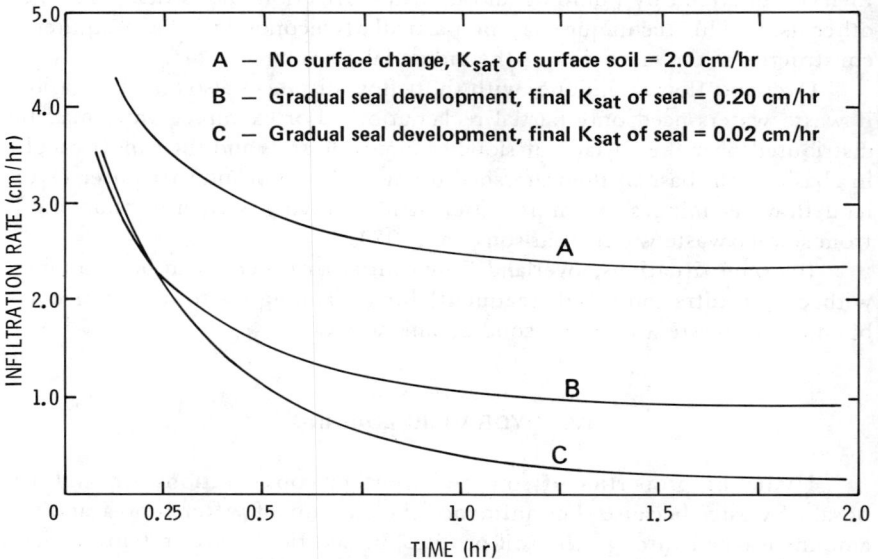

Fig. 4—Estimated infiltration rate for Ida silt loam as influenced by simulated surface conditions (adapted from Edwards and Larson, 1969).

filtration will not be decreased as much as if the hydraulic conductivity decreases less rapidly with increasing suction (Edwards & Larson, 1969).

The shape of the volumetric water content-suction curve in the underlying soil is important because it bears on the amount of air space present. Under many high water loading conditions where the soil surface seals, suctions of 25 to 100 mbars may occur during or soon after irrigation. If at these suctions the air space equals 10 to 12% of the total volume, oxygen exchange between the soil and atmosphere will probably be sufficient for maximum plant growth.

V. DESIRABLE MIXTURES OF ORGANIC WASTES

In many wastes major plant nutrients are not balanced for plant growth. However, different wastes can sometimes be mixed to make a more suitable material for application at high rates to the soil. Sewage sludges and farm manures are often high in N, with C/N ratios of 10 or less being common. When incorporated in the soil at high rates they may release more N than a crop can use, and the excess may lead into the ground water. On the other hand, garbage, paper products, and sawdusts often have low N contents, with C/N ratios of over 100 being common.

Carbon/nitrogen ratios of organic material can be used as a general guide for predicting whether the soil-waste system initially will result in net immobilization or mineralization of inorganic N. Wastes with C/N ratios of less than about 20 to 25 usually result in net mineralization while those with C/N ratios above 20 to 25 result in net immobilization (Harmsen and Kolenbrander, 1965). The C/N ratio of 20 to 25 applies only if all of the C and N are readily available for decomposition. In some wastes appreciable quantities of C and N are quite resistant to oxidation. Approximate C/N ratios for various wastes and plant materials are given in Chapter 2.

Rothwell and Hortenstine (1969) found that, when chicken manure (2.64% N), sewage sludge (5.46% N), and cow manure (1.32% N) were added to an Arrendondo fine sand (Grossarenic Paleudalfs), NO_3-N within the soil decreased upon incubation. However, when equal quantities of cow manure and garbage ($\approx$ 1.20% N) were added, the NO_3-N level remained constant. Equal quantities of garbage and sewage sludge ($\approx$ 3.28% N) and of garbage and chicken manure ($\approx$ 1.96% N) increased the NO_3-N content of the soil during a 64-day incubation period. Perhaps if the sludge and chicken manure had been added at rates of one part to two or three parts of garbage, the mineralized N would have remained nearly constant throughout the incubation. Where the land is cropped, some N mineralization is desired of course.

Sewage sludges may contain large amounts of potentially toxic elements. In high-rate waste utilization systems where different kinds of waste are available, it may be desirable to mix different kinds of wastes to give a desirable chemical composition.

VI. COMPOSTING

Composting sewage sludge prior to land application may be a good way to utilize municipal and animal wastes in or near metropolitan areas (Poincelot, 1972). Large amounts of sludge can be stored in composts with minimum undesirable effects on the environment. During composting some organic matter and N are reduced and (some authors say) pathogens are eliminated. If sewage and other municipal wastes can be composted so that objectionable bacteria and viruses are eliminated and so that the chemical composition is satisfactory, large amounts of the compost might be used for fertilizer on recreational and greenbelt areas within or near metropolitan areas.

Composting of municipal wastes is practiced more commonly in western Europe than in the United States, with many operations using a combination of city garbage and sewage sludge (both slurry and dewatered). Some composts use only garbage (Hart, 1968).

The composted material in Europe is commonly used in horticultural crop operations and on hillside vineyards where erosion control is necessary. Composting is most widely practiced in The Netherlands; 17% of the nation's refuse is made into compost. The sales distribution of the compost in 1965 is given in Table 2 (Hart, 1968).

Composting systems generally fall into three categories: (i) windrow, (ii) pile, and (iii) mechanized or enclosed systems. Most composting systems include separation of ferrous metal and large nonbiodegradable materials, grinding, and storage that favors elevated temperatures for pathogen die-back. The windrow system, which uses filtered sewage sludge, was studied extensively at Beltsville, Md., by Epstein and Wilson (1974). In the present Beltsville system, wood chips are used as a bulking agent during digestion and the chips are later recovered by screening. The project covers 32 ha, of which 5 ha are used for the composting operation. The remaining area is for isolation, screening, and drainage water disposal. Projected capacity of the site is from 18 to 27 metric tons per day (dry). The only environmental problem experienced has been odor emissions during one period which was believed to have been due to overloading, climatic conditions, use of raw rather than

Table 2—Sales distribution of compost in 1965 from municipal refuse in The Netherlands (from Hart, 1968)

Outlet	% of total
Forestland improvement	0.6
Agriculture (field and row crops)	16.4
Fruit farming	6.3
Hotbed vegetable farming	13.0
Greenhouse vegetable farming	8.4
Flower and flower bulb production	17.6
City park, sportfield, and recreational use	37.7

digested sludge, and possibly to chemicals added in the waste water treatment plant.

Composting has been practiced at Largo and Gainesville, Fla., by grinding municipal refuse, adding water or sewage sludge to about 50% water content, and allowing the mixture to digest anaerobically for about 6 days in large bins. After bin digestion, the compost was stockpiled in the open for several weeks where further degradation occurred (Hortenstein and Rothwell, 1969). When the compost was added to Leon fine sand (Aeric Haplaquods) in the greenhouse, turnip, oats, radish, and pearl millet foliage yields were increased at compost addition levels above 32 metric tons/ha. Applications up to 512 metric tons/ha did not reduce yields, although radishes showed phytotoxic symptoms at the 512- metric tons/ha rate. Seed germination was reduced initially, perhaps by excessive soluble salts.

VII. TRENCH INCORPORATION

Walker (1974) has studied trench incorporation of organic waste as a feasible method for simultaneously disposing of large amounts of sewage sludge and improving marginal agricultural land. It seems particularly well suited to dewatered (20% solids) raw-limed sludge and undoubtedly could be used for other municipal organic wastes or mixtures of wastes that are not high in N.

In Walker's (1974) experiments sewage sludge (5 and 20% solids) was placed in 60-cm-wide trenches of different depths and spacings. Dewatered sludge application rates were 800 and 1,200 ton/ha dry solids, respectively, in trenches 60 cm wide by 60 cm deep by 60 cm apart and 60 cm wide by 120 cm deep by 120 cm apart.

Nineteen months after sludge entrenchment, fecal coliform and *Salmonella* were not found in soil more than a few centimeters from the entrenched sludge. No downward movement of heavy metals was detected and heavy metal uptake by crops was moderate. Downward movement of NO_3-N through the soil was found.

VIII. CROPS

Crops species vary in their uptake of toxic metal elements from wastes. Because toxic elements usually accumulate in the leaves and vegetative tissues more than in grains, crops whose vegetative tissues are eaten by humans may not be suitable for production on land receiving wastes containing large amounts of heavy metals.

If high rates of wastes with significant quantities of toxic elements are applied, perhaps crops whose products are nonedible should be grown. Some examples are forests, short rotation forests, kenaf, guayule, jute, castorgeans, ornamentals, flowers, and sod. Crops whose principal products are processed and contain insignificant toxic elements are cotton, sugar, seed, and oil crops. It may be possible to grow these crops in soils containing large amounts of toxic elements if the secondary products are not used for livestock feed. Corn, sorghum, soybeans, and cereals may be suitable crops for production on land that has received waste containing moderate amounts of metals.

IX. HIGH APPLICATION RATES OF WASTES

In this section, we will review briefly examples of high utilization rates of organic wastes in the field. The review will be divided into (A) solid wastes, including liquid sewage sludges, and (B) effluents.

A. Solid Wastes

A number of examples can be given of use of large amounts of sewage sludge to develop otherwise unproductive land in urban centers. Sand dunes in San Francisco were developed into Golden Gate State Park by long-term irrigation with raw sewage followed by fertilization with dried digested sludge (Dalton et al., 1968). The New York City Park Department filled a marshland with garbage and covered it with sand followed by sewage sludge. A grass cover was then established after mixing the sand and sludge by disking (Burd, 1966). A sand dune at Northwestern University in Evanston, Ill., was stabilized against wind erosion by applying sludge solids and planting the area to grass (Hinesly & Sosewitz, 1969). A 55-ha silty clay loam basin filled with assorted wastes was covered with liquid sewage sludge in Chicago. Part of the area received 161 metric tons/ha of sludge and was later seeded to wheat, which yielded 2,860 kg grain/ha (Peterson et al., 1971). An alkali-silica sand waste lagoon at Ottawa, Ill., was successfully vegetated with grass by addition of 280 metric tons/ha of sewage sludge solids. Before sludge application the lagoon had been filled with about 18 m of fine silica material ($< 10~\mu$) and finely ground glass, plus other waste products from a soda-glass factory. The area was successfully vegetated and erosion was stopped. After vegetation was established, the area was fenced and is used as a wildlife sanctuary (Peterson et al., 1971). Other examples of use of sludge for reclaiming urban areas at San Diego, Calif., Las Vegas, Nev., and Miami, Fla., are given by Burd (1966).

Large amounts of sewage sludge per unit area have been applied to unproductive agricultural soils and large increases in plant growth and yield have resulted. Applications of up to 424 ton/ha per year of a dried sewage sludge to an infertile Hubbard sandy loam (Udorthentic Haploborrolls) dramatically increased growth of vegetable crops (Dowdy & Larson, 1975). This relatively low-metal-containing sludge caused no toxicity symptoms, although the metal content of the plants increased slightly. Possible movement of elements out of the root zone was not assessed.

In Illinois, corn grain yields were increased by application of liquid sewage sludge as compared with a fertilized check on Blount silt loam (Aeric Ochraqualfs). Up to 264 tons/ha (dry solids) was applied over a 4-year period by furrow irrigation during the corn growing season. The corn plants showed no toxicity symptoms, even though the sludge contained rather large amounts of heavy metals and the soil's pH was about 5.0. Metal concentration in the plant leaves did increase (Hinesly et al., 1972).

King et al. (1974) applied pulverized municipal refuse at 188 tons/ha singly or in combination with 2.3 cm (5.1 dry tons of organic matter) of liquid sewage sludge. Yields of rye forage planted the same year as waste application and corn grain planted the year after application were not different from yields on the control treatment. Levels of Zn, Cu, Cd, and Pb were slightly higher in the corn and rye grown on the waste-treated area, and soil NO_3-N levels were increased in the root zone.

Much research has been initiated recently in the United States to determine the effects of high rates of animal wastes applied to land on crop growth and soil and water changes. Up to 2,240 metric tons/ha have been applied. Generally, when manure is applied at rates exceeding 135 tons/ha, soluble salts and ammonium toxicity interfere with germination, plant growth, and water uptake.

Mathers and Stewart (1971) applied beef cattle manure at rates of 0, 22, 67, 134, 269, and 538 tons/ha per year for 2 years to grain sorghum. The 538 tons/ha rate was also applied a single time. Annual manure applications up to 134 tons/ha did not reduce yields, but applications of 269 and 538 tons/ha decreased yields in both years. The 538 tons/ha single manure application decreased yields in the application year, but had much less effect the second year. The electrical conductivity of the saturated soil paste increased to 11.7 mmhos/cm on this treatment after 1 year and then decreased in subsequent years. Mathers and Stewart's (1971) data showed that when N in manure was added in excess of crop needs, NO_3-N accumulated and moved downward in the soil profile. Later studies by Mathers et al. (1975) showed that alfalfa could be grown on the areas where high amounts of manure had been applied, and that the NO_3-N accumulated could be removed effectively by the alfalfa at the 1.8- and 3.6-m depths at the 45-ton/ha rate and less. Alfalfa did not effectively remove the NO_3-N at the 3.6- to 6.0-m depths from plots treated at 112 and 224 tons/ha.

Mathers and Stewart (1974) found that a manure application of 22 tons/ha was optimum for the production of corn silage. Higher manure rates caused NO_3-N and salt accumulation in the soil and increased the NO_3-N content of the forage.

Vitosh et al. (1973) applied cattle manure at rates of 22.4, 44.8, and 67.2 tons/ha per year for 9 years to plots on which corn was grown. The 44.8- and 67.2-metric ton/ha per year rates caused a buildup of exchangeable K and resulted in inefficient use of nutrients; hence, increasing the potential for leaching of mobile ions. Cross et al. (1973) in Nebraska disked high rates of manure (0, 90, 269, 582 tons/ha) into the soil as a means of utilization and/or disposal. The 90- and 269-ton/ha manure rate increased the yield of forage sorghum, whereas the 582-ton/ha rate significantly decreased crop yield. Murphy et al. (1972) in Kansas applied manure at rates up to 720 tons/ha (dry weight) on plots planted to corn silage. Silage yields were decreased at all levels above 225 tons/ha.

Reddell et al. (1971) in Texas applied manure at rates of 672, 1,344, and 2,016 tons/ha (wet weight) and incorporated it into the soil with a 46-cm moldboard plow, 76-cm moldboard plow, 127-cm disk, and a 68-cm trencher. Power requirements for these large implements were high. Incorporation with the 46-cm plow and 127-cm disk was not satisfactory, but the 76-cm plow and 68-cm trencher worked well on the high manure rate. The trenching technique shows promise as a means of disposal of large amounts of manure.

B. Effluents

Research results have shown that feedlot effluent can be used to increase production. Generally, no problems result if effluent is applied to meet crop moisture and nutrient requirements.

Sukovaty et al. (1974) applied up to 55 cm of livestock effluent from a runoff-control retention pond to forage sorghum by furrow irrigation and found only minor effects on the sorghum yield. In some instances, the sorghum's NO_3-N content was too high for the sorghum to be used as the sole feed source. They concluded that effluent would not create problems in crop production if used at the level of crop nutrient requirements. Wittmuss and Ellis (1974) concluded that 51 cm of feedlot effluent applied with sprinkler irrigation was optimum for corn.

Lehman et al. (1970) used a natural playa lake bed in Texas as a disposal and/or storage site for feedlot effluent. They found that NO_3-N accumulating in the surface of the playa was minimal and probably was denitrified when the playa was wet and anaerobic conditions prevailed.

Research concerning use of sewage effluents in crop irrigation (Kardos & Sopper, 1973; Tomlinson, 1974) and rapid infiltration (Bouwer, 1973) has recently been reviewed and will not be covered here.

X. SUMMARY

High-rate systems for use of wastes can be developed if the site design is properly matched with the characteristics of the waste. With a proper match-up, plant growth can be enhanced, soils improved, and quality water can be reclaimed without environmental degradation.

LITERATURE CITED

Bauer, W. J., and J. R. Sheaffer. 1974. Water and crops monitoring following applications of sludge. Am. Soc. Agric. Eng. paper no. 74-2542, Dec. 1974.

Bouwer, Herman. 1970. Ground water recharge design for renovating waste water. J. Sanit. Eng. Div., Amer. Soc. Civ. Eng. 96:59–74.

Bouwer, Herman. 1973. Renovating secondary effluent by groundwater recharge with infiltration basins. p. 164–175. *In* William E. Sopper and Louis T. Kardos (ed.) Recycling treated municipal wastewater and sludge through forest and cropland. Pennsylvania State Univ., University Park, Pa.

Burd, R. S. 1966. A study of sludge handling and disposal. Pub. WP-20-4, FWPCA, USDI, Cincinnati, Ohio.

Carlson, Charles A., Patrick G. Hunt, and Thomas B. Delaney, Jr. 1974. Overland flow treatment of wastewater. U. S. Army Corps of Eng., Misc. Paper Y-74-3, Vicksburg, Miss.

Cross, O. E., A. P. Mazurak, and L. Chesnin. 1973. Animal waste utilization for pollution abatement. Amer. Soc. Agric. Eng., Trans. 16(1):160–163.

Dalton, F. E., J. E. Stein, and B. T. Lyman. 1968. Land reclamation as a complete solution of the solid disposal problem. J. Water Pollut. Control Fed. 40:789–804.

Dowdy, R. H., and W. E. Larson. 1975. The availability of sludge-borne metals to various vegetable crops. J. Environ. Qual. 4(2):278–282.

Edwards, W. M., and W. E. Larson. 1969. Infiltration of water into soils as influenced by surface seal development. Amer. Soc. Agric. Eng., Trans. 12(4):463–465, 470.

Epstein, E., and G. B. Wilson. 1974. Composting sewage sludge. Natl. Conference on Municipal Sludge Management, Proc., June 11–13, 1974. Pittsburgh, Pa.

Hanway, J. J., and John M. Laflen. 1974. Plant nutrient losses from tile-outlet terraces. J. Environ. Qual. 3(4):351–356.

Harmsen, G. W., and G. J. Kolenbrander. 1965. Soil inorganic nitrogen. *In* W. V. Bartholomew and F. E. Clark (ed.) Soil nitrogen. Agronomy 10:285–306. Amer. Soc. of Agron., Madison, Wis.

Hart, Samuel A. 1968. Solid waste management/composting European activity and American potential. U. S. Dep. HEW, Public Health Serv., Consumer Protection and Environ. Health Serv., Environ. Control Admin., Solid Wastes Program, Cincinnati, Ohio, Report SW-2c.

Hinesly, T. D., Robert L. Jones, and E. L. Ziegler. 1972. Effects on corn by applications of heated anaerobically digested sludge. Compost Sci. 13(4):26–30.

Hinesly, T. D., and B. Sosewitz. 1969. Digested sludge disposal on crop land. J. Water Pollut. Control Fed. 41:822–830.

Hortenstine, C. C., and D. F. Rothwell. 1969. Evaluation of composted municipal refuse as a plant nutrient source and soil amendment on Leon fine sand. Soil Crop Sci., Fla., Proc. 29:312–319.

Kardos, Louis T., and William E. Sopper. 1973. Renovation of municipal wastewater through land disposal by spray irrigation. p. 148–163. *In* William E. Sopper and Louis T. Kardos (ed.) Recycling treated municipal wastewater and sludge through forest and cropland. Pennsylvania State Univ., University Park, Pa.

King, Larry D., L. A. Rudgers, and L. R. Webber. 1974. Application of municipal refuse and liquid sewage sludge to agricultural land: I. Field study. J. Environ. Qual. 3: 361–366.

Larson, W. E., C. E. Clapp, and R. H. Dowdy. 1973. Research efforts and needs in using sewage wastes on land. p. 142–147. *In* Proc. 28th Annual Meeting, Soil Conserv. Soc. Am., Ankeny, Iowa.

Lehman, O. R., B. A. Stewart, and A. C. Mathers. 1970. Seepage of feedyard runoff water impounded in playas. Pub. No. MP-944. Texas A&M Univ., College Station, Tx.

Mathers, A. C., and B. A. Stewart. 1971. Crop production and soil analysis as affected by applications of cattle feedlot waste. p. 229–231. *In* Livestock waste and pollution abatement. Am. Soc. Agric. Eng., St. Joseph, Mo.

Mathers, A. C., and B. A. Stewart. 1974. Corn silage yield and soil chemical properties as affected by cattle feedlot manure. J. Environ. Qual. 3(2):143–147.

Mathers, A. C., B. A. Stewart, and Betty Blair. 1975. Nitrate-nitrogen removal from soil profiles by alfalfa. J. Environ. Qual. 4(3):403–405.

Murphy, L. S., G. W. Wallingford, W. L. Powers, and H. L. Manges. 1972. Effects of solid beef feedlot wastes on soil conditions and plant growth. p. 449–464. In Proc., Cornell Agric. Waste Management Conference, Syracuse, N. Y.

Peterson, J. R., T. M. McCalla, and George E. Smith. 1971. Human and animal wastes as fertilizers. p. 557–596. In R. A. Olson et al. (ed.) Fertilizer technology and use. 2nd ed. Soil Sci. Soc. Am., Inc., Madison, Wis.

Phillips, R. L. 1969. Tile-outlet terraces: History and development. Am. Soc. Agric. Eng., Trans. 12:517–518.

Poincelot, Raymond P. 1972. The biochemistry and methodology of composting. Conn. Agric. Exp. Stn. Bull. 727.

Reddell, D. L., W. H. Johnson, P. J. Lyerly, and P. Hobgood. 1971. Disposal of beef manure by deep plowing. p. 235–238. In Livestock waste and pollution abatement. Am. Soc. Agric. Eng., St. Joseph, Mo.

Riddell, M. D. R., and J. W. Cormack. 1966. Ultimate disposal of sludge in inland areas. 39th Annual Meeting of Central States Water Pollut. Control Assoc., Eau Claire, Wis.

Rothwell, D. F., and C. C. Hortenstine. 1969. Composted municipal refuse: Its effects on carbon dioxide, nitrate, fungi, and bacteria in Arrendondo fine sand. Agron. J. 61:837–840.

Sukovaty, J. E., L. F. Elliott, and N. P. Swanson. 1974. Some effects of beef-feedlot effluent applied to forage sorghum grown on a Colo silty clay loam soil. J. Environ. Qual. 3(4):381–388.

Tomlinson, J. (ed.). 1974. Land for waste management. p. 388. Proc. Int. Conference on Land for Waste Management, Ottawa, Canada.

Vitosh, M. L., J. F. Davis, and B. D. Knezek. 1973. Long-term effects of manure, fertilizer, and plow depth on chemical properties of soils and nutrient movement in a monoculture corn system. J. Environ. Qual. 2(2):296–299.

Walker, John M. 1974. Trench incorporation of sewage sludge. Proc. of Natl. Conference on Municipal Sludge Management. June 11–13, 1974. Pittsburgh, Pa.

Wittmuss, H. D., and J. R. Ellis. 1974. Sprinkler application of liquid wastes from holding ponds. p. 110–116. In Proc., Nebraska Irrigation Short Course, Lincoln, Neb.

chapter 24

Motor vehicles crowd into New York City, pouring exhaust fumes into the atmosphere and clouding the skyline (Photo courtesy of the U. S. Public Health Service).

Soil Treatment of Organic Waste Gases

H. L. BOHN, University of Arizona, Tucson, Arizona

I. INTRODUCTION

Inasmuch as soil is a major sink for air pollutants, several investigators have proposed using soils as filters for removal of pollutants from waste gases before, rather than after, release to the atmosphere (Dupont, 1964; Carlson & Leiser, 1966; Bohn & Miyamoto, 1973). Man uses soils almost instinctively to cover odor sources and thus prevent diffusion to the atmosphere. He also has used soil directly to avoid breathing hazardous gases. Soldiers in World War I were advised that if their gas mask sorbent became exhausted, they should fill the canister with soil; another suggested technique was to lie face down, cup one's hand around the nose and mouth, and breathe through the soil.

Smell is often the first sense to be assailed by waste disposal sites and probably the most difficult to ignore. Mosier et al. in Chapter 21 of this book list most of the organic substances which affect olfactory organs at low concentrations. Because these gases must be reduced to extremely low levels to be inoffensive or nonhazardous, their treatment by conventional techniques is difficult and expensive (Turk et al., 1972; Beathea et al., 1973).

Soil is a sorbent and reaction medium porous to both liquids and gases. Soil properties important in treating waste waters are also important in treating waste gases. The soil's volumetric air permeability is about 100-fold faster than its water permeability. Water is roughly 1,000-fold denser than air so the soil's air permeability on a mass basis is only one-tenth that of its water permeability. The mass of waste gas from pollution sources, however, is generally much smaller than that of waste water. Many of the offensive components of waste waters and waste gases are similar chemically and are equally good substrates for microbial decomposition.

To my knowledge, the only treatment of air pollutant gases by forced (mass or convective) flow through soil is in Geneva, Switzerland (Dupont, 1964) and Seattle, Wash. (Carlson & Leiser, 1966). The Geneva filter is simply a perforated concrete pipe laid 3 m below the surface. Above the pipe is a layer of impermeable soil to increase horizontal spreading of the gas. The noisome air from a compost plant is removed by a fan and forced out of the perforated pipe into the soil. This simple filter, without breakdown and virtually without maintenance, removes all sensible odors from the air stream. The soil's capacity for retaining organic gases is continually self-regenerative; apparently soil microorganisms oxidize the gases completely.

Any decision to use soils to treat waste gases must consider reaction rates, cost, ultimate disposal of the waste, and degree to which the offensive component must be removed. Direct substitution of soil for synthetic solid sorbents in conventional industrial air filters is usually unsatisfactory. Soil reaction rates are slower than those of synthetic sorbents, and soils are less permeable to air. Soil conditioners might sufficiently improve air permeability in some cases but likely would not improve reaction rates.

Synthetic sorbents only remove and concentrate undesirable gases from the waste stream. The sorbed gases must be removed and disposed of in a second step. This gas disposal and sorbent regeneration requires energy and time. Furthermore, the sorbent's performance often diminishes with each regeneration since the sorbent breaks down chemically and physically.

Soil, on the other hand, sorbs and spontaneously reacts with pollutant gases to form generally innocuous products which constitutes complete disposal of the pollutants (Bohn, 1972). This generalization holds quite well except for some pesticides whose intermediate reaction products in soils are as toxic as the parent compounds (Alexander, 1969) and require much longer times for inactivation and disposal. These reactions simultaneously restore the soil's capacity for more pollution treatment. The spontaneous disposal and regeneration reactions after sorption set soils apart from synthetic sorbents and offset the soil's lower sorption rate, capacity, and air permeability.

The availability and economy of soil can best be utilized in most cases by transporting waste gases to the field rather than vice versa. Relatively small areas are necessary for waste gas treatment. The area required will depend upon air flow, gas permeability, and method of contact with the soil. Since successful air pollution control often increases the value of surrounding land, there may be an economic incentive to set aside land for waste gas treatment. The only limitation on simultaneous use of the land for other purposes is that the soil surface must be open to the atmosphere.

II. CHEMICAL AND MICROBIOLOGICAL CONSIDERATIONS

A. Oxidation and Removal Rates

In contrast to waste waters, waste gases are usually oxygen-rich. The oxygen concentrations necessary to maintain oxidation under the convective flow conditions of a soil filter are uncertain but should be related to the concentration and sorption-oxidation rates of the gases being removed. The O_2/oxidizable gas ratio would appear to be more important than the oxidizable gas concentration per se. Allowing for the slow kinetics of oxygen reduction, ratios of 100 or more on a mass or volume basis should maintain oxidation and are easily attained by diluting the waste gas with air. Lower ratios are probably rare except in waste gases from fermentation reactions. The course of soil reactions with waste gases is therefore oxidative and for organic gases is almost exclusively microbiologically catalyzed.

Removal rates of organic waste gases by soil filters may change from the onset of operation depending upon the nature of the gas and the response of soil microorganisms. The removal of weakly sorbed gases, usually highly volatile and nonpolar, or of gases initially toxic to microorganisms, is limited by population growth and metabolism rates of the microorganisms responsible. The removal rate of methyl mercaptan (CH_3SH), for example, increases slowly over 6 weeks and is then still unstable (Carlson, 1970). Easily sorbed gases, on the other hand, characterized by higher molecular weights or increasing polarity, are initially sorbed more rapidly than the microorganisms can metabolize them. The hourly removal rate of butanol from air, for example, decreases from 500 to a steady value of 100 mg butanol/kg of compost in a compost filter after 50 hours of operation (Helmer, 1972). Compost filters are analogous to soil filters discussed here but should react more rapidly because of the higher volume density of their microbial population.

B. Soil Fertility and Acidity

Soil fertility influences the size, activity, and metabolic rate of the microbial population and consequently the rate of gas sorption and oxidation. The fertility conditions adequate for high microbial activity are probably similar to those which permit good growth of higher plants. Fertilization and pH control may be necessary for high initial and continuing rates of gas treatment by soils. Increasing the natural soil fertility increases the rate of hydrocarbon breakdown (Schwendinger, 1968), even though the microbial population which slowly builds up in response to oil spills and gas main leaks includes nitrogen fixers (Ellis & Adams, 1961). Nitrogen fixation in soil filters, on the other hand, is probably low because of the oxidative conditions. A pH of near neutrality seems best for the removal of most gases, with the possible exception of CO.

C. Carbon–Hydrogen Gases

Volatile aliphatic (saturated carbon–hydrogen) gases are adsorbed weakly and react slowly in soils. The prospects of using soils to lower the concentrations of these gases in air streams seems remote (Carlson, 1970) unless residence times are long. Where a concentrated source such as a leaking gas main emits methane, the soil microbial population increases over a period of months to oxidize part of the methane (Ellis & Adams, 1961). The rates of methane oxidation ranged from 14 to 45 g CH_4/hour-m^3 of soil when oxygen was supplied by diffusion from the soil surface against a slow convective flow of methane (Hoeks, 1972). The microbial population had been exposed for several years to the gas leaks. The better oxidizing conditions in the forced air flow of a soil filter might not permit such a buildup of microorganisms. Whether the abundant supply of oxygen in a soil filter and the greater activity per unit population would offset the higher microbial population in soil affected by a gas leak is unknown.

Higher molecular-weight hydrocarbons such as diesel oil have half lives of 1 to 2 years in soils (Suess et al., 1969). Unsaturated hydrocarbons apparently react much more rapidly. Ethylene and acetylene, for example, are removed from the atmosphere by soils (Abeles et al., 1971) and presumably oxidized to CO_2. Assuming that soil microorganisms can adapt eventually to virtually any organic compound as an energy source, they should utilize aromatic hydrocarbons. Removal rates of these gases are unknown but should be influenced by the composition and number of their functional groups.

D. Carbon–Oxygen Gases

Carbon dioxide and carbon monoxide are emitted to the atmosphere from natural and anthropogenic sources in far greater amounts than other carbonaceous gases. Carbon monoxide and particularly CO_2 react more slowly with soil than the other gases discussed here. Soil treatment to remove CO_2 from waste gases is out of the question.

The compost filters analogous to soil filters are operated at air residence times less than 60 sec and do not remove CO (Lachnit, 1971). Some CO removal may be possible under the longer residence times of soil filters. G. H. Heichel (personal communication) suggests that, since the ratio of total surface area in soil 1 m deep to the area of the soil-air interface is about 10^6, potential CO removal by soil filters may be 10^6 greater than his measurements of diffusional removal from the atmosphere (Heichel, 1973). This means a theoretical CO removal rate of 20 mg CO/hour-m^3 of soil in soil filters. The CO removal rate was faster in an acid forest soil (pH 4.3, 29% organic matter) than in a field soil (pH 5.9, 4.3% organic matter).

E. Carbon–Hydrogen–Oxygen Gases

The increasing structural polarity and chemical substitution of C–H–O gases enormously increases their sorption and transformation rates in soils compared to aliphatic hydrocarbons. As with the hydrocarbons and C–O gases above, the end products are CO_2 and H_2O. The soil's role is wholly catalytic via microorganisms so that its ability to filter these gases can continue indefinitely.

Organic acids decompose rapidly in soil filters (Carlson & Leiser, 1966). Acetoin (H_3C-CO-CHOH-CH_3) is the major component of the gaseous emissions from compost plants (Boninsegni et al., 1974) and is removed below sensible limits in the 9-minute residence time of the Geneva soil filter.

F. Organonitrogen Gases

Unlike the above gases, the oxidation of organic N gases leaves a N residue and a net production of H^+ in the soil. The N end products are unknown but would probably include N_2, NO_3^-, and ammonia NH_3. Based on our knowledge of the fate of organic N compounds in waste waters under oxidizing conditions, a substantial fraction of the N in waste gases will be converted to HNO_3. The mass of N input from waste gas treatment is small enough that ground water pollution by NO_3^- and appreciable changes of soil pH appear unlikely.

The sorption rates of organic N gases by soils are unknown but are most likely high. Even thin layers of soil prevent odorous emanations such as putrescine $(H_2N-[CH_2]_4-NH_2)$ and cadaverine $(H_2N-[CH_2]_5-NH_2)$ from decomposing organic matter. Cyanides are rapidly sorbed by soils and probably degraded inorganically (Fuhr et al., 1948). Microbial degradation of cyanide in soils yields carbonate and NH_3 (Strobel, 1967).

G. Organosulfur Gases

The fate of S in sorbed organic S gases is better understood than that of N. The S of CH_3SH in air is sorbed and oxidized completely to H_2SO_4 (Carlson, 1970). The sorption of large amounts of CH_3SH caused sufficient soil acidity in poorly-buffered soils to hinder further sorption. Liming would restore this activity. In the absence of oxygen, H_2S sorbed by the soil oxidizes to both SO_4^{2-} and elemental S and reduces iron oxides to ferrous sulfide FeS_2 (Carlson, 1970). Organosulfur gases would probably behave similarly but more slowly because as reducing agents they are not as active as H_2S. Carbon disulfide, CS_2, is sorbed only slowly by soils (Fuhr et al., 1948).

H. Organohalogen and Phosphorus Gases

The removal of halogenated gases by soil filters may be of only marginal practicality. Despite their high molecular weight and polarity, these gases apparently are not retained strongly by soils. DDT and its residues, for example, evaporate rather freely from soil, especially under aerobic conditions (Spencer et al., 1974). These gases would also seem to be too resistant to biodegradation for complete treatment in soil filters. At concentrations of about 30 ppm pentachlorophenol (C_6Cl_5OH) in soil perfusion studies, Watanabe (1973) found that a microbial population evolved which utilized even this exceedingly toxic and stable compound as a substrate. This should also happen to some extent in soil filtration of other organohalogens, but its effectiveness is unknown. If the only other alternative for toxic organohalogens is release to water or the atmosphere, the choice of soil treatment is clearly superior.

The chlorofluoromethanes, such as CCl_2F_2, probably react very slowly with soils and may also be untreatable by soil filtration. They are water-insoluble, nonpolar, volatile, and chemically stable which result in lower sorptivity in soils and inertness to microbial decay. Phosgene $(COCl_2)$, on the other hand, is rapidly sorbed by soil and presumably degraded chemically (Fuhr et al., 1948).

Organophosphorous molecules should be sorbed avidly by soils and rapidly degraded. If they hydrolyze in water at appreciable rates, their breakdown rates in soils should be much faster.

III. PHYSICAL CONSIDERATIONS

A. Entrained Particulates and Aerosols in the Gas Stream

Soil treatment would completely remove entrained particles, but soil pores are quickly plugged by particulates. Although removal of particulates from gas streams is advantageous, the soil's primary role would be to remove objectionable gas phases. Particulate removal is easily accomplished by cyclones and other relatively simple physical methods. Liquid aerosols should not present problems if they wet the soil surface and can be conducted away rapidly enough to prevent plugging. Nonwetting aerosols would plug soil pores much like solid particulates.

B. Water Content

Waste gas treatment by soils requires water management. The oxidation rate of organic gases is probably maximal in the same moisture range which satisfies higher plants, water contents equivalent to about 0.3 to 1 bar suction.

Although the polluted air in the Geneva soil filter is apparently saturated with water, the air flow and grass growing on the surface dry the soil in the summer. Unless the filter is watered by sprinklers to restore the water content, removal of the malodorous gases is incomplete. Watering quickly restores the soil's filter function.

C. Temperature

Microbial activity is paramount for the removal of organic gases. The activity rate of most microorganisms roughly doubles with each 10 degree rise in temperature up to about 37C but decreases rapidly at higher temperatures. Thermophilic bacteria are active up to about 65C so that Helmer (1974) suggests 60C as a working maximum for compost air filters. Lachnit (1971) more conservatively suggests 35C. Gases can be cooled by watering the soil or by water injection into the gas stream prior to contact with the soil.

The Geneva soil filter treats the waste gas satisfactorily during the winter. The waste gas and microbial respiration warm the soil sufficiently for microbial activity. The increased sorption of gases at lower temperatures should offset somewhat the lower microbial activity.

D. Residence Time

The time required for sorption and transformation of organic gases in soils depends on the rate of microbial activity and biodegradability of the

Table 1—Gas permeabilities of dry soils and soil materials.

Soil	Permeability μm^2	Material	Permeability μm^2
Cave loam	2.8	Loamy sand	5.6
Edina silt loam	4	Fine sand	56
Sonoita sandy loam	4.7	Coarse sand	240
Karro loam	16	Calcic soil,	
Stewart clay loam	30	0.2-0.4 mm fraction	44
Webster silt loam	44	0.1-1.0 mm fraction	220
		1.0-2.0 mm fraction	2,300

gas. Under conditions of room temperature, adequate moisture, and the natural nutrient level of compost, Helmer (1974) recommends air flow rates equivalent to a minimum residence time of 30 sec for compost filters. Soil reaction rates should be slower because of a lower volume density of microorganisms. The 9-minute residence time of the Geneva soil filter removes odorous gases, predominantly aldehydes, to below detectable concentrations except when the soil is dry.

E. Gas Flow in Soils

Table 1 shows the air permeability of various soils and soil separates. Gas permeability K is the ratio of the macroscopic flow velocity to the pressure gradient, where the pressure gradient is expressed in water head per unit length of macroscopic flow. The gas permeability is independent of the texture (Elgabaly & Elghamry, 1970) among the medium-textured soils shown. The differences can be accounted for by soil structure. Since resistance to gas flow decreases with the fourth power of increasing pore diameter in smooth pores, gas should flow mostly through large soil pores and soil water should impede air flow appreciably only at high water contents (Aljibury & Evans, 1965). Gas permeability is essentially independent of temperature.

The resistance immediately surrounding the distribution pipe largely determines total resistance of a soil filter. This resistance is usually diminished by enclosing the pipe in a gravel envelope. The cost of modifying the soil structure would probably be greater than increasing the pipe length or increasing the diameter of the gravel envelope around the pipe.

Waste gas treatment by the Geneva soil filter is better in the coarser, sandy portion of the filter. The finer-textured soil is more susceptible to cracking during drought in addition to having a lower gas permeability.

F. Distribution Network and Pipe

Theoretical investigations of vertical wells, horizontal trenches, and horizontal buried porous pipes showed that the horizontal pipes required the least power for gas flow and most uniformly distributed the gas in the soil

(Miyamoto et al., 1974). The power requirement decreases some 10% and the uniformity of gas distribution improves if the pipe is beneath the ridges of a corrugated rather than beneath a flat soil surface (Warrick & Miyamoto, 1974).

The soil and compost filters now in operation use horizontal perforated concrete pipe to distribute the gas. The pipe is expensive and its perforations may be too few and too small for optimum air flow. Plastic pipe with many perforations is now available. It is of a smaller diameter but can be laid at low cost. This should offset the necessarily larger distribution network and distribution losses.

A still cheaper alternative might be an open-bottomed arch formed by a plastic strip laid in an inverted U shape. The cross section could be sufficiently large to handle the waste gas flow without constriction yet would expose a large soil area. This configuration forces the gas flow downward and outward so that only a shallow trench would provide an adequate path length through the soil.

G. Power Requirement

The power requirement for treating waste gases by soil in situ is inversely proportional to the air permeability of the soil. The permeability is determined by both soil texture and structure, and secondarily by the soil water content. Low power usage is attainable in soil treatment of gases compared to other methods. A compost filter, for example, requires one-sixth as much power as the water-hypochlorite scrubber which it replaced (A. A. Ernst, Stadtreiningungsamt, Duisburg, West Germany). The power saving was fortuitous; the hypochlorite scrubber was discarded because of unreliability, water pollution, and inadequate odor control.

IV. SOIL FILTER-SCRUBBERS OF WASTE GASES

Soil scrubbers for the treatment of organic waste gases are adaptable to the particular requirements of the problem: volume of gas to be treated, biodegradability of the gas, amount and cost of land, and operating cost which is essentially the cost of power. The equation relating the land and power requirements to the air permeability of the soil and the mass of gas is (Miyamoto et al., 1974)

$$H_p = \mu Q^2 L/\rho^2 k E_p \qquad [1]$$

where
  H_p = power requirement, watts,
   μ = gas viscosity, kg/sec-m,
   Q = mass flow rate of gas per unit length of pipe, kg/sec-m,
   L = pipe length, m,

ρ = gas density, kg/m^3,

k = intrinsic permeability, m^2, and

E_p = power efficiency of gas distribution in the soil.

The value of E_p depends upon the depth, diameter, and spacing of the pipes. Miyamoto et al. (1974) and Warrick and Miyamoto (1974) calculated E_p values for several pipe configurations. In terms of volumetric flow Q_v, Eq. [1] becomes

$$H_p = \mu Q_v^2/k \ E_pL. \qquad [2]$$

When the dimensions of the Geneva filter are substituted into Eq. [2] (E_p = 1.86 for a single buried pipe), the resistance of the whole system corresponds to a gas permeability of the soil of k = 14 μm^2. Some of the flow resistance may be due to inadequate perforation of the conduit pipe. The Geneva scrubber's insensitivity to soil water content suggests that some power losses may occur in the conduit. The power requirement increases only 10% when water is standing on the surface (J. Kröpfli, Dep. Travaux Publics, Geneva, personal communication). The capture of entrained dust in the conduit and soil has had no effect on the flow resistance of the filter-scrubber.

Table 2 also shows four hypothetical soil scrubber-filters calculated from Eq. [2] to yield a range of air residence times. The air residence time required is determined by the sorptivity and biodegradability of the undesirable gases. For most gases, unfortunately, these properties can as yet only be estimated. Since power is virtually the sole operating expense of soil filter-

Table 2—Properties and estimated costs of the Geneva, Switzerland, and four model soil filters. The models were calculated for a low power requirement over a wide variation in the mean residence time of waste gas in the soil

	Geneva soil filter (Dupont, 1964)	Model soil filters			
		A	B	C	D
Pressure drop, cm H$_2$O	100		10		
Power requirement, kW	20		1		
Air flow, m^3/sec	1		1		
Air permeability, μm^2	14		10		
Depth to center of pipe, m	3	2	1	1	0.5
Pipe diameter, m	0.4	0.4	0.2	0.1	0.2
Distance between pipes, m	--	5	2	1	2
Pipe length, m	36	1,100	1,230	1,840	710
Area, m^2	120	5,600	2,460	1,840	1,420
Residence time, min	9	110	28	26	5
Capital costs $24,700/ha, $5/m to $20/m	$1,100	$36,000	$12,200	$13,600	$7,000
Annual power costs (3¢ kW-hour)	5,260	--	--	263	--
Maintenance	--		negligible		--
Total annual cost (assumed 20-year lifetime)	$5,300	$2,100	$900	$950	$600

scrubbers, a much lower power requirement (1 kW per m^3/sec air flow) than that of the Geneva filter was chosen for these models. The air conductivity of the soil was assumed to be 10 μm^2 so that the filter dimensions are probably conservative yet realistic. The power requirement will rise above 1 kW temporarily when the soil moisture content is high.

Table 2 also shows estimated costs for these filters. Trenching and pipe laying were taken as $5 to $20/m depending on pipe depth and diameter. Land cost attributable to filtration was taken as $24,700/ha. Assigning a land value is difficult because land within the existing boundaries of the polluting facility can often be used, because the value of land near pollution sources increases after satisfactory control, and because the filter area can be used simultaneously for some other purposes. The 20-year lifetime is decidedly conservative; the Geneva filter has been operating for 11 years without any indication of diminished capabilities.

Models A through D (Table 2) are various combinations of pipe depth, diameter, and spacing to achieve various residence times at a low power requirement. Long residence times would be necessary for the removal of slowly biodegradable gases such as CO and the hydrocarbons. Less than 15 minutes is sufficient for most odorous gases but depths shallower than 25 cm seem unwise. The residence time for some molecules in such shallow soil filters would be only several minutes and channeling and cracks are more likely at shallow depths. The calculations assume that the distribution pipe is highly porous. To approach this in practice, perforated pipes would have to be surrounded by a gravel envelope of several centimeters thickness.

Taking B as the reference, Model A illustrates the effect of larger pipe, deeply and widely spaced. Filter A requires almost 2.5 times as much area and pipe length, but due to its depth and spacing has almost 4 times the residence time.

The pipe of Model C, on the other hand, is half as deep as B. The pipe length and area are 60% of those of B but the residence time in C is only one-sixth that of B.

Model D illustrates the effect of smaller pipe at shallow depth and narrow spacing. This reduces the land area to one-third that of B and the residence time to one-twelfth. This 15-min average residence time is probably a reasonable but conservative minimum for malodorous organic gases.

Models A through D were calculated for a corrugated surface whose slopes are 18°. Miyamoto et al. (1974) found this to be a better compromise between minimal power requirement and maximal soil contact than a planar surface. A planar soil surface would require about 10% more area and pipe length to achieve the same power requirement and would reduce the air residence time by the same amount.

The annual operating cost of a conventional water–hypochlorite scrubber is approximately $4,000 for a 1 m^3/sec flow and 92% malodorant removal of waste gases from a rendering plant (E. J. Wooldridge, EPA, Research Triangle Park, N. C., personal communication). The operating cost of the model filters in Table 2 is much lower than this hypochlorite scrubber as are the total costs of filters B through D. In my experience, soil filters also remove a higher fraction of the pollutants.

Compared to conventional, high-technology gas cleaning methods, the advantages of soil filters for organic gases are best realized under any combination of the following conditions: (i) waste gases and, if necessary, simple cooling techniques keep the soil temperature below 40C, (ii) particulate concentration in the gas stream is low, (iii) high reliability and low maintenance are important, (iv) the pollutant is resistant to chemical breakdown but is sorbed and degraded in soils, (v) land is cheap, (vi) power costs are high, (vii) input concentration is already low, and (viii) removal of air pollutants must be complete.

ACKNOWLEDGMENT

Part of this work was carried out at the Institut für Bodenkunde und Waldernährung der Georg-August Universität, Göttingen, West Germany, thanks to a U. S. Senior Scientist Award from the Alexander von Humboldt Foundation, Bonn-Bad Godesberg.

LITERATURE CITED

Abeles, F. B., L. E. Craker, L. E. Forrence, and G. R. Leather. 1971. Fate of air pollutants: removal of ethylene, sulfur dioxide and nitrogen dioxide by soil. Science 173:914–916.

Alexander, M. 1969. Microbial degradation and biological effects of pesticides in soils. In Soil biology, Reviews of Research. UNESCO Nat. Resour. Res. 9:209–240.

Aljibury, F. K., and D. D. Evans. 1965. Water permeability of saturated soils as related to air permeability at different moisture tensions. Soil Sci. Soc. Am. Proc. 29:306–369.

Bethea, R. M., B. N. Murthy, and D. R. Carey. 1973. Odor controls for rendering plants. Environ. Sci. Tech. 7:405–510.

Bohn, H. L. 1972. Soil absorption of air pollutants. J. Environ. Qual. 1:372–377.

Bohn, H. L., and S. Miyamoto. 1973. Soil as a sorbent and filter of waste gases. p. 104–114. In J. Tomlinson (ed.) Proc. Int. Conference Land for Waste Management, National Research Council, Ottawa, Canada.

Boninsegni, Ch., A. Deuber, and H.-U Wanner. 1974. Geruchsanalysen bei der Kehrichtkompostierung. Int. Solid Wastes and Public Cleansing Assoc. (Dubendorf, Switzerland) Info. Bull. 13:17–24.

Carlson, D. A., and C. P. Leiser. 1966. Soil beds for the control of sewage odors. J. Water Pollut. Control Fed. 38:829–840.

Carlson, D. A. 1970. The soil filter: a treatment process for removal of odorous gases. Fed. Water Pollut. Control Assoc. Report WP 00883-03.

Dupont, Gerard. 1964. La desodorisation des gas de fermentation des ordures menageres, dans les usines de compostage. Div. de l'assainessement de Dept. des Travaux Publics, Geneva, Switzerland.

Elgabaly, M. M., and W. M. Elghamry. 1970. Air permeability as related to particle size and bulk density in sand systems. Soil Sci. 110:10–12.

Ellis, R., Jr., and R. S. Adams, Jr. 1961. Contamination of soils by petroleum hydrocarbons. Adv. Agron. 13:197–216.

Fuhr, I., A. V. Bransford, and S. D. Silver. 1948. Sorption of fumigant vapors by soils. Science 107:274–275.

Heichel, G. H. 1973. Removal of carbon monoxide by field and forest soils. J. Environ. Qual. 2:419–422.

Helmer, Richard. 1972. Sorption and mikrobieller Abbau in Bodenfiltern bei der Desodorisierung von Luftströmen. Stuttgarter Berichte für Siedlungswasserwirtschaft 49, Inst. fur Siedlungswasserbau und Wassergütewirtschaft, 7 Stuttgart-Büsnau, Bandtale 1, West Germany.

Helmer, Richard. 1974. Desodorisierung von geruchsbeladener Abluft in Bodenfiltern. Gesundheits-Ingenieur 95(1):21–26.

Hoeks, J. 1972. Changes in composition of soil air near leaks in natural gas mains. Soil Sci. 113:46–54.

Lachnit, F. 1971. Abbau geruchsintensiver Stoffe durch Biofilter. p. 65. In F. Meinek (ed.) Schriftenreihe der Vereins fur Wasser-, Boden-, Lufthygiene #35. Commission für Umweltgefahren beim Bundesgesundheitsamt, Berlin, West Germany.

Miyamoto, S., A. W. Warrick, and H. L. Bohn. 1974. Land disposal of waste gases: I. Flow analysis of gas injection systems. J. Environ. Qual. 3:49–54.

Miyamoto, S., A. W. Warrick, and R. J. Prather. 1974. Land disposal of waste gases: III. Sorption patterns from buried gas injection pipes. J. Environ. Qual. 3:161–166.

Schwendinger, R. B. 1968. Reclamation of soils contaminated by oil. J. Inst. Pet. 54: 181–197.

Spencer, W. F., M. M. Cliath, W. J. Farmer, and R. A. Shepard. 1974. Volatility of DDT residues in soil as affected by flooding and organic matter applications. J. Environ. Qual. 3:126–129.

Strobel, G. A. 1967. Cyanide utilization in soil. Soil Sci. 103:299–302.

Suess, A., A. Netzsch-Lehner, and W. Nowak. 1969. Veränderung von Dieselölkomponenten in zwei Boeden. Z. Pflanzenernähr. Düng. Bodenk. 122:4–18.

Turk, A., R. C. Haring, and R. W. Okey. 1972. Odor control technology. Environ. Sci. Tech. 6:602–607.

Warrick, A. W., and S. Miyamoto. 1974. Land disposal of waste gases: II. Gas flow from buried pipes. J. Environ. Qual. 3:55–60.

Watanabe, I. 1973. Isolation of pentachlorophenol decomposing bacteria in soil. Soil Sci. Plant Nutr. 19:109–116.

chapter 25

Future Direction of Waste Utilization

P. F. PRATT, University of California, Riverside, California

M. D. THORNE, University of Illinois, Urbana, Illinois

FRANK WIERSMA, University of Arizona, Tucson, Arizona

I. INTRODUCTION

Waste waters from industries and municipalities, organic sludges, and animal manures should be considered as resources rather than wastes that need to be put "out of sight, out of mind." The ease with which waste utilization can replace waste disposal varies with the nature of the waste, local geographic features, and socio-economic and political situations or systems. However, in a world with limited resources, our efforts in research, technoloigcal development, and application and formulation of private as well as public policy should be directed toward maximum utilization. Minerals should be recovered for reuse and organics should be reused directly or converted to products that can be used. Even in cases where disposal is practiced, serious considerations should be given to disposal under conditions that would make future mining of the materials possible. For example, sewage sludges put into separate and specifically designed landfills might be mined for heavy metals at some future date when scarcities will make recovery economically profitable.

The average amount of fresh water that falls as precipitation in the United States is about 15.5 billion m^3/day or about 4,100 billion gal/day. Seventy percent of this, about 10.8 billion m^3/day, is returned directly to the atmosphere by evaporation and transpiration. About half of this amount is from areas having vegetation of negligible value; only one-fifth is from cropland. Thirty percent of precipitation (4.7 billion m^3/day) goes into streams and underground reservoirs. Only about one-fourth of this (1.17 billion m^3/day) is subsequently withdrawn for use by industries, municipalities, and agriculture, including irrigation. Less than one-third of the amount withdrawn (0.4 billion m^3/day) is actually consumed or lost such that it cannot be reused. Thus, somewhat more than 0.72 billion m^3 (200 billion gallons) of water are withdrawn each day and used in such a manner that it becomes available for possible reuse. This totals over 253 billion m^3 per year (over 200 million acre feet/year). This potential supply for reuse is considerably more water than is used by agriculture (including that for irrigation) and is more than is estimated to be needed by agriculture by the year 2000. Piper (1965) estimated that agriculture will withdraw 0.66 billion m^3/day and consume 0.45 billion m^3/day by the year 2000.

If a significant portion of waste water in the United States were to be used on agricultural land, irrigated acreage would be increased dramatically. Alternatively, waste water could be substituted for much ground, stream or reservoir water, thereby permitting such waters to be used for other purposes.

There is, of course, much reuse now taking place. Municipalities draw water from adjacent streams and discharge used water with varying degrees of treatment where it becomes available for reuse by other municipalities downstream. Utilization of municipal waste water on land will decrease supplies available downstream unless trade-offs are effected or other sources of water are developed.

Present production of sewage sludge in the United States is estimated at about 6 million dry metric tons/year. Implementation of the Water Quality Act of 1972 (Public Law 92-500) will probably produce a three- to fivefold increase in sludge production in the next 5 to 10 years. Of the presently produced sewage sludge, about 40% goes into landfills, 20% is applied to land, 25% is incinerated, and 15% is discharged into lakes and oceans. Challenges and opportunities to develop economically feasible methods for utilization of sewage sludges are great.

The annual production of manure by farm animals is estimated to be 310 million dry metric tons. About half of this manure is distributed on croplands, pastures, and rangelands in a diffuse manner, leaving 155 million metric tons that are deposited in various types of animal confinement facilities where it can be recovered and used in various ways. Perhaps 90% of the recoverable animal manure is presently used on croplands. The remaining 10% goes into specialty uses such as landscaping, nurseries, and home gardens. Some is accumulating and a small amount is used as feed ingredients. If an increase in efficiency of use of 15% could be obtained, about 23 million metric tons would be released for use on croplands that are now receiving chemical fertilizers.

A report of the Council of Agricultural Science and Technology (CAST Report, 1975) dealing with use of sewage sludges and animal manures in agriculture summarized the problems that limit their use as follows:

1) Animal manures and sewage sludges, as sources of plant nutrient elements, are bulky low-grade fertilizers of variable composition.

2) As low-grade fertilizers, these materials cannot be transported more than a few tens of miles before the costs of transportation exceed their fertilizer value.

3) Difficult on-farm management problems are created by the physical properties of these materials. Application techniques are inefficient and time-consuming. The production of animal manures and sewage sludges are continuous, whereas fertilizer needs are seasonal.

4) Animal manures and sewage sludges contain soluble salts that can limit their use in arid irrigated areas.

5) The undesirability of leaching of soluble salts, particularly NO_3^- to ground waters, may limit their use on croplands in some areas.

6) Sewage sludges contain heavy metals that are retained in soils and might accumulate to the point where they are toxic to some plants, and thus restrict the types of crops that can be grown.

7) Incompletely treated sewage sludges contain pathogenic bacteria, viruses, and parasites that represent a public health risk to farm workers and the public via the food chain.

8) Odors and associated nuisances, both real and imagined, create conflicts between urban and rural residents and the farmers of adjacent croplands where sludges and manures are used.

9) The use of sewage sludges on cropland is inhibited by sociological resistance, that stems from fear of pathogens, odors, and nuisances,

and possible general environmental deterioration.

10) Government agencies require monitoring of both crop and water quality if sewage sludges are to be used.

11) The practice of refeeding animal manures is inhibited by the concern for possible transmission of pathogens and undesirable organic residues into foods.

12) There are no meaningful guidelines, for using sewage sludges on cropland or manures as feed ingredients, that can be interpreted on a regional basis or in terms of local situations and needs.

The CAST Report outlined the following actions that can promote the use of sewage sludges and animal manures:

1) An increase in the quality of the products would make sewage sludges and manures more competitive with chemical fertilizers.

2) Development of new management systems, that would not delay other farm operations, would create a better image for these materials in the minds of farm managers.

3) Appropriate guidelines would enhance beneficial uses. Guidelines that are easily interpretable and based on fact and acceptable risks are needed to protect public health and environmental quality in terms of local and regional conditions.

4) An effective educational program stressing the advantages to society of use of sewage sludges and animal manures would help remove sociological resistance.

5) Research and demonstration projects of high visibility could serve not only to develop improved management practices but also to convince local populations that utilization is desirable and feasible.

In any utilization scheme some residue exists that must be put into waters, in the atmosphere, or added to lands via landfills or spread on land surfaces. Use on cropland thus possesses the advantage of land application for both use and disposal in a single operation.

II. FACTORS THAT INFLUENCE TRENDS IN UTILIZATION

A. Economics

Economics exert strong influences on all presently practiced and potential methods of waste utilization. Economics are not the only factor and perhaps should not be the principal factor in decisions on waste utilization, but they cannot be disregarded in the development of utilization schemes nor in predictions of future directions of utilization.

Use of waste waters for irrigation must be economically competitive with other sources of water and must give economic return in terms of increase in crop yields. The use of land-spreading for water treatment and ground-water recharge must also be within economic capabilities of the government agency or entity responsible for its operation.

There are a number of beneficial effects of the application of organic materials to croplands. Soil structure is frequently improved so that water infiltration is more rapid, the soil is less erosive and is easier to till. Organics serve as slow release fertilizers for a number of nutrient elements, provide a fairly balanced nutrient supply for many, but not all, soil-crop combinations, and in some soil conditions increase the availability of mineral elements already in the soil. However, the main value of organics for most situations is their contents of N, P, and K. Thus, the economics of using sewage sludges and animal manures as fertilizers depends largely on the prices of N, P, and K in chemical fertilizers. The logistic and aesthetic advantages of chemical products as compared to these waste products will favor the use of chemical sources unless strong economic incentives make the waste more acceptable to the farm manager.

The use of animal manures as feed ingredients must be economically competitive with forages and grains that make up the bulk of the presently used feeds. The recovery of metals or organics from sludges and/or animal manures for use in any form cannot proceed unless the products compete with the same or similar products from other sources. Of course, external financial support to make them compete can be economically advantageous if the overall costs are less than other alternative methods of utilization or disposal.

B. Public Policies

Public policies can have direct or indirect effects on waste utilization. Guidelines or standards designed to protect water and air quality, or for protection of public health, can be so restrictive that waste water and sewage sludge utilization on croplands will be unacceptable to farm operators for other than economic reasons. Thus, if waste utilization is to be encouraged, guidelines and standards controlling utilization must be such that the expected utilization can occur.

Absolute environmental and public health risks of waste utilization have little meaning by themselves. Obviously, zero risks are preferred, but because zero risks are unattainable, the only meanful procedure is to develop relative or comparative risks for alternative methods of utilization or disposal. Only when we have these relative risks can we make risk-benefit ratios and thus support actions based on fact. Even though the relative risks of utilization of municipal waste waters and sewage sludges on cropland compared to other methods of disposal are unknown for many situations, the authors feel that with proper site and crop selection and with adequate agronomic management, utilization on croplands can give relatively low risks in many areas.

C. Results of Research and Development

The future of waste utilization can be changed dramatically by the development of new processes for waste treatment. If the use of animal manures for energy to convert N into NH_3 proves to be economically sound, large amounts of manures would be channelled away from croplands. However, the production of N fertilizer can also be increased, which in turn might reduce the economic incentives for using the remaining manures on croplands.

If processing of sewage sludges for recovery of N, P, heavy metals, and perhaps other products proves to be economically sound, this can reduce pressures for use of sewage sludges on croplands and for disposal by landfills or by incineration. Recovery of heavy metals without removing N and P would eliminate a major deterrent to use on croplands and would greatly favor use as a fertilizer.

Of course, possibilities currently unknown can be developed into applicable technologies that can revolutionize present waste management schemes. If a low-pressure, low-temperature process of converting N to NH_3 were to be developed which would require less energy than the presently used process, use of manures as N sources would probably decrease. The advantages of inorganic commercial fertilizers in comparison to manures are sufficiently large on most crops and soils that manure use would be reduced unless P and/or K supplies decrease or become so expensive that manures become more valuable for their P and/or K contents. But regardless of price relationships, manures and other organics will continue to be used on croplands because of benefits from improvement in soil physical condition.

D. Quality Control

The acceptance of waste products for any utilization depends on the value of the product which is dependent not only on the concentrations of desired constituents, but also on the variabilities in these concentrations. Thus, a quality product has consistently high concentrations of valuable constituents. Management of wastes for quality control would enhance their value for present users who could then depend on a product of standard quality.

A quality product generated from waste materials is easier to achieve with animal waste than with most sewage sludges because of the nature of the product. Consequently, a high degree of utilization for profit will likely occur from animal wastes much sooner than with municipal wastes.

The concentration of animals into high density units played an important role in generating a pollution problem. With this trend toward concentration, however, came added potential for improved control of the quality of the waste products. Where the animals are concentrated on dirt lots, quality control is somewhat limited. That is, moisture control is nonexistent and foreign matter, especially soil, is invariably integrated into the manure by animal traffic and collection processes.

The trend to concentration has progressed in many areas to include concrete surfaces or slatted floors with collection pits. In either, the constituents are essentially limited to wastes with no foreign matter or debris. Moisture control on concrete surfaces is limited but in collection pits under slatted floors, it can be controlled. The next step is to modify the composition of wastes through management of feed ingredients and feeding methods. This is especially important where recycling directly through animals as an animal feed is involved.

Livestock management systems currently exist with a level of control wherein quality of animal wastes can be maintained for recycling as a feed for animals. Management decisions affecting waste quality in an animal production system are made by the same individual who makes the other decisions in the operation of the system. Since profit is the ultimate goal, and a quality product generates profit, the incentive and the authority for maintaining quality is retained by one individual or management team.

The same is not true for municipal systems. Consequently, quality control and the opportunity for profitable utilization are limited and likely not feasible in the near future. An example of quality control measures is that imposed upon the garbage detail on a number of United States military bases during the past few decades. Garbage from the mess hall was separated and placed in specific containers, each clearly labeled and destined for various uses, including animal feed. It is unlikely that the level of regimentation and control available in that management system will be possible in the near future for control of wastes from the general populace.

E. Public Acceptance and Sociological Problems

If a waste utilization process or product is not acceptable to the public, the operation will not be feasible. Public acceptance can be influenced by economics, public policy, and by educational and cultural changes. Of course, solutions to some environmental problems such as odors or quality of water and food, whether real or imagined, would greatly help in public acceptance of waste utilization proposals.

Too frequently, individuals in charge of utilization projects tend to alienate local residents unnecessarily. This may happen because of unguided enthusiasm for the project, or inability to recognize other viewpoints. "Benefits" of the project which are enumerated and publicized may not appear as benefits to people concerned about or affected by the project. Local groups may be largely ignored—or at least feel they are—and become frustrated and irritated when their complaints and questions go unanswered.

Frequently, project leaders have little agricultural background or little realization of the impact of their statements in a rural community.

If utilization projects are to proceed with a minimum of friction, local groups must be involved from the start. Their concerns must be dealt with honestly and seriously, and possible alternatives discussed and considered. Goals and objectives of the program must be stated and discussed. Consequences which may result should be presented without value judgment as to whether or not they are beneficial. If a candid presentation is made of what is to be done and what consequences are likely to result, people can and will make their own value judgments. Researchers are not immune to the pitfalls mentioned above and have the same need for caution.

III. USE ON CROPLANDS

A. Water for Crop Production

Waste waters can supply both plant nutrients and water for crop production. However, the concentrations of nutrients in reclaimed sewage effluents are so low that it is generally not feasible to use them to supply all of the nutrients needed for crops of high nutrient demands unless high rates of irrigation are practiced. This is particularly true for humid areas where irrigation is used only as a supplement to natural rainfall during short periods of drought. For example, to provide 250 kg of N/ha (223 lbs/acre) about 120 surface cm (about 4 ft) of water containing 20 mg N/liter would be needed. Few crops, even in most arid regions, require this much water. On the other hand, some crops in arid areas can be overfertilized with N if reclaimed effluents containing 20 mg N/liter are used as the only source of water. These crops include tomatoes (*Lycopersicon esculentum* Mill.) (which do not set fruit if N supplies are too high), sugar beets (*Beta vulgaris* L.) (which need low N supplies during the last part of the season so that beet sugar content will not be decreased), and Valencia oranges (*Citrus sinensis* L. Osbeck) (which re-green during the summer if available N is excessive). In each of these crops a source of irrigation waste water containing N can reduce quality and/or quantity of produce.

There are numerous reported crop yield increases resulting from irrigation of crops with treated municipal waste water (Sopper & Kardos, 1973). Irrigated acreages are steadily increasing in the humid areas of the United States. Farmers are realizing that their per-acre-costs of production are so high they cannot afford the risk of decreased yields due to drought. Improved yield potential and price increases for commodities have made the cost/return ratio for irrigating many crops much more favorable than it was in the past. Since neither ground water nor surface water supplies are available for irrigation in many areas, waste water is appearing to be an increasingly attractive source of irrigation water. If the fertilizer costs can be reduced as a consequence of use of such water, this is an added incentive.

B. Animal Manures and Sewage Sludges

Both animal manures and sewage sludges are highly variable in composition. The balance of N-P-K relative to most crop needs is usually much better in animal manures than in sewage sludges. Relative to crop needs, sewage sludges have high ratios of P to K and in cases where most of the NH_4^+-N has been lost by volatilization of NH_3, ratios of P to available N are also high. Thus, one might suggest that sewage sludges could be improved by mixing them with animal wastes to provide a better N-P-K balance.

From the point of view of resource conservation over a long period, maximum use of manures and sewage sludges might be achieved by application at rates sufficient to provide the P needs of crops and supplement with N and K from chemical fertilizers as needed. As long as fuels are available for fixations of N from the atmosphere, an indefinite supply of this element will be available. The North American continent is well supplied with K. However, because the most easily mined and processed P sources in the United States are in danger of being exhausted, costs of P fertilizers in the future will likely increase. Thus, emphasis on organic wastes as P sources could develop.

If a maintenance application of 50 kg of P/ha per year is required for a given cropping pattern on a given soil, sewage sludge containing 2% P should be added at a rate of 2.5 tons of dry material/ha per year. If sludge is used, much larger amounts of N and K will be needed from commercial fertilizers than if animal wastes are used. The amounts of supplemental N needed could be reduced dramatically by preserving all the N in the sludges and manures until they are incorporated into the soil. The use of H_2SO_4 to acidify and to stabilize NH_4, as $(NH_2)_2SO_4$, might be economically feasible if large amounts of S are to be removed from high-S fuels before they are burned. Of course, other acidifying agents might be useful and means other than acidification might be developed.

The use of manures and sewage sludges to supply P to crops would maximize the distribution of these materials over large areas and also maximize distribution and application costs. Application costs could be reduced by making proportionally large applications every fourth or fifth year and relying on starter P fertilizers for crops planted during the years between waste applications.

Wide distributions of animal manures over large agricultural areas by using small rates of application as compared to concentrating them on small areas at higher rates is probably environmentally desirable in most situations. The efficiency of nutrient recycling would increase and soluble salts would be widely distributed with minimal effects on soil productivity and local ground waters. A possible exception might be an irrigated valley in which all the manure produced is used in the valley. Concentrating the manure on a small area and management to concentrate the salts from the manure into a small volume of drainage water, would maximize salt precipitation and minimize mass salt emissions to the ground water or to surface waters via tile or open drains. On the other hand, if the manure is distributed widely over the irrigated valley, salt concentrations in the drainage water will be lower, but the mass emission of salt and accumulation in the ground waters will be increased.

The environmental desirability of low rates and a wide distribution of sewage sludges versus high rates on a smaller area is perhaps debatable. Keeping any undesirable effects, such as accumulations of heavy metals and pathological agents, on small areas that can be well-managed and controlled has some appeal, whereas wide and uncontrolled distribution might increase total risks without much chance for control. If organic wastes containing P, heavy metals, and other mineral elements are used at high rates over a long period of time, the surface soil might accumulate sufficient amounts of these elements to justify extraction from the soil for industrial and/or agricultural uses.

Sewage sludges produced in population centers and animal manures in areas with large feedlots may require transportation over long distances if they are to be utilized on agricultural land. These materials can be handled easily as liquids if solids content is below 10%. Transport of these materials is usually cheapest by pumping through pipelines. Field application of liquid sludge or manure can be by soil injection or by surface or sprinkler irrigation. Special precautions may be necessary with surface application to prevent contamination of surface waters. In addition to providing moisture, this practice adds nutrient and soil organic matter which is especially beneficial in sands or other soils of low organic matter.

Wastes in the solid form must be transported in some type of mobil equipment and distributed on the soil surface. Although the benefits in the form of nutrients and organic matter are obtained, the economic benefits are limited to relatively short hauling distances.

IV. OTHER METHODS OF UTILIZATION

A. Use of Animal Manure as Feed

Although application to cropland is the historically proven technique for the utilization of animal manures, the feeding of manure to ruminants, and to a lesser extent to nonruminants is feasible. In many respects, poultry wastes possess the greatest potential for processing into a feed or feed supplement. This potential has been recognized by the poultry industry, which has taken leadership in this practice. Circumstances provided the incentive. Poultry operations are often complex systems which include high concentrations of large numbers of birds frequently located in areas where little land is available for use of manure as fertilizer. The trend toward complex manure collection systems for all animals provides opportunity for a level of quality control so that waste products are readily adaptable to refeeding.

As economical methods of transforming excreta to a nutritionally valuable feed are developed, additional nutrients will be recovered from the original feed. This practice can increase the value of the animal manure and usually reduce the quantity eventually identified as waste for disposal.

From an economics viewpoint, wastes should be reused within a closed system with each animal recycling its own wastes, thus reducing handling and transportation costs. However, feeding of manures from one type of animal to other types can have advantages of more complete utilization and of reduced health risks. The advantages of a multispecie cycle must be weighed against the added complexity of the system.

Most waste-processing systems have utilized microbiological organisms which are present without inoculation and can exist in the environment generated by the process. One notable exception to this is the aerobic process requiring the injection of oxygen. We have only tapped the surface in what must be a vast potential for developing cultures and management techniques for maintaining cultures which are effective in transforming the raw waste into a more readily utilizable material. An appropriate culture for converting wastes to a high quality animal feed currently looms as the system with the most attractive large-scale economic potential.

Public resistance, a current deterrent to refeeding, will gradually change to acceptance. "Manure" has been successfully transformed into "wastes," a psychologically more palatable substance. Perhaps as wastes become resource materials, an evolutionary change to an even better term can somewhat disguise or at least improve the image of the origin of the materials.

If acceptance can precede the knowledge of what is being accepted, the barriers will be weakened substantially. People enjoyed hot dogs before they knew the skin of the weenies were made from intestines. Because they had learned to enjoy weenies before they could build up a psychological resistance to them, they continued eating hot dogs. Deodorized liquid plant food is poured into flower pots and planters in ornate living rooms by fastidious urban ladies who probably would revolt at the thought of handling manure.

B. Use of Effluents in Ground-water Recharge

The use of reclaimed water for ground-water recharge, using the "living filter" concept of Pennsylvania State University (Sopper & Kardos, 1973) and the tertiary treatment of large volumes of water per unit area of land as developed by the Flushing Meadows Project near Phoenix, Ariz. (Bouwer et al., 1974), are excellent examples of viable alternatives to irrigation of croplands. In the "living filter" concept, water is added to crop, grass, or forested lands at relatively low loading rates so that the usual vegetation in the area is maintained. In the Flushing Meadows Project, 50 to 70 surface meters are infiltrated into the soil each year. The City of Phoenix is now planning a 40-acre project in which ground water from direct land spreading will be pumped and used for unrestricted irrigation. In the Los Angeles area, reclaimed waters from the Whittier Narrows and San Jose Creek Water Treatment Plants are used for ground-water recharge. In these cases, dilutions with other sources of water are used so that reclaimed waters represent 20% or less of the total recharge.

C. Use of Animal Manure for Energy or Fuel

The production of combustible gases from manure continues to be a topic more exciting to the public than practical in fact. By appropriate selection of assumed values, a theoretical analysis of the feasibility of converting wastes to fuel can be made to look favorable or discouraging. That there is a vast quantity of energy in organic wastes cannot be denied. The most feasible means of extraction and utilization remains to be determined.

The economics of bio-gas production has and will continue to be unfavorable as raw materials become more valuable for other uses. Its appeal reached an apex when wastes were synonomous with pollution and the cost of raw material was free or less.

The energy in manure or sludge is represented primarily by its carbon content. The simplest method of extraction is by direct combustion, and we have the technology for extraction by this method. By overcoming a few obstacles associated with pollution, efficiency and transport from source to consumer, this method could perhaps be developed with promise, particularly in arid areas where drying without use of other fuel is possible.

Refeeding of wastes to animals could be an efficient and potentially effective method of extracting the energy. Crop fertilization is another efficient energy-extraction process wherein the carbon provides the metabolic needs of soil microorganisms to transform the nutrients into forms utilized by plants.

Energy extraction by transformation into a usable gas is not economically feasible under present technology and circumstances. However, the subject should by no means be abandoned. The feasibility could take on an entirely new perspective if (i) costs of current fuels increase substantially, (ii) solar energy can be utilized in the gasification process to increase net energy output, (iii) new developments in the conversion process reduces production costs, (iv) techniques for collecting, storage, and utilization are improved, (v) gas production is part of a more complete system which has other marketable products as outputs, and (vi) we gain understanding of the microbiological conversions and are able to control them for gas and protein production.

D. Sewage Sludge as a Source of Metals

The concentrations of such elements as Zn, Cu, Cr, and Ni in many sewage sludges (Page, 1974; Fassell, 1974) suggests that they be used as sources for these elements. In some cases, mineral contents exceed those found in ores that are commercially mined. However, most ores are inorganic, for which well-developed extraction processes are known. Technologies for extraction and recovery of metals from sewage sludges have not been developed.

This lack of technologies leaves an opportunity for profitable research and development.

One approach to the recovery of heavy metals from sewage sludges has been reported by Fassell (1974), who used a wet oxidation process at 232–240C (450–465F) in mildly acid conditions to oxidize the organic matter. The process reduced chemical oxygen demand by 75 to 85% and converted the metals to extractable forms. Also the N was converted to the NH_3 form and the main organic products were simple organic acids. Fassell reported that the extrapolation of data from his small pilot plant to a full-scale plant showed that the recovery of waste heat, NH_3, and heavy metals could completely pay processing costs for some municipal sludges.

Fassell's approach, or others similar to it, offers a number of advantages. The recovered metals are available for commercial markets, and expensive systems now used for recovery from industrial plant waters before discharge into sewage systems can be avoided. If the process can recover N and P as inorganic salts suitable for commercial fertilizer markets, the processed material would be a source of plant nutrients, but with the undesirable metals removed. In the inorganic form, they can be stored and used to meet the seasonal demands of farming systems, and many disadvantages of the use of sewage sludges directly on cropland can be avoided.

Another savings that might be important is prevention of N loss as NH_3 which now occurs during drying of sewage sludges, during dewatering at the treatment plant, and/or by air-drying in the field. If the recovery process receives liquid sludges before NH_3 volatilization, the 50% or more of the N that is usually lost can be prevented.

In many places where use or disposal of sewage sludges is an expense to the city, the use of a process similar to that developed by Fassell, even if it did not pay for itself, would be advantageous. The resources could be recycled or used rather than lost in a sanitary land fill or in the case of use on croplands the benefits could be obtained without the negative effects.

The future use of processes that can recover metals, N, and P and substantially reduce the organic load is unpredictable at this time. However, the many potential advantages suggest that research and development efforts should be directed into these processes. Perhaps this is one activity that will bloom in the next few decades, particularly if government agencies place severe restrictions on direct application to croplands and on disposal in landfills, surface waters and oceans.

V. PREDICTED TRENDS

The following summarization on the future of waste utilization in the United States is presented as the concensus of response to requests that some predictions be made. We hope that the readers of the future will forgive us for inaccurate predictions and give some credit where predictions were correct.

1) Use of reclaimed water for supplemental irrigation in humid regions will increase. The need to eliminate discharge of these reclaimed waters into

streams and lakes without tertiary treatment will favor disposal or use on land. This pressure along with favorable economics for supplemental irrigation during critical periods of drought will promote greater use on croplands in humid areas.

2) The use of soil systems for tertiary treatments of reclaimed waters will increase. Costs of large structures and associated refined or complex tertiary treatment processes will favor use of soil systems except in areas where suitable land is unavailable.

3) In arid areas, use of reclaimed water for irrigation will expand, not only for crop production but for golf courses, parks and ornamental plantings. In some cases, tertiary treatment by passing effluents through soils before recovery will make these waters suitable for a number of uses, including irrigation, with little or no restrictions on crops that can be grown.

4) In arid and hot regions where irrigation can be practiced and where multiple cropping can continue on a 12-month basis, the use of raw or primary treatment sewage for irrigation of specific crops under carefully controlled conditions might increase. The potential savings of direct irrigation are sufficiently high that this system of use likely will become accepted. World-wide experience with direct irrigation with untreated sewage has shown that such practices have low risks. Experiences, in Australia, Mexico, South America, and Europe where highly productive croplands or pastures are being irrigated with sewage waters of various degrees of primary treatment, will further convince us that this practice can be used with relatively low risks in some areas of the United States where careful controls in effluent quality and on cropping practices can be imposed and maintained.

5) As more processes are developed to reduce the heavy metal contents of municipal sludges, to remove these metals after the sludges are produced, or to keep their assimilation by crops within acceptable limits, the amounts of these sludges used on croplands will increase where land that can be properly managed is available.

6) Processes will be developed for recovery of N, P, and heavy metals from a major portion of the municipal sludges produced in the United States. The N and P will be used as inorganic fertilizers or for other commercial processes and the heavy metals and trace elements will go into industrial channels. Residue wastes from these recovery processes will have low environmental hazard and can be easily discharged on land or put into landfills. Recovery will be most applicable in large metropolitan areas where metal concentrations in industrial water are high and where croplands are not available for direct disposal of sludge. In other areas, use of these sludges on cropland will increase as we learn more about relative risks to public health and to the environment in general, as we reduce the heavy metals concentrations, and as public acceptance of recycling as a desirable process increases.

7) The amounts of animal manures that go into uses other than for fertilizing croplands or as soil amendments will increase. These will include (i) use as an energy source by direct combustion and eventually by the production of combustible gases, and (ii) processing for use as ingredients in

animal rations. Use as a fuel will be developed in arid, warm climates where drying can be achieved easily. Processing for feed will not require dry products so it can be used in any area. Either use will require quality control and will change the nature of manure handling for those animal production units that wish to meet this market for their manures.

8) Despite the increasing amounts of animal manures that will go into other uses, most manures will be used as fertilizers and soil amendments for croplands. Because of their many advantages, inorganic fertilizers, when in adequate supply and at reasonable price, will be used instead of manure which will then accumulate near large centers of animal production. When supplies of inorganics decrease, the use of manures will increase, creating cycles of demand for organics. However, with world-wide increases in demands for agricultural products, increasing fuel costs for the production of NH_3 and increasing costs of production of P and K fertilizer materials, we can expect the demand for organics to increase and prevent a repeat of the vast accumulation of manures that occurred in the decade prior to the Arab oil embargo in 1973. On the other hand, large-scale marketing of animal manures as fertilizers will likely remain unprofitable and difficult, and areas with high concentrations of animals will continue to have local excesses of manures.

9) There are sufficient reserves of combustible fuels for N fixation from the atmosphere to preclude any permanent shortage of N fertilizers. However, supplies of P and K are not as inexhaustible and the need for P from municipal sludges and of P and K from animal manures will become increasingly important. Thus, in the long term, we can expect animal wastes and municipal sludges to be used widely as sources of P and K for maintenance applications of these two elements. However, for the immediate future, supplies of these two elements are sufficient to prevent widespread demands for organic sources to replace inorganic sources.

LITERATURE CITED

Bouwer, H., J. C. Lance, and M. S. Riggs. 1974. High-rate land treatment. II. Water quality and economics of the Flushing Meadows Report. J. Water Poll. Contr. Fed. 45:844–859.

CAST Report. 1975. Utilization of animal manures and sewage sludges in food and fiber production. (P. F. Pratt, Task Force Chairman). Council for Agricultural Science and Technology. Report no. 41. Iowa State University, Ames, IO 50010.

Fassell, W. M. 1974. Sludge disposal at a profit? p. 195–204. In Municipal sludge management. Proc. Nat'l. Conference on Municipal Sludge Management. Pittsburgh, Pa. June 1974. Information Transfer Inc., 1625 Eye Street N. W., Wash., DC 20006.

Page, A. L. 1974. Fate and effects of trace elements in sewage sludge when applied to agricultural lands. Environ. Prot. Technol. Ser. EPA-670/2-74-005. U. S. Government Printing Office. 757-580/5306.

Piper, A. M. 1965. Has the United States enough water? U. S. Geol. Surv. Paper 1797. U. S. Government Printing Office, Washington, D. C.

Sopper, W. E., and L. T. Kardos (ed.). 1973. Recycling treated municipal wastewaters and sludge through forest and cropland. Pennsylvania State University Press, University Park, Pa.

SUBJECT INDEX

Acidity. *See* pH
Actinomycetes, 118–19, 542, 550
Activated sludge, carbon losses from, 175
Adsorption. *See also* each element
  anion, 75–77, 80–82
  bonding mechanisms, 75–89, 90–91
  cation, 75–80, 83–84, 86
  chelation, 83–84, 87–88
  chemical, in soil, 105–06
  complexation, 75, 83, 86–94
  ion. *See* cation, anion
  isotherms, 86, 105–06
    Freundlich, 79, 81
    Langmuir, 79–82, 86
  of waste gases, 78, 550, 607–18
  specific, 75, 83–84
Aeration
  denitrification and, 185, 189–90
  drainage
    microbial degradation, 459
    fate of nitrogen, 459
    fate of phosphorus, 459
  nitrification and, 185
  peat and organic deposits, 459–60
  relation to
    earthworm activity, 128
    metal chemistry, 84–85
    metabolic products, 92
    microbial activity, 138
    organic matter, 189–90
    phosphorus chemistry, 81
    porosity, 106–10
    soil acidity, 77
Aerobic lagoons, 410, 560, 595
  effluents, properties of, 53–57
Aerosols
  microbial, 537–38
  odor transport, 547–49
Aggregates, soil
  breakdown of, 104
  effect of organic residues 138–41
  effect of soil organisms, 128, 141
  formation of, 101, 138–41
  stability of, 101, 139–41
Agricultural wastes
  cannery, 25–27
  crop plants, 15, 16, 20–22, 183
  dairy, 12, 13, 441
  forest, 22–25
  livestock, 12–15, 183–85, 220, 403–09, 440–42, 583, 601–02
    waste waters, properties of, 65–70
  packing plant, 25–27, 351
  poultry, 441
  processing, 185–86
  properties of, 11–15, 185–86, 405, 601

vegetable processing, 18, 351, 442
Air permeability, 607, 613
Air pollution
  in arid regions, 480
  odor, 531–71
  pathogen considerations, 578
  treatment
    aerosols, 612
    carbon monoxide, 610
    hydrocarbons, 609
    organonitrogen, 610
    organophosphorus, 611
    organosulfur, 611
    particulates, 612
Air scrubbers
  compost, 609, 614
  hypochlorite, 614, 616
  soil, 607–618
    model, 615
Airborne pollutants, control of at waste disposal sites, 360
Algae, 121–23, 476
Alkalinity. *See also* Salts, Sodium, pH
  soil, 475–76
Aluminum, relation to
  cation exchange, 75–76
  ion retention, 80–82, 84, 86, 89–90
  soil acidity, 77
Ammonia. *See also* Ammonium
  absorption
    plants, 549
    water, 549
  fixation by organic matter, 191–92
  movement of, 547–50
  nitrification of, 150–51, 190, 549
  in primary effluent, 53
  in raw sewage, 49
  toxicity, 205
  volatilization from manure, 304, 314, 547–50
Ammonium. *See also* Ammonia
  in manure, 304, 314
  in sewage effluents, 304, 306–07, 321
  nutrition to plants, 601
Anaerobic conditions
  composting, 599
  effect on denitrification, 189, 602
  effect on gas flow in soils, 109–10
  effect on odor production, 350, 540–46
  effect on P mobilization, 459
  mobilization
    metabolic products, 92
    metal chemistry, 84–85
    phosphorus chemistry, 81
    soil acidity, 77

635